Common Sense Mathematics

Second Edition

AMS/MAA | TEXTBOOKS

VOL **63**

Common Sense Mathematics

Second Edition

Ethan D. Bolker
Maura B. Mast

Providence, Rhode Island

For additional information and updates on this book, visit
www.ams.org/bookpages/text-63

Library of Congress Cataloging-in-Publication Data

Names: Bolker, Ethan D., author. | Mast, Maura B., author.
Title: Common sense mathematics / Ethan D. Bolker, Maura B. Mast.
Description: Second edition. | Providence, Rhode Island : American Mathematical Society, 2020. | Series:
   AMS/MAA textbooks, 2577-1205 ; volume 63 | Includes bibliographical references and index.
Identifiers: LCCN 2020030460 | ISBN 9781470461348 (paperback) | ISBN 9781470462895 (ebook)
Subjects: LCSH: Problem solving–Problems, exercises, etc. | Mathematics–Problems, exercises, etc. | AMS:
   General – Instructional exposition (textbooks, tutorial papers, etc.). | General – General and miscella-
   neous specific topics – General mathematics. | General – General and miscellaneous specific topics –
   Mathematics for nonmathematicians (engineering, social sciences, etc.).
Classification: LCC QA63 .B65 2020 | DDC 510–dc23
LC record available at https://lccn.loc.gov/2020030460

Contents

1   Calculating on the Back of an Envelope   1

In this first chapter we learn how to think about questions that need only good enough answers. We find those answers with quick estimates that start with reasonable assumptions and information you have at your fingertips. To make the arithmetic easy we round numbers drastically and count zeroes when we have to multiply.

2   Units and Unit Conversions   27

In real life there are few naked numbers. Numbers usually measure something like cost, population, time, speed, distance, weight, energy or power. Often what's measured is a rate, like miles per hour, gallons per mile, miles per gallon, dollars per gallon, dollars per euro or centimeters per inch.

3 Percentages, Sales Tax and Discounts   65

The focus of this chapter is the study of relative change, often expressed as a percent. We augment an often much needed review in two ways — stressing quick
paperless estimation for approximate answers and, for precision, a new technique:
multiplying by 1+ (percent change).

4 Inflation   101

We mine the internet for data about inflation and use the 1+ technique from Chapter 3 to understand that data.

5 Average Values   117

We start by remembering that to compute an average you add the values and divide
by the count. We quickly move on to weighted averages, which are more common
and more useful. They're a little harder to understand, but worth the effort. They
help explain some interesting apparent paradoxes.

6 Income Distribution — Spreadsheets, Charts and Statistics   131

This chapter covers a lot of ground — two new kinds of average (median and mode)
and ways to understand numbers when they come in large quantities rather than
just a few at a time: bar charts, histograms, percentiles and the bell curve. To do
that we introduce spreadsheets as a tool.

Contents

vii

7 Electricity Bills and Income Taxes — Linear Functions

We use an electricity bill as a hook on which to hang an introduction to functions in general and linear functions in particular, in algebra and in spreadsheets. Then we apply what we've learned to study taxes — sales, income and Social Security. You'll also find here a general discussion of energy and power.

8 Climate Change — Linear Models

Complicated physical and social phenomena rarely behave linearly, but sometimes data points lie close to a straight line. When that happens you can use a spreadsheet to construct a linear approximation. Sometimes that's useful and informative. Sometimes it's misleading. Common sense can help you understand which.

9 Compound Interest — Exponential Growth

In this chapter we explore how investments and populations grow and how radioactivity decays — exponentially.

10 Borrowing and Saving   243

When you borrow money — on your credit card, for tuition, for a mortgage — you pay it back in installments. Otherwise what you owe would grow exponentially. In this chapter we explore the mathematics that describes paying off your debt.

11 Probability — Counting, Betting, Insurance   257

Pierre de Fermat and Blaise Pascal invented the mathematics of probability to answer gambling questions posed by a French nobleman in the seventeenth century. We follow history by starting this chapter with simple examples involving cards and dice. Then we discuss raffles and lotteries, fair payoffs and the house advantage, insurance, and risks where quantitative reasoning doesn't help at all.

12 Break the Bank — Independent Events   275

Unlikely things happen — just rarely! Here we calculate probabilities for combinations like runs of heads and tails. Then we think about luck and coincidences.

13 How Good Is That Test?   291

In Chapter 12 we looked at probabilities of independent events — things that had nothing to do with one another. Here we think about probabilities in situations where we expect to see connections, such as in screening tests for diseases or DNA evidence for guilt in a criminal trial.

Contents

Preface to the Second Edition

We are delighted that the reception of the first edition makes this second edition possible. What we wrote in the preface there still resonates: we hope to change the way our students' minds work.

A text that regularly refers to current news ages rapidly. The latest stories in the first edition date from 2015. Our primary task for this second edition was to update the exercises and examples. Along the way we revised several chapters and rearranged some sections and words but kept the overall structure. An instructor should be able to design a course that works even when some students have this edition and others the first.

What is different?

- A new first section in Chapter 1, looking at a 2018 report on the number of Uber and Lyft rides in Boston in 2017.

- A complete rewrite of Chapter 3, on percentage calculations, stressing the value of the 1+ trick.

- A new section on inflation history in Chapter 4.

- Updated Excel screenshots and help with alternatives to Excel in Chapter 6.

- 2019 tax rates and brackets in Chapter 7.

- Reworked sections in Chapter 8 to deal more subtly with the correlations among climate change, greenhouse gas emission and economic growth.

- More on debit cards in Chapter 10.

- Many new exercises, some based on material appearing as late as the spring of 2020. Some old ones moved (with their original numbers) to the collection of extra exercises linked from the *Common Sense Mathematics* website www.ams.org/bookpages/text-63 .

- Corrections for the few errata found in the first edition. We hope we haven't introduced too many new ones.

Internet links age. Pages disappear, perhaps to reappear at a new addresses. Federal government bureaus seem to reorganize their websites often. We checked and repaired all the links in August 2019.

As always, we welcome your questions, corrections and ideas — please let us know what works and what doesn't.

There are a few acknowledgments to add to the long list from the first edition. Cathy Gorini (Maharishi University of Management) corrected answers to several exercises. Ben Bolker wrote the script to check the links. Sam Feuer updated Excel screenshots. Eleanor Bolker tried out some of the new exercises. Christine Thivierge at the American Mathematical Society helped track down permissions and shepherded the manuscript through production. Arlene O'Sean copyedited with a graceful light touch. Maura field-tested this new edition at Fordham with her Interdisciplinary Honors STEM I students in fall 2019. Special thanks to these students — Lydia Abraham, Tabitha Anderson, Laura Babiak, Santiago Baena, Chase Behar, Erik Brown, Lily Coilparampil, Phineas Donohue, Macie Grisemer, Amelia Medved, Christian Morales, Jack Moses, Alyssa Nanfro, Julian Navarro, Gabriel Quiroga, Evan Rubenstein, Samantha Sayre, Liam Sisk, Miguel Sutedjo, Nicholas Urbin, Pilar Valdes, Patrick Vivoda, Brooke Warner and Kaitlynn Wilson — for testing new exercises, spotting typos and providing feedback on the draft.

The dedications stay the same.

Newton, MA, and The Bronx, NY
April 2020

Preface to the First Edition

Philosophy. One of the most important questions we ask ourselves as teachers is, "What do we want our students to remember about this course ten years from now?"

Our answer is sobering. From a ten year perspective most thoughts about the syllabus — "what should be covered" — seem irrelevant. What matters more is our wish to change the way our students' minds work — the way they approach a problem, or, more generally, the way they approach the world. Most people "skip the numbers" in newspapers, magazines, on the web and (more importantly) in financial information. We hope that in ten years our students will follow the news, confident in their ability to make sense of the numbers they find there.

To help them, we built this book around problems suggested by the news of the day as we were writing. We also consider issues that are common — and important — such as student loans. Common sense guides the analysis; we introduce new mathematics only when it's really needed. In particular, you'll find here very few problems invented just to teach particular mathematical techniques.

This preface is meant for students. There's also an instructor's manual that offers more detail about how we carry out our intentions. Students are free to read that too.

Organization. Most quantitative reasoning texts are arranged by topic — the table of contents reads like a list of mathematics to be mastered. Since the mathematics is only a part of what we hope to teach, we've chosen another strategy. Each chapter starts with a real story that can be best understood with careful reading and a little mathematics. The stories involve (among other things):

- Back of an envelope estimation.
- Discounts, inflation and compound interest.
- Income distribution in the United States.
- Reading an electricity bill.
- The graduated income tax.
- Reading a credit card bill.
- Paying off a mortgage or a student loan.
- Lotteries, gambling, insurance and the house advantage.
- False positives and the prosecutor's fallacy.

The best tool for understanding is common sense. We start there. When more is called for we practice "just in time" mathematics. The mathematics you need to understand a question appears when we ask the question. We don't ask you to learn something now because you'll need it later. You always see the mathematics in actual use.

This book differs from many at its level because we focus on how to *consume* numbers more than on how to *produce* them.

Paying attention to the numbers. We hope that when you've finished this course you will routinely look critically at the numbers you encounter every day. Questions like these should occur to you naturally:

- What do the numbers really mean?
- What makes them interesting (or not)?
- Are they consistent? Distorted?
- Do I believe them? Where do they come from?
- How might I check them?
- What conclusions can I draw from them?

To help you answer these kinds of questions we will think about:

- Relative and absolute change.
- Percentages.
- Units: all interesting numbers are numbers *of something*.
- Estimation skills. Counting zeroes: million, billions, trillions and beyond.
- Significant digits, orders of magnitude, quick and dirty mental arithmetic.
- Using a spreadsheet to ask "what-if" questions.
- Using a spreadsheet to display data.
- Models: when simple mathematics can clarify how data may be related.
- Probability and randomness.

Common sense and common knowledge. You can understand the numbers in a paragraph from the newspaper only if you understand something about the subject it addresses. Many of the discussions in the text and the exercises provide opportunities to explore — to learn things you might not know about economics or history or psychology or sociology or science or literature. When words or concepts are unfamiliar, or you're unsure of their meanings, look them up. Explore the ideas before you focus on the numbers.

The exercises.

> In guessing a conundrum, or in catching a flea, we do not expect the breathless victor to give us afterwards, in cold blood, a history of the mental or muscular efforts by which he achieved success; but a mathematical calculation is another thing.
>
> Lewis Carroll
> *A Tangled Tale*
> Answers to Knot 4 [**R1**]

Many textbooks give you a head start on the problems because they occur at the end of the section in which the relevant mathematics is taught. You need only look back a few pages to find a sample problem like the one you're working on. Often all you have to do is change the numbers.

There are few like that here. Most of ours call for a more extended solution — at least several sentences, sometimes several paragraphs. You can't simply calculate and then circle the answer on the page. On exams and on homework assignments we frequently remind you of what we expect with boilerplate like this:

Be sure to write complete sentences. Show how you reached the answer you did. Identify any sources you used. When you refer to a website, you should indicate why you think it is a *reliable* source — there's lots posted online that's just plain wrong.

Some exercises have hints at the back of the book. Try not to look at them until you've thought about the problem yourself for a while. If you invent or solve a problem and are particularly pleased with what you've done, send it to us and we'll consider incorporating it in a later edition of the book, with credit to you, of course. If you find an error, please let us know.

When you think you've finished an exercise, read your answer carefully just to see that it makes sense. If you estimate an average lawyer's annual income as $10,000 you have probably made a mistake somewhere. It's better to write "I know this is wrong, but I can't figure out why — please help me" than to submit an answer you know is wrong, hoping no one will notice.

We've annotated some of the exercises. Here's what the tags mean:

S: The solution manual contains an answer.
U: This exercise is untested. We think it might be a good one but haven't yet tried it out in a class.
C: This exercise is complex, or difficult, or ambiguous. Many problems in the real world are like that, so this book has a few too.
W: This is a worthy exercise. It's particularly instructive, perhaps worth taking up in class.
R: A routine exercise.
A: An exercise with artificial numbers. Sometimes problems like this are good for emphasizing particular points, but we try to avoid them when we can.
N: The idea for an exercise, with no details yet.
Goal $x.y$: Contributes to mastery of Goal y of Chapter x.
Section $x.y$: Depends on or adds to material in Section y of Chapter x.

The world is a messy place. When you're reading the newspaper or come across a webpage or see an ad on television you're not told which chapter of the book will help you understand the numbers there. You're on your own.

When we ask open-ended questions like those triggered by the news of the day our students are often uncomfortable. Here are comments expressing that discomfort, from two students, part way through the course:

•   I still don't understand sometimes how we are given questions that are almost meant to confuse the reader.

•   The only improvements I would make would be that we have more structured problems for homework, occasionally they can be broad.

But at the end of the semester students wrote anonymously in answer to the question "What are the strong points of this course?"

•   In Quantitative Reasoning we are learning how to look into numbers instead of just looking at them.

•   Use math in everyday settings instead of thinking, "When will I ever use this?"

- It covers a real-world perspective of math.

- This course taught material that will be extremely helpful in the future.

- It teaches math that can be used every day and skills in Excel that are useful and that I will definitely be using down the line.

- This course is very useful for me in the outside world and I feel that I will benefit from the education I received from this class and I will be able to apply my new knowledge to situations outside of the classroom.

- This course has taught me a lot about obtaining information and using it in ways that I had not before.

- Very applicable subject matter for other areas of academics and professions. The course was something of a blend of refresher mathematics and new ways to apply them to everyday life.

- The math was more interesting, relevant. Good examples, news articles employed.

- I think the hardest part about this class was thinking. When you usually enter a math class the only thinking that you have to do is remember equations, but in this math class I had to do research and find things on my own to help me out to answer a question.

- For the love of all that is good, why is an English major/poet/musician forced to take math all these years? I am not well-rounded or more comfortable with math, it has just drawn out my college career, costing me time and money that I don't have. I will never use math in my life, the types that I will employ I learned in elementary school. This was the best math class I have ever taken though.

Real and up to date. Our philosophy demands that the examples and exercises in *Common Sense Mathematics* pose real questions of genuine interest. Therefore they usually come from the news of the day at the time we wrote them. You don't have to go to the original sources to answer the questions, but if you're curious you can. You will find the bibliographic details in the references section at the end of the book.

One problem with our philosophy is that the text is out of date as soon as it's printed. Our remedy is to rewrite the course and add to the exercises on the fly each time we teach it. We hope other instructors will do that too. That way nothing is ever stale, which is good. What's less good is that the discussions here may not correspond to what actually happens in the course you are taking.

***Common Sense Mathematics* on the web.** The home page for this text is www.ams.org/bookpages/text-63. There you will find the spreadsheets we refer to, errata we have discovered (or you tell us about) and other information teachers and students of quantitative reasoning might find useful.

Technology. We wrote this text to help students understand questions where quantitative reasoning plays a part. To that end, we take advantage of any tools that will reduce drudgery and prevent careless errors.

For most applications, an ordinary four function calculator will do — and these are ubiquitous. You probably have one on your cell phone. When more advanced arithmetic is called for you can use a spreadsheet or the calculator on your computer or the internet.

You'll find references to websites in the exercises and elsewhere throughout the book. The references were accurate when we wrote the book, but we know that the web changes. If a link doesn't work, don't give up. Most likely it's moved somewhere else. A broken link isn't an excuse to skip a homework problem. Instead, be resourceful. Look around on the web or email us or your instructor.

We think an educated citizen these days should be able to refer to the internet wisely and effectively and be comfortable using a spreadsheet. In *Common Sense Mathematics* we use Excel, not because we are particularly fond of Microsoft, but because it is the most common spreadsheet in use today. But almost all our spreadsheets can be recreated in any spreadsheet program now or (we imagine) in the near future.

A spreadsheet program is good for data analysis, for asking "what-if" questions and for drawing graphs. It also helps make mathematical abstractions like "function" real, rather than formal. We introduce Excel in Chapters 6 and 7. From then on we ask you to create spreadsheets and to use simple ones we've built for you — more complex than ones you could write, but not too hard to read and understand as well as use.

We regularly refer to searching with Google, because it is the most commonly used search engine. But any other should do; use your favorite.

One search feature turns out to be particularly useful. Both Google and Bing will do arithmetic for you when you type a numerical calculation in the search field.

Old vs. new. If you know one way to do a problem should you learn another? That depends. ("That depends" is the answer to most interesting questions. If the question calls for a straightforward "yes" or "no" or just a number or something you can discover in one step with a web search the question is probably not very interesting.)

If you rarely encounter similar problems it's not worth the effort needed to understand and remember a new way to do them. But if you expect to see many, then it may pay to learn that new method.

For example, if you plan to spend just a day or so in a foreign country, get a phrase book with the common words you'll need to communicate. But if you plan to live there half the year, learn the language.

Here's a second example. When using a computer, there are many things you can do with either the mouse or the keyboard. The mouse is intuitive. You can see just what's happening, and there's nothing to remember. Just pull down the menu and click. But the keyboard is faster. So if you're going to do something just once or twice, use the mouse, but if you're going to do it a lot, learn the keyboard shortcut. In particular, in computer applications these days you often copy text from one place to paste it in another, whether that's from a webpage, or in your word processor or spreadsheet. You can do that from the edit menu, or you can use the keyboard shortcuts `control-C` and `control-V`. Learn the shortcuts!

We have tried in this book to teach you new ways to do things when we think those new ways will serve you well in the future. We've resisted the temptation when those new ways are just clever tricks mathematicians are fond of that don't really help you in the long run.

Truth and beauty. We've worked to limit the mathematics we cover to just what you need, along with common sense and common knowledge, to help you deal with the quantitative parts of a complex world. But there is another important reason to study mathematics.

You read not only because it's useful, but because reading gives you access to poetry. You cook not only because you must eat to live, but because there can be pleasure in preparing tasty meals and sitting down in good company to enjoy them. We became mathematicians not only because mathematics is useful, but because (for us and some other people like us) it's beautiful, too.

This passage from Henry Wadsworth Longfellow's 1849 novel *Kavanagh* captures both the truth and the beauty of mathematics (as we hope we have).

> "For my part," [says Mary Churchill] "I do not see how you can make mathematics poetical. There is no poetry in them."
>
> "Ah, that is a very great mistake! There is something divine in the science of numbers. Like God, it holds the sea in the hollow of its hand. It measures the earth; it weighs the stars; it illumines the universe; it is law, it is order, it is beauty. And yet we imagine — that is, most of us — that its highest end and culminating point is book-keeping by double entry. It is our way of teaching it that makes it so prosaic." [**R2**]

Contact us. We welcome questions, feedback, suggested problems (and solutions) and notes about errors. You may contact us by email at `ebolker@gmail.com` (Ethan Bolker) or `mmast@fordham.edu` (Maura Mast).

Acknowledgements. We owe much to many for help with *Common Sense Mathematics*.

Many years ago Linda Kime shepherded the first quantitative reasoning requirement at UMass Boston. In 2007 then Mathematics Chair Dennis Wortman allowed us to coteach Math 114 in hopes of reinventing the course. In early years Mark Pawlak pilot tested early versions of this text. He was the first to believe that we were onto a good thing — soon he was scouring the newspaper for examples to use in class, on exams, and in the exercises. His input, drawn from his long involvement in quantitative reasoning and his deep engagement with student learning, has made this a better book. When *Common Sense Mathematics* became the official textbook for quantitative reasoning at UMass Boston, Mark was the course administrator, trained tutors, developed new approaches to assessing student learning and recruited instructors: George Collison, Karen Crounse, Dennis DeBay, Monique Fuguet, Matt Lehman, Nancy Levy, John Lutts, Robert Rosenfeld, Jeremiah Russell, Mette Schwartz, Joseph Sheppeck, Mitchell Silver, Karen Terrell, Charles Wibiralske and Michael Theodore Williams.

We benefited from feedback from colleagues at other schools who asked to use the text: Margot Black (Lewis & Clark), Samuel Cook (Wheelock College), Grace Coulombe (Bates College), Mike Cullinane (Keene State College), Timothy Delworth (Purdue University), Richard Eells (Roxbury Community College), Marc Egeth (Pennsylvania Academy of the Fine Arts), Ken Gauvreau (Keene State College), Krisan Geary

(Saint Michael's College), David Kung (St. Mary's College of Maryland), Donna LaLonde (Washburn University), Carl Lee (University of Kentucky), Alex Meadows (St. Mary's College of Maryland), Wesley Rich (Saginaw Chippewa Tribal College), Rachel Roe-Dale (Skidmore College), Rob Root (Lafayette College), Barbara Savage (Roxbury Community College), Q. Charles Su (Illinois State University), Joseph Witkowski (Keene State College) and several anonymous reviewers.

Students caught typos, suggested rewordings and provided answers to exercises. We promised to credit them here: Courtney Allen, Matt Anthony, Vladimir Altenor, Theresa Aluise, Selene Bataille, Kelsey Bodor, Quonedell Brown, Katerina Budrys, Candace Carroll, Jillian Christensen, Katie Corey, Molly Cusano, Sam Daitsman, Hella Dijsselbloem-Gron, Michelle DiMenna, Shirley Elliot, Lea Ferone, Solomon Fine, Murray Gudesblat, Frances Harangozo, Irene Hartford, Katilyn Healey, Anna Hodges, Anthony Holt, Laura Keegan, Jennifer Kunze, Kevin Lockwood, Jacob Looney, Ashley McClintock, Edward McConaghy, Nicole McKenna, Amanda Miner, Antonio de las Morenas, Daniel Murano, Hannah Myers, Matt Nickerson, Rodrigo Nunez, Gabby Phillips, Jaqueline Ramirez, Hailey Rector, Taylor Spencer, Jaran Stallbaum, Melinda Stein, Willow Smith, Nick Sullivan, Robert Tagliani, Julia Tran, Marcus Zotter, … and many others.

Cong Liu worked on the index. Monica Gonzalez and Alissa Pellegrino tested the links to the web. Paul Mason ferreted out the newspaper headlines for the cover.

The National Science Foundation provided support from grant DUE-0942186. Any opinions, findings and conclusions or recommendations expressed in this material are those of the contributors and do not necessarily reflect the views of the National Science Foundation. We hope they approve of what we've done with their generosity.

The Boston Globe graciously gave us permission to reproduce here the quotes from its journalists we found in our morning paper and brought to class.

Wizards at `tex.stackexchange.com` were always quick to answer TeXnical questions.

Carol Baxter, Stephen Kennedy, Beverly Ruedi and Stanley Seltzer at the Mathematical Association of America were enthusiastic about our book and brought wisdom and competence to design and production.

Ethan: My wife, Joan's, ongoing contribution began 56 years ago when she asked me how I'd feel if I went to medical school and did no more mathematics. It continued with constant support of all kinds — most of the details would be inappropriate here. I will say that I recommend living with a writing coach to hone writing strategies. I've talked for years with my professor children about mathematics and teaching — Jess and Ben make cameo appearances in the text. I dedicate *Common Sense Mathematics* to the next generation: Solomon Bixby and Eleanor Bolker.

Maura: I owe a greater debt than I could ever express to my husband, Jack Reynolds. We met in Boston over 21 years ago (thanks in part to a National Science Foundation grant). Living in Iowa three years later, we saw that UMass Boston wanted to hire a mathematician to work on quantitative reasoning. With Jack's support, I left a tenured position to accept that challenge. I couldn't have done that and my other work at UMass as well as I did without his faith in me and his support. There's more. Jack has been my conscience as well as my partner. He fundamentally believes that each person can make a difference in the world. Because of that, I now see my work in quantitative reasoning, and mathematics, as a way to change the world. I thank my children, Brendan, Maeve and Nuala Reynolds, for their patience and support, especially for the times when I turned their questions into homework problems, as in the tooth fairy exercise. I dedicate *Common Sense Mathematics* to the memory of my parents, Cecil and Mary Mast, who set high ideals grounded in reality. I miss them dearly.

Newton, MA, and The Bronx, NY
October 2015

1

Calculating on the Back of an Envelope

In this first chapter we learn how to think about questions that need only good enough answers. We find those answers with quick estimates that start with reasonable assumptions and information you have at your fingertips. To make the arithmetic easy we round numbers drastically and count zeroes when we have to multiply.

Chapter goals:

Goal 1.1. Verify quantities found in the media, by checking calculations and with independent web searches.

Goal 1.2. Estimate using common sense and common knowledge.

Goal 1.3. Learn about the Google calculator (or another internet calculator).

Goal 1.4. Round quantities to report only an appropriate number of significant digits.

Goal 1.5. Learn when not to use a calculator — become comfortable with quick approximate mental arithmetic.

Goal 1.6. Work with large numbers.

Goal 1.7. Work with (large) metric prefixes

Goal 1.8. Practice with straightforward unit conversions.

1.1 Hailing a ride

On May 1, 2018, a headline in *The Boston Globe* read "There were nearly 100,000 Uber and Lyft rides per day in Boston last year." The article began:

> It's not your imagination: There are an awful lot of Uber and Lyft cars in Boston traffic.
>
> The ride-hailing companies provided nearly 35 million trips in Boston in 2017, or an average of about 96,000 every day, according to data released by Massachusetts officials Tuesday.
>
> Boston accounted for more than half of the 65 million rides Uber and Lyft provided across all of Massachusetts last year, according to data the Department of Public Utilities collected from the ride-hailing companies under a new state law regulating the industry. Every minute in the city there were, on average, 67 Uber and Lyft rides underway. [R3]

There are several numbers here: 100,000, 35 million, 96,000, 65 million, and 67. Do these numbers make sense?

In particular, should you believe Uber and Lyft accounted for "35 million trips in Boston in 2017"?

Note the question. "Should you believe?" not "Do you believe?". To decide whether or not to believe we will think about what that number says in a context where you can see whether (or not) it makes common sense. Imagining a year's worth of trips is hard. Trips per day or per hour or per minute might be easier.

Start by checking the arithmetic in the second paragraph. If we work with 350 days in a year it's easy to divide the 35 million rides among the days. Cancelling the 35 and counting zeroes gives 100,000 rides per day. That's exactly what's in the headline and is for all common sense purposes the same as the 96,000 daily rides in the quotation.

We could have used a calculator to discover that

$$\frac{35,000,000}{365} = 95890.4109589.$$

But why waste the time? All we know to start with is that the numerator is "nearly 35,000,000". That number has just two significant digits, the 3 and the 5. The six zeroes just tell us where the decimal point goes. So the answer should be rounded to two significant digits: about 96,000 daily rides. That is just what the author of the article did. All the other digits are correct arithmetically but make no sense in the discussion.

Should you believe 100,000 rides per day? Let's suppose that the people who use Uber or Lyft use it twice a day. That's two rides per person, so that the 100,000 rides per day are taken by 50,000 people. About 1,000,000 people live in the Boston metropolitan area. Perhaps one in twenty takes a ride-hail round trip. This seems reasonable.

You can argue about some of the assumptions. 50,000 is probably not a very good estimate for the number of riders (each twice). But it does tell us that 100,000 rides per day is in the right ballpark. It has the right number of zeroes. 10,000 rides per day would clearly be too small while 1,000,000 rides per day would be too large.

Another way to look at 100,000 rides per day is to think about the number of rides per hour, and then per minute. There are about 25 hours in a day, so about 4,000 rides per hour. There are 60 minutes in an hour, and 40/6 is about 7, so there are about 70 rides per minute. That's believable if you imagine people all over Boston looking for rides.

We've checked that the 35 million rides per year is a reasonable number. Should we believe 67 rides underway at any particular time? It feels too small. We just estimated about 70 rides starting per minute. If each ride lasted just one minute there would be 70 rides underway at any time. So 67 can't be right.

The last paragraph of the article provides some data to confirm our suspicion.

> The average ride-hailing trip in Massachusetts lasts about 15 minutes and travels about 17 miles an hour, according to the state data. That's heavily influenced by the clustering of trips in Boston, where the average ride-hailing trip travels at 16 miles an hour.

A 15 minute ride at 16 miles per hour is a 4 mile ride. That's probably about right for a trip in the city, or to or from a suburb.

Since the average ride lasts 15 minutes and 70 rides start each minute there must be about $15 \times 70 \approx 1,000$ rides underway (on average) at any time. That makes much more sense than 67. Should we believe it? Let's look for more evidence.

Googling "how many Uber drivers in Boston" finds this quote from 2015 at `www.americaninno.com/boston/google-ride-hailing-service-google-uber-may-be-competitors-2-2/`:

> There are nearly 10,000 Uber drivers in Boston, according to data the ridesharing company released Thursday morning. [R4]

You can't believe everything you read on the internet, but this site says the 10,000 driver figure comes from Uber. Uber might want to exaggerate, but it's probably an OK estimate. That was several years before the 2017 report. Updating and adding Lyft drivers suggests there might have been 20,000 drivers in 2017. If 5 percent of them had passengers at any moment, that would account for a believable 1,000 rides on the road.

Where might the 67 have come from? A clue is how close it is to our estimate that about 70 rides start each minute. In fact it's just what you get when you carefully convert 96,000 rides per day into

$$\frac{96,000 \text{ rides}}{\text{day}} \times \frac{1 \text{ day}}{24 \text{ hours}} \times \frac{1 \text{ hour}}{60 \text{ minutes}} = \frac{66.666\ldots \text{ rides}}{\text{minute}}.$$

So the author meant to say

> Every minute there were, on average, 67 Uber and Lyft rides *starting*.

rather than

> Every minute there were, on average, 67 Uber and Lyft rides *underway*.

If we had discovered this error when the article first appeared we could have written the author asking for a correction in the online story.

What have we learned? First, that almost all the numbers in this article are consistent with one another: they fit together. Only the 67 rides on the road is wrong. Second, though 35,000,000 Boston rides per year might at first seem unbelievably large, it is in fact reasonable. Finally, thinking with the numbers rather than skipping over them led us to discover and correct a mistake and helped us focus on the ideas that the article is meant to convey.

1.2 How many seconds?

Have you been alive for a thousand seconds? A million? A billion? A trillion?

Before we estimate, what's your guess? Write it down, then read on.

To turn a guess into an estimate you have to do some arithmetic. There are two ways to go about the job. You can start with seconds and work up through hours, days and years, or start with thousands, millions and billions of seconds and work backwards to hours, days and years. We'll do it both ways.

How many seconds in an hour? Easy: $60 \times 60 = 3{,}600$. So we've all been alive much more than thousands of seconds.

Before we continue, we're going to change the rules for arithmetic so that we can do all the multiplication in our heads, without calculators or pencil and paper. We will round numbers so that they start with just one nonzero digit, so 60×60 becomes 4,000. Of course we can't say $60 \times 60 = 4{,}000$; the right symbol is $\approx$, which means "is approximately." We call this "curly arithmetic".

Then an hour is

$$60 \times 60 \approx 4{,}000$$

seconds.

There are 24 hours in a day. $4 \times 24 \approx 100$, so there are

$$4{,}000 \times 24 \approx 100{,}000$$

seconds in a day.

Or we could approximate a day as 20 hours, since we overestimated the number of seconds in an hour. That would mean (approximately) 80,000 seconds per day. We'd end up with the same (approximate) answer.

Since there are about a hundred thousand seconds in a day, there are about a million seconds in just 10 days. That's not even close to a lifetime, so we'll skip working on days, weeks or months and move on to years.

How many seconds in a year? Since there are (approximately) 100,000 in a day and (approximately) 400 days in a year there are about 40,000,000 (forty million) seconds in a year.

If we multiply that by 25 the 4 becomes 100, so a 25 year old has lived for about 1,000,000,000 (one billion) seconds.

Does this match the estimate you wrote down for your lifetime in seconds?

A second way to estimate seconds alive is to work backwards. We'll write the time units using fractions — that's looking ahead to the next chapter — and round the numbers whenever that makes the arithmetic easy. Let's start with 1,000 seconds.

$$1{,}000 \text{ seconds} \times \frac{1 \text{ minute}}{60 \text{ seconds}} = \frac{1{,}000}{60} \text{ minutes}$$

$$= \frac{100}{6} \text{ minutes (cancel a 0)}$$

$$= \frac{50}{3} \text{ minutes (cancel a 2)}$$

$$\approx \frac{60}{3} \text{ minutes (change 50 to 60 — make division easy)}$$

$$= 20 \text{ minutes.}$$

We're all older than that.

How about a million seconds? A million has six zeroes — three more than 1,000, so a million seconds is about 20,000 minutes. Still too many zeroes to make sense of, so convert to something we can understand — try hours:

$$20,000 \text{ minutes} \times \frac{1 \text{ hour}}{60 \text{ minutes}} = \frac{20,000}{60} \text{ hours} = \frac{1,000}{3} \text{ hours} \approx 300 \text{ hours}.$$

There are 24 hours in a day. To do the arithmetic approximately, use 25. Then $300/25 = 12$ so 300 hours is about 12 days. We've all been alive that long.

How about a billion seconds? A billion is a thousand million, so we need three more zeroes. We can make sense of that in years:

$$12,000 \text{ days} \times \frac{1 \text{ year}}{365 \text{ days}} = \frac{12,000}{365} \text{ years} \approx \frac{12,000}{400} \text{ years} \approx 30 \text{ years}.$$

Since a billion seconds is about 30 years, it's in the right ballpark for the age of most students. It's in the same ballpark as the 25 year estimate we found doing the arithmetic the other way around.

A trillion is a thousand billion — three more zeroes. So a trillion seconds is about 30,000 years. Longer than recorded history.

1.3 Heartbeats

In *The Canadian Encyclopedia* a blogger noted that

> The human heart expands and contracts roughly 100,000 times a day, pumping about 8,000 liters of blood. Over a lifetime of 70 years, the heart beats more than 2.5 billion times, with no pit stops for lube jobs or repairs. [**R5**]

Should we believe "100,000 times a day" and "2.5 billion times in a lifetime"?

If you think about the arithmetic in the previous section in a new way, you may realize we just answered this question. Since your pulse rate is about 1 heartbeat per second, counting seconds and counting heartbeats are different versions of the same problem. We discovered that there are about 100,000 seconds in a day, so the heartbeat count is about right. We discovered that 30 years was about a billion seconds, and 70 is about two and a half times 30, so 70 years is about 2.5 billion seconds. Both the numbers in the article make sense.

Even if we didn't know whether 100,000 heartbeats in a day was the right number, we could check to see if that number was consistent with 2.5 billion in a lifetime.

To do that, we want to calculate

$$100,000 \frac{\text{beats}}{\text{day}} \times 365 \frac{\text{days}}{\text{year}} \times 70 \frac{\text{years}}{\text{lifetime}}.$$

Since we only need an approximate answer, we can simplify the numbers and do the arithmetic in our heads. If we round the 365 up to 400 then the only real multiplication is $4 \times 7 = 28$. The rest is counting zeroes. There are eight of them, so the answer is approximately $2,800,000,000 = 2.8$ billion. That means the 2.5 billion in the article is about right. Our answer is larger because we rounded up.

The problems we've tackled so far don't have exact numerical answers of the sort you are used to. The estimation and rounding that goes into solving them means that when you're done you can rely on just a few *significant digits* (the digits at the beginning of a number) and the number of zeroes. Often, and in these examples in particular, that's all you need. Problems like these are called "Fermi problems" after Enrico Fermi (1901–1954), an Italian physicist famous for (among other things) his ability to estimate the answers to physical questions using very little information.

1.4 Calculators

The thrust of our work so far has been on mental arithmetic. You can always check yours with the calculator on your phone. But that requires clicking number and operator icons. If you have internet access, Google's is easier to use — simply type

$$\boxed{100,000 * 365 * 70}$$

into the search box to find the number of heartbeats in a lifetime, assuming 100,000 in a day. Google displays a calculator showing

$$\boxed{2555000000} \ .$$

That 2.555 billion answer is even closer than our first estimate to the 2.5 billion approximation in the article. The Bing search engine offers the same feature.

You can click on the number and operation keys in the Google calculator on display to do more arithmetic. Please don't. Use the keyboard rather than the mouse. It's faster, and you can fix typing mistakes easily.

The Google calculator can do more than just arithmetic — it can keep track of units. Although it doesn't count heartbeats, it does know about miles and about speeds like miles per day and miles per year. We can make it do our work for us by asking about miles instead of heartbeats. Search for

$$\boxed{100,000 \text{ miles per day in miles per 70 years}}$$

and Google rewards you with

$$\boxed{100\,000 \text{ (miles per day)} = 2.55669539 \times 10^9 \text{ miles per (70 years)}} \ .$$

The "$\times 10^9$" means "add nine zeroes" or, in this case, "move the decimal point nine places to the right", so

$$100,000 \text{ (miles per day)} = 2.55669539 \times 10^9 \text{ miles per (70 years)}$$
$$\approx 2.6 \text{ billion miles per (70 years)}.$$

That is again "more than 2.5 billion."

The exact answer from Google is even a little more than the 2,555,000,000 we found when we did just the arithmetic since Google knows a year is a little longer than 365 days — that's why we have leap years.

So 100,000 heartbeats per day does add up to about 2.5 billion in 70 years. We've checked that the numbers are consistent — they fit together.

But are they correct? Does your heart beat 100,000 times per day? To think sensibly about a number with lots of zeroes we can convert it to a number of something equivalent with fewer zeroes — in this case, heartbeats per minute. That calls for division rather than multiplication:

$$100,000 \frac{\text{beats}}{\text{day}} \times \frac{1 \text{ day}}{24 \text{ hours}} \times \frac{1 \text{ hour}}{60 \text{ minutes}}.$$

To do the arithmetic in your head, round the 24 to 25. Then $25 \times 6 = 150$ — there are about 1,500 minutes in a day. Then $100,000/1,500 = 1,000/15$. Since $100/15$ is about 7, we can say that $1,000/15$ is about 70. 70 beats per minute is a reasonable estimate for your pulse rate, so 100,000 heartbeats per day is about right.

Google tells us

$$\boxed{100\,000 \text{ (miles per day)} = 69.4444444 \text{ miles per minute}} \quad .$$

The nine and all the fours in that 69.4444444 are much too precise. The only sensible thing to do with that number is to round it to 70 — which is what we discovered without using a calculator.

Sometimes even the significant digits can be wrong and the answer right, as long as the number of zeroes is correct. Informally, that's what we mean when we say the answer is "*in the right ballpark.*" The fancy way to say the same thing is "the *order of magnitude* is correct." For example, it's right to say there are hundreds of days in a year — not thousands, not tens. There are billions (nine zeroes) of heartbeats in a lifetime, not hundreds of millions (eight zeroes), nor tens of billions (ten zeroes).

1.5 Millions of trees?

On May 4, 2010, Olivia Judson wrote in *The New York Times* [R6] about Baba Brinkman, who describes himself on his webpage as

> ...a Canadian rap artist, award-winning playwright, and former tree-planter who has personally planted more than one million trees. [R7]

How long would it take to personally plant a million trees? Is Brinkman's claim reasonable?

To answer that question you need two estimates — the time it takes to plant one tree and the time Brinkman may have spent planting.

To plant a tree you have to dig a hole, put in a seedling and fill in around the root ball. It's hard to imagine you can do that in less than half an hour.

If Brinkman worked eight hours a day he would plant 16 trees per day. Round that up to 20 trees per day to make the arithmetic easier and give him the benefit of the doubt. At that rate it would take him $1,000,000/20 = 50,000$ days to plant a million trees. If he planted trees 100 days each year, it would take him 500 years; if he planted trees for 200 days out of the year, it would take him 250 years. So his claim looks unreasonable.

What if we change our estimates? Suppose he took just ten minutes to plant each tree and worked fifteen hour days. Then he could plant nearly 100 trees per day. At that rate it would take him 10,000 days to plant a million trees. If he worked 100 days each summer he'd still need about 100 years. So on balance we believe he's planted lots of trees, but not "personally ... more than one million."

It's the "personally" that makes this very unlikely. We can believe the million trees if he organized tree-planting parties, perhaps with people manning power diggers of some kind, or if planting acorns counted as planting trees.

This section, first written in 2010, ended with that unfunny joke until 2013, when Charles Wibiralske, teaching from this text, wondered if we might be overestimating the time it takes to plant a tree. To satisfy his curiosity, he found Brinkman's email address and asked. The answer was a surprising (to him and to us) ten seconds! So our estimate of 10 minutes was 60 times too big. That means our 100 year estimate should really have been only about two years! That's certainly possible. If it took him a minute per tree rather than 10 seconds he could still have planted a million trees in several summers.

Brinkman tells the story of Wibiralske's question and this new ending in his blog at `www.bababrinkman.com/insult-to-injury/`. When you visit you can listen to "The Tree Planter's Waltz" (`www.youtube.com/watch?v=jk-jifbpcww`).

The moral of the story: healthy skepticism about what you read is a good thing, as long as you're explicit and open minded about the assumptions you make when you try to check. That's a key part of using common sense.

Brinkman's blog ends this way:

> Hurray! In the end it's a classic example of ... the drunkard's walk towards knowledge. When our views are self-correcting and open to revision based on new evidence, they will continue to hone in on increasingly accurate representations of the real world. That's good honest skepticism, and when it wins over bad, knee-jerk, "it's hard to imagine" skepticism, that's a beautiful thing.

1.6 Carbon footprints

Discussions about global warming and climate change sometimes talk about the *carbon footprint* of an item or an activity. That's the total amount of carbon dioxide (CO_2) the item or activity releases into the atmosphere. An article in *The Boston Globe* on October 14, 2010, listed estimates of carbon footprints for some common activities. Among those was the 210 gram carbon footprint of a glass of orange juice. That includes the carbon dioxide cost of fertilizing the orange trees in Florida and harvesting the oranges and the carbon dioxide generated burning oil or coal to provide energy to squeeze the oranges, concentrate and freeze the juice and then ship it to its destination. It's just an estimate, like the ones we're learning to make, but much too complex to ask you to reproduce. We drew Figure 1.1 using the rest of the data from the article.

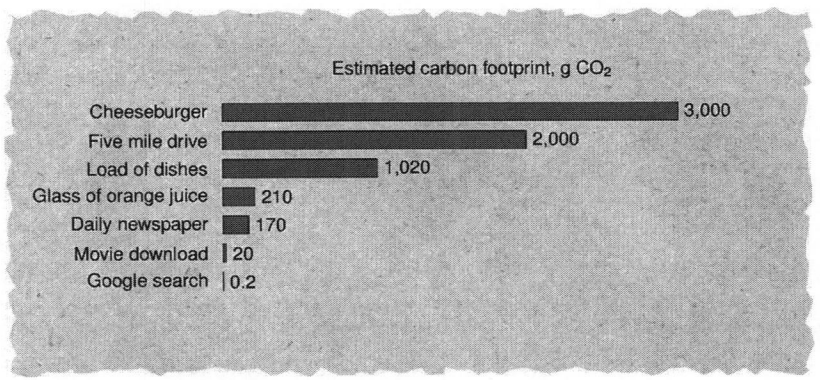

Figure 1.1. Carbon footprints [**R8**]

Let's look at the orange juice. How many glasses are consumed in the United States each day? The population of the United States is about 330 million. To make the arithmetic easy, round that down to 300 million. If we estimate that about 5% of the population has orange juice for breakfast that means 15 million glasses. So the ballpark answer is on the order of 10 to 20 million glasses of orange juice. Since each glass contributes 200 grams, 10 to 20 million glasses contribute 2 to 4 billion grams each day.

It's hard to imagine 2 billion grams. You may have heard of kilograms — a kilogram is about two pounds. In the metric system *kilo* means "multiply by 1,000" so a kilogram is 1,000 grams. Then 2 billion grams of carbon is just 2 million kilograms. That's about 4 million pounds, or 2 thousand tons.

Since there are seven activities listed in the graphic, there are six other Fermi problems like this one for you to work on:

- Google search
- Movie download
- Daily newspaper printed
- Dishwasher run
- Five miles driven
- Cheeseburger consumed

For each you can estimate the total number of daily occurrences and then the total daily carbon contribution. We won't provide answers here because we don't want to spoil a wonderful class exercise.

1.7 Kilo, mega, giga

Counting zeroes is often best done three at a time. That's why we separate groups of three digits by commas. Each step from thousands to millions to billions adds three zeroes. The metric system has prefixes for that job.

We've seen that *kilo* means "multiply by 1,000". Similarly, *mega* means "multiply by 1,000,000". A Megabucks lottery prize is millions of dollars. A megathing is 1,000,000 things, whatever kind of thing you are interested in.

New Hampshire's Seabrook Nuclear Power Plant is rated at 1,270 megawatts. So although you may not know what a watt is, you know this power plant can generate 1,270,000,000 of them. The symbols for "mega" and for "watt" are "M" and "W" so you can write 1,270 megawatts as 1,270 MW.

How many 100 watt bulbs can Seabrook light up? Just take the hundred's two zeroes from the mega's six, leaving four following the 1,270. Putting the commas in the right places, that means 12,700,000 bulbs. Or, if you want to be cute, about 13 megabulbs.

Next after mega is *giga*: nine zeroes. The symbol is "G". When you say "giga" out loud the G is hard, even though it's soft in the word "gigantic".

You could describe Seabrook as a 1.27 gigawatt power plant.

Table 1.2 describes the metric prefixes bigger than giga. There's no need to memorize it. You will rarely need the really big ones. You can look them up when you do.

The metric system also has prefixes for shrinking things as well as these for growing them. Since division is harder than multiplication, we'll postpone discussing those prefixes until we need them, in Section 2.7.

Name	Symbol	Meaning	English name	Zeroes
Kilo	K	$\times 10^3$	thousand	3
Mega	M	$\times 10^6$	million	6
Giga	G	$\times 10^9$	billion	9
Tera	T	$\times 10^{12}$	trillion	12
Peta	P	$\times 10^{15}$	quadrillion	15
Exa	E	$\times 10^{18}$	quintillion	18
Zetta	Z	$\times 10^{21}$	sextillion	21
Yotta	Y	$\times 10^{24}$	septillion	24

Table 1.2. Metric prefixes

1.8 Exercises

Notes about the exercises:

- The Preface to the First Edition has information about the exercises and the solutions, including explanations of the bracketed notations like [S] that accompany each exercise.

- One of the ways to improve your quantitative reasoning skills is to write about what you figure out. The exercises give you many opportunities to practice that. The answer to a question should be more than just a circle around a number or a simple "yes" or "no". Write complete paragraphs that show your reasoning. Your answer should be complete enough so that you can use a corrected homework paper to study for an exam without having to go back to the text to remember the questions.

- It often helps to write about doubt and confusion rather than guessing at what you hope will turn out to be a right answer.

- To solve Fermi problems you make assumptions and estimates. It's those skills we are helping you develop. The web can help, but you should not spend a lot of time searching for answers to the particular questions we ask. That's particularly true since many of the problems don't have a single right answer.

- Be sure to make your estimates and assumptions explicit. Then you can change them easily if necessary, as we did in Section 1.5.

- In some of the chapters the interesting exercises are followed by a few routine exercises where you can practice just the arithmetic, independent of any real applications.

   Your instructor has a Solutions Manual with answers to the exercises, written the way we hope you will write them. Ask him or her to provide some for you.

Exercise 1.8.1. [S][Section 1.3][Goal 1.6] Warren Buffet is very rich.

In *The Boston Globe* on September 30, 2015, you could read that Warren Buffet is worth $62 billion, and that

> ...if [he] gave up on aggressive investing and put his money into a simple savings account, with returns at a bare 1 percent, he'd earn more in interest each hour than the average American earns in a year. [**R9**]

Use the data in this story to estimate the annual earnings of the average American. Do you think the estimate is reasonable?

Exercise 1.8.2. Dropping out.

Moved to Extra Exercises at www.ams.org/bookpages/text-63.

Exercise 1.8.3. [S][Section 1.3][Goal 1.1] [Goal 1.2][Goal 1.5][Goal 1.8] Bumper sticker politics.

In the fall of 2015 at donnellycolt.com you could buy a button that said

> Every Minute 30 Children Die of Hunger and Inadequate Health Care
> While the World Spends $1,700,000 on War

or a small vinyl sticker with the claim

> Every Minute the World Spends $700,000 on War While 30 children
> Die of Hunger & Inadequate Health Care. [**R10**]

What are these items trying to say? Do the numbers make sense?

Your answer should be a few paragraphs combining information you find on the web (cite your sources — how do you know they are reliable?) and a little arithmetic.

Exercise 1.8.4. [S][Section 1.3][Goal 1.1] [Goal 1.2][Goal 1.4][Goal 1.8] Two million matzoh balls.

In 2012 a modest restaurant in Newton Centre, MA, advertised

> *Johnny's Luncheonette*
> *Over 2 Million*
> *Matzoh Balls*
> *served!*

A year or so later students thinking about whether it was reasonable found that Johnny's website at www.johnnysluncheonette.com/ claimed "Over 1 Million Matzoh Balls Served!"

What would you believe?

[See the back of the book for a hint.]

Exercise 1.8.5. [S][Section 1.3][Goal 1.2][Goal 1.5] *Writing Your Dissertation in Fifteen Minutes a Day.*

Joan Bolker's book with that title sold about 120,000 copies in the first fifteen years since its publication in 1998.

Estimate the fraction of doctoral students who bought this book.

Exercise 1.8.6. [S][W][Section 1.3][Goal 1.1] [Goal 1.2][Goal 1.4][Goal 1.5] Smartphone apps may help retail scanning catch on.

Some grocery stores are experimenting with a new technology that allows customers to scan items as they shop. Once the customer is done, he or she completes the
transaction online and never has to stand in the checkout line. On March 11, 2012, *The
Boston Globe* reported that

> Modiv Media's scan-it-yourself technology [is installed] in about 350
> Stop & Shop and Giant stores in the United States. Many consumers
> have embraced the system; Stop & Shop spokeswoman Suzi Robinson
> said the service handles about one million transactions per month.
> [**R11**]

(a) Estimate the number of customers per day per store who use this self-scanning
technology.

(b) Estimate the number of customers per day per store.

(c) Estimate the percentage of customers who use the technology.

Exercise 1.8.7. Health care costs for the uninsured.
Moved to Extra Exercises at www.ams.org/bookpages/text-63 .

Exercise 1.8.8. [S][Section 1.2][Goal 1.1][Goal 1.3] Is 25 the same as 30?
In Section 1.2 we showed that a 25 year old has lived for about a billion seconds.
Then we estimated that a billion seconds is about 30 years.

(a) Explain why we got two different answers — 25 and 30 years.

(b) Are the answers really different?

(c) Compare them to what the Google calculator says about a billion seconds.

Exercise 1.8.9. [S][Section 1.2][Goal 1.1] [Goal 1.2][Goal 1.4][Goal 1.5] Spoons around
the world.
In January 2011 *Whole Foods Magazine* reported that

> According to the Clean Air Council, enough paper and plastic utensils
> are thrown away every year to circle the equator 300 times. [**R12**]

(a) Estimate the number of utensils it would take to circle the Earth 300 times.

(b) Is the assertion in the quote reasonable?

(c) This 2011 assertion is viral on the internet. Someone writes it once, then it's copied
from website to website. What are the earliest and latest versions you can find?

Exercise 1.8.10. [S][Section 1.2][Goal 1.1] [Goal 1.2][Goal 1.3]
1,000,000,000,000,000,000,000,000,000.
On July 25, 2010, Christopher Shea wrote in *The Boston Globe* about "hella", a new
metric prefix popular among geeks in northern California.

Austin Sendek, a physics major at the University of California Davis, wants to take "hella" from the streets and into the lab. With the help of a Facebook-driven public relations campaign, he's petitioning the Consultative Committee on Units, a division of the very serious Bureau International des Poids et Mesures, to anoint "hella" as the official term for a previously unnamed, rather large number: 10 to the 27th power. (The diameter of the universe, by Sendek's reckoning, is 1.4 hellameters.) [**R13**]

(a) Does the Google calculator know about hella?

(b) Does the Bing search engine know about hella?

Exercise 1.8.11. [S][Section 1.2][Goal 1.2] [Goal 1.4][Goal 1.5] Counting fish.

In *To the Top of the Continent* Frederick Cook wrote about this incident in his Alaska travels.

The run of the hulligans was very exciting ... Mr. Porter's thoughts ran to mathematics, he figured that the train of hulligans was twelve inches wide and six inches deep and that it probably extended a hundred miles. Estimating the number of fish in a cubic foot at ninety-one and one half, he went on to so many millions that he gave it up, suggesting that we try and catch some. [**R14**]

(a) How many millions of hulligans did Mr. Porter try to count?

(b) What's wrong with the precision in this paragraph?

(c) What's a hulligan? Cook's book was published in 1908. Are there any hulligans around today?

Exercise 1.8.12. [S][Section 1.2][Goal 1.1][Goal 1.8] Millions jam street-level crime map website.

In early 2011, the British government introduced a crime-mapping website that allows people to see crimes reported by entering a street name. The launch of the site, however, was problematic. The BBC reported that the website was jammed by up to five million hits per hour, about 75 thousand a minute. [**R15**]

(a) Are the figures "five million an hour" and "75,000 a minute" consistent?

(b) Estimate the fraction of the population of London trying to look at that website. Does your answer make sense?

[See the back of the book for a hint.]

Exercise 1.8.13. The popularity of social networks.

Moved to Extra Exercises at www.ams.org/bookpages/text-63 .

Exercise 1.8.14. [S][Section 1.2][Goal 1.1][Goal 1.2][Goal 1.5][Goal 1.8] How rich is rich?

On September 19th, 2011, Aaron S. from Florida posted a comment at *The New York Times* in which he claims that if you left one of the 400 wealthy people with just

a billion dollars he could not spend his fortune in 30 years at a rate of $100,000 a day.
[**R16**]

Is Aaron's arithmetic right? Can you do this calculation without a calculator? Without pencil and paper?

Exercise 1.8.15. Greek debt.

Moved to Extra Exercises at `www.ams.org/bookpages/text-63`.

Exercise 1.8.16. [S][Section 1.3][Goal 1.1] [Goal 1.4][Goal 1.6] No Lunch Left Behind.

From *The New York Times*, Feb 20, 2009, in a column by Alice Waters and Katrina Heron with that headline:

> How much would it cost to feed 30 million American schoolchildren
> a wholesome meal? It could be done for about $5 per child, or roughly
> $27 billion a year, plus a one-time investment in real kitchens. [**R17**]

There are three numbers in the paragraph. Are they reasonable? Are they consistent with each other and with other numbers you know?

Exercise 1.8.17. [S][Section 1.3][Goal 1.2] [Goal 1.4][Goal 1.5] Brush your teeth twice a day — but turn off the water.

The Environmental Protection Agency says on its website that

> You can save up to 8 gallons of water by turning off the faucet when
> you brush your teeth in the morning and before bedtime. [**R18**]

(a) Estimate how much water a family of four would use each week, assuming they left the water running while brushing.

(b) Estimate how much water would be saved in one day if the entire United States turned off the faucet while brushing.

(c) Put your answer to the previous question in context (compare it to the volume of water in a lake or a swimming pool, for example).

(d) Realistically, you need some water to brush your teeth because you need to get the toothbrush wet and you need to rinse the brush and your teeth. Estimate how much water that involves, per brushing, then redo the estimate in part (b).

Exercise 1.8.18. [W][S][Section 1.3][Goal 1.1][Goal 1.2][Goal 1.5] Lady Liberty.

On May 9, 2009, *The Boston Globe* reported that the Statue of Liberty's crown will reopen and that

> 50,000 people, 10 at a time, will get to visit the 265-foot-high crown.
> [**R19**]

Estimate how long each visitor will have in the crown to enjoy the view.

Exercise 1.8.19. [S][Section 1.3][Goal 1.2] [Goal 1.5] Look ma! No zipper!

The bag of Lundberg Zipper Free California White Basmati Rice advertises

> We've removed the re-closable zipper from our two pound bags, which
> will save about 15% of the material used to make the bag, which will
> save 35,000 lbs. of plastic from landfills every year.

(a) How much plastic is still ending up in landfills?

(b) What fact would you need to figure out how many bags of rice Lundberg sells each year? Estimate that number, and then estimate the answer.

Exercise 1.8.20. [S][Section 1.3][Goal 1.4][Goal 1.5][Goal 1.8] Leisure in Peru.

On page 32 in *The New Yorker* on December 7, 2009, Lauren Collins wrote that late arrivals in Peru are said to amount to three billion hours each year. [**R20**]

We suspect that the source of Collins's assertion is an article in the July 1, 2007, edition of *Psychology Today* that commented on the campaign for punctuality and included a feature called "Tardiness by the Numbers" that provided the data:

- 107 hours: annual tardiness per Peruvian
- $5 billion: cost to the country
- 84%: Peruvians who think their compatriots are punctual only "sometimes" or "never"
- 15%: think tardiness is a local custom that doesn't need fixing [**R21**]

(a) According to Collins, how late are Peruvians, in hours per person per day?

(b) Is your answer to the previous question consistent with the numbers in the *Psychology Today* article?

(c) Is the $5 billion "cost to the country" a reasonable estimate?

[See the back of the book for a hint.]

Exercise 1.8.21. [R][S][Section 1.3][Goal 1.1] The white cliffs of Dover.

In his essay "Season on the Chalk" in the March 12, 2007, issue of *The New Yorker* John McPhee wrote:

> The chalk accumulated at the rate of about one millimetre in a century, and the thickness got past three hundred metres in some thirty-five million years. [**R22**]

Check McPhee's arithmetic.

Exercise 1.8.22. [S][C][Section 1.3][Goal 1.1][Goal 1.4] [Goal 1.5] Social media and internet statistics.

In January 2009 Adam Singer blogged:

> I thought it might be fun to take a step back and look at some interesting/amazing social media, Web 2.0, crowdsourcing and internet statistics. I tried to find stats that are the most up-to-date as possible at the time of publishing this post. [**R23**]

(a) Read that blog entry, choose a few numbers you find interesting, and make sense of them. Are they reasonable? Are they consistent?

(b) Estimate (or research) what those numbers might be now (when you are answering this question.)

(c) We saw this information on the blog: in March 2008, there were 70 million videos on YouTube. It would take 412.3 years to view all of that YouTube content. Thirteen hours of video are uploaded to YouTube every minute. Can you make sense of these numbers? Are they reasonable? Are they consistent?

(d) Can you locate the source of the statistics above, or other sources that confirm them?

Exercise 1.8.23. [S][Section 1.3][Goal 1.1] [Goal 1.3][Goal 1.8] Bottle deposits.

A headline in *The Boston Globe* on July 15, 2010, read: "State panel OKs expansion of nickel deposit to bottled water." At the time, Massachusetts required a 5 cent bottle deposit for all bottles containing carbonated liquids. There was a debate about extending the deposit law to other liquids, including bottled water. In the article you could read that

> The Patrick administration, which supports the bottle bill, has estimated the state would raise about $58 million by allowing the redemption of an additional 1.5 billion containers a year, or about $20 million more than the state earns from the current law, and that municipalities would save as much as $7 million in disposal costs. [**R24**]

(a) Is it reasonable to estimate that 1.5 billion water bottles would be recycled in a year if users paid a nickel deposit on each?

(b) Is $7 million a reasonable estimate of the cost of disposing of 1.5 billion bottles (probably in a landfill) rather than recycling them?

(c) Use the data to estimate the number of water bottles potentially redeemed relative to the number of bottles and cans currently being redeemed. Does the result of the comparison seem reasonable?

(d) The article says the administration estimates that the state will collect $58 million by keeping the deposits paid by the people who don't return the bottles. Use that information to estimate the percentage of bottles that they expect will be recycled.

Note: This bill was defeated in the state legislature.

Exercise 1.8.24. Drivers curb habits as cost of gas soars.
Moved to Extra Exercises at www.ams.org/bookpages/text-63 .

Exercise 1.8.25. [S][Section 1.3][Goal 1.1] [Goal 1.2][Goal 1.4][Goal 1.5] The Homemade Cafe.

Figure 1.3 appeared on the back of Berkeley California's Homemade Cafe tenth anniversary tee shirt in 1989.

(a) Check that the numbers there make sense.

(b) Assume that the Homemade Cafe is still in business when you are working on this exercise. What numbers would go on this year's tee shirt?

Exercise 1.8.26. [S][Section 1.3][Goal 1.1] [Goal 1.2][Goal 1.3][Goal 1.5] So Many Books, So Little Time.

On Sunday, March 4, 2012, Anthony Doerr calculated in *The Boston Globe* that reading one book a week for 70 years would get you to 3,640 books. Then he wrote:

> If you consider that the Harvard University Library system's collection is counted in the tens of millions, or that a new book of fiction is published every 30 minutes, 3,640 doesn't seem like so many. [**R26**]

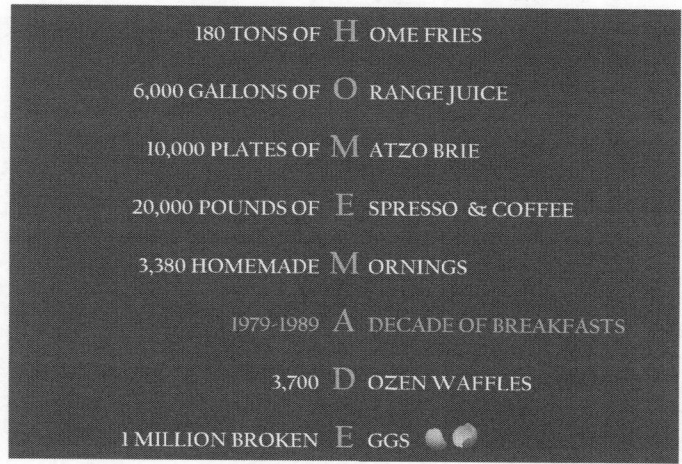

180 TONS OF H OME FRIES

6,000 GALLONS OF O RANGE JUICE

10,000 PLATES OF M ATZO BRIE

20,000 POUNDS OF E SPRESSO & COFFEE

3,380 HOMEMADE M ORNINGS

1979-1989 A DECADE OF BREAKFASTS

3,700 D OZEN WAFFLES

1 MILLION BROKEN E GGS

Figure 1.3. The Homemade Cafe [**R25**]

(a) Confirm that 70 years of reading one book per week would amount to 3,640 books read in a lifetime.

(b) How many people reading one book per week during their lifetime would it take to read all the books in the Harvard University Library system?

(c) If you read one book of fiction each week this year, what percent of all the fiction published this year will you have read?

Exercise 1.8.27. Counting car crashes.
Moved to Extra Exercises at www.ams.org/bookpages/text-63.

Exercise 1.8.28. [R][S][Goal 1.3][Section 1.4] Do parentheses matter?
What does the Google calculator tell you if you accidentally leave out the parentheses in the computation $12/(2 * 3)$?

Exercise 1.8.29. [U][R][S][Section 1.4][Goal 1.4] Should the U.S. Really Try to Host Another World Cup?

> The proposed budget for the 2010 [Soccer World Cup] games was about $225 million for stadiums and $421 million overall. Expenses have far exceeded those numbers. Reported stadium expenses jumped from the planned level of $225 million to $2.13 billion, and overall expenses jumped similarly from $421 million to over $5 billion. [**R27**]

How many orders of magnitude off were these estimates? That is, how many places were the decimal points away from where they should have been?

Exercise 1.8.30. [U][Section 1.5][Goal 1.5][Goal 1.1] Check our arithmetic, please.
When we found out in Section 1.5 that Baba Brinkman needed just ten seconds to plant a tree, we decided that his claim was possible.

(a) Verify the statement in the section that "... our 100 year estimate should really have been only about two years!"

(b) Verify the next statement: "If it took him a minute per tree rather than 10 seconds he could still have planted a million trees in ten summers."

Exercise 1.8.31. [S][Section 1.6][Goal 1.2][Goal 1.6] [Goal 1.8] The tooth fairy.

How many visits per day does the tooth fairy make in the United States? What's the daily transaction volume (in dollars) in the tooth fairy sector of the economy?

Before you turn to the internet for data, use what you know and some common sense to estimate answers. Write down your assumptions. Only then search the web if you wish.

Exercise 1.8.32. [S][Section 1.6] [Goal 1.2][Goal 1.4][Goal 1.8] Paying for college.

In 2006, the ABC program 20/20 told the story of a couple on Los Angeles who put their children through college by collecting and redeeming soda cans and bottles. Their oldest son went to MIT and their two other children attended California state schools. According to the article, the Garcias collected cans and bottles for 21 years with the goal of saving for their children's college tuition. [**R28**]

Is this possible?

[See the back of the book for a hint.]

Exercise 1.8.33. [S][Section 1.6][Goal 1.2] [Goal 1.4][Goal 1.8] Low flow toilets.

In 1994, a U.S. federal law went into effect that required all new residential toilets to be "low-flow", using just 1.6 gallons of water per flush instead of the five gallons per flush of older toilets.

Estimate how much water a household could save in one year by switching to low flow toilets.

Exercise 1.8.34. [S][W][Section 1.6][Goal 1.1] [Goal 1.5] Vet bills add up.

The inside back cover of the September/October 2008 issue of *BARk* magazine carried an ad for pet insurance asserting that every ten seconds a pet owner faced a $1,000 vet bill.

Is this claim reasonable?

Exercise 1.8.35. [U][Section 1.6][Goal 1.2] [Goal 1.6][Goal 1.8] Americans love animals.

In the November 29, 2009, issue of *The New Yorker* Elizabeth Kolbert wrote in a review of Jonathan Safran Foer's *Eating Animals* that there were 46 million dog-owning households, 38 million with cats and 13 million aquariums with more than 170 million fish.

> Collectively, these creatures cost Americans some forty billion dollars annually. (Seventeen billion goes to food and another twelve billion to veterinary bills.) [**R29**]

Is the twelve billion dollar figure she quotes for veterinary bills consistent with the numbers in the previous problem?

Exercise 1.8.36. [U][Section 1.6][Goal 1.2] [Goal 1.4][Goal 1.5][Goal 1.8] Total carbon footprint.

Use the estimates for the seven tasks discussed in Section 1.6 to rank those tasks in order of *total* daily carbon footprint.

Exercise 1.8.37. [S][Section 1.7][Goal 1.1][Goal 1.5][Goal 1.6] How may internet ads? An employee from Akamai claimed that

> There are 4.5×10^{12} internet advertisements annually. That's two thousand ads per person per year.

(a) Are the figures for the total number of ads and the number per person consistent?

(b) Do you think two thousand ads per person per year is a good estimate?

Exercise 1.8.38. [R][S][Section 1.7][Goal 1.7] [Goal 1.8] Metric ton.
A *metric ton*, also known as a *tonne*, is 1,000 kilograms.

(a) Is a metric ton a megagram or a gigagram?

(b) How many grams are there in a kilotonne?

(c) How many grams are there in a megatonne?

Exercise 1.8.39. e-reading.
Moved to Extra Exercises at `www.ams.org/bookpages/text-63` .

Exercise 1.8.40. Personal storage.
Moved to Extra Exercises at `www.ams.org/bookpages/text-63` .

Exercise 1.8.41. Backing up the Library of Congress.
Moved to Extra Exercises at `www.ams.org/bookpages/text-63` .

Exercise 1.8.42. [U][Section 1.7][Goal 1.6][Goal 1.7] giga-usa.
The website `www.giga-usa.com/` advertises itself as an

> Extensive collection of 100,000+ ancient and modern quotations, aphorisms, maxims, proverbs, sayings, truisms, mottoes, book excerpts, poems and the like browsable by 6,000+ authors or 3,500+ cross-referenced topics. [**R30**]

Is the website properly named?

Exercise 1.8.43. Data glut.
Moved to Extra Exercises at `www.ams.org/bookpages/text-63` .

Exercise 1.8.44. Zettabytes.
Moved to Extra Exercises at `www.ams.org/bookpages/text-63` .

Exercise 1.8.45. Zettabytes redux.
Moved to Extra Exercises at `www.ams.org/bookpages/text-63` .

Exercise 1.8.46. [W][S][Section 1.3][Goal 1.1] Waiting for the light to change.
In the Pooch Cafe comic strip on August 27, 2012, Poncho the dog is sitting in the car with his master Chazz. He says, "Did you know the average person spends six months of their life waiting at red lights?" (You can see the strip at `www.gocomics.com/poochcafe/2012/08/27`)
What do you think of Poncho's estimation skills?

Exercise 1.8.47. Killer cats.

Moved to Extra Exercises at www.ams.org/bookpages/text-63 .

Exercise 1.8.48. [S][Section 1.5][Goal 1.1][Goal 1.6] Viagra, anyone?

A 2014 Viagra ad on TV stated that more than 20 million men already use Viagra. Use the 2018 U.S. population pyramid in Figure 1.4 to argue whether or not this claim seems reasonable. Be explicit about any assumptions you make about the age groups of men who might typically use this drug.

Does using the 2018 population pyramid rather than one from 2014 affect your conclusion?

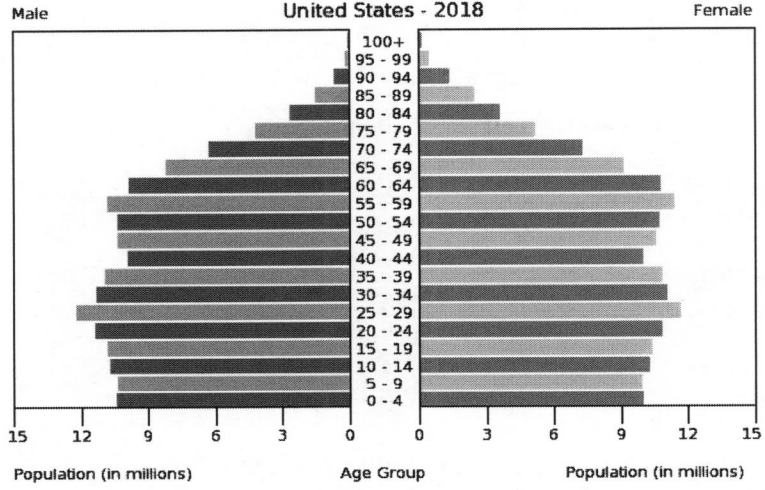

Figure 1.4. U.S. Population pyramid (2018) [**R31**]

Exercise 1.8.49. [U][Section 1.1][Goal 1.1][Goal 1.2] Lots of olives?

On August 23, 2014, the Associated Press reported that in the 1980's Robert Crandall, the CEO of American Airlines decided that offering one less olive in every salad would save $40,000 a year. [**R32**]

(a) About how many olives could Crandall buy for $40,000?

(b) Is your answer to the previous question in the same ballpark as the number of American Airlines passengers in 1980?

Exercise 1.8.50. [U][C][Section 1.3][Goal 1.1] [Goal 1.2][Goal 1.5][Goal 1.8] Nuclear bombs.

"Kiloton" and "megaton" are terms you commonly hear when nuclear bombs are being discussed. In that context the "ton" refers not to 2,000 pounds, but to the explosive yield of a ton of TNT.

(a) What was the explosive yield of the only (two) atomic bombs ever used in war?

(b) How does the explosive yield of the hydrogen bombs in the current arsenals of the United States and Russia (and other countries) compare to that of those first atomic bombs?

(c) Estimate the destructive power of the world's current stockpile of nuclear weapons, in terms easier to grasp than kilotons or megatons or gigatons or ....

Review exercises.

Exercise 1.8.51. [A] Do each of these calculations by counting zeroes. Use the Google or Bing calculator (or another calculator) to check your answers.

(a) One million times one billion.

(b) Four hundred times three thousand.

(c) Two billion divided by two hundred.

(d) One-tenth times five thousand.

(e) Four thousand divided by two hundred.

(f) Twelve thousand times two hundred.

(g) $10,000 \times \frac{2}{1,000}$.

(h) $450,000 \times 100$.

(i) $\frac{50,000}{200}$.

(j) $\frac{4,000,000,000,000}{2,000,000}$.

Exercise 1.8.52. [A] Use rounding to estimate each of these quantities. Then check your answers.

(a) The number of feet in ten miles.

(b) The number of minutes in a week.

(c) The number of inches in a mile.

(d) The number of yards in a mile.

(e) The number of seconds in a month.

Exercise 1.8.53. [A] Answer each of these questions without using pencil and paper (or a calculator).

(a) There are twelve cans of soda in a case. If I buy five cases of soda, how many cans will I have in total?

(b) A case of bottled water contains 24 bottles. About how many bottles are there in four cases?

(c) At the market my cart contains three bags of cereal at $2.19 each, a gallon of milk for $2.99, a pound of potatoes that costs $3.50 and a $3.99 bag of apples. Will my total be more than twenty dollars?

(d) At our local deli, a bagel costs $1.19. How much would five bagels cost? If the deli offers six bagels for $6.99, is that a better deal?

(e) It costs $2.50 to ride the New York City subway. A seven-day unlimited pass costs $29. How many rides in a seven day period should you take before it's a better deal to buy the pass?

(f) I have a $100 gift card for a department store. Can I buy a pair of shoes for $19.99, two shirts that are $12.59 each and three pairs of pants that each cost $20.50?

Exercise 1.8.54. [A] Convert the following measurements:

(a) One kilometer into meters.

(b) One megameter into kilometers.

(c) One terabyte into megabytes.

(d) Three megawatts into kilowatts.

(e) Five gigabytes into kilobytes.

(f) One thousand kilograms into megagrams.

Exercises added for the second edition.

Exercise 1.8.55. [S][Section 1.1][Goal 1.1][Goal 1.2][Goal 1.4][Goal 1.6] Chicken salad all around.

On May Day 2019 *Bloomberg* reported from the Milken Institute Global Conference that

> In Beverly Hills, the chicken Caesar salad costs $25.95. Stephen Schwarzman, with a net worth of $14.3 billion, could buy one for all 329 million people in the U.S. today — and then do so again tomorrow for 222 million of them. [**R33**]

(a) How would you verify this estimate using just simple arithmetic, without a calculator?

(b) Check the estimate with a calculator.

Exercise 1.8.56. [U][S][Section 1.2][Goal 1.1][Goal 1.6][Goal 1.8] Jeff Bezos is really rich.

In October 2018 the website QUARTZ*at*WORK posted a story about Amazon's plan to pay its 250,000 full time and 100,000 part time workers $15 an hour. Amazon claims that their lowest paid employee would make about $30,000 a year.

The article then notes that

> ...in the past 12 months, [founder Jeff Bezos's] net worth increased by $82.6 billion. ...Presuming his wealth creation continues at a similar pace, Bezos will "make" the annual salary of one of Amazon's newly minted $15/hour employees every 11.5 seconds. [**R34**]

(a) Check that at $15/hour "Amazon's lowest-paid full-time employee in the U.S. will make around $30,000 a year."

(b) Compare what Amazon distributes yearly in wages for those employees to the projected increase in Bezos's wealth.

(c) Check the claim that Bezos will collect $30,000 every 11.5 seconds is in the right ballpark.

Exercise 1.8.57. [U][S][Section 1.1][Goal 1.2] A million pitches.
In *The Boston Globe* on June 16, 2015, Stan Grossfeld wrote that

> In the 105-year history of Fenway, hitters have faced more than a million pitches. [**R35**]

Is that million pitches a reasonable estimate?
[See the back of the book for a hint.]

Exercise 1.8.58. [S][Section 1.3][Goal 1.1][Goal 1.2] [Goal 1.4] Dropping out.
On the website DoSomething.org you can read that

> Every year, over 1.2 million students drop out of high school in the United States alone. That's a student every 26 seconds — or 7,000 a day. [**R36**]

Are the three numbers in this quotation consistent? If not, can you explain the discrepancy?

Exercise 1.8.59. [S][Section 1.7][Goal 1.6][Goal 1.2][Goal 1.7] Gone phishin'
On January 13, 2019, *The Boston Globe* reported on a small company helping to counter internet phishing scams.

> [T]he data it produced on the [recent Netflix] attack were among 532,765,897 million fields of data it provided last month alone. [**R37**]

The "fields" referred to are the boxes you fill in on a form in your browser. Figure 1.5 shows a Netflix registration form with two fields, one for an email address, one for a password. Later in the process you would be asked for your name and your credit card information. We estimate that the average length of the fields is at least 10 characters, which you can think of as at least 10 bytes.

Figure 1.5. Netflix account creation form

(a) What is "phishing"?

(b) Write the number in this quotation as a number of bytes, using the appropriate metric prefix and an appropriate number of significant digits.

(c) Estimate the number of Netflix requests to open an account or order a movie it would take to generate that much data.

(d) Argue convincingly that the number is much too large to have been correctly reported.

Exercise 1.8.60. [S][Section 1.2][Goal 1.1][Goal 1.2][Goal 1.6] Counting galaxies.

Until recently astrophysicists estimated that the universe contained 100 to 200 billion galaxies. Then an international team proposed a minimum tenfold increase to 1 to 2 trillion. The AP report on the study concluded with this observation from Professor Conselice of Nottingham University, the lead researcher on the team, on how hard it is to imagine such large numbers:

> 2 trillion is equivalent to the number of seconds in 1,000 average lifetimes. [**R38**]

(a) Is the new estimate really a minimum tenfold increase?

(b) Is Conselice's count of seconds right?

Exercise 1.8.61. [U][S][Section 1.2][Goal 1.2][Goal 1.6] So many bitcoins, so many Benjamins.

In his *The New York Times* op-ed on January 29, 2018, Paul Krugman wrote:

> Like Bitcoins, $100 bills aren't much use for ordinary transactions: Most shops won't accept them. But "Benjamins" are popular with thieves, drug dealers and tax evaders. And while most of us can go years without seeing a $100 bill, there are a lot of those bills out there — more than a trillion dollars' worth, accounting for 78 percent of the value of U.S. currency in circulation. [**R39**]

(a) How many Benjamins are there in circulation?

(b) Use the data in this exercise to estimate the value of U.S. currency in circulation.

(c) Confirm that estimate with a web search.

(d) Why is the $100 bill a "Benjamin"?

Exercise 1.8.62. [S][Section 1.1][Goal 1.6] Don't answer the phone.

In April 2019, *The New York Times* reported that

> The seemingly endless stream of robocalls reached a new monthly high of 5.23 billion nationwide in March, according to the call-blocking service YouMail. [**R40**]

Write a short paragraph in which you convert the 5.23 billion robocalls into a number that you can grasp. Then discuss why you think it's too small, too large, or just right.

Exercise 1.8.63. [U][S][Section 1.1][Goal 1.1][Goal 1.2][Goal 1.6] One hundred billion bottles.

On October 29, 2019, the Portland Maine *PressHerald* featured an Associated Press report that said

> Every year, an estimated 100 billion plastic bottles are produced in the U.S. [**R41**]

(a) Show that this claim is about one bottle per person per day. You can do that with common knowledge and a little elementary arithmetic. No need for a calculator or the internet.

(b) Write a sentence or two about whether you think this claim is correct.

2

Units and Unit Conversions

In real life there are few naked numbers. Numbers usually measure something like cost, population, time, speed, distance, weight, energy or power. Often what's measured is a rate, like miles per hour, gallons per mile, miles per gallon, dollars per gallon, dollars per euro or centimeters per inch.

Chapter goals:

Goal 2.1. Explicitly manipulate units in expressions.

Goal 2.2. Explore metric prefixes.

Goal 2.3. Read and write scientific notation.

Goal 2.4. Understand how units of length determine units of area and volume.

Goal 2.5. Learn to use the Google calculator to make unit conversions.

Goal 2.6. Compare information by converting to similar units (dollars per ounce, for example, or births per one million people).

2.1 Rate times time equals distance

If you drive at 60 miles per hour for an hour and a half how far do you travel?

Since $60 \times 1.5 = 90$ the answer is 90 miles. You don't need the machinery we're inventing to get it right, but this easy problem is a good place to learn that machinery, which shows how to multiply the units along with the numbers:

$$60 \, \frac{\text{miles}}{\text{hour}} \times 1.5 \, \text{hours} = 90 \, \text{miles}.$$

When the same factor appears on the top and the bottom of a fraction you can cancel it, as in these calculations we made up to help you remember that fact.

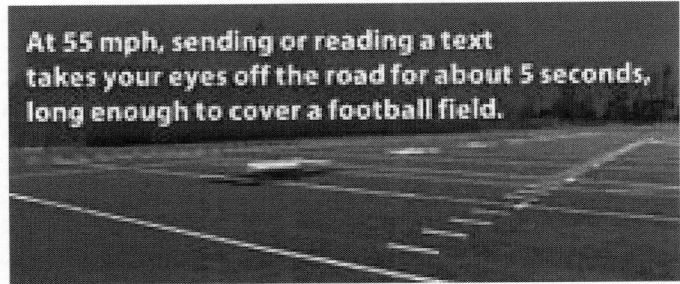

Figure 2.1. Texting while driving [**R42**]

With numbers:

$$3 \times 11 \times \frac{7}{5 \times 11} = 3 \times \cancel{11} \times \frac{7}{5 \times \cancel{11}} = 3 \times \frac{7}{5} = \frac{3 \times 7}{5},$$

particularly for zeroes:

$$\frac{1,000,000}{100} = \frac{1,000,0\cancel{0}0}{1\cancel{0}0} = 10,000$$

where cancelling a zero corresponds to cancelling a factor of 10, and in algebraic expressions:

$$3x \frac{y}{wx} = 3 \cancel{x} \frac{y}{w\cancel{x}} = 3\frac{y}{w} = \frac{3y}{w}.$$

When you write numbers with their units you can treat the words expressing the units just as you treat the numbers and letters — cancel them when they match. Revisiting our original example, we see that the hours cancel, leaving just miles as the units of the answer:

$$60 \frac{\text{miles}}{\cancel{\text{hour}}} \times 1.5 \, \cancel{\text{hours}} = 90 \text{ miles}, \qquad (2.1)$$

which illustrates an important habit to develop:

> Always write the units along with the numbers.

Let's see how this plays out in a real situation. It's common knowledge and common sense that you should not text while driving. Figure 2.1 from the Centers for Disease Control explains why. It says that driving at 55 miles per hour for 5 seconds you travel more than 100 yards. To check that, we want to calculate

$$\frac{55 \text{ miles}}{\text{hour}} \times 5 \text{ seconds},$$

but we can't simply cancel the units "hours" and "seconds". Since 1 hour = 3,600 seconds

$$\frac{1 \text{ hour}}{3,600 \text{ seconds}} = 1$$

so

$$\frac{55 \text{ miles}}{\text{hour}} \times 5 \text{ seconds} = \frac{55 \text{ miles}}{\cancel{\text{hour}}} \times 5 \, \cancel{\text{seconds}} \times \frac{1 \, \cancel{\text{hour}}}{3600 \, \cancel{\text{seconds}}} = 0.076 \text{ miles}.$$

How does that number compare to the length of a 100 yard football field? Since there are 1,760 yards in a mile,

$$0.076 \text{ miles} \times \frac{1,760 \text{ yards}}{\text{mile}} \approx 134 \text{ yards}$$

which is indeed more than a football field.

It's important to know how to do this kind of calculation, but also important to know how to ask someone to do it for you — in this case, Google (or Bing) for

$$\boxed{55 \text{ miles per hour} * 5 \text{ seconds in yards}}$$

and see

$$\boxed{134.444444 \text{ yards}} \, .$$

Think about that football field so you remember:

$$\boxed{\text{Don't text while driving.}}$$

2.2 The MPG illusion

What's a better gas saving choice — adding a second electric power source to a gas guzzling SUV to increase its fuel efficiency from 12 to 14 miles per gallon (MPG), or replacing an ordinary sedan that gets 28 miles per gallon by a hybrid that gets 40 miles per gallon?

Before you read on, jot down your first guess.

In June 2008 Richard Larrick and Jack Soll wrote in *Science* that

> Many people consider fuel efficiency when purchasing a car, hoping to reduce gas consumption and carbon emissions. However, an accurate understanding of fuel efficiency is critical to making an informed decision. We will show that there is a systematic misperception in judging fuel efficiency when it is expressed as miles per gallon (MPG), which is the measure used in the U.S.A. [R43]

In their article they calculated that to drive 10,000 miles you'd need 833 gallons at 12 miles per gallon but only 714 at 14 miles per gallon. Then they showed that saves more gas than replacing a car that gets 28 miles per gallon with an efficient hybrid getting 40 miles per gallon.

Let's check the calculations. To figure out how much gas is used driving 10,000 miles in a car that gets 12 miles per gallon we must do something with the numbers 10,000 and 12. Don't just guess whether to multiply or divide, write the equation with just the units first. We're starting with miles and want to end up with gallons, so we need to multiply by a fraction with units gallons/mile:

$$\text{miles} \times \frac{\text{gallons}}{\text{mile}} = \text{miles} \times \frac{\text{gallons}}{\text{mile}} = \text{gallons.}$$

That equation tells us that to fill in the numbers we need the fuel efficiency in gallons/mile rather than in miles/gallon. 12 miles/gallon means driving 12 miles uses one gallon of gas. That says one gallon takes us 12 miles, which is a rate of (1 gallon) per (12 miles). Now we know where to put the numbers:

$$10,000 \text{ miles} \times \frac{1 \text{ gallon}}{12 \text{ miles}} = \frac{10,000}{12} \text{gallons} = 833 \text{ gallons},$$

just as Larrick and Soll say.

Replacing 12 by 14 in this equation tells us that at 14 miles per gallon you will use about 714 gallons to drive 10,000 miles. So retrofitting the SUV saves $833 - 714 = 119 \approx$ 120 gallons.

To see the savings when you trade in the sedan for a hybrid, make the same calculation for 28 and 40 miles per gallon. That saves $357 - 250 = 107$ gallons — less than the 120 gallons saved by the SUV fix.

The paper in *Science* reported on the authors' survey showing that most people didn't know this. The reason is psychological: the change in MPG from 28 to 40 looks much more significant than the change from 12 to 14. But looks can be deceiving. We found the truth by using units carefully to think through the arithmetic.

Larrick and Soll go on to argue that the psychological problem can't be solved by asking people to do the work we just did. They conclude instead that

> These studies have demonstrated a systematic misunderstanding of MPG as a measure of fuel efficiency. Relying on linear reasoning about MPG leads people to undervalue small improvements on inefficient vehicles. We believe this general misunderstanding of MPG has implications for both public policy and research on environmental decision-making. From a policy perspective, these results imply that the United States should express fuel efficiency as a ratio of volume of consumption to a unit of distance. Although MPG is useful for estimating the range of a car's gas tank, gallons per mile (GPM) allows consumers to understand exactly how much gas they are using on a given car trip or in a given year and, with additional information, how much carbon they are releasing. GPM also makes cost savings from reduced gas consumption easier to calculate.

To convert MPG to GPM, we simply turn the fraction upside down:

$$\frac{1}{\frac{12 \text{ miles}}{\text{gallon}}} = \frac{1}{12} \frac{\text{gallons}}{\text{mile}} = 0.083 \frac{\text{gallons}}{\text{mile}}.$$

The old-fashioned fraction rule $\frac{1}{a/b} = b/a$ handles the units just fine.

If (like most people) you're not comfortable with small numbers like 0.083, write this as

$$8.3 \frac{\text{gallons}}{100 \text{ miles}}.$$

Larrick and Soll argue persuasively that this figure should appear on the fact sheets for new cars, as it does in Europe (in liters per 100 km). The EPA was convinced. Figure 2.2 shows the car window sticker design as of 2013. Item 5 gives the fuel efficiency in gallons per hundred miles. Too bad it's in smaller type than item 2, the more familiar miles per gallon.

Visit `www.epa.gov/fueleconomy/interactive-version-gasoline-vehicle-label` for an interactive graphic that explains what the other numbers mean.

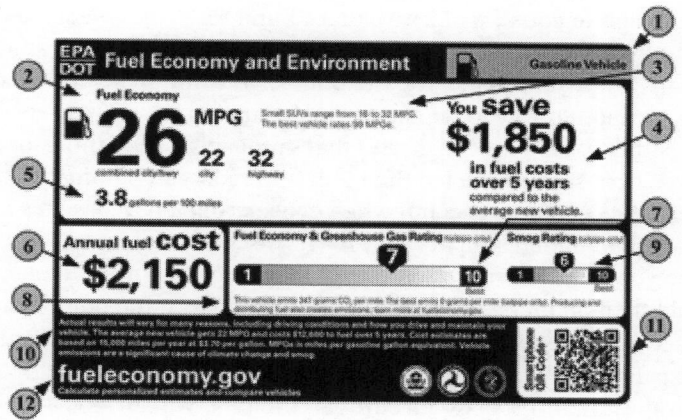

Figure 2.2. Gasoline vehicle label (2015) [**R44**]

2.3 Converting currency

When you visit a foreign country you will have to exchange your dollars for that country's currency. There are websites that will do the arithmetic for you, but you may not always have access to the internet. And you'll find it easier to travel if you can make quick estimates without consulting your smart cell phone (which may not work abroad). Here's how to use units to understand the conversions.

Suppose your trip is to France, where they use the euro. The value of the euro (in dollars) varies from day to day — in fact, from minute to minute. When we wrote this section the going rate was

$$€1 = \$1.38 .\tag{2.2}$$

That means you could buy 1 euro for 1.38 dollars. In fraction form, with units, the conversion rate is

$$\frac{1.38\ \$}{1\ €} = 1.38\ \frac{\$}{€} .\tag{2.3}$$

If you always write the conversion factor with its units you will never have to worry about whether to multiply or divide by 1.38.

Suppose you are tempted by a pair of shoes that costs €49. You want to know the dollar cost. To estimate the answer, think "49 is close to 50" and visualize the conversion factor with its units. That tells you that if you know the number of euros, you must multiply by (dollars per euro) to make the euros cancel, leaving you with dollars. Since the conversion factor is 1.38 \$/€ ≈ 1.4 \$/€, the cost is about 50 × 1.4 = 70 dollars.

If you're comfortable with percentage calculations you can read the conversion factor of 1.38 \$/€ as telling you that the cost in dollars will be about 40% larger than the cost in euros. Since 40% more than 50 is 70, the €49 pair of shoes will cost about \$70.

To make the conversion precisely, with units carefully displayed, first write

$$€ \times \frac{\$}{€} = \$.\tag{2.4}$$

Then fill in the numbers:

$$49\ € \times 1.38\frac{\$}{€} = 49 \times 1.38\ \$ = 68.05\ \$.\tag{2.5}$$

The equals sign in equation 2.2 says that €1 and \$1.38 are *the same*. Since dividing something by itself always gives the number 1, the expressions on both sides of equation 2.3 are just two ways to write the number 1. Multiplying by 1 doesn't change anything. That's another way to see the truth of equation 2.5

Our \$70 estimate was a little high. But that's probably a good thing for two reasons. First, we haven't considered the fee the bank (or credit card company) charges each time you exchange currency. Second, when considering a purchase it's always better to overestimate than underestimate what it will cost.

In France you may be driving as well as shopping. You need to plan for the cost of gas. It's sold there in liters, not gallons. The unit price at the pump when we wrote this section was $1.30\frac{€}{\text{liter}}$. Looks like a bargain — let's find out, by converting the units to dollars per gallon.

We know how to convert between euros and dollars. We can look up the conversion between gallons and liters: there are 0.264 gallons in 1 liter.

As usual, set up the computation first just with units:

$$\frac{€}{\text{liter}} \times \frac{\$}{€} \times \frac{\text{liters}}{\text{gallon}} = \frac{\$}{\text{gallon}}.$$

Then fill in the numbers and do the arithmetic:

$$1.30\,\frac{€}{\text{liter}} \times \frac{1.38\ \$}{€} \times \frac{1\ \text{liter}}{0.264\ \text{gallon}} = \frac{1.30 \times 1.38}{0.264}\,\frac{\$}{\text{gallon}}$$

$$= 6.79545455\,\frac{\$}{\text{gallon}}$$

$$\approx 6.80\,\frac{\$}{\text{gallon}}.$$

Not such a bargain after all. When we wrote this section gas in France cost more than twice what it cost here.

When a question has a surprising answer first check your work. We did. The answer is right. Gas was really twice as expensive in France. Why?

The first two answers that many people try are

- It's because dollars are worth less than euros.
- It's because liters are smaller than gallons.

Both of these statements are correct, but neither explains the high cost of gas in Europe. Those facts were *precisely what we took into account* when we converted $1.30\,\frac{€}{\text{liter}}$ to $6.80\,\frac{\$}{\text{gallon}}$. You need to search further for the real reason: gasoline taxes in Europe are much higher than they are in the United States. In Exercise 2.9.50 we ask you to research that question further.

Once you understand how the units work you can ask the Google calculator to do the work for you. For this calculation it will tell you something like

$$\boxed{\text{1.3 (Euros per liter)} = 6.60550571 \text{ U.S. dollars per US gallon}}\ .$$

The \$6.60 per gallon here does not match the \$6.80 per gallon in the text above. They differ because the Google calculator looks up the current exchange rate for the euro. We wrote this paragraph a month or so after we found the \$6.80. The dollar value of the euro changed in the interim.

Google will tell you the current exchange rate (rounded to the nearest penny). We searched for

$$\boxed{\text{1 euro in dollars}}$$

and were told

$$\boxed{\text{1 Euro equals 1.33 US Dollar}}$$

which is a little less than the \$1.38/€ we used above.

2.4 Unit pricing and crime rates

Grocery shopping is full of decisions. Suppose you're staring at two boxes of crackers — the brand you like. The 16 ounce box costs \$1.99; the 24 ounce box costs \$2.69. Which is the better deal?

To decide, you need to know the *unit price*. The units of the unit price will be $\frac{\$}{\text{ounce}}$.

A quick estimate helps: 24 ounces is 50% more than 16 ounces. The 16 ounce box costs \$2, so if the larger box costs less than \$3 then it's a better buy. It is in this case.

Writing out the calculation with the units, we see that for the small box the unit price is

$$\frac{1.99\ \$}{16\ \text{ounces}} = 0.124\ \frac{\$}{\text{ounce}}$$

while for the big box it's

$$\frac{2.69\ \$}{24\ \text{ounces}} = 0.112\ \frac{\$}{\text{ounce}}.$$

Our estimate gave us the right answer: the larger box is the better buy, ounce for ounce.

Some states require stores to list unit prices on the shelves. But they don't require consistency in the units used, so you may see unit prices in dollars per ounce next to those in dollars per pound or dollars per serving. You still need to think.

The supermarket isn't the only place where converting absolute quantities to rates (*something per something*) helps make sense of numbers. The FBI compiles and reports crime statistics for cities across the United States. The 2011 data recorded 3,354 violent crimes for Sacramento, CA, and 3,206 for San Jose, CA. [R45]

Which city had more crime?

They seem to be about equally dangerous. But for a true comparison we should take into account the sizes of the two cities. Sacramento's population was 471,972 while San Jose's was 957,062: San Jose is about twice the size of Sacramento. Since the number of violent crimes in each city was about the same, the crime *rate* in San Jose was about half the rate in Sacramento.

Formally, with units and numbers, for Sacramento

$$\frac{3,354\ \text{crimes}}{471,972\ \text{people}} = 0.00711\ \text{crimes per person.}$$

If we make the same calculation for San Jose we find a rate of 0.00335 crimes per person.

It's hard to think about small numbers like those with zeroes after the decimal point. For that reason it's better to report the crime rate with units crimes per 1,000 people rather than crimes per person. We do that by extending our calculation using units this way:

$$\frac{3,354 \text{ crimes}}{471,972 \text{ people}} \times \frac{1,000 \text{ people}}{\text{thousand people}} = 7.11 \text{ crimes per thousand people}.$$

The second fraction in the computation, $\frac{1,000 \text{ people}}{\text{thousand people}}$, is just another way to write the number 1. Multiplying by 1 doesn't change the value of the number, just its appearance. Multiplying by 1,000 in the numerator moves the decimal point three places, turning the clumsy 0.00711 into the easy to read 7.11.

The crime rate for San Jose is just 3.35 crimes per thousand people. Now we have easy to read numbers to compare. San Jose's crime rate is just under half of Sacramento's.

2.5 The metric system

Most of the world uses the *metric system*. Although it's not common in the United States, sometimes it pops up in unexpected places.

> *NASA's metric confusion caused Mars orbiter loss*
> September 30, 1999
>   (CNN) — NASA lost a $125 million Mars orbiter because one engineering team used metric units while another used English units for a key spacecraft operation, according to a review finding released Thursday. [**R46**]

On a less serious note, when we used the Google calculator to check our solution to the problem in Section 2.1 the answer was weird. Asking for

$$\boxed{1.5 \text{ hours} * 60 \text{ miles per hour}}$$

we were surprised to see

$$\boxed{1.5 \text{ hours times } (60 \text{ miles per hour}) = 144.84096 \text{ kilometers}} \;.$$

We've no idea why Google decided to use kilometers instead of miles. We did make it tell us what we really wanted to know this way:

$$\boxed{1.5 \text{ hours} * 60 \text{ miles per hour in miles}}$$

leads to

$$\boxed{(1.5 \text{ hours}) * 60 \text{ miles per hour} = 90 \text{ miles}} \;.$$

We'll spend the rest of this section learning just a little about some of the metric units and how to convert among them and between them and English units.

To carry on everyday quantitative life we use small, medium and large units so that we don't have to deal with numbers that are very small or very large. For example, we measure short lengths (the size of your waist) in inches, medium lengths (the size of a room) in feet or yards and large lengths (the distance you commute) in miles. In the metric system the small, medium and large units for length are the centimeter, the meter and the kilometer.

In the metric system the units of different sizes are always related by a conversion factor that's a power of 10. Since it's easy to multiply or divide by a power of 10 by moving the decimal point, it's easy to move among the units. In contrast, converting units in the English system means knowing (or looking up) awkward conversion factors like 12 inches/foot or 5,280 feet/mile, and then actually multiplying or dividing by them.

A *centimeter* is about the width of your fingernail. There are about two and a half centimeters in an inch. Your index finger is about 10 centimeters (four inches) long. The metric system has nothing like the foot for intermediate lengths. It relies on the *meter*: like a yard, but 10% longer. For longer distances the metric equivalent of the mile is the *kilometer* — about 60% of a mile.

If you lived in a metric world you would think in metric terms rather than in our English system (even England has mostly abandoned the English system).

You would understand "meter" directly, not as "about a yard." You'd never need to convert meters to yards or kilometers to miles any more than you'd need to translate French to English if you lived in France and spoke the language like a native.

But in our world you may occasionally need to convert, with the following factors. Remember the approximations; look up the exact values up when you need them.

$$2.54 \ \frac{\text{centimeters}}{\text{inch}} \approx 2.5 \ \frac{\text{centimeters}}{\text{inch}},$$

$$0.393700787 \ \frac{\text{inches}}{\text{centimeter}} \approx 0.4 \ \frac{\text{inches}}{\text{centimeter}},$$

so 10 centimeters is about four inches.

$$0.9144 \ \frac{\text{meters}}{\text{yard}} \approx 0.9 \ \frac{\text{meters}}{\text{yard}}.$$

$$1.0936133 \ \frac{\text{yards}}{\text{meter}} \approx 1.1 \ \frac{\text{yards}}{\text{meter}}.$$

$$1.609344 \ \frac{\text{kilometers}}{\text{mile}} \approx 1.6 \ \frac{\text{kilometers}}{\text{mile}}.$$

$$0.621371192 \ \frac{\text{miles}}{\text{kilometer}} \approx 0.6 \ \frac{\text{miles}}{\text{kilometer}}.$$

The last of these approximations leads to an easy answer to the question asked implicitly at the start of this chapter. 100 kilometers is about 60 miles, so 100 kilometers per hour is about 60 miles per hour.

The small, medium and large units for weight in the metric system are the gram, kilogram and metric ton.

A paperclip weighs about a *gram* — about a thirtieth of an ounce. So there are about 30 grams in an ounce.

A *kilogram* is just over two pounds — about 2.2, or 10% more. So a pound is just under half a kilogram (about 10%).

A *metric ton* (1,000 kilograms) is about 10% larger than a ton (2,000 pounds); a ton is about 10% smaller than a metric ton.

The small and medium metric units for volume are the cubic centimeter and the liter.

It takes about 30 cubic centimeters to fill an ounce, so the small volume unit in the metric system is a lot smaller than the corresponding unit in our English system. (One of many annoying feature of the English system is that there are two different kinds of ounces — one for weight and one for volume. You can't convert between them.)

The name "cubic centimeter" is a clue to a nice feature of the metric system. The different kinds of units are related in as sensible a way as possible. The unit used for small volumes was created from the unit for small lengths. The gram (the unit for small weights) is just the weight of a cubic centimeter of water.

The metric unit for volumes of medium size is the liter. It's just about 5% larger than a quart. A quart is about 95% of a liter. We encountered liters earlier in the chapter when thinking about buying gasoline in France.

2.6 Working on the railroad

On April 28, 2010, *The Boston Globe* reported that "Urgent fixes will disrupt rail lines — T to spend $91.5m to repair crumbling Old Colony ties":

> Starting in August, T officials say, they plan to tear up 150,000 concrete ties along 57 miles of track and to replace them with wooden ties. [**R47**]

There are three numbers in this quotation and the headline: 150,000 *ties*, 57 *miles* and 91.5 million *dollars*.

Using them two at a time we can figure out both how far apart the ties are and how much each tie costs.

Railroad ties are fairly close together, so we should measure the distance between them using feet, not miles: we want an answer with units feet/tie. Since we have miles of track, the units equation will be

$$\frac{\text{miles (of track)}}{\text{(number of) ties}} \times \frac{\text{feet}}{\text{mile}} = \frac{\text{feet}}{\text{tie}} \,.$$

Putting in the numbers and doing the arithmetic:

$$\frac{57 \text{ miles}}{150,000 \text{ ties}} \times \frac{5,280 \text{ feet}}{\text{mile}} = \frac{57 \times 5,280}{150,000} \frac{\text{feet}}{\text{tie}}$$

$$= 2.0064 \frac{\text{feet}}{\text{tie}}$$

$$\approx 2 \frac{\text{feet}}{\text{tie}}$$

so the ties are about two feet apart.

Here's a way to trick the Google calculator into doing the unit calculations as well as the arithmetic. It doesn't know about railroad ties, but we can pretend ties are gallons. Ask it for

> 57 miles per 150000 gallons in feet per gallon

and it tells us

> (57 miles) per (150 000 US gallons) = 2.0064 feet per US gallon .

That answer is suspiciously close to the round number of 2 feet. That suggests that the T estimated the number of ties they'd need by assuming the two foot separation. They knew that they were laying 57 miles of new track. They converted 57 miles to $57 \times 5,280 = 300,960$ feet. With ties two feet apart they need half that many ties, which rounds nicely to 150,000.

To compute the cost in dollars per tie, the units tell you to find

$$\frac{91.5 \text{ million \$}}{150,000 \text{ ties}} \, .$$

If you ask Google for

$$\boxed{91.5 \text{ million} / 150000}$$

you find out that

$$\boxed{91.5 \text{ million} / 150\,000 = \text{six hundred ten}}$$

so the ties cost about \$600 each. That includes material and labor.

We were surprised and interested to see Google write 610 as "six hundred ten". (That's the right way to say "610" aloud. "Six hundred and ten" is common, but wrong.) We realized that was because our search used the word "million". Had we asked instead for

$$\boxed{91.5 * 1,000,000 / 150000}$$

or

$$\boxed{91,500,000 / 150000}$$

we'd have been told

$$\boxed{91\,500\,000 / 150\,000 = 610} \, .$$

2.7 Scientific notation, milli and micro

In the exercises in Chapter 1 we found out that 70 years was about 2.5 billion seconds. Since 70 is about two and a half times 30, 30 years is about a billion seconds. If we search Google for

$$\boxed{30 \text{ years in seconds}}$$

we're told

$$\boxed{30 \text{ years} = 946\,707\,779 \text{ seconds}}$$

so our estimate of one billion is quite good.

If we ask for the number of seconds in 60 years we expect an answer just twice as big:

$$\boxed{60 \text{ years} = 1\,893\,415\,558 \text{ seconds}}$$

but see instead

$$\boxed{60 \text{ years} = 1.89341556 \times 10^9 \text{ seconds}} \, .$$

What is going on?

Google chose to express the answer in scientific notation. It's telling you to multiply the number $1.89\ldots56$ by 10, nine times. Multiplying by 10 moves the decimal point to the right. If we follow those instructions,

$$1.89341556 \times 10^9 = 1,893,415,600$$

which is quite close to the 1,893,415,558 we expected to see.

Figure 2.3. Stressing the calculator [**R48**]

When you read "1.893 ... × 10⁹" aloud, you say

"one point eight nine three ... times ten to the ninth" .

A common error (which we hope you will never make) is to say instead "one point eight nine three ... to the ninth". To help you remember that the exponent 9 is a power of 10 you have to *say* the 10.

The Google calculator can read powers of 10 as well as write them. To enter an exponent, use the caret (^) on your keyboard. It's the uppercase 6. The 91.5 million we worked with in the previous section is 91.5×10^6. Ask Google for

$$91.5 * 10 \char94 6$$

and you will see

$$91.5 * (10 \char94 6) = 91\,500\,000$$.

Google's isn't the only calculator that switches to scientific notation for really large numbers. Figure 2.3 shows what happens at www.online-calculator.com/ when you ask for

$$123,456,789 * 1,000,000,000,000,000.$$

Why? 24 digits won't fit on its screen, so it can't show

$$123,456,789,000,000,000,000,000$$

(even without the commas). Moreover, it hasn't the height to display exponents, so it can't show

$$1.23456789 \times 10^{23}.$$

Instead it uses the symbols "e+23" to tell you the exponent that it can't write is (positive) 23. If you do need to say this aloud, you put the "ten" with its power back: "1 point 23 ...89 times 10 to the 23rd", *not* "1.12...89 to the 23rd".

You don't often encounter scientific notation in daily life. But we want to show you what it looks like so that you can read it if you stumble on it.

A number written in scientific notation always has just one nonzero digit before the decimal point, as many after the decimal point as appropriate for the precision desired, and then instructions about how many places to move the decimal point right or left. The power of 10 tells you the number of places.

Name	Symbol	English name	Power of 10
deci	d	one tenth	−1
centi	c	one hundredth	−2
milli	m	one thousandth	−3
micro	μ	one millionth	−6
nano	n	one billionth	−9
pico	p	one trillionth	−12
femto	f	one quadrillionth	−15

Table 2.4. Metric prefixes for shrinking

You can make numbers bigger by adding zeroes but you can't always make them smaller by removing zeroes. Scientific notation provides a way. For example,

$$2.5 \times 10^{-4} = 0.00025;$$

the decimal point has been moved four places to the *left* because

$$10^{-4} = \frac{1}{10^4}.$$

It's no surprise that 2.5×10^{-4} is a very small number.

We've seen the metric prefixes kilo, mega, giga, ...that make units larger. There are also prefixes for fractions of a unit. The most common is *milli*, which means "divide by 1,000" or "multiply by 10^{-3}" or "move the decimal point three spaces to the left". The symbol is "m", in lower case. We've already seen it at work: a milliliter is one one-thousandth of a liter. The thickness of lead in a pencil is measured in millimeters.

In Greek mythology, Helen's beauty led to the Trojan War — hers was "the face that launch'd a thousand ships" [R49] so a milliHelen is the unit of beauty sufficient to launch one ship.

For microscopic things you need the next prefix, micro. It tells you to multiply by 10^{-6} — one one-millionth. The symbol is the Greek letter *mu*: "μ". There are few calls for micro in everyday use.

Although most metric prefixes move the decimal point three places, one common prefix moves it just two: *centi* means "one one-hundredth of" or "divide by 100." So there are 100 centimeters in a meter and 10 millimeters in a centimeter.

Table 2.4 shows the metric prefixes for shrinking things. These days "nano" sometimes means just "very small" rather than "one billionth", as in "nanotechnology" or even "ipod nano".

2.8 Carpeting and paint

How much carpeting does it take to cover the floor of a 12 foot by 15 foot room?

To find the floor area of the room you multiply the length by the width. Multiply the units along with the numbers:

$$12 \text{ feet} \times 15 \text{ feet} = 180 \text{ feet} \times \text{feet} = 180 \text{ feet}^2$$

which we read as "180 square feet".

Carpeting is sold by the square yard. How many square yards will you need? How do you convert from square feet to square yards? We know

$$1 \text{ yard} = 3 \text{ feet}$$

so

$$(1 \text{ yard})^2 = (3 \text{ feet})^2 = 9 \text{ feet}^2.$$

In other words, the fraction

$$\frac{1 \text{ yard}^2}{9 \text{ feet}^2}$$

is just another way to write the number 1, so multiplying by it doesn't change the value, just the way the answer looks:

$$180 \text{ feet}^2 \times \frac{1 \text{ yard}^2}{9 \text{ feet}^2} = 20 \text{ yards}^2.$$

Once you see that simple answer you may realize that the problem would have been even easier if you'd worked in yards from the start: the room is 4 yards wide by 5 yards long, so it has an area of 20 square yards.

Before you lay the new carpeting you want to paint the room. How much paint will you need?

To continue computing, you must know the height of the ceiling — suppose it's nine feet. To find the area to cover, think four walls: two are 12 feet long and 9 feet high, two are 15 feet long and 9 feet high, so the total area to be covered is

$$(2 \times 12 \text{ feet} + 2 \times 15 \text{ feet}) \times 9 \text{ feet} = 486 \text{ feet}^2.$$

A gallon of paint covers 400 square feet, so you will need about a gallon and a quarter. That's convenient, because you can buy a gallon and a quart.

If you want to paint the ceiling too you'll need more paint. The area of the ceiling is the same as the area of the floor to be carpeted: 180 square feet. So buy two quarts of ceiling paint. (You can't simply get two gallons of the wall paint and use two quarts of that for the ceiling since the ceiling requires a different kind of paint even if it's the same color.)

Finally, if you plan to paint a real room the real computation is more complicated. Some colors (particularly yellow) require two coats, or a primer. The doors and windows represent wall area you don't have to cover, but you do need to paint the woodwork that surrounds them — that's yet another kind of paint. And you can buy a gallon of paint for the same price as three quarts — be sure to get enough. When you're done, carefully label and close the partially full paint cans. Be sure to paint the ceiling before you paint the walls, and do all the painting before you install the carpet.

We've used this discussion of paint and carpet to show how units for length determine square units for area. Next we look at cubic units for volume.

You happen to have a nice wooden box one foot on each side. You want to fill it for your grandchildren with those nice kids' alphabet blocks each one inch on a side. How many blocks will you need? A lot more than you might think.

A row of 12 blocks will be a foot long. 12 of those rows, or 144 blocks, will fill one layer on the bottom of the box. Then you need 12 layers of 144, or 1,728 blocks in all, to fill the box.

The size of that answer — nearly 2,000 blocks — surprises many people. Now that you know it you will remember that volumes grow quickly. The volume of the box is one cubic foot: 1 foot^3. The volume of each block is one cubic inch. Then

$$1 \text{ foot}^3 = (12 \text{ inches})^3 = 12^3 \text{ inches}^3 = 1,728 \text{ inches}^3 .$$

2.9 Exercises

Exercise 2.9.1. [Section 2.1][Goal 2.1] Units in the news.

Find a current news or magazine article that uses unit conversions either explicitly or implicitly. Verify the calculation. Include a copy of the article (or a link to it on the web) when you turn in your work.

[See the back of the book for a hint.]

Exercise 2.9.2. [S][Section 2.1][Goal 2.1] Drive carefully!

Suppose you drive from Here to There at a speed of 30 miles/hour and then drive back from There to Here at 60 miles/hour.

What is your average speed?

[See the back of the book for a hint.]

Exercise 2.9.3. [R][S][Section 2.1][Goal 2.1] Airline ticket taxes.

On July 22, 2011, the Associated Press quoted Transportation Secretary Ray La-Hood saying that a partial shutdown of the Federal Aviation Administration (FAA) would cost the government about $200 million a week in airline ticket taxes. [**R50**]

(a) Calculate the total amount of lost revenue from the partial shutdown of the FAA that began on July 23, 2011, and ended on August 4, 2011.

(b) One of the issues at stake in the debate over the shutdown was the $16.5 million the FAA provided in federal subsidies to rural airports. House Republicans wanted to eliminate these subsidies. Estimate how many hours of collecting federal aviation taxes would be needed to support these subsidies.

Exercise 2.9.4. [S][Section 2.1][Goal 2.1] A race through the tree of life.

Jessica Bolker teaches biology at the University of New Hampshire. She writes in an email:

```
If I were going to cover all the animal groups in
proportion to their extant (living) species diversity
in a single 50-minute lecture, how long would I spend
on each?

Start animals at 10:10  (50 minutes to cover ~1,255,000
species; ~0.0024 seconds each)

Talk about arthropods (1,000,000) for 40 minutes,
until 10:50 am

mollusks (110,000) from 10:50 till about 10:54
```

```
all the other non-chordates, combined (94,000, but
perhaps really a lot more) for about 3.75 minutes,
until almost 10:58

chordates beginning at 10:58 (remember class ends at 11)

fish (31,000) for a minute and a quarter (75 seconds)

reptiles (7,000) for 17 seconds

amphibians (5,500) for 13 seconds

mammals (4,000) for almost 10 seconds

birds (3,000) for a bit over 7 seconds

- and we're done.
```

For the first few questions, assume that her data about species counts are correct.

(a) Is her estimate of about 0.0024 seconds per species correct?

(b) Do her species counts by group add up to approximately 1,255,000?

(c) Does the way she apportions lecture time match the apportionment of species?

Now think about the data.

(d) The website www.biologicaldiversity.org/species/birds/ says there are in fact about 10,000 bird species, not the 3,000 in Professor Bolker's letter. How would you rearrange the last two minutes of her class to take that new information into account?

(e) You probably know about reptiles, amphibians, mammals and birds. What are arthropods, mollusks, chordates? Don't just copy and paste a definition from the internet — provide one that shows that you understand what you've written. Did you learn anything doing this part of the problem?

(f) While you were answering the previous question you probably found websites for each of those groups. Do the number of species of each kind match Professor Bolker's assumptions? (Full sentences, and citations, please, not just "yes" or "no".)

(g) How is Professor Bolker related to one of the authors of this book?

Exercise 2.9.5. The penny stops here.
Moved to Extra Exercises at www.ams.org/bookpages/text-63 .

Exercise 2.9.6. [S][Section 2.1][Goal 2.1] Penny wise, pound foolish. [**R51**]
The graphs in Figure 2.5 display data from the U.S. Mint on the number of pennies produced each year and the cost to make them.

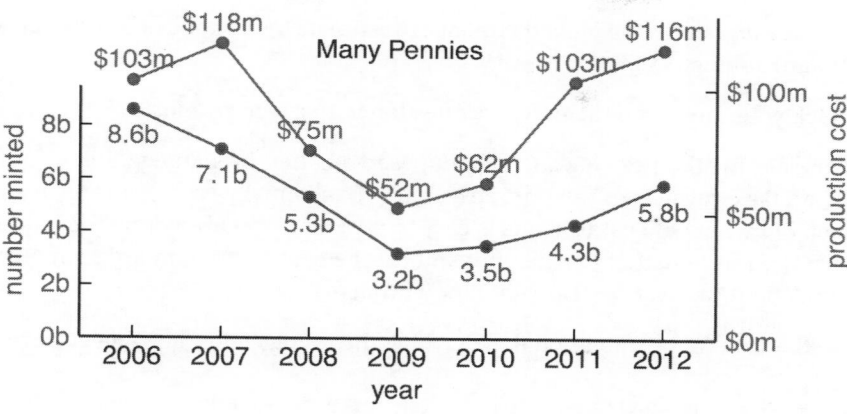

Figure 2.5. What's a penny worth these days? [R52]

(a) How much did it cost to manufacture one penny in 2012?

(b) How much money did the United States lose for each penny minted in 2012?

(c) How can you tell from the graphs without doing any arithmetic that it cost more to make a penny in 2007 than in 2006?

(d) In which of the years 2006–2012 was the total cost of producing pennies the largest?

(e) In which of the years was the cost per penny the largest?

(f) What does the phrase "penny wise, pound foolish" mean? What is its origin?

Exercise 2.9.7. [S][Section 2.1][Goal 2.1] A literal ballpark estimate.
The Boston Globe reported on June 21, 2012, that

> It was so hot at Fenway that Dave Mellor, the Red Sox groundskeeper, said he could probably fry enough eggs on the Green Monster to feed the entire city. Not possible, mathematically, but he could probably fry enough for everyone in the stadium. [R53]

Verify the writer's "Not possible, mathematically, but he could probably …". Write up your argument in the form of a letter to the editor.

There's enough information on the web to estimate the area of the Green Monster.

Exercise 2.9.8. [S][Section 2.1][Goal 2.1][Goal 2.5] Counting red blood cells.
On page 240 of the April 2011 issue of the *Bulletin of the American Mathematical Society* Misha Gromov wrote that

> Erythrocytes are continuously produced in the red bone marrow of large bones …at the rate ≈ 2.5 million/second or ≈ 200 billion/day. Adult humans have 20–30 trillion erythrocytes, ≈ 5 million/ cubic millimeter of blood. [R54]

(a) What is an erythrocyte?

(b) Check that 2.5 million/second is about 200 billion/day.

(c) Use the numbers in the second sentence to estimate the volume of blood in an adult human, in liters.

(d) Check your answer to the previous question with a web search.

(e) (Warmup for the next question.) Suppose your hen lays an egg a day, and you always have seven eggs in your refrigerator. Explain why each egg is on average a week old when you eat it. (This is easy to explain if you always eat the oldest egg first. It's a little harder to explain if you eat the eggs in random order and have to think about the average. Don't bother with that.)

(f) Use the data in the quotation to determine the average lifetime of an erythrocyte.

Exercise 2.9.9. Global warming opens Arctic for Tokyo-London undersea cable.
Moved to Extra Exercises at `www.ams.org/bookpages/text-63` .

Exercise 2.9.10. [S][Section 2.1][Goal 2.1][Goal 2.6] Maple syrup.
The March page of the 2008 Massachusetts "Celebrating the Seasons of Agriculture" calendar contains the quote

> Farm Fact #3
> In 2007, Massachusetts maple producers set 230,000 taps to collect sap, which, after boiling produced 30,000 gallons of maple syrup. It takes about 40 gallons of sap to produce one gallon of syrup.

And take a look at `www.gocomics.com/frazz/2015/09/25` .

(a) How many gallons of syrup did each tap yield?

(b) How many gallons of sap did each tap yield?

(c) What extra information would you need to estimate the number of sugar maple trees that were tapped?

(d) Suppose the sap is collected in gallon buckets that hang under each tap and that the maple syrup season lasts one month. How many times a day will the buckets need to be emptied?

(e) Visit a local supermarket and compare the price of 100% maple syrup to the price of some "pancake syrup." Read the label on the pancake syrup to find out how much maple syrup it contains. (If you can find this information on the web without visiting a supermarket, more power to you.)

What fraction of the cost of the generic syrup can you attribute to the actual maple syrup it contains?

Exercise 2.9.11. [W][Section 2.2][Goal 2.1] The MPG illusion.

(a) Take the quiz at `www.fuqua.duke.edu/news/mpg/mpg.html` . Then ask two friends to take the quiz, and watch over their shoulder as they do. If they are willing to think out loud for you, so much the better. If they get the wrong answer, explain the correct one.

Write a brief summary of how you did on the quiz and how your friends did.

(b) When you ask Google for

> 12 miles per gallon in gallons per 10,000 miles

you get more than the answer we used in the text — you will see many links. Follow some that you find particularly interesting — please not just the first two — and write about what you find.

Exercise 2.9.12. [R][S][Section 2.2][Goal 2.1][Goal 2.5] Fuel economy.

The government's 2011 *Fuel Economy Guide* (`www.fueleconomy.gov/feg/pdfs/guides/FEG2011.pdf`) lists the fuel economy for a Chevrolet Malibu as 22 MPG for city driving and 33 MPG for highway driving, with an annual fuel cost of $2,114. The *Guide* says that figure is computed this way:

> Combined city and highway MPG estimates ... assume you will drive 55% in the city and 45% on the highway. Annual fuel costs assume you travel 15,000 miles each year and fuel costs $3.66/gallon for regular unleaded gasoline and $3.90/gallon for premium.

(a) Convert these ratings to gallons per mile and gallons per hundred miles, using a calculator just for the arithmetic, not Google's for the units.

(b) Check your answers with the Google calculator.

(c) Verify the annual fuel cost figure. (The Malibu does not require premium gasoline.)

(d) Explain why the computation in the previous part of the exercise would have been easier if the fuel economy was reported in GPM instead of (or in addition to) MPG.

(e) What would the annual fuel cost be at current gas prices?

(f) (Optional) Look at the *Guide*, find something interesting there, and write briefly about it.

Exercise 2.9.13. [S][Section 2.2][Goal 2.1] Sticker shock.

An editorial in *The New York Times* on Sunday, June 4, 2011, discussed the redesigned fuel economy stickers to be required on new cars starting in 2013.

> Consider a new midsize Ford Fusion S model. At $4-a-gallon gas, annual fuel costs would be about $2,200 for 15,000 miles driven. ... And that's for a car with a rated mileage of 34.9 m.p.g. (actual highway mileage, as with all vehicles, is lower, in this case about 27 m.p.g.). Industry clearly needs to do better. [R55]

Did the *Times* use the rated or the actual mileage in its computation?

Exercise 2.9.14. [W][S][Section 2.2][Goal 2.1] Hybrids vs. nonhybrids: The 5 year equation.

On February 23, 2011, Matthew Wald blogged at *The New York Times* about a study in *Consumer Reports* saying that

> A car buyer who lays out an extra $6,200 extra [sic] to buy the hybrid version of the Lexus RX will get the money back in gas savings within five years, according to Consumer Reports magazine, but only if gasoline averages $8.77 a gallon. Otherwise, the nonhybrid RX 350 is a better buy than the Hybrid 450h. [R56]

Wald notes that the study assumes

- the car will be driven 12,000 miles a year,

- gas will cost $2.80 a gallon,

- the hybrid gets 26 miles per gallon, the nonhybrid, 21.

(a) Show that the computation is wrong — that at $8.77 per gallon of gas you can't save $6,200 in 60,000 miles of driving.

(b) Show that you can save that much with that much driving if gas costs $8.77 per gallon more than $2.80 per gallon.

(c) The *Times* blogger was reporting on a study from *Consumer Reports* magazine. Do you think the error was the blogger's or the magazine's? What would you have to do to find out which?

(d) Write a response to post as a comment on the blog.

Exercise 2.9.15. [S][Section 2.2][Goal 2.1][Goal 2.6] Lots of gasoline.

> In [the U.S. in] 2011, the weighted average combined fuel economy of cars and light trucks combined was 21.4 miles per gallon (FHWA 2013). The average vehicle miles traveled in 2011 was 11,318 miles per year. [**R57**]

(a) Convert the weighted average combined fuel economy to gallons per hundred miles.

(b) Compute the amount of gasoline the average vehicle used in 2007.

(c) Estimate the number of cars and light trucks on the road in the U.S. in 2007.

(d) Estimate the total amount of gasoline used by cars and light trucks in the U.S. in 2011.

(e) At `www.americanfuels.net/2014/03/2013-gasoline-consumption.html` you could read that the U.S. consumed 134,179,668,000 gallons of gasoline in 2011. Is that number consistent with your answer to the previous part of the problem?

Exercise 2.9.16. [S][Section 2.2][Goal 2.1] Yet another problem on gas mileage.
    On February 18, 2012, ScottW of Chapel Hill, NC, responded to an article in *The New York Times* saying that a car that averaged 35 mpg instead of 20 would cut your gas bill almost in half. [**R58**]
    Is ScottW's arithmetic right?

Exercise 2.9.17. [S][Section 2.3][Goal 2.1] Refining oil.
    When a barrel of oil is refined it produces 19.5 gallons of gasoline, 9 gallons of fuel oil, 4 gallons of jet fuel and 11 gallons of other products (kerosene, etc.).

(a) How many gallons of oil are there in a barrel before any of it is refined?

(b) Determine the fraction (or percentage) of the barrel that gasoline represents.

(c) Find the current price for a barrel of oil and the current price for a gallon of gasoline at the pump.

(d) What percentage of the price of a gallon of gas at the pump is accounted for by the cost of the oil from which it is made? What do you think accounts for the rest?

(e) In 2005, the United States consumed oil at an average rate of about 22 million barrels per day. What is the corresponding consumption of gasoline in gallons per day?

(f) Suppose each gallon can is about one foot tall. If we stacked them up would those cans of gasoline reach the moon?

Exercise 2.9.18. [S][Section 2.3][Goal 2.1] Harry Potter.
The last Harry Potter book went on sale at 12:01 a.m. on July 21, 2007.

> Early reports estimate that Scholastic broke all publishing records, selling an unprecedented 8.3 million copies of *Harry Potter and the Deathly Hallows* in its first 24 hours on sale. [**R59**]

(a) On the average, how many books per minute were sold during that first day?

(b) Estimate how many bookstores were selling the book.

(c) In Harry Potter's wizard world currency is measured in galleons, sickles and knuts, where

$$1 \text{ galleon} = 17 \text{ sickles}, \quad 1 \text{ sickle} = 29 \text{ knuts}.$$

Scholastic Books, the publisher of the Harry Potter series in the United States, has issued paperback copies of books that (they claim) Harry Potter used at Hogwarts. The price is listed as

$5.99 US (14 Sickles 3 Knuts)

Use this information to figure out the number of dollars per galleon.

(d) The list price of the hardback version of *Harry Potter and the Deathly Hallows* was $34.99. Convert this price to galleons. (You may write your answer with decimal fractions of a galleon. No need to convert the fractional part to sickles and knuts — but you can if you want to.)

Exercise 2.9.19. The national debt.
Moved to Extra Exercises at www.ams.org/bookpages/text-63 .

Exercise 2.9.20. [S][Section 2.4][Goal 2.6] [Goal 2.3] Medicare fraud.
An Associated Press story in Easton, Maryland's, *The Star Democrat* on June 3, 2010, reported that

> All told, scam artists are believed to have stolen about $47 billion from Medicare in the 2009 fiscal year, nearly triple the toll a year earlier. Medicare spokesman Peter Ashkanaz said that ... charges have been filed against 103 defendants in cases involving more than $100 million in Medicare fraud. [**R60**]

(a) What percentage of the Medicare fraud has been targeted by filed charges?

(b) What is the average claim in each fraud charge?

(c) How many of these average size claims would need to be filed to recover the entire $47 billion?

(d) The article suggests that the administration is vigorously pursuing Medicare fraud. Do the numbers support that suggestion?

Exercise 2.9.21. [S][Section 2.4][Goal 2.6] [Goal 2.5] Mom blogs and unit prices.
carrotsncake.com/2011/01/grocery-shopping-101-unit-price.html is just one of many "mom blogs" that discuss shopping wisely using unit prices. There you will find the following quote:

> In the photo [on the website], the unit price is listed in orange on the left side of the label. In this case, it's how much you pay per gallon of olive oil. The cost per gallon is $50.67, but you're only buying 16.9 ounces of olive oil, so you pay $6.69. It might seem like a deal because you're only paying $6.69, but the cost per gallon is high. [**R61**]

(a) Verify the unit price calculation in the quotation.

(b) Visit the blog. There you will find some comments in each of the categories

> "I've always looked at unit prices."
> "This is new to me, and cool. I'll do this from now on."
> "My husband/boyfriend told me about this."

Estimate the percentage or fraction of comments in each category — perhaps "About a third of the comments are from people who learned about unit pricing on this blog."

(c) If you were to post a comment, what would it be? (Optional: post it.)

Exercise 2.9.22. [S][Section 2.4][Goal 2.6] Greenies.
A 12 ounce box of Greenies costs $12.99. A 30 ounce box costs $26.99.

(a) What is the cost in dollars per ounce in each box? Which size should you buy?

(b) Suppose you have a coupon worth $5 toward the price of any box of Greenies. If you use the coupon, which size should you buy? Is it always better to buy the large economy size?

(c) Answer the previous question if your coupon gives you a 10% discount rather than $5 off.

(d) What's a Greenie? Are these prices reasonable?

Exercise 2.9.23. [S][R][Section 2.4][Goal 2.6] [Goal 2.5] Race to the moon.
On July 22, 2011, *The Seattle Times* reported that Astrobiotics Technology planned to charge $820,000 a pound to send scientific experiments to the moon in a spacecraft that could carry 240 pounds of cargo. They would collect two hundred million dollars for the delivery. [**R62**]

In August 2014 the company advertised on its website that the cost to deliver a payload to the lunar surface was $1.2M per kilo. [**R63**]

(a) Verify the arithmetic in the 2011 article from *The Seattle Times*.

(b) How did the quoted price change between then and 2015?

Exercise 2.9.24. [S][Section 2.4][Goal 2.6] When's the beef?
    Use the graph in Figure 2.6 to answer the following questions. You may need to supplement the data with information from the web, but don't ask the web for answers to the questions, except (if you wish) to check your computations.

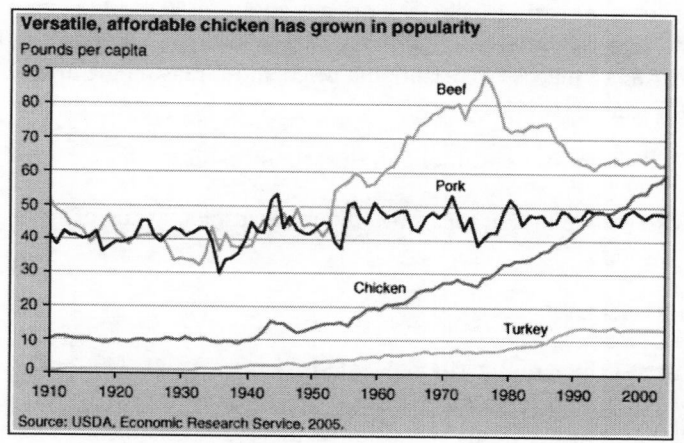

Figure 2.6. Beef, chicken, pork, turkey [**R64**]

(a) What does "per capita" mean?

(b) Estimate the number of pounds per capita of beef, chicken, pork and turkey consumed in the year 2000.

(c) The graph stops at about 2004. Use the trends it shows to estimate the number of pounds per capita of beef, chicken, pork and turkey that are being consumed this year.

(d) The peak year for per capita beef consumption was about 1975. Was more total beef consumed that year than at any other time in the past century?

Exercise 2.9.25. Apple's app store.
    Moved to Extra Exercises at www.ams.org/bookpages/text-63 .

Exercise 2.9.26. [S][Section 2.4][Goal 2.6] [Goal 2.5] Seven thousand dollars for sushi?
    On April 5, 2012, in *The New York Times* Gail Collins wondered "Why sushi rolls for 300 people cost $7,000" at a "four-day gathering [that] wound up costing more than $820,000. ...Among the most notable excesses was $6,325 worth of commemorative coins in velvet boxes."

Several of the online commenters did some arithmetic:

Truth Will Out (Olympia, WA)
No biggie. This is the misleading tyranny of large aggregate totals. Break it down, and it is only $683 per person per day. Not ex-
actly off the wall for conference expenses, especially if this included
room (Gail doesn't say). Even those commemorative coins only break
down to a $20 chachka per person. Pretty cheap as far as conference
mementos go.

P Desenex (Tokyo) I agree. $7000 for sushi for 300 people is
$23.33 per person. At a kaiten sushi place (one of the more economi-
cal venues for eating sushi) the price is about $1.50 to $3.00 per piece
depending on the kind of fish. We can then figure each person had
from 7 to 15 pieces. A reasonable price and a reasonable amount to
eat for lunch. [**R65**]

(a) Check the unit pricing arithmetic in these comments.

(b) What does Truth Will Out mean by "the misleading tyranny of large aggregate totals"?

(c) What's a "chachka"?

Exercise 2.9.27. [S][Section 2.4][Goal 2.6] Splash!
The Tournament Players Club at Sawgrass is a golf course in Florida and is considered one of the most difficult in the world. Its most famous hole is the 17th, known as
the "Island Green", which is completely surrounded by water. The Wikipedia page for
the course notes that

It is estimated that more than 100,000 balls are retrieved from the
surrounding water every year .... [**R66**]

(a) Estimate the number of hours per year golf is played at that course in Florida.

(b) If the estimate in the article is correct, how many golf balls enter the water per
minute?

(c) Do you think the estimate in the article makes sense?

(d) How much water would 100,000 golf balls displace if they were all left in the water?

(e) The lake surrounding the island green is roughly rectangular, about 100 feet wide
and 200 feet long. How much do the golf balls raise the water level?

(f) What if they were put in a box after they were taken out of the water. How big
would the box have to be?

(g) What is the monetary value of the golf balls?

(h) Do some people collect golf balls as a job? If yes, how does that pay?

Exercise 2.9.28. [S][Section 2.4][Goal 2.6] Automobile accident statistics.
Here are two quotations from the web:

> According to the National Highway Traffic Administration, car acci-
> dents happen every minute of the day. Motor vehicle accidents occur
> in any part of the world every 60 seconds. And if it's all summed up in
> a yearly basis, there are 5.25 million driving accidents that take place
> per year. [**R67**]

> Car Crash Stats: There were nearly 6,420,000 auto accidents in the
> United States in 2005. The financial cost of these crashes is more than
> 230 billion dollars. 2.9 million people were injured and 42,636 people
> killed. About 115 people die every day in vehicle crashes in the United
> States — one death every 13 minutes. [**R68**]

(a) The first quotation gives accident rates in two ways, with two different time units.
Are the rates the same when you convert one to the other? If not, what mistake
might the writer have made?

(b) Have the approximate death rates in the second quotation been computed correctly
from the total number of deaths?

(c) Are the accident rates in the two quotations consistent? Can you find other evidence on the internet to confirm them?

[See the back of the book for a hint.]

Exercise 2.9.29. [S][Section 2.4][Goal 2.6] [Goal 2.5] Gas is cheap.
B. P. Phillips of Broomfield, Colorado, conducted "An Impartial Price Survey of
Various Household Liquids as Compared to a Gallon of Gasoline". You can find it in
the *Annals of Improbable Research*.

> With the price of oil surging past $100 per barrel recently, consumers
> and the media in the United States have been in an uproar over the
> outrageous price of gasoline. In my opinion, however, they have failed
> to place this cost in context to other common, household liquids.
> Thus, I decided to perform an informal price survey.
>   I visited my local grocery store, where I recorded the prices and
> respective volumes of 22 off-the-shelf fluids. Then, after returning
> home, I computed the price per gallon of each item, and arranged
> them in ascending order by cost, as shown in Table 2.7.
>   Clearly, all of the fluids I recorded are at least as expensive, if
> not more so, than gasoline, which is currently hovering around $3.25
> per gallon in Broomfield, Colorado. In fact, the only two products
> with prices that are even comparable are Coca-Cola and milk, each
> costing less than four dollars per gallon. All of the other products
> range anywhere from $6.38 per gallon of orange juice all the way up
> to a staggering $400.46 per gallon of Wite-Out Correction Fluid.

Product	Price	Price per Gallon
Coca-Cola	$1.89 / 2 liters	$3.58
Lucerne 2% Milk	$3.79 / 1 gal.	$3.79
Tropicana Orange Juice	$3.19 / 64 fl. oz.	$6.38
Tree Top Apple Juice	$3.55 / 64 fl. oz.	$7.10
Evian Water	$1.89 / 1 liter	$7.16
Bud Light	$1.99 / 24 fl. oz.	$10.61
Heinz Ketchup	$4.19 / 46 fl. oz.	$11.66
Crisco Vegetable Oil	$5.69 / 48 fl. oz.	$15.17
Tide Laundry Detergent	$14.49 / 100 fl. oz.	$18.55
Windex Glass Cleaner	$3.99 / 26 fl. oz.	$19.64
Aunt Jemima Maple Syrup	$4.39 / 24 fl. oz.	$23.41
Starbucks Frappuccino	$7.29 / 38 fl. oz.	$24.56
Hidden Valley Ranch Dressing	$3.75 / 16 fl. oz.	$30.00
Red Bull Energy Drink	$1.99 / 8.3 fl. oz.	$30.69
Pert Shampoo	$3.49 / 13.5 fl. oz.	$33.09
Lucky Clover Honey	$4.45 / 12 fl. oz.	$47.47
Pepto-Bismol	$5.99 / 16 fl. oz.	$47.92
A1 Steak Sauce	$7.34 / 15 fl. oz.	$62.63
Listerine Mouthwash	$3.59 / 5.8 fl. oz.	$79.23
Bertolli Extra Virgin Olive Oil	$11.49 / 17 fl. oz.	$86.51
Vick's Nyquil	$7.49 / 10 fl. oz.	$95.87
Wite-Out Correction Fluid	$2.19 / 0.7 fl. oz.	$400.46

Table 2.7. Unit prices for common fluids

After being made aware of my results, it should be painfully obvious to consumers and the media that gasoline is still relatively inexpensive compared to other liquids. In closing, I ask readers to provide us with additional examples of outstandingly expensive fluids from the U.S. or other countries. [R69]

(a) Verify at least four of the numbers in Table 2.7.

(b) Do some research (either on the web at an online grocery store or at a local store) and see if you can find items less expensive than Coca-Cola and more expensive than Wite-Out Correction Fluid.

(c) Investigate the cost of bottled water in dollars per gallon. The answer will depend on where you buy it and in what quantity. Discuss where your answers fit in this list.

Exercise 2.9.30. [S][R][A][Section 2.5] [Goal 2.2][Goal 2.5] Mph and kph.

(a) Convert 100 km/hour to miles per hour, so you'll know how fast you may drive in Canada.

(b) Convert 65 miles/hour to km/hour.

Write your answers using just two significant digits, not lots of decimal places.

Exercise 2.9.31. [U][Section 2.5] [Goal 2.1][Goal 2.2] Roll your own.
Invent a problem based on something that catches your fancy on Wikipedia's list of humorous units of measurement: `en.wikipedia.org/wiki/List_of_humorous_units_of_measurement`.

Exercise 2.9.32. [S][Section 2.5][Goal 2.1][Goal 2.2] The chron.
The World Time Organization is proposing that we convert to a metric time measurement system. The basic unit would be the chron; a day would be 10 chrons long. Use this information to answer the following questions.

(a) How many kilochrons are there in a year?

(b) How many seconds are there in a millichron?

(c) If we switch to this new system, the Department of Transportation will need to replace its speed limit signs. What would the following sign look like if miles per hour were converted to kilometers per chron?

How would the Federal Highway Administration change this regulation?

> This sign is used to display the limit established by law, ordinance, regulation, or as adopted by the authorized agency. The speed limits shown shall be in multiples of 10 km/h or 5 mph. [**R70**]

(d) French revolutionary metric time (optional)

The chron isn't entirely our own invention. The metric system of measurement for lengths and weights was a byproduct of the French Revolution. There were proposals then for a metric time system too:

- 10 metric hours in a day
- 100 metric minutes in a metric hour
- 100 metric seconds in a metric minute
- 10 days in a metric week (called a dekade)

There are several websites devoted to debating whether this or something like it would be a good idea. Find some and write about the question.

Exercise 2.9.33. [R][S][Section 2.5] [Goal 2.1][Goal 2.2][Goal 2.5] How old is the United States?
The United States was created in 1776. How old is it in seconds? Write your answer in metric terms with the proper prefix. (Kiloseconds? Petaseconds? Decide what works best here.) Remember not to use too many significant digits.

Exercise 2.9.34. [U][Section 2.5][Goal 2.2] Metric prefixes as metaphors.

The metric prefixes are often used to suggest that quantities are large or small without any particular quantitative meaning. What is microfinance? Micromanagement? Nanotechnology? Megalomania? What about the Mega Society at megasociety.org/about.html ? Find other similar usages.

Exercise 2.9.35. [S][Section 2.4][Goal 2.1] World vital events per time unit.

The data in Table 2.8 are from the U.S. Census website.

Time unit	Births	Deaths	Natural increase
Year	134,176,254	56,605,700	77,570,553
Month	11,181,355	4,717,142	6,464,213
Day	367,606	155,084	212,522
Minute	255	108	148
Second	4.3	1.8	2.5

Table 2.8. World vital events per time unit: 2014 (figures may not add to totals due to rounding)

(a) Check that the rates in each column are consistent.

(b) Check that the rates in each row are consistent.

(c) Which of the rates is the easiest to understand the impact of? Why do you think so?

Exercise 2.9.36. [S][C][S][Section 2.5] [Goal 2.2] Pumping blood.

We haven't done anything with the "8,000 liters of blood" in Section 1.3. What does it mean? Clearly it's not the amount of blood in the body — the number is way too large for that. It's the amount pumped by the heart, in liters per day.

(a) Verify the assertion that the human body contains about 5 liters of blood.

(b) Use that fact to estimate how many times a day (on average) each red blood cell passes through the heart on its way to the rest of your body, and how long the round trip takes.

Exercise 2.9.37. [S][W][Section 2.6] [Goal 2.1][Goal 2.4] Arizona using donations for border fence.

On November 24, 2011, *The Washington Times* published an article from the Associated Press with that headline. There you could read that the state was seeking donations to run a fence along the border with Mexico.

> ...[State Senator] Smith acknowledges he has a long way to go to make the fence a reality. The $255,000 collected will barely cover a half mile of fencing. Mr. Smith estimates that the total supplies alone will cost $34 million, or about $426,000 a mile. Much of the work is expected to be done by prisoners at 50 cents an hour. [**R71**]

(a) Use the information in the quotation above to estimate the length of Arizona's border with Mexico, in miles.

(b) Find another estimate of the length of Arizona's border with Mexico. Is the answer you find consistent with your answer to the previous question?

(c) If $255,000 pays for labor and supplies for half a mile of fence, how much does the labor cost?

(d) How many hours will it take prisoners to construct that half mile? Does your answer make sense?

Exercise 2.9.38. [S][Section 2.6][Goal 2.5] U.S. gallons.
Explain why Google says

$$\boxed{12 \text{ miles per gallon} = 833.333333 \text{ US gallons per } (10\,000 \text{ miles})}$$

instead of simply

$$\boxed{12 \text{ miles per gallon} = 833.333333 \text{ gallons per } (10\,000 \text{ miles})} \; .$$

Exercise 2.9.39. [S][Section 2.8][Goal 2.4] Comet 67P/Churyumov-Gerasimenko.
On August 5, 2014, *The New York Times* reported that

> In June, the [Rosetta] spacecraft measured the flow of water vapor streaming off the comet [67P/Churyumov-Gerasimenko] at a rate of about two cups a second, which would fill an Olympic-size swimming pool in about 100 days. [**R72**]

Check this volume calculation.

Exercise 2.9.40. [U][Section 2.8][Goal 2.4] A table for your puzzle.
You are about to start work on your new 1,000 piece jigsaw puzzle and you need to know how much space to clear for it on the table. Each piece is approximately square, about 3/4 of an inch on each side. The finished puzzle is a rectangle about one and a half times as wide as it is high. How big is the puzzle?
(Optional) Answer the same question if the finished puzzle is circular.

Exercise 2.9.41. [U][Section 2.8][Goal 2.2][Goal 2.4] A pint's a pound the whole world round.
Discuss the accuracy of that well-known saying.
[See the back of the book for a hint.]

Exercise 2.9.42. [S][Section 2.8][Goal 2.1] [Goal 2.4] An ark full of books.
In March of 2012 *The New York Times* reported on a warehouse in California storing forty-foot shipping containers full of books. Each week twenty thousand new books arrive. There were 500,000 stored at the time the article was written. The site plans to expand to store 10 million. [**R73**]

(a) How long had the collection been accumulating when the article appeared?

(b) Estimate how many volumes are in each shipping container (you can assume that these are standard forty-foot shipping containers, which means you will have to look up the dimensions of this type of container).

(c) How many shipping containers would be needed to store 10 million volumes?

Exercise 2.9.43. [S][Section 2.8][Goal 2.1][Goal 2.4] Floating trash threatens Three Gorges Dam.

A story with that headline in the August 2, 2010, *China Daily* reported about a flood in China piling garbage 60 cm thick over an area of 50,000 square meters every day. To protect the Three Gorges Dam 3,000 tons of garbage were being collected each day. [**R74**]

(a) Check that 60 cm is approximately two feet and that 50,000 square meters is approximately half a million square feet.

(b) The article suggests that collecting 3,000 tons a day will keep the pile under control. Is that reasonable?

(c) Estimate how long it would take to clear the garbage at the rate of 3,000 tons a day.

(d) How long would it take the city you live in to produce a layer of garbage covering half a million square feet two feet deep?

Exercise 2.9.44. Stressing your calculator.
Moved to Extra Exercises at www.ams.org/bookpages/text-63 .

Exercise 2.9.45. Gold.
Moved to Extra Exercises at www.ams.org/bookpages/text-63 .

Exercise 2.9.46. Expensive solar energy.
Moved to Extra Exercises at www.ams.org/bookpages/text-63 .

Exercise 2.9.47. [S][Goal 2.1] Boca negra.
Charlotte Seeley's boca negra cake recipe calls for 12 ounces of chocolate. She has a 17.6 ounce bar of Trader Joe's Pound Plus Belgian Chocolate that's divided into 40 squares.

(a) How many of the squares should she use?

(b) Why do you think the package comes in such a peculiar size?

(c) What's a boca negra cake?

Exercise 2.9.48. [S][Section 2.6][Goal 2.1] Smoots.
If you follow Mass Ave from Boston to Cambridge you will cross the Charles River via the Harvard Bridge. Looking down you may notice that the bridge is marked off in *smoots*. From one end to the other the length of the bridge is 364.4 smoots. Wikipedia, reliable in this case, says a smoot is "a nonstandard unit of length created as part of an MIT fraternity prank. It is named after Oliver R. Smoot, a fraternity pledge to Lambda Chi Alpha, who in October 1958 lay on the Harvard Bridge ... and was used by his fraternity brothers to measure the length of the bridge." [**R75**]

(a) The length of the Harvard Bridge in standard English measure is 2,035 feet. How tall was Oliver R. Smoot, in feet and inches?

(b) How tall was Oliver R. Smoot, in meters? (Use only the right number of decimal places in your answer.)

(c) In 2014 Kelly Olynyk was the 7 foot tall starting center for the Boston Celtics. How long would the Harvard Bridge be in olynyks?

(d) What is the conversion rate between smoots and olynyks?

Exercise 2.9.49. [S][Section 2.5][Goal 2.2] Looking into the nanoworld.
   In October 2014 Eric Betzig, Stefan W. Hell and William E. Moerner won the 2014 Nobel Prize in Chemistry for inventing a microscope using fluorescence to see things as tiny as the creation of synapses between brain cells.
   In the Reuter's report of the award you could read that

> In 1873, Ernst Abbe [a German scientist] stipulated that resolution could never be better than 0.2 micrometers, or around 500 times smaller than the width of a human hair. [**R76**]

(a) Use the data in this quote to estimate the width of a human hair. Choose the appropriate metric unit of length: meter, millimeter, micrometer or nanometer.

(b) Use the web to confirm the accuracy of your estimate.

(c) The Nobel Prize award statement said that "Due to their achievements the optical microscope can now peer into the nanoworld."

  • Express the width of a human hair in nanometers.
  • How much smaller than a human hair are the things scientists can now see?

Exercise 2.9.50. [R][Section 2.3][Goal 2.1][Goal 2.2] Gasoline elsewhere.
   Compare the cost of gasoline in a foreign country of your choice with the cost where you live.
   You will need to find the price at the pump and the tax rates in each place. Then use appropriate unit conversions and subtract the tax from the price at the pump to find the cost of the gasoline itself.

Exercise 2.9.51. [S][Section 2.8][Goal 2.1] How thick is paint?
   In Section 2.8 you can read that a gallon of paint covers about 400 square feet.
   How thick is the layer of paint on the wall? What are good units for the answer?

Exercise 2.9.52. [S][Section 2.8][Goal 2.4] Molasses!
   The story headlined in the *The Boston Post* on the cover of this book tells of a January 15, 1919, disaster in Boston's North End. Chuck Lyons retold the story in *History Today* in 2009, describing how a tank ruptured, "releasing two million gallons of molasses." [**R77**]
   Wikipedia adds that "Several blocks were flooded to a depth of 2 to 3 feet." [**R78**]
   Lyons goes on to say that several years later the auditor appointed to decide on compensation recommended

> . . . around $300,000 in damages, equivalent to around $30 million today, with about $6,000 going to the families of those killed, $25,000 to the City of Boston, and $42,000 to the Boston Elevated Railway Company.

(a) Estimate the number of blocks that could be covered two to three feet deep by that much molasses.

(b) Is Lyons's inflation calculation correct?

(c) Estimate the number of families that lost loved ones. How much would each receive, in today's dollars? How does that amount compare to damage awards in similar cases today?

[See the back of the book for a hint.]

Review exercises.

Exercise 2.9.53. [A]

(a) A car travels for 3 hours at an average speed of 50 miles per hour. How far has it gone?

(b) A car travels for 2 hours at an average speed of 55 miles per hour, then travels for 30 minutes at an average speed of 40 miles per hour, then travels for 30 minutes at an average speed of 55 miles per hour. How far has it gone?

(c) If you plan to drive between two cities that are 250 miles apart, and you expect to go about 55 miles per hour, how long will it take you?

(d) If you earn $9.10 per hour and you work a 37 hour week, what are your gross (before tax) earnings?

(e) Apples sell for $1.39 per pound. If you buy 4.5 pounds, how much will you have to pay?

Exercise 2.9.54. [A] The exchange rate between euros and U.S. dollars is 1 € = $1.23.

(a) How much will it cost to buy one hundred euros?

(b) How many euros can you buy with one hundred dollars?

(c) The price of gas in a town I visited in France was 1.75 euros per liter. Convert this to dollars per gallon.

(d) A nice restaurant meal in Germany cost 32.50 euros. Convert this to dollars.

Exercise 2.9.55. [A] Suppose the exchange rate between U.S. dollars and GBP is £1=$2. If a meal in a London pub costs £13 and a pint costs £5, how much would a meal and 3 pints cost in U.S. dollars?

Exercise 2.9.56. [A] Write each of the following expressions in standard notation (that is, using all of the zeros).

(a) 3.2×10^5

(b) 4.666×10^8

(c) -3.3×10^6

(d) 5.3×10^{-3}

(e) 6.2×10^{-5}

(f) -3.222×10^{-6}

Exercise 2.9.57. [A]

(a) If my favorite cereal costs $2.99 for a 14 ounce box, what is the price per ounce?

(b) A gallon of milk costs $2.89. What is the price per ounce?

(c) A three pound bag of apples is on sale for $2.49. What is the price per pound?

(d) In one store, a 36-pack of diapers costs $12.99. Another store is selling bonus packs, containing 42 diapers, for $14.59. Which is the better deal?

Exercise 2.9.58. [A] Do each conversion.

(a) One kilogram into centigrams.

(b) Two millimeters into nanometers.

(c) Three inches into centimeters.

(d) Ten feet into decimeters.

(e) Forty gallons into liters.

(f) Forty gallons into milliliters.

(g) Forty gallons into cubic centimeters.

(h) Forty gallons into picoliters.

(i) 1,000 nanometers into micrometers.

(j) 100 milliliters into liters.

(k) 40 centiliters into liters.

Exercise 2.9.59. [A] Find each of these areas:

(a) A 12 foot by 18 foot carpet.

(b) A small rug measuring 4 feet by 18 inches.

(c) A deck measuring 5 meters by 7 meters.

(d) A football field in the United States measuring 120 yards by 110 yards.

(e) An NBA basketball court measuring 94 feet by 50 feet.

Exercise 2.9.60. [A] Find each of these volumes:

(a) A room measuring 12 feet by 18 feet by 9 feet.

(b) A bathtub measuring 6 feet by 2 feet by 3 feet.

(c) A wooden block measuring 4 centimeters by 2 centimeters by 1.5 centimeters.

Exercise 2.9.61. [A] Provide sensible answers with just one or two digits.

(a) One yard per second in miles per hour.

(b) One foot per second in miles per hour.

(c) One meter per second in kilometers per hour.

Exercises added for the second edition.

Exercise 2.9.62. [U][S][Section 2.5][Goal 2.1][Goal 2.2] A trillion gigabytes.

On May 23, 2016, the wireless industry trade group CTIA released its annual survey of internet usage, reporting that

> Americans used 9.6 trillion megabytes (MB) of data in 2015, three times the 3.2 trillion MB in 2013. This is the equivalent of consumers streaming 59,219 videos every minute or roughly 18 million MB. [**R79**]

The total was reported in the *Baltimore Sun* as "nearly 1 trillion gigabytes of mobile data". [**R80**]

(a) Is the *Baltimore Sun* quote correct?

(b) Is this "roughly 18 million MB" per minute?

(c) Criticise the precision in "59,219 videos every minute". Can you make sense of that number?

Exercise 2.9.63. [U][W][S][Section 2.5][Goal 2.1] Walmart goes green.

On April 19, 2017, a Walmart press release described an attempt to help the company's suppliers reduce greenhouse gas emission.

> Dubbed Project Gigaton, this initiative will provide an emissions reduction toolkit to a broad network of suppliers seeking to eliminate one gigaton of emissions, focusing on areas such as manufacturing, materials and use of products by 2030. That's the equivalent to taking more than 211 million passenger vehicles off of U.S. roads and highways for a year. [**R81**]

A year later, on April 18, 2018,

> Walmart announced that suppliers have reported reducing more than 20 million metric tons (MMT) of greenhouse gas emissions in the global value chain, as part of the company's Project Gigaton initiative. [**R82**]

The EPA says

> A typical passenger vehicle emits about 4.7 metric tons of carbon dioxide per year. This assumes the average gasoline vehicle on the road today has a fuel economy of about 21.6 miles per gallon and drives around 11,400 miles per year. Every gallon of gasoline burned creates about 8,887 grams of CO_2. www.epa.gov/greenvehicles/greenhouse-gas-emissions-typical-passenger-vehicle-0

(a) Check that Walgreens and the EPA agree about the amount of carbon dioxide the average gasoline vehicle emits in a year.

(b) Where do you think Mr. Krupp got his 211 million car estimate? Critique it.

(c) Estimate, or look up, the number of cars on the road in the United States. Compare that number to the 211 million that the gigaton project would take off the road. What does your comparison tell you?

(d) The second press release reports the first year's progress. Is Walmart on track to reach its gigaton target?

Exercise 2.9.64. [U][S][Section 2.8][Goal 2.1][Goal 2.4] Termites.

On November 20, 2018, *The New York Times* published a story on the discovery of a field of 200 million termite mounds in Brazil. They estimated that

> ...to build 200 million mounds, the termites excavated 2.4 cubic miles of dirt — a volume equal to about 4,000 great pyramids of Giza. [**R83**]

(a) What is the volume of each termite mound? Write your answer in appropriate units.

(b) The article says elsewhere that the mounds are about 10 feet high. Use that fact and the rest of the data to estimate the height of the great pyramid of Giza. Check your estimate with a web search.

[See the back of the book for a hint.]

Exercise 2.9.65. [U][S][Section 2.8][Goal 2.1][Goal 2.4] Not a drop to drink.

On January 30, 2019, *The New York Times* reported that

> The United States Bureau of Reclamation estimates that the Colorado River's watershed could face an annual shortfall of 3.2 million acre-feet by midcentury. That's 1.04 trillion gallons, almost half what Arizona uses per year. [**R84**]

The article links to www.arizonawaterfacts.com/water-your-facts where you can see Figure 2.9.

The website worldpopulationreview.com/states/arizona-population/ reports that the January 2019 population of Arizona was about 7.2 million.

(a) Verify the volume conversion between acre-feet and gallons.

(b) Use the data provided to calculate how much water a typical Arizonan uses in their home in a year.

(c) Find a trustworthy estimate on the web for domestic daily per capita water consumption. Use that estimate to critique your answer to the previous question.

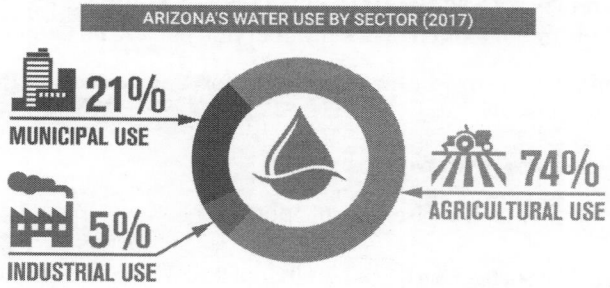

Figure 2.9. Arizona Water Use [**R85**]

Exercise 2.9.66. [U][S][Section 2.6][Goal 2.1][Goal 2.2] Ultima Thule.

In January 2019 User S. Kohn posted this query on an Astronomy question and answer site:

> I just got the news that the New Horizons has passed by some remote planet on the edge of the solar system.
>
> I was surprised that the guy from NASA says that it might take 24 months from us to get the photo of that planet.
>
> The solar system is not that big, right? It is slow because the signal transmission is slow, right? But why is the transmission so slow? [**R86**]

One of the answers to that question notes that

> During the encounter with Ultima Thule, all of the 7 instruments on New Horizons were gathering data (although not all at the same time) and the total data collected is expected to be about 50 gigabits of data.
> …
>
> Since New Horizons is about another billion miles further out than Pluto was and 3 more years have elapsed, there is less power for the (tiny) transmitter and the signals are much weaker. The bit rate is about 1000 bits per second.

(a) Why did astronomers name that object "Ultima Thule"?

(b) Verify the estimate of two years to transmit the data from New Horizons.

(c) Look up the size in bytes of a two hour movie (standard, not HD). How many two hour movies could you store in the 50 gigabit memory on New Horizons.

(d) Compare the rate in bits per second to stream that movie to the rate at which data downloads from New Horizons.

Exercise 2.9.67. [U][S][Section 2.6][Goal 2.1] Double double.

On February 6, 2019, *The New York Times* published an obituary of Ron Joyce, who was instrumental in a Canadian success story:

> Tim Hortons, which started as a doughnut and coffee shop but later expanded into other fare, dominates Canada's fast-food business to an extraordinary degree. There is one Timmies, as the shops are popularly known, for every 9,800 Canadians. By contrast, McDonald's restaurants in the United States number one per 23,100 people. [**R87**]

(a) Were there more Tim Hortons in Canada or more McDonalds in the United States when Ron Joyce died?

(b) Check your estimates with web searches.

(c) Why did we name this exercise "double double"?

Exercise 2.9.68. [U][Section 2.4][Goal 2.1][Goal 2.6] The cost of war.

According to estimates by the Costs of War Project at Brown University's Watson Institute for International and Public Affairs, the war on terror has cost Americans a staggering $5.6 trillion since October 7, 2001, when the U.S. invaded Afghanistan. This figure includes not just the Pentagon's war fund, but also future obligations such as social services for an ever-growing number of post-9/11 veterans.

It's hard for most of us to even begin to grasp such an enormous number.

It means Americans spend $32 million per hour, according to a counter by the National Priorities Project at the Institute for Policy Studies.

Put another way: Since 2001, every American taxpayer has spent almost $24,000 on the wars — equal to the average down payment on a house, a new Honda Accord or a year at a public university.

(a) Verify that $5.6 trillion spent since October 7, 2001, is approximately $32 million spent every hour since the invasion of Afghanistan began.

(b) Verify that if every American bought a Honda Accord for $24,000 the sum of all their purchases would be equal to $5.6 trillion.

Exercise 2.9.69. [U][S][Section 2.4][Goal 2.1][Goal 2.6] Equal pay for equal work. On April 14, 2016, *Metro Boston* reported that

> A 35-year-old woman working in architecture and engineering earns an average of $34 to a man's $42 [per hour], a gap of $16,000 per year — or the equivalent of 181 months of rent over 30 years. according to the calculator. [**R88**]

(a) Confirm that the earnings gap is $16,000 a year.

(b) Can you make sense of the claim that $16,000 per year is equivalent to 181 months of rent over 30 years? What might the author have been thinking?

3

Percentages, Sales Tax and Discounts

The focus of this chapter is the study of relative change, often expressed as a percent. We augment an often much needed review in two ways — stressing quick paperless estimation for approximate answers and, for precision, a new technique: multiplying by 1+ (percent change).

Chapter goals:

Goal 3.1. Understand absolute and relative change.

Goal 3.2. Work with relative change expressed as a percentage.

Goal 3.3. Master strategies for arranging percentage calculations.

Goal 3.4. Learn (and appreciate) the 1+ trick for computing with percentages.

Goal 3.5. Calculate successive percentage changes.

Goal 3.6. Calculate percentage discounts.

Goal 3.7. Understand the difference between percentages and percentage points.

Goal 3.8. Understand what a percentile is and how to interpret percentile information.

3.1 The federal budget

Figure 3.1 shows how the federal government spent $4.0 trillion in 2017. In that pie chart each wedge represents a category of expenditure. The numbers in the wedges report both the actual dollar expenditure and that amount as a percentage of the gross domestic product (GDP). We will start by comparing the dollar amounts to the total

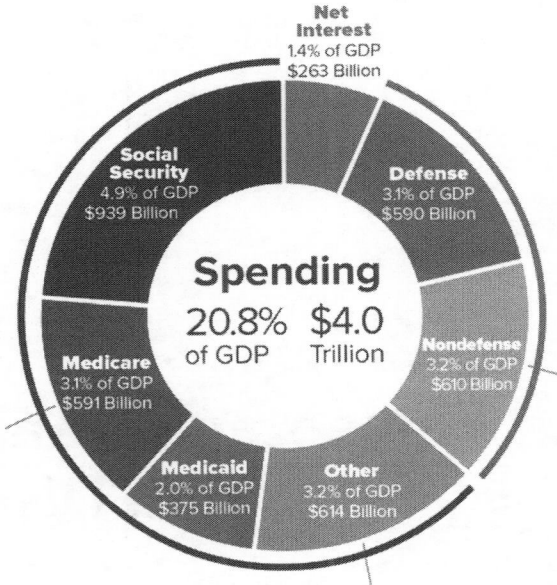

Figure 3.1. The 2017 federal budget [**R89**]

budget. Defense accounted for $590 billion. The best way to compare that amount to the total is with the fraction

$$\frac{\$590 \text{ billion}}{\$4.0 \text{ trillion}} = 0.1475 \approx 0.15.$$

The leading zero (before the decimal point) in 0.1475 does not change the value of that number. Writing .1475 would be just as good. But the extra zero helps you read the number. Without it you might not see the decimal point. We will always use it. You should learn to.

Note that the dollar units in the fraction cancel. The answer is just a number. Since defense spending is just a part of total spending, that number is less than 1. Numbers less than 1 are hard to think about so it's traditional to multiply them by 100 to make them friendlier. We report this as a percent:

$$0.15 = 15\%,$$

where the percent sign means "divide by 100". The word itself tells you that: "per" means "divide" and "cent" is Latin for "one hundred".

So federal spending on defense is about 15 percent of total spending. 15 percent — 15 out of 100 — is about 1/7, so about one dollar of every seven the government spends is for defense.

The two rightmost wedges in the figure, Defense and Nondefense, make up the total *discretionary spending*. Congress decides how much to spend and how to spend it. They are about the same size and account for about 30 percent of the total. The remaining 70 percent represents *mandatory* spending (spending required by law on things like Social Security) and interest on the federal debt.

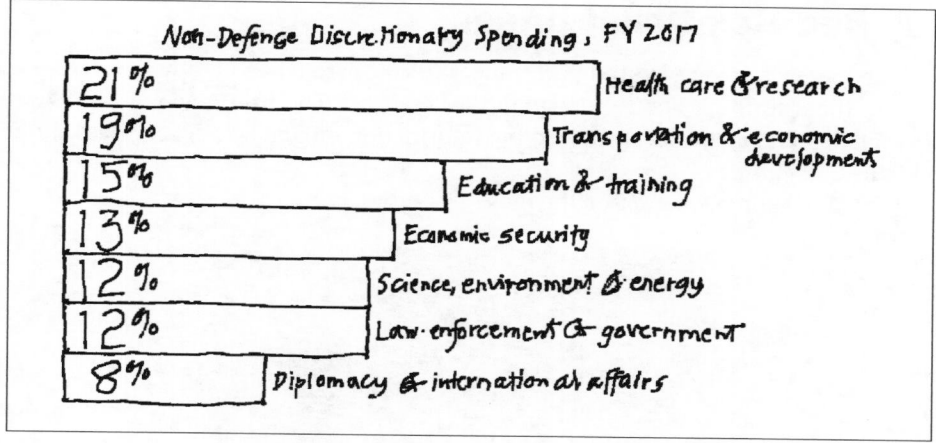

Non-Defense Discretionary Spending, FY 2017

21%	Health care & research
19%	Transportation & economic development
15%	Education & training
13%	Economic security
12%	Science, environment & energy
12%	Law-enforcement & government
8%	Diplomacy & international affairs

Figure 3.2. The 2017 federal budget [**R90**]

Figure 3.2 shows that 12 percent of nondefense discretionary spending is for Science, Environment and Energy. To compare that to the total federal budget we calculate 12% of 15%:

$$0.12 \times 0.15 = 0.018 = 1.8\%.$$

Science, the environment and energy receive less than 2 cents out of every dollar the government spends. That statistic is a useful part of an argument that we should spend more. If you would rather argue for less, find the dollar amount:

$$0.018 \times \$4.0 \text{ trillion} = \$72 \text{ billion}.$$

Then you can complain about how much we spend on those three things.

There is another percent in the center of the chart in Figure 3.1: 20.8% of GDP. That says the $4 trillion the government spends is 20.8 percent of the gross domestic product, the the total value of goods and services produced in 2017. We can use that fact to find the 2017 GDP. Since 20.8% is close to 20% = 0.20 = 1/5, the 2017 GDP was about 5 × $4 trillion = $20 trillion. To see that result another way, start with

$$0.208 \times \text{ GDP} = \$4.0 \text{ trillion}$$

so

$$\text{GDP} = \frac{\$4.0 \text{ trillion}}{0.208} = \$19.9 \text{ trillion}.$$

We end this section with three morals:

- Use the fraction part/whole to compare a part to a whole.

- Working with percentages calls for multiplication and division, not addition and subtraction.

- We spend a lot more on defense than on education, science and the environment combined.

3.2 Red Sox ticket prices

In the previous section we used percentages to compare a part to a whole. Here we will use them to think about how quantities change. The front page of *The Boston Globe* on September 9, 2008, provided the information on Red Sox ticket prices shown in Figure 3.3.

Which ticket price increased the most?

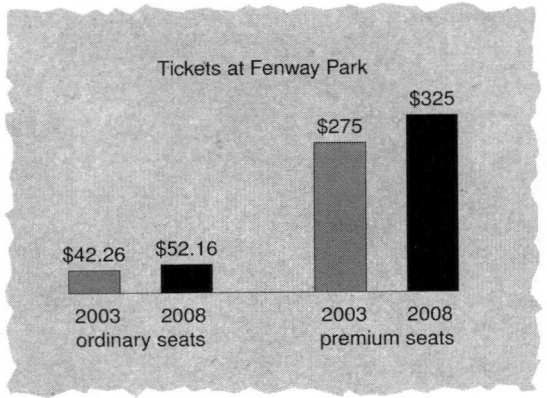

Figure 3.3. Red Sox ticket prices [**R91**]

The natural way to compare these ticket prices five years apart is with the fraction

$$\frac{2008 \text{ premium seat}}{2003 \text{ premium seat}} = \frac{\$325}{\$275} \approx 1.18.$$

That fraction is greater than 1 because the ticket price went up. The 0.18 in

$$1.18 = 1.00 + 0.18$$

tells you just how much: the premium seat price increased by 18 percent.

A similar calculation for ordinary seats,

$$\frac{2008 \text{ ordinary seat}}{2003 \text{ ordinary seat}} = \frac{\$52.16}{\$42.26} \approx 1.23,$$

shows that those seats increased 23 percent in price.

The seat price changes in dollars would tell another story. Premium seats cost $325-$275 = $50 more while the ordinary seat price rose only $52.16-$42.26 = $9.90.

These two ways to compare have names. The 18 percent increase for the premium seats is the *relative change*. The $50 increase is the *absolute change*. Most of the time the relative change is more useful.

3.3 The 1+ trick

States and towns often collect sales tax — a percentage of the sales price on most things you buy. You can find out the tax rate and exemptions for nontaxable goods for most places in the United States at www.salestaxstates.com/ . In March 2019 the sales tax rate in Berkeley, California, was 9.25 percent. How much would you pay there for a widget costing $114.95?

To work with the sales tax of 9.25%, we convert it to a decimal: 0.0925. Add that to the 100% = 1 representing the purchase price and multiply by the purchase price:

$$1.0925 \times \$114.95 = \$125.58.$$

The price with tax included is $125.58.

Here is another way to reach the same answer. First find 9.25 percent of $114.95:

$$\frac{9.25 \times \$114.95}{100} = \$10.63.$$

Then add

$$\$114.95 + \$10.63 = \$125.58.$$

Both ways work, but the first, multiplying by 1.0925, is faster and more reliable. We have a name for it: the *one plus trick*. Sometimes we write "1+ trick".

The 1+ trick helps solve a harder problem: Working backward from a known percentage increase. Suppose that widget is on sale for $115, tax included. Find the before tax sale price. Because

$$1.0925 \times \text{(sale price)} = \$115$$

we know

$$\text{sale price} = \frac{\$115}{1.0925} = \$105.26.$$

If you try to find the before tax sale price without the 1+ trick your intuition may lead you astray. If you calculate 9.25% of $115 and subtract you get $104.36, which is about a dollar too small.

Here's an even more dramatic example. In March 2019 TV station WKBN in Youngstown, Ohio, reported that

> Governor Mike DeWine has proposed a 95-percent increase in Ohio's investment to children services.
>
> ...
>
> DeWine's $74 million proposal would give financially-strapped county agencies the ability to pay the rising costs of serving children, help support struggling relatives who unexpectedly find themselves caring for children, invest in recruiting foster parents and help caseworkers be more efficient and productive in the field. [**R92**]

To find the investment in children's services in the then-current budget, calculate

$$\frac{\$74 \text{ million}}{1.95} = \$37.9 \text{ million.}$$

That makes sense: since 95 percent is nearly 100 percent, the proposed budget is almost twice the then-current budget. If we mistakenly subtracted 95% of $74 million from $74 million we'd get $3.7 million, which is just 5 percent of $74 million. Common sense would tell us we made a mistake somewhere.

The moral of the story:

> *Don't subtract to undo a percentage increase. Use the 1+ trick and divide.*

Perhaps we shouldn't say this, but when the percentage increase is small the error undoing it by subtracting instead of dividing by 1+ the increase may not matter. A two percent increase in $1000 is $20. Undoing that increase by subtracting two percent of $1,020 is $999.60. You may not care about the 40 cent error.

Sometimes solving a problem using an incorrect method gets you close enough to the correct answer. Using a correct method always works.

3.4 Exploiting the 1+ trick

In the previous section we encountered the 1+ trick and used it to study percentage increases. You may have learned and may even remember other ways to answer the questions posed there. If so, why take the time and effort to learn this new way? In this section we'll tackle several kinds of questions involving percentages that are easy with the 1+ trick and hard without it. When it becomes familiar it will no longer be a trick. In mathematician and teacher George Pólya's words, "What is the difference between method and device? A method is a device which you used twice." [R93]

Successive increases. Lucky you. You've just received a 20% salary increase. Last year your raise was 10%. How far ahead of the game are you now?

Your first guess might add the two percentages to get 30%. The 1+ trick will tell us the correct answer and see why the 30% guess is wrong.

If working with naked percentages makes you uncomfortable, here's a general strategy that may help you through this kind of problem. Imagine a particular starting salary and work with that actual number to find out what happens. Suppose that two years ago you were making $100,000. (Since it's your salary, why not imagine a nice large one that's easy to work with?) After the first raise the 1+ trick tells you your salary was

$$1.10 \times \$100,000 = \$110,000.$$

Then the second raise kicks in, and you're earning

$$1.20 \times \$110,000 = \$132,000.$$

That's a 32% increase overall — more than the first (wrong) answer of 30%. The extra 2% is because the second raise was applied to your salary after the first raise took effect.

Even better, you can write just one equation and do the rate increase multiplications first:

$$\begin{aligned} \text{new salary} &= 1.20 \times (1.10 \times \text{ old salary}) \\ &= (1.20 \times 1.10) \times \text{ old salary} \\ &= 1.32 \times \text{ old salary}. \end{aligned}$$

So your combined increase is 32%, and you didn't even have to imagine an actual dollar value to start with.

When combining percentage increases it helps to remember first what *not* to do:

> *Don't add percentage increases. Use the 1+ trick and multiply.*

Discounts. Suppose you have a coupon that offers you a 20% discount on the $114.95 widget you want to buy in Berkeley. What is the discounted price? You could calculate that 20% and subtract, but the 1+ trick is easier. A 20% discount is a change of $-20\% = -0.2$ so

$$1 + (\text{percent change}) = 1 + (-0.2) = 1 - 0.2 = 0.8.$$

That's no surprise. A 20% discount means you pay 80%. Then

$$\text{discount price} = 0.8 \times \$114.95 = \$91.96.$$

Without the 1+ trick you need two steps: find 20% of $114.95, then subtract. With the trick it's easy to calculate the final price including tax:

$$1.095 \times 0.80 \times \$114.95 = 0.876 \times \$114.95 = \$100.79.$$

If you also have a store coupon for 10% off that you may use on a discounted item then your net discount for the widget will be 28% (not 30%) because

$$0.9 \times 0.8 = 0.72 = 1 - 0.28.$$

1+ trick summary examples.

- To *increase* a number by 15% you multiply by

$$1 + 0.15 = 1.15.$$

- To *decrease* a number by 15% you multiply by

$$1 - 0.15 = 0.85.$$

- To *undo* a 15% increase you have to "unmultiply" by 1.15. So you *divide by* 1.15. Since

$$\frac{1}{1.15} = 0.8695652173 \approx 0.87 = 1 - 0.13,$$

undoing a 15% increase requires a 13% decrease.

- To calculate *successive* increases of 15% and 10% you multiply by

$$1.15 \times 1.1 = 1.265.$$

That 26.5% change is larger than what you get by adding the percentages.

3.5 Large and small percentages

Large percentages. A 2018 staff report from the Permanent Subcommittee on Investigations of the United States Senate described how

> On November 18, Portman and Carper released the findings of a new investigative report, which were also highlighted on CBS News' 60 Minutes, detailing how drug manufacturer kalèo exploited the opioid crisis by increasing the price of its naloxone drug EVZIO by more than 600 percent by 2016 (from the initial price of $575 per unit in July 2014 to $3,750 in February 2016 and then to $4,100 11 months later in January 2017), launching a new distribution model planning to "capitalize on the opportunity" of "opioid overdose at epidemic levels." The company's sales force focused on ensuring doctors' offices signed necessary paperwork indicating that EVZIO was medically necessary, ensuring the drug would be covered by government programs like Medicare and Medicaid. The plan worked, resulting in increased costs to taxpayers, to date, totaling more than $142 million in just the last four years, despite the fact that less costly versions of naloxone exist. [R94]

It's easy to check the arithmetic:

$$\frac{\$4,100}{\$575} = 7.13 = 1 + 6.13.$$

Since $6.13 = 613/100 = 613$ percent the report correcty says "more than 600 percent".

It's often clearer to describe increases much greater than 100 percent using multiplication instead of percentage change. A 100 percent increase doubles. A 1,000 percent increase means multiplying by 999, so nearly a thousand times as much.

We would have preferred to read in the Senate committee report that

> ... increasing the price of its naloxone drug EVZIO more than seven-fold (from an initial price of \$575 per unit to \$3,750 and then \$4,100 eleven months later) ...

The report's authors might want to keep "more than 600%" since it sounds more dramatic. Then rewriting the next sentence to point out the multiplication could double the drama, by concluding that "charges to taxpayers increased sevenfold from \$20 million to \$142 million in just four years."

Small percentages. We said at the start of this chapter that we imagine people invented percentages to avoid numbers: 17 percent is easier to think about than 0.17. But sometimes the percent itself is less than 1 — often a lot less. In the April 18, 2016, issue of *The New Yorker* Elizabeth Kolbert wrote:

> Powell wants to produce ten thousand genetically modified chestnut trees that can be made available to the public. If all these trees get planted and survive, they will represent [0].00025 per cent of the chestnuts that grew in America before the blight. [**R95**]

How many chestnut trees were there before the blight struck at the start of the twentieth century?

There are already three zeroes after the decimal point in 0.00025. Since it's a percent we have to insert two more to write it as a number. The most reliable way to count the zeroes in the answer is with powers of 10:

$$\text{number of pre-blight chestnuts} = \frac{10,000}{0.0000025} = \frac{100 \times 10^2}{25 \times 10^{-7}} = 4 \times 10^9 = 4 \text{ billion.}$$

To put that large number in perspective, it's more than 50 per person in 1900. In the chestnut's natural range east of the Mississippi River 4 billion trees is an average of more than 10,000 per square mile. So the new crop would be less than one square mile's worth.

The quote continues:

> "This is a century-long project," Powell said. "That's why I tell people, 'You've got to get your children, you've got to get your grandchildren involved in this.' "

3.6 Percentage points

When you want to compare percentages or discuss a change in percentage it's easy to be confused and confusing.

In Section 3.2 we saw that premium Red Sox ticket prices increased 18% between 2003 and 2008. In the same period, ordinary ticket prices increased 23%. The ordinary ticket price increase was 5 *percentage points* greater. It's important to describe the increase in percentage points. To find a 5 *percent* increase in 18% you would calculate $1.05 \times 18\% = 18.9\%$.

The Federal Reserve Board sets the prime interest rate (in part to try to control inflation, but that's a complicated story), which in turn determines interest rates for savings accounts and for loans such as mortgages. To describe a rise in the prime rate from 2.00% to 2.25% you could say, "The interest rate has risen a quarter of a *percentage point.*" This *percent* increase in the prime rate would be 12.5 percent because $2.25\%/2.00\% = 1.125$.

To avoid confusion, the Federal Reserve Board calls that percentage point change an increase of 25 *basis points* or a quarter point.

Figure 3.4 shows poor Senator Grayson pondering a 19% drop in his polling from a previous dismal 20%.

Figure 3.4. Percentage points [R96]

If that 19 is a *percentage* then the senator is polling at 81% of what he was before, so at

$$0.81 \times 20\% \approx 16\%.$$

If it's 19 *percentage points* he's in even more trouble, since

$$20\% - 19\% = 1\%.$$

3.7 Percentiles

On February 19, 2019, we visited the website www.infantchart.com/ to collect the data displayed in Table 3.5. Reading down the first column, you can see that a one year

Weight (pounds)	Percentile	Weight (pounds)	Percentile
15.5	0.2	23.5	82.2
16.5	1.1	24.5	90.3
17.5	3.9	25.5	95.3
18.5	10.2	26.5	97.9
19.5	21.5	27.5	99.1
20.5	36.8	28.5	99.7
21.5	54.1	29.5	99.9
22.5	69.9		

Table 3.5. Year old male baby weights [R97]

old male baby weighing 20.5 pounds would be in the 36.8th *percentile*. That means that he would weigh more than about 37 percent of the babies, less than 63 percent. A one year old weighing less than 21.5 pounds would be in the 54.1 percentile, so about 54% of one year old babies weigh less than he weighs.

The difference in the percentiles for 21.5 and 20.5 pounds is 54.1%−36.8% = 17.3%. That's the percentage of babies that weigh about 21 pounds.

With guess-and-check on the website we found that a weight of 21.25 pounds was as close as we could get to the 50th percentile. Half the one year olds weigh more than that, half less. That's the *median* baby weight.

Some other commonly used percentiles are the

- tenth: any baby weighing less than about 18.5 pounds is in the tenth percentile,
- ninetieth: about 90% of the babies weigh less than about 24.5 pounds,
- the quartiles — the 25th, 50th and 75th percentiles.

3.8 Exercises

Exercise 3.8.1. [S][Goal 3.2][Section 3.2] Ordinary vs. premium.
Figure 3.3 gives some data about prices of different levels of seats at Fenway Park.

(a) Compare the cost of an ordinary seat to the cost of a premium seat in 2003. You need to decide which comparison is the most informative. Consider absolute and relative change, perhaps with percentages.

(b) Compare the cost of an ordinary seat to the cost of a premium seat in 2008.

(c) Use your calculations to make a statement about the cost of ordinary seats compared to premium seats between 2003 and 2008.

Exercise 3.8.2. [S][Section 3.1][Goal 3.1][Goal 3.2] The shrinking rainy day fund.
On October 5, 2015, *The Boston Globe* published a story headlined "State's rainy day fund has dwindled over past decade" that said (in part):

> In the summer of 2007, before the massive recession began, the rainy day fund had $2.3 billion — a cushion of about 7.8 percent of total state spending, according to the Taxpayers Foundation. This summer,

the rainy day balance stood at $1.1 billion — about 2.7 percent of total state spending, a fraction small enough to raise a red flag for analysts. [**R98**]

(a) Calculate total state spending as of the summer of 2007 and the summer of 2015.

(b) What is the percentage change in total state spending between the summer of 2007 and the summer of 2015.

(c) Write a short argument supporting the statement "The 2015 rainy day fund is only about half what it was in 2007". (One good sentence will do the job.)

(d) Write a short argument supporting the statement "The 2015 rainy day fund is only about a third of what it was in 2007".

(e) Which of the previous two statements is a better description of the situation? (Why do you think so? Don't just say it's one or the other.)

Exercise 3.8.3. [S][R][Goal 3.2][Section 3.2] The New York marathon.
46,795 runners finished the 2011 New York marathon. The 1981 marathon had 13,203 finishers.
What was the percentage increase in runners who finished?

Exercise 3.8.4. [R][A][S][Section 3.1][Goal 3.4][Goal 3.2] A raise at last!
In a recent negotiation, the union negotiated a 1.5% raise for all staff. If a staff member's annual salary is $35,000, what is her salary after the raise takes effect?

Exercise 3.8.5. [S][R][Section 3.1] [Goal 3.3] The cell phone market.
In *The New Yorker* on March 29, 2010, James Surowicki wrote that the 25 million iPhones Apple sold in 2009 represented 2.2 percent of world cell phone purchases. [**R99**]

(a) Use Surowicki's numbers to estimate how many people bought a cell phone in 2009.

(b) Estimate the percentage of the world population that bought a cell phone in 2009.

(c) Estimate the percentage of the United States population that bought a cell phone in 2009.

Exercise 3.8.6. [S][Section 3.1][Goal 3.3] The 2010 oil spill.
Reuters, reporting in June on the April 2010 Gulf of Mexico oil spill, noted that

BP continues to siphon more oil from the blown-out deep-sea well. It said it collected or burned off 23,290 barrels (978,180 gallons/3.7 million liters) of crude on Sunday, still well below the 35,000 to 60,000 barrels a day that government scientists estimate are gushing from the well. [**R100**]

(a) Check that the reporters are describing the same amount of oil independent of the units (barrels, gallons or liters) used.

(b) Criticize the article for its inconsistent use of significant digits (precision) in reporting these numbers.

(c) What percentage of the oil is being collected or burned off? Your answer should be a range, not a number, expressed with just the appropriate amount of precision.

Exercise 3.8.7. New taxes?
Moved to Extra Exercises at `www.ams.org/bookpages/text-63` .

Exercise 3.8.8. [W][S][Section 3.1] [Goal 3.3] Youth sports head injuries.
The Massachusetts Department of Public Health collects annual data on brain injuries in school athletics. In 2011, the department reported that in their survey of middle and high school students, about 18 percent of students who played on a team in the previous 12 months reported symptoms of a traumatic brain injury while playing sports. These symptoms include losing consciousness, having memory problems, double or blurry vision, headaches or nausea. During that year, about 200,000 Massachusetts high school students participated in extracurricular sports. [**R101**]

(a) How many reported injuries were there?

(b) The survey reports the number injured as a percentage of the number of students participating in extracurricular sports. What is the number injured as a percentage of the population of Massachusetts?

(c) How many injured high school students would you expect in the town in which you live?

(d) Discuss the reliability of the reported 18% figure. What factors might make the true value greater? What factors might make it less?

Exercise 3.8.9. [W][S][Section 3.1] [Goal 3.3] Listen up on public broadcasting.
In 2011, the United States government was facing a $1.5 trillion federal budget deficit. Numerous cuts were proposed, including cutting federal funding of the Corporation for Public Broadcasting (CPB). The Western Reserve Public Media website argued against the cuts, saying:

> Public television is America's largest classroom, ...all at the cost of about $1.35 per person per year....
>
> Eliminating the ...investment in CPB would only reduce the $1.5 trillion federal budget deficit by less than 3 ten-thousandths of one percent, but it would have a devastating impact on local communities nationwide. [**R102**]

(a) Use the information in the first paragraph to estimate annual federal spending on public broadcasting.

(b) Find evidence online showing that your answer to the previous question is in the right ballpark.

(c) Show that the information in the last paragraph of the quotation leads to an estimate of annual federal spending on public broadcasting that is wrong by two orders of magnitude.

(d) How would you rewrite the last paragraph so that it was correct?

(e) The article discusses the spending on public broadcasting as a percentage of the federal *deficit*. That's unusual — most expenditures are reported as a percentage of the federal *budget*.

What percentage of the federal budget is that spending?

(f) Find current data on federal spending on public broadcasting, the U.S. population and the deficit, and update the statistics in the quotation.

(g) (Optional, no credit) Do you listen to public radio or public television? If so, do you contribute when they ask for money?

Exercise 3.8.10. [R][S][Section 3.1] [Goal 3.3] Coming as tourists, leaving with American babies.

From *The Arizona Republic* on August 27, 2011:

> In 2008, slightly more than 7,400 children were born in the U.S. to non-citizens who said they lived outside the country, according to the National Center for Health Statistics. ...
>
> Although up nearly 50 percent since 2000, the 7,462 children are still just a tiny fraction of the 4,255,156 babies born in the U.S. that year. [**R103**]

(a) What percentage of the total births were to foreign residents? Report your answer with the proper number of decimal places.

(b) Criticize the precision of the numbers in the article.

(c) About how many children were born in 2000 in the U.S. to non-citizens living elsewhere?

Exercise 3.8.11. [S][Section 3.1][Goal 3.3] Mercury in fog.

A researcher at the University of California at Santa Cruz studied the level of methyl mercury in coastal fog. The UCSC student newspaper reported that

> methyl mercury concentrations ranged from about 1.5 parts per trillion to 10 parts per trillion, averaging at 3.4 parts per trillion — five times higher than concentrations formerly observed in rainwater. [**R104**]

The Green Blog at *The New York Times* notes that fish is safe to eat if it has less than 0.3 parts per million of methyl mercury. [**R105**]

(a) What are the highest levels of methyl mercury recorded in rainwater?

(b) Express the methyl mercury content of the fog in parts per million.

(c) Express the methyl mercury content of the fog as a percent.

(d) Compare the methyl mercury content of the fog to the safe threshold for methyl mercury in fish.

[See the back of the book for a hint.]

Exercise 3.8.12. [S][Section 3.1][Goal 3.3] For-profit colleges could be banned from using taxpayer money for ads.

In April 2012, Senators Tom Harkin and Kay Hagan introduced a Senate bill to prohibit colleges from using federal education dollars for advertising or marketing. In March 2013 they reintroduced the bill. The Senate committee on Health, Education, Labor and Pensions reported that

- Fifteen of the largest for-profit education companies received 86 percent of their revenues from federal student aid programs — such as the G.I. Bill and Pell grants.

- In Fiscal Year 2009, these for-profit education companies spent $3.7 billion dollars, or 23 percent of their budgets, on advertising, marketing and recruitment, which was often very aggressive and deceptive. [**R106**]

(a) What were the total revenues of those fifteen for-profit colleges in fiscal year 2009?

(b) How much did those colleges receive in Federal student aid in fiscal year 2009?

(c) How much of the money they spent on advertising, marketing and recruitment could be considered as coming from the Federal government?

(d) Could the colleges maintain the same level of advertising, marketing and recruiting expenses if they did not use any Federal dollars for those purposes?

(e) What was the fate of Harkin and Hagan's bill?

Check out *Doonesbury* for February 19, 2014, at www.gocomics.com/doonesbury/2014/02/19 .

Exercise 3.8.13. [S][R][Section 3.1] [Goal 3.3] Paper jams.

A box of 20 reams of computer printer paper advertises "99.99% jam free." How many paper jams would you expect from that box?

Exercise 3.8.14. [R][S][Section 3.2] [Goal 3.1] 300 million?

We regularly use 300 million as an estimate for the population of the United States.

(a) Find the absolute and relative errors when you use that figure instead of the correct one.

(b) Answer the previous question for the years 2000, 2005 and 2010.

Exercise 3.8.15. [R][S][Section 3.2] [Goal 3.1] The minimum wage.

The federal minimum wage increased from $5.85 per hour to $7.25 per hour in 2009. Describe this increase in absolute and relative terms.

Exercise 3.8.16. [S][R][Section 3.2][Goal 3.1] Health care spending.

The January 2010 National Health Expenditures report stated that overall health care spending in the United States rose from $7,421 per person in 2007 to $7,681 per person in 2008.

(a) Calculate the absolute change in health care spending per person from 2007 to 2008.

(b) Calculate the percentage change in health care spending per person from 2007 to 2008.

(c) Estimate the total amount spent on health care in the United States in 2007 and in 2008.

Exercise 3.8.17. [S][Section 3.2] [Goal 3.1][Goal 3.2] AMG.

On October 6, 2010, you could find a graph like the one in Figure 3.6 in *The Boston Globe*, along with the assertion that stock for Affiliated Managers jumped 4.4 percent to close at its highest point since May 4. **[R107]**

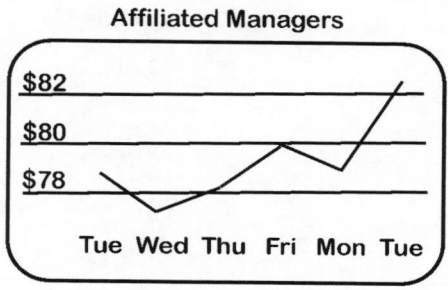

Figure 3.6. AMG stock

(a) What were the absolute and relative changes in the AMG share price between Monday and Tuesday of the week?

(b) Is the "4.4 percent jump" in the article consistent with the data in the graph?

(c) Why does the graph seem to show the AMG share price more than doubling between Monday and Tuesday even though the change was just 4.4 percent?

(d) What can you say about the AMG share price on May 4 (without trying to look up the exact value)?

Exercise 3.8.18. [S][Section 3.2][Goal 3.2] Five nines.

The posting titled "Five nines: chasing the dream?" at www.continuitycentral. com/feature0267.htm discusses the cost to industries when their computer systems are down. The "five nines" in the title refers to 99.999% availability.

Tables 3.7 and 3.8 appear there along with this quotation:

> Let us examine the math of it, first. …[T]he maximum downtime permitted per year may be calculated as reflected in Table 3.7. Please do not debate with me leap years, lost seconds or even changes to Gregorian Calendar. Equally let us not debate time travel! To quote a Cypriot saying: "I am from a village: I know nothing." The figures [in the table] are sufficiently accurate to make the points this article is trying to get across. **[R108]**

Uptime	Uptime (%)	Maximum Downtime per Year
Six nines	99.9999	31.5 seconds
Five nines	99.999	5 minutes 35 seconds
Four nines	99.99	52 minutes 33 seconds
Three nines	99.9	8 hours 46 minutes
Two nines	99	87 hours 36 minutes
One nine	90	36 days 12 hours

Table 3.7. Uptime and maximum downtime

Industry Sector	$000's Revenue / hour
Energy	2,818
Telecom	2,066
Manufacturing	1,611
Finance	1,495
IT	1,344
Insurance	1,202
Retail	1,107

Table 3.8. Cost of downtime

(a) Check the arithmetic in Table 3.7.

(b) Why do the numbers in the second column increase by a factor of 10 from each line to the next?

(c) How much would an energy sector company save if it increased its computer system availability from four nines to five? From five nines to six?

(d) Compare the savings in these two scenarios (move from four nines to five, or from five nines to six). Which upgrade is likely to cost more? Which is more likely to be worth the cost? (The web posting discusses the estimated costs of these upgrades and compares them to the savings, but that's more complex quantitatively than what we can do in this book.)

Exercise 3.8.19. [R][W][S][Section 3.1] [Goal 3.2][Goal 3.3] How many master's degrees?

The National Center for Educational Statistics reported that in 2008–2009, 178,564 master's degrees in education were awarded in the United States and that this represented 27% of all master's degrees awarded. [**R109**]

How many master's degrees were awarded in 2008–2009?

(a) Make a quick back-of-an-envelope estimate for the answer.

(b) Calculate to find a more precise result, appropriately rounded.

(c) Compare your estimate and your calculation.

Exercise 3.8.20. [S][Section 3.3] [Goal 3.3] Rent in Boston.

According to a survey conducted in spring 2006 by Northeast Apartment Advisors, an Acton, MA, research firm, the average monthly rent in Greater Boston was $1,355, an increase of 3.6 percent from 2005. What was the average monthly rent in 2005?

Exercise 3.8.21. [R][S][Section 3.1] [Goal 3.3] Counting birds.

In *The Washington Post* on April 25, 2010, Juliet Eilperin wrote that birder Timothy Boucher said he had 4,257 species on his "life list" of birds he'd seen; 43 percent of all the bird species there are. [**R110**]

(a) If this report is correct, about how many bird species are there?

(b) Can you find independent evidence that there are that many species?

Exercise 3.8.22. [S][Section 3.1] [Goal 3.1][Goal 3.2] 60 years in seconds.

In Chapter 2 we calculated the number of seconds in 60 years two ways and found 1.89341556×10^9 and $1,893,415,558$.

(a) What is the absolute difference between these two figures?

(b) What is the relative change, expressed as a decimal and as a percentage?

(c) Which way of describing the difference is likely to be more useful?

Exercise 3.8.23. [S][Section 3.1][Goal 3.2] How long is a year?

In Fermi problems we often use 400 days/year to make the arithmetic easier.

(a) What is the relative error when you use 400 days/year instead of 365 days/year?

(b) What does the Google calculator tell you when you ask it for

$$\boxed{\text{one year in days}} \ ?$$

Find a website that tells you the number of days in a year accurate to at least four decimal places.

(c) What are the relative and absolute errors when you use Google's answer instead of the four decimal place answer?

(d) A leap year every four years would correspond to a year with 365.25. How does our current leap year schedule take the discrepancy into account?

Exercise 3.8.24. [C][W][S][Section 3.3] [Goal 3.3] Improving fuel economy.

If your car's fuel economy increases by 20% will you use 20% less gas?

(a) Show that the answer to this question is "no" if you measure fuel economy in miles per gallon.

(b) Show that the answer to this question is "yes" if you measure fuel economy in gallons per mile.

[See the back of the book for a hint.]

Exercise 3.8.25. [S][Section 3.3][Goal 3.2] Bytes.

In Chapter 1 we saw that a kilobyte is 1,024 bytes, not the 1,000 bytes you expect.

(a) What is the percentage error if you use 1,000 bytes instead of 1 kilobyte?

(b) How many bytes in a megabyte (exact answer, please)? In a gigabyte?

(c) What is the percentage error if you work with 1,000,000 bytes per megabyte?

Exercise 3.8.26. [S][W][Goal 3.5] [Section 3.3][Goal 3.4][Goal 3.3] Hard times.

The boss says, "Hard times. In order to avoid layoffs, everyone takes a 10% pay cut." The next day he says, "Things aren't as bad as I thought. Everyone gets a 10% pay raise, so we're all even."

(a) Is the boss right?

(b) What if the 10% pay raise comes first, followed by a 10% cut?

Exercise 3.8.27. [S][Section 3.3][Goal 3.3] [Goal 2.3] Oil spill consequences.

In a June 19, 2010, article headlined "Gulf oil spill could lead to drop in global output" *The Denver Post* quoted the Dow Jones Newswire:

> Global oil output could decline up to 900,000 barrels a day from pro-
> jected levels for 2015 if oil-producing countries follow the U.S. lead
> and impose moratoriums on development of new offshore oil reserves.
> ...[that] would represent a mere 1 percent or so of global oil output.
> [**R111**]

(We consider other numbers related to the Gulf oil spill in Exercise 3.8.6.)

(a) Use the figures in the quotation to estimate the projected global output for 2015.

(b) Do enough research to verify the order of magnitude of your answer.

Exercise 3.8.28. [S][Section 3.3] [Goal 3.3][Goal 3.2] Taking the fifth.

Several years ago the liquor industry in the United States started selling wine bottles containing 0.75 liters instead of bottles containing 1/5 of a gallon ("fifths"). They charged the same amount for the new bottle as the old. What was the percentage change in the cost of wine?

Exercise 3.8.29. [S][Section 3.3][Goal 3.3] Tax holiday.

Occasionally a state designates one weekend as a tax-free holiday. That is, consumers can purchase some items without paying the sales tax. Many stores advertise the savings in the days before the weekend, to encourage customers to shop with them. The sales tax in South Carolina (in 2013) was 6%. Would it be right for an ad to read, "No sales tax this weekend — save 6% on your purchases!"

Exercise 3.8.30. [S][W][Section 3.3] [Goal 3.3][Goal 3.6] Measuring markups.

The standard markup in the book industry is 50%: the retail price of a book is one and one half times the wholesale price.

(a) What percentage of the retail price of a book is the bookstore's markup?

(b) The New England Mobile Book Fair advertises its bestsellers as "30% off retail." What is their markup on bestsellers, as a percentage of the wholesale price?

(c) Answer the previous question if the Book Fair discounts bestsellers by 40%. Would they ever do that?

[See the back of the book for a hint.]

Exercise 3.8.31. [S][C][Section 3.3][Goal 3.2][Goal 3.3] Goldman Sachs bonuses.

On January 21, 2010, the *Tampa Bay Times* carried an Associated Press report headlined "Goldman Sachs limits pay, earns $4.79 billion in fourth quarter" reporting that in 2009 the bank gave out salaries and bonuses worth $16.2 billion. That figure was 47 percent more than the year before.

> In all, compensation accounted for 36 percent of Goldman's $45.17 billion in 2009 revenue. ...In 2008, Goldman set aside 48 percent of its revenue to pay employees. [**R112**]

(a) Are the numbers $16.2 billion, 36% and $45.17 billion in the report consistent?

(b) How much did Goldman Sachs pay out in bonuses in 2008?

(c) What were Goldman Sachs's revenues in 2008?

(d) What do you make of the discrepancy between the headline and the text in the article? Read the article to find out.

Exercise 3.8.32. [S][C][Section 3.3] [Goal 3.2][Goal 3.3][Goal 3.7] More than 40m now use food stamps.

In October 2010 Bloomberg News quoted a government report that

> The number of Americans receiving food stamps rose to a record 41.8 million in July, [**R113**]

noting that was an 18 percent increase over the previous year.

The White House estimated that in the following year food stamps will go each month to "43.3 million people, more than an eighth of the population".

(a) About how many Americans were receiving food stamps in 2009, the year before this article appeared?

(b) How did the percentage of the population receiving food stamps in July change between 2009 and 2010?

(c) Was 43.3 million people more than an eighth of the population?

Exercise 3.8.33. [S][C][Section 3.3] [Goal 3.3][Goal 3.2][Goal 3.6] Gamblers spending less time, money in AC casinos.

In December 2010 NBC News reported that compared to 2006 gamblers were spending 22 percent less time in Atlantic City casinos and spending almost 30 percent less money. As a result, revenue in 2009 was just $3.9 billion — down from $5.2 billion in 2006.

Gross gaming revenue fell from $9.13 per hour in 2006 to $6.42 [in 2010].

Gross operating profit per visitor hour went from $2.74 in 2006 to $1.05 in the third quarter of this year. [**R114**]

(a) The units "per hour" in the first quoted line are wrong. What should they be?

(b) Calculate the percentage change in gross operating profit per visitor hour from 2006 to 2009. Does this match the assertion in the article?

(c) Calculate the percentage change in gaming revenue per visitor hour from 2006 to 2009. Does this match the assertion in the article?

(d) The article says the total number of hours gamblers spent in the third quarter is down 22% and the gross revenue per hour is down 30%. What percentage decrease in revenue would this combination lead to?

(e) Calculate the percentage change in revenue. Is your answer consistent with the numbers in the article?

Exercise 3.8.34. [S][Section 3.3][Goal 3.1][Goal 3.2] Tracking Harvard's endowment.

On September 23, 2011, *The Boston Globe* published data from Harvard University on the performance of its endowment from 2007 to 2011. We've displayed that information in Figure 3.9. [**R115**]

The graph on the left shows the return — the overall interest — on the endowment from year to year; the one on the right the actual value.

(a) Between 2007 and 2008 the graph on the left goes down while the one on the right goes up. How is that possible?

(b) For the years 2008, 2009, 2010 and 2011 use the second graph to compute the absolute and relative changes in endowment assets from the previous year.

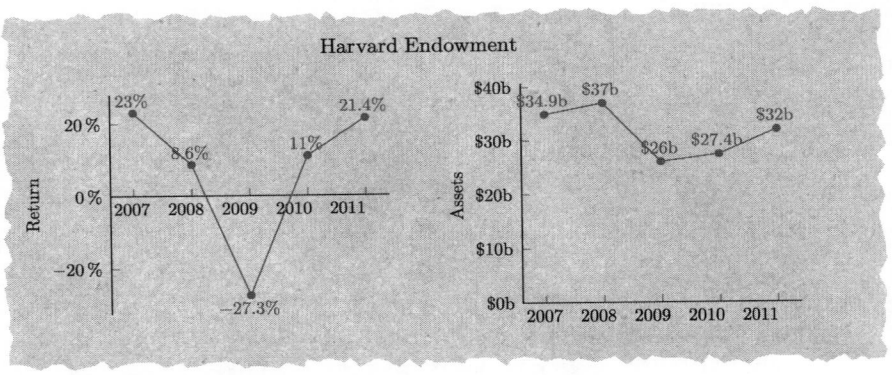

Figure 3.9. Tracking Harvard's endowment

Exercise 3.8.35. [S][Section 3.3][Goal 3.2][Goal 3.3][Goal 3.7] Banks to make customers pay fee for using debit cards.

From *The New York Times* , Friday, September 30, 2011:

> Until now, [debit card] fees have been 44 cents a transaction, on average. The Federal Reserve in June agreed to cut the fees to a maximum of about 24 cents. While the fee amounts to pennies per swipe, it rapidly adds up across millions of transactions. The new limit is expected to cost the banks about $6.6 billion in revenue a year. [**R116**]

(a) Use the data in the article to estimate the number of debit card transactions in a year.

[See the back of the book for a hint.]

(b) What assumptions did you make in arriving at your estimate?

(c) Is the phrase "millions of transactions" a good way to describe the order of magnitude of your answer?

(d) What percentage of a $10 debit card transaction does a merchant have to pay to the bank that issued the card? How will that percentage change when the new rule takes effect?

(e) Rewrite your estimate in (a) in transactions per day.

(f) Estimate the number of people in the United States who use a debit card. Then rewrite your answer to the previous problem in transactions per person per day.

(g) (Optional) Do you use a debit card? Did you know how much your use costs the merchant? Will you change your behavior if your bank charges $5/month for debit card use?

Exercise 3.8.36. [S][Section 3.3] [Goal 3.1][Goal 3.3] Taxing gasoline.

The federal tax on gasoline at the end of 2011 was 18.4 cents per gallon. That hadn't changed since 1993, when gasoline cost just $1.16 per gallon (tax included).

(a) What percentage of the 1993 cost of gasoline was the federal tax in 1993?

(b) If the average cost of gas in 2011 was $3.40 per gallon, what percentage of the cost of gasoline then was the federal tax?

(c) What would the cost of a gallon of gasoline have been at the end of 2011 if the percentage rather than the amount of federal tax was the same then as in 1993?

(d) Gasoline consumption in the United States was estimated to be about 175 million gallons per day in 2011. How much revenue was generated by the federal gas tax at its then-current rate? How much would have been generated if tax were computed using your answer to part (c)?

Exercise 3.8.37. [S][Section 3.3] [Goal 3.1][Goal 3.3] How long is a microcentury?

The mathematician John von Neumann is often identified as the source of the fact that a 50 minute lecture lasts about as long as a microcentury.

(a) Check this fact by doing your own arithmetic.

(b) What are the absolute and relative errors in the "about"?

(c) If you could listen to lectures one after another, day in and day out for a century, about how many would you hear?

(d) You can find several answers on the web to the question, "How long is a microcentury?" Critique them.

(e) Estimate how many years it would take for professors at your school to have delivered a century's worth of classes. (Don't assume the classes were back to back year round.)

Exercise 3.8.38. [S][W][Section 3.3][Goal 3.2] [Goal 3.3][Goal 3.4] Currency conversion.

When you read this question you will see that you need at least two days to answer it.

(a) Suppose your company is sending you to France for business and asks you to convert $1,250 (U.S. dollars) to euros in preparation for your trip. Find a currency conversion calculator on the web and use it to find out how many euros your $1,250 will buy. Your answer should clearly identify the website, the date and time and the conversion factor used (in euros/dollar) as well as the amount of euros you would get. Imagine that you have done the exchange, so that you now have euros instead of dollars.

(If you'd rather your company arranged your business trip to some other country, feel free to change the problem accordingly.)

(b) Two days after you changed your money into euros, your company calls off the trip. You need to change your money back into dollars. Go back to the web to figure out what you would get back in dollars for the euros that you have. Again identify the website, date and time, and conversion factor, this time in dollars/euro. Once you've done the conversion, compare the amount of money you now have with the original amount. Calculate the percentage gain (or loss) due to the change in the exchange rate.

(c) If you actually converted dollars to euros and back, the bank would charge a fee each time. Suppose that fee is 2% of the amount converted. Go through the problem again, but this time account for the 2% fee. That is, figure out how many euros you would have gotten in the first conversion, with the 2% fee included, then how many dollars back in the second, with another 2% fee. Finally, calculate the total percentage gain (or loss).

(d) If the conversion rate was exactly the same each time (unlikely, but imagine it) would your loss in the conversion to euros and back be 4%?

Travel note: If you're planning a trip to a foreign country, explore the cheapest way to convert your money. Getting cash at a local bank is probably not the answer. Your debit and credit cards will probably work worldwide — ask your credit card company about the fees. And tell them that you will be using the card elsewhere so they don't think it's been stolen.

Exercise 3.8.39. [S][Section 3.3][Goal 3.4] Town weighs future with fossil fuel.
   The *Times* reported on April 18, 2012, that Boardman, Oregon, is one of several port cities thinking about shipping coal mined in Wyoming and Montana to Asia.
   If all the cities decided to do this

> ... as much as 150 million tons of coal per year could be exported from the Northwest, nearly 50 percent more than the nation's entire coal export output last year. [**R117**]

(a) Use the data in the quotation (not a web search) to estimate how many tons of coal the United States exported in 2011.

(b) Use the web to confirm (or not) your estimate.

Exercise 3.8.40. [W][S][Section 3.3][Goal 3.1] [Goal 3.4][Goal 3.6] Brand loyalty.
   An article headlined "Seeking savings, some ditch brand loyalty" appeared in *The Boston Globe* on January 29, 2010, saying (among other things) that

> Unit sales of private label goods have jumped 8 percent since 2007, while brand names have declined roughly 4 percent, according to Nielsen Co. [**R118**]

You can see the data in Figure 3.10.

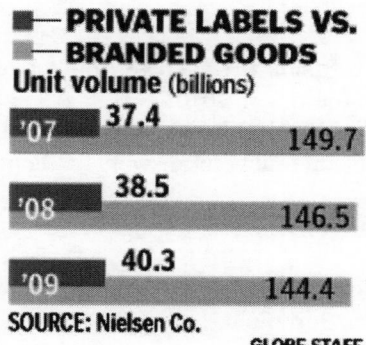

PRIVATE LABELS VS. BRANDED GOODS
Unit volume (billions)

'07 37.4   149.7

'08 38.5   146.5

'09 40.3   144.4

SOURCE: Nielsen Co.
                    GLOBE STAFF

Figure 3.10. Private labels, branded goods

(a) Find the absolute and relative changes in unit sales of private label and brand name goods in the years since 2007.

(b) Are the percentages in the article consistent with the numbers in the graphic?

(c) How have total sales changed between 2007 and 2009?

(d) How has the percentage of private label sales changed in this period?

Exercise 3.8.41. [S][Section 3.3][Goal 3.2] [Goal 3.3][Goal 3.4] Research funding.
   On February 10, 2011, *The Boston Globe* carried an article that said

> The University of Massachusetts received a record amount of funding for research in fiscal year 2010, taking in $536 million, according to preliminary figures. That was an increase of $47 million, or 9.5 percent, over the previous year. [**R119**]

Use the information in the article to calculate the total research funding for the UMass system in fiscal year 2009 two ways, by subtracting, and using the "1 + " idea. Do you get the same answer each way?

Exercise 3.8.42. [S][Section 3.3][Goal 3.4] [Goal 3.3] Long gone?
    On March 4, 2011, columnist Alex Beam wrote in *The Boston Globe*:

> PBooks? No one has read a "p-book" — that's industry jargon for "print books" — in years, according to nice Mr. Bezos of Amazon, who would like to sell you e-books on his Kindle gizmo. How counterintuitive, then, that total U.S. books sales increased 3.9 percent to almost $12 billion in 2010, according to the Association of American Publishers.
>
> I just read two p-books: "The Last Crossing," by Guy Vanderhaeghe, and Mordecai Richler's hilarious "Barney's Version," which I am told is 20 times better than the movie. And I read them for free, thanks to my local library.
>
> Wait — aren't libraries obsolete? The subject of my next column, perhaps. [**R120**]

(a) What were book sales in 2009?

(b) Use the data given to estimate the average number of dollars spent and books bought by each adult in the U.S. in 2010.

(c) Discuss how reasonable you find your answers to the previous question.

(d) Can you make quantitative sense of "20 times better" in the second paragraph?

Exercise 3.8.43. [S][Section 3.3][Goal 3.3] [Goal 3.4] Student loan growth.
    On March 21, 2011, John Sununu wrote in *The Boston Globe* that in 2010 for-profit schools had nearly 2 million students — 10 percent of American higher education.

> Since 2003, the total debt burden from education loans to private and public institutions has grown 18 percent per year and now stands at over $500 billion. [**R121**]

(a) Use this information to calculate the total number of students in American higher education in 2010.

(b) What was the average student debt in 2010? (Assume the $500 billion figure is for 2010.)

(c) Use the information in the quotation to calculate total student loan debt burden for each of the years between 2003 and 2009.

(d) Estimate the average student debt for each of the years between 2003 and 2010.

    [See the back of the book for a hint.]

Exercise 3.8.44. [U][Section 3.3][Goal 3.3] [Goal 3.4] One thousand three hundred percent.

On September 29, 2011, Hiawatha Bray wrote in *The Boston Globe* that

> The social networking team at Google Inc. has a great sense of timing. They picked last week to tear down the gates at their new Google+ service, which was previously accessible only by invitation. It worked. Google+ visits surged by nearly 1,300 percent on the week of September 24, with 15 million users coming to the site. [**R122**]

How many users were there before the gates fell?

Exercise 3.8.45. [S][W][Section 3.4] [Goal 3.4] What's a good basketball player worth these days?

In *The Boston Globe* on June 30, 2010, you could read that

> [A] maximum deal for [LeBron] James would start at $16.56 million (at the minimum) and increase by 8 percent each season over the five-year period. [**R123**]

(a) Use the 1+ trick four times to calculate James's salary in the fifth season.

(b) In the five seasons, James will get four eight percent salary increases. Compare the result of your calculation to a single $4 \times 8\% = 32\%$ increase.

Exercise 3.8.46. [S][R][Section 3.4][Goal 3.3] [Goal 3.4][Goal 3.6] An unexpected bill.

On December 17, 2010, the Wilmington, North Carolina, *StarNews* carried an Associated Press report headlined "13 million get unexpected tax bill from Obama tax credit". The report said that

> The Internal Revenue Service reported that the average tax refund was $2,892 in the 2010 filing season, up from $2,663 in 2009. However, the number of refunds dropped by 3.5 percent, to 93.3 million. [**R124**]

(a) What was the percentage increase in the average tax refund from 2009 to 2010?

(b) How many refunds were there in 2009 and in 2010?

(c) The number of refunds decreased while the average refund increased, so you need some arithmetic to decide how the total amount refunded changed from 2009 to 2010. Find that change, in both absolute and percentage terms.

(d) Does the headline match the quote in accuracy? In tone?

Exercise 3.8.47. [S][R][Section 3.4] [Goal 3.2][Goal 3.6] Bargain bread.

A local supermarket offers

> Special
> Pain au Levain
> 2 for $5
> Regularly $3.99
> Over 30% off.

(a) Is the claim true?

(b) How much more than 30% is the actual savings?

Exercise 3.8.48. [S][A][Section 3.4] [Goal 3.3][Goal 3.4][Goal 3.6] Tax and discount.
 Suppose sales tax is 5% and a $117 item is discounted 40%. Compute the final price and the sales tax two ways:

(a) First find the discounted price, then the sales tax, then the final price.

(b) First find the sales tax, then the discounted (final) price.

(c) Does it matter to you which way the computation is made? Does it matter to the state? (Remember: the state collects the sales tax!)

Exercise 3.8.49. [S][Section 3.4] [Goal 3.6][Goal 3.4] Free!?
 If a shirt on the clearance rack is marked "50% off original price" and you have a preferred customer coupon giving you 50% off the sale price of any item, do you get the shirt free?

Exercise 3.8.50. [S][Section 3.6] [Goal 3.7][Goal 3.2] Paying off your student loan.
 We found this information about student loans on the website for Colorado's Red Rocks Communtity College.

> Direct debit is the most convenient way to make your student loan payments — on time, every month. Direct debit is a free service. You may qualify for a 0.25% interest rate reduction when actively making payments on Direct Debit. [**R125**]

(a) Explain how direct debit works. (You may need to look this up. You may want to look it up even if you think you know.)

(b) If you have a student loan on which you pay interest at a rate of 6.8% what will your interest rate be if you sign up for direct debit?

(c) Why should the quotation say "0.25 percentage point interest rate reduction" instead of "0.25% interest rate reduction"?

(d) If the quotation really means what it says, what will the interest rate be on your 6.8% loan if you sign up for direct debit?

Exercise 3.8.51. [S][W][Section 3.6][Goal 3.7] [Goal 3.2] Benefits take hit in Patrick budget.
 In *The Boston Globe* on January 13, 2008, reporter Matt Viser wrote:

> Looking for ways to trim a looming $1.3 billion state budget gap, Governor Deval Patrick will propose shifting more of the cost of health insurance premiums onto tens of thousands of state employees.
>
> Under his plan, about 37,000 employees would see their monthly premiums increase by 10 percent.
>
> ...
>
> Right now, most employees pay 15 percent, and the state covers 85 percent. [In the plan] those making more than $50,000 would pay 25 percent.

...

For the 37,000 employees who would face the 10 percent increase, that would mean additional monthly costs of $51 for an individual plan and $120 for a family plan. [**R126**]

(The Commonwealth of Massachusetts went ahead and made this change in 2009.)

(a) Explain why this increase of 10 percentage points is really a 67 percent increase.

(b) Why might Governor Patrick prefer to publicize the 10 instead of the 67?

(c) Write a letter to the editor pointing out the error in the article.

Your letter should be short, accurate and pointed or funny in some way if possible.

Exercise 3.8.52. [S][C][Section 3.3][Goal 3.2][Goal 3.6] Slow down; save gas. A U.S. government webpage on fuel economy said in June 2015 that

While each vehicle reaches its optimal fuel economy at a different speed (or range of speeds), gas mileage usually decreases rapidly at speeds above 50 mph.

You can assume that each 5 mph you drive over 50 mph is like paying an additional $0.19 per gallon for gas.

Observing the speed limit is also safer. [**R127**]

Figure 3.11 is from the same webpage.

(a) Use the graph to estimate the percentage decrease in fuel economy that results when you increase your average speed from 50 mph to 55 mph. (Think about whether you should measure fuel economy in miles per gallon or in gallons per mile. You might want to look at Section 2.2.)

(b) Use the graph to estimate the percentage decrease in fuel economy that results when you increase your average speed from 55 mph to 60 mph.

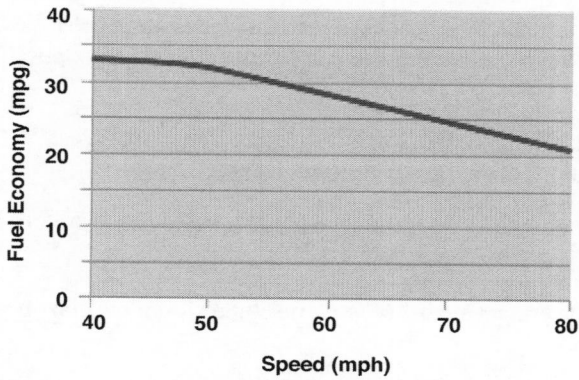

Figure 3.11. Slow down, save gas [**R128**]

(c) Use the graph to estimate the percentage decrease in fuel economy that results when you increase your average speed from 60 mph to 65 mph.

(d) Are these percentages the same? Do your calculations support the common assumption that (once you are at a high enough speed) every time you increase your speed by 5 mph you decrease your fuel efficiency by about 5%?

(e) What assumption is being made about the price of gas in these calculations?

Exercise 3.8.53. [S][Section 3.3][Goal 3.2][Goal 3.3] Holiday shows bringing box office joy.

The Boston Globe published the data in Figure 3.12 on December 27, 2012.

(a) Verify the four percentage calculations.

(b) For both venues, revenue increased faster than attendance between 2009 and 2012. That means the average ticket price must have increased. Compute the average ticket price in each year for each venue, and the percentage change in each case.

BSO Holiday Pops		Christmas Revels	
Attendance		**Attendance**	
2009	68,771	2009	15,400
2012	78,916	2012	17,200
	+15%		+12%
Revenue		**Revenue**	
2009	$4.9m	2009	$662k
2012	$5.8m	2012	$770k
	+18%		+16%

Figure 3.12. Good news at the box office [**R129**]

Exercise 3.8.54. [U][Section 3.3][Goal 3.2][Goal 3.4] Taxing online purchases.

On October 28, 2013, The Boston Globe reported that on the following Friday Massachusetts would begin collecting the 6.25% sales tax on items purchased from Amazon.

> The state Department of Revenue expects the tax to raise $36.7 million before the fiscal year ends June 30. [**R130**]

(a) What total taxable sales does the Department of Revenue predict for the rest of the fiscal year?

(b) Estimate the average (mean) Massachusetts resident's Amazon purchases during that time period.

(c) Estimate the number of Massachusetts residents who shop online. Then estimate their Amazon purchases during that time period.

Exercise 3.8.55. [U][S][Section 3.3][Goal 3.4] How hot was it?

On Friday, March 23, 2012, *The Boston Globe* claimed that the 82 degree record temperature on March 22 was about 35% higher than the historical average of 61 degrees.

(a) Convert these Boston record and average temperatures to their equivalents in degrees Celsius.

(b) Ben Bolker lives in Hamilton, Ontario, Canada, where they measure temperature in degrees Celsius. What percentage change would his newspaper, the *Hamilton Spectator*, have reported for those temperatures?

(c) The different answers to the previous two questions show that talking about a percentage change in a temperature makes no sense. Write a letter to the editor explaining what's wrong with the claim.

Exercise 3.8.56. [S] [Section 3.4][Goal 3.2][Goal 3.4][Goal 3.6] What should you do after you graduate?

According to Vox.com, "the top 25 hedge fund managers earned a collective $21.1 billion this year." [**R131**]

Vox.com put this figure into context, saying "it's about 0.13 percent of total national income for 2013 being earned by something like 0.00000008 percent of the American population."

(a) Check the calculation that the 25 hedge fund managers are the percent of the population claimed.

(b) Use the data from Vox.com to estimate the total national income in 2013.

(c) The same Vox.com article stated that the earnings of the 25 hedge fund managers were "about 2.5 times the income of every kindergarten teacher in the country combined." Can you verify this income comparison of hedge fund managers and kindergarten teachers?

Exercise 3.8.57. [S][Section 3.6][Goal 3.7] Payday loans in the military.

In their study "In Harm's Way? Payday Loan Access and Military Personnel Performance" in the *Review of Financial Studies* S. Carrell and J. Zinman report that

> Access [to payday loans] significantly increases the likelihood that an airman is ineligible to reenlist by 1.1 percentage points (i.e., by 3.9%). We find a comparable decline in reenlistment. Payday loan access also significantly increases the likelihood that an airman is sanctioned for critically poor readiness by 0.2 percentage points (5.3%). [**R132**]

(a) What is the likelihood (as a percent) that an airman is ineligible to reenlist when no payday loans are available? What is the percentage when they are?

(b) What is the likelihood (as a percent) that an airman is sanctioned for critically poor readiness when no payday loans are available? What is the percentage when they are?

(c) What is a payday loan? Should servicemen and women be paid so poorly that they need them?

Exercise 3.8.58. [S][R][Section 3.3][Goal 3.2] Who's ahead?

On September 10, 2014, the race for the American League Central Division was close. Both the Kansas City Royals and the Detroit Tigers had a winning percentage of 55.2%. Their won-lost records were

team	won	lost
Royals	79	64
Tigers	80	65

(a) Check the percentage calculations.

(b) Which team was really ahead? By how much?

Exercise 3.8.59. [S][Section 3.2][Goal 3.2] Doubling down on education.

On October 4, 2014, Josh Boak of the Associated Press reported that in the 2007 recession families behaved differently. Those in the top 10%, earning $253,146 annually, spent more on education for their children. That was a 35% increase to $5,210 per year over the previous two years.

> For the remaining 90 percent of households, such spending averaged about a flat $1,000, according to research by Emory University sociologist Sabino Kornrich. [**R133**]

Use the numbers in the quotation as much as you can to answer the questions that follow. When you need numbers that aren't there, estimate them, with or without the web. Be clear about any assumptions you make. If you do look things up, be sure to cite your sources.

(a) What were the wealthiest households spending per child on education before the increase?

(b) What percentage of their income were the wealthiest households spending on education after the increase?

(c) How did the percentage of their income devoted to education change when the recession started?

(d) What was the average education spending per child for all households?

(e) Estimate the 2007 total national household annual education spending on which this study is based.

(f) What does "double down" mean? Did the wealthiest households double down on education spending?

Exercise 3.8.60. [S][Section 3.3][Goal 3.2][Goal 3.4] Billions more for nukes.

On November 14, 2014, Secretary of Defense Chuck Hagel announced that

> the Defense Department will boost spending on the nuclear forces by about 10 percent a year for the next five years. ...That would be a total increase of about $10 billion over the five years. [**R134**]

(a) What percentage increase will five years at 10 percent per year amount to?

(b) Use your answer to the previous question to estimate 2014 Defense Department spending on nuclear forces.

(c) Compare your answer to the amount spent subsidizing school lunches. (You can do that with a web search or a Fermi problem estimate.)

Exercise 3.8.61. [S][Section 3.3][Goal 3.2][Goal 3.4][Goal 3.6] Top 400.
On November 25, 2014, a blogger at *The Washington Post* wrote about an IRS report noting that in 2010 the top 400 households

> ... or the top 0.0003 percent ... took home 16 percent of all capital gains. That's right: one out of every six dollars that Americans made selling stocks, bonds, and real estate ... went to the top-third of the top-thousandth percent of households. [**R135**]

At www.irs.gov/pub/irs-soi/10intop400.pdf you can find out that the total capital gains reported by those 400 households was about $59 million. (That website seems to be the source for the data in the quote.)

(a) Is 0.0003 percent one-third of one-thousandth of a percent?

(b) Use the data in the quotation to estimate the number of households in the U.S. in 2010. Report only a reasonable number of significant digits in your answer.

(c) Estimate the number of households some other way to show that this answer is the right order of magnitude.

(d) Is 16% the same as "one out of every six"?

(e) How much total capital gains were reported by all households in 2010?

(f) When this blog post appeared, tax in the top bracket for ordinary income was 39.6 percent while the tax on capital gains was just 15 percent. How much more in tax would the government have collected if the capital gains had been taxed at the ordinary income rate?

Exercise 3.8.62. [S][Section 3.1][Goal 3.2][Goal 3.4] Batting averages.
In 1941 Ted Williams's batting average was .406. (We would recommend writing that as 0.406, but baseball doesn't do it that way.) He had 185 hits in 456 official at bats.
According to Wikipedia,

> Before the game on September 28, Williams was batting .39955, which would have been rounded up to a .400 average. Williams, who had the chance to sit out the final, decided to play a doubleheader against the Philadelphia Athletics. Williams explained that he didn't really deserve the .400 average if he did sit out. Williams went 6-for-8 on the day, finishing the baseball season at .406. [**R136**]

(a) Check the computation of Williams's final batting average.

(b) Check the computation of Williams's batting average before the final day of the season.

(c) Williams had six hits in eight at bats on that last day. Would he have batted .400 for the season if he'd gotten only five? What if only four? Three?

Exercise 3.8.63. [S][Section 3.3][Goal 3.2][Goal 3.4] Boston city councilors vote for $20,000-a-year raise.

That was the headline in an October 2014 story in *The Boston Globe*. The article said the new salary would be $107,500. [**R137**]

(a) What percent increase in salary does this represent?

(b) If the federal minimum wage of $7.25 per hour were increased by the same percentage, how much more per year would someone earn who worked full time at the minimum wage? (Be sure to explain your definition of "full-time".)

Exercise 3.8.64. [S][Section 3.1][Goal 3.2] Dogfish.

The dogfish shark is one of the few fish still abundant in New England coastal waters. In a September 28, 2014, AP story reported in *The Washington Times* you could read that

> Maine fishermen caught a little more than 100,000 pounds of dogfish in 2013 at a total value of $17,945, barely a tenth the price per pound of haddock, and less than 7 percent of the price per pound of cod. The total value of the cod catch was $736,154, while for haddock it was $211,279. [**R138**]

(a) What was the 2013 price per pound of dogfish?

(b) What were the approximate prices per pound of haddock and of cod?

(c) Estimate the size in pounds of both the haddock and cod catch in Maine in 2013.

Exercise 3.8.65. [U][Section 3.1][Goal 3.3] Use first gear!

Figure 3.13 shows a road sign you might see at the bottom of a hill on a road in New Zealand, warning that a steep grade lies ahead.

(a) If the grade is 4 km long, how many meters higher is the top of the hill than the bottom?

(b) If the grade is 2.5 miles long, how many feet higher is the top of the hill than the bottom?

Figure 3.13. Steep grade [**R139**]

(c) Why are these questions a little easier to answer in the metric system than in the English system?

(d) Is the grade in the figure really about 12%?

Exercise 3.8.66. [U][Section 3.1][Goal 3.3] Writing your dissertation.

Author Joan Bolker's book *Writing Your Dissertation in Fifteen Minutes a Day* was published in 1998. The hundred thousandth copy sold in 2011.

Approximately what percentage of the graduate students writing dissertations in those years bought the book?

Review exercises.

Exercise 3.8.67. [A] Some routine percentage calculations.

(a) Calculate 40% of 250.

(b) Calculate 130% of 79.

(c) 85 is what percentage of 140?

(d) 62 is what percentage of 30?

(e) 30% of what number is 211?

(f) 115% of what number is 52?

(g) After a 25% increase a number is 240. What was the number before the increase?

(h) After a 20% decrease a number is 192. What was the number before the decrease?

Exercise 3.8.68. [A] For each problem, calculate the 6.25% sales tax and also the final price. Redo the calculation using the 1+ trick.

(a) A book with sticker price $12.99.

(b) A computer priced at $499.99.

(c) A necklace priced at $32.00.

(d) A DVD selling for $5.50.

Exercises added for the second edition.

Exercise 3.8.69. [S][Section 3.3][Goal 3.2][Goal 3.4] Miniscule meals tax.

Maria Mellone, an instructor at UMass Boston, encountered Figure 3.14 at Walnut Street Cafe in Lynn, MA, and wrote

I'm out getting coffee and spotted this … I'm cringing.

(a) Calculate the tax and the total purchase price for the egg and cheese.

(b) What would the total purchase price be if the tax were written correctly?

(c) How did the restaurant arrive at the weird price $3.53 for the egg and cheese sandwich?

Figure 3.14. What was that tax, again?

(d) What would the restaurant charge for a sandwich they planned to sell for $5 including tax?

Exercise 3.8.70. [U][S][R][Section 3.3][Goal 3.4] Federal workers deserve a raise.
On February 8, 2019, the *Government Executive* website featured a story headlined

Democrats Propose 3.6 Percent Raise for Feds in 2020. [**R140**]

How much would a federal worker making $46,000 per year earn after such a raise?

Exercise 3.8.71. [S][R][Section 3.3][Goal 3.4] Town taxes on the increase.
In March 2019, the *Hartford Courant* reported that

Newington Town Manager Tanya Lane proposed a $124.2 million budget Monday night — a 4.7 percent increase over the previous year. [**R141**]

What was Newington's budget in the "previous year"?

Exercise 3.8.72. [U][S][Section 3.7][Goal 3.8] Who's gifted?
The blog giftedissues.davidsongifted.org/BB/ubbthreads.php/topics/152941/Re_Innumeracy_in_Gifted_Educat.html offers two very interesting quotations: [**R142**]
From *The Everything Parent's Guide to Raising a Gifted Child* by Sarah Robbins (p. 125):

Unfortunately, highly gifted children (those in the 95th percentile) only occur in approximately 1 out of 1,000 preschoolers, and profoundly gifted children (those in the 99.9th percentile) are as rare as 1 in 10,000 preschoolers.

From *Giftedness 101* by Linda Silverman (p. 87):

> In our mushrooming populace, over 3 million Americans and approx-
> imately 70 million global citizens are highly gifted or beyond (99.9th
> percentile).

What is wrong with the arithmetic?

Exercise 3.8.73. [S][Section 3.2][Goal 3.2][Goal 3.4] Bigger as well as faster.
The New York Times published Figure 3.15 on March 4, 2019.

	portion size (grams)	**calorie count**
1986	100	234
1991	112	266
2016	171	420

Figure 3.15. Fast food entrees [**R143**]

(a) Calculate the percent changes in portion size between 1986 and 1991, and between 1991 and 2016.

(b) Use the 1+ trick to find the combined percentage increase, and check the result with a direct calculation.

(c) How did the number of calories per gram change between 1986 and 2016?

Exercise 3.8.74. [S][Section 3.1][Goal 3.4][Goal 3.6] Ice cream!
Figure 3.16 shows the ice cream rewards card from Truly Yogurt, in Wellesley, MA.

Figure 3.16. Ice cream rewards

(a) Kiddie cones cost $3.00. How much will you pay for 11 of them if you use your rewards card?

(b) What is the absolute cost of each of the 11 kiddie cones? What is the percentage discount?

(c) Can you calculate the percentage discount on the regular cones without knowing what each costs?

(d) Suppose I used the rewards card filled with 10 kiddie cone purchases to get a $6 large cone free. Calculate the percentage discount on all my ice cream purchases.

Exercise 3.8.75. [S][Section 3.3][Goal 3.4] It will cost more to see Bryce Harper.

On March 1, 2019, Fox Business reported that the cheapest seat for the Philadelphia Phillies home opener increased to $117 right after Bryce Harber agreed to a $330 million 13 year contract.

> The jump marked a 166 percent increase compared to the price before Harper's signing, when the cheapest available ticket could be purchased for just $44. [**R144**]

(a) Use the 1+ trick to confirm the percentage increase in the cheap tickets.

(b) How many cheap tickets would the Phillies have to sell to pay Harper?

(c) Can ticket sales support Harper's 13 year contract?

Exercise 3.8.76. [S][Section 3.4][Goal 3.2][Goal 3.4][Goal 3.5][Goal 3.5][Goal 3.6] Batteries are cheaper.

In April 2019 *Politico* reported that

> The cost of lithium-ion batteries has plunged 85 percent in a decade, and 30 percent in just the past year, [**R145**]

What was the percent drop in battery cost for the first nine years of the decade reported on?

Exercise 3.8.77. [S][U][Section 3.4][Goal 3.2][Goal 3.3][Goal 3.4] Driving down, deaths down?

On May 20, 2020, an article in *The Washington Post* reported an 8 percent drop in highway fatalities compared to a year ago. According to the National Safety Council,

> The number of miles driven dropped 18.6 percent in March compared with the same month last year . . . [b]ut the death rate per 100 million vehicle miles driven was 1.22 in March, up from 1.07 in March 2019. [**R146**]

(a) Elsewhere in the article you can read that the fatality rate per miles driven increased by 14 percent. Verify that calculation.

(b) Is the combination of an increased fatality rate per mile driven and a decrease in the number of miles driven consistent with the claim that deaths decreased by 8 percent?

4

Inflation

Our everyday experience tells us that pretty much everything costs more each year than the year before. That's known as *inflation*. A dollar today is somehow not the same as a dollar last year. Common sense says it's important to understand that quantitatively in order to understand changes in your salary, the prices you pay and the value of what you own. In this chapter we explore inflation in some familiar contexts. We mine the internet for data about inflation and use the 1+ technique from Chapter 3 to understand that data.

Chapter goals:

Goal 4.1. Use an online inflation calculator.

Goal 4.2. Adjust prices for the effect of inflation.

Goal 4.3. Understand the historic value of money, using the terms *current dollars* and *constant dollars*.

Goal 4.4. Reinforce techniques for dealing with percentage changes. Understand the effective change of a price over time.

4.1 Red Sox ticket prices

In Section 3.2 we studied the increase in Red Sox ticket prices from 2003 to 2008 illustrated in Figure 3.3. We found that ordinary ticket prices seemed to have grown by a staggering 23% from $42.26 to $52.16.

To understand what that increase means, we need to take inflation into account. How does the 23% increase in the Red Sox ticket price compare to the average increase of pretty much everything from 2003 to 2008?

Fortunately, the federal Bureau of Labor Statistics provides an Inflation Calculator on line at `www.bls.gov/data/inflation_calculator.htm` that you can use to do

101

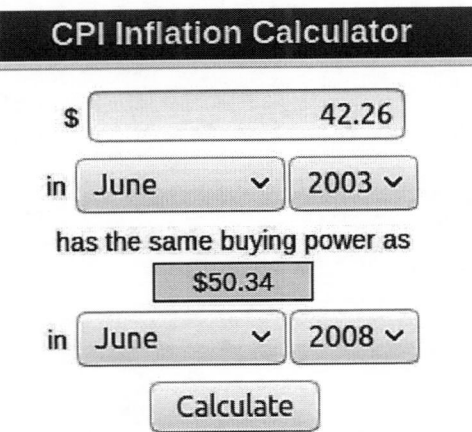

Figure 4.1. Bureau of Labor Statistics inflation calculator [R147]

just that. The screen shot in Figure 4.1 shows that you'd need $50.34 in June 2008 to buy what cost $42.26 in June 2003. That means ordinary ticket prices at Fenway Park increased by $52.16 − $50.34 = $1.82 when measured in "2008 dollars". The relative change was $52.16/$50.34 = 1.036, an increase of just over 3.5% — a lot less than 23%.

The $275 you paid for a premium seat in 2003 would be worth $327.57 in 2008. That is more than the $325 price so, adjusting for inflation, the price actually decreased by $2.57 — a little less than 1 percent.

These calculations show that the increases in Red Sox ticket prices just about matched the average price increase for everything. But that's just an average. Some costs increase faster than the average inflation rate. For example, health care spending per person increased by 4.6% from 2010 to 2011 according to a report issued by the Health Care Cost Institute. [R148]

Since the inflation rate from December 2010 to December 2011 was 2.96%, health care costs increased faster than inflation.

4.2 Inflation is a rate

The government collects data on inflation every month, which is why their calculator offers that level of granularity. Sometimes that much precision makes sense but often the months don't matter. The calculator www.usinflationcalculator.com/ (Figure 4.2), works with the same data to provide a yearly inflation rate. You can see there that on average you would have to pay $102.85 in 2007 for stuff that cost $100 in 2006.

There are several ways to think about the increase from $100 to $102.85. The absolute change is $2.85. But the relative change is more informative and more traditional:

$$\frac{\$102.85}{\$100} = 1.0285,$$

which we recognize as an increase of 2.85% — the *annual inflation rate*. The calculator display reasonably rounds that to 2.8%.

Inflation means the "value of a dollar" changes from year to year. That's often confusing. Sometimes it helps to think of adjusting for inflation as currency conversion.

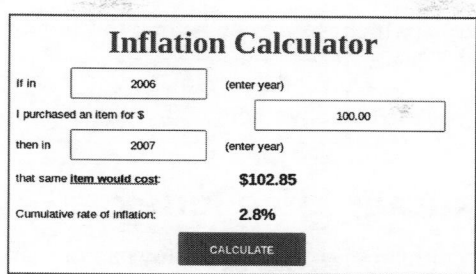

Figure 4.2. US inflation calculator [**R149**]

To make that point clearer, suppose 2006 dollar bills have faded to yellow, while 2007 dollars are the original green. The conversion rate is

$$1.0285 \; \frac{\text{green dollars}}{\text{yellow dollar}} = 1.0285 \; \frac{\text{2007 dollars}}{\text{2006 dollar}}$$

or

$$0.9723 \; \frac{\text{yellow dollars}}{\text{green dollar}} = 0.9723 \; \frac{\text{2006 dollars}}{\text{2007 dollar}}.$$

Yellow dollars are worth more than green dollars.

What would be the cost in 2007 of an item that cost $10,000 in 2006? Since the inflation rate is 2.85% we need just one step with the 1+ trick:

$$1.0285 \times \$10,000 = \$10,285.$$

If you ask the inflation calculator to do the work, it tells you the answer is $10,284.82. That's essentially the same 2.85% inflation rate. The pennies add too much precision to the approximate answer.

The 1+ trick lets us combine inflation rates over two years. The inflation rate from 2006 to 2007 was 2.85%. The inflation calculator tells us that the inflation rate from 2007 to 2008 was 3.84%. It's tempting to guess that the inflation rate from 2005 to 2007 was 2.85% + 3.84% = 6.69% but we know better. The inflation computation from 2007 to 2008 should start with the already inflated prices from 2007. Here's how to work out the right answer using the trick:

$$1.0285 \times 1.0384 = 1.0679944 \approx 1.0680.$$

Then read off the combined inflation rate as 6.80% by subtracting the 1.

If you're still uncomfortable with the 1+ trick all by itself, check the answer the old-fashioned way. Stuff that cost $100 in 2006 would cost $100 × 1.0285 = $102.85 in 2007. That stuff would cost $102.85 × 1.0384 = $106.79944 ≈ $106.80 in 2008: an inflation rate of 6.80%.

The trailing zero in 6.80 doesn't change the value of the number. Writing 6.8 might seem to be just as good. But that zero at the end does matter. It tells you how precise the figure is. In this example 6.8 could be anything between 6.75 and 6.85, while 6.80 must be between 6.795 and 6.805. Here we want the second decimal place in the answer since there are two decimal places of precision in the data we started with. So don't throw away trailing zeroes.

The inflation calculator confirms our analysis.

The moral of the story (which we've seen before, in Chapter 3):

> Don't add percentages.
> Use the 1+ trick to combine them with a single multiplication.

4.3 The Consumer Price Index

An inflation calculator compares prices between any pair of years. The *Consumer Price Index* (CPI) is a single number that captures inflation in a different way. The CPI compares prices with those in 1982. The table at `inflationdata.com/Inflation/Consumer_Price_Index/HistoricalCPI.aspx` says that the average CPI for 2018 was 251.107. In 2018 you needed $251.12 to buy what you paid $100 for in 1982.

When we check that with the inflation calculator using the base year 1982 we find $260.21, not 251.107, the 2018 CPI. The numbers don't match exactly because

> The Consumer Price Index (CPI-U) is said to be based upon a 1982 Base for ease of thought. But in actuality the BLS set the index to an average for the period from 1982 through 1984 (inclusive) equal to 100. [**R150**]

The difference is not likely to matter in any context you encounter.

To find the inflation rate between two years when you know the CPI for each, the ratio does the job. For example, the 2016 CPI was 236.916. Then

$$\frac{\text{2018 CPI}}{\text{2016 CPI}} = \frac{251.107}{236.916} \approx 1.06$$

tells us that the inflation rate from 2016 to 2018 was six percent. In other words, the percentage change in the Consumer Price Index is just the inflation rate.

4.4 More than 100 %

Since the 2018 CPI was $251.107 \approx 250$, prices between 1982 and 2018 increased on average by a factor of 2.5: they are about two and a half times what they were. To use the 1+ trick to express the increase as a percentage we have to find the missing "1" that represents the whole:

$$\text{2018 price} = 2.5 \times (\text{1982 price})$$
$$= (\mathbf{1 + 1.5}) \times (\text{1982 price})$$
$$= \text{1982 price} + 1.5 \times (\text{1982 price})$$

so the inflation rate for the 36 year time period was 150%.

For large changes — more than 100% — it's often better to report how much you multiply by rather than the percent change: the factor of 2.5 for adjusting prices from 1982 to 2018 is more informative than the 150% inflation rate.

Here's a more dramatic example.

On Sunday, June 19, 2011, *The Boston Globe* reported that "The United States spends around $30 billion a year on the National Institutes of Health" and later in the article that "NIH funding in 1939 totaled less than $500,000 a year, a sum that supported just one institute. Adjusting for inflation, the budget has since increased nearly 4,000-fold." [**R151**]

The numbers seem to say that the NIH budget was ($30 billion / $500,000) = 60,000 times as large in 2011 as it was in 1939. But that's not right once you adjust for inflation. Let's check that the author did that correctly. The inflation calculator shows that you'd need just over $8 million in 2011 to buy what cost $500,000 in 1939. Therefore the actual purchasing power of the NIH budget grew by a factor of about ($30 billion / $8 million) = 3,750. That's the same order of magnitude as the "nearly 4,000" in the article. It's a lot, but a lot less than 60,000 times as much.

4.5 How much is your raise worth?

Suppose your salary in 2017 was $40,000 and your contract stated that you would receive a yearly cost of living raise equal to the inflation rate. The inflation rate from 2017 to 2018 was 2.5%, so your 2018 salary would be

$$1.025 \times \$40,000 = \$41,000.$$

Your salary increases. You see it in your paycheck each month. But you're not really any better off. Because of inflation, you would need all of the $41,000 in 2018 to buy the same things you bought with $40,000 in 2017. Since your raise is equal to the inflation rate, your buying power has not changed at all. Effectively, your salary has not increased.

Of course you were clever enough to know this, so you negotiated a 5% raise from 2017 to 2018 so that you would actually be earning more after taking inflation into account. With that raise your 2018 salary would be $42,000.

You are better off. By how much? The inflation rate was 2.5%, so the difference predicts an increase of 5% − 2.5% = 2.5%. But you should be suspicious. Subtracting percentages is as unreliable as adding them.

To find the actual change in your buying power you should find the buying power of your 5% higher 2018 salary in 2017 dollars, by undoing the inflation. That requires division:

$$1.05/1.025 = 1.0243902439 \approx 1.024.$$

That is, your buying power has increased only by 2.4%. The subtraction estimate was 2.5% — incorrect, but not different enough to matter since the actual percentages involved are small.

This calculation is pretty subtle (though you should be used to it by now). We can check the answer the old-fashioned way. Remember that we just figured out that you would need $41,000 in 2018 to maintain the buying power you had in 2017. To see the effective change in your buying power, calculate the relative change:

$$\text{relative change} = \frac{\text{new value}}{\text{reference value}} = \frac{42,000}{41,000} = 1.0243902439 \approx 1.024.$$

That's the same conclusion arrived at in a different way.

4.6 The minimum wage

The federal minimum wage is the legal minimum employers must offer workers paid by the hour. Table 4.3 shows its value at each increase since it was first set at $0.25 per hour in 1938.

Month / Year	Minimum Wage	Month / Year	Minimum Wage
October 1938	$0.25	January 1978	$2.65
October 1939	$0.30	January 1979	$2.90
October 1945	$0.40	January 1980	$3.10
January 1950	$0.75	January 1981	$3.35
March 1956	$1.00	April 1990	$3.80
September 1961	$1.15	April 1991	$4.25
September 1963	$1.25	October 1996	$4.75
February 1967	$1.40	September 1997	$5.15
February 1968	$1.60	July, 2007	$5.85
May 1974	$2.00	July, 2008	$6.55
January 1975	$2.10	July, 2009	$7.25
January 1976	$2.30		

Table 4.3. Federal hourly minimum wage history [R152]

Figure 4.4. The minimum wage, in actual and 2019 dollars [R153]

If we take inflation into account we see that history in perspective. Figure 4.4 traces the minimum wage in both actual and real 2019 dollars. Note the jumps in the real value in the years when Congress passed an increase, followed by gradual decline as inflation ate away at the gain. That is particularly clear for the periods from 1981 to 1990 and from the last increase to $7.25 per hour in 2009 to the time we wrote this in 2019, when minimum wage earners were worse off than in almost all the years since 1950.

4.7 Inflation history

The New York Times printed Figure 4.5 in time for Presidents' Day 2017. It shows part of an advertisement that appeared in *Claypoole's American Daily Advertiser* on May 26, 1796.

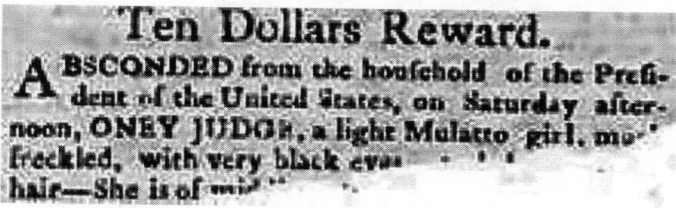

Figure 4.5. George Washington wants his slave back [**R154**]

How much would the $10 reward be in 2017 dollars?

The bad news is that the calculators using Bureau of Labor Statistics data go back only to 1913. The good news is that a Google search for

$10 in 1796 inflation calculator

finds the website www.in2013dollars.com/1796-dollars-in-2016. The calculator there says that $10 in 1796 is equivalent to $194.36 in 2019, along with the assertion that the dollar experienced an average inflation rate of 1.34% per year during this period.

We found that annual increase surprisingly small. An average inflation rate of only 1.3% per year does not match what we see in recent history.

Figure 4.6 explains the contradiction. The green spikes below the axis show that until the late 1930's there were many periods of significant *deflation* alternating with the inflation we're used to.

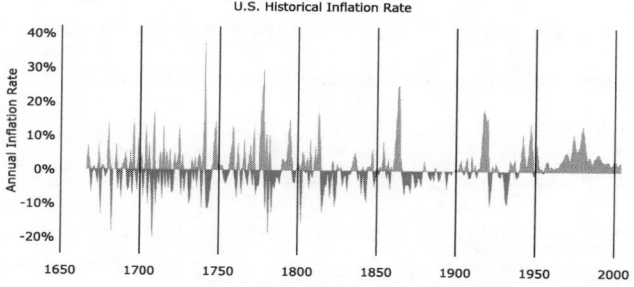

Figure 4.6. Inflation through the years [**R155**]

Source: The Bureau of Labor Statistics' annual Consumer Price Index (CPI), established in 1913. Inflation data from 1665 to 1912 is sourced from a historical study conducted by political science professor Robert Sahr at Oregon State University.

Since the financial reforms enacted in response to the Great Depression of 1929 there has been no significant deflation — just a little at the end of the 1930's and then more recently in 2008 and 2015. Figure 4.7 shows a more recent tiny temporary deflation for January of that year.

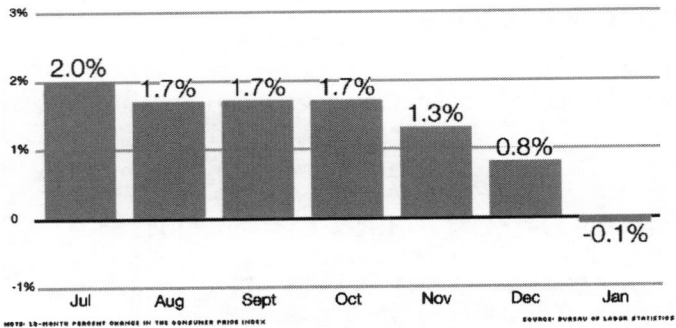

Figure 4.7. Deflation in January 2015 [**R156**]

4.8 Exercises

Exercise 4.8.1. [S][R][W][Section 4.1][Goal 4.1] Inflation calculator practice.

Use an inflation calculator to find the inflation rate (as a percent) from 2003 to 2007, using purchase amounts of $1, $100 and $10,000. Explain why the answers differ. Which number is the most appropriate?

Exercise 4.8.2. [S][Section 4.1][Goal 4.1] [Goal 4.2] A million dollar Apple computer.

On October 22, 2014, the website *appleinsider* reported that an antique Apple 1 computer

> ...was sold at an auction on Wednesday, with the nearly 40-year-old machine crushing expectations to fetch a record-breaking $905,000. [**R157**]

The 200 Apple 1's sold for $666.66 each when they were introduced in 1976.

(a) Compare the auction price to the appropriately inflated value of the original price.

(b) Was the auction price a bargain?

(c) (Optional) Why is 666 called "the number of the beast"?

Exercise 4.8.3. [S][Section 4.1][Goal 4.1] [Goal 4.2] Good to the last penny.

On the Commonwealth of Massachusetts webpage at www.sec.state.ma.us/trs/trsbok/mod.htm you can read that

> The Massachusetts State House cornerstone was laid on the Fourth of July, 1795, by Governor Sam Adams and Paul Revere, Grand Master of the Masons. The stone was drawn by fifteen white horses, one for each of the states of the Union at that time. The cost of the original building? $133,333.33
>
> ...
>
> Paul Revere & Sons coppered the dome in 1802 to prevent water leakage. Some seventy years later the dome was gilded with 23 carat gold leaf for the first time. The cost was $2862.50; the most recent gilding, in 1997, cost $300,000. [**R158**]

(a) What's strange about the number $133,333.33 for the original cost of the State House? How do you think the person who wrote these paragraphs came up with that number?

(b) Adjust the three dollar amounts in this quotation to take inflation into account. Have building and gilding gotten more or less expensive over the years?

[See the back of the book for a hint.]

Exercise 4.8.4. [S][Section 4.1][Goal 4.1] Several ways to skin a cat.

Find another inflation calculator on the internet and compare it to the one provided by the government. Do they give the same answers? If not, why not? Do they cover the same range of years? Are they equally easy to use? Make sure you give a clear citation for the calculator you find.

If it's easy for you, make screen shots comparing your inflation calculator to the one at the Bureau of Labor Statistics.

Exercise 4.8.5. [S][Section 4.1][Goal 4.1] [Goal 4.2] When was a dime a nickel?

In 2011 Massachusetts was considering raising the nickel bottle deposit to a dime. On December 29 Chris Lohmann wrote in a letter to *The Boston Globe*:

> Let's not forget that a dime now is worth less than a nickel was when the original bottle bill was passed. [R159]

(a) Use the inflation calculator to estimate when the Massachusetts law calling for a nickel bottle deposit was passed.

(b) If you can, check your estimate by finding the actual year.

[See the back of the book for a hint.]

Exercise 4.8.6. [S][R][Section 4.2] [Goal 4.1][Goal 4.4] Twenty years as two decades.

(a) Use the inflation rate calculator to find the inflation rate (as a percent) from 1990 to 2000, from 2000 to 2010, and from 1990 to 2010.

(b) Explain why the inflation rate from 1990 to 2010 is not just the sum of the rate from 1990 to 2000 and the rate from 2000 to 2010.

(c) Use the 1+ trick to compute the 1990 to 2010 rate correctly from the rates for the two previous decades.

Exercise 4.8.7. [S][Section 4.2][Goal 4.1] [Goal 4.3] Piecework.

In the 1880's many young women in large cities worked at home sewing clothing. They were paid by the piece, hence the name "piecework" for this type of labor. A pants stitcher would finish a pair of pants, putting in canvas for the pockets and waistband linings, and might be paid 12.5 cents per pair of pants finished. She could generally finish 16 pairs of pants a week, working from 8 am until dark. (You can read more about this on the web at `historymatters.gmu.edu/d/5753/` .)

(a) What was her hourly rate of pay in 1880?

(b) Convert her 1880 wages to current wages.

(c) Compare her hourly rate to the current minimum wage.

Exercise 4.8.8. [S][Section 4.2][Goal 4.1][Goal 4.3][Goal 4.2][Goal 4.4] Hollywood math: bad to worse.

   A short article in the January 10, 2011, edition of *The New York Times* discussed the amount spent on DVDs over the past six years:

> [In 2004] consumers spent about $21.8 billion to rent and buy DVDs, Blu-ray discs, digital downloads and other forms of home entertainment ... [T]he number has fallen every year since, for a total drop of about 13.8 percent, to $18.8 billion in 2010. [**R160**]

The reporter noted that the actual drop was about double what it seemed to be when the figures were adjusted for inflation: the $21.8 billion figure from 2004 would amount to $25.3 billion in current dollars.

(a) Verify the reporter's inflation calculation.

(b) If we adjust for inflation, how dramatic is the drop in spending over the six year period?

(c) Explain how adjusting for inflation doubles the reported drop in spending.

(d) The article also noted that box office revenues rose from $9.3 billion in 2004 to $10.6 billion in 2010. Calculate the percentage increase in box office revenues, then recalculate the percentage increase using inflation-adjusted dollars. Have box office revenues increased or decreased over this six year time period?

Exercise 4.8.9. [S][Section 4.2][Goal 4.1] [Goal 4.4] Deflation.
   What was the inflation rate from 2008 to 2009?
   [See the back of the book for a hint.]

Exercise 4.8.10. Holiday Pops.
   Moved to Extra Exercises at `www.ams.org/bookpages/text-63` .

Exercise 4.8.11. [S][Section 4.2][Goal 4.1][Goal 4.2][Goal 4.4] The Jollity Building.
   In 1941 A. J. Liebling wrote in *The New Yorker* that Mr. Ormont, the manager of the Jollity Building, was paid $50 a week plus a commission of two percent of the rents. That commission earned him an extra two thousand dollars a year. [**R161**]

(a) What was the average weekly rent on which Mr. Ormont's commission was based?

(b) What would Mr. Ormont's annual income be today, adjusted for inflation?

(c) Mr. Ormont managed the Jollity Building, a rather seedy low class establishment for "the petty nomads of Broadway — chiefly orchestra leaders, theatrical agents, bookmakers, and miscellaneous promoters." Was he reasonably well paid for this job?

(d) Sometimes we find quantitative reasoning questions in our casual reading (not the daily paper). This one comes from a reprint of "The Jollity Building" in *Just Enough Liebling*, a collection of the author's *New Yorker* pieces, where it was dated 1938. [**R162**] Redo the previous questions with this date. How much of a difference does the three years make?

Exercise 4.8.12. [S][C][Section 4.2] [Goal 4.2][Goal 4.4] Firefighters' pay.

On June 10, 2010, *The Boston Globe* reported that in the five years between 2006 and 2011 Boston firefighters received raises of 2%, 2.5%, 3%, 3.5%, 2.5% and 1.5%.

(a) Explain why the overall raise for this five year period is not just the sum of the six percentage increases.

(b) What was the firefighters' overall raise for the five year period?

(c) What was the firefighters' overall raise when you take inflation into account?

[See the back of the book for a hint.]

Exercise 4.8.13. [S][Section 4.2][Goal 4.2] [Goal 4.4] A raise tied to the inflation rate.

Consider these two different ways to calculate a raise that's "10% over the rate of inflation":

(a) What is your new salary if you add the inflation rate to your 10% raise and use that as your percentage raise?

(b) What will you earn if you first compute your 10% increase and then increase that to take inflation into account?

(c) Which is the best approach (for you — since it's your salary)?

[See the back of the book for a hint.]

Exercise 4.8.14. [S][Section 4.5] [Goal 4.2][Goal 4.4] Committee approves 1 percent pay raise ....

On June 26, 2013, the Madison, Wisconsin, *Star Tribune* reported that most state employees in Wisconsin will get a one percent pay raise — the first since 2008.

> "One percent doesn't even hold the workers even," said Marty Bell, executive director of the Wisconsin State Employees Union. [**R163**]

(a) What is the effective raise for these state workers?

(b) The reporter noted that the cost of paying the salary increases for University of Wisconsin workers would be $52.4 million over two years. What is the total salary pool of these employees (before the raises)?

(c) Is your answer to the previous question a reasonable estimate for total employee salaries at the University of Wisconsin?

[See the back of the book for a hint.]

Exercise 4.8.15. When is a raise not a raise?

Moved to Extra Exercises at www.ams.org/bookpages/text-63 .

Exercise 4.8.16. [S][R][Section 4.5][Goal 4.1][Goal 4.2][Goal 4.3] Private colleges vastly outspent public peers.

The Fort Wayne, IN, *Journal Gazette* reported on July 10, 2010, on a study from the Delta Project on Postsecondary Education Costs, Productivity, and Accountability, a Washington-based nonprofit, that said

> Private institutions, on average, laid out $19,520 per student for in-
> struction [in 2007–2008], a 22 percent increase from a decade earlier,
> ...Public universities spent $9,732 for each student, up 10 percent in
> the decade. [**R164**]

Calculate these increases in spending when inflation is taken into account.

Exercise 4.8.17. [S][Section 4.5] [Goal 4.1][Goal 4.2][Goal 4.3] DPS, teachers' union reach accord.

EdNews Colorado reported on June 19, 2010, that Denver school teachers may be about to get their first cost-of-living adjustment in three years.

> Under the terms of the tentative agreement, Denver teachers would
> receive ...a 1 percent cost-of-living raise if a proposed $49 million tax
> increase for operating dollars is ...approved by voters. ...In addi-
> tion, if the increase passes, teachers would receive a .5 percent raise
> in 2013–14 and a .5 percent raise in 2014–15. [**R165**]

(a) Assuming all of the increases pass, what is the total cost of living increase that a teacher would receive by the end of the 2015 school year?

(b) If a teacher did not receive any merit increases over the past three years (some did not), what is the effective percentage decrease in that teacher's salary during this time period?

Exercise 4.8.18. [U][S] [Section 4.4] [Goal 4.1][Goal 4.2] Wages down 14 percent.

On June 6, 2012, the Oregon *Portland Tribune* reported on a study from the Oregon Employment Department and the Bureau of Economic Analysis. The article noted that

> ...between 2000 and 2011, the average wages for ...jobs in Columbia
> County rose by 13 percent. But county wages actually decreased 13.8
> percent during that time when inflation is taken into account.
>   In neighboring Clatsop County, ...(t)he inflation adjusted wages
> ...rose 0.4 percent from 2000 to 2011. [**R166**]

(a) Use the CPI inflation calculator to make sense of the numbers for workers in Columbia County.

(b) What was the average increase (without taking inflation into account) from 2000 to 2011 for workers in Clatsop County?

Exercise 4.8.19. [S][Section 4.4] [Goal 4.1][Goal 4.2][Goal 4.3] Newspaper sales slid to 1984 levels in 2011.

A post to the blog "Reflections of a newsosaur" reported on March 15, 2012, on figures released by the Newspaper Association of America:

> In the poorest showing since 1984, advertising revenues at newspa-
> pers last year fell 7.3% to $23.9 billion.
>   ...
>   The combined print and digital sales ...for last year are less than
> half of the all-time sales peak of $49.4 billion achieved as recently as
> 2005.

...

The last time sales were this low was 1984, when they totaled $23.5 billion. Adjusted for inflation, the 1984 sum would be worth nearly $50 billion today. [**R167**]

(a) Use the CPI calculator to verify the claim in the third paragraph.

(b) A comment to the blog post stated, "Shouldn't the headline be 'Newspaper sales slid to half of 1984 level in 2011?' The comparison seems meaningless when not adjusted for inflation." Do you agree? Explain your answer.

(c) The claim in the second paragraph has not been adjusted for inflation. Would the statement change if you adjusted that figure for inflation?

(d) The blog *Carpe Diem* gave updated information on newspaper revenues in a post on September 6, 2012. In that post, the author predicted that print newspaper advertising will be lower in 2012 than in 1950, when adjusted for inflation. [**R168**]

In fact total advertising in 2012 was $22.3 million. In 1950, total advertising was $2.07 million (in 1950 dollars). Use the CPI calculator to see if he was right.

Exercise 4.8.20. [S][R][Section 4.5] [Goal 4.1][Goal 4.2][Goal 4.3] The MAA.
Annual dues for the Mathematical Association of America were $3 in 1916. They were $175 in 2019.

(a) How much would the $3 dues in 1916 be in 2019 dollars?

(b) How much have dues gone up (or down) between 1916 and 2019, in 2019 dollars, in absolute and relative terms?

Exercise 4.8.21. [R][S][Section 4.6] [Goal 4.1][Goal 4.3][Goal 4.4] Working for the minimum wage.

(a) Figure out the annual income in 2007 dollars for someone who worked at a minimum wage job 40 hours/week for 50 weeks in 1975.

(b) Compare that income to the annual income of someone working the same amount at the minimum wage in 2007.

You'll have to look up the minimum wage for each year — or look back in this chapter to find that information.

Exercise 4.8.22. [S][Section 4.6][Goal 4.2] [Goal 4.3] Penny dreadful.
On page 60 of the March 31, 2008, issue of *The New Yorker* David Owen wrote that you'd earn less than the federal minimum wage if it took you longer than 6.15 seconds to pick up a penny. [**R169**]

(a) Use the information in the quotation to figure out the minimum wage when Owen wrote his article.

(b) Check Owen's arithmetic by comparing your answer to the actual federal minimum wage at that time. (This information is available on the web and in the text.)

(c) How much time could you spend picking up a penny if you wanted to earn the minimum hourly wage today for your work?

(d) What is the origin of the phrase "penny dreadful"?

Exercise 4.8.23. [S][A][Section 4.2][Goal 4.1][Goal 4.4] A decade of inflation.

(a) Suppose the annual inflation rate was 3% for 10 years in a row. What was the inflation rate for the decade?

(b) Find the inflation rate for the decade from 1960 to 1970. If the inflation rate had been the same for each year in that decade, what would it have been?

[See the back of the book for a hint.]

Exercise 4.8.24. [S][Section 4.2][Goal 4.1] Running to keep up with inflation.

(a) Use the data in Exercise 3.8.3 to compare the growth in the number of finishers in the New York Marathon to the inflation rate for the corresponding years.

(b) Why is this a ridiculous question?

Review exercises.

Exercise 4.8.25. [A] For each of the following, use the online CPI inflation calculator.

(a) Find the buying power in 2010 of $12.50 in 2004.

(b) Find the buying power in 2011 of $42.99 in 2000.

(c) Find the buying power in 2008 of $20.00 in 2010.

(d) Find the buying power in 2010 of $100 in 1992.

Exercise 4.8.26. [A] Rewrite each sentence, expressing the change as a percentage.

(a) The cost of gasoline doubled during this time period.

(b) The value of the painting tripled over the past decade.

(c) The price of milk is two-thirds of what it cost last year.

(d) The CEO's salary is ten times the salary of the lowest-paid worker in the company.

(e) I cut my bills by a third last year.

(f) His salary is one and a half times more than her salary.

(g) Our budget this year is half of what it was last year.

Exercises added for the second edition.

Exercise 4.8.27. [S][Section 4.2][Goal 4.2][Goal 4.3] When is a raise not a raise?
    On March 7, 2018, you could read in *The New York Times* that the Governor of West Virginia offered the state's teachers "a 1 percent a year raise for the next five years." The article noted that "if inflation averages 2 percent a year for that period, this translates into an effective 5 percent pay cut." [**R170**]
    Is the "5 percent pay cut" claim correct?

Exercise 4.8.28. [S][Section 4.7][Goal 4.1][Goal 4.2] What is Babe Ruth worth?

One of the headlines on the cover of this book announces the 1920 sale of Babe Ruth from the Boston Red Sox to the New York Yankees for $100,000.

(a) What does that amount represent in 2020 dollars?

(b) If you care about baseball, speculate about whether that sale price would be reasonable.

5

Average Values

We start by remembering that to compute an average you add the values and divide by the count. We quickly move on to weighted averages, which are more common and more useful. They're a little harder to understand, but worth the effort. They help explain some interesting apparent paradoxes.

Chapter goals:

Goal 5.1. Compute means using weighted averages.

Goal 5.2. Investigate what it takes to change a weighted average.

Goal 5.3. Understand paradoxes resulting from weighted averages.

Goal 5.4. Study the Consumer Price Index.

5.1 Average test score

Suppose a student has taken ten quizzes and earned scores of

$$90, 90, 80, 90, 60, 90, 90, 70, 90, 80.$$

To keep the computations simple and to focus on the ideas, we've made up this short unrealistic example. Later we will return to our prejudice in favor of real data.

To find her average score you add the ten scores and divide by ten:

$$\frac{90 + 90 + 80 + 90 + 60 + 90 + 90 + 70 + 90 + 80}{10} = \frac{830}{10} = 83.$$

There are only four different values in this list: $60, 70, 80$ and 90. But her average score isn't just $(60 + 70 + 80 + 90)/4 = 75$. A correct calculation must take into account the fact that she had lots of grades of 90 but just one 60. Here is a way to do that explicitly:

$$\frac{(1 \times 60) + (1 \times 70) + (2 \times 80) + (6 \times 90)}{10} = \frac{830}{10} = 83.$$

To work with the fraction of times each score occurs rather than the number of times just divide the 10 in the denominator into each of the terms in the numerator:

$$\frac{1}{10} \times 60 + \frac{1}{10} \times 70 + \frac{2}{10} \times 80 + \frac{6}{10} \times 90 = 83.$$

Viewed that way, we see exactly how each of the four different quiz grades contributes to the average with its proper weight.

We can rewrite that weighted average showing the weights as decimal fractions:

$$0.1 \times 60 + 0.1 \times 70 + 0.2 \times 80 + 0.6 \times 90 = 83$$

or percentages:

$$10\% \times 60 + 10\% \times 70 + 20\% \times 80 + 60\% \times 90 = 83.$$

In each case all the tests are accounted for so the weights expressed as fractions or as decimals sum to 1. As percents they sum to 100%.

The same strategy finds the average value of a card when (as in blackjack) the value of the face cards (Jack, Queen and King) is 10. Imagine that you choose a card at random, write down its value, return it to the deck, shuffle and do it again — many times. What will the average value be? It's sure to be greater than the simple average of the numbers 1 to 10 (which is 5.5) since there are more cards with value 10 than any other. This is a weighted average, where the weight of each value is its probability. There are 4 chances in 52 that you see (say) a four, and 16 chances in 52 that you see a card with a value of 10. The average value is

$$\frac{4}{52} \times 1 + \frac{4}{52} \times 2 + \cdots + \frac{4}{52} \times 9 + \frac{16}{52} \times 10 = \frac{1 + 2 + \cdots + 9 + (4 \times 10)}{13}$$

$$\approx 6.54.$$

The first step in the computation puts all the fractions over a common denominator and cancels a 4. We could have started there by thinking about the cards one suit at a time.

5.2 Grade point average

The UMass Boston registrar posts the rules used to compute student grade point averages at `www.umb.edu/registrar/grades_transcripts/grading_system`. Here is what that webpage said in 2019.

> **Grading System**
> Each letter grade has a grade point equivalent. List your grades in a column, then each grade point equivalent next to the letter grade. Multiply each grade point equivalent by the number of credits for each class. Total all products and divide by the total number of credits. The answer will be your grade point average for that semester.
> [R171]

The site then displays Table 5.1 showing a sample computation for a student who took 12 courses for a total of 33 credits and earned one grade of each kind.

The "average" in "grade point average" is a weighted average of the numerical grades, with credits as the weights.

GRADE	GRADE POINT	EQUIVALENT CREDITS	QUALITY POINT
A	4.00	3	12.00
A-	3.70	3	11.10
B+	3.30	3	9.90
B	3.00	2	6.00
B-	2.70	3	8.10
C+	2.30	3	6.90
C	2.00	1	2.00
C-	1.70	2	3.40
D+	1.30	3	3.90
D	1.00	4	4.00
D-	0.70	3	2.10
F or IF	0.00	3	0.00
*TOTAL		33	69.40

* Example: Divide total quality points (69.40) by total number of registered credits (33) = 2.103 (grade point average).

Table 5.1. Computing a grade point average [R172]

Suppose a student has taken four courses worth 4, 3, 3 and 2 credits and earned grades of B+, A, B+ and C-, respectively. Since she has earned a total of 12 credits, her GPA is

$$\frac{4 \times 3.3 + 3 \times 4 + 3 \times 3.3 + 2 \times 1.7}{12} = \frac{38.5}{12} = 3.21.$$

That's a solid B average in spite of the C-.

To see the weights more clearly, rewrite the computation showing the fraction of credits for each course:

$$\frac{4}{12} \times 3.3 + \frac{3}{12} \times 4 + \frac{3}{12} \times 3.3 + \frac{2}{12} \times 1.7 = 3.21.$$

The Registrar's website tells you to multiply the grade equivalent by the number of credits. We do the multiplication in the other order — fraction of credits times grade equivalent — to make the weights more visible.

5.3 Improving averages

Suppose a student at the end of her junior year has a GPA of 2.8 for the 90 credits she has taken so far. What GPA must she earn for the 30 credits she will take as a senior so that she can graduate with a GPA of 3.0?

The ordinary average of 2.8 and 3.2 is 3.0, so she might think at first that a 3.2 as a senior will do the trick. We know that can't be right, since there are more credits in her first three years than in her last. She will need more than a 3.2 to bring her GPA up to 3.0. We need to figure out how much more.

If her senior year GPA is G then her combined GPA is the weighted average

$$\frac{90 \times 2.8 + 30\,G}{120}. \tag{5.1}$$

What value of G will make the arithmetic in this expression come out at least 3.0?

We'll answer this question three ways. Each method has its advantages and disadvantages.

If you remember even a little bit of algebra you can solve the equation

$$\frac{90 \times 2.8 + 30\,G}{120} = 3.0$$

for the unknown G. Multiply both sides by 120 to get

$$90 \times 2.8 + 30\,G = 360$$

so

$$30\,G = 360 - 90 \times 2.8 = 360 - 252 = 108$$

so

$$G = \frac{108}{30} = 3.6.$$

Once you see the answer you can see why it is right. The 3.0 GPA she wants will be just one fourth of the way from the 2.8 GPA she has so far to what she needs as a senior, since she has already earned three fourths of her credits — the 2.8 carries 75% of the weight. That leads to another way to solve the problem. If you noticed the one fourth at the start you could do it in your head: G must be three times as far from 3.0 as 2.8 is, so it must be 3.6.

Finally, you can answer the question even if you've forgotten your algebra and don't see the answer right away. Just guess, check your guess, and guess again as long as you have to.

Try a first guess of $G = 3.0$ in (5.1):

$$\frac{90 \times 2.8 + 30 \times 3.0}{120} = 2.85,$$

so 3.0 is too small. How about 4.0 (straight A work as a senior)?

$$\frac{90 \times 2.8 + 30 \times 4.0}{120} = 3.1,$$

which is more than she needs. The answer is somewhere between 3.0 and 4.0. Try 3.5:

$$\frac{90 \times 2.8 + 30 \times 3.5}{120} = 2.975.$$

That's almost enough. Guess 3.6 next:

$$\frac{90 \times 2.8 + 30 \times 3.6}{120} = 3.0.$$

Bingo! Got it!

Guess-and-check isn't as efficient as algebra, but it's easy to remember, and it works in places where algebra won't help. That makes it a better life skill.

5.4 The Consumer Price Index

The CPI calculator says that the inflation rate from 2016 to 2017 was 2.50%. That does not mean every price increased by that percent. The inflation rate is a weighted average of the changes in costs of various items. The Bureau of Labor Statistics surveys the population to find out how much we spend on various goods and services to discover what weights to use when averaging the changes in prices:

> The CPI market basket is developed from detailed expenditure information provided by families and individuals on what they actually bought. There is a time lag between the expenditure survey and its use in the CPI. For example, CPI data in 2016 and 2017 was based on data collected from the Consumer Expenditure Surveys for 2013 and 2014. In each of those years, about 24,000 consumers from around the country provided information each quarter on their spending habits in the interview survey. To collect information on frequently purchased items, such as food and personal care products, another 12,000 consumers in each of these years kept diaries listing everything they bought during a 2-week period.
>
> Over the 2 year period, then, expenditure information came from approximately 24,000 weekly diaries and 48,000 quarterly interviews used to determine the importance, or weight, of the item categories in the CPI index structure. [R173]

Table 5.2 shows the weights and the percentage changes for the several categories for Pittsburgh, PA, for the period 2016–2017. In fact, each of these changes is a weighted average of subcategories within each category. For example, food and beverages are broken down into those consumed at home and those consumed away from home.

Category	Weight (%)	2016–2017 change (%)
Food and beverages	14.314	1.9
Housing	42.202	1.9
Apparel	2.959	−1.5
Transportation	16.348	1.4
Medical care	8.682	10.8
Recreation	5.694	−1.6
Education and communication	6.596	−0.6
Other goods and services	3.204	4.3
All items	100	2.2

Table 5.2. Consumer Price Index market basket [R174]

The percentages in the second column add to 100, as they should. It's interesting to note from the third column that average prices in Apparel and Recreation actually decreased. The 2.2 in the last row, last column is the weighted average of the changes in each category, calculated this way:

$$0.14314 \times 1.9 + 0.42202 \times 1.9 - 0.02959 \times 1.5$$
$$+ 0.16348 \times 1.4 + 0.08682 \times 10.8$$
$$- 0.05694 \times 1.6 + 0.06596 \times 0.6$$
$$+ 0.03204 \times 4.3$$
$$\approx 2.20.$$

The 2.2% inflation rate for Pittsburgh in 2016–2017 doesn't exactly match the nationwide average of 2.50% from the inflation calculator. That's because the nationwide average is itself a weighted average of the regional averages, and prices in the Pittsbugh area increased less than those elsewhere.

5.5 New car prices fall …

On September 5, 2008, the business section in *The Columbus Dispatch* carried an Associated Press story headlined "New-vehicle prices plunge, report says". The article said that in the second quarter the average cost of a new vehicle was $25,632, 2.3 percent less than a year earlier. [R175]

Since average costs (the Consumer Price Index) increased in that year, we were puzzled. We were pretty sure car prices had gone up too. When we read further we found that

> …Truck-based vehicles such as pickup trucks, minivans, and SUVs accounted for less than half of all sales in the second quarter for the first time since 2001.…

Aha! That means vehicle prices could all have risen even while the average fell! We've made up some numbers to show how the arithmetic might work. Suppose truck-based vehicles accounted for 45% of the sales in 2008 — that's "less than half" — and that the average prices for truck-based vehicles and cars were $32.2K and $20.2K, respectively, as shown in Table 5.3.

Model	Average price ($K)	Percent of market
car	20.2	55
truck	32.2	45

Table 5.3. Vehicle sales: second quarter 2008

Then the average price for a vehicle would be

$$0.55 \times \$20.2K + 0.45 \times \$32.2K = \$25,600 = \$25.6K,$$

which is close to the reported average of $25,632.

Now imagine what the numbers might have been a year earlier, when, perhaps, truck-based vehicles outsold cars 55% to 45%, as in Table 5.4.

Model	Average price ($K)	Percent of market
car	19.6	45
truck	31.6	55

Table 5.4. Vehicle sales: second quarter 2007

The average price for a vehicle would have been

$$0.45 \times \$19.6K + 0.55 \times \$31.6K = \$26,200 \approx \$26.2K.$$

With these assumptions, the average price fell by $600, a relative decrease of

$$\$600/\$26{,}200 = 0.0229 \approx 2.3\%,$$

as the article reports. But average prices for cars and trucks separately both increased.

This is really a warning about averaging averages. The average vehicle cost is an average of the average car cost and the average truck cost. The weights in that second average matter.

5.6 An averaging paradox

The professor's average class size is smaller than the student's average class size.

How can that be? Sometimes the best way to understand a paradox is to imagine a simple extreme case rather than trying to untangle complex real data.

Suppose a small department (with just one professor) offers just two classes. One is a large lecture, with 100 students. The other is a seminar on a topic so narrow that no students sign up. Then the average class size (from the professor's point of view) is $(100 + 0)/2 = 50$ students, while each student's average class size is 100.

5.7 Exercises

Exercise 5.7.1. [S][A][Section 5.1][Goal 5.1] Can she earn a B?

Suppose a student's final grade in a biology course is determined using the following weights:

    quizzes are worth 5%,
    exam 1 is worth 20%,
    exam 2 is worth 20%,
    lab reports are worth 15%,
    research paper is worth 15%,
    final exam is worth 25%.

Just before the final, she has earned the following grades (all out of 100):

Lab report grades:	75, 90, 85, 69, 70, 75, 80, 75
Quiz grades:	85, 80, 0, 60, 70, 80, 80, 75
Exam 1:	80
Exam 2:	70
Paper:	85

(a) What is her lab report average?

(b) What is her quiz average?

(c) What is her course average just before the final?

(d) For a B she needs a course average of at least 82%. What is the lowest grade she can get on the final and achieve that goal?

Exercise 5.7.2. Fundraising.

Moved to Extra Exercises at `www.ams.org/bookpages/text-63`.

Exercise 5.7.3. [R][W][Section 5.2][Goal 5.1] A grade point average that matters.

Check that your own GPA has been computed correctly using the rules in effect at your school.

If you are in your first semester and don't yet have a GPA, imagine the grades you expect at the end of the semester and figure out what your GPA would be.

If you are in your last semester then this exercise will take you a long time. Instead you may check your GPA for just one semester, or for the courses in your major.

If all the courses you took carry the same number of credits, check that you get the correct GPA if you compute the average the old-fashioned way, without using any weights.

Exercise 5.7.4. [W][S][Section 5.3][Goal 5.2] Ways to raise your GPA.

A UMass Boston student has earned 55 credits toward his degree but has a GPA of just 1.80. The registrar has informed him that he is on probation and will be suspended unless his overall GPA is at least 2.0 after one more semester.

(a) If he takes 12 credits, what is the minimum GPA he must earn in that semester to avoid being suspended?

(b) Answer the same question if he takes 9 credits.

(c) What if he took just 6 credits?

(d) (Optional) What would you do if you found yourself in a similar situation — take fewer courses in hopes of doing really well in them or take more courses so that you could afford to do not quite so well in each?

Exercise 5.7.5. [S][Section 5.3][Goal 5.1] [Goal 5.2] Gaming the system.

In the *Chess Notes* column in *The Boston Globe* on Tuesday, March 10, 2008, Harold Dondis and Patrick Wolff wrote that

> ...the sensation of the [Amateur East] tourney was the team with the highest score, GGGg (no relation to the song Gigi but standing for the three Grandmasters and one future Grandmaster.) The players were our Eugene Perelshteyn, Roman Dzindzichasvili, Zviad Izoria, all Grandmasters, and 5-year-old Stephen Fanning who rounded out the team. Was this a valid lineup? Well, yes, it was. The rules of the Amateur provide that the average rating of a team could not exceed 2200. GGGg's three Grandmasters were well above that, but Stephen Fanning ...had a current rating of 178, which brought the average rating to 2017.
>
> The Grandmasters of GGGg ...delivered wins, while their fourth board, who would have won the prize as the cutest chess player (if the sponsors had had the foresight to establish such a prize) struggled to make legal moves, sometimes failing to do so. Naturally, there followed an extensive debate as to whether the victorious ensemble had gamed the system. [**R176**]

(a) If the three grandmasters had the same rating, what would it have been?

(b) Did GGGg game the system?

(c) How might the tourney organizers change the rules to prevent this kind of team from winning?

Exercise 5.7.6. [S][Section 5.3] [Goal 5.1][Goal 5.2] Good day, sunshine.
On June 23, 2009, *The Boston Globe* reported that

> June 2009 in Boston might turn out to be the dimmest on record. So far in the month, the sun was shining only 32% of the time. The record low was in 1903, when the sun shone only 25% of the time. [**R177**]

(a) What percentage of sunshine for the remaining days of the month would make 2009 at least a tie for the dimmest June?

(b) Did that happen?

Exercise 5.7.7. [S][Section 5.3][Goal 5.1][Goal 5.2] Five million unemployed.
In *The Hightower Lowdown* (Volume 12, Number 5, May 2010) you could read:

- **5 MILLION PEOPLE** (about 10% of the workforce are out of work).
- **UNEMPLOYMENT IS HEAVILY SKEWED BY CLASS.** Among the **wealthiest 10%** of American families (incomes above $150,000), only **3% are unemployed** — a jobless rate that rises as you go down the income scale. Among the **bottom 10%, more than 30% are out of work.** [**R178**]

What average unemployment rate for the middle 80% of families fits with the given values for the top and bottom 10% to work out to the overall (weighted) average unemployment rate of 10%?

Exercise 5.7.8. [S][A][W][Section 5.3][Goal 5.1][Goal 5.3] Who wins?
Alice and Bob are both students at ESU. In September they start a friendly competition. In June they compare transcripts. Alice had a higher GPA for both the fall and spring semesters. Bob had a higher GPA for the full year.

(a) Explain how this can happen, by imagining their transcripts — number of credits and GPA for each, for the two semesters and for the full year, as in this table:

	fall credits	fall GPA	spring credits	spring GPA	year GPA
Alice					
Bob					

(b) Who wins?

Exercise 5.7.9. [C][U][Section 5.4][Goal 5.4][Goal 5.1][Goal 5.2] Your rate may vary.

(a) Estimate the weights in your life now for the various categories used to compute the CPI in Table 5.2.

Then figure out how much your cost of living would have increased from 2016–2017 if you'd been living in Pittsburgh then with the same lifestyle you have now.

(b) Compare your answer in (a) to the 2.2 percent Pittsburg average. Explain why it's higher or lower.

Exercise 5.7.10. Regional differences in the CPI.
Moved to Extra Exercises at www.ams.org/bookpages/text-63 .

Exercise 5.7.11. [S][Section 5.4][Goal 5.4][Goal 5.1] [Goal 5.2] Eat out or in?
The overall change in the Consumer Price Index is a weighted average of changes in various categories. On November 19, 2011, *The Boston Globe* reported the data displayed in Figure 5.5.

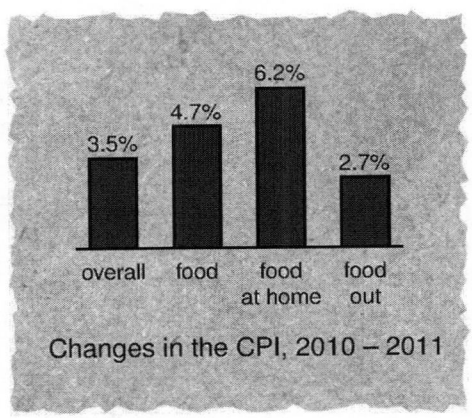

Figure 5.5. The cost of everything and the cost of food [**R179**]

(a) The change in the CPI for food is a weighted average of the change in the cost of food at home and the change in the cost of food away from home. Find the weights.

(b) Is it reasonable to say that about 40% of food expenses are for food away from home?

(c) The overall change in the CPI is a weighted average of the change for food and the change for everything else. Do you have enough information to find the weights?

Exercise 5.7.12. [S][C][Section 5.5][Goal 5.1] Projected cost of F-35 program up to $382B.
On June 2, 2010, *USA Today* reported a story from Bloomberg News that said that the total price tag for the project to build the new F-35 Joint Strike Fighter would be 65% higher than the original $232 billion estimate. The cost now is estimated to be $382 billion.
The total cost of each plane would be $112.4 million, up 81% from the original $62 million. The cost to produce each plane would be $92.4 million, about 85% higher than the original estimate of $50 million. [**R180**]

(a) Check the arithmetic to see that the three percentage increases reported are correct.

(b) Can you explain how the project cost can have increased by just 65 percent when the overall total per plane increased by 81 percent and the construction cost alone increased by almost 85 percent?

(c) How would the description of the increase in project cost change if you took into account inflation between 2002 and 2010?

Exercise 5.7.13. [S][Section 5.6][Goal 5.3] [Goal 5.1] Your car is more crowded than you think.

Table 5.6 reports results from a 1969 Personal Transportation Survey on "home-to-work" trips in metropolitan areas.

Number of riders	Percentage of cars
1	73.5
2	18.2
3	4.7
4	1.9
5	1.1
6	0.5
7	0.1

Table 5.6. Rush hour car occupancy [**R181**]

(a) The survey stated that the average car occupancy was 1.4 people. Check that calculation.

(b) Show that the average number of riders in the car of a typical commuter is 1.9.

(c) (Optional and tricky) Suppose you could persuade enough people who drive alone to switch to five person car pools in order to increase the average number of riders per car from 1.4 to 2.

What would the percentage of single occupant cars be then? What percentage of the people would be driving alone? How many people would there be in the car of a typical commuter?

[See the back of the book for a hint.]

Exercise 5.7.14. [S][Section 5.6][Goal 5.1][Goal 5.3] Why your friends have more friends than you do.

Imagine a small social network — perhaps Facebook when it was just starting out — with 100 people. One of them is friends with the other 99, but none of those 99 is a friend of any of the others.

(a) What is the average number of friends in this network?

(b) Explain why the statement "your friends have more friends than you do" is true for almost everybody in this network.

This network is just a toy one for making easy computations. But the paradox is true for real social networks, whenever there are some people with many friends and some with few. It's surely true for Facebook.

For more on this interesting paradox, see the article "Why Your Friends Have More Friends Than You Do" by Scott L. Feld, *American Journal of Sociology*, Vol. 96, No. 6 (May 1991), pp. 1464–1477, written long before Facebook.

Review exercises.

Exercise 5.7.15. [A] Find the average of each set of numbers.

(a) 40, 20, 30, 50, 40, 40, 55, 45, 60, 30

(b) 0, 0, 0, 0, 5, 10, 10, 10, 10

(c) 5, 5, 5, 5, 5, 5, 5, 5, 5, 5

(d) 0, 0, 0, 100, 100, 100

(e) 0.2, 0.4, 0.33, 0.45, 0.2, 0.1, 0.1, 0.1, 0.2, 0.4

(f) 5, 3, 6, 1, 4, 4, 4, 7, 6, 6, 4, 5, 1

(g) 86, 72, 86, 90, 91, 86, 75, 88, 42, 89, 90

Exercise 5.7.16. [A] Find each weighted average.

(a) Three test scores were 75, five test scores were 88 and two test scores were 90.

(b) Over the past month, I took three 5 mile runs, four 7 mile runs and ten 3 mile runs.

(c) Last week, the store sold 25 DVDs at $14.99 each, 13 DVDs at $12.99 each and 19 DVDs at $7.99 each.

Exercise 5.7.17. [A] Find the grade point average for each student, using Table 5.1.

(a) Bob earned an A in a 3 credit course, a B- in a 3 credit course, a B+ in a 4 credit course and an A- in a 3 credit course.

(b) Mary earned a C+ in a 3 credit course, an A- in a 4 credit course, a B+ in a 2 credit course, a B+ in a 3 credit course and a B in a 3 credit course.

(c) Alice earned a D+ in a 2 credit course, an A in a 3 credit course, an A- in a 4 credit course and a B+ in a 4 credit course.

Exercise 5.7.18. [A] Find the semester and cumulative GPA for each student, using Table 5.1.

(a) Mike has a 2.9 cumulative GPA and 45 credits. This semester he took three 4 credit courses and earned grades of B-, B+ and B+. He also took a 3 credit course and earned a C grade.

(b) Hilda has a 2.5 cumulative GPA and 16 credits. This semester she took a 3 credit course and earned a C+, a 4 credit course and earned a B- and a 5 credit course and earned an A-.

Exercises added for the second edition.

Exercise 5.7.19. [U][S] Pennies cost more.

Pennies are 2.5% copper with the balance zinc. Between 2016 and 2017 the cost to produce, administer and distribute a penny increased from 1.50 to 1.82 cents.

> "Compared to last year, FY 2017 … average copper prices rose 21.3 percent to $5,782.69, and average zinc prices increased 45.4 percent to $2,715.62," the U.S. Mint's annual report noted. [R182]

(a) What was the percentage increase in the cost to produce, administer and deliver a penny?

(b) How do you know that the increase in the price of the metal was about 45%?

(c) The quotation says, "Higher metal prices were one factor in the increases." The other factors were possible increases in administration and distribution. If those other factors did not increase, what fraction of the cost of producing a penny is the cost of the metal?

Exercise 5.7.20. [U][S][Section 5.2][Goal 5.1] Vaccination.

Figure 5.7 shows the percentage of California counties with various measles vaccination coverage (as a percent).

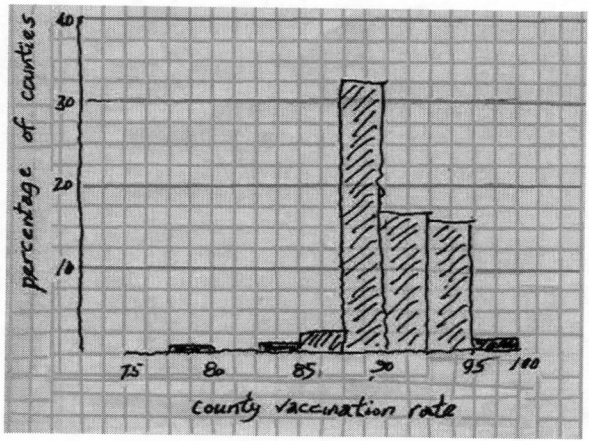

Figure 5.7. Measles vaccination coverage [R183]

(a) Check the assertion that about one in three students lived in a county with a vaccination rate below 90%. What assumption are you making about the number of students in each county?

(b) Estimate the weighted average county vaccination rate.

(c) Is your answer to the previous question consistent with the assertion elsewhere in the article that "In 2014, for California over all, about 93 percent of entering kindergartners were vaccinated for measles." ?

(d) What is "herd immunity"? What fraction of the population should be vaccinated for measles in order for herd immunity to take effect? Is California safe?

6

Income Distribution — Spreadsheets, Charts and Statistics

This chapter covers a lot of ground — two new kinds of average (median and mode) and ways to understand numbers when they come in large quantities rather than just a few at a time: bar charts, histograms, percentiles and the bell curve. To do that we introduce spreadsheets as a tool.

Chapter goals:

Goal 6.1. Work with mean, median and mode of a dataset.

Goal 6.2. Introduce the normal distribution, with its mean and standard deviation.

Goal 6.3. Understand how skewed distributions lead to inequalities among mean, median and mode.

Goal 6.4. Make routine calculations in Excel.

Goal 6.5. Use a spreadsheet to ask and answer "what-if" questions.

Goal 6.6. Create bar charts and other types of charts with a spreadsheet.

Goal 6.7. Use histograms to group and explore data.

Goal 6.8. Calculate averages for grouped data.

Goal 6.9. Understand the basics of descriptive statistics, including bell curve, bimodal data and margin of error.

6.1 Salaries at Wing Aero

Table 6.1 shows the distribution of workers' salaries at Wing Aero, a small hypothetical company. In keeping with our *Common Sense Mathematics* philosophy we should work with real data. But most companies keep this kind of information private. Any similarity between our imagined Wing Aero and any real company is purely coincidental. In Exercise 6.12.18 we'll apply the lessons we learn to look at income distribution in society at large.

The company has about 30 employees. We want to understand the salary distribution. A natural place to begin is with the average. But adding thirty numbers by hand (even with a calculator) is tedious and error-prone. A spreadsheet on your computer can do the arithmetic faster and more accurately. We have worked the examples in this text with Microsoft's Excel and with the free spreadsheet LibreCalc, available from `www.libreoffice.org/` . Online applications like Google sheets offer most of the features you will need. Use the examples here to work out how to access them.

We've organized this chapter as a spreadsheet tutorial — you can follow it step by step in Excel as you read it. If you are using LibreCalc you will find the same features. Sometimes we will provide screenshots from both.

See Section 6.3 for some general software tips and information about alternatives to Excel.

If you're online you can save typing time by downloading the Wing Aero spreadsheet `WingAero.xlsx`. That spreadsheet and all the others you'll need live at `www.ams.org/bookpages/text-63`. If you build it for yourself, use column `A` for the employees and column `B` for the salaries. Put the labels in row 7 and the data in rows `8:37`, not side by side as in the table. You should see Figure 6.2.

Employee	Salary (thousands of $)	Employee	Salary (thousands of $)
CEO	299	Supervisor	43
CTO	250	Supervisor	51
CIO	250	Supervisor	38
CFO	290	Supervisor	33
Manager	77	Supervisor	42
Manager	123	Supervisor	49
Manager	84	Worker	25
Manager	63	Worker	19
Manager	68	Worker	41
Manager	49	Worker	17
Manager	82	Worker	26
Manager	87	Worker	25
Supervisor	42	Worker	21
Supervisor	37	Worker	28
Supervisor	29	Worker	27

Table 6.1. Wing Aero salary distribution

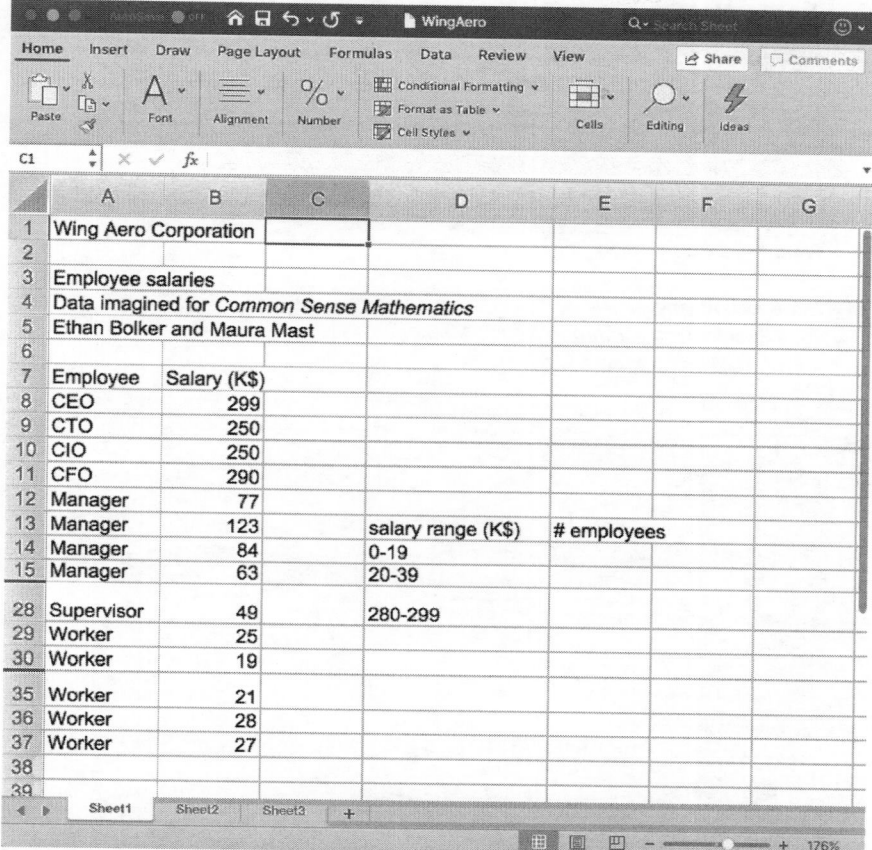

Figure 6.2. Wing Aero spreadsheet (some hidden rows)

What do you notice?

The employees are listed in decreasing order of importance (or prestige), but only approximately in decreasing order of salary. Some supervisors earn more than some managers, and some workers more than some supervisors. We can make those discrepancies visible by sorting the data.

Select the rectangular block of data in rows 8 through 37, columns A and B. Be sure to select both columns so they will be sorted together. Choose the Sort dialog box from the Data tab, select sorting by Salary, Largest to Smallest, as in Figure 6.3.

Often Excel offers you more than one way to do a job. This is one way to sort, in Excel 2013. There are others. Other versions of Excel may use different menus. But the ability to sort will be available in any spreadsheet program you use. Figure 6.4 shows how to do it in LibreCalc.

If you sort the data again alphabetically (by Employee, A to Z) the table returns (nearly) to its original state, because the employee categories were alphabetical at the start. But each category is now sorted by salary.

To find the average salary at Wing Aero we tell Excel to add the entries in column B and divide by the number of employees.

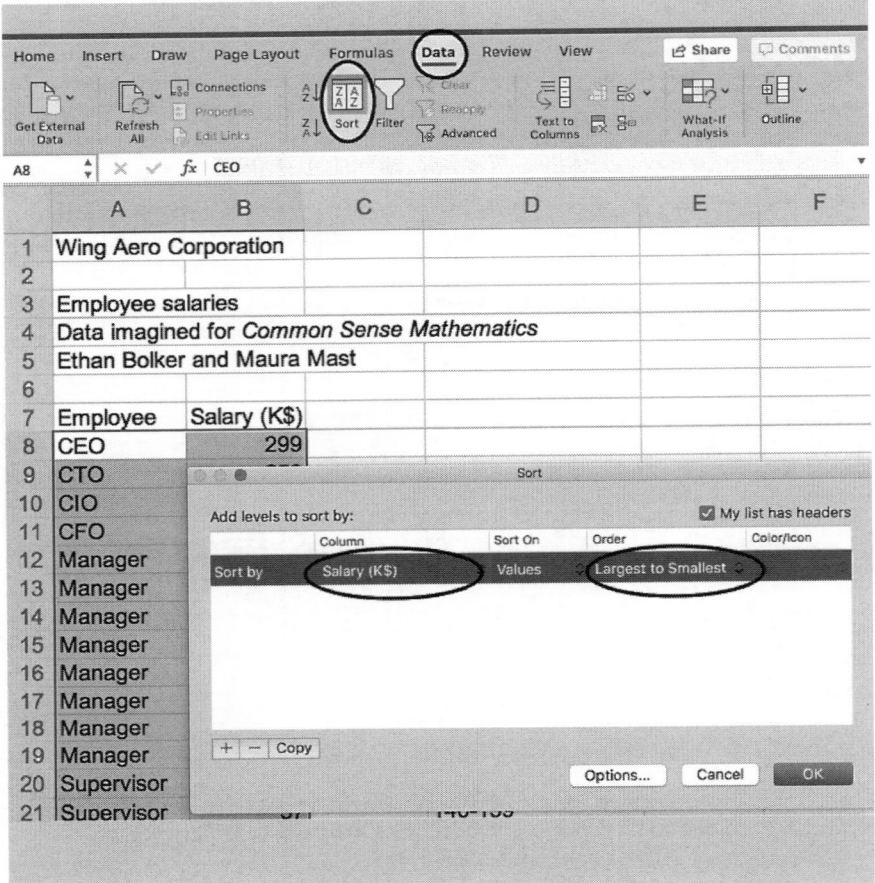

Figure 6.3. Sorting Wing Aero salaries (Excel)

Enter the label `Total` in cell A38. Then go to cell B38. Make sure the `Formula Bar` is visible. (Use the `View` menu to find it if it's not.) In the formula box type an equals sign =, to tell Excel you want it to do some arithmetic, and then the name of the operation,

$$=\text{SUM}($$

since you are about to add up some numbers. The open parenthesis asks Excel to prompt you for information. It suggests

$$\text{SUM}([\text{number1}], [\text{number2}], \ldots)$$

as in Figure 6.5.

Select cells B8:B37, close the parentheses and type `enter` or click the check icon on the `Formula Bar`. You should see `Total 2315` in cells A38 and B38. Wing Aero's total annual payroll is \$2,315K — about \$2.3 million.

To find the average salary we must divide the total \$2,315K payroll by the number of employees. Rows 8 through 37 contain employee records so there are $37-8+1 = 30$ employees. But it's better to ask Excel to count the rows for you. Type the label `Count`

Figure 6.4. Sorting Wing Aero salaries (LibreCalc)

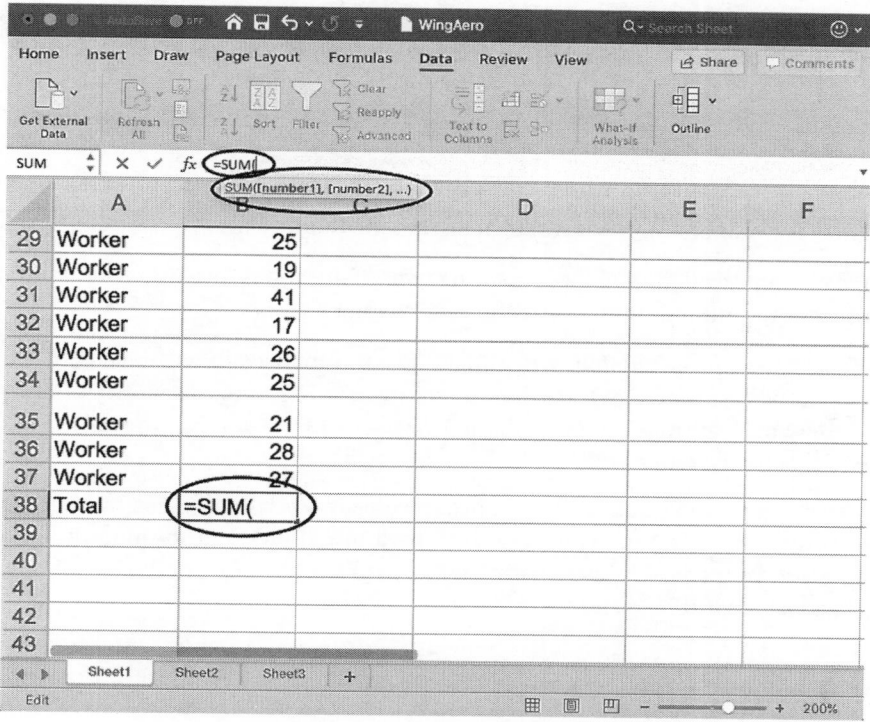

Figure 6.5. Summing Wing Aero salaries

in cell A39 and begin formula =COUNT(in cell B39. Finish the formula by selecting cells B8:B37 or by typing the addresses of those cells and closing the parentheses.

You should see

Count   30

Finally, type the label Average in cell A40 and put formula =B38/B39 in cell B40. You should see

Average   77.16667

so the average annual salary at Wing Aero is about $77,000. Excel rounded the exact 77.1666... to 77.16667 when it ran out of space. We rounded to two significant digits because that's all the precision we have in almost all the data. In Exercise 6.12.3 you'll learn how to tell Excel to round for you.

Excel's built-in functions SUM and COUNT are separately useful, which is part of why we showed them to you, but if it's the average you want, Excel can find it in one step.

Enter =AVERAGE(B8:B37) in cell B41, click the check icon and Excel tells you again that the average is 77.16667. Put "computed using SUM/COUNT" in cell C40 and "computed using AVERAGE function" in cell C41.

6.2 What if?

Suppose the CEO convinced the Board of Directors to double his salary, to $598K (even though the company lost money). To see how that would affect the payroll statistics, go to cell B8 and change the 299 there to 598. Excel automatically updates all your computations, increasing the total annual payroll to $2,614 thousand and the average annual salary by about $10,000 to more than $87,000.

We can learn several useful lessons from our work so far.

- Excel is faster and more accurate at arithmetic than we are.

- Excel is an excellent tool for answering what-if questions, because it automatically updates its computations when the data change.

- The average we calculated with SUM/COUNT and then again with AVERAGE is a terrible way to summarize the salary structure at Wing Aero. The CEO's 100% raise increases the average salary by about $10,000, or 13% — but he's the only employee who's actually better off!

After you've tried out these changes, restore the original values by clicking the Undo icon on the toolbar. This very useful feature allows you to undo the last few changes you've made. Use it to fix mistakes or to get back to where you were before you asked a "what-if" question.

6.3 Using software

Why use a spreadsheet at all? There are several reasons:

- to carry out large tedious calculations rapidly and correctly,

- to answer "what-if?" questions without having to redo arithmetic,

- to draw charts,

- to learn a tool you may use long after you've finished this course.

  Here are some tips for working with Excel and with software packages in general.

- To figure out something new, you can use the application's built-in help, search the web, ask a friend or teacher, or just play around. Which you try first depends on your personality.

- Many applications provide several ways to do the same task. That means you may get different advice or instructions from different sources. Choose a way that suits your style.

- Use the menus for things you do rarely, but learn the keyboard shortcuts for things you do often — in particular, `control-C` for copy and `control-V` for paste can save time.

- Learn about undo. **Save your work often**.

- When you are about to make significant changes to a spreadsheet or a document make a copy of what you have, so that you can return to it if you change your mind about what you should have done. In the Wing Aero study we created several, calling them `WingAero1.xlsx`, `WingAero2.xlsx` and so on.

- Create backup copies of important documents often — off your computer. Use a thumb drive (flash drive), an external hardrive, or the cloud.

- Some things work the same way in different applications (browsers, Word, Excel) — for example, selecting with the mouse, cut and paste with keyboard shortcuts.

- In many software applications placing the mouse over a feature you are interested in and right clicking often lets you view and change the properties of that feature. "Do the right thing" is a good mnemonic.

- Applications often try to guess what you intend to do. That can be good or bad. We'll see soon that Excel can adjust cell references automatically — that's usually, but not always, what you want. Word processors may try to fix your spelling — perhaps correctly, perhaps not.

Excel and LibreCalc are full function spreadsheet programs. With Google Sheets (`www.google.com/sheets/about/`) you can create spreadsheets in the cloud. That software is powerful enough to do the arithmetic we need for *Common Sense Mathematics* but it has far fewer chart formatting features than full-fledged programs. Spreadsheet applications on tablet computers lack those feature too. Don't even think of trying to do spreadsheet work on your phone.

6.4 Median

In Section 6.1 we found that the average salary at Wing Aero was $77,000. This is a pretty good annual salary. If you saw that in a job advertisement you'd think it was a pretty good company to work for. Maybe, maybe not. In Section 6.2 we saw how it's skewed by the CEO's earnings. When his (or her) salary increased from $299,000 to $598,000, the average salary jumped by $10,000 to $87,000. But no one else's salary changed!

The $77,000 "average" is misleading in other ways too. Most of the employees — 26 out of 30 — have salaries less than the average. That contradicts what we like to think "average" means. To find a salary that's "in the middle", sort the 30 line table again so that salaries are increasing. Since the table starts in row 8 and has 30 entries, rows 22 and 23 are the middle rows and the entries in cells B22 and B23 are the middle salaries. That means half the employees make $42,000 or less (the entry in cell B22) and half make $43,000 or more. So we might want to say that the "average" salary is $42,500. There's a name for this kind of "average" — it's the *median*. The first "average" we computed above is called the *mean*.

On the spreadsheet, change the Average label in cell A41 to Mean.

Then put

```
Median   42.5   computed by finding middle of sorted list
```

into cells A45, B45 and C45.

In some ways the median is a fairer "average" than the mean for describing the Wing Aero salary structure. It tells you more about the way salaries are distributed. In particular, the median salary isn't affected by the CEO's big raise. Try changing that salary again in Excel: the mean changes, as it did before, but the median stays the same.

Excel knows how to compute medians. Enter =MEDIAN(B8:B37) in cell B46 and check that you get the same value: 42.5. Enter "computed using MEDIAN function" in C46.

Finding the median with the MEDIAN function is better than finding the middle of the sorted list yourself because it works even when the data aren't sorted. Suppose the supervisor making $43K gets a raise to $50,000. Enter that new value as 50 in the spreadsheet. Then see that Excel has recalculated the median in cell B46: it's now 45.5.

There's a third kind of average, the *mode*. We'll return to that after we've summarized the salary data in a different way.

6.5 Bar charts

Often pictures are better than numbers when we wish to convey information convincingly. A *bar chart* is one such picture.

We use bar charts when we have data that fall naturally into categories. The height of each bar represents the value for that category. When you want general understanding rather than numerical detail it's easier to compare the heights of bars visually (in both relative and absolute terms) than the values of numbers.

In order to understand the Wing Aero income distribution better we will use Excel to draw two bar charts with four columns each, one showing the number of employees in each of the four job categories executive, manager, supervisor and worker, the other the total earnings in each category. We will use the original Wing Aero salary data from Section 6.1.

You can find the data from Table 6.6 in the range D7:F11 in the copy of the Wing Aero salary distribution spreadsheet at WingAeroBarCharts.xlsx. Excel calculated the values in columns E and F using the COUNT and SUM functions. For Figure 6.7 we asked Excel to show us how it made those calculations.

Figure 6.8 shows the first two pictures from that spreadsheet.

These side-by-side bar charts dramatically demonstrate that the executives make up a small part of the workforce but enjoy a large part of the salary expenses!

Job	Number	Total salary ($K)
Executive	4	1,089
Manager	8	633
Supervisor	9	364
Worker	9	229

Table 6.6. Wing Aero salary distribution by job category

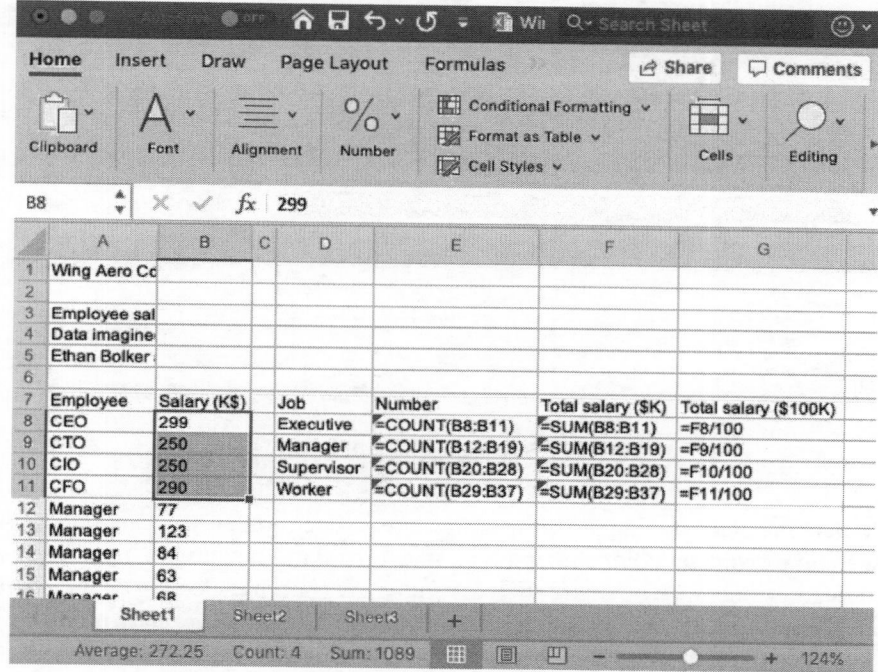

Figure 6.7. Showing the formulas used in a spreadsheet

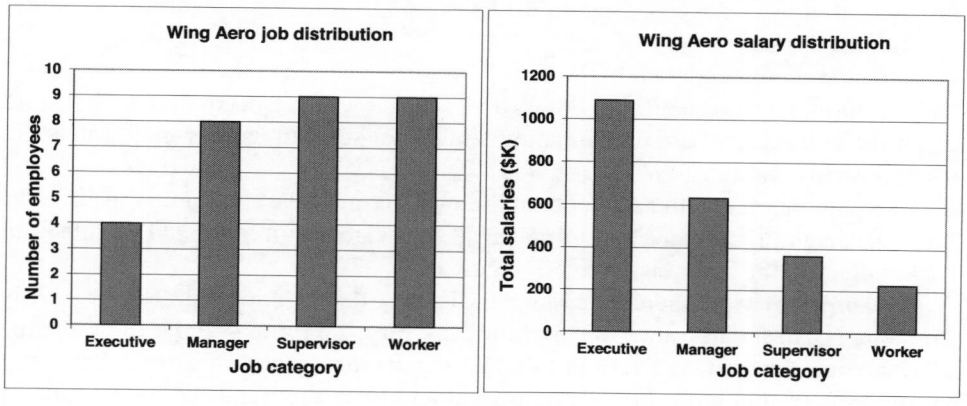

Figure 6.8. Wing Aero employee information by category

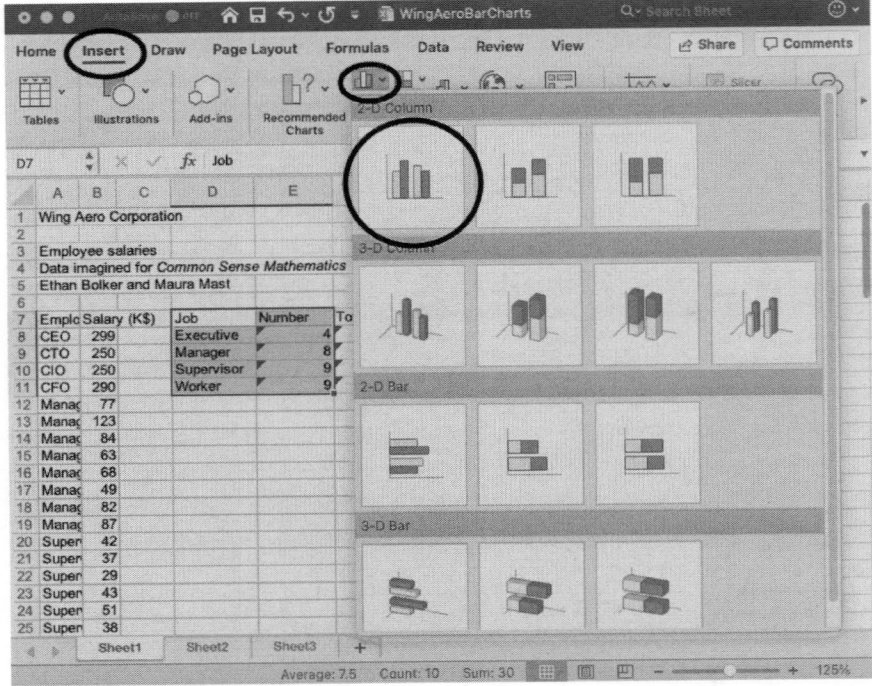

Figure 6.9. Inserting a chart in a spreadsheet

Delete the charts from your copy of the spreadsheet, so that you can learn how to build them for yourself. For the first one, select the data in columns D and E, rows 8 through 12 — the range D8:E12. Then ask for a new chart by selecting Column and 2D Column from the Insert tab, as in Figure 6.9. (Don't ask for a bar chart.)

Use the mouse to move the chart so that it does not hide any of the data. Then add the appropriate labels and change the colors so that the result is suitable for black and white printing. There are several ways to go about these tasks; experiment until you find ones that work for you. Section 6.3 has a tip about how the right mouse button can help with these tasks.

To build the second picture the same way you must select the data in columns D and F without selecting column E. To do that, select rows 8:12 in column D. Then hold down the control (PC) or apple or command (Mac) key and use the mouse to select those rows in column F.

It would be even better to combine the two pictures. We can do that in Excel by building a column chart for the full range D8:F12. The result (after adding titles and fixing colors) is the chart on the left in Figure 6.10.

The problem with that picture is that the bars for the employee numbers are nearly invisible, because data values for the numbers vary from 4 to 9 while those for the total salaries vary from $200K to $1200K. We can fix that by reporting the salary totals in hundreds of thousands of dollars rather than just in thousands of dollars (that is, by changing the units for measuring salary). Column G contains those numbers; we used it to draw the second picture. There you see clearly the opposing trends in the categories: total wages decrease as the number of employees increases.

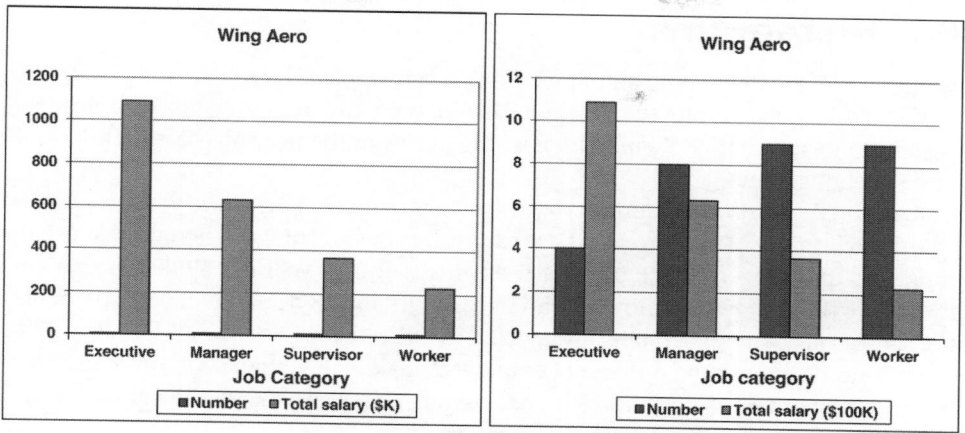

Figure 6.10. Wing Aero: side-by-side bar charts

6.6 Pie charts

Excel allows you to change the chart type on the fly. Select the column chart showing the total salary by job category. Right click on that chart. Select Change Chart Type . . . and then the first Pie. Adjust labels and colors to create the first of the charts in Figure 6.11.

The pie chart shows clearly in yet another way that the executives are the winners at Wing Aero. They take home nearly half the salary total. If you wanted to make that look a little less dramatic, you could omit the percentages and ask Excel to show a three dimensional version of the chart, as in the second picture. There we've rotated the picture so that the executive wedge is at the back, so it looks even smaller in perspective.

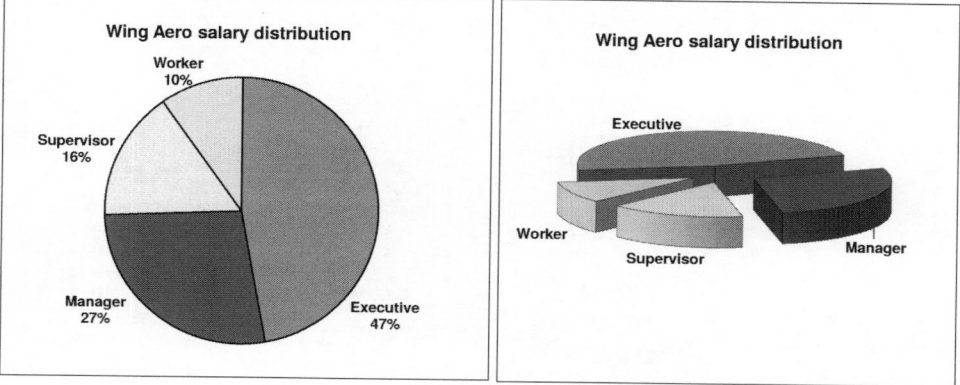

Figure 6.11. Wing Aero salary distribution pie charts

6.7 Histograms

Wing Aero is small enough so that we can see the whole salary table at a glance. But if there were 1,000 employees that wouldn't be possible. To understand the numbers we'd have to summarize them. The four categories in the previous section might not provide enough detail.

Another useful way to summarize the data is to divide the salaries into ranges and count the number of employees whose salary falls in each range. Then we can use the ranges as the categories in a bar chart. In this example we'll use $20K ranges. That means we think of two employees who make $29K and $33K as having approximately the same salary, since each falls in the $20K–$39K category.

We need to count the number of employees making less than $20K, then the number making between $20K and $39K, and so on. That's easy when the data are sorted in increasing order. Table 6.12 shows what we found.

To save you typing, we've listed the categories in cells `D15:D29` in `WingAero.xlsx`. You should check our work and enter the data in column E.

To see that you haven't missed anyone, `SUM` the range `E15:E29` to make sure the answer is 30, the known number of employees. (The sum isn't a perfect check. Although the total is correct, we might have put some employees into the wrong categories.)

We can use the data to draw a *histogram* — a bar chart where the categories on the x-axis specify data ranges and the y-axis counts or percentages for each range. You can see the resulting histogram in Figure 6.14.

Start as usual by building a column chart from the data in cells `D13:E28`. In a histogram it's conventional to make adjacent bars touch. To do that in Excel, right or

Salary range ($K)	Number of employees
0–19	2
20–39	10
40–59	7
60–79	3
80–99	3
100–119	0
120–139	1
140–159	0
160–179	0
180–199	0
200–219	0
220–239	0
240–259	2
260–279	0
280–299	2

Table 6.12. Wing Aero salary distribution by salary range

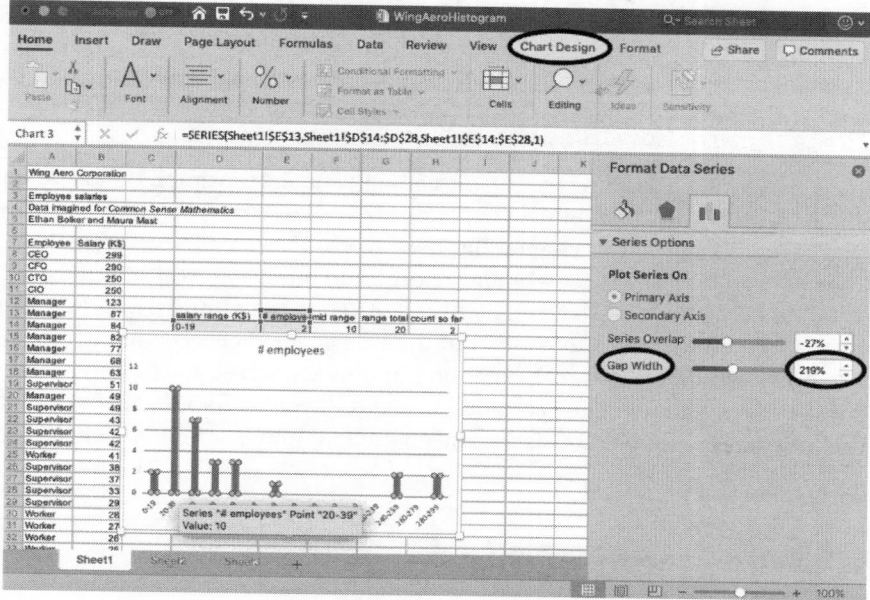

Figure 6.13. Formatting a histogram

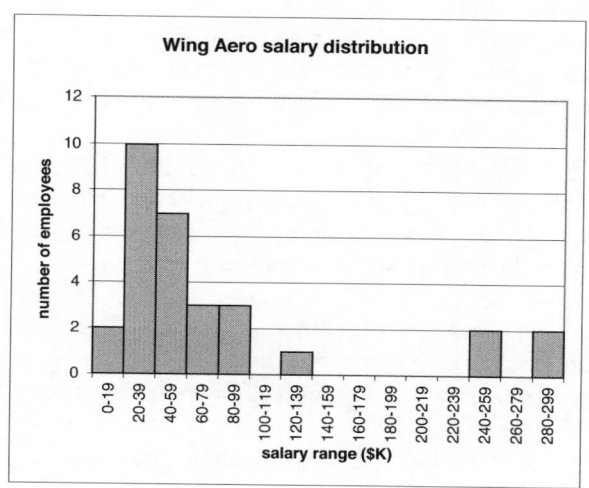

Figure 6.14. Wing Aero salary histogram

double click on one of the columns and look at the menu that appears. There you can change the Gap Width to 0, as in Figure 6.13. While you're there, figure out how to change the colors of the bars, and fix the labels. You can see the full spreadsheet with all the computations we've done so far at WingAeroHistogram.xlsx.

It's worth taking some time to study this histogram, which shows how the data are distributed. Most of the salaries are less than $100K and there's a large gap in salaries between $139K and the executives who make more than $200K. Although all this information is in the table, the histogram makes it visible and dramatic.

We chose $20K for the size of the salary ranges so that we would have enough categories to show what was going on but few enough to make the graph understandable. In Exercise 6.12.14 you can explore what happens with different choices.

One of the themes of this chapter has been to use Excel for tedious repetitive calculations. So you might wonder whether Excel could build Table 6.12 for the grouped data if we told it the ranges we were interested in. Then the numbers would automatically update when we asked "what-if" questions. The good news is that Excel can do this job. The bad news is that its histogram building tools are rather clumsy. We will content ourselves with doing the counting by hand. If you're ambitious, try Excel help or search the internet for "excel histogram" to find out how to group data for an Excel histogram.

6.8 Mean, median, mode

There's a third kind of "average" — the mode — that's sometimes informative. The *mode* is the most common value. The histogram from the previous section shows that there are more employees with salaries in the $20K–$40K range than any other. So the mode is about $30K. It's the category with the highest bar.

The mode is most useful for data aggregated into ranges, as in a histogram. In the raw Wing Aero data there is no well-defined mode. Each of the values $250K, $49K, $42K and $25K appears twice. In Excel the function MODE(B8:B37) reports $250K when column B is sorted in descending order and $25K when the column is sorted the other way.

Each of "mean," "median" and "mode" can legitimately be called an "average." That ambiguity makes it easy to lie with statistics without actually lying. The CEO at Wing Aero may brag that workers at his company earn an average $77K per year, while the union argues that the average salary is $30K per year.

A cynic would advise you to use the "average" that tells the story your way and hope your listener won't know the difference.

When distributions are symmetric, the mean, median and mode are in the same place. The Wing Aero salary distribution isn't symmetric; it's skewed to the right. That's the fancy way to say that the bulk of the data cluster toward the left of the histogram with a long tail off to the right. For data that's right skewed, as this is:

$$\begin{array}{ccccc} \text{mode} & < & \text{median} & < & \text{mean} \\ 30 & < & 42.5 & < & 77.1 \end{array}$$

In a histogram the mode is the peak, the median splits the area in half and the mean is where the graph would balance if it were a cardboard cutout.

If you're learning about these several kinds of averages for the first time you may want mnemonics to help you remember which is which. The "med" in median suggests correctly that it's in the middle. You can remember mode because there are "mo' of them than anything else."

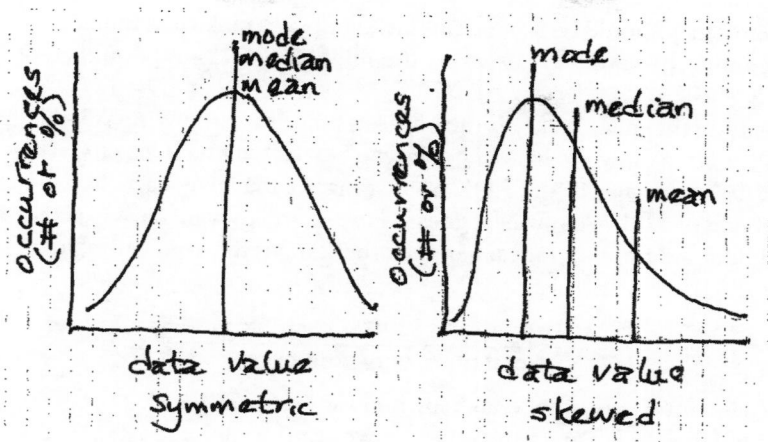

Figure 6.15. Symmetric and skewed distributions

6.9 Computing averages from histograms

Often all we know about data is a summary like that presented in a histogram. We'll see here how to estimate the mode, median and mean from that information. We will use the Wing Aero histogram in Figure 6.14 as an example. Since we know the correct averages we can see how good our estimates are.

We've already found the mode. It's the highest bar in the histogram — the range with the largest entry: $20,000–$39,000. Note carefully — the mode is the range, not the height of the bar. We could report that range, or report the mode as the middle of the range: about $30,000. The raw data do not have a mode that makes sense, so there's nothing to compare this estimate to.

To estimate the median salary from the histogram we need to find the salary such that half the employees make less and half more. The first bar tells us that two make less than $19K. Adding the number of employees in the first two ranges tells us that 2 + 10 = 12 make less than $39K. Looking at the next range we see that 2 + 10 + 7 = 12 + 7 = 19 make less than $59K. Since there are 30 employees, it's the 15th and 16th whose salaries are closest to the median. They are clearly pretty much in the middle of the third category, so we can estimate the median as the midpoint of that category, say $50,000. The correct value (from the raw data) is $42,500.

Since there are just eight ranges with data and the median occurred in the third one, we didn't need Excel to do the arithmetic. In a more complicated example things might not be so easy, so let's see how to make Excel do the work. Label column H in cell H15 as "count so far". Then copy the value from E15 to H15 by typing =E15 in H15. Then enter =H15+E16 in H16, to add the number of employees counted so far (in H15) to the number in this range (in E16).

Now select that formula and copy it to H17:H29. A miracle has happened! Cell H29 contains the value 30. If you click on that cell you can see that it's the result of the formula =H28+E29. Excel read your mind and automatically updated the row references to cells in columns H and E with each copy down the column. It knew you wanted to add the value in the cell above to the value in the cell three over to the left.

The last entry should be 30 since all the employees make less than \$299K. Now it's easy to see that the count so far passes the midpoint of 15 in the middle of the third range.

Estimating the mean salary is the hardest. Since the only Wing Aero data we have is what we used to draw the histogram we can't expect to find it exactly. All we can say about the 10 employees in the \$20K–\$39K range is that they earn somewhere in the neighborhood of \$30K. So our best guess is to assume they all earn exactly that. If we make a similar assumption for each of the other ranges then our estimate for the mean is the weighted average

$$\frac{2 \times \$10K + 10 \times \$30K + \cdots + 2 \times \$290K}{\text{total number of employees}}.$$

That's too much arithmetic to do by hand so we'll use Excel.

Put the label `mid range` in cell F13. We want to use cells F14:F28 to hold the values 10, 30, ..., 290 that are the middles of the ranges in cells D14:D28. There's a quick trick for that. Enter the 10 in cell F14 and enter the formula

$$\texttt{=F14+20}$$

in cell F15 . Excel will display 30 there. That's because it reads the formula as

add 20 to the contents of cell F14.

Now we want to add 20 each time you move down a row. To do that, copy the formula in F15 and paste it into cells F16:F28. (This takes advantage again of Excel's correct guess about what we are trying to do.)

Next label column G by typing `range total` in cell G13. Then put the formula

$$\texttt{=E14*F14}$$

in cell G14. That asks Excel to multiply the numbers in cells E14 and F14. You should see 20. That's the first number to add in the weighted average computation we're working on.

When you copy that formula to cells G15:G28 you should see 580 at the end of the list. That's the miracle yet again.

To compute the weighted average you must sum the values in column G. Since cell E29 contains the sum of the values in column E, just copy the formula from that cell to cell G29. Excel will automatically change the column reference, turning the formula =SUM(E14:E28) into =SUM(G14:G28). The sum is 2360. To find the mean, enter the formula =G20/E29 in cell G30. Excel shows you 78.66667. Label that value as the mean.

Our estimate of the mean salary from the histogram is \$79,000. We shouldn't report more precision than that, since we made many approximations along the way. Do note that the estimate is not very far from the true mean of \$77,167 computed from the raw data.

6.10 The bell curve

We used the data on baby weights in Table 3.5 to study percentiles. Figure 6.16 shows the histogram we constructed from that table. The highest bar shows that the mode — the most common weight for male babies one year old in 2019 was about 21.5 pounds.

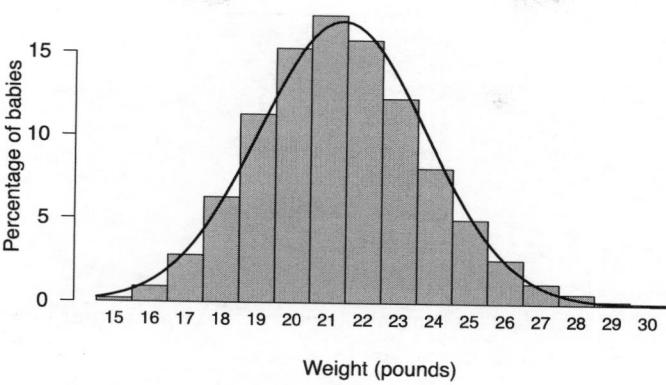

Figure 6.16. Baby weights

In Section 3.6 we found the median to be about the same. Using the techniques from Section 6.9 we computed the mean as a weighted average. It is 21.4 pounds.

That suggests that the distribution is just little bit skewed to the right.

Many factors contribute to a baby's weight at one year: birth weight, heredity, nutrition, ... In general, when many small effects combine to give a total, the distribution of values forms a *bell curve*. The one shown in the figure is a mathematically correct bell curve that approximates the real data. The mathematically correct name for the bell curve is *normal distribution*.

A normal distribution is always symmetrical. Its mean, median and mode are in the same place. To construct the one in the figure we need one more number besides the mean — a measure of how fast the curve spreads out. That measure is called the *standard deviation*. It's usually written with the Greek letter sigma: σ; the mean is usually written with the Greek letter mu: μ. For the baby weight data the standard deviation is about 2.6 pounds.

There's a nice way to use the standard deviation to describe how a bell curve spreads out:

- 2/3 of the values are less than one standard deviation away from the mean,
- 95% of the values are less than two standard deviations away,
- 99.7% are less than three standard deviations away.

Figure 6.18 illustrates these percentages. Table 6.17 summarizes them in terms of percentiles.

Since the baby body weight distribution is very close to normal, with mean μ about 21.4 pounds and standard deviation σ about 2.4 pounds, we know about 2/3 of the babies weighed between $\mu - \sigma = 21.4 - 2.4 = 19.0$ and $\mu + \sigma = 21.4 + 2.4 = 23.8$ pounds. Approximately 95% weighed between 16.6 and 26.2 pounds. 2.5% weigh more than 28.6 pounds and 2.5% weighed less than 14.2 pounds.

Figure 6.19 shows three bell curves with the same mean $\mu = 21.4$, but three different standard deviations, $\sigma = 1.2$, 2.4 and 4.8. The middle one matches the baby weight data.

The mathematics needed to calculate standard deviations and draw bell curves is more than we will present here. You will learn about it if you take an introductory course in statistics. All you should remember now is the rough relationship between

value	percentile
$\mu - 3\sigma$	0.1
$\mu - 2\sigma$	2.3
$\mu - \sigma$	15.9
μ	50.0
$\mu + \sigma$	84.1
$\mu + 2\sigma$	97.7
$\mu + 3\sigma$	99.9

Table 6.17. Percentiles for the normal curve, mean μ, standard deviation σ

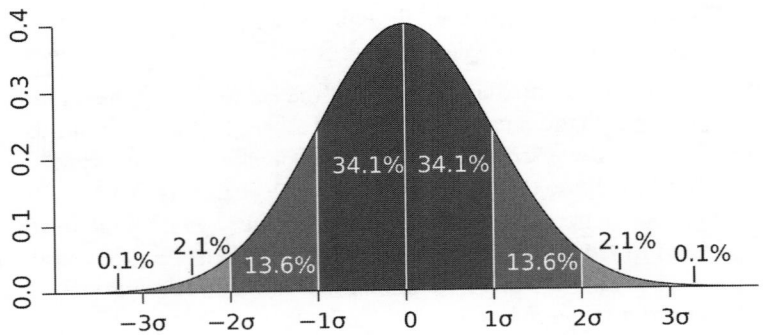

Figure 6.18. How the normal distribution spreads out [R184]

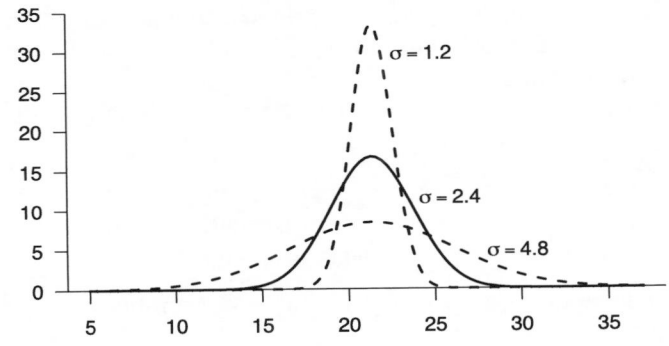

Figure 6.19. Three bell curves

standard deviation and spread we sketched above and that the official name for the bell curve is *normal distribution*.

Sometimes normal distributions are hidden in data. The black points in Figure 6.20 plot how the rate of diagnosis (in cases per 100,000 people) of Hodgkin's lymphoma (a kind of cancer) for white females depends on the age of the woman diagnosed. This graph is *bimodal*: there are two peaks, one at about 20 years, the other at 75 years. There is no single value that can legitimately be called the mode. The mean and the median would each be about 45 years, but they make no sense at all. Even though they

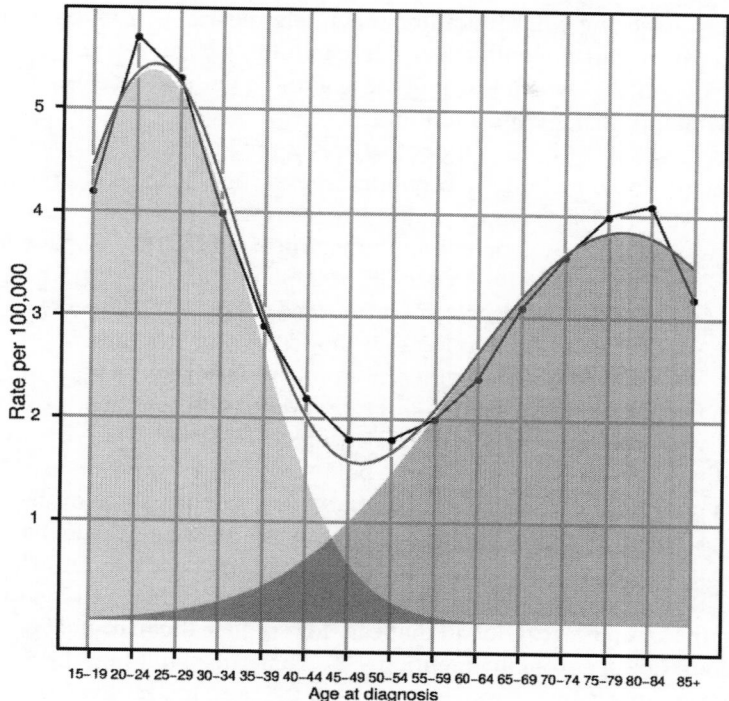

Figure 6.20. Hodgkin's lymphoma incidence (white females) [**R185**]

are "averages" there are few 45 year olds with the disease. The disease probably has two very different causes, one of which occurs more often in young people, the other in old people.

We can understand this distribution as a combination of two normal distributions. The left bell curve for the early onset of Hodgkin's lymphoma has a mean of about 24 years with a standard deviation of 11 years. The right bell curve for late onset has a mean of about 79 years with a standard deviation of 19 years — it spreads out more slowly. The smooth curve is the sum of the two normal distributions. It matches the data very well.

6.11 Margin of error

The Pew Research Center conducted a study in July of 2012 that asked about support for President Obama's tax position. Their report said (in part):

> By two-to-one (44% to 22%), the public says that raising taxes on incomes above $250,000 would help the economy rather than hurt it, while 24% say this would not make a difference.
>
> [The poll reached 1,015 adults and has a margin of sampling error of plus or minus 3.6 percentage points.] [**R186**]

The first paragraph quoted above reports the results of a survey. The second tells you something about how reliable those results are. It's clear that the smaller the margin of error the more you can trust the results. Understanding the margin of error

quantitatively — seeing what the number actually means — is much more complicated. A statistics course would cover that carefully; we can't here. Since the term
occurs so frequently, it's worth learning the beginning of the story. Even that is a little
hard to understand, so pay close attention.

The survey was conducted in order to discover the number of people who thought
the tax increase would benefit the economy. If everyone in the country offered their
opinion then we would know that number exactly. If we gave the survey to just three or
four people we could hardly conclude anything. The people at the Pew Research Center
decided to survey the opinions of a *sample* of the population — 1,015 people chosen
at random. Of the particular people surveyed, $0.44 \times 1,015 = 447$ people thought the
tax increase would benefit the economy. If they'd surveyed a different group of 1,015
people, they would probably see a different number, so a different percent.

The 3.6 percentage point margin of error says that if they carried out the survey
many times with different samples of 1,015 people, 95% of those surveys would report
an answer that was within 3.6 percentage points of the true value.

There's no way to know whether this particular sample is one of the 95% or one
of the others. About five of every 100 surveys you see in the news are likely to be bad
ones where the margin of error surrounding the reported answer doesn't include the
true value. Survey designers can reduce the margin of error by asking more people
(increasing the sample size). But all that can do is reduce the margin of error.

The report doesn't explicitly mention 95%. That's just built into the mathematical
formula that computes the margin of error from the sample size. Even that conclusion
may be too optimistic. The margin of error computation only works if the sample is
chosen in a fair way, so that everyone is equally likely to be included. If they asked 1,015
people at random from an area where most people were Democrats (or Republicans)
or rich (or poor) the result would be even less reliable. The report describes the efforts
taken to get a representative sample.

6.12 Exercises

A spreadsheet is just a tool. It doesn't answer questions; it provides numbers and pictures you use to answer questions. So keep on writing complete sentences explaining
the meaning and context of the numbers you report. The numbers in cells in a spreadsheet are of no more use by themselves than the numbers in a calculator display.

If an exercise asks for hard copy of a spreadsheet, format it well. Be sure to preview
before you print to make sure that charts fit on one page and don't cover important
numbers. Label the data columns, cells containing important computations, axes and
legends in charts. Numbers are useless when they can't be understood.

Exercise 6.12.1. [S][R][Section 6.1] CXO.
   What do "CEO", "CTO", "CIO" and "CFO" stand for in the Wing Aero salary table
in Section 6.1?

Exercise 6.12.2. [S] [Section 6.2][Section 6.4] [Goal 6.1][Goal 6.5] What if...?
   Open up the original Wing Aero spreadsheet and use your spreadsheet to calculate
the mean and median, as we did in Section 6.1. For each exercise that follows, write a
clear statement — using the numbers — that summarizes what you found.

(a) Suppose all the managers get a $10K raise. Change their salaries and see how Excel updates the mean and median.

(b) Go back to the original spreadsheet (use the Undo feature if you can). Experiment with changing salaries (of any of the workers — your choices) so that the mean and the median increase. Explain how you made your choices.

(c) Reset back to the original spreadsheet. How would you change salaries so that the mean decreases but the median stays the same?

(d) Reset back to the original spreadsheet again. How would you change salaries so that the mean stays the same but the median decreases? Which salaries did you change?

Exercise 6.12.3. [S][R][Section 6.3] [Goal 6.4] Formatting in Excel.
Format the cells containing averages in the Wing Aero spreadsheet so that the numbers displayed for the various averages are rounded to the nearest thousand dollars (no decimal places).

Exercise 6.12.4. [S][Section 6.4][Goal 6.1][Goal 6.4] [Goal 6.6] Practice finding the median.
Open up the original Wing Aero spreadsheet from WingAero.xlsx.

(a) Use Excel to find the median salaries for each category of employees (workers, managers, etc.).

(b) Show the data in a properly labeled bar chart.

(c) Do you think the median is a representative "average" for each category? Explain.

Exercise 6.12.5. [S][Section 6.5][Goal 6.6][Goal 6.1] Averaging averages.

(a) Find the average (mean) salary at Wing Aero for each of the four categories of employees (use the original data set, from Section 6.1).

(b) Show the data in a properly labeled bar chart.

(c) Compute the weighted average of these averages to check that you get the correct mean for the whole payroll.

[See the back of the book for a hint.]

Exercise 6.12.6. [S][Section 6.5][Goal 6.1] [Goal 6.6] Cash-strapped T proposes 23 percent fare increase.
On March 28, 2012, the Massachusetts Bay Transportation Authority (MBTA) provided the fare data in Table 6.21. Riders can pay with a stored-value Charlie Card or with a Charlie Ticket bought on the spot.

(a) What are the relative and absolute changes in the Charlie Card bus fare?
You don't have to do this in your head, without a calculator, but you should be able to.

(b) Create a spreadsheet for these data, with eight rows (one for each of the eight categories) and three columns, for the category name, the existing fare and the proposed fare. Create a properly labeled bar graph to display the data.

Fare Category	Current	Proposed
Charlie Card		
Bus	$1.25	$1.50
Subway	$1.70	$2.00
Senior Bus	$0.40	$0.75
Senior Subway	$0.60	$1.00
Student Bus	$0.60	$0.75
Student Subway	$0.85	$1.00
Charlie Ticket		
Bus	$1.50	$2.00
Subway	$2.00	$2.50

Table 6.21. MBTA fare increases [R187]

(c) Label the next two columns appropriately to hold the relative and absolute changes. Fill those columns with Excel formulas to compute the correct values. Do not compute the values elsewhere and enter them in the spreadsheet as numbers.

[See the back of the book for a hint.]

(d) Imagine that you are addressing a public meeting about these fare increases. How would you argue that an unfair burden is being placed on people who pay using a Charlie Ticket? How would you argue that senior citizens are most hard hit?

(e) Find the mean of the relative percent increases. Explain why the answer is not the 23 percent quoted in the headline.

(f) What extra information would you need to check that the correct mean is 23 percent?

(g) (Optional) Find out why the stored-value card is called a Charlie Card.

(h) (Optional and difficult) Create a column for the percentage of MBTA revenue for each of the categories and fill in some values that sum to 100% and give a weighted average fare increase of about 23%.

There's an important and subtle distinction here between weights as a percentage of revenue and weights as a percentage of trips.

Exercise 6.12.7. [S][Section 6.6][Goal 6.6] Why not pie charts?
Do some internet research to discover why bar charts are usually better than pie charts. Write the reasons in your own words (don't just cut and paste). Identify the sources of your information and comment on why you think those sources are reliable.

Exercise 6.12.8. [A][U][Section 6.6][Goal 6.6] Misleading pie charts.
Table 6.22 shows some student enrollment data at a college. It's clearly incomplete: Juniors and Seniors are missing.

Class	Percentage
Freshman	40
Sophomore	25

Table 6.22. Freshman and sophomore enrollments

Flashy pie charts are common in the media, but staid and boring bar charts are usually more informative and less deceiving. Because Excel makes it so easy to switch to a pie from a bar, designers may be tempted by the glitz. You should not succumb to that temptation. Here's an exercise where you can find out why.

(a) Construct a bar chart displaying these data, with columns for Freshman and Sophomore enrollments.

(b) Change the chart type to pie in your spreadsheet.

(c) Explain what is wrong with the new chart.

Exercise 6.12.9. [S][W][Section 6.8][Goal 6.5] [Goal 6.1] What if?
Add five workers each earning $18K to the original Wing Aero payroll by inserting some rows in the table. Comment on what happens to the mean, median and modal incomes.

Note that Excel automatically recomputes these averages, but not the charts for which you created data by hand.

When you've done this exercise, undo your changes and check that the three kinds of averages revert to their old values.

Exercise 6.12.10. [S][Section 6.8][Goal 6.3][Goal 6.7] Population pyramids.
At www.census.gov/data-tools/demo/idb/informationGateway.php you can choose a country and a year, then ask for a kind of bar chart known as a *population pyramid*. You can also download the data used to build the chart.

(a) Construct population pyramids for the United States and for Sudan for the year 2010.

(b) Estimate the number of people in the United States in 2010 between 0 and 9 years old. Do the same for Sudan. What fractions of the populations do these numbers represent?

(c) Find the modal age range for the United States population. Do the same for Sudan.

(d) Find the age range for the United States that has the smallest population. Do the same for Sudan.

(e) Compare the population distributions of the United States and Sudan. Write several sentences that highlight aspects of each distribution that you think are quantitatively significant. Use the results of the previous part of the problem.

Exercise 6.12.11. [S][Section 6.8][Goal 6.1] Lake Wobegon.

(a) Look back at the Wing Aero data and answer the following questions:

- What percentage of the employees make more than the mean salary?
- What percentage of the employees make more than the median salary?
- What percentage of the employees make more than the mode salary?

(b) In his radio show *A Prairie Home Companion* host Garrison Keillor regularly tells his audience about Lake Wobegon, where "all the children are above average".

Is that possible, with any of the meanings of "average"?

Exercise 6.12.12. [S][C][Section 6.8][Goal 6.1] Working for Walmart.

On December 2, 2009, Bloomberg News reported on a settlement in which Walmart agreed to pay $40 million to up to 87,500 employees because the company had failed to pay overtime, allow rest and meal breaks and, in addition, manipulated time cards.

Eligible present and former employees would receive $400 to $2,500 each — on average $734.

The lawyers for the employees asked for fees of $15.2 million from the $40 million.
[R188]

(a) Compute the mean compensation, assuming that there are 87,500 eligible employees and that the lawyers have taken their cut.

(b) Compare your answer to the reported minimum compensation of $400 and the reported $734 the average worker will receive.

(c) Draft a letter to the editor or the reporter, politely pointing out that both the reported "averages" made no sense and asking for more detail or a correction.

Exercise 6.12.13. [R][A][S][Section 6.8][Goal 6.1] Is it discrimination?

Table 6.23 shows the salary structure of two departments in a hypothetical university.

(a) What is the average (mean) salary of the professors? Of the women professors? Of the men?

(b) Answer the same questions for the median.

(c) Answer the same questions for the mode.

(d) Write a few sentences to convince someone that men in this university are paid better than women. Then write a few sentences to convince someone of just the opposite. Explain the contradiction.

	Physics		English	
	professors	salary	professors	salary
Women	1	$100K	8	$50K
Men	9	$90K	2	$40K

Table 6.23. Salary structure at a university

These calculations are so straightforward that they're easier with pencil and paper (maybe a calculator) than with Excel.

Exercise 6.12.14. [S][Section 6.9] [Goal 6.8][Goal 6.7] Choosing data ranges for a histogram.

When you change the widths of the intervals in a histogram you get a (slightly) different picture of the data.

(a) Redo the Wing Aero histograms in the text using salary ranges of size $10K and then of size $50K.

[See the back of the book for a hint.]

(b) In each case use the techniques from Section 6.9 to estimate the mean, median and mode.

(c) Discuss the advantages and disadvantages of these possible choices for the salary range, comparing them to our choice of $20K.

Exercise 6.12.15. [S][Section 5.1][Goal 5.1] Texting teens.

We drew the chart in Figure 6.24 using data from *The Boston Globe* on April 15, 2012.

(a) Estimate the mode, median and mean number of text messages sent by teenagers each day.

(b) In total, approximately how many text messages are sent by the 23 million American teens each day?

(c) The percentages don't add up to 100%. Why might that have happened?

(d) If you asked a random teenager how many text messages she sent yesterday what are the chances (what is the probability) that it was more than 50? More than 100? More than 25?

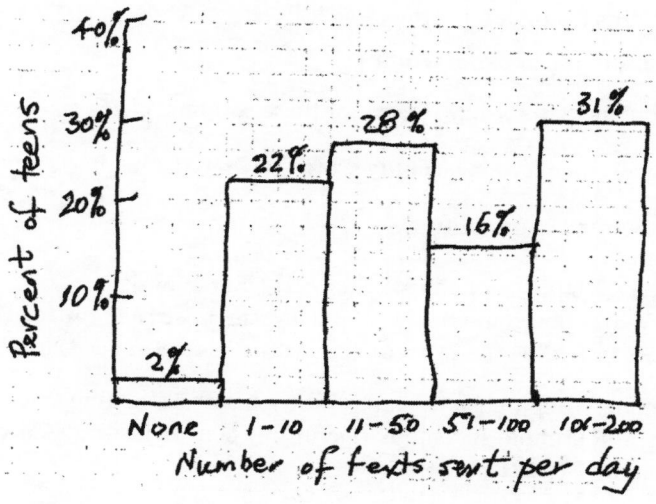

Figure 6.24. Teen texting [**R189**]

(e) What percent of teenagers text more than the median amount?

(f) Does the figure display a histogram?

(g) Create an Excel chart that reproduces the figure.

Exercise 6.12.16. [S][Section 6.9][Goal 6.8] [Goal 6.7] Websites are often confusing.

Jakob Nielsen evaluated the usability of voter information websites for the 2008 election for each of the fifty states and the District of Columbia. You can read his analysis at `www.nngroup.com/articles/aspects-of-design-quality/`. His article includes the histogram in Figure 6.25.

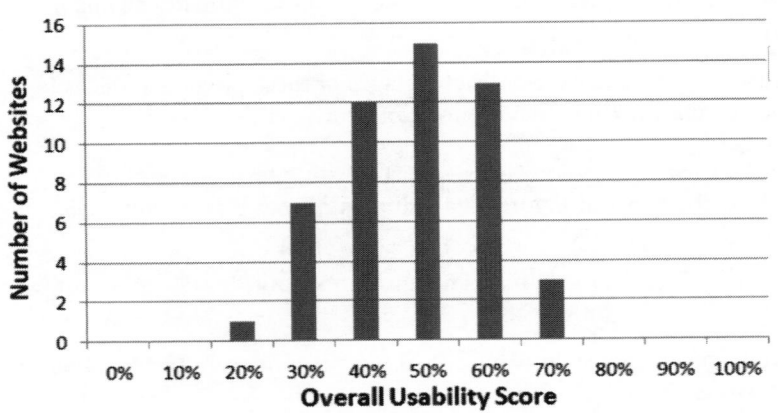

Figure 6.25. Website usability [**R190**]

(a) What is the modal usability score for these 51 home pages?

(b) Reproduce this histogram in Excel.

   [See the back of the book for a hint.]

(c) Estimate the median usability score for these 51 home pages.

(d) How many of them have a usability score less than the median score?

(e) Estimate their mean usability score.

Exercise 6.12.17. [U][S][Section 6.4][Goal 6.1] Doublethink.

In January 2012 one could read this at `www.businessinsider.com/where-the-one-percent-live-the-15-richest-counties-in-america-2012-2`:

> Living in Arlington isn't cheap, so you'd better be making at least the median household income to live in this county just outside Washington, D.C. [**R191**]

The "median household income" in the quote is the median income in Arlington County. How does this statement contradict itself?

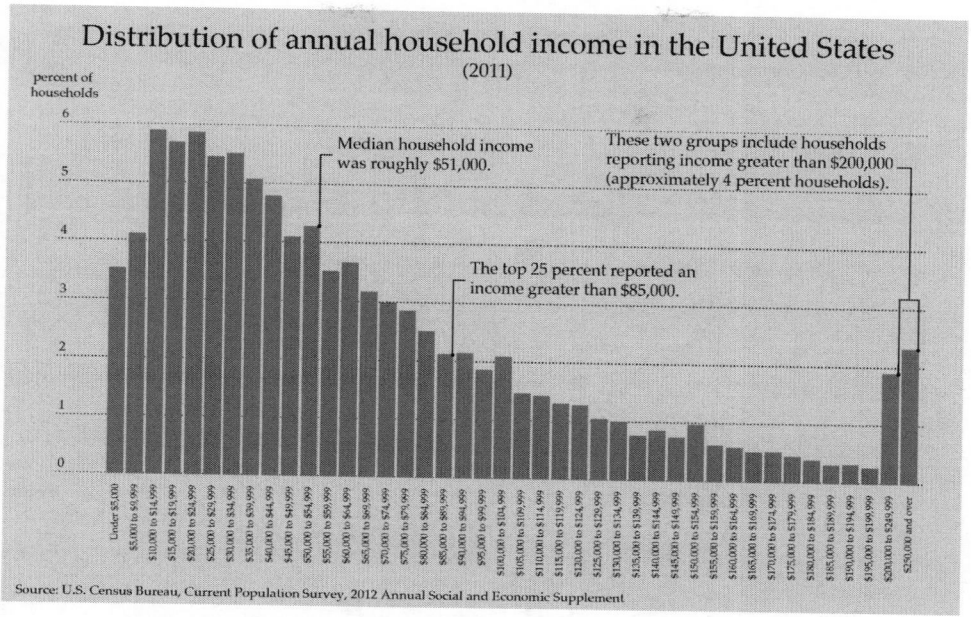

Figure 6.26. United States household income — 2012 Estimate [**R192**]

Exercise 6.12.18. [S][C][W] [Section 6.9][Goal 6.8][Goal 6.7] Household income in the United States.

The histogram in Figure 6.26 shows the estimated percentages of households in income groups $5,000 increments apart, except for the two farthest right columns.

We've put the data (and a copy of the figure) in the spreadsheet `Households2012.xlsx`.

(a) Check the quantitative assertions in the text in the Wikipedia chart.

(b) Build a histogram in Excel that comes as close as possible to matching the one from Wikipedia. Create the same chart and axis titles. Change the grid lines. Put in the comments as text boxes. Match the fonts.

(c) Do the percentages sum to 100%? If not, what might explain the discrepancy?

(d) To estimate the mean household income you will need an estimate of the mean for the households with incomes greater than $250,000. There's no top to this range, so you can't use the middle of the range.

What value for the mean for the last category makes the mean for the whole population equal to the median?

(e) Search for an estimate of the mean household income for the whole population. What mean for the last category results in this overall mean?

Exercise 6.12.19. [S][Section 6.9] [Goal 6.8][Goal 6.7] Fight for the Senate.

A graph like the one in Figure 6.27 appeared in Nate Silver's Five Thirty Eight column in *The New York Times* on October 31, 2012. The *x*-axis displays the number of

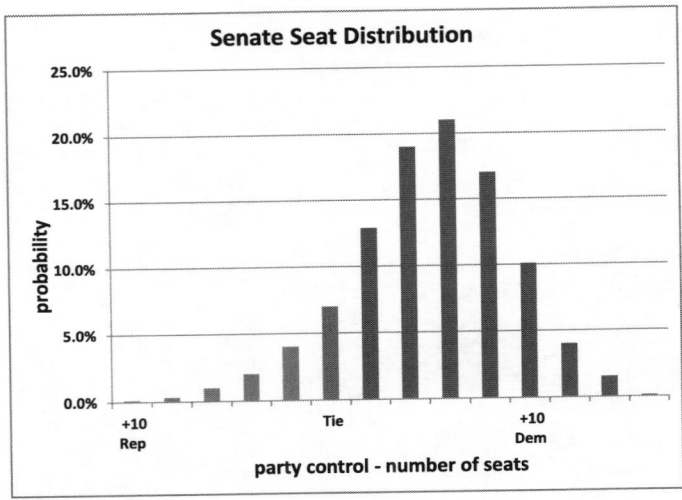

Figure 6.27. The fight for the Senate [**R193**]

seats held by each party: the Tie in the middle is 50 Democrats, 50 Republicans. The +10 Dem corresponds to 55 Democrats, 45 Republicans.

Nate Silver constructed this histogram by imagining (simulating) many thousands of elections and recording the percentage of time each Democratic/Republican split occurred. We estimated the percentages in his chart and entered them in the spreadsheet `Oct31SenateProjection.xlsx` so you don't have to type them yourself. (We rounded the really tiny percentages to zero.) Use Excel whenever it's most convenient for you.

(a) What is the most likely number of Democratic senators?

(b) What number of Democratic senators represents the mode of this distribution?

(c) What is the probability that there are more than 50 Democratic senators?

(d) What number of Democratic senators is the median of this distribution?

(e) If you had the complete list of all Nate Silver's imagined elections and sorted it by the number of Democratic senators, how many Democratic Senators would there be in the middle election on that list?

(f) Use Excel to compute the (weighted) average number of Democratic senators for these imagined elections.

(g) What actually happened in the election?

Exercise 6.12.20. [S][W][Section 6.9] [Goal 6.1][Goal 6.8] What cities pay for fire protection.

On Monday, March 30, 2009, *The Boston Globe* published an article comparing the amount various cities spent on Fire and EMS services. Figure 6.28 is a screenshot of a spreadsheet where we entered some of the data, along with population figures. You can download that spreadsheet from `FireSpending.xlsx`. Use the data to answer these questions.

Spending on Fire/EMS services

Study from the Boston Globe, March 30, 2009
2007 population figures from Wikipedia

City	Population	Fire/EMS spending per resident	Fire/EMS personnel per 1000 residents
Boston	608,352	$452.15	3.4
San Francisco	799,183	$315.81	2.2
Columbus, OH	747,755	$255.70	2.1
Seattle	594,210	$247.75	1.8
Baltimore	637,455	$225.98	2.7
Memphis	674,028	$220.22	2.5
Detroit	916,952	$201.54	1.6
Nashville	590,807	$194.43	1.9
Philadelphia	1,449,634	$187.63	1.6
Jacksonville	805,605	$179.99	1.5
New York	8,274,527	$157.56	1.7
Los Angeles	3,834,340	$137.80	0.9

Figure 6.28. Fire protection spending [R194]

(i) Population

   (a) What is the mean population of the twelve cities for which data are presented?
   (b) What is the median population of the twelve cities for which data are presented?
   (c) Create Table 6.29 in Excel. Fill in the second column there. Then create a properly labeled histogram for the data.
   (d) Use your histogram to estimate the mode population for these cities.
   (e) What percent of the U.S. population lives in these twelve cities?

(ii) Fire/EMS spending per person

   (a) What is the mean amount spent for Fire/EMS services per person in these twelve cities?
   (b) Estimate the median amount spent for Fire/EMS services per person in these twelve cities.

Population range	Number of cities
500K–600K	
600K–700K	
700K–800K	
800K–900K	
900K–1000K	
1000K–2000K	
2000K–3000K	
3000K–4000K	
> 4000K	

Table 6.29. City populations

(c) Estimate the mode amount spent for Fire/EMS services per person in these twelve cities.

(iii) What do firefighters earn?

There is enough information in the spreadsheet to calculate the average (mean) earnings of Fire/EMS personnel in each of the twelve cities. Do that, in a fresh column in your spreadsheet.

(a) In which city do Fire/EMS personnel have the highest average salary? How much is it?

(b) In which city do Fire/EMS personnel have the lowest average salary? How much is it?

(c) Where does Boston rank in the list of Fire/EMS personnel salaries?

(d) Explain how Boston can be at the top of the list in Fire/EMS expenses per resident although it does not pay the highest salaries.

(iv) Correction the next day!

On Tuesday, the next day, *The Boston Globe* published a correction, which said that Boston's fire department expenses were $285 per resident in the last fiscal year. [**R195**]

Look at the answers to the questions above and indicate which have changed (and how) and which stayed the same.

[See the back of the book for a hint.]

Exercise 6.12.21. [S][Section 6.9] [Goal 6.8] College presidents' pay.
Figure 6.30 is a histogram showing the total compensation for the 100 best paid presidents of public universities. The data are from an article in the April 3, 2011, issue of *The Chronicle of Higher Education* (chronicle.com/article/Presidents-Defend-Their/126971 .)
To save you staring at the picture, the number of presidents in each of the ranges (reading from left to right) is 51, 25, 15, 8 and then 1 in the last range.

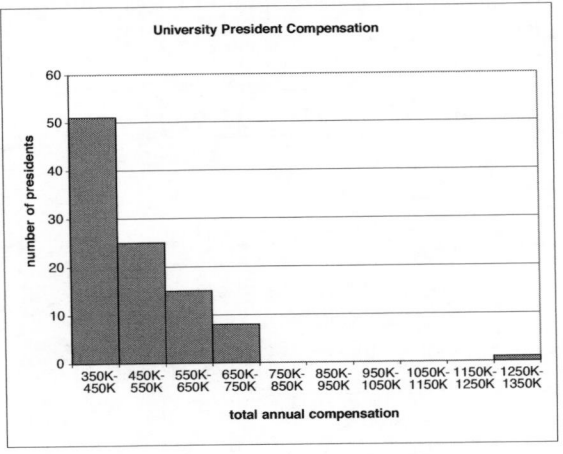

Figure 6.30. College presidents' pay [**R196**]

You may use Excel for this exercise if you wish, but you don't have to.

(a) What is the mode of this distribution?

(b) Estimate the median compensation.

(c) Estimate the mean compensation.

(d) Write two arguments one of which indicates that the average president is paid appropriately, one that presidential pay is too high.

Exercise 6.12.22. [A][S][W][Section 6.9] [Goal 6.8][Goal 6.7][Goal 6.9] Xorlon fleegs.
    Table 6.31 shows the distribution of weights of a sample of fleegs from the planet Xorlon, where weight is measured in frams.
    We made up the numbers in this artificial problem so that the arithmetic would be easy. We made up the words, too. Professor Bolker could have used "frams" in a Scrabble game with his wife — for 57 points on a triple word score.

weight range (frams)	percent of sample
20–40	10
40–60	10
60–80	20
80–100	10
100–120	50

Table 6.31. Xorlon fleegs

(a) Sketch a neat properly labeled histogram that displays the data.

(b) Create a properly labeled histogram in Excel that matches the one you just drew. You can start by downloading XorlonFleegs.xlsx.

    Answer the following questions. You may work in Excel or with a calculator or do mental arithmetic.

(c) Estimate the mode fleeg weight.

(d) Estimate the median fleeg weight.

(e) Estimate the mean fleeg weight.

(f) Estimate the percentage of fleegs that weigh more than the median.

(g) Estimate the percentage of fleegs that weigh more than the mean.

(h) Estimate the percentage of fleegs that weigh more than the mode.

Exercise 6.12.23. [S][Section 6.9] [Goal 6.9] Erdős numbers.
    Paul Erdős (1913–1996) was the most prolific mathematician of the twentieth century. He was famous (in mathematical circles) for the way he worked — he traveled from school to school, writing joint papers with mathematicians at each.

Erdős number	mathematicians
0	1
1	504
2	6,593
3	33,605
4	83,642
5	87,760
6	40,014
7	11,591
8	3,146
9	819
10	244
11	68
12	23
13	5

Table 6.32. Erdős numbers [**R198**]

From Wikipedia:

> An Erdős number describes a person's degree of separation from Erdős himself, based on their collaboration with him, or with another who has their own Erdős number. Erdős alone was assigned the Erdős number of 0 (for being himself), while his immediate collaborators could claim an Erdős number of 1, their collaborators have Erdős number at most 2, and so on. [**R197**]

You can think of the Erdős Number Project as a description of the social network of mathematicians. Its home page is `www.oakland.edu/enp` ; there you can find out that

> . . . the median Erdős number is 5; the mean is 4.65, and the standard deviation is 1.21.

Also, see the raw data in Table 6.32.

> [T]he lists of coauthors and the various other statistics on this site are updated about once every five years. The current version was posted on July 14, 2015.

(a) Use Excel to draw a histogram for the distribution of Erdős numbers.

(b) What is the mode of this distribution?

(c) Verify the claims for the median and mean.

(d) Professor Bolker (one of the authors of this book) has an Erdős number of 2 (he wrote a paper with Patrick O'Neil, who wrote a paper with Erdős).

Professor Bolker also wrote a paper with his granddaughter, Eleanor, and his son, her uncle Benjamin. So their Erdős numbers and that of Professor Mast (the other author of this book) are at most 3. Might any of them be less? Might they decrease in time?

(e) How many mathematicians have a finite Erdős number?

(f) There are some mathematicians whose Erdős number is infinite. How can that be?

(g) (Optional) Check to see whether the statistics at www.oakland.edu/enp have been updated, and update this question if they have been.

Exercise 6.12.24. [S][A][Section 6.9] [Goal 6.8] Ruritania.

Find the mean, median and mode age for male residents of Ruritania using the histogram or the data in the spreadsheet Ruritania.xlsx. (Ruritania is a fictional country in central Europe which forms the setting for *The Prisoner of Zenda*, a fantasy novel written by Anthony Hope.)

Exercise 6.12.25. [A][S][Section 6.7][Goal 6.7] Wing Aero percentiles.

In a histogram, data are grouped into ranges of the same numerical width. When finding percentiles the ranges contain known percentages of the data items — the widths vary. Use your sorted list of Wing Aero salaries to answer the following questions.

(a) How many Wing Aero employees are in the bottom tenth percentile in salary?

(b) What is the salary cutoff for the bottom tenth percentile?

(c) Answer the same questions for the top tenth percentile.

Exercise 6.12.26. [U][Section 6.11][Goal 6.2] [Goal 6.9] Sick-leave proposal.

An article in the *Orlando Sentinel* on August 6, 2012, discussed a ballot initiative that would require employers with 15 or more workers to provide paid time off for employees for illness-related issues. The article polled voters to gauge support for placing this question on the ballot for the November election and noted that among likely voters, 67 percent supported the initiative while 26 percent opposed it.

The article reported that the poll surveyed 500 people and had a margin of error of 4.4 percentage points. [**R199**]

(a) Why don't the percentages from this poll add up to 100%?

(b) Explain why this statement is not true: "67% of the residents likely to vote in November support the measure."

(c) Explain what the 4.4 percentage point margin of error means for this poll.

Exercise 6.12.27. [S][Section 6.9][Goal 6.7][Goal 6.8] The Boston Marathon.

Table 6.33 contains data for the numbers of men and women who finished the 2012 Boston Marathon, grouped by finishing times. For example, 26 men and one woman finished with a time between two and two and a half hours. (That one woman was a wheelchair racer.)

We've entered the data in the spreadsheet Marathon2012.xlsx.

Answer the following questions. Do as much of the arithmetic in Excel as possible.

(a) Sketch a neat histogram for this data.

(b) Draw your histogram with Excel. Does it match your sketch?

Finishing time	Men	Women
2:00–2:30	26	1
2:30–3:00	444	27
3:00–3:30	1,844	260
3:30–4:00	3,389	1,714
4:00–4:30	2,819	2,833
4:30–5:00	1,861	1,966
5:00–5:30	1,068	1,013
5:30–6:00	607	609
6:00–6:30	323	339
6:30–7:00	160	162

Table 6.33. The 2012 Boston Marathon

(c) How many men finished the marathon? How many women?

(d) Use the data to estimate the mode, median and mean for the men's finishing times. Mark these times on the handwritten histogram sketch.

(e) Suppose my friend ran the marathon and finished ahead of half the men. What was his finishing time (approximately)?

(f) About what percentage of the women finished ahead of half the men?

[See the back of the book for a hint.]

Exercise 6.12.28. Income growth.
 Moved to Extra Exercises at www.ams.org/bookpages/text-63 .

Exercise 6.12.29. [S][Section 6.5][Goal 6.1][Goal 6.6] Scrabble.
 Wikipedia page en.wikipedia.org/wiki/Scrabble_letter_distributions shows the point value of each of the 100 Scrabble tiles.

(a) Draw a bar chart illustrating the number of tiles with each of the point values from 0 to 10. The x-axis labels should be

$$0\ 1\ 2\ 3\ 4\ 5\ 6\ 7\ 8\ 9\ 10$$

The heights of the bars should correspond to the number of tiles with each value.

This is a difficult chart to create in Excel. Before you try, draw it by hand, so you know what you want the end result to look like. What Excel shows you first is likely to be far from your goal.

(b) What are the mode, median and mean point values? Show them on your (hand-written) chart.

(c) What percentage of the tiles are worth more than 1 point?

(d) What percentage of the tiles are worth less than the median number of points?

(e) Answer these questions for Scrabble in some other language (your choice) and discuss the differences between that language and English.

Exercise 6.12.30. [S][Section 6.9][Goal 6.7][Goal 6.8][Goal 6.3] Many flights arrive early!

The spreadsheet at `ArrivalDelays.xlsx` contains data on how many minutes late American Airlines flights to Boston's Logan Airport were in January 2014.

(1) What does a "negative delay" mean?

(2) Later you'll be asked to draw a histogram of this data in Excel. Sketch a neat approximate version first, with proper titles and reasonable scales for both axes and a proper title for the whole chart. You don't need to draw all the bars!

(3) Draw your histogram with Excel. Does it match your sketch?

(4) How many flights were counted in this data?

(5) What percentage of the flights arrived on time?

(6) Use the data to estimate the mode, median and mean arrival delay. Show these values on your histogram sketch.

(7) Flights that are more than two hours late are *outliers* — the delay is probably not American Airlines' fault. Estimate the mode, median and mean arrival delays if you don't include the outliers.

Exercise 6.12.31. [S][Section 6.9][Goal 6.7][Goal 6.8] Quotes in *Common Sense Mathematics*.

The spreadsheet `CSMquotes.xlsx` contains data on the number of words in quotes used in an early draft of this text.

(a) Create a properly labeled histogram displaying the data. You may sketch the histogram with pencil and paper, or use Excel.

(b) Calculate the total number of quotes.

(c) Estimate the total number of words in the quotes.

(d) Estimate mode, median and mean quote sizes, and mark them on your histogram.

(e) Explain why the mean is the largest of the three averages.

(f) Estimate the total number of words in the text.

(g) Estimate the percentage of words in the text that are in quotes.

Exercise 6.12.32. [S][W][Section 6.2][Goal 3.1][Goal 6.4][Goal 6.5] Ricky's tacos.

A story in *The Boston Globe* on February 6, 2015, stated that

> Food prices over the past year have increased at four times the rate of overall inflation, with fresh products, such as meat, vegetables, and dairy, soaring even faster. Ground beef prices, for example, are up about 20 percent from a year ago. Shoppers at local grocery stores have felt the sharp rise in prices, but for Ricky Reyes, owner of the taqueria on Dorchester Avenue, costlier ingredients mean it is getting harder to keep the price of his signature beef taco down. [**R200**]

Ingredient	2013 cost per pound	2014 cost per pound	percent increase in cost	percent of taco filling
Beef	$3.46	$4.16		45
Tomato	$1.73	$2.19		20
Cheese	$5.39	$5.44		20
Lettuce	$0.99	$1.11		15

Table 6.34. Taco costs

Use Table 6.34 to answer the following questions. We've entered the data in the spreadsheet tacos.xlsx.

(a) Is the *Globe* correct about the percent increase in the cost of beef?

(b) Fill in the column showing the percent increase in cost of each of the ingredients.

[See the back of the book for a hint.]

(c) Find the cost of a pound of taco filling in 2013 and 2014. Then find the percent increase in the cost of the filling.

(d) One way for Mr. Reyes to reduce the cost increase would be to change the percentages of meat and cheese, keeping the lettuce and tomato the same. What would the percent of each be if he wanted to keep the increase in a pound of filling to just 10%?

(e) Do you think customers would notice if Mr. Reyes changed the recipe using your answer to (d)?

[See the back of the book for a hint.]

Exercise 6.12.33. [S][Section 6.9][Goal 6.7][Goal 6.8] State populations.
Figure 6.35 shows the U.S. population distribution among the 50 states based on the 2010 U.S. Census. The data are in StatePopulations2010.xlsx.

(a) Recreate the histogram in Excel.

(b) What is the modal population of states?

(c) Estimate the median population of states from the histogram. Compare that to the median population Excel calculates.

(d) Estimate the mean population of states from the histogram. Compare that to the mean population Excel calculates.

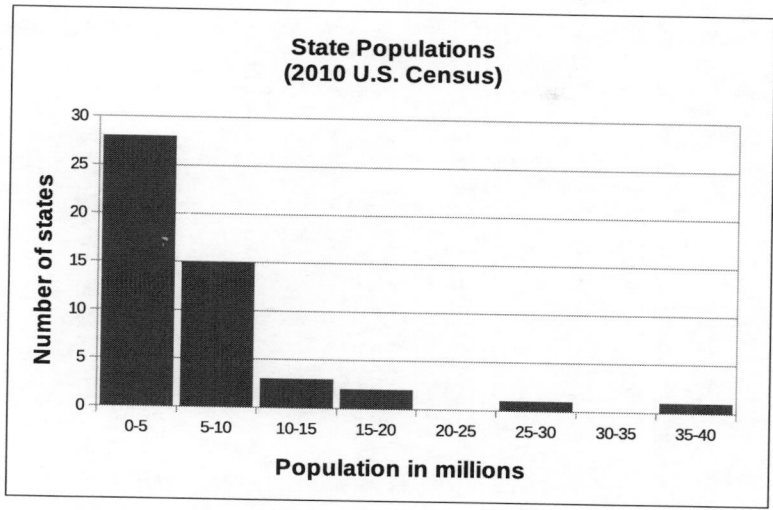

Figure 6.35. State populations

Review exercises.

Exercise 6.12.34. [A][R][Section 6.1][Goal 6.1][Goal 6.4]
Create an Excel spreadsheet and put the following numbers in the first column:

14, 15, 22, 50, 0, 33, 16, 18, 23, 40, 47.

(a) Use Excel to find the mean, median and mode of these numbers

(b) Change the first number from 14 to 23. How do the Excel calculations change?

(c) Click the "undo" button and confirm that Excel reverts back to the original set of numbers.

(d) Change the last four numbers to 0, so that the data now read

14, 15, 22, 50, 0, 33, 16, 0, 0, 0, 0.

How do the different averages change? Explain how the data are skewed.

Exercise added for the second edition.

Exercise 6.12.35. [S][Section 6.1][Goal 6.3] Jonathan Dushoff's beer.
Jonathan Dushoff says he averages two beers a week but drinks just one beer in an average week.
Invent a month's worth of data that explains this seeming contradiction, with one of the averages the mean and the other the mode.

7

Electricity Bills and Income Taxes — Linear Functions

We use an electricity bill as a hook on which to hang an introduction to functions in general and linear functions in particular, in algebra and in spreadsheets. Then we apply what we've learned to study taxes — sales, income and Social Security. You'll also find here a general discussion of energy and power.

Chapter goals:

Goal 7.1. Understand direct proportion as a linear equation with intercept 0.

Goal 7.2. Study situations governed by linear equations.

Goal 7.3. Stress the meaning and units of the slope and intercept.

Goal 7.4. View functions as tables, graphs and formulas.

Goal 7.5. Construct flexible spreadsheets to model linear equations.

Goal 7.6. Understand the piecewise linear income tax computations.

Goal 7.7. Sort out the confusing distinction between energy and power.

7.1 Rates

We began Chapter 2 with a discussion of the relationship

$$\text{distance} = \text{rate} \times \text{time},$$

or, with units,

$$\text{distance (miles)} = \text{rate (miles/hour)} \times \text{time (hours)}. \tag{7.1}$$

169

There we worked with particular numbers; now we want to look at that relationship a little more generally. If the rate is fixed (say 60 miles/hour), then we can find the distance traveled whenever we know the time. To say that with some algebra, write D for distance and T for time. Then

$$D = 60 \times T,$$

or, more generally,

$$D = r \times T,$$

where r is the rate of travel, in miles/hour.

That formula says that distance traveled is *proportional* to travel time. The rate in miles per hour is the *proportionality constant*. If you drive for twice as long you go twice as far. If you drive ten times as long you go ten times as far.

In Section 2.2 we introduced the units (gallons per hundred miles) as a useful way to measure automobile fuel economy: the amount of gas you use is proportional to the distance you drive. Let F be the amount of gasoline you use, in gallons, to drive D hundred miles. Then

$$F = r \times D$$

where the proportionality constant r is the fuel use rate, with units gallons per hundred miles. If you drive 10 times as far you use 10 times as much gas. You don't use any gas at all just sitting in the driveway (unless you're idling to warm up the car).

The proportionality constant is always a rate: it appears with units. In these examples the units are miles per hour and gallons per 100 miles.

The unit pricing discussion in Section 2.4 provides more examples of proportionality.

Finally, sales tax is computed as a proportion. If you spend twice as much you pay twice as much tax. The tax rate is the proportionality constant. If it's 5% then you pay (five dollars of tax) per (hundred dollars of purchase).

7.2 Reading your electricity bill

The more electricity you use at home, the more you pay. But the relationship isn't quite proportional. You don't pay twice as much to use twice as much. Figure 7.1 shows a simple sample electricity bill based on one we found on a British website. [**R201**]

This bill explains itself. We'll study it before we look at a real one. Since it comes from Great Britain the costs are expressed in pounds and pence rather than dollars and cents, and it comes once a quarter (every three months) rather than once a month, but you can ignore that while you read it — from the bottom up.

The last line is the total bill, computed as

$$\boxed{\texttt{Cost of electricity + fixed charge}}.$$

Checking the arithmetic:

$$£34.62 + £9.49 = £44.11.$$

The third line from the bottom explains the £34.62:

$$\boxed{\texttt{Number of units} \times \texttt{cost per unit}}.$$

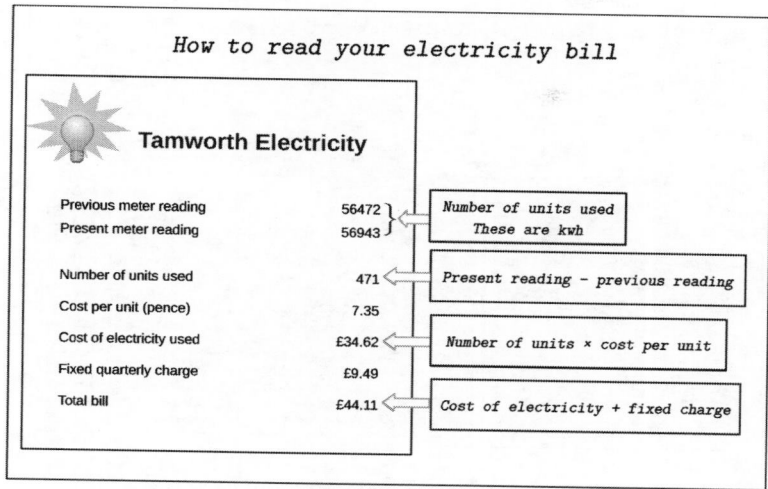

Figure 7.1. Tamworth electricity bill

That's our old friend proportionality. The previous line gives the proportionality constant: the cost per unit as 7.35 pence per unit. Later on in the document we're told that a unit is just a kilowatt-hour, abbreviated "kwh". (It's too bad the bill talks about "units" instead of just "kwh" since for us "units" has a more general meaning.)

There are 100 pence in a pound, so the cost per unit is

$$0.0735\frac{£}{\text{kwh}}.$$

(The computation would have been much more complicated before February 15, 1971 — the day England converted from pounds/shillings/pence to decimal currency. See `news.bbc.co.uk/onthisday/hi/dates/stories/february/15/newsid_2543000/2543665.stm`.)

In the current quarter this customer used 471 kwh of electricity — the difference between the meter reading before and after the quarter.

Here's all the arithmetic, with units:

$$£44.11 = 0.0735\frac{£}{\text{kwh}} \times 471\ \text{kwh} + £9.49.$$

That English bill is easy to read. Figure 7.2 shows a real one that's a little more complex, from NStar, in Boston.

We can identify the same two components. The fixed charge is the $6.43 labeled "Customer Charge." It's the part of the $145.26 total that does not depend on the amount of electricity used — in this case, 813 kwh. The six lines on the bill that do depend on that contribute

$$(0.04432 + 0.01039 + 0.00468 + 0.00050 + 0.00250 + 0.10838) \times 813$$

$$= 0.17077 \times 813$$

$$= 138.83601$$

to the total bill, which is

$$\$145.26 = 0.17077\frac{\$}{\text{kwh}} \times 813\ \text{kwh} + \$6.43.$$

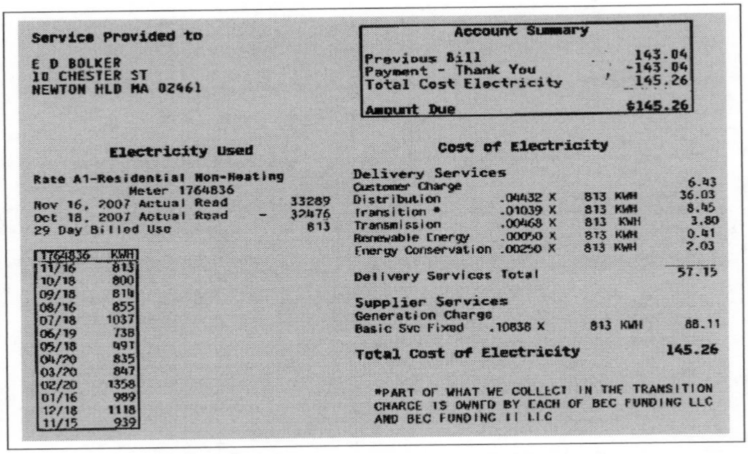

Service Provided to

E D BOLKER
10 CHESTER ST
NEWTON HLD MA 02461

Account Summary

Previous Bill	143.04
Payment - Thank You	-143.04
Total Cost Electricity	145.26
Amount Due	**$145.26**

Electricity Used

Rate A1-Residential Non-Heating
Meter 1764836

Nov 16, 2007 Actual Read		33289
Oct 18, 2007 Actual Read	-	32476
29 Day Billed Use		813

1764836	KWH
11/16	813
10/18	800
09/18	814
08/16	855
07/18	1037
06/19	738
05/18	491
04/20	835
03/20	847
02/20	1358
01/16	989
12/18	1118
11/15	939

Cost of Electricity

Delivery Services			
Customer Charge			6.43
Distribution	.04432 X	813 KWH	36.03
Transition *	.01039 X	813 KWH	8.45
Transmission	.00468 X	813 KWH	3.80
Renewable Energy	.00050 X	813 KWH	0.41
Energy Conservation	.00250 X	813 KWH	2.03
Delivery Services Total			57.15
Supplier Services			
Generation Charge			
Basic Svc Fixed	.10838 X	813 KWH	88.11
Total Cost of Electricity			**145.26**

*PART OF WHAT WE COLLECT IN THE TRANSITION CHARGE IS OWNED BY EACH OF BEC FUNDING LLC AND BEC FUNDING II LLC

Figure 7.2. NStar electricity bill (2007)

Note that NStar rounded \$138.83601 down to \$138.83 rather than up to the nearest penny. We should be grateful for small favors.

7.3 Linear functions

So far *Common Sense Mathematics* has called for hardly any algebra. Now a little bit will come in handy.

Suppose you buy your electricity from NStar as in the example above and want to study how your bill changes when you use different amounts of electricity. The monthly \$6.43 Customer Charge does not change. The rest of your bill is proportional to the amount of electricity. The proportionality constant is 0.17077 \$/kwh in the sample bill. We will assume that it does not change, although in fact it does change slightly when the electric company changes its rates.

If in a given month you use E kwh of electricity your total bill B can be computed with the formula

$$B\$ = 0.17077\frac{\$}{\cancel{\text{kwh}}} \times E\,\cancel{\text{kwh}} + \$6.43.$$

That formula captures how the dollar amount of your electricity bill depends on the amount of electricity you use, measured in kwh. The first term is the part that's proportional to the amount of electricity used. The second term (the amount \$6.43) is fixed. It represents the electric company's fixed costs: things like generating the bill and mailing it to you and maintaining the power lines on the street in front of your house. Those are expenses they must cover even if you're on vacation and have turned off all the appliances.

You probably encountered a similar formula once in an algebra class — it may look more familiar without the units

$$B = 0.17077 \times E + 6.43.$$

It may look even more familiar if we call the variables by the traditional names x (for the independent variable) and y (for the dependent) instead of E and B:

$$y = 0.17077x + 6.43.$$

This is a *linear function*, which standard algebra texts write in *slope-intercept* form

$$y = mx + b.$$

In this example the *slope m* is $0.17077 \frac{\$}{\text{kwh}}$ and the *intercept b* is \$6.43. For the English bill the slope is $0.0735 \frac{£}{\text{kwh}}$ and the intercept is £9.49.

There are many everyday examples where a linear equation describes how a total is computed by adding a fixed amount to a varying part that's a proportion. The fixed amount is the intercept. The proportionality constant is the slope.

- The most familiar examples are the ones where the intercept is 0: all the ones in Section 7.1.

- When renting a truck, the amount you pay is

(rate in dollars/mile) × (miles driven) + (fixed charge) .

- Your monthly cell phone bill might be

(rate in dollars per minute) × (number of minutes) + (fixed fee) .

(A real cell phone bill will probably be more complicated, perhaps with separate charges for phone minutes, text messages and data transfer, perhaps with some of each kind of use built into the fixed fee.)

- If you work as a salesperson and your commission is 15% of total sales your total wages are

0.15 × (total sales) + (your base salary) .

The pattern is

total = (rate) × (amount of some quantity) + (fixed constant).

In each case the slope is the rate and the intercept is the fixed constant. The units of a slope are always those of a rate. In the truck example, the slope is the rate in dollars per mile; in the cell phone example, the units of the slope are dollars per minute; in the salesperson example, the slope is 0.15 dollars of commission per dollar of total sales.

Think of the intercept as an initial or starting value; it's what happens when the input is zero. It has proper units too — in each of these examples that unit is dollars. If you rent a truck but don't drive it anywhere, you still pay the fixed charge. If you make no cell phone calls you still pay the fixed fee. If you don't sell anything in a month, your commission is \$0 but you still earn your base salary.

7.4 Linear functions in a spreadsheet

In Section 7.2 we saw that the amount you pay for electricity in a month is a linear function of the amount you use. In this section we'll use Excel to calculate electricity bills and to draw a picture of the results.

Figure 7.3 is a screen shot of the Excel spreadsheet `TamworthElectric.xlsx`. We put the slope 0.735 in cell C4, with its units £/kwh in cell D4. We put the intercept 9.49 in cell C5 and the units (£) in cell D5.

Then we entered column labels in cells B7:C8 and a few values in rows 9 through 13 in column B. Finally, we asked Excel to calculate the electricity bills in column C. To

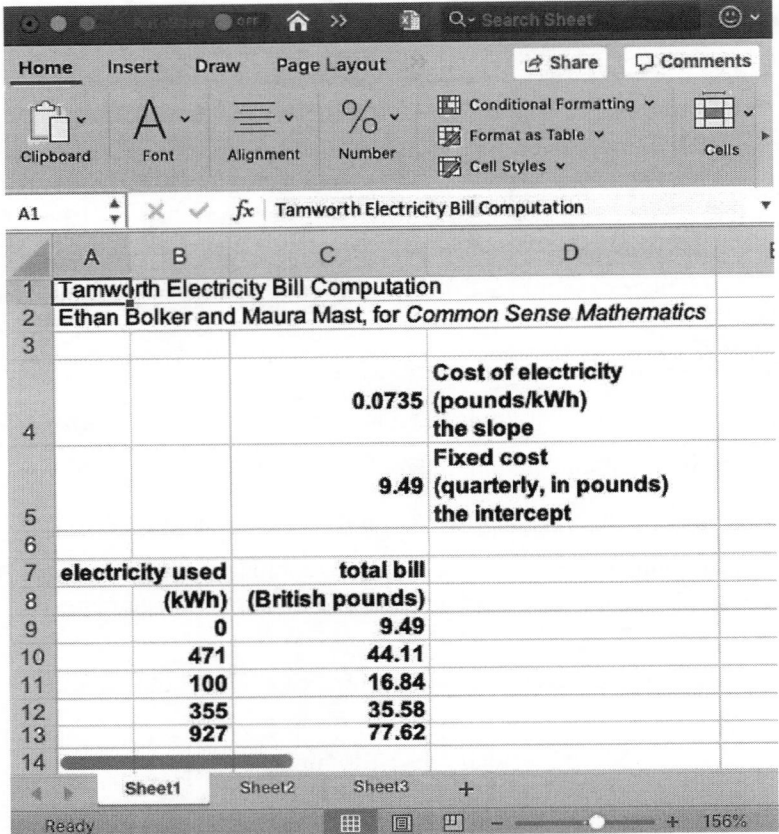

Figure 7.3. Tamworth electricity bill

do that, we started with the formula

$$\texttt{=C4*B9+C5}$$

in cell C9. The = sign tells Excel to multiply the numbers in cells C4 and B9 and add the number in C5. The result is 9.49, as expected.

 The next step is to copy that formula from cell C9 to cell C10. With any luck Excel will guess what we want to do, change B9 to B10, compute

$$\texttt{=C4*B10+C5}$$

and display 44.11. But that's not what happens! Excel shows 4469.79 instead! If you look at the contents of cell C10 you will find

$$\texttt{=C5*B10+C6}$$

so Excel added 1 to the row numbers for cells C4 and C5 as well as for B9. There's nothing in cell C6. Excel treats that as a zero and adds it to 471×9.49 to get 4469.79.

 That's not what we want. Changing B9 to B10 is right, but we want Excel to leave the references to cells C5 and C6 alone. The trick that makes that happen is to put a $ in front of the 5 and the 6. This is not something you could have figured out. There's no particular reason why this trick should work. Just remember it. The right formula to use in cell C9 is

$$=C\$4*B9+C\$5 .$$

When we copy that formula from C9 to cells C10:C13 we get Figure 7.3.

Figure 7.4 is a screen shot of the same spreadsheet — after we asked Excel to show the formulas for each cell instead of the values.

In Chapter 6 we learned how to use Excel to draw bar charts and histograms so that we could visualize data organized into categories. The x-axis displayed category names, with corresponding values on the y-axis. That won't work for the data in Figure 7.3, since there both the x- and y-axes have numerical values. Instead, after selecting cells B7:C13 we must ask Excel for a chart of type XY(Scatter). Figure 7.5 shows the result.

The graph is a straight line — that's why the function is called "linear". The slope tells us how steep the line is and the intercept tells us where it crosses the vertical axis — in this case at the value £9.49, the total bill when you use no electricity at all.

Excel will let you change the type of a chart once it's built. If you change the chart in Figure 7.5 to a Line Chart Excel will use the data in the column B as category labels rather than as the numbers of kilowatt-hours. It will space them evenly along the x-axis, whatever their values, and draw the nonsense you see in Figure 7.6.

If you change the chart type to scatter you get two scatters, one for each column. You can get Figure 7.5 only if you start with a scatter plot. If you select the two columns of data and build a line chart first, things are even worse. Excel thinks each row is a category for which you have two pieces of data. It labels the categories 1, 2, ... and shows a line for each. Try this and see what happens.

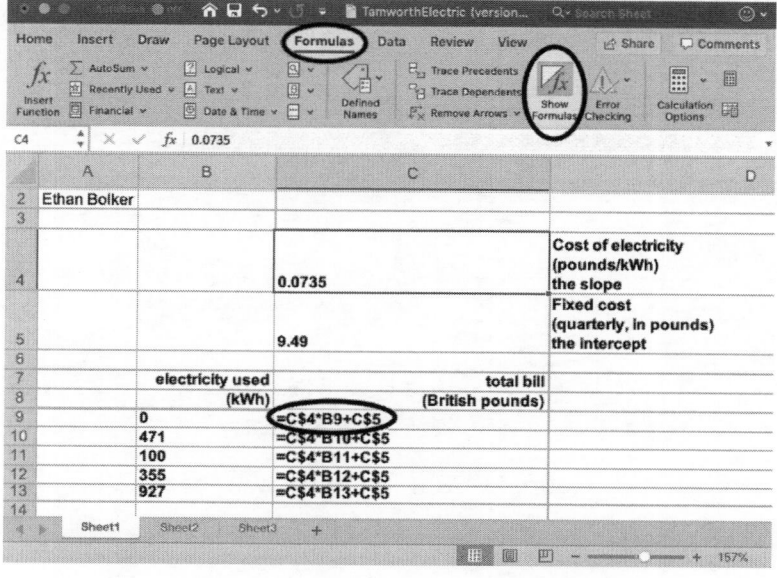

Figure 7.4. Tamworth electricity bill — formulas

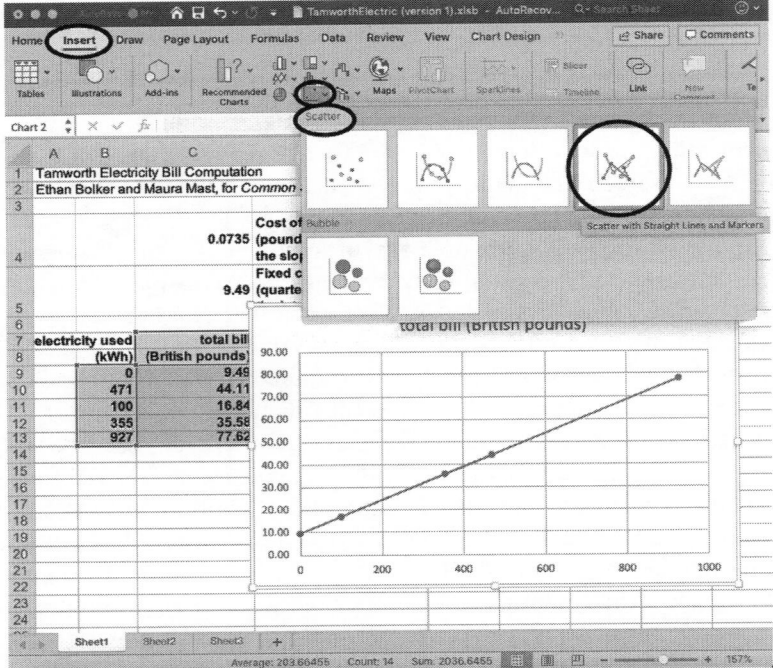

Figure 7.5. Tamworth electricity bill — chart

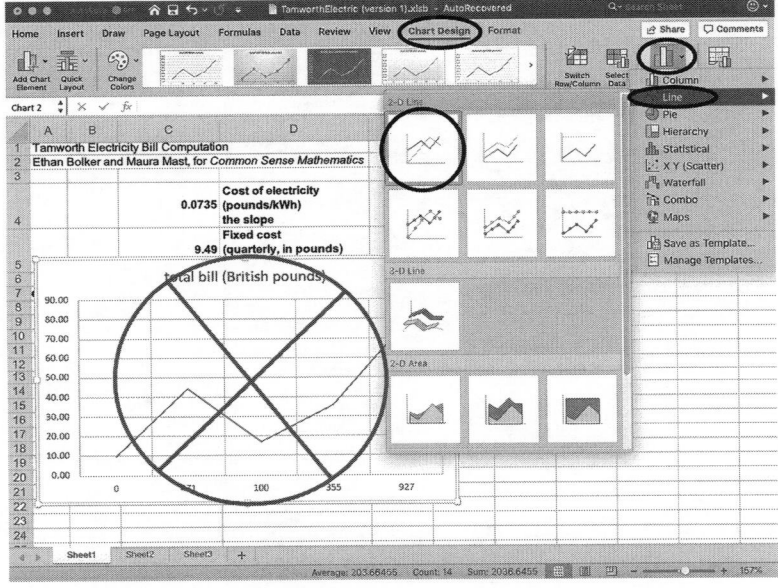

Figure 7.6. How NOT to draw a line

7.5 Which truck to rent?

Table 7.7 shows the cost of renting a truck for one day in Boston in March of 2015. Four companies offer equivalent models. Which one should you choose?

	Watertown	U-Haul	Budget	Enterprise
fixed cost	79	29.95	29.95	59.99
$/mile	0	1.39	0.99	0.59

Table 7.7. One day truck rental costs

It's clear from the numbers that for a very few miles Budget will be cheapest, since it's tied with U-Haul for the lowest fixed cost and charges less per mile. For a really long move Watertown will be best since there is no mileage charge. The Excel chart in Figure 7.8 tells the whole story. Budget is cheapest up to about 50 miles. For longer trips, choose Watertown. It never makes financial sense to rent from U-Haul or Enterprise.

The figure is a good reminder of the meaning of slope and intercept for straight line graphs. A line crosses the vertical axis at the intercept. In this example intercepts represent the fixed costs. The units of the intercept are dollars — the units on the y-axis. The slope of a line measures how steep it is. That's particularly visible when you compare U-Haul and Budget, which have the same intercept but different slopes — 1.39 dollars/mile and 0.99 dollars/mile. The units of the slope are always (units on y-axis)/(units on x-axis). The line for Watertown is horizontal since its slope is 0 dollars/mile.

The spreadsheet with that figure is at `TruckRental.xlsx`. Figure 7.9 shows how we arranged the formulas in the spreadsheet to compute the total cost for each company. Column A lists the mileages we're considering, from 0 in cell A11 to 100. The

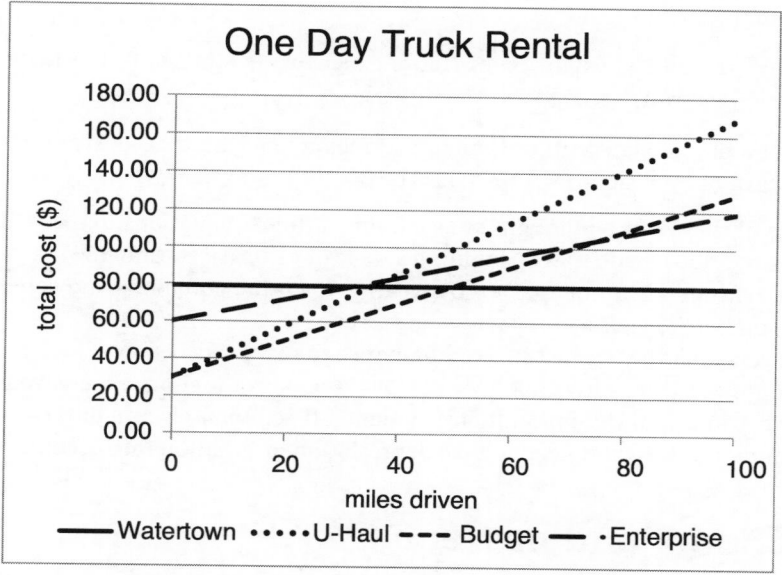

Figure 7.8. Comparing truck rental costs

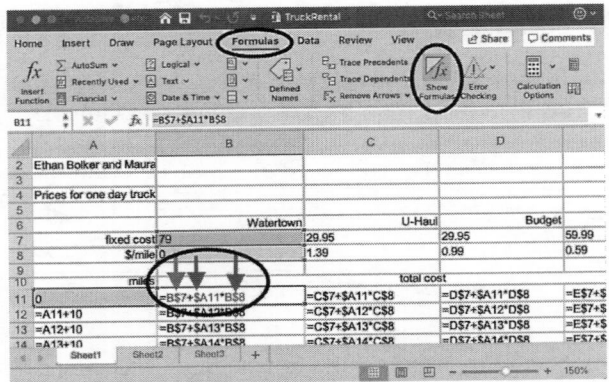

Figure 7.9. Computing truck rental costs

formula =A11+10 in cell A12 fills column A when we copy it to cells A13:A21. The formula in cell B11 is

$$=B\$7+\$A11*B\$8 .$$

It uses three $ signs to keep Excel from adjusting references for rows 7 and 8 and for column A. That allowed us to copy it to all of the range B11:E21.

The problem we've just solved is typical of situations where you have to decide among options, some with small startup cost but a high ongoing rate, others the reverse. Here are some examples:

- Insulate your house (high initial investment compared to doing nothing) in order to pay less for heat in the winter (lower rate for use).

- Buy a hybrid instead of a conventional car (higher initial cost, lower rate of fuel consumption).

- Buy energy efficient light bulbs (more expensive to start with, but they use less electricity to run),

- Select a phone plan with unlimited text messaging (more expensive than pay-as-you-text, but the slope ($ per text) is zero).

Each of these can be thought of as looking at linear equations to see where their graphs cross. You can do that by building a table of values or by drawing the graphs (in Excel, or with pencil and paper) or by writing down the equations and solving them with remembered algebra, or by guess-and-check.

But you can't rely on just this mathematics to make a decision. There are always other important things to think about. How long does the more expensive purchase last? Can you afford the initial high payment? If so, what else might you rather do with that money? Do you need to take depreciation or inflation into account?

7.6 Energy and power

How much electrical energy does a 100 watt light bulb use? That depends on how long it's on. When it's switched off, it doesn't use any at all. If it's on for two hours it must

Energy transferred	=	power	×	time
(kilowatt-hour, kWh)		(kilowatt, kW)		(hour, h)

Figure 7.10. How much electrical energy?

use twice as much as it does in one hour. Figure 7.10 (from the second page of the Tamworth bill) shows the proportion lurking there.

The second line in that figure displays the units for the quantities in the first line. Time is measured in hours, of course. Power is measured in kilowatts. Energy is measured in kilowatt-hours: the product of the units for power and for time.

That tells us right away that energy and power are not the same thing. Comparing the figure to equation (7.1), you can see that energy is like distance — a thing that's consumed or traveled, while power is like speed — the rate at which energy is used or distance covered.

If you turn a 100 watt light bulb on for 2 hours the formula in Figure 7.10 tells you how much electrical energy you use:

$$\begin{aligned} 100 \text{ watts} \times 2 \text{ hours} &= 200 \text{ watt-hours} && (7.2) \\ &= 0.2 \text{ kilowatt-hours.} && (7.3) \end{aligned}$$

(7.2) is just multiplication. (7.3) changes watt-hours to kilowatt-hours.

How much does it cost to leave a 40 watt bulb on all the time in your basement for a year? There are about 9,000 hours in a year. That is 9 kilo-hours, so you'll use about

$$40 \text{ watts} \times 9 \text{ kilo-hours} = 360 \text{ kilowatt-hours.}$$

If you pay \$0.10 per kwh for electricity it will cost you about \$36 per year to guarantee that you don't fall down the basement steps in the dark.

The electrical energy that flows through the wires in your house to your appliances probably comes to you from a power plant, which might be burning coal or natural gas or extracting energy from nuclear fuel. (You might have a wind turbine in your neighborhood, or a hot spring, or solar panels on your roof, but these are unlikely power sources for most people.) So power plants produce energy, not power. The power of a power plant is the rate at which it can produce energy, so "energy plant" would be a better name than "power plant".

The website for Chicago's Cook Nuclear Plant says that

> The 1,048 net megawatt (MW) Unit 1 and 1,107 net MW Unit 2 combined produce enough electricity for more than one and one half million average homes. [**R202**]

Let's check this. The combined total power is 2,155 megawatts. This is the rate at which that plant produces electrical energy when it is running at full power. (When it's not running it's still just as powerful, but not producing any energy.) When it's running, how many average homes could it produce electricity for?

If the Cook plant ran all year (about 9,000 hours) it would produce

$$\begin{aligned} 2{,}155 \text{ megawatts} \times 9{,}000 \text{ hours} &\approx 18{,}000{,}000 \text{ megawatt-hours} \\ &= 18{,}000{,}000{,}000 \text{ kilowatt-hours} \end{aligned}$$

of electrical energy. Googling "average household electricity usage" finds

> 6,000 kwh per household per year for 3 residents average per household

from `www.physics.uci.edu/~silverma/actions/HouseholdEnergy.html` . The source is a physics professor's website, so it's probably reliable. At 6,000 kwh per household per year Cook could power 3 million homes. The quotation claims half that, so it's clearly in the right ballpark. The 6,000 kwh per household per year is a southern California average — households in northern Illinois might well use more electricity.

Energy comes in many forms besides electric. The Cook plant converts the energy in its nuclear fuel to electricity. Driving a car uses the energy stored in the gasoline. Running a marathon uses the energy in the food you eat. Each form of energy has its own units. We've seen that electrical energy is measured in kilowatt-hours. If you cook on a gas stove, the energy in the gas is measured in *therms*. The energy in the oil that heats your house is measured in *British Thermal Units* or Btus. The energy in the food you eat is measured in *calories*. Physicists measure energy in *ergs* or *joules*; you rarely see those units in everyday life. You can look up conversion factors for these units — for example, the energy in a barrel (42 gallons) of oil is about 5.8 million Btu, which is equivalent to 1,700 kilowatt-hours. So it would take about a fifth of a barrel to keep that 40 watt light bulb burning for a year.

Converting among the units for energy is just like converting among the units for length (meters, feet, yard, miles, …). You can use a table, an online calculator like the one at the National Institute of Standards and Technology (`physics.nist.gov/cuu/Constants/energy.html`) or the Google calculator.

Possibly the most interesting energy conversion is the one that Einstein discovered in 1905: mass and energy are the same thing, measured in different units. The conversion factor is the square of the speed of light — hence the famous equation

$$e = mc^2.$$

To see that at work, look again at the yearly energy output of the Cook plant. The Google calculator tells us that

$$\boxed{18\,000\,000\,000 \text{ kilowatt-hours} = 6.48 \times 10^{16} \text{ joules}} \ .$$

The National Institute of Standards and Technology website says that corresponds to a mass of about 0.72 kg, which is 720 grams. That means just about 1.6 pounds of matter must be converted to energy to power millions of Chicago homes for a year. The Google calculator does not know Einstein's equation, so it wouldn't convert kwh to grams directly!

7.7 Federal payroll taxes

Income tax. Taxes are a part of life (the only other certainty is death), so it's only common sense to learn how they work. In Section 7.1 we studied sales taxes. Cities and states collect them; they are computed as a percentage of the purchase price. In this section we'll explain two important federal taxes that depend on your income, not on how you spend it.

Federal income tax is not simply a proportion of your income. It's a *progressive graduated tax*. When you make more money you not only pay more tax, some of your

Bracket ($)	Marginal Tax Rate (%)
0 – 9,700	10
9,701 – 39,475	12
39,476 – 84,200	22
84,201 – 160,725	24
160,726 – 204,100	32
204,101 – 510,300	35
510,301 –	37

Table 7.11. 2019 single taxpayer brackets and rates

income may be taxed at a higher rate. Table 7.11 shows the 2019 *tax brackets* for single taxpayers.

That tells you that the first $9,700 of your income is taxed at 10%. If you make exactly that much, you pay $970 in tax. If you make more, the extra income is taxed at a higher rate — you have moved to a higher *tax bracket*. For example, if you make between $9,700 and $39,475 you will pay $970 for the first $9,700 and 12% of the amount you earn over $9,700. If you earn more than $39,475 you start paying at a 22% rate on the extra.

Let's try an example. If your taxable income is $50,000, then your total tax is

$$\begin{aligned} \text{total tax} &= 0.10 \times \$9,700 + 0.12 \times (\$39,475 - \$9,700) \quad (7.4) \\ &\quad + 0.22 \times (\$50,000 - \$39,475) \\ &= \$970.00 + \$3,573.00 + \$2,315.50 \\ &= \$6,858.50 \,. \end{aligned}$$

Note carefully that when you are in a higher tax bracket the higher rate applies only to the extra income. The taxpayer in this example is in the 22% bracket, but that rate applies only to her earnings in that bracket.

Figure 7.12 explains this rule in another way.

Individual Taxpayers

If Taxable Income Is Between:	The Tax Due Is:
0 - $9,700	10% of taxable income
$9,701 - $39,475	$970 + 12% of the amount over $9,700
$39,476 - $84,200	$4,543 + 22% of the amount over $39,475
$84,201 - $160,725	$14,382.50 + 24% of the amount over $84,200
$160,726 - $204,100	$32,748.50 + 32% of the amount over $160,725
$204,101 - $510,300	$46,628.50 + 35% of the amount over $204,100
$510,301 +	$153,798.50 + 37% of the amount over $510,300

Figure 7.12. 2019 tax brackets [R203]

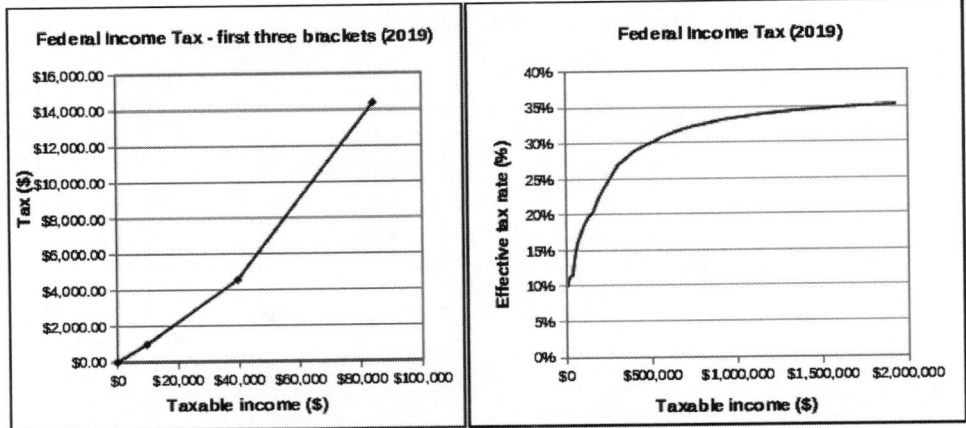

Figure 7.13. 2019 single taxpayer tax liability

The first graph in Figure 7.13 from the spreadsheet `GraduatedTax2019.xlsx` shows that the dependence of tax on income is *piecewise linear* — built from pieces of straight lines that become steeper as income increases.

The second graph shows the effective tax rate — the percentage of your income you pay in federal income tax. In equation (7.4) we found that the total tax on a taxable income of $50,000 was $6,858.50. The effective tax rate is $6,858.50/$50,000 = 13.72%. This is a weighted average of the three bracket rates 10%, 12% and 22%, with weights the amount of income taxed at each rate. The effective tax rate is less than the rate in your top bracket because you pay at a lower rate on the first part of your income. The effective rate does not reach 35% until about $2 million in income — well into the top 37% bracket.

In fact, the actual effective tax rate is lower than this for wealthier households because income tax is collected only on income from wages and earnings. Income from capital gains — returns on investment — is taxed at a lower rate. Figure 7.14 shows the actual effective federal tax rate by total household income for the year 2007. The rate for the wealthiest households was just 16.6% — less than half the 35% rate for the top income tax bracket that year. It's reasonable to assume that a similar discrepancy is still true.

Social Security. Social Security tax payroll deductions show up labeled "FICA" on your pay stub. That acronym is from the "Federal Insurance Contributions Act". Those taxes pay for Social Security and Medicare.

In 2019 the starting tax rate was 6.2% for Social Security and 1.45% for Medicare. The Social Security tax is collected only on the first $132,900 of your earnings. Up to that income level the combined rate is 7.65%.

When your earnings exceed 132, 900 you pay no more Social Security tax, but you continue to pay Medicare tax at the 1.45% rate. When your income reaches $200,000 the Medicare rate increases to 2.35% on the amount over $200,000.

The actual rules are a little more complicated. First, the tax applies only to wages. Other income (like stock dividends or interest) are not subject to this tax. Second, the

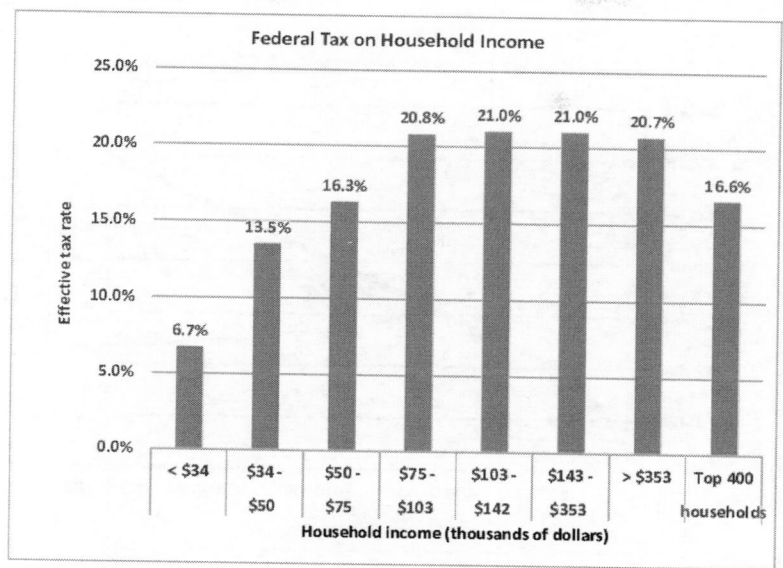

Figure 7.14. Actual effective federal tax rate by income, 2007 [**R204**]

real rates are twice the quoted amounts, but your employer is required to pay half. If you're self-employed you pay it all.

If you earn $500,000 your FICA tax is

$$0.062 \times \$132,9000 + 0.0145 \times \$200,000 + 0.0235 \times (\$500,000 - \$200,000)$$
$$= \$18,190.$$

Since you pay no Social Security on wages over $132,900 the percentage of your earnings collected for FICA taxes decreases as your earnings increase even though you still pay for Medicare. So FICA taxes are *regressive*. Up to $139,700 the effective rate is 7.45%. For $500,000 the effective rate is just $18,190/$500,000 = 3.64%. For higher incomes, the effective rate is even smaller. At huge incomes it levels off at the top Medicare rate of 2.35%. Figure 7.15 from spreadsheet `SocialSecurityTax2019.xlsx` shows the amount of FICA tax paid and the decreasing effective tax rate as a function of FICA earnings.

Income tax history. For complex historical, legal and political reasons the Supreme Court rejected the first attempts to collect an income tax. This Constitutional amendment, ratified in 1909, made that kind of tax legal.

> Amendment XVI
>
> The Congress shall have power to lay and collect taxes on incomes, from whatever source derived, without apportionment among the several States, and without regard to any census or enumeration. [**R205**]

The first federal income taxes were collected in 1913. Figure 7.16 shows the significant fluctuations in the rate for the top bracket through 2019. They have been near their historic lows since the late 1980's.

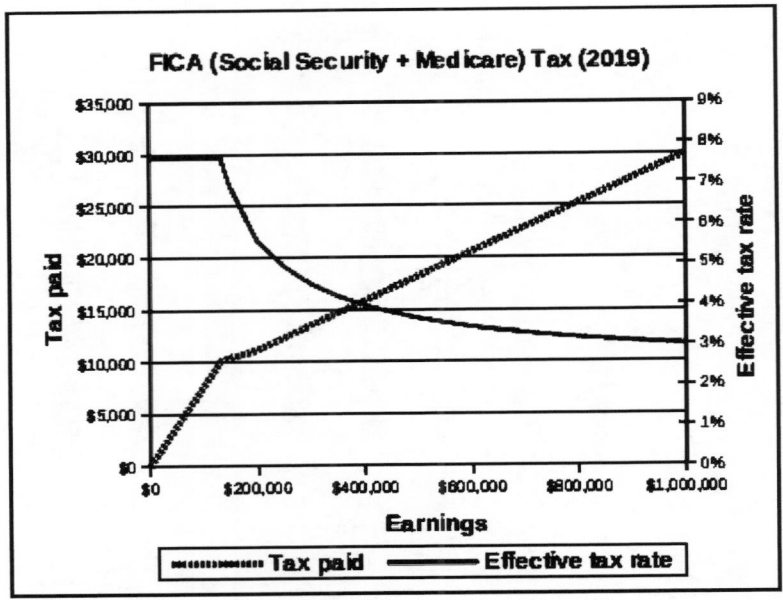

Figure 7.15. FICA (Social Security and Medicare) tax

Top Federal Tax Rates

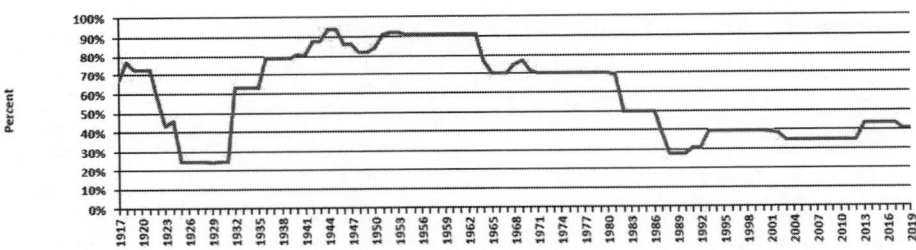

Figure 7.16. Historical top tax bracket rate [**R206**]

To assess the social and economic impact of the changes in income tax rates would require much more time and knowledge than we can offer here. The spreadsheet Federalindividualratehistory.xlsx contains a complete history of income tax brackets and rates from the inception of the income tax in 1913 through its hundredth anniversary in 2013, in both dollars current in each year and adjusted for inflation (2012 dollars).

Even in years when Congress does not revise the tax code, the IRS routinely adjusts the brackets (not the rates) to take inflation into account. If that were not done then salaries increased by inflation would move people into higher brackets even when their increased wages did not correspond to increased purchasing power.

Figure 7.17 shows the brackets for the tax year 2018. You can check that the 2019 brackets are all just about 2% larger than these.

Individual Taxpayers

If Taxable Income Is Between:	The Tax Due Is:
0 - $9,525	10% of taxable income
$9,526 - $38,700	$952.50 + 12% of the amount over $9,525
$38,701 - $82,500	$4,453.50 + 22% of the amount over $38,700
$82,501 - $157,500	$14,089.50 + 24% of the amount over $82,500
$157,501 - $200,000	$32,089.50 + 32% of the amount over $157,500
$200,001 - $500,000	$45,689.50 + 35% of the amount over $200,000
$500,001 +	$150,689.50 + 37% of the amount over $500,000

Figure 7.17. 2018 tax brackets [**R207**]

7.8 Exercises

Exercise 7.8.1. [U][Section 7.2][Goal 7.2] [Goal 7.3] Your electricity bill.

Verify the computations on your current electricity bill, either with a calculator or by modifying the Tamworth bill spreadsheet at `TamworthElectric.xlsx`.

If you don't have a current electricity bill, check the website for your local electric company, which probably provides a sample bill you can use instead.

Exercise 7.8.2. [U][Section 7.2][Goal 4.2] [Goal 7.2] Electricity costs now and then and here and there.

Compare residential electricity cost in Boston in 2007 (the date of the NStar bill in Figure 7.2) to the cost where you live today. Your answer should take inflation into account.

If you have a current electricity bill, use it. If you don't, try to find the fixed monthly cost and the cost of electricity in $/kwh from your local electric company. Perhaps their website has that information.

Exercise 7.8.3. [R][S][Section 7.2][Goal 7.2] [Goal 7.3] How much electricity does it use ...?

The document containing the Tamworth bill asks how much electricity various appliances use.

(1) Calculate how much electricity is consumed by a

  (a) 100 watt lamp on for 2 hours,
  (b) 500 watt TV on for 5 hours,
  (c) 2 kilowatt kettle on for half an hour,
  (d) 10 watt electric blanket on for 15 minutes.

(2) Compare the electricity used by

  (a) 2 kilowatt heater on for 2 hours or 3 kilowatt heater on for 3 hours,
  (b) 900 watt toaster on for 15 minutes or 2 kilowatt grill on for 10 minutes,

(c) 100 watt radio on for 2 hours or 500 watt radio on for 45 minutes.

(d) Which appliance would be cheaper to use in each case?

Exercise 7.8.4. [S][Section 7.2] [Goal 7.1][Goal 7.7] Direct current.
    In an article from *The New York Times* on November 17, 2011, headlined "From Edison's Trunk, Direct Current Gets Another Look" you can read that

> In a data center redesigned to use more direct current, monthly utility bills can be cut by 10 to 20 percent, according to Trent Waterhouse, vice president of marketing for power electronics at General Electric. Verizon Communications, a G.E. customer, expects to save 1 billion kilowatt-hours a year from a nationwide retrofit of its data centers, which translates to roughly enough to power 77,000 homes. [**R208**]

(a) How many watt-hours is a billion kilowatt-hours?

(b) About how much electricity was Verizon using in 2011, given that they hoped to save a billion kilowatt-hours per year?

(c) Verify the estimate that those billion kwh are "roughly enough to power 77,000 homes". (You won't find anything useful in the original article. How else will you search?)

Exercise 7.8.5. [S][Section 7.3][Goal 7.2] How hot was it?
    Figure 7.18 shows a weather forecast for Hamilton, Ontario, Canada. The temperature there is displayed in degrees Celsius, marked °C.

(a) Look up the linear relationship for converting temperatures measured on the Celsius scale to temperatures on the Fahrenheit scale.

(b) Use the formula you found to calculate the temperature in degrees Fahrenheit in Hamilton, Ontario, on Thursday, August 13, 2015, at 12:45.

(c) Check your answer using the Google calculator.

(d) What does "Wind: SW 13 km/h" in the figure mean?

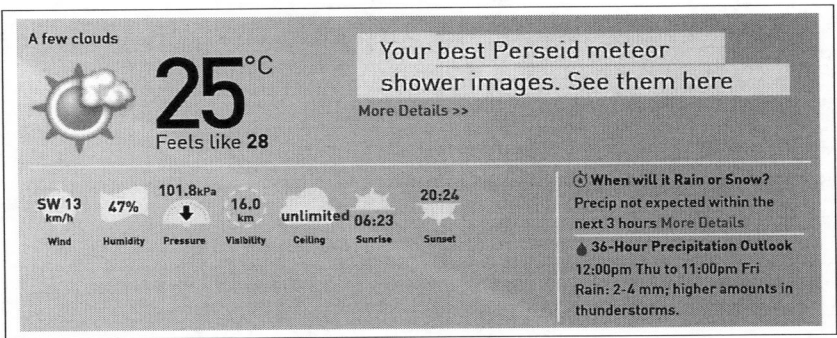

Figure 7.18. Weather in Canada

(e) What does "Pressure: 101.8 kPa" in the figure mean?

(f) What is the Perseid meteor shower?

Exercise 7.8.6. [S][Section 7.3][Goal 7.2] [Goal 7.3][Goal 7.4] The Jollity Building.
    In Exercise 4.8.11 there's an implicit linear model for Mr. Ormont's weekly income. Write the linear equation for that model. Clearly identify the independent and dependent variables and the units for the slope and the intercept.

Exercise 7.8.7. [S][Section 7.3] [Goal 7.2][Goal 7.3] Newton trees.
    In an article in the April 2012 issue of the Newton Conservators newsletter you can read that

> In the early 1970s there were approximately 40,000 trees lining the streets of Newton. Today, that number is about 26,000 — a 35% loss. The current annual rate of decline is about 650 trees per year. At this rate, if unchecked, public street trees would diminish to approximately 10,000 within a generation (25 years), and in 40 years, public street trees would no longer be part of the Newton landscape. [**R209**]

(a) Check the arithmetic that leads to the claimed "35% loss".

(b) Check the arithmetic that leads to a "current annual rate of 650 trees per year".

(c) Check the predictions in the last sentence. Are they likely to come to pass?

(d) Write the equation for the linear model implicit in this quotation (use years since 2012 as the independent variable). Identify the slope and the intercept, with their units.

Exercise 7.8.8. [R][S][A][Section 7.4] [Goal 7.2][Goal 7.3][Goal 7.5] Apple time.
    Kuipers Family Farm (www.kuipersfamilyfarm.com/apple-orchard/), in Maple Park, IL, charged $10 per person for admission to the apple orchard during the 2018 apple season. This included a $\frac{1}{2}$ peck bag of apples and a hayride to the orchard. Visitors could pick additional peck bags of apples for $15 each. (A bag of apples in the grocery store usually contains $\frac{1}{2}$ peck.)

(a) If you pick that additional peck of apples, how much do you pay?

(b) If you pick two additional pecks of apples, how much do you pay?

(c) If you just enjoy the free cider, your $\frac{1}{2}$ peck of apples and the sunshine, what do you pay?

(d) Write a linear function that shows how the total cost of the farm visit depends on how many pecks of apples you pick. Identify the slope and intercept, with their units.

(e) Build an Excel spreadsheet to compute this function and use it to check the values you worked out by hand above. Include a chart in your spreadsheet.

(f) Use your spreadsheet to calculate your apple cost (in $/peck) when you pick no extra pecks, then when you pick 1, 2 or 10 extra pecks.

Exercise 7.8.9. [S][A][Section 7.5] [Goal 7.2][Goal 7.5] Comparing telephone calling plans.

A cell phone company has introduced a pay-as-you-go price structure, with three possibilities.

Plan 1	$10 a month	10 cents per minute
Plan 2	$15 a month	7.5 cents per minute
Plan 3	$30 a month	5 cents per minute

(a) For each plan, find a linear function that describes how the total cost for one month depends on the number of minutes used.

(b) Construct a table in Excel showing the total cost for one month for each of the three plans. Organize your data this way:

   - Create a sequence of cells in column A for the various possible numbers of minutes. Label that column. Start with 0 minutes. What's a good step to use? What's a reasonable place to stop?

   - Use columns B, C and D for each of the three plans. The fixed charge and charge per call should be in cells in those columns too, so you can use the same formula everywhere in the data table. (That will call for clever use of the $ to keep Excel from changing row numbers and column letters when you don't want it to.)

(c) Use Excel to draw one chart showing how the monthly bill (y-axis) depends on the number of minutes you use the phone (x-axis) for all three plans.

(d) Write a paragraph explaining to your friend how she should go about choosing the plan that's best for her.

[See the back of the book for a hint.]

Exercise 7.8.10. [S][Section 7.5][Goal 7.2] [Goal 7.5] Prepaid phones.

Until summer 2012, if you wanted an iPhone you needed to lock into a two year contract. Then some mobile companies started selling the iPhone and letting you choose your own plan, with no contract. You can use the ideas from this chapter to make a quantitative comparison (we'll let you decide what other factors, such as paying a lot up front, matter in your decision).

Virgin Mobile began selling a 16 GB iPhone 4S for $649.99, with no plan or contract. They offered a $55 per month "unlimited" data plan (in fact it was not unlimited, as once you go past 2.5GB of data they slowed the phone speed down considerably). If you purchased the phone through Sprint, it cost just $149.99. However, you had to sign up for a two year contract. The least expensive option was $79.99 with what they called unlimited data.

(a) For each plan, find a linear function that shows how the total cost of the phone depends on the number of months you have it.

(b) Build a spreadsheet in Excel using your functions and fill in the cost of the two different options over several months.

(c) Graph the data in your spreadsheet.

(d) Write a short statement comparing the two plans. Clearly the Virgin Mobile plan is more expensive at first. When does it become the less expensive plan? If you had to choose one of the plans, which one would you choose and why? What other factors would you consider?

Exercise 7.8.11. [S][Section 7.5][Goal 7.3] [Goal 7.5] Hybrid payback.

The "Best & Worst Cars 2011" issue of *Consumer Reports* provides the following data for new Toyota Camrys:

	conventional	hybrid
cost	$19,720	$26,575
fuel economy	26 MPG	34 MPG

Assume gasoline costs $3.50/gallon.

(a) Questions about the conventional Camry.

    (i) Once you own the car, how much does it cost to run, in dollars per mile? Does your answer make sense?

    (ii) Calculate the total cost (purchase plus gasoline) to drive the conventional Camry 10,000 miles.

    (iii) Write the linear equation that computes the total cost C of driving the conventional Camry M miles.

    (iv) Identify the slope and the intercept of this equation, with their units.

(b) Open the spreadsheet `ConventionalvsHybrid.xlsx`. Enter the numerical data from the table and the cost of gasoline in the appropriate cells. What formula should you enter in cell `C15` to check your answer to part (i)?

(c) Copy your formula to cells `B14:D29` to fill in the table. Where must you add $ signs to keep Excel from changing row and column references?

Create a properly formatted and labeled chart in Excel showing how the cost of driving each car depends on the number of miles driven. Use your graph along with the table to answer the following questions.

(d) If you drive 120,000 miles will you recover in gas savings the extra initial cost of the hybrid? Write a complete sentence or two and use appropriate precision for the numbers you use to make your argument.

(e) When will you recover the extra initial cost in gas savings if the government (re)instates a $3,000 tax rebate for hybrid purchases?

(f) With the original initial costs, how much would the price of gasoline have to be in order for the breakeven point to occur at 30,000 miles?

(g) Restore all the inputs to their original values. Arrange your spreadsheet so that it will print on one page, with the chart below the data table.

(If you're thinking of buying a car, remember that there's a lot more that goes into the cost of driving one car or another (or any car at all) than just these simple computations using initial cost and miles driven.)

Exercise 7.8.12. [S][Section 7.5][Goal 7.3] [Goal 7.5] Contract or not?

Table 7.19 provides data that appeared in a story in *The Boston Globe* on June 14, 2012, headlined "Pay full price for iPhone, avoid contract".

Plan	Phone cost	Monthly charge	Two year cost
Cricket Wireless	499.99	55	1,819.99
Virgin Mobile	649	30	1,369

Table 7.19. Comparing cell phone plans [**R210**]

(a) How much would it cost (in total) to buy the Cricket phone and use it for two months?

(b) Write an equation for the total cost to buy and use the Virgin Mobile phone for *M* months.

(c) Identify the slope and the intercept of your equation, with proper units for each.

(d) Create an Excel spreadsheet and use Excel formulas to complete a table like this:

Months	Cricket	Virgin
0		
1		
...		
24		

(e) Check that your spreadsheet produces the answers in the table for 24 months.

(f) Create a properly labeled and formatted chart displaying the data in your table.

(g) When (in terms of months of use) would it be better to choose the Cricket phone?

(h) Suppose the monthly charge for the Cricket phone was just $45 while that for the Virgin phone increased to $35/month. Answer the previous question with this new data.

Exercise 7.8.13. [S][Section 7.6][Goal 7.2] [Goal 7.1][Goal 7.7] Regenerative braking.

When you apply the brakes in a Toyota Prius the car uses some of the energy of the forward motion to recharge the battery. The dashboard displays a little car icon each time that recharging has collected 50 watt-hours.

(a) Estimate the energy equivalent of each icon in gallons of gasoline.

(b) Estimate the dollar value of that gasoline.

(c) Compare your estimate to the dollar value of 50 watt-hours of electricity in your house (what it would cost to keep a 100 watt bulb on for half an hour).

(d) Discuss the value of the display.

[See the back of the book for a hint.]

Exercise 7.8.14. [S][Section 7.6] [Goal 7.1][Goal 7.7] Computers don't sleep soundly.

The website michaelbluejay.com/electricity/computers.html gives information about how much energy a computer uses while asleep, in standby mode, or in use. The iMac G5, for example, uses 97 watts while "doing nothing", compared to 3.5 watts while asleep.

(a) What do you think "doing nothing" means?

(b) If this type of computer is doing nothing all day (24 hours), how much electricity does it use? Express your answer in kilowatt-hours.

(c) Now suppose the computer goes to sleep after 15 minutes of doing nothing. How much electricity does it use in an idle day?

(d) If a kilowatt-hour of electricity costs 20 cents, how much money is saved in a day because the computer is smart enough to go to sleep?

Exercise 7.8.15. [S][C][Section 7.6] [Goal 7.1][Goal 7.7] How Much Water Does Pasta Really Need?

On February 24, 2009, *The New York Times* published an article by Harold McGee addressing that question.

McGee's kitchen experiments convinced him that he could cook pasta in far less water than is customary. Since (he says) we consume about a billion pounds of pasta a year:

> My rough figuring indicates an energy savings at the stove top of several trillion B.T.U.s. At the power plant, that would mean saving 250,000 to 500,000 barrels of oil, or $10 million to $20 million at current prices. Significant numbers, though these days they sound like small drops in a very large pot. [**R211**]

(a) Verify the author's conversion of "several trillion B.T.U.s" to barrels of oil and then to dollars.

(b) Does McGee's estimate of a billion pounds of pasta per year make sense?

(c) How much water do Americans use cooking pasta? How much would they use if they followed McGee's advice?

(d) Does not boiling the extra water really save the amount of energy McGee claims?

Exercise 7.8.16. [S][Section 7.6][Goal 7.7] Solar energy.

On May 1, 2013, *The Arizona Republic* reported on the $500 million Arlington Valley Solar Energy II project near Phoenix. The article said

> [the project] will have 127 megawatts of capacity when finished. One megawatt is enough electricity to supply about 250 Arizona homes at once, when the sun is shining on the solar panels. [**R212**]

(a) How many watts of power does the average Arizona home need to run its appliances when the sun is shining?

(b) Research the power requirements of several typical home appliances: air conditioners, stoves, television sets, … Then decide whether the article's claim about home power requirement on a sunny day in Arizona is reasonable.

Exercise 7.8.17. [S][C][Section 7.6] [Goal 7.1][Goal 7.7] Solar power at Wellesley College.

The sign on a solar panel array at Wellesley College reads:

> Solar Photovoltaic Array
> This 10-kilowatt Solar PV Array is composed
> of 48 panels, each 210 watts. It will generate approximately
> 13,000 kilowatt hours of electricity per year, enough to
> power 2 homes, 32 metal halide street lights
> or 85 LED street lights for an entire year.
> For real time electrical output please go to:
> `www.sunwatchmeter.com/home/day/wellesley-college`
> PLEASE KEEP OFF THE PANELS

You can see a picture at `www.theswellesleyreport.com/2010/09/wellesley-college-saves-the-planet/solar-panel/` .

(a) Check the consistency of some of the numbers.

(b) How many hours of sunshine per day do the designers expect the installation to see?

(c) Visit the website on the sign and write about it. What do the graph and the meters represent? What is happening there now?

Exercise 7.8.18. [C][Section 7.6][Goal 7.7] Wind power.

From *The Los Angeles Times*, March 1, 2009:

> The U.S. last year surpassed Germany as the world's No. 1 wind-powered nation, with more than 25,000 megawatts in place. Wind could supply 20% of America's electricity needs by 2030, up from less than 1% now, according to a recent Energy Department report. [**R213**]

(a) What do these data say when you calculate wind power in megawatts per person, or as a percentage of the total power available?

(b) Explain why it might or might not be true to say that when this article appeared the U.S. now produced more wind energy than Germany?

Exercise 7.8.19. [S][Section 7.6][Goal 7.7] [Goal 7.1] World solar power.

In a posting on their website on July 31, 2013, the Earth Policy Institute reported that

> The world installed 31,100 megawatts of solar photovoltaics (PV) in 2012 — an all-time annual high that pushed global PV capacity above 100,000 megawatts. There is now enough PV operating to meet the household electricity needs of nearly 70 million people at the European level of use. [**R214**]

At `www.wec-indicators.enerdata.eu/household-electricity-use.html` a graphic shows that average household electricity consumption in Europe in 2013 was about 4,000 kwh/year. [**R215**]

(a) Use the data in this exercise to estimate the average number of hours per day that these solar panels are producing electricity.

(b) How does average household electricity consumption in the U.S. compare to that in Europe?

[See the back of the book for a hint.]

Exercise 7.8.20. [S][C][Section 7.6] [Goal 7.1][Goal 7.7] Chilling out by the quarry.
On August 16, 2010, *The Boston Globe* described a local business's plan to cool its corporate facility with water from a nearby quarry rather than with conventional air conditioning.

> [Director of facilities] Dondero estimated that the cooling system, which eliminates the need for any type of refrigerant in the building, saves about $75,000 a year, reduces annual water use by one million gallons, and cuts yearly energy use by about 300,000 kilowatt hours — enough to power about 30 homes. [**R216**]

(a) What rate in dollars per kwh is Dondero using to support his assertion that this change will save $75,000 a year?

(b) Is the claim that 300,000 kilowatt-hours would power 30 homes for a year reasonable?

Exercise 7.8.21. [S][Section 7.6][Goal 7.1] [Goal 7.7] Energy savings at MIT.
On March 26, 2011, Jon Coifman wrote in *The Boston Globe* that

> In just 36 months, [MIT and NStar] plan to cut the university's energy use 15 percent — enough to power 4,500 Massachusetts homes for a year. [**R217**]

(a) If all of MIT's energy use were devoted to powering Massachusetts homes, how many homes would that be?

(b) Compare your answer in part (a) to the number of homes in Cambridge.

(c) Estimate MIT's total annual energy use, in Btus.

(d) Convert your answer to the previous question from Btus to watt-hours.

Exercise 7.8.22. [S][Section 7.6][Goal 7.1] [Goal 7.7] The governor gets the units wrong.
On June 7, 2011, *The Norwich Bulletin* reported that

> Connecticut Governor Daniel Malloy recently signed off on a deal to tax electricity generators one quarter of one cent per kilowatt hour, or 25 cents per $100. [**R218**]

(a) What is wrong with the units in this quotation?

(b) Estimate the percentage change in the cost of electricity that would result from a one-quarter of one cent increase per kilowatt-hour.

(c) What do you think Governor Malloy intended to say?

[See the back of the book for a hint.]

Exercise 7.8.23. Your total federal tax bill.
Moved to Extra Exercises at `www.ams.org/bookpages/text-63`.

Exercise 7.8.24. [S][C][Section 7.7][Goal 7.1] [Goal 7.6] President Obama's income tax.
According to the White House website

> [The President] and the First Lady filed their [2013] income tax re-
> turns jointly and reported adjusted gross income of $481,098. The
> Obamas paid $98,169 in total tax.
>   The President and First Lady also reported donating $59,251 —
> or about 12.3 percent of their adjusted gross income — to 32 differ-
> ent charities. The largest reported gift to charity was $8,751 to the
> Fisher House Foundation. The President's effective federal income
> tax rate is 20.4 percent. ... The President and First Lady also released
> their Illinois income tax return and reported paying $23,328 in state
> income tax. [**R219**]

The President itemized deductions, so he could deduct charitable contributions and state tax from his adjusted gross income:

> In the United States income tax system, adjusted gross income (AGI)
> is an individual's total gross income minus specific deductions. Tax-
> able income is adjusted gross income minus allowances for personal
> exemptions and itemized deductions. [**R220**]

(a) With the information given, what is the largest possible value for the President's taxable income? What tax bracket would he be in?

(b) Use the Married Filing Jointly brackets and rates in the spreadsheet at `Federalindividualratehistory.xlsx` to compute the Obamas' 2013 federal income tax bill for your answer to part (a). If your result does not match the reported figure, what might explain the difference?

(c) If the Obamas had not made those charitable contributions the money would be part of their taxable income. Use your answers to the previous questions to answer these.

   (1) What would their taxable income have been? What bracket would that have put them in? What would their tax have been?

   (2) What fraction of the contribution was (essentially) made by the government?

   (3) What fraction of the Obamas' income did they contribute to charity?

   (4) Did they tithe?

   (5) How does their contribution compare to the national average?

Exercise 7.8.25. [S][Section 7.7][Goal 7.1][Goal 7.6] Using the tax table.
Use Table 7.11 to answer the following questions.

(a) Compute the tax due in 2019 on a net taxable income of $80K. Show your work.

(b) Compute the effective tax rate for that income.

(c) Check your answers with those in the spreadsheet `GraduatedTax.xlsx`.

Exercise 7.8.26. [S][Section 7.7][Goal 7.1][Goal 7.6] Taxes and inflation.
The spreadsheet `Federalindividualratehistory.xlsx` contains a complete history of income tax brackets and rates from the inception of the income tax in 1913 through 2013.

(a) Compute the tax due in 2003 for a single taxpayer with a net taxable income of $30K. What is her effective tax rate?

(b) Suppose that taxpayer received raises each year that kept up with inflation. Use an inflation calculator to calculate her net taxable income in 2013.

(c) Use the 2013 tax tables to compute her tax in 2013. What is her effective tax rate?

(d) Compare her 2003 tax and effective tax rate with her 2013 tax and effective tax rate, taking inflation into account. Has her tax gone up or down or stayed the same?

Exercise 7.8.27. [S][Section 7.3][Goal 7.2] [Goal 7.3] Pandora growing fast!
On June 10, 2011, CNN Money reported that the internet music site Pandora is adding new users at the rate of one per second. Between February and April the number of users grew from 80 to 90 million. [**R221**]
Is the slope of one user per second correct based on the February and April numbers of users?

Exercise 7.8.28. [S][Section 7.6][Goal 7.7] Bicycle power in Times Square.
On December 30, 2012, *The Boston Globe* reported on an Associated Press story about six bicycles that would help illuminate the famed falling ball in Times Square on New Year's Eve.

> Each bike will generate an average of 75 watts an hour. It takes 50,000 watts to light up the ball's LEDs. [**R222**]

Unfortunately, the Associated Press reporter is quite confused about the difference between energy and power. The "generate …75 watts an hour" in the quote makes no sense. We think what he or she is trying to say is that while someone is actually pedaling it, each bike could power a 75 watt light bulb. All six bikes together could light up just 450 watts worth of LEDs.

(a) How many bikes would have to be pedaled simultaneously to light up all the ball's LEDs?

(b) Since there are only six bikes, people pedaling during the day will store the energy they generate in batteries, which will then be used to light the ball. Suppose the lights need to be on for two minutes while the ball drops at midnight.

How many hours of pedaling will it take to generate (and save) the electrical energy needed?

Exercise 7.8.29. [S][Section 7.3][Goal 7.2][Goal 7.3] Flying twice as far.

A curious traveler asked this question on Stack Exchange:

> A flight from Los Angeles to Albuquerque is about 2 hours but is ≈ 670.2 miles.
>
> A flight from San Jose to Chicago is 4 hours but is ≈ 1859.0 miles.
>
> Can anyone explain why the travel time from San Jose to Chicago is not longer and closer to 5.75 hours?
>
> If the distance increases by 2, shouldn't the time increase by a factor of 2 as well? [**R223**]

(a) Write a linear model for this question. Takeoff and landing will take a fixed amount of time. Actual travel in the air will take time proportional to the distance traveled. Think about which of the variables (time and distance) is the independent variable, and identify the slope and intercept with their units.

(b) Use the data in the quotation to estimate the two constants in your linear model.

(c) Compare your answer to those at the link to the quotation.

Exercise 7.8.30. [S][Section 7.3][Goal 7.2][Goal 7.3] Express lane?

In *The Boston Globe* on November 27, 2015, you could read that

> Amid the holiday grocery shopping madness, every line feels like the wrong one. And yet, some are wronger than others. Given equally capable cashiers, you are often better off bypassing the express lane. Research conducted at a large, unnamed, California grocery store found that while each item adds 3 seconds to the check-out time, it takes 41 seconds for a person to move through the line even before their items are added to the tally. Bottom line: The big time-consumers are not the items, but the small talk and the paying, says Dan Meyer, who has a doctorate in math education from Stanford University. [**R224**]

Suppose you have 10 items in your cart, so you are allowed to use the express lane. How much longer must the line there be (compared to the regular lane) to make the wait in the regular lane less?

You can answer that question with any strategy that makes sense to you, as long as you explain what you're thinking. If you need a starting place, one way is to use these steps:

(a) Write the linear equations showing how the time it takes a shopper to check out depends on the number of items in her cart. What are the slope and intercept, with their units?

(b) Suppose shoppers in the express lane buy 6 items (on average), while those in the regular lane buy about 20. Write the linear equations showing that your waiting time in each line depends on the number of shoppers ahead of you.

(c) Now work on the main question — which line should you join when you have 10 items in your cart? How much longer must the express lane line be to make the wait on the regular lane line less?

[See the back of the book for a hint.]

Review exercises.

Exercise 7.8.31. [A] If you drive at a rate of 50 miles per hour for 3 hours, how far have you driven? Identify each piece of this proportion: the quantities being measured and the proportionality constant, with the appropriate units.

Exercise 7.8.32. [A] You may remember from geometry that the circumference of a circle is directly proportional to the diameter of that circle. The relationship is

$$c = \pi d,$$

where c represents the circumference and d represents the diameter and $\pi \approx 3.14$ is the proportionality constant. If the diameter of a circle is doubled, how does the circumference change?

Exercise 7.8.33. [A] The cost of potatoes is proportional to the weight (in pounds) you buy. If potatoes cost $0.69 per pound, what is the cost for 3 pounds of potatoes?

Exercise 7.8.34. [A] The conversion from £ to U.S. $ is 1.53 $/£. How much is £200 worth in U.S.$?

Exercise 7.8.35. [A] Suppose that y is directly proportional to x. When $x = 16$, then $y = 4$.

(a) What is the proportionality constant?

(b) If $x = 32$, what is y?

(c) If $y = 32$, what is x?

Exercise 7.8.36. [A] Identify the slope and intercept in each of the following. When appropriate, state the units.

(a) $y = 2.5x + 6$.

(b) $y = -5x + 20$.

(c) $y = 300 + 40x$.

(d) $Q = 0.004E - 300$.

(e) He earns $9.25 per hour.

(f) To rent a car for one day, the cost is $25 plus $0.15 per mile.

(g) My new phone cost $25, plus a monthly charge of $15.

(h) The conversion from £ to U.S.$ is 1.53 $/£.

(i) The salesperson worked only on commission, earning 20% of the total amount sold.

Exercise 7.8.37. [A] Solve each problem.

(a) If $y = 2.5x + 6$ and $x = 4$, what is y?

(b) If $y = -5x + 20$ and $y = 0$, what is x?

(c) If $y = 300 + 40x$ and $x = -10$, what is y?

(d) If his salary is $9.25 per hour and he works 5 hours, how much does he earn?

(e) If the conversion from U.S. dollars to pounds sterling is 1.80 $/£, how much money would you get by changing $100 to £?

(f) If my new phone cost $25 and I pay a monthly charge of $15, what is my total cost after 10 months? When does my total cost reach $250?

Exercises added for the second edition.

Exercise 7.8.38. [U][S][Section 7.7][Goal 7.2][Goal 7.6] Instructions from the IRS.
Figure 7.20 is the form the Internal Revenue Service provides for computing the 2018 tax on incomes over $100,000.

(a) Calculate the tax on an income of $150,000 using this form. (That income is the amount on line 10.)

(b) Calculate the tax on an income of $150,000 using the procedure in Figure 7.17. Check that you get the same answer.

(c) Explain how the IRS arrived at the figure $5,710.50 in column (d) in the first row of the form.

Section A—Use if your filing status is **Single**. Complete the row below that applies to you.

Taxable income. If line 10 is—	(a) Enter the amount from line 10	(b) Multiplication amount	(c) Multiply (a) by (b)	(d) Subtraction amount	Tax. Subtract (d) from (c). Enter the result here and on the entry space on line 11a.
At least $100,000 but not over $157,500	$	× 24% (0.24)	$	$ 5,710.50	$
Over $157,500 but not over $200,000	$	× 32% (0.32)	$	$18,310.50	$
Over $200,000 but not over $500,000	$	× 35% (0.35)	$	$24,310.50	$
Over $500,000	$	× 37% (0.37)	$	$34,310.50	$

Figure 7.20. Tax calculation (2018) [**R225**]

8

Climate Change — Linear Models

Complicated physical and social phenomena rarely behave linearly, but sometimes data points lie close to a straight line. When that happens you can use a spreadsheet to construct a linear approximation. Sometimes that's useful and informative. Sometimes it's misleading. Common sense can help you understand which.

Chapter goals:

Goal 8.1. Draw regression lines using Excel. Interpret regression lines.

Goal 8.2. Recognize when rounding too much distorts conclusions.

Goal 8.3. Think about causation vs correlation.

8.1 Climate change

Climate change (global warming) is a current hot topic. Is the Earth's average temperature increasing? If so, how fast? What might the consequences be? What is the cause? What might we do about it? Should we try? The science is complex and the politics even more so. In a book like this we can barely begin to unravel those complexities. To start we will look briefly at the recent average temperature of the Earth. The spreadsheet EarthDataRegression2019.xlsx has data we downloaded from www.earth-policy.org/data_center/C23 .

The chart on the left in Figure 8.1 shows the average global temperature in degrees Celsius for the years 1960–2014. The temperature seems to be increasing, with jagged small ups and downs. Spreadsheets know how to find the *regression line* that best captures the average increase in the average temperature. The chart on the right shows the regression line, along with its equation:

$$T = 0.0145 \, Y - 14.5750 \, .$$

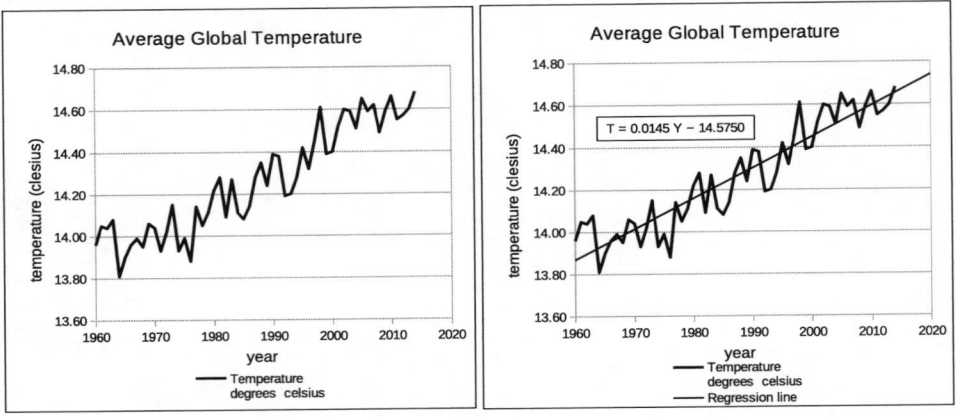

Figure 8.1. Average global temperature, 1960–2014

The important number in that equation is the slope. It tells us that the temperature increased at an average rate of

$$0.0145 \; \frac{\text{degrees Celsius}}{\text{year}}$$

between 1960 and 2014.

We extended the regression line on the chart to estimate the average global temperature in 2020: that's 6 years after 2014 so the expected increase would be about $6 \times 0.0145 = 0.087$ degrees Celsius — not quite a tenth of a degree.

The intercept for the regression line, with its units, is

$$-14.5750 \text{ degrees Celsius.}$$

Supposedly, that is the temperature predicted (retroactively) by the regression line for year 0. That's nonsense, of course.

We end this section with instructions on how to tell your spreadsheet to construct and format the regression line in Figure 8.1.

Excel. Click on the chart to open the `Chart Design` menu. Select `Add Chart Element` and hover over `Trendline`. (Spreadsheet software uses the term "trend line" for the more official "regression line".) Select `Linear` from the dropdown menu as shown in Figure 8.2. The trendline should appear. Double click on it to open the `Format Trendline` sidebar shown in Figure 8.3.

Alternatively, right click on any data point and select `Add  Trendline` from the menu that appears.

Then you can rename the trendline "Regression Line". You can also forecast forward 6 periods. Finally, display both the equation and R-squared value on the chart.

In Figure 8.1 the variable names in the trendline equation are Y for year and T for temperature. That figure was drawn with LibreCalc, where you can change the default names x and y. We don't know how to do that in Excel.

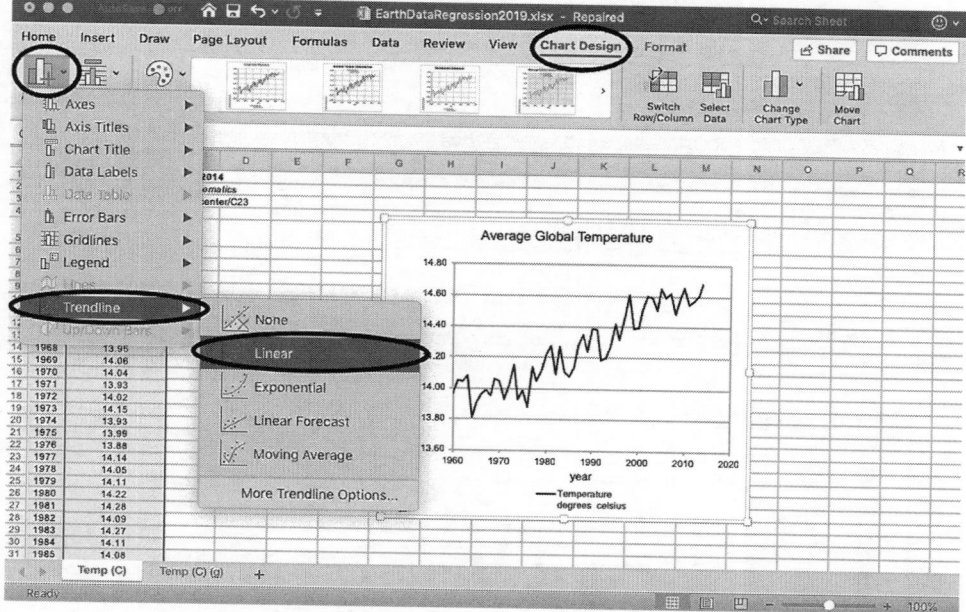

Figure 8.2. Adding a trendline to a chart (Excel)

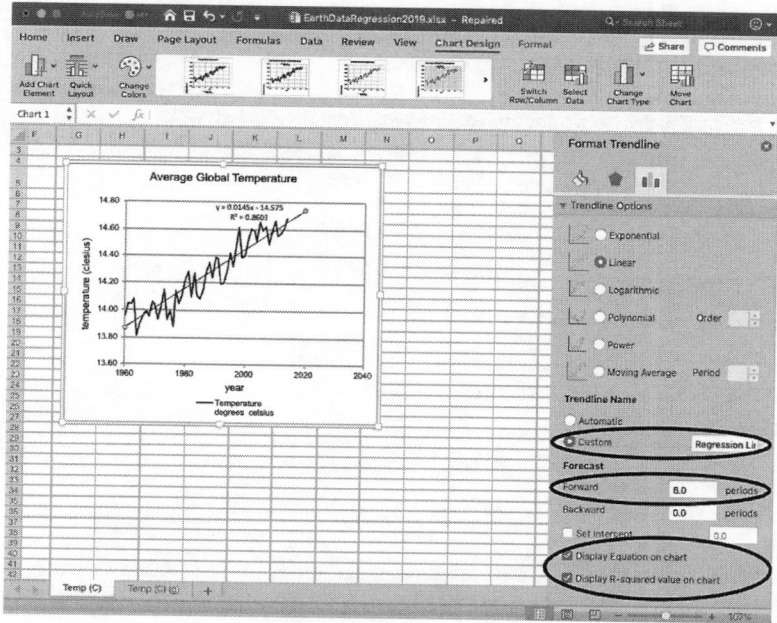

Figure 8.3. Specifying a trendline (Excel)

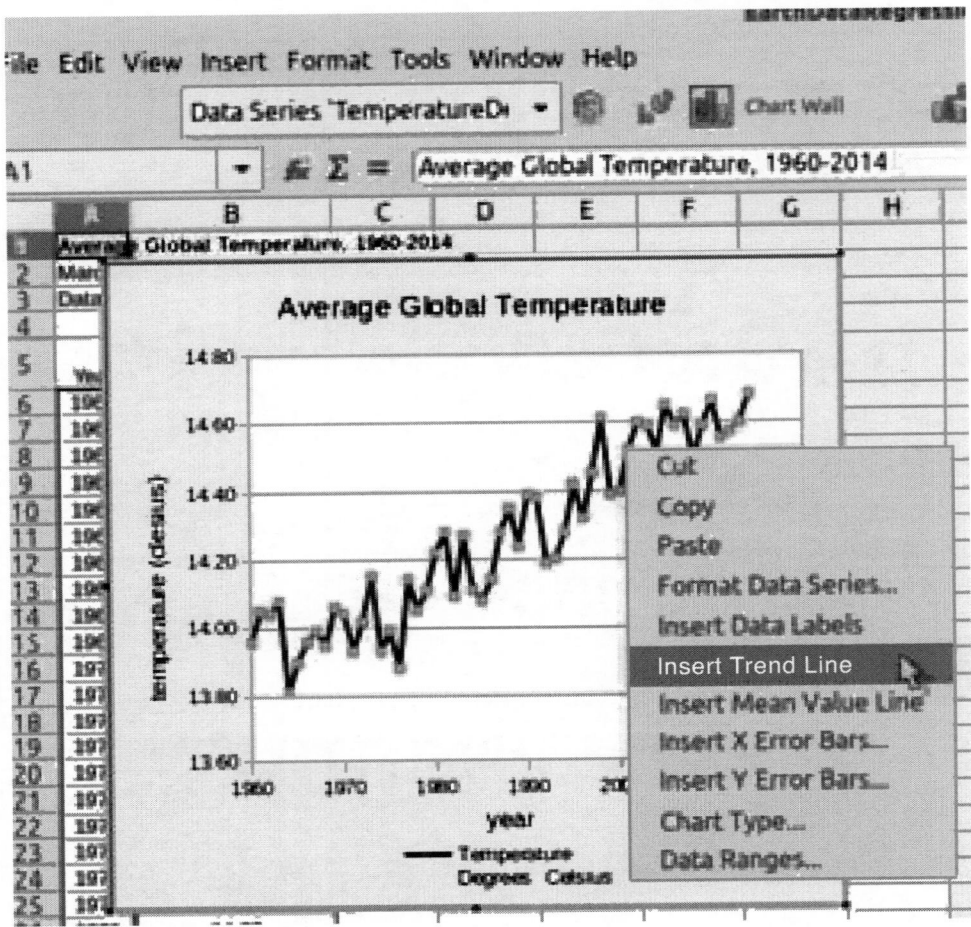

Figure 8.4. Adding a trendline to a chart (LibreCalc)

LibreCalc. Click on the chart to select it, right click and select Edit from the menu.
Then click on any data point on the chart, right click and select Insert Trend Line
. . . as shown in Figure 8.4.

Select the Type tab to see the options in Figure 8.5. Choose the Linear option.
Name the Trendline Regression line. Extrapolate forward 6 to draw the line as far
as year 2000. Click to show the equation and the R-squared value, and name the X
variable "Y" for "year" and the Y variable "T" for "temperature". Finally, click OK. You
should see Figure 8.4.

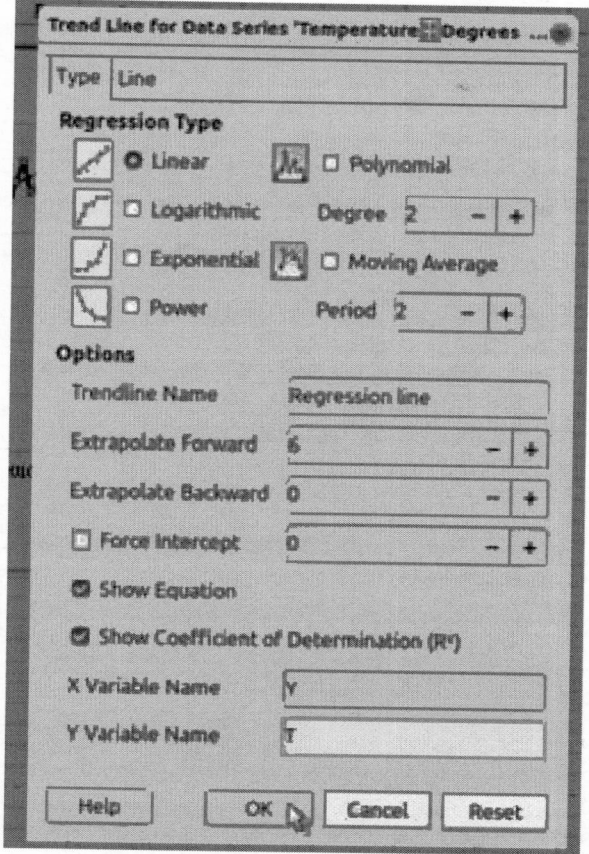

Figure 8.5. Specifying a trendline (LibreCalc)

8.2 The greenhouse effect

The fact that average global temperature has increased at more than a tenth of a degree Celsius per decade for the last half century does not mean the increase will continue. To make a prediction we need a reason for the increase, not just a pattern in the data.

The *greenhouse effect* provides a clue. A greenhouse is warm in the winter because sunlight entering through the glass roof warms the air inside. The roof prevents the warm inside air from escaping. Carbon dioxide (CO_2) in the atmosphere behaves similarly — it lets sunlight in but doesn't let heat out. That is why it's called a "greenhouse gas".

NASA reports that

> Multiple studies published in peer-reviewed scientific journals show that 97 percent or more of actively publishing climate scientists agree: Climate-warming trends over the past century are extremely likely due to human activities. [**R226**]

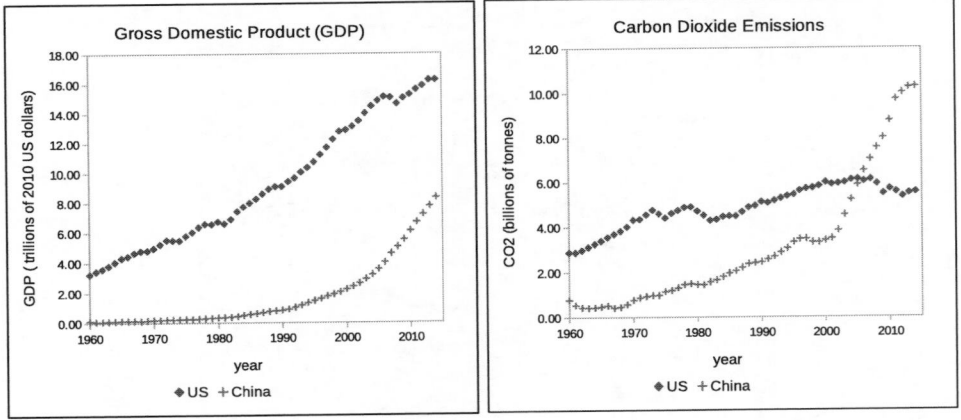

Figure 8.6. GDP and CO_2 emission, US and China, 1960–2014

In short, the Earth is warming because an increasingly prosperous population is using energy produced by burning fossil fuels that add carbon dioxide to the atmosphere. The charts in Figure 8.6 show how the gross domestic product (GDP) and total carbon dioxide emissions have changed since 1960 in the world's two largest economies.

The chart on the left shows that the U.S. economy is larger than China's, but growing more slowly. The chart on the right shows that CO_2 emissions from China passed those from the U.S. in about 2006 and by 2014 were nearly twice as much. We can understand these data better by studying how CO_2 emissions compare to GDP rather than how each varies over time.

We've done that in Figure 8.7. The chart on the left shows that in both economies a larger GDP correlates with more CO_2 emission. That makes sense: more economic activity requires more energy. At comparable levels of gross domestic product, China emits much more CO_2 than the U.S. That suggests that we have a more energy efficient economy. But the second chart in the figure seems to tell a different story. It shows that both *per capita* GDP and CO_2 consumption in the U.S. were dramatically larger than the corresponding values for China. As China modernizes those gaps may narrow.

In this section and the last we offered reasons for accepting the argument that human economic activity leads to CO_2 emissions which in turn warm the planet, using data from the years 1960–2014. The argument is sound but needs a little clarification. Adding carbon dioxide to the atmosphere does not warm the Earth instantaneously. There is about a 40 year delay. The change we see between 1960 and 2014 was affected by emissions between 1920 and 1974. More recent emissions will affect the climate years into the future even if we somehow cap greenhouse gas emissions today.

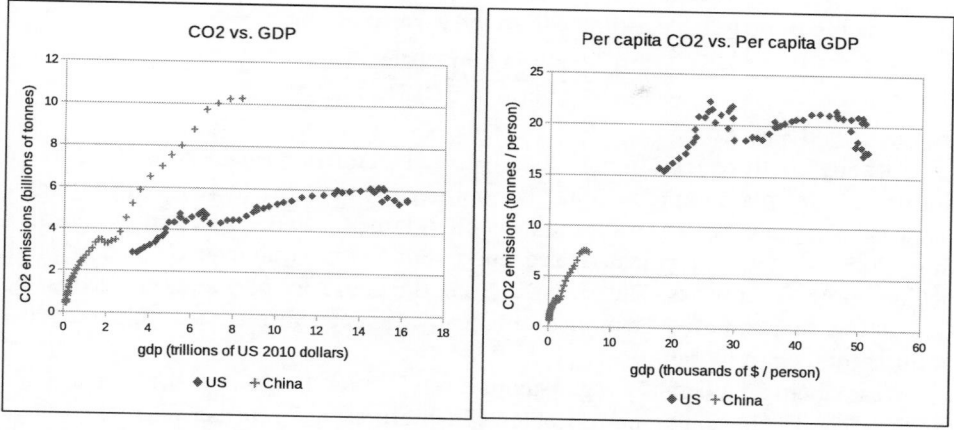

Figure 8.7. CO_2 emission vs GDP, US and China, 1960–2014

8.3 How good is the linear model?

In Figure 8.8 we added regression lines to the plots of CO_2 emissions as a function of gross domestic product. How much a regression line helps understand the data depends in part on how close the data points are to the line.

The official statistical measure of "close to the line" is a number between zero and one called "R-squared", written R^2. The closer R-squared is to 1 the better the regression line fits the data. In this example the R^2 values in the first chart are 0.98 and 0.80. That reflects the obvious conclusion from the graph: the average growth rate of

$$\frac{1.27 \text{ billion tonnes of } CO_2}{\text{trillion \$ of GDP}}$$

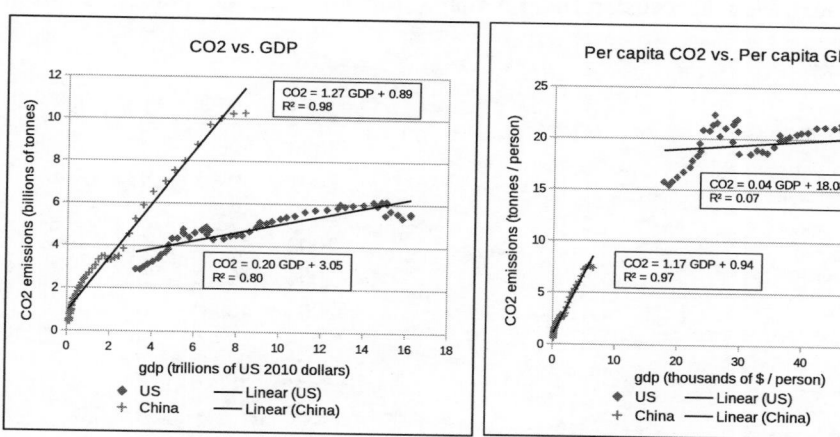

Figure 8.8. CO_2 emission vs GDP, US and China, 1960–2014

for China is a more reliable estimate than the corresponding

$$\frac{0.20 \text{ billion tonnes of } CO_2}{\text{trillion \$ of GDP}}$$

for the United States.

Although both regression lines look reasonable, the data suggest that they might not be good for making predictions. For both economies, the increase in CO_2 emissions as GDP grows seems to start to level off at large GDP. Perhaps that represents a transition from a manufacturing-based, energy-intensive economy to one whose products are more like services. The second chart in the figure, for per capita emissions and GDP, shows that pattern even more dramatically. It's even visible for the United States in the second chart in Figure 8.6,

We are being deliberately vague about how close R^2 should be to 1 to declare that the fit is "good". One of the many problems in interpreting R^2 correctly is that it is near 0 when the regression line is horizontal even if the points are scattered, You can see that in the second chart, where its value for the United States is just 0.07.

8.4 Regression nonsense

The graphic in Figure 8.9 resembles one that appeared in *The Boston Globe* on January 14, 2010, in a story headlined "Imaginary fiends", which began

> In 2009, crime went down. In fact it's been going down for a decade.
> But more and more Americans believe it's getting worse. [R227]

The data are from the FBI and the Gallup Poll. The FBI measures the crime rate in violent crimes per 100,000 people. The fear index is the percentage of people who say crime is going up.

The headline seems to announce a juicy story. The graph is drawn to accentuate the apparent contradiction, since the scales on both y-axes don't start at 0. We will use these numbers to illustrate the kinds of nonsense arguments you can make with regression lines. There are three variables to play with: the year, the crime rate, and the fear index. We will focus on them two at a time and imagine different kinds of conclusions.

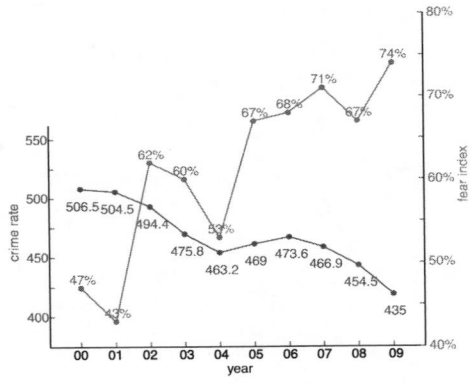

Year	Crime	Fear
2000	506.5	47
2001	504.5	43
2002	494.4	62
2003	475.8	60
2004	463.2	53
2005	469.0	67
2006	473.6	68
2007	466.9	71
2008	454.5	67
2009	435.0	74

Figure 8.9. Crime down, fear up [R228]

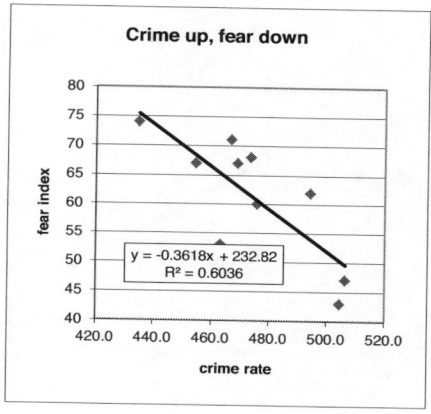
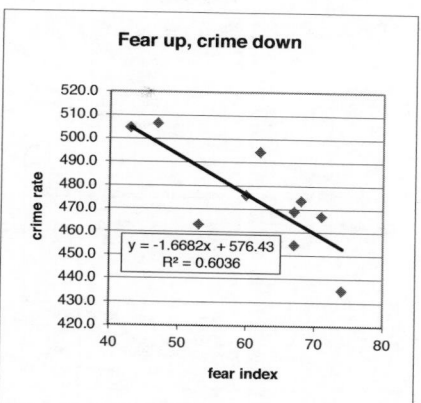

Figure 8.10. Crime vs fear regressions

Our work is in the spreadsheet crimeDropsFearsRise.xlsx.

The first graph in Figure 8.10 shows a scatterplot and trendline for the last two columns in the table. There we asked Excel to construct a graph with crime rate as the independent variable.

Since we chose crime rate as the independent variable it's easy to look at the graph — and the trend line — and conclude that the increase in crime rate is closely related to the decrease in the fear index. The regression line slopes down — high crime rates seem to come along with decreased fear of crime. The R-squared value is 0.60 — perhaps not compellingly high, but we won't let that stop us from thinking about the data. What might the correlation mean? Could an increase in crime (the independent variable on the x-axis) cause people to be less afraid? Here's an attempt at an explanation: perhaps when crime is rare it's reported spectacularly in the news and people are frightened, while when it's common it gets less press and most people don't notice it as much because it isn't happening to them.

Does that make sense? Not to us, but it's the kind of argument you frequently see or hear — a simpleminded attempt to explain what seems to be a real "this is true because of that" connection, or perhaps what a politician would like you to believe is a real connection.

The second graph in Figure 8.10 shows the same data with the fear index as the independent variable. That changes our view of the data. Now we see crime dropping as fear increases. How might we explain that? Perhaps we'd argue that increasing fear of crime leads to more pressure on the police to arrest criminals, thus reducing the amount of crime. That's more plausible than the other way around, but still a shallow unconvincing analysis of complex social phenomena. Both the crime rate and the fear of crime are changing over time, one decreasing while the other increases, but just because we can find a trendline doesn't mean either change causes the other.

We can see the two trends separately if we plot each with time as the independent variable, as in Figure 8.11. With these charts we can create other nonsense arguments. The slope of the fear index regression line is about 3 percentage points per year. Since the index was at 74% in 2009, if the trend continues then in about 8 more years, in 2017, 98% of the population would have believed that crime was getting worse every

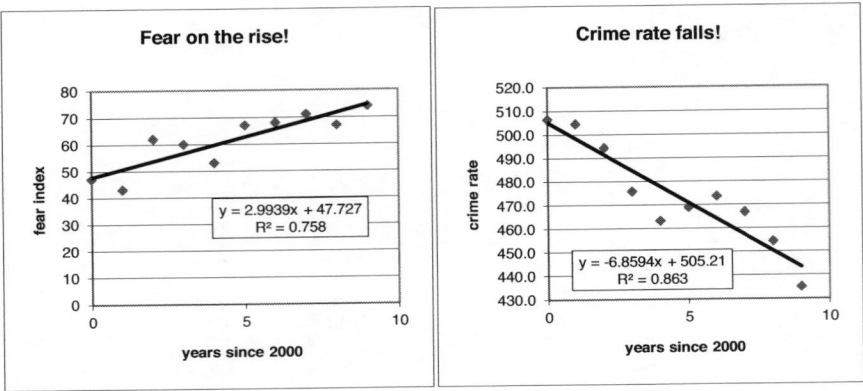

Figure 8.11. Fear index and crime rate over time

year. The second regression line says the crime rate is actually falling each year by about 7 violent crimes per 100,000 people, so it predicted that in 2017 when everyone believed things were getting worse it would be down from 435 to about 380. Neither of these predictions carries much conviction. Neither came about.

The news story that prompted this discussion is misleading in another way. When we found the data on which it is based we discovered that in the previous decade, from 1990 to 2000, the crime rate and the fear index were both decreasing. The author of the article chose not to tell us that. He *cherry-picked* the data to make his point (whatever it is) more dramatic. You can find all the numbers in our spreadsheet at `crimeDropsFearsRise.xlsx`.

The moral of this story:

> Correlation is not causation.

This cautionary quote drives home the point:

> [I]n psychological and sociological investigations involving very large numbers of subjects, it is regularly found that almost all correlations or differences between means are statistically significant. ... Data currently being analyzed by Dr. David Lykken and myself, derived from a huge sample of over 55,000 Minnesota high school seniors, reveal statistically significant relationships in 91% of pairwise associations among a congeries of 45 miscellaneous variables such as sex, birth order, religious preference, number of siblings, vocational choice, club membership, college choice, mother's education, dancing, interest in woodworking, liking for school, and the like. [**R229**]

It's very easy to use regression to link variables (crime rate and fear index, as in this example) to suggest trends and to make predictions or interpret correlation as explanation. Just because you can doesn't mean you should. It's often wrong.

8.5 Exercises

Exercise 8.5.1. [S][W][Section 8.1][Goal 8.1] A trendline for linear data.

(a) What values would you expect to see for the slope, the intercept and R-squared if you were to add a trendline to the Tamworth electricity bill in the spreadsheet TamworthElectric.xlsx?

(b) What would the trendline look like on the graph in Figure 7.5?

(c) Add the trendline and verify your predictions.

Exercise 8.5.2. [S][Section 8.3][Goal 8.1] Anscombe's quartet.

> Anscombe's quartet comprises four datasets that have nearly identical simple statistical properties, yet appear very different when graphed. Each dataset consists of eleven (x, y) points. They were constructed in 1973 by the statistician Francis Anscombe to demonstrate both the importance of graphing data before analyzing it and the effect of outliers on statistical properties. [**R230**]

Use the data in AnscombesQuartet.xlsx for the tasks that follow.

(a) For each data set, use a spreadsheet to find the mean of the x and y values. Label them in your spreadsheet.

(b) Do the mean values describe these four data sets very well? Explain.

(c) Graph each set of (x, y) values. Label each graph ("data set 1", etc.). Write a sentence or two describing the relationship between the x and y values, using what you see on the graph. Talk about how strong that relationship is (but don't calculate the R-squared value yet).

(d) Display the trendline, the trendline equation and the R^2 value on each graph.

(e) Round the slope and intercept to two decimal places. Write a sentence comparing the slope, intercept and R-squared value for each of the data sets.

(f) Explain in your own words how these examples demonstrate the importance of graphing data before analyzing it.

(g) The short description at the beginning of this problem also talked about the effect of "outliers" on statistical properties. In this context, an outlier is a number that lies outside most of the numbers in the data set. Does each of the data sets contain an outlier? If so, how does that outlier influence the basic statistics for each data set?

Exercise 8.5.3. [S][Section 8.1][Section 8.2][Goal 8.1] Faster than a speeding bullet.
The spreadsheet at MarathonWinningTimes.xlsx shows the history of the winning time in the Boston Marathon for men and women from 1966 (when women first ran) through 2013.

(a) Graph the men's and women's winning times depending on the year, properly label the axes and add a trendline for each data column.

(b) What is the average rate at which the men's finishing time changed from year to year?

(c) Use the trendline to predict when the men's winner will finish in two hours. How confident are you in that prediction?

(d) Use the trendline to predict when the men's winner will finish in one hour. How confident are you in that prediction?

(e) The trendlines suggest that in about six years the fastest woman will be as fast as the fastest man and will be faster thereafter. Explain why the lines say that and why it's nonsense.

(f) Make a better prediction about the long run relation between men's and women's winner finishing times.

[See the back of the book for a hint.]

Exercise 8.5.4. [U][Section 8.1][Goal 8.1] The leaning tower of Pisa.

The famous "Leaning Tower of Pisa" began to lean even while it was under construction in the 1170s. The table in Figure 8.12 shows the measured lean for the years 1975 through 1987: the distance in meters between where a particular point on the tower would be if the tower were straight and where it actually was.

(a) Construct the regression line for this data and estimate (visually) what the lean was in the year 2000.

(b) How good is that estimate likely to be?

(c) What is the slope of the regression line? What are its units? What does it mean?

(d) Check your estimate using the equation of the regression line. Can you use the formula as it appears in the chart or do you need more decimal places?

Year	Lean (m)
1975	2.9642
1976	2.9644
1977	2.9656
1978	2.9667
1979	2.9673
1980	2.9688
1981	2.9696
1982	2.9698
1983	2.9713
1984	2.9717
1985	2.9725
1986	2.9742
1987	2.9757

Figure 8.12. The Tower of Pisa [**R231**]

(e) Explain why the actual numbers in the data table for the Tower of Pisa depend on the height of the "particular point" at which measurements were taken. What would the numbers be if the point were twice as high? Would the linear regression line be just as good?

(f) What has happened to the Tower of Pisa since 1987?

Exercise 8.5.5. [S][Section 8.1][Goal 8.1] Beverage consumption.
The spreadsheet at BeverageConsumption.xlsx contains data on the amounts of milk, bottled water and soft drinks consumed in the United States between 1980 and 2004.

(a) Use Excel to create a scatter plot of this data. Label the data series and the axes correctly.

(b) Explore correlations among the various categories (for example, between milk and water). Write about what you discover. In particular, which kinds of consumption are most closely correlated?

(c) Use the regression lines to make some predictions for years following 2004.

(d) Find the source of the data in BeverageConsumption.xlsx. If you find data for other years there, discuss the validity of your predictions.

[See the back of the book for a hint.]

Exercise 8.5.6. [S][Section 8.1][Goal 8.1] Energy consumption.
The Excel spreadsheet EnergyConsumption.xlsx contains a table showing the annual United States energy consumption, measured in terawatt-hours, between 1949 and 2005.

(a) Insert a new column labeled "years since 1949" in between the years column and the consumption column. Use Excel to fill in the cells for this column.

(b) Use Excel to find a linear trendline for this data. Include the equation and R^2 value for the trendline on the graph.

(c) Is this trendline a good fit for the data?

(d) What is the slope of this line? Include the units in your answer. Use your answer for the slope to complete the sentence: "For every additional year that passes, total energy consumption ...."

(e) Estimate total energy consumption in the years from 2006 to the present.

(f) Look for data with which to check the estimates from the previous part of the exercise.

Exercise 8.5.7. [S][Section 8.1][Goal 8.1] Supply and demand for office space.
   The data in Table 8.13 appeared on page B5 in *The Boston Globe* on April 3, 2010.

quarter	vacancy rate	rent ($/ft^2)
Q1 '06	11.8%	38.76
Q1 '07	7.5%	47.54
Q1 '08	6.0%	62.20
Q1 '09	9.0%	49.24
Q1 '10	11.1%	42.46

Table 8.13. Lower rent, more vacancy

(a) Build and then discuss a linear regression line for the dependence of rent per square
   foot on vacancy rate.

(b) How do your conclusions change when you adjust rents to take inflation into account?

Exercise 8.5.8. [S][Section 8.1][Goal 8.1] [Goal 8.3] Office rents.
   On February 22, 2008, *The Boston Globe* ran a story under the headline "Office
rents reach dizzying heights" that featured graphs like those in Figure 8.14.
   The shapes of the curves illustrate the law of supply and demand — the more space
is available the less you have to pay for it.
   You can find the data in the spreadsheet `BostonOfficeRents.xlsx`.

(a) Show how rental cost depends on the percent of space available by creating a scatter
   plot using columns D and F and a regression line for that scatter plot. Identify the
   slope and its units. How good is the correlation?

(b) Use the graph and the formula to estimate office rent when the availability rate
   is 8%.

(c) The spreadsheet contains data on the vacancy rate as well as the availability rate.
   Create a scatterplot illustrating how the vacancy rate depends on the availability
   rate. Add a regression line and discuss what it tells you.

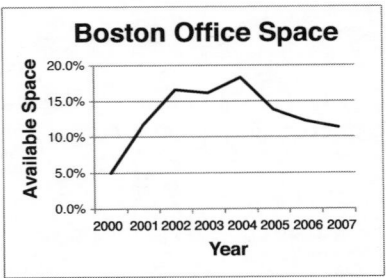 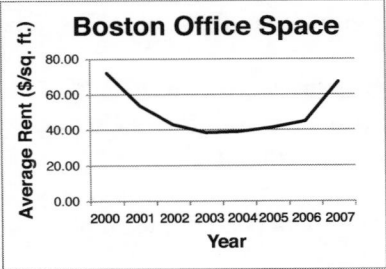

Figure 8.14. Boston office rental rates

Exercise 8.5.9. [U][Section 8.3][Goal 8.1] First class mail.

Table 8.15 shows the cost of sending first class mail weighing up to one ounce.

Year	cents/oz	Year	cents/oz
1885	2	1991	29
1917	3	1995	32
1919	2	1999	22
1932	3	2001	34
1958	4	2002	37
1963	5	2006	39
1968	6	2007	41
1971	8	2008	42
1974	10	2009	44
1975	13	2012	45
1978	15	2013	46
1981	18	2014	49
1982	20	2018	50
1985	22	2019	55
1988	25		

Table 8.15. First class mail [R232]

(a) Enter the data in a spreadsheet; then draw a graph of the data.

(b) Insert the trendline and display the trendline equation and the R-squared value on the graph.

(c) Write a sentence interpreting the slope of the trendline.

(d) Is this a strong correlation? Explain.

(e) Answer the same questions if you restrict your trendline to the years since 1970.

Exercise 8.5.10. [S][Section 8.1][Goal 8.1] College costs.

The spreadsheet CollegeCosts2010.xlsx shows the annual mean cost for tuition and fees at private and public four year colleges in the U.S. between 1999 and 2010.

(a) Create a properly labeled graph showing how mean private and public education costs changed in the years 1999–2010.

Insert a linear trendline for each set of data. Use Excel to forecast the trendline out to 2015 (that is, 16 years past 1999).

(b) Interpret the numerical value of the slope in each trendline equation. That is, write a sentence explaining what the slope represents.

(c) Use your trendline equations to determine the projected mean tuition cost at both private and public four year colleges for 2015.

(d) Compare your answers from the previous questions with the graph. Are the answers consistent or do you need to use more digits in your calculation?

(e) (Optional) Confirm your prediction for the year in which you are answering the question.

Exercise 8.5.11. [U][Section 8.1][Goal 8.1] Speed vs. MPG, revisited.
Exercise 3.8.52 looked at the relationship between speed and fuel consumption. You can do this problem even if you didn't do that one.

(a) Read data from the graph in Figure 3.11 and enter it in Excel.

(b) The information cited in Exercise 3.8.52 states that for each 5 mph you drive over 50 mph, your decrease in fuel economy means that you pay an additional $0.25 for gas. Use Excel to graph the data corresponding to speeds above 50 mph. Construct a regression line for this data. What does the slope of the regression line tell you about how fuel economy changes as speed increases? If your speed increases by 5 mph, how does your fuel economy change, on average?

(c) Use Excel to convert the data in your table from mpg to gallons per 100 miles. Graph the data again and insert the regression line. What does the slope of the regression line tell you about how fuel economy changes as speed increases? Is it easier to explain how fuel economy changes when your speed increases by 5 mph?

Exercise 8.5.12. [S][Section 8.4][Goal 8.1] Playing with regression lines.
Use the spreadsheet PlayWithRegression.xlsx to explore the following questions.

(a) What happens when all the y-values are the same?

(b) What if all but one of the y-values are the same and you vary that one?

(c) What if y decreases as x increases?

(d) What if the x and y values match?

Exercise 8.5.13. [S][Section 8.4][Goal 8.1] Should businesses use private jets?
On May 26, 2012, *The Boston Globe* published a letter to the editor from David V. Dineen, Executive Director of the Massachusetts Airport Management Association. He observed that companies using their own private jets had earnings 434 percent higher than those using commercial airlines. [R233]

Explain how and why Dineen is using the statistic he quotes to encourage readers to confuse correlation with causation.

Exercise 8.5.14. [S][Section 8.4][Goal 8.3] Cherry-picking.
In Section 8.4, we discovered that the author had "cherry-picked" the data. Find out what "cherry-picking" means and where the phrase comes from. Find and discuss some examples.

Exercise 8.5.15. [S][Section 8.4][Goal 8.1] [Goal 8.3] Watch TV! Live Longer!
The data in the spreadsheet TVData.xlsx show the life expectancy in years for several countries, along with the number of people per television set in those countries. (The idea (and the data) for this problem come from the article www.amstat. org/publications/jse/v2n2/datasets.rossman.html.)

(a) Which countries have the highest and lowest life expectancy at birth? Which have the highest and lowest number of people per television set?

(b) Use Excel to create a properly labeled scatter plot of the life expectancy and people per television data. Find the trendline and display the equation and the R-squared value on your graph.

(c) What is the slope of the trendline (with its units)? Explain its meaning in a sentence.

(d) Does a small number of people per television set improve health? Would people in countries with low life expectancy live longer if we sent them shiploads of television sets?

(e) Does living longer increase the number of television sets? If we improved the life expectancy in a country by providing better medical care would that cause there to be fewer people per television set?

(f) What else could be going on here? Why might high life expectancy be strongly correlated with a low ratio of people per TV set?

Exercise 8.5.16. [S][W][Section 8.4][Goal 8.1][Goal 8.3] Crime rates revisited.

(a) Use the data in `crimeDropsFearsRise.xlsx` to redo the analysis for the entire period from 1990 to 2009.

(b) Are the crime rates in this exercise consistent with those in the example we studied in Chapter 2?

[See the back of the book for a hint.]

Exercise 8.5.17. [S][Section 8.4] [Goal 8.1][Goal 8.3] The Mississippi River.

> In the space of one hundred and seventy-six years the Lower Mississippi has shortened itself two hundred and forty-two miles. That is an average of a trifle over one mile and a third per year. Therefore, any calm person, who is not blind or idiotic, can see that in the Old Oolitic Silurian Period, just a million years ago next November, the Lower Mississippi River was upwards of one million three hundred thousand miles long, and stuck out over the Gulf of Mexico like a fishing-rod. And by the same token any person can see that seven hundred and forty-two years from now the Lower Mississippi will be only a mile and three-quarters long, and Cairo and New Orleans will have joined their streets together, and be plodding comfortably along under a single mayor and a mutual board of aldermen. There is something fascinating about science. One gets such wholesale returns of conjecture out of such a trifling investment of fact.
>
> Mark Twain
> Life on the Mississippi [**R234**]

Discuss this linear model for the length of the Mississippi River. What's the slope? Can you verify Twain's arithmetic?

Exercise 8.5.18. [W][Section 8.4][Goal 8.3] Well, maybe.

Explain the joke in the cartoon in Figure 8.16 reproduced from xkcd.com/ .

Figure 8.16. Well, maybe. [**R235**]

9

Compound Interest — Exponential Growth

In this chapter we explore how investments and populations grow and how radio-activity decays — exponentially.

Chapter goals:

Goal 9.1. Understand that exponential growth (or decay) is constant relative change.

Goal 9.2. Understand how compound interest is calculated.

Goal 9.3. Work with exponential decay.

Goal 9.4. Reason using doubling times, half-lives, rule of 70.

Goal 9.5. Fit exponential models to data.

9.1 Money earns money

Imagine that you have $1,000 to invest. Would you rather earn $100 per year in interest or 8% per year in interest? In each case the interest is added into your principal (the balance in your account) each year, and you never make any withdrawals.

The first scenario is called simple interest. You find the new balance by adding $100 each year to the previous balance. After one year the balance would be $1,100, after two years $1,200, and so on.

The second scenario is a little more complicated. The interest each year is a fixed percentage of the balance. The one-plus trick finds the new balance in one step: after one year you would have $1.08 \times \$1,000 = \$1,080$. After two years your balance would be $1.08 \times \$1,080 = \$1,166.40$. This pattern is called compound interest.

Table 9.1 shows your balance in each case for the first four years.

	Balance	
Year	Simple interest	Compound interest
now	$1,000	$1,000.00
1	$1,100	$1,080.00
2	$1,200	$1,166.40
3	$1,300	$1,259.71
4	$1,400	$1,360.49

Table 9.1. Simple and compound interest

So far simple interest offers a better return on your investment. What would the numbers be in 10 years? We could continue building the table a year at a time by hand (which is tedious), we could have a spreadsheet calculate for us (we'll do that in a minute) or we could look for a pattern and find a formula for each scenario, so that we can compute for any year we like without having to do the work for all the years in between. We'll do that first.

Simple interest leads to a linear equation. Each year the balance increases by a fixed amount, $100, so the slope is $100/year. The intercept is the starting value, $1,000. The linear function is

$$B = 1,000 + 100 \times T$$

where B represents the balance, in dollars, and T the number of years. If you leave your money growing until $T = 10$ years, your balance will be $1,000 + 100 \times 10 = 2,000$ dollars.

The function describing compound interest isn't linear. The percentage increase is constant but the amount of interest changes from year to year. In the first year you earn $80, while in the second you earn $86.40. To see what kind of function to use, we unwind the arithmetic in the compound interest column of Table 9.1:

Year 1: $1,080.00 = 1,000.00 \times 1.08$

Year 2: $1,166.40 = 1,080.00 \times 1.08 = (1,000 \times 1.08) \times 1.08 = 1,000 \times 1.08^2$

Year 3: $1,259.71 = 1,166.40 \times 1.08 = (1,080 \times 1.08^2) \times 1.08 = 1,000 \times 1.08^3$

It's clear that the function describing this growth is

$$B = 1,000 \times (1.08)^T \tag{9.1}$$

where, as before, B represents the balance, in dollars, and T the time, in years. It's an exponential function, because the independent variable T is the exponent of 1.08. The 0.08 in $1.08 = 1 + 0.08$ is the constant relative change for each additional year. The 1,000 is where we start: the value of B when $T = 0$. (You may remember but not have enjoyed the fact that $1.08^0 = 1$. If so, perhaps it makes a little more sense in this context. After 0 years you've received no interest, so your balance should be multiplied just by 1.)

Suppose you want to compare the balances at simple and compound interest after year 10. With simple interest you will have $1,000 + 10 \times 100 = 2,000$ dollars. But how can you compute $1,000 \times 1.08^{10}$ without boringly multiplying by 1.08 ten times?

The calculator in Figure 2.3 isn't powerful enough. For that job you need a *scientific calculator*, one with a key labeled $\boxed{y^x}$ or $\boxed{x^y}$.

There are many online. Here are two: `www.math.com/students/calculators/source/scientific.html` and `web2.0calc.com/`. Each will tell you that at the end of year 10 the balance will be about $2159, so the exponential growth has caught up with the linear.

You can do the computation with the Google calculator's buttons but to use the search bar or a spreadsheet you need to know how to enter the exponent from the keyboard without a $\boxed{y^x}$ key. Both use the caret character "^" to raise a number to a power. That's meant to suggest literally "raising" the exponent. You just type

$$1000 * 1.08 \wedge 10$$

into the Google search bar or as a formula (preceded by an equal sign) in a cell in a spreadsheet to check the arithmetic in the previous paragraphs.

In *Common Sense Mathematics* we rarely put things to remember in boxes, but the moral of this discussion deserves that treatment:

> In linear growth, the absolute change is constant.
> In exponential growth, the relative change is constant.

Interest isn't the only place exponential growth happens. In Exercise 9.7.2 we ask you to think about others.

9.2 Exploring exponential growth with a spreadsheet

You can answer "what-if" questions about exponential growth by changing the initial investment and interest rate in the the spreadsheet `exponentialGrowth.xlsx`. Figure 9.2 shows two examples, for the equations

$$B = 1{,}000 \times (1.08)^T$$

and

$$B = 1{,}000 \times (1.16)^T.$$

Each swoops upward at an increasing rate. That shape is the signature for exponential growth. The graphs look similar, but the scales on the vertical axes tell a different story. The spreadsheet has a second tab (labeled "Compare two growth trajectories") that plots two exponential curves on the same set of axes. Figure 9.3 clearly shows how much faster growth is at 16% than at 8%.

Let's take some time to see how Excel updates the calculations when we change the constants in the equation. Click on one of the cells in the exponential column, say B17. The formula bar reads

$$= \text{START} * \text{RELCHANGE}^{\wedge}\text{A17}, \tag{9.2}$$

which is Excel's version of (9.1). We labeled cells A10 and A11 as START and RELCHANGE so that we could use (9.2) instead of

$$= \$A\$10 * \$A\$11^{\wedge}\text{A19}.$$

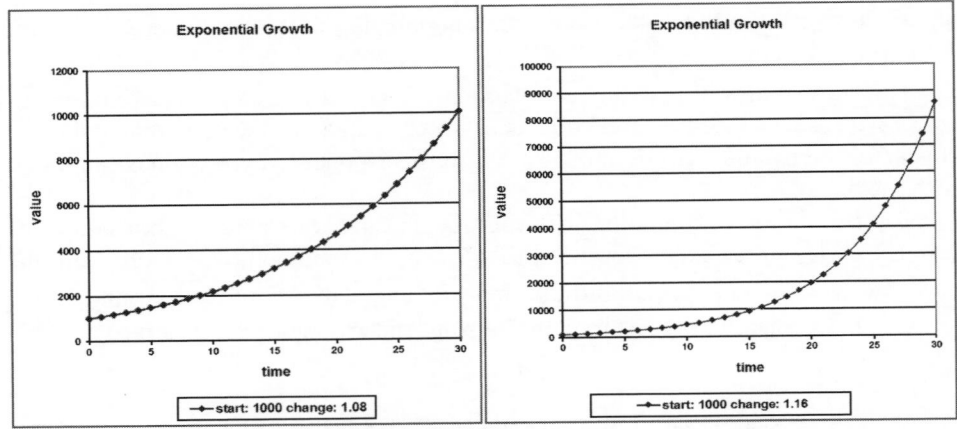

Figure 9.2. Two exponential graphs

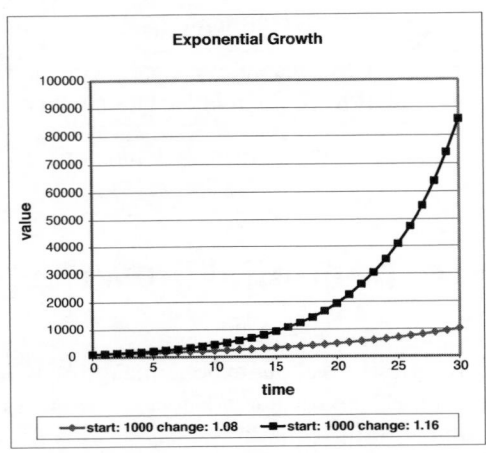

Figure 9.3. Two exponential graphs on the same set of axes

The version using cell labels is much easier to understand than the one with cell references, and it doesn't need the dollar signs to tell Excel not to change those references
when we copy from one row to another.

To label a cell, click on it. The Name Box at the left of the Formatting Toolbar will
contain the address of the cell, so if you click on cell H4 you will see H4 there. You can
highlight the contents of the box and type in your own name.

Figure 9.4 shows two screen shots of our spreadsheet, the first with cell values, the
second with cell formulas.

The numbers in columns B and C are the same. Excel computes them in different
ways. We've seen how B17 uses the algebra in (9.2). The value in cell C17 comes from
the previous value in C18 instead:

$$= \text{C16*RELCHANGE}.$$

In the hypothetical investment comparison at the start of this chapter, linear
growth starts out better but by year 10 exponential growth leads to a higher balance. To

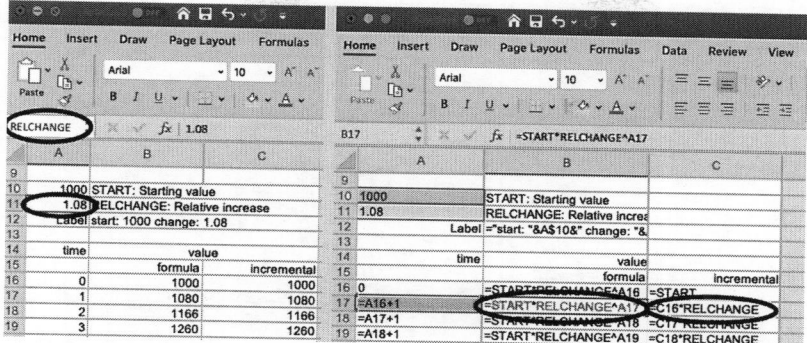

Figure 9.4. Screenshots showing Excel formulas using named cells

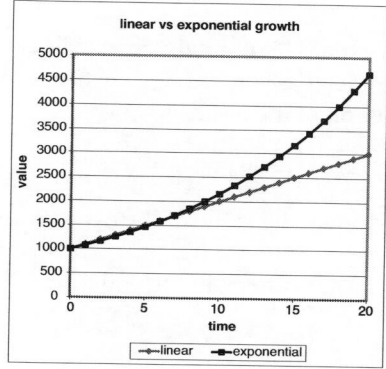

Figure 9.5. Linear vs exponential growth

explore what happens in more detail, use the spreadsheet `linearExponential.xlsx`. It extends Table 9.1 to cover 15 years. The graph in Figure 9.5 shows that starting at year 7, the value of the exponential function is larger than the linear.

Now we can answer "what-if" questions. Suppose, for example, our money earned 7% interest instead of 8% interest. To redo the calculations we need to change just one number: replace the 1.08 in cell A9 with 1.07. Excel recomputes the values of the exponential function in column C and redraws the graph. Then you can see that with this lower interest rate, we have to wait 11 years before the exponential growth of compound interest gives us a better return.

9.3 Depreciation

It's always easier to think about increases (adding and multiplying) than decreases (subtracting and dividing) but sometimes things do decrease.

Suppose you buy a new car for $20,000. As soon as you drive it out of the dealer's lot it's worth less. In fact it's worth less each year: it *depreciates*. Its value depends on its age.

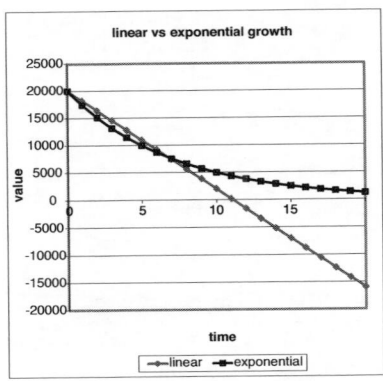

Figure 9.6. Linear vs exponential depreciation

If the car is a business expense you might choose linear depreciation for tax purposes — suppose the value decreases by $1,800 each year. The equation that deter-
mines the value V as a function of the age A is

$$V = 20{,}000 - 1{,}800A.$$

But a more realistic way to model the value of the car is to assume that the percentage
decrease is the same each year. Suppose it's 13%. Then each year its value is 87% of
what it was the year before. The corresponding equation is

$$V = 20{,}000 \times 0.87^A.$$

We can use our old friend linearExponential.xlsx to draw Figure 9.6 showing what
the car is worth over time in each case. Set START to 20,000, ABSCHANGE to -1,800, and
RELCHANGE to $1 - 0.13 = 0.87$. (The relative change is still positive. It's a decrease
rather than an increase because it's less than 1.)

When the depreciation is linear the car is worthless (at least on paper) after about
11 years. Excel doesn't know that, so it continues the graph on into negative values. If
we wanted to use this graph in a more formal presentation we'd have to prevent that
and change the labels. Leaving it this way exhibits the power of thinking abstractly in
Excel — the original spreadsheet can manage shrinking just as easily as growth.

9.4 Doubling times and half-lives

How long will it take to double your money? (We'll assume you're clever enough to
insist on compound interest.) The answer depends on the interest rate and the initial balance. The exponentialGrowth.xlsx spreadsheet shows that at 8% interest
with an initial investment of $1,000 the balance is $2,000 after 9 years (the table shows
$1,999.004627, which is quite close enough to double).

If you change the initial investment to $100 then Excel shows a balance of $200
after the same 9 years. Experimenting with many different initial investments always
shows the same doubling time. So the time it takes to double your money does not
depend on the amount you start with.

interest rate (%)	approximate doubling time	70/rate
2	35	35.0
3	24	23.3
5	14.5	14.0
7	10.5	10.0
8	9	8.8
10	7.5	7.0
15	5	4.7
20	4	3.5
50	1.7	1.4
100	1	0.7

Table 9.7. Double your money

What about a different interest rate? If you use 5% interest in the spreadsheet the doubling time seems to be between 14 and 15 years. We can do a calculation: $1.05^{14.5} = 2.028826162$, so 14.5 is a good guess.

At 2% interest it takes more than 30 years to double your money, so the spreadsheet doesn't give us the answer. We could find it by adding some rows, but we'll use another method instead. We'll try to guess the value of T in the equation $1.02^T = 2$ and adjust our guess until we're close enough. Perhaps the answer is $T = 40$ years:

$$1.02^{40} = 2.208039664.$$

Too big, so we need less time. Try 35:

$$1.02^{35} = 1.999889553.$$

Bingo!

We've collected these results and a few more in the second column of Table 9.7.

The third column in that table shows the results from the "Rule of 70", which says that you can estimate the compound interest doubling time by dividing the magic number 70 by the annual interest rate as a percent. The approximation is better when the interest rate isn't too large; those are just the cases that matter most in everyday investing. The most commonly quoted consequence of the Rule of 70 is that money invested at 7% will double in 10 years.

Figure 9.8 shows how good the Rule of 70 is for interest rates up to 20%.

When a relative increase occurs repeatedly the doubling time is independent of the initial value. So if inflation is 5% per year, all prices will double in 14 years.

Knowing the doubling time helps you make quick calculations. Since 5% inflation doubles prices in 14 years it will quadruple them in 28 years. In 42 years they will be eight times as large. The Bureau of Labor Statistics inflation calculator says that inflation in the 42 years from 1968 to 2010 increased the cost of a $100 item to $626. That's not quite eight times as much, so the average inflation rate for those years was not quite 5% per year.

The Rule of 70 applies to depreciation as well — it tells you the *half-life*. That's the equivalent for depreciation of the doubling time — the time until half the original value has disappeared. Like doubling time, the half-life depends on the depreciation

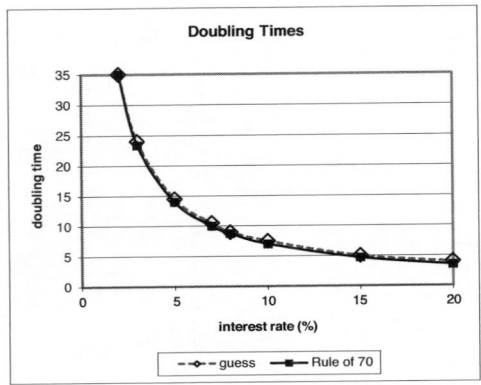

Figure 9.8. Doubling times

rate, but not on the original value. For 13% annual depreciation it's approximately $70/13 = 5.38461538 \approx 5.4$ years.

The term half-life comes from atomic physics, where it describes the way the quantity of a radioactive element decreases over time. The following quotation from `http://archives.nirs.us/factsheets/hlwfcst.htm` provides food for quantitative thought.

> After ten half-lives, one-thousandth of the original concentration [of a radioactive substance] is left; after 20 half-lives, one millionth. Generally 10–20 half-lives is called the hazardous life of the waste. Example: plutonium-239, which is in irradiated fuel [from a nuclear power plant], has a half-life of 24,400 years. It is dangerous for a quarter million years, or 12,000 human generations. [**R236**]

This is the kind of quotation that begs to have its numbers checked.

First let's look at "ten half-lives". After one half-life the concentration is half what it was at the start. After two half-lives it's half of a half, or 1/4. After three half-lives it's 1/8. So after 10 half-lives it's

$$\frac{1}{2} \times \frac{1}{2} \times \cdots \times \frac{1}{2} = \frac{1}{2^{10}}$$

of what it was.

We saw when we studied the metric prefixes that $2^{10} = 1,024 \approx 1,000$. That's why "kilo" means "1,000" most of the time but "1,024" when computers are involved. Now we use the same fact to see why $1/2^{10}$, which is exactly 1/1,024, is approximately 1/1,000 — one one-thousandth.

What about twenty half-lives? In that time the original concentration will be reduced to 1/1,000 of 1/1,000 of what it was at the start. Since a thousand thousand is a million, that's one one-millionth.

According the quotation, plutonium-239 will be dangerous for at least ten 24,000 year half-lives. That's 240,000 years, which is indeed about a quarter of a million years. Is it 12,000 generations? Yes, if you calculate with $240,000/12,000 = 20$ years per generation. That's perhaps a little low for the developed world, but good enough to highlight the danger of nuclear waste.

9.5 Exponential models

For compound interest and radioactive decay the equation for exponential change gives exact answers, just as the linear equation gives exact answers for simple interest and electricity bills.

When change is approximately linear a regression line may be useful. When it's approximately exponential, we can construct an *exponential trendline*. Most elementary texts discuss the reproduction of bacteria as a toy example of exponential growth. Bacteria reproduce by dividing, so each individual gives rise to 2, then 4, then 8 descendants, and so on. The number of bacteria grows exponentially. But that can't go on forever. Since exponential growth curves quickly become steep, it can't go on for very long. Eventually, crowding or diminishing resources cause growth to slow, perhaps even to reverse as organisms die faster than new ones are born.

Table 9.9 records the population of three different strains of the *Burkholderia ceno-cepacia* bacterium in a one day experiment conducted by Professor Vaughn Cooper at the University of New Hampshire.

time	population		
	W	R	S
4	7.9	8.5	17.5
8	17.3	13.5	42.1
12	44.7	48.9	225.2
17	119.3	268.7	407.3
20	41.3	98.0	149.3
24	41.3	64.0	160.0

Table 9.9. Bacteria growth [R237]

Time: hours
Population: millions of organisms per milliliter

The raw data points in the graph on the left in Figure 9.10 (built in the spreadsheet `bacteriaGrowth.xlsx`) suggest that the growth of each strain was exponential until about hour 17. To construct the graph on the right we plotted the data for each strain for that period. In the resulting chart we right-clicked on a data point for each strain, selected `Add Trendline . . .`, and chose `Exponential` on the `Type` tab. As usual, we asked for the equations and the *R*-squared values. Those are all pretty close to 1, best for strain W, worst for strain S.

Let's look at the equation for the exponential trendline for strain W, which you can see in the figure:

$$y = 3.3663 \, e^{0.2109x}. \tag{9.3}$$

What is the "*e*" in this equation? The complete answer to that question calls for much more mathematics than you need to know to apply common sense to quantitative arguments. But since Excel uses it you may encounter it somewhere so we'll discuss it briefly.

The simplest explanation is that *e* is just a particular number — approximately 2.7183. Like $\pi \approx 3.1416$ it's one of those numbers whose decimal expansion "goes on forever", so the first few decimal places give just an approximation. The number e appears naturally in discussions of exponential growth just as π appears in discussions of circles. It was named by the prolific mathematician Leonhard Euler (1707–1783) who was the first to recognize its importance.

Back to our exponential growth function for the strain W. We see that

$$e^{0.2109} \approx (2.7183)^{0.2109} \approx 1.2348.$$

You don't have to remember the value of e to compute with it, since Excel and LibreCalc and the Google calculator provide the built-in function EXP to do the job. Entering =EXP(0.2109) in a spreadsheet cell or the same thing without the equal sign in the Google search bar will give you the same answer.

Now we can substitute 1.2348 for $e^{0.2109}$ in (9.3) and rewrite it as

$$y = 3.3663 \times 1.2348^{x}.$$

In that form we recognize this behavior as exponential growth at a constant rate of about 23% per hour. The Rule of 70 tells us to expect a doubling time of about $70/23 \approx 3$ hours. That matches the data, which does indeed show the strain W population doubling about every three hours for most of the day.

If you experiment with EXP you will find that $e^{0.7} = 2.0137527 \ldots \approx 2$. It's the 0.7 in the exponent that leads to the Rule of 70. To learn just how, go on to take a course in calculus.

Using the exponential trendlines for constant growth rate to predict the future populations would fail in this experiment. About two thirds of the way through the 24 hour day a different reality appears. The gray bullets in Figure 9.10 show that all three populations drop dramatically. (A newspaper reporter wanting to emphasize the drama might say, incorrectly, that the populations dropped "exponentially.")

It's the biologist's job to understand why. The analysis of the exponential growth at the start has told him only how the bacteria grow when there's lots of food, lots of room and no competition.

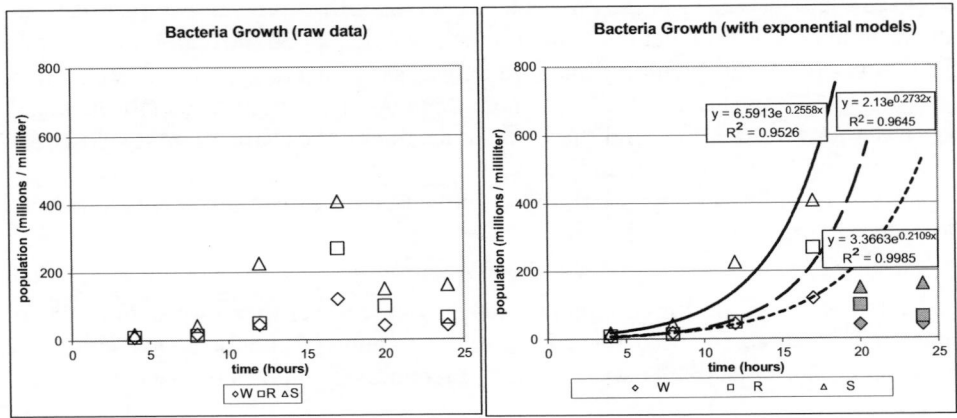

Figure 9.10. Bacteria growth

9.6 "Exponentially"

A Google search for a definition of *exponentially* finds this first meaning:

> 1. (with reference to an increase) more and more rapidly.
> "our business has been growing exponentially"

This is from *The New York Times* in April 2019:

> "The way ads are targeted today is radically different from the way it was done 10 or 15 years ago," said Frederike Kaltheuner, who heads the corporate exploitation program at Privacy International. "It's become exponentially more invasive, and most people are completely unaware of what kinds of data feeds into the targeting." [**R238**]

The *Daily Beast* seems particularly fond of the word:

> "[I]n 2009, the [Sketchbook Museum] project moved to New York and exponentially grew across the globe." [**R239**]

> "Exponentially powerful technologies are transforming our sphere of possibilities." [**R240**]

It's sad to see "exponentially" reduced to a bland adverb when we understand its precise mathematical meaning — the kind of growth captured in the formula for compound interest:

$$\text{balance} = \text{start} \times (1 + \text{rate})^{\text{time}}.$$

Even writers who know that "exponentially" must somehow involve an exponent can get the mathematics wrong. Victoria Markovitz wrote in a *National Geographic* article that

> [t]he amount of power you can produce [with a wind turbine] is determined by the square of the blade radius. That means increasing the size of the turbine has an exponential effect on power. [**R241**]

To think about this claim we put the power output equation from the linked post at www.nationalgeographic.com/environment/great-energy-challenge/2012/worlds-largest-wind-turbines-is-bigger-always-better/ in spreadsheet `turbinepower.xlsx`. Then we drew Figure 9.11 to see how power production P depends on blade length R for lengths up to 35 feet. The dotted curve predicts values using the square of the blade radius. The solid curve that grows much faster predicts values using an exponential trendline. The two equations are essentially

$$\text{quadratic} : \quad \text{power} = 0.0016\text{radius}^2$$

and

$$\text{exponential} : \quad \text{power} = 0.19e^{0.07\text{radius}} \approx 0.19(1.07)^{\text{radius}}.$$

When Markovitz writes that blade length has an "exponential effect on power" she is misusing the word.

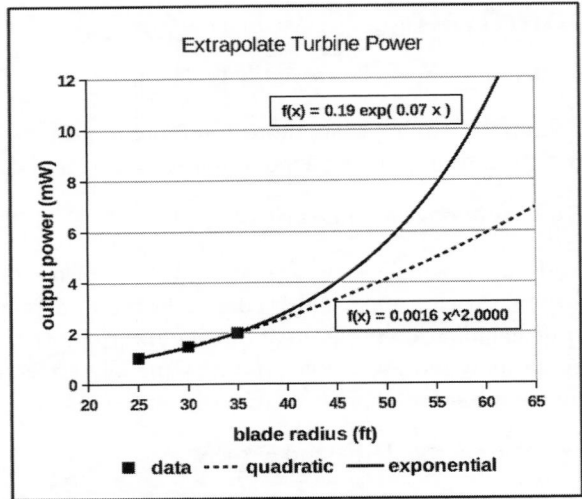

Figure 9.11. Turbine power

9.7 Exercises

Exercise 9.7.1. [S][R][Section 9.1] [Goal 9.1][Goal 9.2] Compound interest computations.

If you invest $1,500 at 7% interest compounded every year, how much will you have at the end of 10 years? 15 years? 20 years? Use the formula for exponential growth; then check your answers with the exponentialGrowth.xlsx spreadsheet.

Exercise 9.7.2. [S][Section 9.1][Goal 9.1] When do you expect exponential growth?

In each of the following situations, explain why you would expect linear or exponential growth.

Think about whether the change is best described as an absolute rate (like dollars per hour or gallons per mile) or a percentage (like percent per year or percent per washing).

Write the units for the kind of rate you decide on.

(a) Price increases from year to year due to inflation.

(b) How the amount of gas you use depends on how far you drive.

(c) The amount of money left on your public transportation debit card as the days go by and you commute to school or work.

(d) The amount of sales tax you pay, depending on how much you buy.

(e) The amount of dirt left in your kid's filthy jeans when you wash them over and over again.

(f) The population of the world as the years go by.

(g) Your credit card balance if you stop making payments. (We will study credit cards in the next chapter.)

(h) The height of the snow as it accumulates in a big storm.

(i) The number of people sick in the first weeks of the flu season.

(j) The number of subscribers to a hot new social network in its first days.

Think about your answers before you look at the hints.
[See the back of the book for a hint.]

Exercise 9.7.3. [S][Section 9.6][Goal 9.1] Is it really exponential?
In everyday usage the phrase "growing exponentially" is just a vibrant synonym for "growing rapidly". It's rare that it really means a constant relative change.
Find instances of "exponential" growth in the media where what's meant is just very rapid growth.

Exercise 9.7.4. [S][Section 9.1][Goal 9.1] Health care spending.
In Chapter 3, Exercise 3.8.16, we used data from the 2010 National Health Expenditures report to compute the absolute and relative changes in health care spending per person from 2007 to 2008.

(a) Use the results of those calculations to build linear and exponential models for the growth of health care spending per person.

(b) Use each model to predict when health care spending will reach $10,000 per person per year.

Exercise 9.7.5. [S][Section 9.1][Goal 9.1] Malthus.
In 1798 Thomas Malthus, an English economist and clergyman, wrote "An Essay on the Principle of Population". He said there:

I think I may fairly make two postulata.
First, That food is necessary to the existence of man.
Secondly, That the passion between the sexes is necessary and will remain nearly in its present state.
These two laws, ever since we have had any knowledge of mankind, appear to have been fixed laws of our nature, and, as we have not hitherto seen any alteration in them, we have no right to conclude that they will ever cease to be what they now are, without an immediate act of power in that Being who first arranged the system of the universe, and for the advantage of his creatures, still executes, according to fixed laws, all its various operations.
...
Assuming then my postulata as granted, I say, that the power of population is indefinitely greater than the power in the earth to produce subsistence for man.
Population, when unchecked, increases in a geometrical ratio. Subsistence increases only in an arithmetical ratio. A slight acquaintance with numbers will shew the immensity of the first power in comparison of the second.
...

The power of population is so superior to the power in the earth to produce subsistence for man, that premature death must in some shape or other visit the human race. The vices of mankind are active and able ministers of depopulation. They are the precursors in the great army of destruction; and often finish the dreadful work themselves. But should they fail in this war of extermination, sickly seasons, epidemics, pestilence, and plague, advance in terrific array, and sweep off their thousands and ten thousands. Should success be still incomplete, gigantic inevitable famine stalks in the rear, and with one mighty blow levels the population with the food of the world. [R242]

Malthus claimed that the food supply grows in a linear fashion. As a unit of food supply he used the amount of food needed for one person for one year. He estimated food production in Britain in 1798 as 7,000,000 food units and that food production might increase by a constant 280,000 units each year.

Malthus also believed that the population of Britain was growing at a rate of 2.8% each year. In 1798, the population was about 7,000,000.

(a) Write a linear function that models food production.

(b) Write an exponential function that models population growth.

(c) Was there enough food for each individual in Britain in 1798?

(d) Using Malthus's models, determine whether there would be enough food for each individual in Britain in 1800.

(e) Malthus claimed that the population in Britain would eventually outstrip the food supply — a prediction we now call "the Malthusian dilemma". He didn't have Excel to do the arithmetic for him, but we do. Use it to estimate when Malthus's predicted disaster would occur. Was Malthus right?

Exercise 9.7.6. [S][Section 9.1][Goal 9.1] [Goal 9.2] The pawn shop business model.
On April 9, 2011, *The New York Times* reported on a pawn shop that opened in an ex-Blockbuster store:

The borrowers are given 60 days to pay back the loan, and La Familia charges a 20 percent interest rate per month. (So for a $100 loan, the borrower would need to pay back $140 after 60 days.) [R243]

(a) Explain why 20% interest per month on a $100 loan for two months would actually require repayment of a little more than $140.

(b) What is the annual interest rate when this business lends money?

Exercise 9.7.7. [S][W][Section 9.2][Goal 9.1] [Goal 9.2] Playing with exponential growth.
Open the spreadsheet `exponentialGrowth.xlsx` and describe what happens to the graph when you make each of the following experiments. If you can see easily what happens to the numbers, describe that too.

(a) Change the value of START from 1,000 to 10, then 100, then 10,000. Change it to some other random positive values that aren't as nice.

(b) Change the value of START from 1,000 to $-1,000$.

(c) Change the value of RELCHANGE to 1.

(d) Change the value of RELCHANGE to 1.01 (1% growth). Fit a linear trendline to the data. What is the R-squared value? What does it tell you?

(e) Change the value of RELCHANGE to 2. Why does the graph look flat at 0 as far as $T = 20$? Is it really flat?

(f) Change the value of RELCHANGE to 10.

(g) Change the value of RELCHANGE to 0.9 (a 10% decrease).

(h) (Optional) Can you figure out how we got the label on the graph to incorporate the values of START and RELCHANGE?

Exercise 9.7.8. [S][Section 9.2][Goal 9.1] Five percent.

If you try to use linearExponential.xlsx to see when exponential growth at 5% catches up to linear growth you see that it's still behind at 20 years, which is as far as the table goes.

Modify the spreadsheet to determine when it catches up.

Exercise 9.7.9. [U][Section 9.2][Goal 9.1] [Goal 9.2] Playing with linear vs exponential growth.

Use the spreadsheet linearExponential.xlsx to answer these questions.

(a) How does changing the initial investment change the time it takes for the expo-nential function to catch up with the linear function?

(b) Does doubling or tripling both the initial investment and the absolute change affect the time it takes for the exponential function to catch up to the linear function?

Exercise 9.7.10. [S][Section 9.2][Goal 9.1] [Goal 9.2] Deals you can't believe.

The data in this problem aren't real. But the problem is interesting and instructive, so it's worth spending time on.

Suppose you are shopping for a car and find three deals advertised:

• Make a $10,000 down payment and pay only $100 per month for two years.

• Just $5,000 down, monthly payments start at a low $50 and increase by $50 each month for two years.

• Give me $1.00 today and take the car home! Pay 1 penny for the first month. Then double your payment each month. After two years, the car is yours.

(a) Before you do any calculating, which deal do you think is best? Why?

(b) What would your monthly payments be in the second and tenth months if you take the second dealer's offer?

(c) What would your monthly payments be in the second and tenth months if you take the third dealer's offer?

(d) For each deal, write an algebraic expression that gives the monthly payment.

Month	Payment		
	Deal 1	Deal 2	Deal 3
(down) 0	10,000	5,000	1.00
1	100	50	0.01
2			
⋮			
24			
Total			

Table 9.12. Three car deals

(e) Use Excel to calculate your total payments for the 24 months. Set up four columns as in Table 9.12. Then tell Excel how to fill in the columns to 24 months. Finally, use the SUM function to add up the payments.

(f) Now use what your calculations tell you to compare the three deals. Which is best? Which is worst?

Exercise 9.7.11. [S][C][Section 9.2][Goal 9.1] Green energy in China.

In the December 21 & 28, 2009, issue of *The New Yorker* Evan Osnos wrote in his essay "Green Giant: Beijing's crash program for clean energy" that China's spending on R&D, now seventy billion dollars a year, has been growing at an annual rate of about twenty percent for two decades. [**R244**]

(a) What does "R&D" stand for?

(b) Use Excel to build a chart of annual Chinese R&D expenditures for the years 1989–2008.

(c) Add a data column showing the annual expenditures adjusted for inflation (use the United States cost of living index) and display that data on your chart.

Exercise 9.7.12. [S][Section 9.3] [Goal 9.1] Car excise tax.

In Massachusetts you pay excise tax each year on the current value of your automobile. Assume for the sake of this problem that the rate is 3%, so you would pay $600 in excise tax in the first year you owned a new $20,000 car.

Use Excel to answer the following questions.

(a) Suppose the car depreciates linearly at a rate of $1,800 per year. What formula calculates the amount of excise tax you pay as a function of the age of the car?

(b) If you own the car for ten years, what will the car be worth then and how much total excise tax will you have paid?

(c) Answer the same questions if it depreciates at the rate of 13% per year.

(d) Find real data on the way a new car depreciates in value. Is an exponential model a good approximation?

Exercise 9.7.13. [S][Section 9.4][Goal 9.4] Iodine 131.

An article in *The New York Times* on April 6, 2011, soon after the Fukushima disaster discussed levels of radioactive iodine (iodine 131) in fish caught near Japan. The article noted that Japan recently revised the safety limit for iodine 131 in fish to 2,000 becquerels per kilogram. (A becquerel is a measure of radiation.)

Radioactive iodine has a half-life of about 8 days.

If a fish contained 10,000 becquerels of iodine 131 per kilogram, how long would it take for the iodine to decay to a "safe" level?

Exercise 9.7.14. [U][Section 9.4][Goal 9.1][Goal 9.4] Quadrupling time.

(a) Explain why the quadrupling time in exponential growth is just twice the doubling time.

(b) Show that quadrupling time is given by a "Rule of 140" analogous to the rule of 70.

Exercise 9.7.15. [S][Section 9.4][Goal 9.1][Goal 9.4][Goal 9.2] Tripling time.

Suppose you invest $1,000 at 10% interest compounded every year. (That's a pretty good rate of return if you can get it — don't trust a Madoff promise!)

(a) How long will it be until your balance is $3,000?

[See the back of the book for a hint.]

(b) Find the tripling time for some other interest rates.

(c) Check that the tripling time in exponential growth is given (approximately) by a "Rule of 110".

(d) Check that $e^{1.1} \approx 3$.

Exercise 9.7.16. [U][Section 9.4][Goal 9.4][Goal 9.2] Compounding very frequently.

(a) Calculate the effective rate for 8% annual interest when it's compounded weekly, daily, hourly and once every second.

(b) Estimate the effective rate if the interest is compounded every instant.

[See the back of the book for a hint.]

(c) Redo the calculations starting with a 25% annual increase. (Not realistic for interest on a bank account!) Show that the Rule of 70 for doubling times is more accurate the more frequently you compound the interest.

Exercise 9.7.17. [S][C][W][Section 9.4] [Goal 9.1][Goal 9.4] How fast does information double?

In the Preface to the Carnegie Corporation report *Writing to Read* Vartan Gregorian wrote that he's been told that the amount of available information doubles every two to three years. [R245]

(a) What growth rate in percent per year would lead to a doubling time of two to three years?

(b) Who is Vartan Gregorian?

(c) Can you verify his assertion?

[See the back of the book for a hint.]

Exercise 9.7.18. [S][Section 9.5] [Goal 9.5][Goal 9.4] Bacteria doubling time.

Find the approximate doubling times for strains R and S in the bacteria growth example in Section 9.5.

Exercise 9.7.19. [S][Section 9.5][Goal 9.1] [Goal 9.5] When will R catch S?

The population of strain S outnumbers that of strain R for the entire first 17 hours of the experiment discussed in Section 9.5. But the exponential trendline equation shows that strain R is growing faster. If the exponential growth were to continue (which it didn't) when would strain W catch up?

[See the back of the book for a hint.]

Exercise 9.7.20. [U][Section 9.5][Goal 9.5] The magic number e.

(a) Find the value of e in Excel using the formula =EXP(1).

(b) Find the value from Google with the same formula (without the equal sign). Check that the answers agree as far as they go together.

(c) Which provides more digits?

(d) Can you get more precision from Excel by formatting the cell in which the number appears?

(e) Find even more digits with an internet search.

Exercise 9.7.21. [S][Section 9.5][Goal 9.1] Educating mothers saves lives, study says.

On September 17, 2010, *The Boston Globe* carried an Associated Press report on a study that found that the death rate for children under 5 dropped by nearly 10 percent for every extra year of their mother's education. That education saved 4.2 million children in the developing countries in 2009.

The story continued with these baseline numbes:

> In 1970, women aged 18 to 44 in developing countries went to school for about two years. That rose to seven years in 2009. [R246]

(a) How much did the death rate for children under 5 decline from 1970 to 2009?

(b) Build as much as you can of the exponential model implicit in this quotation. What are the independent and dependent variables? What is the annual relative change?

Exercise 9.7.22. [S][Section 9.5][Goal 9.1] [Goal 9.5] Email.

On May 29, 2011, Virginia Heffernan blogged in *The New York Times* that the number of email accounts grew from about 15 million in the early 1990's to 569 million by December 1999 and that today [when she was writing] there are [were] more than 3 billion. [R247]

(a) Is this exponential growth?

(b) Can you use these numbers to make predictions?

[See the back of the book for a hint.]

Exercise 9.7.23. [S][Section 9.5][Goal 9.5] When will India pass China?

In an article dated April 1, 2011, on the website About.com you could read that India, the world's second largest country, had a population of 1.21 billion. India was expected to pass China by 2030, when it would have more than 1.53 billion people. China's population then would be just 1.46 billion.

The article noted that India's growth rate of 1.6% per year doubles its population in less than 44 years. [**R248**]

(a) Is the article correct in stating that an annual growth rate of 1.6% means India's population will double in 44 years?

(b) Assuming that India's growth rate remains 1.6% annually, what will its population be in 2030 when it surpasses China's?

(c) Assuming that India's growth rate remains 1.6% annually from 2011 on, what will its population be in the year 2100? Compare that figure to the current population of the world. Do you think India's growth rate can in fact continue at 1.6% for the 89 years from 2011 to 2100?

Exercise 9.7.24. [S][Section 9.5] [Goal 9.5] Health care expenditures grow.

The National Health Expenditures report, released in January 2009, stated that overall health care spending in the United States rose from $7,062 per person in 2006 to $7,421 per person in 2007.

(a) Calculate both the absolute change and percentage change in health care spending per person from 2006 to 2007.

(b) Using 2006 as your starting year (2006 = year 0), determine an exponential equation that calculates the amount of health care spending over time assuming the annual percentage change stays the same. Clearly identify the variable names and symbols in your equation.

(c) Using 2006 as your starting year (2006 = year 0), determine a linear equation that calculates the amount of health care spending over time assuming the annual absolute change stays the same. Clearly identify the variable names and symbols in your equation.

(d) Create an Excel spreadsheet to compare the two growth models' predictions for health care spending through the year 2021. Include a chart showing both models.

(e) Which model first predicts that U.S. health care spending will reach a level of $10,000 per person? In what year will that occur?

Exercise 9.7.25. [S][Section 9.5][Goal 9.1][Goal 9.5] Joe Seeley, *in memoriam*.

Joe Seeley died at age 50 in the fall of 2012.

He was a brave and witty blogger at `joes-blasts.blogspot.com/` throughout his hospitalization, creating virtual lemonade from the sourest of lemons. I think his words helped him; I know they helped those who cared for him to cheer him on. They will help the hospital staff care better for patients who come after him. And they will help you learn a little mathematics.

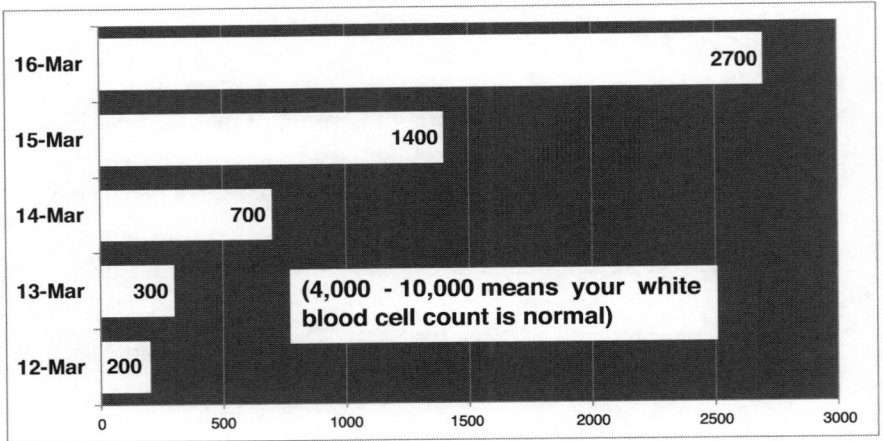

Figure 9.13. Proliferating white blood cells

Figure 9.13 appeared in the blog at a hopeful moment in his odyssey. It shows Joe's white blood cell counts on days following a stem cell transplant. He chose white for the bars, to symbolize white blood cells, and red for the background, for blood in general. I wrote him about it:

> March 18, 2011, 6:58 AM
>
> Ethan Bolker said …
>
> Exponential growth is good! … Will you still have a daily double after the predicted short dip? May I use your data for my quantitative reasoning class at UMass Boston?

> March 18, 2011, 9:48 AM
>
> Joseph Seeley said …
>
> I will not see doubling again, unless something is wrong. Over the next few months, the counts will rise and fall, sometimes for no reason that the doctors can determine.
>
> I hereby authorize the use of my blood count data for any and all educational purposes.

(a) Enter the data in Excel. Reproduce Joe's chart. Match the formatting (labels, colors, sizes, fonts) as well as you can.

(b) Create an exponential trendline for the data.

(c) Use common sense or your trendline to predict when Joe's white blood cell count will be in the normal range.

(d) His white blood cell count on March 17 was 5,100. Does that match your prediction?

(e) Modify your chart to include this new information. Mark it with a suitable exclamation!

Exercise 9.7.26. [S][Section 9.3][Goal 9.3] How impatient are you?
The Boston Globe reported on September 24, 2012, that

> [MIT grad Andy Berkheimer] found that [YouTube] viewers start clos-
> ing out if there's even a two-second delay. Every one-second delay af-
> ter that results in a 5.8 percent increase in the number of people who
> give up. A 40-second wait costs a video nearly a third of its audience.
> [**R249**]

Show that at this rate more than ninety percent of the viewers would give up after
40 seconds — not the "nearly a third" in the quote.

Exercise 9.7.27. [U][Section 9.1][Goal 9.1] Does compounding always matter?
On page 32 of his excellent and highly recommended *The Signal and the Noise* Nate
Silver wrote

> [Over the] 100-year-period from 1896 through 1996 ...sale prices of
> houses had increased by just 6 percent *total* after inflation, or about
> 0.06 percent annually. [**R250**]

Silver clearly divided six percent by 100 to reach his 0.06 percent annually conclusion. But that's not how percent increases work. The 0.06 percent annual increase
must be compounded.

Would taking compounding into account change Silver's fundamental point?

Exercise 9.7.28. [S][Section 9.1][Goal 9.2] Cuba, you owe us $7 billion.
On April 18, 2014, Leon Neyfakh wrote in *The Boston Globe* that property confis-
cated by the Cuban government in the 1959 revolution was

> ...originally valued at $1.8 billion, which at 6 percent simple interest
> translates to nearly $7 billion today. [**R251**]

(a) Is the simple interest calculation in the quotation correct?

(b) What would the value be today at 6 percent compound interest?

(c) What would the value be today simply taking inflation into account?

(d) Discuss which of the three valuations makes the most sense.

Exercise 9.7.29. [S][Section 9.1][Goal 9.2] "As Time Goes By".
On December 13, 2012, you could read in *The New York Times* that the piano from
Rick's place in the 1942 movie *Casablanca* is up for auction.

> Sotheby's expects [it] to sell from $800,000 to $1.2 million in the auc-
> tion on Friday. That is between 34 to 48 times what [Ingrid] Bergman
> was paid for sharing top billing with Humphrey Bogart. [**R252**]

(a) How much was Ingrid Bergman paid for her role in the film? Calculate this two
ways using the data in the quotation and comment on what you discover.

(b) Would adjusting her pay to take inflation into account allow her to bid on the piano
in 2012?

(c) What compound interest rate would she have to have earned on her pay to bid on
the piano in 2012?

(d) Find out what happened at the auction.

Review exercises.

Exercise 9.7.30. [A] You invest $500 in an account that earns $10 in interest each year.

(a) At the end of 24 months, what is the balance?

(b) At the end of 30 months, what is the balance?

(c) At the end of 5 years, what is the balance?

(d) Find the linear equation that gives the balance after t years.

Exercise 9.7.31. [A] You buy a car for $15,000 and for tax purposes you depreciate it at a rate of 11% per year.

(a) At the end of 24 months, what is the value of the car?

(b) At the end of 5 years, what is the value of the car?

(c) Find the exponential equation that gives the value of the car after t years.

(d) Does the value of the car ever reach $0?

Exercise 9.7.32. [A] Calculate the percentage.

(a) What is 8% of $2,000?

(b) What is 108% of $2,000?

(c) What is 3.25% of $800?

(d) What is 103.25% of $800?

Exercise 9.7.33. [A] Use a calculator to evaluate these expressions using exponents. (You may find typing into the Google or Bing calculator much faster than using one that requires you to press keys, either with your fingers or with a mouse.)

(a) 1.03^4

(b) 0.89^5

(c) 140×1.03^4

(d) 80×0.89^5

(e) $\frac{1}{3^8}$

(f) $\left(\frac{1}{3}\right)^8$

(g) 1.25^0

(h) 1.25^1

(i) e^2

(j) e^{15}

Exercise 9.7.34. [A] In the exponential functions below, identify the relative change and the initial amount.

(a) $P = 100 \times (1.05)^T$.

(b) $y = 400 \times (0.88)^x$.

(c) $S = 550 \times (1.22)^Q$.

(d) $P = 96 \times (0.50)^T$.

Exercise 9.7.35. [A] Excel gives the following best-fit exponential function for a set of data: $y = 2.099 \times e^{1.344x}$. Find the constant growth rate and rewrite the function without using e.

Exercises added for the second edition.

Exercise 9.7.36. [U][S][Section 9.1][Goal 9.1][Goal 9.2] There's more next year. Lewis Carroll's Professor talks with Sylvie and Bruno:

> "Come in!" [said the Professor]
>
> "Only the tailor, Sir, with your little bill," said a meek voice outside the door.
>
> "Ah, well, I can soon settle his business," the Professor said to the children, "if you'll just wait a minute. How much is it, this year, my man?" The tailor had come in while he was speaking.
>
> "Well, it's been a doubling so many years, you see," the tailor replied, a little gruffly, "and I think I'd like the money now. It's two thousand pound, it is!"
>
> "Oh, that's nothing!" the Professor carelessly remarked, feeling in his pocket, as if he always carried at least that amount about with him. "But wouldn't you like to wait just another year, and make it four thousand? Just think how rich you'd be! Why, you might be a King, if you liked!"
>
> "I don't know as I'd care about being a King," the man said thoughtfully. "But it dew sound a powerful sight o' money! Well, I think I'll wait —"
>
> "Of course you will!" said the Professor. "There's good sense in you, I see. Good-day to you, my man!"
>
> "Will you ever have to pay him that four thousand pounds?" Sylvie asked as the door closed on the departing creditor.
>
> "Never, my child!" the Professor replied emphatically. "He'll go on doubling it, till he dies. You see it's always worth while waiting another year, to get twice as much money!" [**R253**]

(a) Use the fact that $2^{10} \approx 1{,}000$ to estimate the original tailor's bill if it's been doubling for five years.

(b) Check your estimate with a precise calculation.

Exercise 9.7.37. [U][S][Section 9.4][Goal 9.4] Light pollution.

On August 18, 2019, Kelsey Johnson wrote in *The New York Times* that

> [T]he global amount of artificial light at night has been growing by
> at least 2 percent per year. At this rate the amount of light pollution
> originating from Earth-based sources alone will double in less than
> 50 years. [**R254**]

In a more technical discussion linked from that article you can read Jeff Hecht's
observation that

> A new analysis of satellite data from the past four years shows that
> the total acreage lit by artificial light at night increased by an average
> of 2.2 percent a year. The brightness of the areas lit at the start of
> the study also increased by the same rate — 2.2 percent annually —
> around the globe. [**R255**]

(a) Use the Rule of 70 to find a better estimate of the light pollution doubling time.

(b) How do you think Johnson arrived at the 50 year doubling time?

(c) Use the data in the quote from Hecht to explain why the annual light pollution
increase might be better reported as almost four and a half percent.

Exercise 9.7.38. [S][U][Section 9.4][Goal 9.4] Thirty four years at seven percent.

From *Swan Boats at Four*, a novel by George V. Higgins:

> Rutledge said "In other words, if we'd painted over that damned pic-
> ture in the summer of nineteen seventy-eight, we would've made the
> club, and ourselves individually, liable for a hundred thousand bucks,
> plus interest at, say, an average of seven percent per annum, com-
> pounded for thirty-four years. ..."
>   "Offhand," [David] said, "I can't even imagine how much that
> would've been."
>   "At the time, I couldn't either," Rutledge said, " ...so we looked it
> up — I don't mean we figured it out. ...I don't recall the exact figure,
> but it came out to around a million and a half dollars." [**R256**]

David is a banker. He would know the Rule of 70 and figure it out offhand, without
pencil and paper. Higgins should have known that.

(a) Use the Rule of 70 to decide whether Rutledge was right when he said the figure
was "around a million and a half dollars".

(b) Calculate the liability accurately.

Exercise 9.7.39. [U][S][Section 9.3][Goal 9.1][Goal 9.3] Disrupting the cow.

A November 1919 article in *The Boston Globe* reported that the cost of precision
fermentation (PF), a process for growing meat in a laboratory, was falling "exponentially"

> — from $1 million per kilogram in 2000 to about $100 today. With
> the technologies we have today, we project these costs will fall even
> lower — to $10 per kilogram by 2023–25. [**R257**]

(a) Use the data from the *Globe* to verify the statement

> The cost of PF falls by a factor of about 1/100 per decade.

(b) Rewrite the statement in (a) using an assertion about percentage change.

(c) Explain why the cost of PF is falling by a factor of about 1/10 every five years, to 10% of what it starts at.

(d) Is the *Globe* article right when it projects $10 per kilogram by 2023–25?

10

Borrowing and Saving

When you borrow money — on your credit card, for tuition, for a mortgage — you pay it back in installments. Otherwise what you owe would grow exponentially. In this chapter we explore the mathematics that describes paying off your debt.

Chapter goals:

Goal 10.1. Examine how debit and credit cards work.

Goal 10.2. Study balance and interest when paying off a loan periodically.

Goal 10.3. Calculate monthly mortgage payments and examine the costs and benefits of home ownership.

Goal 10.4. Understand periodic compounding, APR and other interest terms.

Goal 10.5. Understand the basics of saving money with a long-term goal like retirement.

10.1 Debit and credit cards

In the old days people shopped with paper money. Now the answer to "paper or plastic?" is more likely to be "plastic" — or "neither," if you pay with an app.

The infrastructure that supports the convenience of a cash-free economy doesn't come free. In this section we will look at how it is paid for.

Debit cards. You open a bank account and deposit money. The bank gives you a debit card — magic plastic that you present at a coffee shop in exchange for a $3 latte. The bank sends the shop $3 from your account, after subtracting a processing fee.

> [A] merchant who accepts a swiped Visa debit payment from a customer would pay either 0.80% plus $0.15 or 0.05% plus $0.21 in Interchange fees for that transaction. [**R258**]

The fee for your $3 latte would be about 18 or 21 cents, depending on the option, so about 6%. The percentage would be less on a larger purchase.

Merchants take debit card fees into account when setting prices, so you pay them, indirectly. They are a reasonable price to pay for the convenience of the card, as long as you have the money in your account. If you don't, the bank can pay the merchant anyway, collect from you later, and charge you an *overdraft fee* for the extra convenience.

The online bank *Chime* reported that

> in 2014, the Consumer Financial Protection Bureau (CFPB) found that the majority of overdraft fees were charged on transactions of $24 or less. With a median fee of $34 at the time, the same type of charge on a loan for a similar three day period would result in an annual percentage rate (APR) of 17,000%. [**R259**]

Let's check the arithmetic. If the bank honors your $24 purchase even when you have no money in your account they are lending you that money until you pay back $24 + $34 = $58 three days later. The $34 fee corresponds to an interest rate of

$$\frac{\$34}{\$24} \approx 1.42 = 142\%.$$

Since there are 120 three day periods in a year, the annual percentage rate is about

$$120 \times 142\% \approx 17{,}000\%.$$

That agrees with what Chime reported.

You can avoid borrowing money at that interest rate.

> Banks get to decide either to cover or reject a transaction that would make your balance negative, but you can control one thing. Opting out of an overdraft coverage program means that your bank cannot cover one-time debit card or ATM transactions or charge overdraft fees on them. [**R260**]

Forego the latte if you can't pay for it right now:

Opt out of overdraft coverage.

Credit cards. When you pay with a debit card you spend money you already have. When you pay with a credit card you promise to pay later. Your bill explains the interest you're charged for borrowing that money. Figure 10.1 shows a sample credit card statement.

If you have a credit card you get a statement like this once a month. Glyphne Muse[1] (the owner of this card) charged $125.24 in merchandise and services during January. She's decided not to use this card any longer and will settle her debt by paying $20 each month. When will she be debt free, and how much interest will she have paid?

She makes a minimum payment of $20 in February for her January purchases, so her balance is $125.24 − $20 = $105.24. She's paid no interest so far. But now that changes. The credit card company charges interest on the balance she carries in February. The FINANCE CHARGE SUMMARY shows a periodic (that is, monthly)

---

[1] Eleanor Bolker discovered Glyphne, the muse of graffiti.

```
                          BANK OF THE PELOPONNESE
                           CREDIT CARD STATEMENT

  ACCOUNT NUMBER        NAME              STATEMENT DATE    PAYMENT DUE DATE
  314159265359          Glyphne Muse      02/01/20          03/01/20

  CREDIT LINE           CREDIT AVAILABLE  NEW BALANCE       MINIMUM PAYMENT DUE
  $1,200.00             $1,074.76         $125.24           $20.00
```

REFERENCE	SOLD	POSTED	ACTIVITY SINCE LAST STATEMENT	AMOUNT
2P71828182		1/25	PAYMENT THANK YOU	-166.80
1P41421356	1/12	1/16	MERCURY TRANSPORT	14.83
0P69314718	1/13	1/16	HAMMER & CHISEL HARDWARE	30.55
1P73205080	1/18	1/18	MADAME DELPHI	27.50
6P02214E23	1/20	1/22	OLYMPIA DINER	12.26
6P6260E-34	1/28	1/30	CALLIOPE CONSULTING	40.10

```
  Previous Balance   (+)      166.80        Current Amount Due       125.24
  Purchases          (+)      125.24        Amount Past Due
  Cash Advance       (+)                    Amount Over Credit Line
  Payments           (-)      166.80        Minimum Payment Due       20.00
  Credits            (-)
  FINANCE CHARGES    (+)
  Late Chartes       (-)
  NEW BALANCE        (-)      125.24
```

FINANCE CHARGE SUMMARY	PURCHASES	ADVANCES
Periodic Rate	1.65%	0.54%
Annual Percentage Rate	19.80%	6.48%

Figure 10.1. A credit card statement [**R261**]

rate of 1.65% so she will pay $105.24 \times 0.0165 = \$1.74$ in interest. The 1+ trick tell us that at the beginning of March she owes

$$\$105.24 \times 1.0165 = \$106.98.$$

After her $20 payment on March 1 her balance is

$$\$105.24 \times 1.0165 - \$20 = \$86.98.$$

The credit card company used $1.74 of the payment for the February interest. The rest they subtracted from her balance.

Table 10.2 tells the rest of the story. Figure 10.3 shows the Excel formulas we used in PayOffDebt.xlsx to calculate the values in that table, along with a graph showing how the balance decreases each month at a slightly faster rate, until it reaches 0.

Glyphne's last payment is for the $10.86 balance.

In seven months she's paid $5.80 in interest. That doesn't seem too terrible. It's $5.80/\$125.24 = 0.046311 \approx 4.6\%$. But don't be fooled. This is a monthly statement, so that 4.6% isn't the annual interest rate. Glyphne didn't borrow that money for a year. Some of it she had for seven months, some for just one.

The law requires the credit card company to tell you the *APR* (*annual percentage rate*) somewhere on the monthly statement. You can find it on this one in the **FINANCE CHARGE SUMMARY** section: it's 19.80%. You can check their arithmetic: $1.65\% \times 12 = 19.80\%$. We will have more to say about the APR in Section 10.4.

That large interest rate is why the credit card company wants you to pay just the small minimum each month. The full balance appears at the top of the statement labeled NEW BALANCE — but the only payment shown is the MINIMUM PAYMENT DUE. You have to know that it's best for you to pay the full balance at once.

Month	Balance	Interest
Jan	125.24	0.00
Feb	105.24	1.74
Mar	86.98	1.44
Apr	68.41	1.13
May	49.54	0.82
Jun	30.36	0.50
Jul	10.86	0.00
total		5.80

Table 10.2. Seven months to pay it off

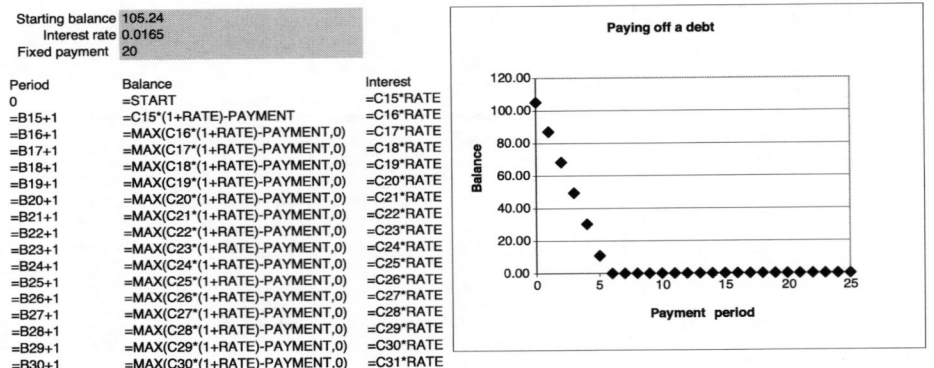

Figure 10.3. Paying off credit card debt

There are other ways credit card companies make you pay for the convenience of borrowing their money. One kicks in if you miss a payment by even a day or so. Then they charge a substantial late payment fee and may also increase the already large interest rate. The law requires credit card companies to print a warning on your statement. Here's what one says:

> **Late Payment Warning:** If we do not receive your minimum payment by the date listed above you may have to pay up to a $39.00 late fee and your APRs will be subject to increase to a maximum Penalty APR of 29.99%.

Moreover, the late payment will show up on your credit report, so when you go to a bank later to take out a mortgage on the condo you want to buy, they may charge you a higher interest rate too.

Does all this mean that using a credit card is a bad idea? No, as long as you're careful. Then you can take advantage of some of the good things credit can do for you:

- Glyphne's statement shows that she paid her last balance of $168.80 on time in full, avoiding all finance charges. So, in fact, she borrowed that money for a month from the credit card company at no cost. If she kept it in a savings account until it was time to pay her bill she'd even have made a few pennies in the meanwhile.

- Some credit cards give you back a reward at the end of the year — perhaps 1% of your total purchase dollars or some frequent flyer airline miles.

- Merchants pay credit card service fees like those for debit cards, but usually larger — perhaps 2%. When making a large purchase you may be able to negotiate a discount for paying by cash or check. Even when you pay the full price you may want to save the local merchant the fee.

- If you miss a payment your credit rating may suffer. But just avoiding credit errors won't get you a good credit rating. For that you have to prove you can manage debt — by having a credit card and paying the balance in full when due. Then when it's time to borrow money for a car or a condo your good credit rating may get you a lower interest rate.

- If you have a balance on an existing card that you can't afford to pay off immediately, consider opening a second card and transferring the balance. The new card company may offer you 0% interest for a while to encourage you to switch. If you do that and then don't use the new card you can pay off the old balance over time without any further interest charges. Be sure to read the small print before you do this — the transferred debt may be interest free, but often there's a charge (perhaps three or four percent) to make the transfer.

- Federal legislation passed in response to the 2009 financial crisis forced credit card companies to change their policies so that "payments above the Minimum Payment Due will be applied first to higher interest rate balances." That notification appeared on one of the authors' statements, along with the kind thought that "This may help you to pay off your highest interest rate balances more quickly and reduce your interest charges." They did not reveal how much money they spent lobbying in Washington against the regulation.

- Finally, you may find a credit card issued by one of your favorite charities. Then the charity collects a small fraction of the fees the merchants pay.

Do remember:

> Pay your full balance on time every month.

You can even arrange to have that happen automatically from your bank account, so you don't have to remember and you save the cost of a stamp. Just make sure there's enough money in the bank.

10.2 Can you afford a mortgage?

There's a $250,000 condominium in Denver you want to buy. You've managed to scrape together $50,000 for the down payment (savings, your parents, ...) but will have to take out a mortgage for the $200,000 balance. Can you afford it? There are many websites that provide a place to start. We visited smartasset.com/mortgage/colorado-mortgage-calculator, filled out the Mortgage Calculator and discovered that on June 3, 2019, in Denver, Colorado you could get a 30 year fixed rate mortgage at 4.38% annual interest with a monthly payment of $999 or a 15 year fixed rate mortgage at 3.88% with a monthly payment of $1,467.

In this section we'll look at what those numbers mean, see how they are calculated and discuss a few important issues (some quantitative, some not) that you should think about when making a decision like this one.

Paying off a mortgage is like paying off a credit card balance when you make no new purchases. There's an annual rate. Your balance at the end of a month includes interest computed at one twelfth of the annual rate. Each month you pay all the current interest and some of the principal. Since the principal is decreasing, there's less interest each month so more of the payment goes toward the principal. One difference is that the credit card company sets the minimum payment; then it takes as long as it takes to pay off the balance, while the mortgage payment is figured out in advance so that everything is paid off at a particular time — usually 15 or 30 years.

The mortgage company uses this formula to calculate the monthly payment:

$$P \times \frac{r/12}{1 - \left(1 + \frac{r}{12}\right)^{-12y}} \tag{10.1}$$

where P is the principal (the amount of your mortgage), r is the annual interest rate, and y is the length of the mortgage, in years.

It is probably the most complicated formula in *Common Sense Mathematics*. We won't explain where it comes from, and you need not memorize it. But you can understand some parts of it. It has the form

$$P \times (\text{complex expression involving } r \text{ and } y)$$

which tells you that your monthly payment is proportional to P. The complex part is the expression in parentheses — the *effective monthly rate*. That's the number of dollars in your payment for each dollar you borrow. There the $r/12$ finds the monthly rate from the annual rate. The product $12y$ is the number of months in y years.

You can use that formula to check that the Wells Fargo calculator finds the right monthly payment of $984 for a 30 year $200,000 loan at 4.250% interest. We did the arithmetic in Excel, with the formula

```
=(STARTBALANCE*INTERESTRATE/12)/(1-(1+INTERESTRATE/12)^(-12*YEARS))
```

in cell C11 on the `mortgage` worksheet in `PayOffDebt.xlsx`. There you can see the principal balance at the end of each year and the total interest paid. On that 30 year mortgage it's $154,196.72.

When you borrow you always pay back more than the amount you borrowed — in this case, $150,000 in interest in addition to the $250,000 principal. Should that frighten you? Maybe or maybe not. Is it worth it? Perhaps, for several reasons.

- It would take a long time to save up the full purchase price (to avoid borrowing). Saving would be difficult because you would be paying rent the whole time. So you can think of the mortgage payments as money spent instead of rent.

- The condo may well be worth more after 15 or 30 years than the total you paid for it — even including the interest on the mortgage.

- Inflation is pretty nearly inevitable over the years. These computations are all made in dollars computed in the year you make the purchase, but the actual value of that money when you pay it to the bank in later years will be less, in then-current dollars. Think of it this way: your salary is likely to increase at least as fast as inflation, so the fixed monthly mortgage payments will be a smaller and smaller percentage of your take-home pay.

- That said, you do want to minimize the amount of interest you pay, by paying attention to the significant difference between a 15 and a 30 year mortgage. The short one has a lower interest rate (2.5% instead of 3.5%) and a much lower total interest cost: about $40,000 instead of $123,000. So you should choose it if you can manage the extra $430 per month in payments.

- You will also get a lower rate if you have established a good credit rating in the years before you apply for your mortgage. So start now to use a credit card wisely.

Words of warning. This discussion shows how, in principle, you pay off a loan by paying some interest and some principal periodically. That's just one of the financial things you'll need to understand when you think about buying a house or condo. Just asking the bank or shopping online for an interest rate isn't sufficient. As with most other topics in this book our hope is to provide a quantitative foundation for further questions. Some of those will address these issues:

- There are other up-front costs: legal fees, title searches, inspections, points.

- The cost of owning is more than just the cost of the mortgage. You must be prepared for expenses that your landlord would cover if you were renting — things like real estate tax, insurance, repairs.

- Variable rate mortgages generally start out with lower rates than fixed rate mortgages — but payments can balloon when the initial rate expires.

There are many books and webpages that may help — here's just one we found with a simple search: `www.ourfamilyplace.com/homebuyer/checklist.html` .

10.3 Saving for college or retirement

In Chapter 9 we studied how money accumulates when you invest a big chunk and let the interest compound. But you rarely save a big chunk of money all at once. A more realistic way to save, for college (for your children) or for retirement, is to put away a fixed amount on a regular basis.

The kind of calculations we made above to study paying off a debt will help now to study how money saved regularly accumulates. Suppose you can invest $1,200 a year and make your payment at the end of the year.

You think you can get a 6% return on your investment, since you're willing to take some short-term risk for the sake of long-term return. At the end of the second year you will have

$$\$1,200 \times 1.06 + \$1,200 = \$2,472$$

and after the third

$$\$2,472 \times 1.06 + \$1,200 = \$3,820.32.$$

These calculations look just like the ones we made for paying down credit card debt, except that in this case we add the periodic payment to the balance rather than subtracting it. That means we can use the debt payment spreadsheet to see how our

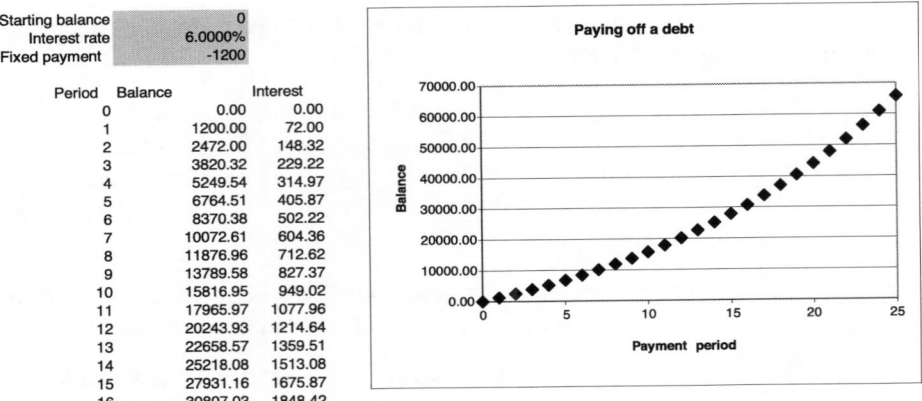

Starting balance	0
Interest rate	6.0000%
Fixed payment	-1200

Period	Balance	Interest
0	0.00	0.00
1	1200.00	72.00
2	2472.00	148.32
3	3820.32	229.22
4	5249.54	314.97
5	6764.51	405.87
6	8370.38	502.22
7	10072.61	604.36
8	11876.96	712.62
9	13789.58	827.37
10	15816.95	949.02
11	17965.97	1077.96
12	20243.93	1214.64
13	22658.57	1359.51
14	25218.08	1513.08
15	27931.16	1675.87
16	30807.03	1848.42

Figure 10.4. Saving for retirement

money accumulates by entering a negative "payment" to be added to the growing balance. Figure 10.4 shows the result (we haven't changed the labels).

In 25 years you will have accumulated nearly $66,000. Of that amount, you contributed just $1,200 × 25 = $30,000. The rest, more than $35,000, is interest. Making precise sense of the total accumulation in cell C40 and the total interest in cell D41 is a little tricky. You have to think carefully about whether these values are computed at the beginning or the end of the year, before or after interest is credited. You don't need to do this detailed analysis to understand the principle.

10.4 Effective interest rate

We discovered in Chapter 9 that compounding is a good thing for your investments. When you have a balance on your credit card it's a good thing for the credit card company. The sample bill in Section 10.1 lists the Annual Percentage Rate (APR) as 19.08% and the Periodic Rate as 19.08%/12 = 1.65%.

At the end of the year interest charged monthly will have been compounded twelve times. Since
$$1.0165^{12} = 1.217,$$

the *effective annual rate* (EAR) is 21.7% — almost 2 full percentage points higher than the already substantial advertised APR of 19.8%. In Europe credit cards must report the EAR where we see the APR.

The effective rate works for you rather than against you when you invest rather than borrow. In the last section we looked at saving for retirement by saving $1,200 a year at 6% interest, paid once a year. You shouldn't be surprised that monthly compounding will earn you interest on your interest if you save $100 each month instead. Then the computation

$$\left(1 + \frac{0.06}{12}\right)^{12} = 1.0616778119 \approx 1.0617 \qquad (10.2)$$

tells you your effective interest rate is about 6.17%. It's not a coincidence that the left side of this equation matches part of the complicated formula in equation (10.1).

The effective annual rate for the already outrageous debit card fee APR of 17,000% is unimaginable:

$$142^{120} \approx 1.88 \times 10^{258},$$

which is 188 followed by 256 zeroes.

Whether you think of the true interest rate as the APR or the EAR the moral is clear. Don't spend small sums of money you don't have. Tell your bank to refuse payment when your account is empty. Go without that particular cup of coffee.

10.5 Instantaneous compounding

We have just spent some time working with expressions like the one in equation (10.2), which quantifies the advantage of montnly compounding: 6% is effectively 6.17%.

If monthly compounding is good then daily compounding must be even better. To see what six percent annual interest compounded daily leads to, compute

$$\left(1 + \frac{0.06}{365}\right)^{365} = 1.06183131.$$

That corresponds to an effective annual rate of 6.183%. Hourly compounding gives

$$\left(1 + \frac{0.06}{8765}\right)^{8765} = 1.06183633,$$

which is just a tiny bit better. Compounding every minute results in 1.06183654, which differs only in the seventh decimal place. These computations suggest that as you compound more and more often you do better and better, but by less and less. There seems to be a limit. In fact there is. You can find it with the magic number e and the Excel function EXP we discussed in Section 9.5. If you compound 6% annual interest *every instant* the computation

$$e^{0.06} = \text{EXP}(0.06) = 1.061836547$$

tells you the effective interest rate to nine decimal places.

To compare monthly and instantaneous compounding in terms that are easier to understand, suppose you invested a thousand dollars. Then the Google calculator tells you

$$\boxed{1\,000 * ((e\char`\^0.06) - ((1 + 0.06/12)\char`\^12)) = 0.158734681}$$

so the difference after one year is about 16 cents. That's chump change for a thousand dollar investment.

Finally, suppose you could find someone to pay you 100% interest annually. Then without compounding, one dollar would double and become two. If you compounded instantaneously your dollar would turn into $e = \text{EXP}(1) = 2.72$ dollars in a year.

10.6 Exercises

Exercise 10.6.1. [W][Section 10.1][Goal 10.1] Your credit report.

> The Fair Credit Reporting Act (FCRA) requires each of the nation-wide credit reporting companies — Equifax, Experian, and TransUnion — to provide you with a free copy of your credit report,

at your request, once every 12 months. The FCRA promotes the accuracy and privacy of information in the files of the nation's credit reporting companies. The Federal Trade Commission (FTC), the nation's consumer protection agency, enforces the FCRA with respect to credit reporting companies. [**R262**]

To get your reports, visit `www.annualcreditreport.com/index.action`. That's not a clickable link. You have to type it into your browser. Here's why, from `www.annualcreditreport.com/aboutThisSite.action`.

AnnualCreditReport.com is the official site to get your free annual credit reports. This right is guaranteed by Federal law. To verify that this is the official site, visit `www.consumerfinance.gov/askcfpb/311/how-do-i-get-a-copy-of-my-credit-report.html`.

Don't be fooled by look-alike sites. You can be sure that you are on the right site if you type `www.annualcreditreport.com` in your browser address line. Don't come to this site by clicking on a link in another site or in an email.

Now you have your credit *report*. That's not the same as credit *score*.

Based on the information in your credit report, lenders calculate your credit score so they can assess the risk you pose to them before they decide whether they will give you credit. The higher your score, the less risk you pose to creditors.

The information in your credit report is used to calculate your FICO (the acronym stands for Fair, Isaac and Company) score. Your score can range anywhere from 300–850. Aiming for a score in the 700s will put you in good standing. A high score, for example, makes it easier for you to obtain a loan, rent an apartment or lower your insurance rate. [**R263**]

Look for a place on the web that will give you an estimate of your credit score. Wherever you go, be sure to read the fine print, and don't pay for anything. Some credit and debit card companies provide an updated score to their customers.

Without revealing any of your personal information, write about how easy or difficult it was to estimate your credit score. What did you learn as you did this research?

Exercise 10.6.2. [S][Section 10.1][Goal 10.1] How long to pay it off?

Starting in 2010 credit card companies were required to provide the information in Table 10.5 each month. The numbers there are for a bill with a balance of $2,020.37, a minimum payment amount of $40.00 and an annual percentage rate of 12.24%.

If you use the `PayOffDebt.xlsx` spreadsheet to work on this exercise you will need the `mortgage` worksheet, since the `plain` worksheet only covers 25 payment periods.

(a) Verify the three year time to pay off the balance at a rate of $67 per month.

(b) Show that a constant monthly payment of $40 is much more than is needed to pay off the balance in 18 years. How can the 18 year claim be correct?

If you make no additional charges using this card and each month you pay ...	You will pay off the balance shown on this statement in about ...	And you will end up paying an estimated total of ...
Only the minimum payment	18 years	$3,843
$67	3 years	$2,426 (Savings = $1,417)

Table 10.5. Paying off a credit card balance

(c) The 2010 Consumer Credit Law allows banks to raise the minimum payment on an account to a constant amount sufficient to pay off the balance in five years. What would that minimum payment be for this bill?

Exercise 10.6.3. [U][R][Section 10.2][Goal 10.3] Build your own mortgage.

Redo the computations in Section 10.2 for a house or condo of your choice in your town. Start with a reasonable cost and down payment. Find rates from at least two separate online sites; check them with the formula and the PayOffDebt.xlsx spreadsheet.

Exercise 10.6.4. [U][Section 10.2][Goal 10.3][Goal 10.2] Using the debt payoff spreadsheet.

The debt payoff spreadsheet can reproduce some of the computations from the exponential growth spreadsheet we introduced in Chapter 9. Test that by setting the monthly payment to 0 and the annual interest rate to 12 times the growth rate you want to study.

In particular, what happens if the annual growth rate is 1,200%, the starting balance is 1 and the monthly payment is 0?

Exercise 10.6.5. [R][S][Section 10.2][Goal 10.2] [Goal 10.3] Jumbo loans.

On November 20, 2010, a story in *The Boston Globe* headlined "Rates for big loans tumble" said that

> Over the past year, the average interest rate for so-called jumbo loans — $523,750 and up in the Boston area — has fallen from 6 percent to about 5 percent for a 30-year, fixed-rate mortgage. That translates into a monthly savings of about $375 on a $600,000 loan. [**R264**]

(a) What monthly payment will retire the loan when the interest rate is 6%?

(b) What monthly payment will retire the loan when the interest rate is 5%?

(c) Is the newspaper's claim of a $375 monthly saving correct?

Exercise 10.6.6. [R][S][Section 10.2][Goal 10.2] [Goal 10.3][Goal 10.4] Mortgages in the news.

A March 3, 2011, article in *The New York Times* headlined "Without Loan Giants, 30-Year Mortgage May Fade Away" claimed that the monthly payment on a 30 year mortgage at six percent interest would be $600 but just $716 for a 20 year mortgage. [**R265**]

On the next day in an article in *The Boston Globe* headlined "The end of 30-year fixed-rate mortgage?" you could read that

> The difference between a 15- and 30-year mortgage amounts to well over $600 per month on a $300,000 loan, a substantial amount that may prevent wide swaths of the middle class from buying homes. [**R266**]

Verify the calculations in each of these quotations.

Exercise 10.6.7. [U][W][Section 10.3][Goal 10.5] [Goal 10.4] Retirement planning.

Find an online retirement income calculator. Use it with data you imagine for yourself. Write down what you do as you proceed. (Screenshots would be nice.) Record what it tells you at the end.

Their calculator is much more sophisticated than the simple one in Excel we introduced in this chapter. See if you can use ours to get answers that match what it told you.

Exercise 10.6.8. [S][Section 10.4][Goal 10.4] Payday loans.

The Boston Globe on New Year's Day 2009 reported that a New Hampshire law will cap the interest rate on payday loans at 36 percent per year.

> Payday lenders typically charge $20 per $100 for two-week loans backed by the borrower's car title or next paycheck. That amounts to 1.43 percent interest per day, an annual rate of 521 percent. [**R267**]

The cap will limit the daily rate to about 0.1 percent, so just $1.38 — a dime a day — on that two week $100 loan.

(a) What is a "payday loan"?

(b) Verify the computation that 1.43% interest per day is 521% interest annually.

(c) If the 1.43% daily interest is compounded daily then the true annual rate of interest is in fact much more than 521%. How much is it?

   [See the back of the book for a hint.]

(d) Verify that paying interest of $1.38 on a two week loan of $100 is just about a "dime a day" and corresponds to a daily interest rate of about a tenth of a percent. What annual rate does that represent?

(e) Visit a payday loan website and report on what you discover there about interest rates.

Exercise 10.6.9. [S][C][Section 10.4][Goal 10.5] [Goal 4.2] Supporting a hospital bed.
The headline "Charity sues R.I. hospital over donation in 1912" accompanied an article in the City and Region section of *The Boston Globe* on February 23, 2008. The article described a gift intended to provide a free bed in perpetuity for needy patients. There you could read that

> Mark E. Swirbalus, a Boston lawyer representing Children's Friend, said that "as far as we know, the hospital never set aside a bed and never set aside the money." The $4,000, if conservatively invested by the hospital in 1912, would be worth about $1.5 million today, he said. [R268]

(a) Is Swirbalus's claim about a "conservative investment" correct?

[See the back of the book for a hint.]

(b) What the hospital should have done was invest the money and use just the interest each year to fund the bed. That would work — if only there were no inflation that made the cost of the bed increase.

Suppose the hospital got a 6.5% percent yearly return on investment and annual inflation was 3.5%. Explain why it would be able to spend about $120 on the bed in 1912 and could keep spending at that rate as the years went on.

(c) If the cost of providing a hospital bed in 1912 was $120, what would it be in 2008 if all you had to do was adjust for inflation?

Exercise 10.6.10. [S][Section 10.2][Goal 10.3] Half the time, more than twice the benefit.
Show that taking out a 15 year mortgage instead of a 30 year mortgage (for the same loan amount at the same annual rate) doesn't double your monthly payment and more than halves the total interest you pay on your loan.
(The advantages are usually even greater since you can usually negotiate a lower interest rate for a shorter mortgage.)

Exercise 10.6.11. [S][Section 10.3][Goal 10.5] Saving $50,000.

(a) Use the spreadsheet PayOffDebt.xlsx to figure out how much you'd have to save per year at 3% interest (compounded annually) to have a balance of $50,000 in your account at the end of 25 years.

(b) Then use the mortgage tab in that spreadsheet to answer the same question if you save a fixed amount each month rather than each year.

Exercise 10.6.12. [S][Section 10.1][Goal 10.1] What merchants pay for credit card services.
In May 2014 *The Nilson Report* said that total spending for credit, debit and prepaid purchases in 2013 was $4.530 trillion, broken down as follows:

credit $2.399 trillion
debit $1.949 trillion
prepaid $0.182 trillion. [R269]

The Boston Globe reported that "merchants in the United States spent $71.7 billion on fees [for these transactions] last year."

(a) What is the average merchant fee, as a percentage?

(b) Make sense of the $4.530 trillion total: think about it in units like dollars per person per day, dollars per transaction, … .

Exercise 10.6.13. [U][Section 10.2][Goal 10.2][Goal 10.3][Goal 10.4] Excel templates from the internet.

At `www.excely.com/template/loan-calculator.shtml` you can download a Loan Calculator Excel Template.

(a) Do the calculations there match those in `PayOffDebt.xlsx`?

(b) Find out where that template uses Excel's built-in PMT function. Compare how it works to the formula in the spreadsheet `PayOffDebt.xlsx`.

Exercises added for the second edition.

Exercise 10.6.14. [U][S][Section 10.1][Goal 10.1] Credit card fees.

On October 16, 2017, Bloomberg News reported on the Supreme Court's decision to take a case on American Express credit card fees. There you could read that "[Merchants pay] $50 billion in fees to credit-card companies each year." Those fees come from "the 'astronomical number' of credit-card transactions each year — 22 billion totaling more than $2 trillion in 2011." [**R270**]

In June of 2018 the Court sided with American Express, ruling that the company could demand that merchants not ask customers to use a different credtit card. [**R271**]

(a) Does the figure 22 billion transactions per year make sense?

(b) What is the average dollar value of a credit card transaction? Does your answer seem reasonable?

(c) The average you computed in the previous question is the mean. Would you expect the mean and mode to be smaller or larger? Why?

(d) What is the average percentage fee charged merchants for credit card transactions?

Exercise 10.6.15. [U][S][Section 10.1][Section 10.4] What does a missed payment really cost?

In Section 10.1 we quoted a late penalty fee with an APR of 29.99%. What is the actual EAR?

Exercise 10.6.16. [U][W][Section 10.1][Goal 10.2][Goal 10.4] Predatory lending.

Take one of the Predatory Lending Awareness Quizzes at `extension.missouri.edu/cfe/wcap/quizzes.htm` . Write about what you discovered. Will that change your behavior?

11

Probability — Counting, Betting, Insurance

Pierre de Fermat and Blaise Pascal invented the mathematics of probability to answer gambling questions posed by a French nobleman in the seventeenth century. We follow history by starting this chapter with simple examples involving cards and dice. Then we discuss raffles and lotteries, fair payoffs and the house advantage, insurance, and risks where quantitative reasoning doesn't help at all.

Chapter goals:

Goal 11.1. Compute probabilities for games of chance by counting outcomes.

Goal 11.2. Calculate fair price of a bet as a weighted average.

Goal 11.3. Calculate house advantage as (payout)/(income).

Goal 11.4. Understand insurance as a lottery.

11.1 Equally likely

In its everyday qualitative meaning "probably" is just a synonym for "likely" or "I think so but I'm not sure." In this chapter we start with simple examples where we can make "probably" quantitative by counting the possibilities.

To think about the chance of some particular event involving coins, dice, cards or raffles happening, count the possible equally likely outcomes, then count how many match what you're looking for and write down the appropriate fraction.

- The probability of heads when tossing a fair coin is $\frac{1}{2}$.

- The probability of rolling a 6 with a fair die is $\frac{1}{6}$.

- The probability of drawing an ace from a well-shuffled deck is $\frac{4}{52}$.

In words,

$$\text{probability of an event} = \frac{\text{number of outcomes that match the event}}{\text{number of possible outcomes}}.$$

Writing probabilities as fractions helps you remember what they mean. But since they're just numbers, we can write them as decimals if we wish. Since they are numbers between 0 and 1, we often express them as percentages.

- The probability of heads tossing a fair coin is $\frac{1}{2} = 0.5 = 50\%$.

- The probability of rolling a 6 with a fair die is $\frac{1}{6} \approx 0.167 \approx 17\%$.

- The probability of drawing an ace from a well-shuffled deck is $\frac{4}{52} \approx 0.077 = 7.7\%$.

Events that can never happen have probability 0. Events with probability 1 are certain to happen.

- The probability of rolling a 7 with a die is $\frac{0}{6} = 0 = 0\%$. It doesn't matter whether the die is fair or not.

- The probability of drawing a heart, a club, a spade or a diamond from a deck of cards is $\frac{52}{52} = 1 = 100\%$. It doesn't matter whether the deck is well shuffled or arranged in some nice order.

There are other probability problems you can solve by counting, as long as you're careful to count the right things.

Many state lotteries offer a prize if you pick the right six numbers in the some range. The numbers must be different, with no repetitions, but the order in which you pick them doesn't matter. To find the probability that your pick will win you have to count how many ways there are to pick six numbers. That's a problem for a math course more advanced than this one: the answer is 20,358,520 when the range is numbers from 1 to 52. So the probability of winning pick-six is about one twenty-millionth. If twenty million people play, expect about one winner.

We will have much more to say about lotteries in Section 11.4.

11.2 Odds

Another way to describe a coin toss is to say "the odds are fifty-fifty." Heads and tails are equally likely — the odds are even.

Here are the odds for some gambling events; we usually write odds with a colon (:) and read the colon out loud as "to".

- The odds for rolling a 6 with a fair die are 1 : 5, or one to five. The odds against are 5 : 1, or five to one.

- The odds for drawing an ace from a well-shuffled deck are 4 : 48, or 1 : 12. The odds against are twelve to one.

- The odds for heads tossing a fair coin are 1 : 1.

These examples illustrate how to find the odds for an event when you can count the equally likely possibilities and decide which ones are favorable. You compute

(number of favorable cases) : (number of unfavorable cases) .

The odds against the event are

(number of unfavorable cases) : (number of favorable cases).

Odds are fractions in disguise, so the odds against drawing a spade from a deck of cards may be expressed as 39 : 13 (counting all the possibilities) or simply as 3 : 1 (three to one).

The odds against a winning pick-six ticket are about 20 million to 1.

You can convert back and forth between odds and probabilities. Since the odds against drawing a spade are 39 : 13, the probability that you won't draw a spade is $\frac{39}{52} = \frac{3}{4}$. In general, if the odds for an event are $a : b$ then its probability is $a/(a + b)$.

If you start out knowing that you will draw a spade with probability 25% you know too that the probability that you'll draw a heart, a diamond or a club is 75%. With both those probabilities it's easy to find the odds: they are 0.25 : 0.75 for drawing a spade. That's just our old friend 1 : 3 in disguise. Gamblers usually describe bets in terms of odds rather than probabilities. We will use odds that way in Section 11.6.

In general, if the probability of an event is p then the odds for that event are $p :$ $(1 - p)$. The odds against are $(1 - p) : p$.

The few formulas in this section are just common sense. If you understand them you won't have to memorize them. If you try to memorize them without understanding them you may end up using them in the wrong places.

11.3 Raffles

Simple raffles are gambles with computable probabilities. Tickets are sold, some are chosen at random and the people who hold those tickets get prizes. You may be familiar with fundraising raffles run by school parent-teacher organizations (PTOs).

Suppose the PTO sells 500 tickets for a raffle with a single prize.

Since each of the 500 tickets has an equal chance of being selected, the odds of a ticket winning are 1 : 499, or 499 : 1 against. The probability that any particular ticket wins is $\frac{1}{500} = 0.002 = 0.2\%$, or two tenths of a percent.

The probability that a particular person wins may be different. If you buy 10 tickets then you win with probability $\frac{10}{500} = 0.02 = 2\%$. If you don't play, the probability is 0. If you buy all of the tickets then you win with probability $\frac{500}{500} = 1 = 100\%$.

Now let's connect probability with money, as the inventors of the mathematics of probability did centuries ago. Suppose the PTO wants to offer a $1,000 prize to the winner. Then the *fair price* of a ticket is what it would cost if all the money collected were distributed as prizes:

$$\text{fair price} = \frac{\text{total prize money}}{\text{number of tickets}} = \frac{\$1,000}{500 \text{ tickets}} = 2\frac{\$}{\text{ticket}}. \tag{11.1}$$

Using what we learned in Chapter 5 we can rewrite this computation as a weighted average. One of the tickets is worth $1,000; the others are worthless, so

$$\text{fair price} = \frac{\text{total value of tickets}}{\text{number of tickets}}$$
$$= \frac{499 \times \$0 + 1 \times \$1,000}{500}$$
$$= \frac{499}{500} \times \$0 + \frac{1}{500} \times \$1,000$$
$$= \text{probability of losing} \times \text{value of losing ticket}$$
$$\quad + \text{probability of winning} \times \text{value of winning ticket}$$
$$= 0.998 \times \$0 + 0.002 \times \$1,000$$
$$= \$2.$$

In the fourth line of the computation the ticket counts disappear. The fair price is the weighted average value of a ticket, weighted by the probabilities for each kind of ticket.

That average is the fair price of a ticket because all the money collected is returned in prizes. That may make for an exciting evening at the PTO meeting, but it won't raise any money. So the PTO decides to charge $3.00 for each ticket, keep the prize at $1,000 and use the other $500 to buy classroom supplies for the kids.

Since the total prize money and the number of tickets have not changed, the fair price is still $2. So on average each ticket loses

$$\text{cost of ticket} - \text{fair price of ticket} = \$3 - \$2 = \$1.$$

Of course you never lose exactly one dollar with one ticket. You either collect $1,000 for a net gain of $997 or get nothing and lose your $3 bet.

Yet another way to calculate the average loss is to see that the prize is just 2/3 of what the PTO collects, so the fair price is 2/3 of the $3 cost, or $2. Then on average each ticket loses the other 1/3, or $1.

Would you buy a $3 ticket when the fair price is just $2, knowing that on average you will lose $1.00? Perhaps. Even though you're very likely to lose your three dollars you can feel good about supporting the school. Maybe the thrill you get anticipating what you will do with the prize if you win despite long odds makes the probable loss more bearable.

11.4 State lotteries

On July 10, 2016, Jeff Jacoby wrote a column in *The Boston Globe* about the possibility that the Massachusetts State Lottery might begin online ticket sales:

> Massachusetts officials boast that the state lottery is the nation's most successful. Some $5 billion in lottery tickets were sold last year, a record high. On a per capita basis, lottery sales in Massachusetts — the amount spent on scratch tickets, the Numbers Game, Megabucks, and all the rest — averaged $740. That is a stunning number. The 43 states and District of Columbia that operate lotteries did about $70 billion worth of business last year, which averages out to $230 in gambling revenue for every man, woman, and child within their borders.

Thus sales in Massachusetts were more than three times the US average.

...

State lotteries are often justified as an effective, yet voluntary, means of raising money to pay for public services. In Massachusetts last year, $945 million in State Lottery profit was disbursed as local aid to the Commonwealth's cities and towns. [R272]

As usual, we start thinking about these numbers with a quick sanity check. Do the per capita amounts match the totals?

Estimating the U.S. population in 2016,

$$320 \text{ million people} \times 230 \frac{\$}{\text{person}} = 70.4 \text{ billion } \$$$

which is close enough to the $70 billion in the article.

For Massachusetts,

$$6.5 \text{ million people} \times 740 \frac{\$}{\text{person}} = 4.8 \text{ billion } \$$$

which is close enough to the $5 billion in the article.

The payoff rules for the various games are very complex and vary widely from game to game. Fortunately, the Census Bureau provides the total amount returned in prizes so we can calculate the average fair price of a dollar ticket. Table 11.1 shows $66.9 billion for total U.S. state lottery revenues and $1.3 million for the net proceeds available in Massachusetts. These don't quite match Jacoby's figures, but they are in the same ballpark.

Massachusetts players spent $5 billion on tickets and received $3.6 billion in prizes. The average return was thus

$$\frac{3.6 \text{ billion prize dollars}}{5 \text{ billion purchase dollars}} = 0.72 \frac{\text{prize dollars}}{\text{purchase dollar}}.$$

The fair price for a one dollar Massachusetts ticket is just about 72 cents.

With that figure we can estimate the probability of winning when we know the prize structure. For example, for a single million dollar payout the Lottery Commision will have to sell

$$\frac{\$1,000,000}{0.72 \text{ \$/ticket}} \approx 1,400,000 \text{ dollar tickets}$$

in order to pay out 72% in winnings. Therefore the odds that a ticket wins are 1:1,400,000.

	Income Ticket sales (excluding commissions)	Apportionment of funds		
		Prizes	Administration	Proceeds available
Massachusetts	5,005,635	3,641,351	100,590	1,263,694
U.S.	66,885,544	42,893,054	3,125,938	20,900,504

Table 11.1. Massachusetts and U.S. state lotteries (2015) [R273]

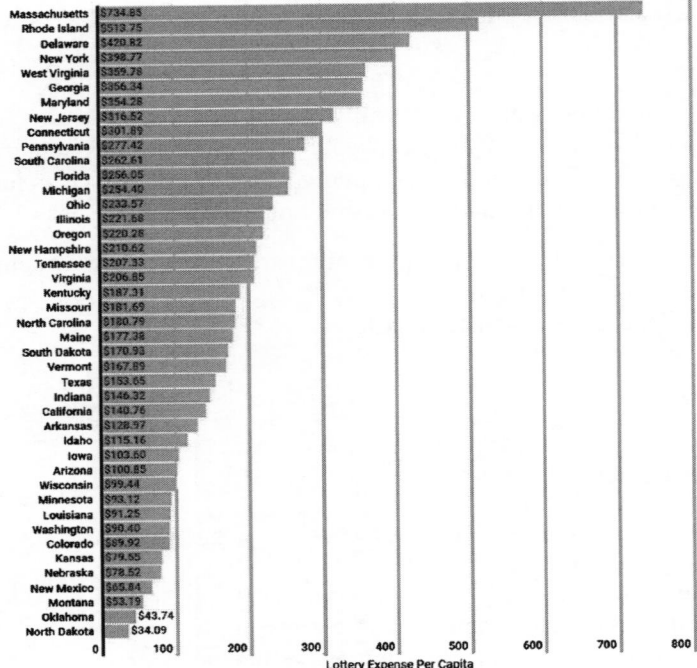

Figure 11.2. State per capita lottery expenses (2016) [**R274**]

You would have to buy 700,000 one dollar tickets for a 50% chance at the million dollar prize. To add insult to injury, if you won you would have to pay federal income tax on your winnings.

Using the government's data, the nationwide average fair price for a dollar ticket was

$$\frac{42.9 \text{ billion prize dollars}}{66.9 \text{ billion purchase dollars}} = 0.64 \frac{\text{prize dollars}}{\text{purchase dollar}},$$

so the Massachusetts gamblers are a little better off, per dollar. But they spend many more dollars. Figure 11.2 breaks down the total $66.9 billion by state.

States use lottery revenue to help balance their budgets, so you can think of lotteries as a kind of tax. What kind? Income and property taxes are roughly proportional to income or property values. Sales taxes are roughly proportional to sales. But prosperous citizens don't play the lottery in proportion to their prosperity, so a lottery is in effect a regressive tax . We close this section with a further quote from Jacoby:

> But it is odious for government to raise money by preying on the poor and the foolish. Everyone knows that lottery tickets are most frequently purchased by people least able to afford them.

11.5 The house advantage

Raffles and lotteries are designed to make money. So is casino gambling — for the casinos. They make a profit, and states tax the proceeds to raise revenue.

Before you lay down your bet at a casino, you should think about how much you will pay to play — the difference between a dollar bet and fair price of that bet (the average amount returned to you for your dollar). That difference is called the *house advantage*.

In Section 11.4 we discovered that the house advantage on state lotteries averages about 37%. At gambling casinos it's much smaller. As in the state lotteries, the house advantage varies from game to game. It's the highest (about 10%) for slot machines — and there is no way to know that when you decide to play. But for roulette we can actually calculate the house advantage.

A fair roulette wheel is a circle divided into 36 equal wedges numbered from 1 to 36, colored alternately red and black. A ball runs around the rim of the wheel, slowing down until it falls into a random wedge. Before the wheel spins you place your bet, perhaps:

- On the number 17 ("straight-up"), with a winning probability of $\frac{1}{36}$. The odds are 35 to 1 against.
- On red, with a winning probability of $\frac{18}{36} = \frac{1}{2}$. Even odds.
- On odd, at even odds.
- On one of the numbers 1 through 12 (a "dozen bet"), with a winning probability of $\frac{12}{36} = \frac{1}{3}$. Two to one against.

  What would be a fair return on a $1 bet?

- If you bet straight-up the payoff should be $36.
- If you bet on red, the payoff should be $2.
- If you bet on odd, the payoff should be $2.
- For a dozen bet the payoff should be $3.

  There are several ways to see that these are fair. We'll work them out with the $36 payoff for the straight-up dollar bet on a single number.

- Imagine the spin of the wheel as a raffle with 36 tickets. A dollar bet on 17 is like buying one of the tickets. Imagine that others have bought the other 35 for $1 each. Then the casino has collected $36. The fair thing to do would be to pay that to the winner — then all the money collected is awarded as prizes.

- Using the technique we learned in Section 11.3, we can check the numbers in this equation:

  winning probability × winning payoff + losing probability × losing payoff.

  For the straight-up bet that equation says

  $$\frac{1}{36} \times \$36 + \frac{35}{36} \times \$0 = \$1$$

  which is a perfectly fair average return on a $1 bet!

- What happens when you play for a long time? Since you pay $1 for each spin of the wheel and win about $\frac{1}{36}$ of the time, you should collect $36 for each win in order to break even in the long run.

Figure 11.3. The house advantage in roulette [**R275**]

- The odds for winning are 1 : 35. Your $1 bet on 17 is a bet against the casino. They put up $35 to match your $1. The winner takes all $36.

The other computations (to check the fair payoffs for bets on red or odd or 1-12) work the same way.

In real life the casino must cover expenses, pay the state its share of the take and still turn a profit, so the average value of a bet must be less than the fair price. The difference is the house advantage.

In the PTO raffle the organization assumes it sells all the tickets and decides how much of what it collects to return as prizes. But the casino can't count on people betting on all the numbers and can't know how many people will bet.

Figure 11.3 shows how they collect the house advantage in roulette. The picture shows a wheel with an extra green wedge numbered 0. An American roulette wheel will have another green wedge numbered 00.

The casino uses the old fair price payoffs: $36 for a winning $1 straight-up bet on 17. But the extra wedges change the probabilities. Here is the calculation for an American wheel with 38 wedges:

$$\frac{1}{38} \times \$36 + \frac{37}{38} \times \$0 = \$0.94736842105 \approx \$0.95$$

which means that on average you lose a little more than a nickel of every dollar you bet. The house advantage is just over 5.25%.

Does this mean you shouldn't play? Not necessarily. You may be willing to pay the house advantage in return for the thrill of the gamble. But before you do, you should understand the odds for the game you choose.

There are casino games in which a skilled player can win — slowly, and with great effort. At the poker table you are competing with other gamblers, not with the house, which pays its expenses and profits by taking a fraction of the ante or pot on each deal. So the house always wins, but a skilled poker player can win too by beating the other players.

In principle, you can also win at blackjack. We'll think about why in the next chapter.

11.6 One-time events

In our discussions so far we've assumed each example is "fair" (even if payoffs weren't) — coins and dice and roulette wheels are properly balanced, decks of cards are properly shuffled, no one peeks when drawing the winning raffle ticket. In each case all possible outcomes are equally likely so we could compute probabilities just by counting cases.

To test for whether a particular coin or die is really "fair" you could imagine repeating an experiment many times. A fair coin should come up heads about half the time (but not exactly half the time, which would be very unlikely). A fair die should show a 5 about 1/6 of the time. We'll return to this topic in Section 12.2.

There are many situations in real life where probabilities and odds appear but can't be computed by simple counting or be checked by repeated experiments. Will the Chicago Cubs win the World Series? Which horse will win the Kentucky Derby? Who will be elected? Will it rain tomorrow?

Suppose you bet your Chicago friend that the odds against the Cubs winning the World Series are 99 : 1. You put up $99, she puts up $1 and the winner takes home $100 when the season is over. That means that (in principle) you believe that probability of that Cubs World Series win is just 1/100. (Those might have been the right odds before 2017, when the Cubs won the World Series for the first time in more than a century.)

When lots of people have an opinion they are willing to bet on, they can decide the probability collectively.

There's a way in which many state lottery payoffs depend on what the bettors think: the total prize money for a winning pick-six combination is divided among the people who bet on that combination. The odds for any particular number combination don't change, but the payoff does. Exercise 11.9.11 pursues this idea.

In horse racing the odds at the track depend on the bets placed, in what's called *parimutuel* betting. Most readers of this book won't be playing the horses, and those who do will (or should) know all about this kind of betting. We discuss it here anyway since it provides an example where we can actually see how the bets determine the odds.

Before the race the punters place their bets at the tote. ("Punter" and "tote" are racing terms. You can look them up if you don't know what they mean.) After the race the track skims its *take* or *commission* — a percentage of the total amount bet. The winning bettors share the remainder in proportion to the amount each bet.

Table 11.4 shows the amount bet on each of six horses in an imaginary race. The horses are real — all winners of the Kentucky Derby — but we made up the numbers.

The favorite horse is Twenty Grand precisely because more people have bet on him to win.

Since the total amount bet is $1,716,500, the collective wisdom at the track says that Apollo will win with probability $155{,}000/1{,}716{,}500 = 0.0903000291 \approx 9\%$. The fair payoff is $1{,}716{,}500/155{,}000 = 11.0741935 \approx 11$ dollars per dollar bet.

That corresponds to odds against of about 10 to 1. If Apollo wins, each dollar bet will collect $11: the original dollar plus the ten the other bettors put up in vain.

We had fun choosing the amounts bet on each of these six horses so that the odds of each are close to their odds in the Derby they won. We've included the longest shot of all, Donerail, and a favorite, Twenty Grand, who ran at less than even odds.

Horse	Bets (K$)
Barbaro	239.0
Spend a Buck	333.2
Donerail	18.6
Twenty Grand	904.4
Apollo	155.0
Dark Star	66.3
total	1,716.5

Table 11.4. Race of champions

The 10 : 1 odds for Apollo do not take into account the race track's commission. We don't know how much that was or even whether it was the same in all six races. Suppose that in this fantasy race it's 10%.

Suppose Apollo wins. The track pays 90% of the take to those who bet on Apollo — $0.9 \times \$1,716,500/\$155,000 = \$9.96677\ldots \approx 10$ dollars per dollar, instead of 11 dollars per dollar. The odds are effectively just 9 : 1 against. The track makes money by lowering the odds, which no longer reflect the probabilities determined by the bets.

The moral of the story: you can win at the race track if you really know better than most people which horse is likely to win. There's that word "likely" again — you need to know a lot about the horses and play a lot for your knowledge to pay off. Perhaps the best way to win is to sell suckers a system ....

11.7 Insurance

When you buy insurance you're gambling. In this case the gamble is one you hope to lose — you don't want to get sick or have your house burn down or total your car. In each of those situations you've made a small advance payment you hope and expect to lose in order to cover your losses when a catastrophic event with small probability happens.

Insurance companies estimate probabilities in order to determine the fair price for their policies, then add what they need to cover their administrative expenses and make a profit — their "house advantage." In the long run, on average, their customers never get all their money back. Therefore you want to think things through when you're deciding whether to buy insurance for more than the fair price. Sometimes you may be better off accepting the risk yourself.

Here's a sample of the kind of advice you can find on the web. It's from Liz Pulliam Weston, writing for *MSN Money*.

> Say you have a 10-year-old Honda that's worth $4,000 in a private-party sale and have a $500 deductible. Your risk is $3,500. If your premiums for collision and comprehensive are more than $350 a year, it may be wiser to bank that money toward a newer car. [R276]

If we make a simple assumption we can think about this using probabilities. Suppose that the only kind of accident to worry about is one that totals the car. Then Weston's advice is reasonable if you think that the probability that you'll have such an accident is less than 10%. Here's why. Imagine that the insurance policy is a lottery ticket, which "wins" if you have an accident. A winning ticket is worth $3,500. If you think you have a 10% chance of winning, then the fair price (for you) is $350. If you think your chance of totaling your car is less than 10% then the fair price is more than $350, so perhaps you shouldn't buy the insurance.

Of course the real decision isn't this easy. You should take into account the fact that your accident might not total the car. You have to think about making this decision every year — sometimes your car will be worth more than $4,000, sometimes less. But the principle is clear. If the premiums are very high compared to your estimate of your risk, you should consider not buying collision and comprehensive insurance. Over the course of a driving lifetime you will probably save money.

However, there are often good reasons to pay more than the fair price for insurance. If you don't have the money to replace a totaled car and you must have one, then you need that insurance. Even if you have the money, the cost to you of a large loss may be more than you can afford or may feel like more than the dollar amount.

For a discussion of answers to the question "Why buy insurance?" visit money. stackexchange.com/questions/54561/why-buy-insurance.

Sometimes you may be required to buy insurance. In order to drive, you must carry liability insurance to cover the cost of injuries to others in an accident you caused. If you have a mortgage on a house the bank will insist on fire insurance to protect their interest in the money they've lent you. The taxes you pay to support the police and fire departments can be considered a kind of insurance. You will probably never need their services, but you want them to be there when you do. Healthy people buy health insurance (and may even be required to do so) to spread the cost of catastrophic medical bills.

George Bernard Shaw wrote about this in "The Vice of Gambling and the Virtue of Insurance" [R277]. There's a section on health insurance that's the clearest argument we've seen for single payer "medicare for all". Too bad it was written a century ago by a socialist.

11.8 Sometimes the numbers don't help at all

About forty years ago Joan Bolker had to decide whether to invest three years of hard work in hopes of earning a clinical psychology license.

Only after more than two thousand hours of clinical internships (which she would have to arrange) could she petition to have her doctorate in education count as appropriate postgraduate preparation for her new career. Only if that petition were granted would she be allowed to take the psychology licensing examination, much of which covered material she had not studied in any course.

Clearly the odds were long. She faced a significant investment of time, energy and lost income, with an unknown and hard to estimate probability of success at the end. She took the risk. She won her gamble, with a combination of talent, persistence and luck.

The moral of the story: sometimes numbers don't help. "Not everything that can be counted counts, and not everything that counts can be counted." (A quote often (wrongly) attributed to Albert Einstein). [R278]

In this case there was no way to quantify the costs, the benefits and the probabilities in order to make what might look like a rational choice. The kind of back-of-an-envelope probability calculations we've studied about playing the lottery or buying insurance are often of little help when making life-changing one-of-a-kind choices.

11.9 Exercises

Exercise 11.9.1. [S][R][Section 11.1][Goal 11.1] What's in a name?

(a) What is the probability that the name of a state (of the United States) chosen at random begins with the letter "A"?

(b) What is the probability that the name of a state (of the United States) chosen at random begins with the letter "Z"?

(c) How much more likely is it that a state name begins with "M" than with "A"?

Exercise 11.9.2. [U][Section 11.1][Goal 11.1] "Probably" in everyday English.

(a) Use the index to this book to find places where we used the words "probably" or "likely" other than in the chapters devoted to studying probability. Discuss the meaning of the word there. When it makes sense, provide a numerical estimate of the probability.

(b) Do the same for two or three occurrences of "probably" or "likely" in the media.

Exercise 11.9.3. [S][Section 11.1][Goal 11.1] Is it safe to swim?
    In an article in *The Boston Globe* reprinted from the *Washington Post* on March 4, 2012, you could read a story headlined "Possible cut to beach testing a health threat, critics say". The story reports on Environmental Protection Agency estimates that say that the average person goes to a beach, lake or river about 10 days a year and that about 3.5 million people get sick from splashing in bacterial contaminated water. [**R279**]
    What is the probability that a visit to the beach will make you sick?

Exercise 11.9.4. [S][Section 11.2][Goal 11.1] It's a horse race.
    Use the data in Table 11.4 to compute

(a) the odds and payoff for Donerail, the long shot,

(b) the odds and payoff for Twenty Grand, the favorite,

(c) the payoff for these two horses if the track takes a 10% commission before paying off any bets.

    [See the back of the book for a hint.]

Exercise 11.9.5. [S][Section 11.2][Goal 11.1] Extended warranties.

The list below from tv.about.com/od/warranties/a/buyexwarranty.htm outlines a set of points to think about when deciding whether to buy an extended warranty for your new TV. We think something important is missing from this list. What is it?

    (1) Value of item being purchased.
    (2) Price of extended warranty..
    (3) Length of manufacturer's warranty
    (4) Length of extended warranty and date coverage begins. **[R280]**

Exercise 11.9.6. [S][R][Section 11.3][Goal 11.2] Which average?

In the raffle discussed in Section 11.3 there are 500 tickets and a $1,000 prize. We found that the average value of a ticket was $2.

(a) Which average is that — mean, median or mode?

(b) Compute the other two "average" ticket values.

Exercise 11.9.7. [S][R][Section 11.3][Goal 11.2] Multiple prizes.

Suppose a lottery with 1,000,000 tickets has a first prize of $200,000, three second prizes of $60,000 each and 100 third prizes of $200 each.

(a) What is the probability that a ticket wins the first prize?

(b) What is the probability that a ticket wins some prize?

(c) What is the fair price of a ticket?

(d) How much should the state charge for a ticket if it needs 10% of the revenue for overhead and wants to make $500,000 profit?

Exercise 11.9.8. 1996 was a long time ago.

Moved to Extra Exercises at www.ams.org/bookpages/text-63.

Exercise 11.9.9. [S][W][Section 11.4] [Goal 11.2] Massachusetts Lottery statistics.

• The Massachusetts Lottery Commission reported that in 2006 they distributed over $761 million in direct local aid to the cities and towns of the Commonwealth.

• From www.masslottery.com/winners/faqs.html:

> What happens to the revenue which the Lottery generates from sales?
>
>   1. A minimum of 45% of revenues stays in the State Lottery Fund to be paid out in prizes. The Lottery's current prize percentage is over 69
>
>   2. A portion of revenues is transferred to the Commonwealth's General Fund for the expenses incurred in administering and operating the Lottery. The administrative and operating expenses of the Lottery are appropriated by the legislature as part of the annual state budget. Operating expenses cannot exceed 15%. Currently, operating expenses are under 8%. These operating expenses include 5.8% in commissions and bonuses paid to

the sales agents who sell the tickets and under 2% in administrative expenses due to Lottery operation.

    3. After prizes and expenses, the remaining Lottery revenues (approximately 23%) is transferred to the Local Aid Fund and returned to the cities and towns of the Commonwealth in the form of local aid. [**R281**]

- Several years later, on January 5, 2011, *The Boston Globe* reported that about $26 million in tickets had been sold in hopes of winning the $355 million Mega Millions jackpot and that "the tickets have raised $11 million for cities and towns". [**R282**]

(a) Sketch a pie chart showing how the money collected by the Lottery Commission was distributed among prizes, overhead and aid to cities and towns. Label each of the three slices with its percentage, and one of the slices with an amount of money.

(b) What was the total dollar amount collected by the Lottery Commission in 2006?

(c) What was the fair price of a $5 ticket?

(d) How much on average did people in Massachusetts spend on lottery tickets in 2006? On average, how much did they get back in prizes? Is this "average" the mean, the median or the mode?

(e) Does the 2011 payout for the 16 draw series match the prize percentage reported in 2006?

Exercise 11.9.10. [S][Section 11.4][Goal 11.2] Megabucks changes the odds.

    The first two paragraphs of an article in *The Boston Globe* on March 21, 2009, said that

> Like anyone who plays the lottery, Dean Thornblad was hoping to get rich quickly. He studied the odds of winning the various games before shelling out $150 for three season tickets that automatically enter him in twice-weekly drawings of Megabucks. At 1 in 5.2 million, the odds of hitting the jackpot, long by any standard, seemed to him at least "somewhat imaginable."
>
> But even his boundless optimism is being stretched by the lottery's latest proposal. The agency, under mounting pressure to return more money to cash-strapped cities and towns, is planning to make the odds of winning even slimmer, reducing them to 1 in 13.9 million beginning May 2, by making players match six numbers between 1 and 49, instead of six between 1 and 42. [**R283**]

What has happened to the expected value of Thornblad's ticket?

Exercise 11.9.11. [S][C][Section 11.4][Goal 11.2] Uncommon numbers.

    In many state lotteries the customer picks the numbers she thinks will win. The prize is then divided among all the people who happened to pick the winning numbers. Much as we try to analyze only real situations, the real Massachusetts Lottery is too complicated for this class. (Many people find it too complicated to choose the numbers they want to bet on and elect "quick picks" instead.) So this question is about an imaginary lottery.

Here is how our lottery works. Tickets cost $1. Each person buying a ticket chooses the number between 1 and 100 that she thinks will win. When all the tickets have been sold, the state picks a number at random between 1 and 100. All the people who have chosen that number divide 70% of the total collected among themselves. (The other 30% the state uses for overhead and local aid.) So the fair price for a $1 ticket is $0.70 or 70 cents.

Of course the winners collect much more than the fair price (since the losers collect nothing). For example, if 1,000 people bought tickets, 39 was the winning number, and 8 people chose 39, each would get ($1,000 × 0.7)/8 = $87.50.

If everyone buying tickets used "quick pick" then the 1,000 tickets would (more or less) consist of 10 for each of the 100 numbers, ten people would have the winning number and the typical payoff would be $1,000 × 0.7/10 = $70.

Now that you've read this far and understood the game, we can ask an interesting question. Suppose you know that people are so afraid of the number 13 that no one ever picks it. You think (correctly), "If I buy a ticket and choose 13, I'm probably not going to win. But if I do win, I will win big because I won't have to share the prize." So every day you buy one of the 1,000 tickets and choose 13, knowing that no one else will. You lose with probability 99/100 = 0.99 = 99% and win with probability 1/100 = 0.01 = 1%.

In the long run, how much money do you win (on the average) each day?

[See the back of the book for a hint.]

At blogs.wsj.com/numbersguy/lottery-math-101-801/ you can read more about this idea:

> Low numbers are particularly popular, some of them because birth-days are a popular source of numbers to play. Research conducted by Tom Holtgraves showed that bettors also avoid numbers with re-peated digits, though these are just as likely to turn up in lotteries as numbers without. [R284]

For still more information, see "Q3.4: Can RANDOM.ORG help me win the lottery?" at www.random.org/faq/\#Q3.4 .

Exercise 11.9.12. [R][Section 11.5][Goal 11.3] It's always 5.26%.

Compute the house advantage with an American wheel for each of the roulette bets in Section 11.5 to show that it's the same for each bet.

Exercise 11.9.13. [S][Section 11.5][Goal 11.3] Single zero roulette.

(a) Compute the house advantage for a roulette wheel with one extra wedge.

(b) Show that the house advantage in single zero roulette is approximately but not exactly half the house advantage in double zero roulette.

Exercise 11.9.14. [S][Section 11.5][Goal 11.3] Help this fellow out, please.

A questioner on the web has posted his roulette strategy. He says he will pick 19 numbers and bet on them. Since there are 38 spaces, he will win half the time, with a 35 to 1 payoff. So when he wins he's ahead 36 − 18 = 18 dollars. Then he asks,

> Am I missing something, or is it really that simple? [R285]

Answer his question.

Exercise 11.9.15. What you're counting counts.
Moved to Extra Exercises at `www.ams.org/bookpages/text-63`.

Exercise 11.9.16. What is wrong with this estimate?
Moved to Extra Exercises at `www.ams.org/bookpages/text-63`.

Exercises added for the second edition.

Exercise 11.9.17. [S][R][Section 11.3][Goal 11.1][Goal 11.2] What was fair in Texas?
The table at www.ncsl.org/research/financial-services-and-commerce/lotteries-and-revenue-by-state-2010.aspx lists the following data for the 2010 Texas lotteries:

Income Ticket sales (excluding commissions)	Apportionment of funds		
	Prizes	Administration	Proceeds available
$3,542,210	$2,300,182	$184,980	$1,057,048

What was the fair price of a $1 ticket?

Exercise 11.9.18. [U][Section 11.4][Goal 11.2] Your state lottery.
Find and analyze the data for the most recent lottery in your state. Calculate the fair price of a dollar ticket and the per capita dollar sales figure.

Exercise 11.9.19. [S][Section 11.1][Goal 11.1] Differ by two
We found this question at `math.stackexchange.com/questions/1716651/roll-two-dice-what-is-the-probability-that-one-die-shows-exactly-two-more-than`:

> Two fair six-sided dice are rolled. What is the probability that one die shows exactly two more than the other die (for example, rolling a 1 and 3, or rolling a 6 and a 4)? **[R286]**

Exercise 11.9.20. [U][S][Section 11.1][Goal 11.1] Stretching for the ball.
Peter Abraham reported on Jackie Bradley's catch in the Red Sox 2017 home opener:

> According to MLB's Statcast system, which tracks plays with high-resolution cameras and radar, Bradley had only a 55 percent chance of catching the ball and had to go nearly 30 yards to get there. **[R287]**

(a) How might the Statcast system come up with the "55 percent" estimate?

(b) Was this a lucky catch?

Exercise 11.9.21. [S][Section 11.2][Section 11.6][Goal 11.1] Spoiler alert?
On May 21, 2019, *The New York Times* reported on betting activity just before the final episode of *Game of Thrones*.

> Bran was no mere favorite: The final wagers for him to rule the kingdom were placed at 2-to-9 odds on one major offshore betting site, which implied that the probability he would prevail was almost 82 percent.

That made for an expensive bet. For every $45 risked, the profit would only be $10. It was the sort of wager to make only by those who are pretty certain they are right. [**R288**]

Note that gambling odds like these are often quoted in the order opposite that we described in [Section 11.2]. There the odds for the unlikely 6 in a die toss are 1 : 5. A gambler would say that as "5 : 1 against". So in this example, the odds of 2 : 9 mean that Bran's success is likely, not a long shot.

(a) Verify the quoted probability.

(b) If you bet $100 on Bran what would you have collected?

(c) Why do you think the author of this article used a $45 bet to illustrate his point?

Exercise 11.9.22. [S][Section 11.2][Section 11.6][Goal 11.1] Football.
In May 2019 *The New York Times* reported on the surprising success of Liverpool's soccer team in a match against Barcelona.

Before the series started, Barcelona were the strong favorite to advance to the final, and the outcome of the first game validated that assessment. After that, someone who wanted to win $100 betting on Barcelona needed to risk $1,800 to do it. [**R289**]

What was the probability bookmakers assigned for the match after the first game?

Exercise 11.9.23. [S][Section 11.1][Goal 11.1] Can you play?
At en.wikipedia.org/wiki/Lotteries_in_the_United_States Wikipedia (probably reliable for these data) says that (as of 2019) 45 states and the District of Columbia have lotteries.
Then
$$\frac{\text{number of states with lotteries}}{\text{number of states}} = \frac{46}{51} = 90\%.$$

(a) Why is it wrong to conclude that there is a 90% probability that a random person chosen from the United States lives in a jurisdiction with a lottery? Is the correct probability more or less than 90%?

(b) How would you compute the probability correctly?

Exercise 11.9.24. [U][Section 11.3][Goal 11.2] Price your raffle tickets.
www.ticketprinting.com/Articles/RaffleTicketPriceCalculator/ provides an online calculator for pricing raffle tickets.
Check that kit does so correctly.

12

Break the Bank — Independent Events

> I've thought that there may be more collisions ... in life than in books. Maybe the element of coincidence is played down in literature because it seems like cheating or can't be made believable. Whereas life itself doesn't have to be fair, or convincing.
>
> Shirley Hazzard
> *The Transit of Venus* [**R290**]

Unlikely things happen — just rarely! Here we calculate probabilities for combinations like runs of heads and tails. Then we think about luck and coincidences.

Chapter goals:

Goal 12.1. Experiment with multiple independent events.

Goal 12.2. Understand that rare events do happen!

Goal 12.3. Understand common cultural references to events and their probabilities.

12.1 A coin and a die

We know that if we flip a fair coin then the probability of getting heads is $\frac{1}{2}$ — that's the very definition of "fair".

Suppose you have a coin in one hand and a die in the other. What's the probability that when you flip the coin and toss the die you see a head and a 4?

In order to calculate

$$\frac{\text{number of desired outcomes}}{\text{number of possible outcomes}}$$

we first list all 12 possible outcomes:

Coin	Die	Coin	Die
head	1	head	4
tail	1	tail	4
head	2	head	5
tail	2	tail	5
head	3	head	6
tail	3	tail	6

Tossing the coin and rolling the die are completely unrelated. If you find out that the coin landed heads up you know nothing about what the die shows. The tosses are *independent* events. Each of the twelve equally likely outcomes occurs with probability 1/12. In particular, that's the probability of a head and a 4.

Figure 12.1 displays those twelve possible outcomes in a square so that you can see the various probabilities as areas. The shaded column for heads takes up half the area, corresponding to a probability of 1/2. The shaded row for 4 takes up one sixth, corresponding to a probability of 1/6. The overlap represents a head and a 4: that doubly shaded rectangle has area $\frac{1}{2} \times \frac{1}{6} = \frac{1}{12}$.

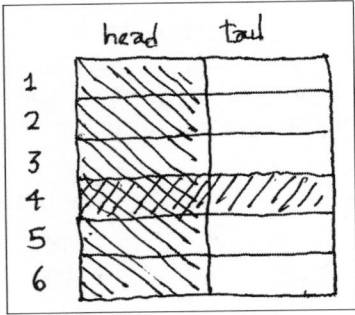

Figure 12.1. A coin and a die

The general principle is:

> When two events are independent the probability that both happen is the product of the probabilities for each one separately.

While we have this example in mind, we can think about a question that sounds similar: what is the probability of seeing a head or a four? Changing "and" to "or" in the question changes the answer dramatically. Of the 12 possible outcomes, in 6 the coin shows heads; in 2 the die shows four. So a first guess is that in 6 + 2 = 8 of the 12 possible outcomes we see a head or a four. That first guess is wrong — it counts the combination "head and four" twice. The correct count is 7, so the probability is 7/12, not 8/12. That's clear in Figure 12.1, where 7 of the 12 small rectangles are shaded.

Suppose the question were asked slightly differently: what is the probability of seeing a head or a four *but not both*? The answer would be slightly different. We'd want to count just the singly shaded areas in the figure, for a probability of $(5 + 1)/12 = 50\%$.

This computation may help you remember several ideas that can help you navigate probability computations for independent events.

- "And" is straightforward. Multiply.

- "Or" is tricky. You can't just add. The sum is too big. To adjust it you have to think about whether in the particular problem you face the "or" means "but not both".

12.2 Repeated coin flips

What is the probability that you will get two heads when you flip two coins? Since neither coin cares about the other one, the two flips are independent. The same is true if you flip one coin twice: the second flip is independent of the first since the coin can't remember how it landed the first time.

That means we should multiply the two probabilities: $\frac{1}{2} \times \frac{1}{2} = \frac{1}{4}$ in order to count just one of the four equally likely outcomes for (first coin, second coin), or, if we have just one coin to play with, outcomes for (first flip, second flip):

$$
\begin{array}{cc}
H & H \\
H & T \\
T & H \\
T & T
\end{array}
$$

The probability of two tails is also $\frac{1}{4}$.

In the other two cases there's one head and one tail, with probability $\frac{2}{4} = \frac{1}{2}$.

If you flip three coins (or one coin three times) there are eight possible outcomes, which we can list by putting first a head and then a tail in front of the four possibilities for two coins:

$$
\begin{array}{cc}
HHH & THH \\
HHT & THT \\
HTH & TTH \\
HTT & TTT
\end{array}
$$

The probability of three heads is clearly 1/8. You can see that by looking at the list of outcomes or by computing

$$
\frac{1}{2} \times \frac{1}{2} \times \frac{1}{2} = \frac{1}{8}.
$$

The same is true for three tails.

To find the probability for two heads and a tail, or for two tails and a head, you can count the cases. The answer is 3/8.

If you flip 10 coins there are

$$
2 \times 2 \times 2 \times 2 \times 2 \times 2 \times 2 \times 2 \times 2 \times 2 = 2^{10} = 1{,}024
$$

possible outcomes. Just one of those is all heads, with probability $(1/2)^{10} = 1/1{,}024$, which is just about one tenth of one percent. (We've seen the approximation $1{,}024 \approx 1{,}000$ before, in Chapter 1 where we discussed the different meanings of "kilo" in kilogram and kilobyte.)

What is the probability of just one head in ten tosses? That head could come from any one of the ten flips, so there are ten ways to get one head:

HTTTTTTTTT

T**H**TTTTTTTT

...

TTTTTTTTT**H**

so the probability is $10/1{,}024 \approx 1/100 = 1\%$.

Any combination of heads and tails can happen. Counting the number of ways for each is harder when there are a few heads and a few tails. You can study that mathematics in a statistics course, not here. The probability for exactly half heads and half tails (5 of each) turns out to be $252/1{,}024 \approx 25\%$. That means that about a quarter of all tosses of 10 coins will show a $50 : 50$ split.

To develop your intuition about what happens in practice we conducted an experiment flipping ten coins at a time, many times, counting the number of heads (between none and ten) each time. Figure 12.2 shows the results. (We didn't actually flip coins — we did the experiment in Excel, with the spreadsheet `flipcoins.xlsx`. You can repeat it if you wish.)

The figure shows that when we flipped our ten coins just ten times we never saw exactly 5 heads but did see 8 and 9 heads once each. Continuing to 100 ten-coin flips the observed values look like a better match to the gray bars, which show the predicted counts, calculated by looking at all the possibilities. But it was still true that five heads

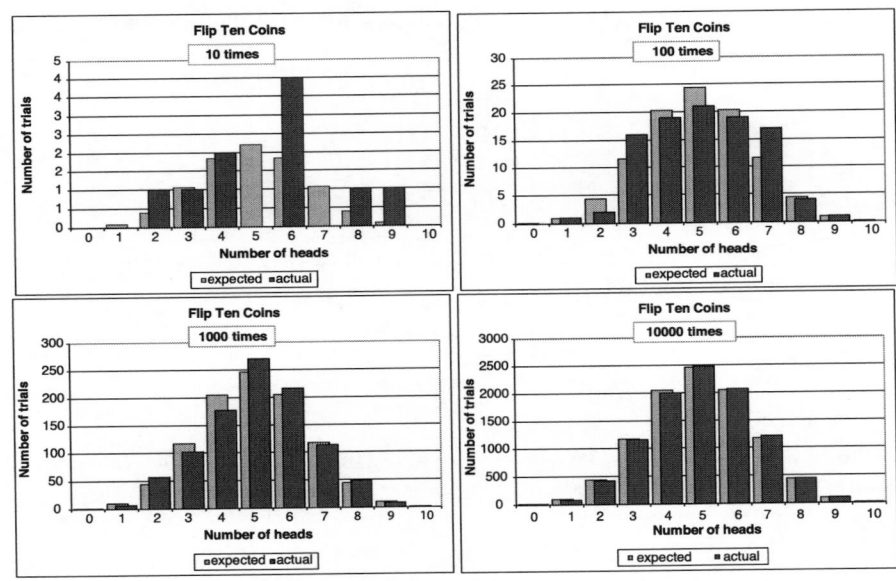

Figure 12.2. Flipping ten coins at once, many times

The categories on the *x*-axis are the number of heads (0 to 10).
The black bars record how many times each number appeared.
The gray bars are what the mathematics predicts.

out of ten coins did not occur quite as often as the mathematics predicts. At 1,000 tries there were noticeably more occurrences of six heads out of ten than of four.

At 10,000 tries the observed values are close to the expected ones, but even there the number of times four heads occurred is visibly less than what it "should be". The outline of both column charts is the common bell curve we studied in Section 6.10. The mean, median and mode are all at the same place: half heads and half tails, with probability about 25% — odds of 1 : 3. The standard deviation is about 1. The probability of four or five or six heads is about 66%. Two thirds of the time that's what you'll see.

The gray bars look the same in the last three charts. That's because Excel chooses the scale on the y-axis so that it conveniently fits the values it needs to graph. The number of tries is ten times as large as you move from one experiment to the next. The scale on the y-axis covers ten times as much as well.

This experiment shows that when you flip several fair coins over and over again then in the long run you will see approximately half heads and half tails most of the time and that the distribution will approach the standard bell curve. Mathematicians have been working for about three hundred years to say precisely what "in the long run", "approximately half heads", "most of the time" and "approach" mean. One result is a formula for the standard deviation of the approximated bell curve. That formula says that if you toss 100 coins rather than just 10 the standard deviation is 5 — about two thirds of the time you will see between 45 and 55 heads in one hundred coin tosses. Some day you may learn the formula in a statistics course. You don't need it now to begin to think sensibly about the probabilities for multiple coin flips.

12.3 Double your bet?

Here's a strategy that's bound to win money tossing a fair coin: whenever you lose, double your bet to cover your loss.

Bet a dollar. If you lose, you're out $1, so bet $2 next time. If you win, you collect $4: your $2 bet and the $2 your opponent put up. You've spent $3 altogether, so you're ahead a dollar. If you lose that second bet, try a third at $4. If you win, you get $8, which is still a dollar more than the $4 + $2 + $1 = $7 you've spent so far. Next time bet $8 if you have to.

It seems clear that you will always win your dollar — eventually. But . . . you should always be suspicious of any "system" that guarantees a gambling win. This doubling scheme is one of the most common. Here are two reasons why it won't work, both related to the fact that long runs, though rare, do occur.

In Section 12.2 we found that the probability of 10 heads in a row is

$$\left(\frac{1}{2}\right)^{10} = \frac{1}{1,024}.$$

That's close to 1,000 : 1 against. It can happen, but it's very unlikely. Ten heads in a row will occur about once every thousand times, so if thousands of people try the system this will happen to some of them. You can't know that it won't happen to you.

That doesn't yet explain why the system must fail.

One reason is that you will run out of money to bet. Losing ten times in a row will cost you $1,023. Your next bet must be $2,048. If you don't have that much left, you're out more than $1,000, which you just risked in hopes of making $1.

You might feel safer with a million dollar bankroll. But if you lose twenty times in a row (which will happen from time to time) then you're out your million dollars and the next bet is more than a million. So if you didn't have two million to begin with you're stuck.

It doesn't matter how much you start with. You will lose it all sooner or later.

You can see the second flaw in the system when you think about the person you're betting against. If it's a casino, they will have a *house limit* — the maximum bet they're willing to accept. If that's set at $1,000 you won't be able to make your tenth bet after losing nine times in a row. If it's $1,000,000 you won't be able to make your twentieth. Whatever it is, sooner or later you'll reach that limit. The higher it is, the more likely it is that you will think the system is working — but the more you'll lose when you encounter it.

Someone using this system will win small amounts frequently and lose more than her total winnings every once in a long while. What makes the system tempting is people's false belief that long runs never happen. We know better.

12.4 Cancer clusters

In an online post titled "Cancer Clusters: Findings Vs Feelings" you can read that

> A variety of factors often work together to create the appearance of a cluster where nothing abnormal is occurring. Looking for clusters is analogous to drawing a bull's eye after you have thrown darts at the wall at random. In this situation, there is possibly a place in which a bull's eye can be drawn that will leave multiple darts in close proximity to some common center. According to the American Cancer Society, cancer was diagnosed in an estimated 1,268,000 Americans in 2001. Finding clusters in cancer data is, thus, something like looking for patterns in the location of more than a million darts thrown at a dartboard the size of the United States. [R291]

To illustrate this phenomenon, we downloaded the populations of the 3,134 counties in the United States in the 2000 Census. The total population then was 279,517,404. The 1,268,000 cancer cases diagnosed in 2001 mean the national rate was

$$\frac{1,268,000}{279,517,404} \approx 0.0045,$$

or just under one half of one percent — about one person in every 200. Then we did an experiment. We randomly assigned the cancer cases to the counties in proportion to their population. (The data and the Excel commands that drive the simulation are in the spreadsheet `cancerclusters.xlsm`.)

Then we sorted the list to see where cancer seemed to be concentrated. Table 12.3 shows the beginning and the end of that list.

You can just imagine the editor of the local paper in MacPherson County, Nebraska, or Roberts County, Texas, writing an angry editorial wondering why the cancer rate there was twice the national average and demanding a federal investigation, while the Chambers of Commerce in Arthur County, Nebraska, and Loving County, Texas, wrote press releases bragging about what healthy places they were to live in.

State	County	Population	Cancer cases	Incidence rate
Nebraska	McPherson County	533	5	0.938 %
Texas	Roberts County	887	8	0.902 %
Idaho	Adams County	3,476	30	0.863 %
Montana	Petroleum County	493	4	0.811 %
Montana	Powder River County	1,858	15	0.807 %
	...			
Texas	Sherman County	3,186	5	0.157 %
Texas	Shackelford County	3,302	5	0.151 %
Colorado	Mineral County	831	1	0.120 %
Nebraska	Arthur County	444	0	0.000 %
Texas	Loving County	67	0	0.000 %

Table 12.3. Simulated cancer distribution

State	County	Population	Cancer cases	Incidence rate
Texas	Kenedy County	414	5	1.208 %
Colorado	Hinsdale County	790	8	1.013 %
Montana	Garfield County	1,279	11	0.860 %
Texas	King County	356	3	0.842,7%
South Dakota	Hyde County	1,671	14	0.837,8%
	...			
Nebraska	Banner County	819	1	0.122 %
Texas	Roberts County	887	1	0.113 %
Montana	Petroleum County	493	0	0.000 %
Nebraska	Hooker County	783	0	0.000 %
Texas	Loving County	67	0	0.000 %

Table 12.4. Simulated cancer distribution (again)

Table 12.4 shows what happened when we repeated the experiment. Note that Petroleum County, Montana, moved from fourth most cancerous to cancer free!

The moral of the story is that events that look surpising from a local perspective might be the result of random large scale happenings.

12.5 The hundred year flood

On May 8, 2019, NPR aired a story about a suburb of St. Louis that experienced three major floods since 2015, two of which were approximately "1-in-100-year events." That prompted the headline

> When "1-In-100-Year" Floods Happen Often, What Should You Call Them?

The story continued:

> "The educated layperson or elected officials, they think, 'Well, you
> scientists and engineers can't get it straight because we had a 100-year
> flood two years ago! Why are we having another one? You guys must
> have your numbers wrong.' It makes people think we don't know
> what we're doing," says Robert Holmes, the national flood hazard co-
> ordinator at the U.S. Geological Survey. [R292]

We know enough to "get it straight". When the U.S. Geological Survey looked at
the historical data for the St. Louis suburb they saw floods that severe about once in
100 years on average, but with no particular pattern. Since the weather in any year
is essentially independent of the weather the year before, predicting a flood in any
particular year is like flipping a coin with a 1% chance of heads (wet) and a 99% chance
of tails (dry). We know that the probability of a run of 100 dry years is

$$0.99^{100} = 0.366032341 \approx 37\%.$$

That means that about 1/3 of the time you will go a whole century in St. Louis without
seeing a hundred year flood and 2/3 of the time you will see at least one. You may well
see two less than 100 years apart. You can't properly argue that if there has been no
100-year flood in a long time you are "due for one" or that if there's a horrendous flood
in 2019 you are "safe for another century".

The NPR report went on to suggest a better way to talk about a "hundred year
flood"

> ... by telling people what their risk of flooding is over time rather than
> each year.
>
> For example, if there is a 1% chance that a home will flood each
> year, that means there's a 26% chance it will flood over the course of
> a 30-year mortgage.

We can check the arithmetic. Since

$$0.99^{30} = 0.73970037338 \approx 74\%$$

there is a $100\% - 74\% = 26\%$ chance for a flood in a 30 year period.

However, there may well be some reason to pay attention to the idea expressed in
"we had a 100-year flood two years ago! Why are we having another one?" The historical data does suggest that the frequency of severe floods is increasing, even though
there is no year-to-year pattern. The weather one year is still independent of the weather
the year before, but perhaps what used to be a 100-year flood is now a 90-year flood.
Since

$$0.9^{30} = 0.04239115827 \approx 4\%$$

that would mean a 96% chance that your home will be flooded before you have paid off
your mortgage.

12.6 Improbable things happen all the time

In August 2017 in the wake of hurricane Harvey *The Washington Post* offered an argument explaining why rare weather events are more common than you think:

> The United States has 3.5 million miles of rivers and streams and 95,000 miles of shoreline, which means lots of potential for flooding — even flooding that's historic, by local or regional standards — in any given year. As it turns out, the country experiences multiple 500-year flood or storm events (that is to say, an event that it had a 1 in 500 chance of occurring in that given place) every single year. Including Harvey, the country has experienced at least 25 such 500-year rain events since 2010, according to the National Weather Service. [R293]

Although it's unlikely that any particular person will win the state lottery, it's certain that someone will win.

Although it's unlikely that the Doonesbury cartoon characters Alex and Toggle will meet (as they did in that comic strip on March 21, 2009), that's no less likely than any other particular pairing. [R294]

When you encounter an unlikely event you're likely to look for a reason. Sometimes there is none — some rare things happen with probability one (someone wins the lottery), some unlikely things happen because there are lots of tries — heads ten or even twenty times in a row, cancer clusters, floods somewhere.

On September 24, 2009, Carl Bialik, "The Numbers Guy," blogged in *The Wall Street Journal* about a Bulgarian lottery that produced the same six number winning combination twice in a row. Since each particular combination has a winning probability less than one in 5.2 million, the government investigated to check for cheating. "It just happened," a spokeswoman for the Bulgarian embassy in Washington, D.C., said. [R295]

Yet again, this time from Andrew Gelman's blog on May 26, 2011:

> It was reported last year that the national lottery of Israel featured the exact same 6 numbers (out of 45) twice in the same month, and statistics professor Isaac Meilijson of Tel Aviv University was quoted as saying that "the incident of six numbers repeating themselves within a month is an event of once in 10,000 years."
>
> …
>
> But wait a second … How many lotteries are there out there? A quick Wikipedia search yields the following:
> - 62 international lotteries. I think I'm undercounting here because it looks like several countries have multiple "Pick m out of n" lotteries but I'm counting each country only once.
> - 46 states or jurisdictions of the United States have lotteries. Some of these appear to be joint between states, however.
> I think a safe approximate guess is 100 major lotteries worldwide.
>
> These lotteries have different rules — some are more frequent than twice a week, some less frequent, some are easier to win than "pick 6 out of 45", some are harder to win. But a quick calculation is

that if the Israeli lottery will have a repeat in a single month, once in 10,000 years, that if there are 100 lotteries out there, you'll see "the incident of six numbers repeating themselves within a month" roughly once in 100 years. [**R296**]

12.7 Exercises

Exercise 12.7.1. [S][R][Section 12.1][Goal 12.1] Craps.

(a) Write down the 36 ways in which two dice can fall.

(b) Find the probability of each of the possible totals 2, 3, …, 12.

(c) Check that the probabilities sum to 1.

(d) What is the probability of throwing doubles?

Exercise 12.7.2. [S][Section 12.1][Goal 12.1] Sicherman dice.

George Sicherman found a strange way to number the faces of a pair of dice that leads to the same probability distribution of totals as an ordinary pair. One die is marked (1, 3, 4, 5, 6, 8) and the other (1, 2, 2, 3, 3, 4). [**R297**]

(a) Write down the 36 ways in which two Sicherman dice can fall.

(b) Check that the totals from a pair of Sicherman dice have the same probabilities as the totals from ordinary dice.

(c) Compare the odds of throwing doubles with Sicherman dice to those of throwing doubles with ordinary dice.

Exercise 12.7.3. [S][Section 12.1][Goal 12.1] Nontransitive dice.

Here are three strangely numbered dice:

$$A \quad — \quad 3, 3, 3, 3, 3, 6,$$
$$B \quad — \quad 2, 2, 2, 5, 5, 5,$$
$$C \quad — \quad 1, 4, 4, 4, 4, 4.$$

(a) Suppose you roll die A and your friend rolls die B. Find the probability that your die shows a higher number.

[See the back of the book for a hint.]

(b) Answer the previous question if you roll B and she rolls C.

(c) Answer the previous question if you roll C and she rolls A.

(d) Which is the most powerful die?

Exercise 12.7.4. [S][Section 12.1][Goal 12.1] [Goal 12.3] Rain, rain go away.

(a) What is the probability that it will rain tomorrow where you live?

(b) What is the probability that it will rain tomorrow in London?

(c) What is the probability that it will rain tomorrow in both places?

(d) What is the probability that it will rain tomorrow in neither place?

(e) The webpage www.weather.gov/ffc/pop explains how the National Weather Service calculates the "probability of precipitation". Read it and write about what you read.

Exercise 12.7.5. [S][Section 12.1][Goal 12.1][Goal 12.3] Who gave the money?
   On April 24, 2009, an Associated Press story in *The Boston Globe* wondered why a dozen colleges with female presidents got large gifts from a single anonymous donor. The writer asked,

> Coincidence? Unlikely. With about 23 percent of US college presidents women, the odds of a dozen randomly selected institutions all having female leaders are 1 in 50 million. [**R298**]

Verify the computation in the quoted paragraph.

Exercise 12.7.6. [S][R][Section 12.1][Goal 12.1] Blackjack.
   Unlike craps and roulette, where each throw or spin is independent of the one before it, in blackjack the probabilities for the next card depend on the cards that have been dealt so far.
   Since the dealer never changes his strategy, the player can overcome the house advantage by carefully counting the cards that have been dealt and betting more when his odds of winning are better. (For more on this, see en.wikipedia.org/wiki/Card_counting.)

(a) What is the probability that the first card dealt will be a face card (a jack, queen or king)?

(b) At what points in the deal can the probability of a face card be the same as it was at the start?

(c) What is the probability that the next card dealt will be a face card if the 12 cards dealt so far have been J829K6Q9A3K4?

(d) When might the probability of a face card be the same as it is for a full deck as the deal in the previous question continues?

Exercise 12.7.7. [S][Section 12.2] [Goal 12.1][Goal 12.2] Impossible?
   What is the probability that a fair coin will come up heads 100 times in a row? [See the back of the book for a hint.]

Exercise 12.7.8. [S][Section 12.2][Goal 12.1] Would you bet?

(a) What is the probability that a fair coin lands tails up eight times in a row?

(b) Suppose you're offered a bet at 250 : 1 odds against a fair coin landing tails up eight times in a row. What would you do — take one side of the bet or the other, or decide not to play at all? (Your answer should depend both on your answer to the previous part of the problem and on your own personal ideas about risk and money.)

(c) Answer the previous question if you are operating a casino. Compute the house advantage and explain how much money you would make in the long run.

(d) Which is more likely, that eight flips will be TTTTTTTT or that they will be THTHTHTH?

Exercise 12.7.9. [U][Section 12.2][Goal 12.1] Simulating dice.

Use the spreadsheet at `flipcoins.xlsx` to roll dice by setting the probability of a head to be 1/6 instead of 1/2. Then the experiment is the same as rolling ten dice and counting the number of one's that come up.

(a) Describe what happens as you increase the number of tries.

(b) Explain why the probability that you see all ten dice showing a one is $(1/6)^{10}$. Did that ever happen in your experiment? Did you try long enough to expect it to happen?

(c) Explain why the probability that none of the ten dice shows a one is $(5/6)^{10}$. Did that ever happen in your experiment? Did you try long enough to expect it to happen?

Exercise 12.7.10. [U][Section 12.2][Goal 12.2] Online experiments.

Play with some of the applications at `www.random.org/`. Write about what you discover. What was interesting? What was surprising?

Exercise 12.7.11. [S][R][Section 12.2][Goal 12.1] Runs of twenty.

If everyone in the United States flipped a coin twenty times about how many people would see twenty heads? Twenty tails? Heads and tails alternating perfectly?

Exercise 12.7.12. [U][Section 12.4][Goal 12.2][Goal 12.3] Cancer clusters.

(a) Why do you think the counties with the highest and lowest (simulated) cancer incidence are all rural?

(b) Redo the experiment when the population is subdivided into fifty states rather than three thousand counties. Why is the result less dramatic?

[See the back of the book for a hint.]

Exercise 12.7.13. [U][Section 12.5][Goal 12.1][Goal 12.3] Misinformation on the web.

Criticize this argument from `www.numbersplanet.com/`:

> Would you play "14 22 38 49 59 16" on your Powerball ticket if you knew they had already hit on Oct 10th 2009? I would think not! While it's entirely possible that the same six numbers can draw more than once, it simply doesn't happen, or if it does, not very often. [**R299**]

Exercise 12.7.14. [S][Section 12.5][Goal 12.1] [Goal 12.3] More misinformation on the web.

Here's an online discussion about using Powerball statistics to beat the system. The website starts with the data in this table summarizing the result of drawing six numbers multiple times:

Evens	Odds	Drawings
6	0	17
5	1	127
4	2	331
3	3	431
2	4	353
1	5	118
0	6	16

and then says

> The important statistic is the "3 Even / 3 Odd: 431" statistic ... [t]elling you that you should be playing six numbers that has a "3 Even and 3 Odd! number combination". [**R300**]

(a) Plot the data and show that it approximates the normal bell curve, as it should.

(b) Criticize the argument suggesting that you should always bet on a three even, three odd combination.

(c) Fix the website's grammar.

Exercise 12.7.15. [S][Section 12.6][Goal 12.2][Goal 12.3] The Tour de France.

On August 6, 2008, London's *Telegraph* featured an article headlined "Olympic Games drug testing means 'cheaters escape and innocents tarnished'". That article featured a comment from Professor Donald Berry discussing doping charges against the cyclist Floyd Landis:

> With 126 samples taken in the Tour de France, assuming 99 per cent specificity, the false-positive rate is 72 per cent. So, an apparently unusual test result may not be unusual at all when viewed from the perspective of multiple tests. ... This is well understood by statisticians, who routinely adjust for multiple testing. [**R301**]

Professor Berry was then head of the Division of Quantitative Sciences and chair of the Department of Biostatistics and Frank T. McGraw Memorial Chair of Cancer Research, MD Anderson Cancer Center, so presumably he knows what he is talking about.

Check his conclusion that a drug test that reports a false positive only 1% of the time will return a false positive with 72% probability when it's repeated 126 times.

(News reports in 2010 suggested that Landis was in fact guilty, even though you could not reliably conclude that if all you had were the results of those particular tests.)

Exercise 12.7.16. [U][Section 12.6][Goal 12.2][Goal 12.3] Old Friends Farm.
An email message from Old Friends Farm (`www.oldfriendsfarm.com/`) in Amherst, MA, notes that

> Two members of our [six person] crew realized that they had met 10 years ago on a farm in France, and not seen each other since. What are the chances of that!? [**R302**]

Discuss the coincidence.

Exercise 12.7.17. [U][Section 12.6][Goal 12.2][Goal 12.3] The law of averages.
Mma Makutsi warns Mma Ramotswe:
In *The Double Comfort Safari Club* Alexander McCall Smith puts these words in Mma Makutsi's mouth:

> There is something called the law of averages — you may have heard of it. It says that if you haven't trodden on a snake yet, then you may tread on one soon-soon. [**R303**]

Discuss how this opinion fits with the ideas in this chapter.

Review exercises.

Exercise 12.7.18. [A] Find the probability of each event.

(a) Tossing two fair coins and getting HT as the result.

(b) Tossing a fair coin twice and getting HT as the result.

(c) Drawing a red card from a deck of playing cards.

(d) Drawing a red ace from a deck of playing cards.

(e) Drawing an ace from a deck of cards and then drawing a king.

(f) Rolling two fair dice and getting snake eyes (1 and 1).

(g) Rolling two fair dice and getting boxcars (6 and 6).

(h) Rolling two fair dice and getting a total of 7.

Exercises added for the second edition.

Exercise 12.7.19. [S][Section 12.6][Goal 12.1][Goal 12.2] Just luck?
On February 16, 2018, David Grinberg asked on stackexchange

> My tax refund is exactly $3000. Did I hit a cap? [**R304**]

(a) Estimate the number of tax returns filed in 2018 for 2017 year taxes.

(b) Check your estimate with a web search.

(c) Estimate the probability of a tax refund that's an even number of thousands of dollars.

(d) Should Grinberg have been surprised?

(e) Should you be surprised by Grinberg's surprise?

Exercise 12.7.20. [U][S][Section 12.5][Goal 12.2] Will I get my feet wet?
The website www.gfdrr.org/en/100-year-flood offers an online calculator for flood probabilities. Visit, ask it for a probability, and check the answer it gives.

Exercise 12.7.21. [U][S][Section 12.6][Goal 12.1][Goal 12.2] Only girls?
In August 2019 this question was posted at skeptics.stackexchange.com:

> German media has reported that a village in Poland has had only girls born for the last decade:
>
> "The head of the parish promises a reward, the Catholic village priest asks for God's help: In the Polish village of 300 souls, Mistitz, no boys have been born for almost ten years."
>
> Source: "Mistitz in Polen — Dorf ohne Jungen", tagesschau Stand: 17.08.2019 11:43
>
> How is that possible that there are no boys born? Statistically, it seems to me to be nearly impossible. Is this real, or is it a publicity gag or something else?
>
> English media has also reported on the village: "Girls only: Tiny Polish village of 300 people waits for first birth of a boy for nearly a DECADE", Daily Mail. [**R305**]

Might this be real?
[See the back of the book for a hint.]

Exercise 12.7.22. [U][S][Section 12.1][Goal 12.1] Cautious optimism.
At the height of the Covid-19 pandemic, Mark Lampert and his coauthors compared the search for a therapeutic agent to the need to take lots of shots on goal in low scoring games like soccer and hockey. Although each shot has low probability, when you take many the laws of probability predict eventual success. They note that there are about 100 drugs now being tested.

> With so many shots on goal, if each of these candidates has even a 5 percent chance of success, the probability that at least one impactful drug emerges is over 99 percent. [**R306**]

(a) Why do the authors believe that the probability that all the drugs fail is $(1-0.05)^{100}$?

(b) Evaluate that expression and confirm that there is more than a 99 percent chance that at least one drug succeeds.

(c) The implicit assumption in the first question is that the individual probabilites of success are independent. Suppose that the 100 drugs represent just five different kinds of drugs and that each kind has a 5 percent chance of success. Calculate the probability that at least one kind succeeds.

13

How Good Is That Test?

In Chapter 12 we looked at probabilities of independent events — things that had nothing to do with one another. Here we think about probabilities in situations where we expect to see connections, such as in screening tests for diseases or DNA evidence for guilt in a criminal trial.

Chapter goals:

Goal 13.1. Interpret and build two way contingency tables.

Goal 13.2. Understand how to compute probabilities for dependent events.

Goal 13.3. Understand the implications of false positives.

13.1 UMass Boston enrollment

Table 13.1 summarizes student enrollment at UMass Boston in 2007 by category two ways: graduate/undergraduate and male/female. We can use the data to answer some probability questions about a random student.

- What is the probability that a student chosen at random is an undergraduate?

	Undergraduate	Graduate	Total
Female	5,680	2,388	8,068
Male	4,328	1,037	5,365
Total	10,008	3,425	13,433

Table 13.1. UMass Boston enrollment, 2007

The last row of the table has the numbers we need:
$$\frac{\text{number of undergraduates}}{\text{number of students}} = \frac{10,008}{13,433}$$
$$= 0.745$$
$$\approx 75\%.$$

Three quarters of the students are undergraduates.

- What is the probability that a student is female?

  For that computation we use the totals in the last column:
  $$\frac{\text{number of females}}{\text{number of students}} = \frac{8,068}{13,433}$$
  $$= 0.600610437$$
  $$\approx 60\%.$$

- What is the probability that a student is a female undergraduate?

  Use the count in the first column of the first row:
  $$\frac{\text{number of female undergraduates}}{\text{number of students}} = \frac{5,680}{13,433}$$
  $$= 0.4228392764$$
  $$\approx 42\%.$$

In each of these probability calculations we used the total number of students (13,433) in the denominator.

Continuing ...

- What is the probability that a female student is an undergraduate?

  Since this is a question about the female students, we need a different denominator:
  $$\frac{\text{number of female undergraduates}}{\text{number of female students}} = \frac{5,680}{8,068}$$
  $$= 0.70401586514$$
  $$\approx 70\%.$$

- What is the probability that an undergraduate is female?

  That's a different question. This time we know the student is an undergraduate. That calls for a different denominator:
  $$\frac{\text{number of female undergraduates}}{\text{number of undergraduates}} = \frac{5,680}{10,008}$$
  $$= 0.56754596322$$
  $$\approx 57\%.$$

The last two questions sound similar but have different answers, because each begins with a different assumption. In the first we know the student is female and wonder whether she's an undergraduate. In the second, we know that the student is an undergraduate and wonder whether it's a she.

We're not finished thinking about these probabilities. We found that there's a 60% probability that a student is female. But if we know the student is an undergraduate then that probability drops to 57%, because the proportion of women is different for undergraduates than for the student body as a whole. This is not what happened when we thought about a coin and a die in Section 12.1. The probability that the die shows a four is the same whether the coin comes up heads or tails. Those events are *independent*. The facts "is female" and "is an undergraduate" are *dependent*. When you know one of them you know something about the probability of the other.

We learned in Section 12.1 that when events are independent you multiply to compute the probability that both happen:

$$\text{probability(coin H and die 4)} = \text{probability(coin H)} \times \text{probability(die 4)}$$
$$= \frac{1}{2} \times \frac{1}{6}$$
$$= \frac{1}{12}.$$

For dependent events that won't work. We found that

$$\text{probability(female and undergraduate)} = 42\%$$

but

$$\text{probability(female)} \times \text{probability(undergraduate)} = 60\% \times 75\%$$
$$= 45\%.$$

Those answers are close, but not the same, because the proportion of females among the undergraduates is close to but not the same as the proportion among the graduate students.

In the rest of this chapter we will look at the probabilities for dependent events, working with displays like Table 13.1 in examples where the consequences matter much more than they do here.

13.2 False positives and false negatives

Many women have periodic mammograms to look for breast cancer. Many men have periodic PSA tests to look for prostate cancer. In each there are four possibilities. We'll spell them out for breast cancer.

- *True positive*: a woman has breast cancer and the mammogram says so.

- *True negative*: a woman does not have breast cancer and the mammogram says she doesn't.

- *False positive*: a woman doesn't have breast cancer but the mammogram mistakenly says she does.

- *False negative*: a woman does have breast cancer but the mammogram doesn't detect it.

If the test were perfect there would be no false positives and no false negatives — but there are very few perfect tests. In order to understand what the test results mean you can build a table like the one in the first section of this chapter.

We'll do that with a real example. Figure 13.2 appeared in the article "False positives, false negatives, and the validity of the diagnosis of major depression in primary care" in the September 1998 *Archives of Family Medicine*.

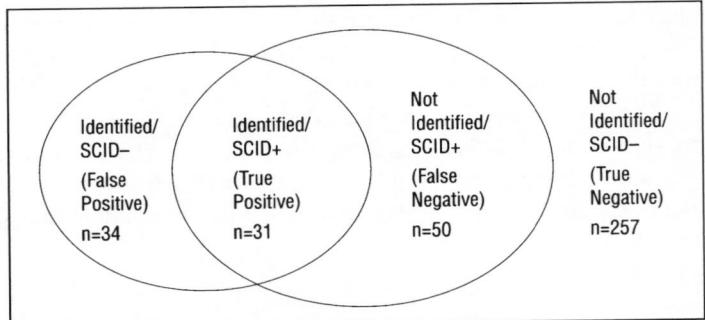

Proportions of patients identified as depressed by physicians, Structured Clinical Interview for Diagnostic and Statistical Manual of Mental Disorders, Revised Third Edition (SCID), both methods, or neither method (N = 372).

Figure 13.2. Diagnosing depression [**R307**]

It summarizes the results of a study of 372 patients who were screened by family physicians for clinical depression.

The numbers in the four categories in the figure are easier to understand when we put them in Table 13.3.

		depressed		total
		yes	no	total
diagnosed	yes	31	34	65
	no	50	257	307
	total	81	291	372

Table 13.3. Diagnosing depression

Two-by-two tables like this are called *contingency tables.* Figure 13.4 shows the standard names for the four cells with raw data: true positive, false positive, false negative and true negative. In this example they have values 31, 34, 50 and 257.

The totals tell us that 65 people in the population of 372 (17%) were diagnosed as depressed and that 81 (22%) were depressed. Those numbers are pretty close. But does that make it a good test? To answer that question we need to look at the columns separately.

- The first column tells us there were 31 true positives and 50 false negatives from the total of 81 subjects who were in fact depressed. So if a subject was depressed

		condition	
		present	absent
screened positive	yes	true positive	false positive
	no	false negative	true negative

Figure 13.4. A two way contingency table

the probability that he or she was diagnosed correctly is only $31/81 \approx 38\%$. That is the true positive rate. There's a 62% chance the condition was missed. That 62% is the false negative rate.

• The second column says that even when a subject was not depressed the chance of a diagnosis of depression was $34/291 = 0.117 \approx 12\%$. That is the false positive rate.

Whether this is a "good" test is a difficult decision.

Although the chance of misdiagnosis of depression when it doesn't exist is fairly low — about 12% — the 62% false negative rate says that test will identify fewer than half the depressed people.

13.3 Screening for a rare disease

A test with a small false positive rate looks like a good candidate for screening large populations for a nasty disease. However, if the disease is rare, the test may not be as good as it looks. In this section we'll study two examples, one made up and one real.

Suppose a drug company has developed a test for the rare disease X. Clinical trials show that the test is 90% accurate at detection, so the false negative rate is 10%. Those trials also show that the false positive rate is only 1%.

These are the important questions:

(1) What is the probability that a person who suffers from X tests positive?

(2) What is the probability that a person who tests positive suffers from X?

If the test were perfect — no false positives, no false negatives — each question would have the same answer: 100%. But the two facts "suffers from X" and "tests positive for X" are not exactly the same. Knowing either one makes the other more likely, but not certain. We want to find out how much more likely in each case.

Question 1 is easy: the drug company's clinical trials found that there is a 90% probability that a person who suffers from X tests positive for X.

Whether that test is as good as it sounds depends in part on the answer to the second question. That answer depends on two things: the false positive rate and the number of people who actually have X. Suppose just one person in every 1,000 suffers from X (one tenth of one percent of the population). Then even though the false positive rate is only 1%, most of the positive results will come from healthy people. We can use a contingency table to find the actual value for "most of".

Since percentages (particularly small percentages) are often confusing, we'll build our table for an imaginary population of 100,000 people that just matches the statistical profile for this test. In a population of 100,000, one out of every 1,000 will have the disease. That's 100 people. Of those 100, 90% (so 90 people) will test positive. The other 10 will be the false negatives. Of the 99,900 healthy people, one percent (999) will test positive. The other 98,901 will be the true negatives. Table 13.5 shows the contingency table.

		suffers from X		
		yes	no	total
test + for X	yes	90	999	1,089
	no	10	98,901	98,911
	total	100	99,900	100,000

Table 13.5. Screening for disease X

Now we can answer the second question. The probability that someone who tests positive is actually ill with X is only 90/1,089 = 8.26%.

Is this acceptable? Maybe, maybe not. If the test is inexpensive and there's a second test (perhaps more expensive) that can weed out the false positives and the disease can be treated successfully if detected, perhaps the screening is a good idea. If all the people who test positive must undergo expensive painful unreliable treatment, which would be unnecessary for more than 90% of them, then the screening is probably a bad investment of scarce health care resources.

For a real application of this technique to the statistics of screening for breast cancer, work Exercise 13.7.7.

13.4 Trisomy 18

In this section we'll work through the numbers when considering whether to call for routine prenatal screening for the rare birth defect trisomy 18.

> K Spencer and colleagues' claims for prenatal detection of trisomy 18 by measurement of maternal serum (alpha) fetoprotein and free β human chorionic gonadotrophin concentrations are impressive. Detection of 50% of cases for a false positive rate of only 1% seems to compare favourably with the detection rate for Down's syndrome when similar techniques are used, which is 70% for a false positive rate of 5%. Unfortunately, the authors fail to emphasise the importance of the relative incidence of the two conditions at birth before concluding that screening for trisomy 18 should be introduced. [**R308**]

Davies notes that there are about 12.6 instances of Down's syndrome per 10,000 births. The incidence for trisomy 18 is just 1.3 per 10,000 births.

A positive test leads to a second procedure, an amniocentesis to check whether the positive is true or false. Davies calculates that screening 10,000 pregnant women for Down's syndrome "would result in 8.8 cases being detected at the cost of 500 amniocenteses (5% of 10,000). This means that one case of Down's syndrome is detected for every 57 amniocenteses performed." For trisomy 18 his figures are 0.65 cases detected, 100 amniocenteses, so 154 amniocenteses to detect one true case.

Let's check his arithmetic for trisomy 18. To build the contingency table we need three numbers:

- The false negative rate. Since the test detects 50% of cases the other 50% are the false negatives.

- The false positive rate. It's only 1%.

Contingency Table Calculations
Ethan Bolker and Maura Mast, for *Common Sense Mathematics*

Set population to 1 to see percentages in the table.
Set population to a large round number like 1000 to see meaningful numbers of cases.

The incidence, false positive and false negative rates are formatted as percentages.
You can use actual values in Excel formulas to set them - for example, =1/1000 for 0.1%.

population:	10000	
incidence:	0.0130%	
false+:	1.00%	
false-:	50.00%	

	actual+	actual-	total
test+	0.65	99.987	100.637
test-	0.65	9898.713	9899.363
total	1.3	9998.7	

0.65% prob(actual+ when test+)
99.99% prob(actual- when test-)

Figure 13.6. Screening for trisomy 18

- The incidence rate. The second paragraph tells us it's 1.3 per 10,000 births.

Figure 13.6 shows a screenshot of the spreadsheet ContingencyTable.xlsx with entries for this problem. Cell B12 is named INCIDENCE; it contains the formula =1.3/10000, formatted as a percent. Cell B17 for the number of true positive results contains the formula

=POPULATION*INCIDENCE*(1-FALSENEG)

which in this example is

$$10,000 \times \frac{1.3}{10,000} \times (1 - 0.5) = 0.65,$$

confirming Davies's "0.65 cases per 10,000 women tested." The spreadsheet shows 100.637 positive tests, which matches Davies's estimate of 100. So it would take 100 amniocenteses to find 0.65 cases of trisomy 18. That works out to $100/0.65 = 154$ amniocenteses to find each case. (Exercise 13.7.3 asks you to check Davies's arithmetic for Down's syndrome.)

You don't have to know what amniocentesis is, but you do have to know that it has some risks: there is a small chance that it will lead to a miscarriage. Davies claims that "a screening programme would cause the abortion of at least as many normal fetuses as it would detect cases of trisomy 18." ("Abortion" is his synonym for "miscarriage," not the politically charged "abortion" so much in the news.)

That would be true if the risk of miscarriage from amniocentesis was about one in 150. It's probably smaller. Several web sources provide statistics like these:

> Miscarriage is the primary risk related to amniocentesis. The risk of miscarriage ranges from 1 in 400 to 1 in 200. In facilities where amniocentesis is performed regularly, the rates are closer to 1 in 400.
> [R309]

If we use one in 300 instead of one in 150 as the probability of miscarriage from an amniocentesis then it costs about one unnecessary miscarriage to detect two cases of trisomy 18. That's still a pretty high risk.

Davies compares his risk estimate to the much lower estimate for similar screening for Down's syndrome, noting that "[i]n many places it is still undecided whether screening for Down's syndrome is worth the disbenefits for the prospective parents."

13.5 The prosecutor's fallacy

The Cornell University Legal Information Institute posted a discussion of *McDaniel v. Brown* when that case was on the docket of the Supreme Court. They wrote:

> Following a state conviction for sexual assault, Troy Brown filed a petition for writ of habeas corpus in the United States District Court for the District of Nevada. The District Court allowed Brown to present new evidence: a report from Dr. Lawrence Mueller. This report detailed a statistical error ("prosecutor's fallacy") made by the prosecution during the presentation of DNA evidence. Based on Dr. Mueller's report, the District Court dismissed the DNA evidence from consideration, found insufficient evidence to convict Brown, and ordered a retrial.
>
> ...
>
> At trial, Renee Romero, a forensic scientist at the Washoe County Crime Lab, testified that the DNA found in the victim's underwear matched Brown's DNA; only one in three million people would match the DNA tested. The prosecutor asked Romero to express this statistic as "the likelihood that the DNA found ... is the same as the DNA found in [Brown's] blood." Romero concluded that the likelihood was 99.999967 percent. Based on this statistic, the prosecutor then asked Romero if it would be fair to conclude that there was a 0.000033 percent chance that the DNA did not belong to Brown. Romero agreed with the prosecutor, stating that this was "not inaccurate." [**R310**]

Romero's arithmetic is right: one in three million is 0.000033 percent. But her thinking is wrong.

The prosecutor's fallacy is the claim that the one in three million probability of a random match is the same as the probability that the defendant is innocent. We can use a contingency table to show why those probabilities are different.

First we need an estimate of the population in which a possible DNA match might be found. To make the arithmetic easier, we'll take that to be 9 million people (Los Angeles is near enough to Nevada). Then the "one in three million" statistic says we should expect three DNA matches from the innocent people in that population. This contingency table summarizes the data:

		truth		
		guilty	innocent	total
DNA	match	1	3	4
	nonmatch	0	8,999,996	8,999,996
	total	1	8,999,999	9,000,000

Make sure you understand the first row of the table. One person is guilty and is a DNA match. The three matches from innocent people are false positives. In other words, the first row of that table tells us that if the only evidence in the case is the DNA match the odds are 3 : 1 that the suspect is innocent! The probability that he's guilty is only 25%. That's a far cry from the "99.999967% guilty" that the prosecutor asked the jury to believe.

The defense didn't make this argument using a hypothetical 9,000,000 population of potential suspects. Instead they questioned the "one in three million" chance of a match. The defendant had near relatives in the area which increased the chances of a match to about one in 6,500, according to a defense specialist. That would reduce the chance of an accidental match to $6499/6500 = 0.999846154 \approx 99.98\%$. We're not surprised that the change from 99.999967 percent to 99.98% did not convince the jury to acquit. 99.98% still sounds very much like a sure thing.

But it's not, because of the prosecutor's fallacy. That was the basis for the appeal. Suppose we reduce the population from which the match might come to just 100,000 — the nearby area where there may be close relatives. Then the 1 in 6,500 chance of a match means there will be about 15 matches in that population in addition to the one match for the guilty party. The numbers in the revised contingency table below show there is now a 15 : 1 chance that the DNA match fingers an innocent person rather than the true criminal:

		truth		total
		guilty	innocent	
DNA	match	1	15	16
	nonmatch	0	99,984	99,984
	total	1	99,999	100,000

Nevertheless, the story did not end well for Brown.

> The Supreme Court [overturning the appeals court order for a retrial] said in a per curiam opinion that overstated estimates of a DNA match at trial did not warrant reversal of a conviction when there is still "convincing evidence of guilt." [R311]

13.6 The boy who cried "Wolf"

After unusual disasters like terrorist attacks, earthquakes, severe storms or airplane crashes you often hear finger-pointing discussions about the incompetence of the agencies charged with predicting (perhaps even preventing) what happened. Those discussions may start with a search that discovers warning signs that were ignored.

Sometimes there were real lapses, and policies and practices must be designed to prevent a recurrence. But often blame is unjustified. Table 13.7 explains why, even without numbers. You might call this *qualitative reasoning*.

With numbers in the first column you can compute the probability that a disaster occurs with no warning at all. That's not good. To guard against it, there should be more warnings. Then with numbers in the first row you can compute the probability that a particular warning actually corresponds to a disaster about to happen. But more warnings don't lead to more disasters, just to more false positives.

		what happens		
		disaster	nothing	total
warning?	yes	rare	usually	infrequent
	no	rare	almost always	almost always
	total	rare	almost always	always

Table 13.7. Should it have been predicted?

That means there are often good reasons for ignoring a warning. State and governmental agencies have to balance the severity of the warning with the cost and inconvenience of asking the public to respond. For example, an earthquake warning may lead to an order to evacuate an entire city. The expense and disruption from repeated evacuations that are not followed by an earthquake may be worse than the consequences in the rare instance when the earthquake happens. Just because after the fact you look back and find clues in the seismic record that suggested an earthquake might be imminent doesn't mean evacuation was the right call.

13.7 Exercises

Exercise 13.7.1. [S][Section 13.2] [Goal 13.1][Goal 13.3] Chronic fatigue syndrome.

On August 24, 2010, a headline in *The Boston Globe* read "Researchers link chronic fatigue syndrome to class of virus". The story reported on a study of 37 patients with the disease. 32 tested positive for a particular suspicious virus. Only 3 of 44 healthy people tested positive. [**R312**]

A 2003 study in the *Archives of Internal Medicine* reported that "The overall ... prevalence of CFS ... was 235 per 100,000 persons." [**R313**]

(a) Construct the contingency table for this diagnostic tool. You may do this by hand or with the spreadsheet ContingencyTable.xlsx.

(b) Explain why this test is potentially important for research on chronic fatigue syndrome but might not be a good screening test.

[See the back of the book for a hint.]

Exercise 13.7.2. [S][W][Section 13.2][Goal 13.1] [Goal 13.3] Pregnancy tests.

An online website on pregnancy testing says that

> Usually, if all care has been taken, [home] pregnancy tests are 97% accurate. [**R314**]

Assume that "97% accurate" means a false positive rate and a false negative rate of 3%. Since a woman is unlikely to use a home pregnancy test unless she thinks she's probably pregnant, assume that 80% of the women who try one are in fact pregnant.

Explain why a positive test indicates a pregnancy more than 99% of the time even though the false positive rate is 3%.

Exercise 13.7.3. [U][Section 13.3] [Goal 13.1][Goal 13.3] Prenatal screening.

Check the calculations for Down's syndrome testing using the data in the quotation in Section 13.3.

Exercise 13.7.4. [U][C][Section 13.2][Goal 13.1] [Goal 13.3] Spam.

Spam is junk email. Most mail systems have a spam filter that tries to decide whether each piece of email you get is spam. When the spam filter finds something it thinks is spam, it may throw it away, or put it in a junk mail folder so that you can decide whether to throw it away without reading it.

Before my university department set up a spam filter I ran my own. (The "I" here is Ethan Bolker, one of the authors, not the generic authorial "we" we use in most of the book.)

I got about 250 emails each day. My spam filter trapped about 175 of them. Of those about five were legitimate and should have been delivered directly to me. My inbox, which should have contained just the legitimate messages was usually about half spam. So (in words) my spam filter is pretty good (but not perfect) at recognizing legitimate email but not very good at calling spam, spam.

(a) Build a two way contingency table with row categories "marked spam" and "not marked spam", column categories "spam" and "legitimate".

(b) Compute and interpret the false positive and false negative rates.

(c) Explain why both the false positives and the false negatives make dealing with my email harder.

(d) I can adjust the settings in my spam filter to reduce the false positive rate. Explain why that would increase the false negative rate.

(e) Is the number of spam emails I received consistent with the claim in the August 6, 2008, issue of *The New Yorker* that there are more than a hundred billion spam emails every day? [**R315**]

(f) What is the original meaning of the word "spam"? Does the company that sells (the real) spam object to the new meaning?

(g) How do you deal with spam? If your email provider does all the filtering for you, you may not even know it's throwing things away before you see them, so you may need to do some research on your email provider's website to find the answers to the following questions.

- Who provides your email service (your university, your internet service provider, Google, Yahoo, ...) ?

- Do you have any say in how your email provider filters spam for you? If so, what do you tell it?

- Estimate the data you need to build the two way table for your spam statistics and compute the false negative and false positive rates.

Here are some websites to look at if you want to find out more about spam.

- www.imediaconnection.com/content/3649.asp . There are some useful tips here about how to keep other people's spam filters from thinking mail from you is spam.

- Tools your system administrator might use: www.spamcop.net/, www.spamhaus.org/

Exercise 13.7.5. [S][Section 13.2][Goal 13.1] [Goal 13.3] Plagiarism.

In 2006 UMass Boston experimented with the plagiarism detection software described at www.turnitin.com that claims it can identify plagiarism in essays students write. UMass did not purchase the software after the experiment. Perhaps the possibility of false positives contributed to that decision.

Suppose that the software can actually detect every cheater and that it's 99% accurate in declaring honest students honest. (We made up "detect every cheater" and "99%" since the company does not advertise them.) Sounds like a pretty good test.

(a) Estimate how many papers are submitted by students at your school each semester.

(b) Suppose that most students are honest. Estimate how many students will be falsely accused of cheating.

(c) What are the advantages and disadvantages of using the software? (There are several arguments on both sides of the question. Think of as many as you can.)

(d) Read and write about this article from *The New York Times*: www.nytimes.com/2010/07/06/education/06cheat.html.

Exercise 13.7.6. [U][Section 13.2][Goal 13.3] Airport screening.

In response to the article "Screening programme evaluation applied to airport security" in the December 10, 2007, issue of the *British Medical Journal*, Ganesan Karthikeyan wrote:

> It is probably true that airport security in its present form is not an efficient screening measure. However, one important difference exists between screening for disease in individual patients and screening for, say, explosives in airports. While one missed cancer on screening can cause the loss of at the most, one life, the number of potential lives lost per missed screening at airports can be substantially larger. This has to be factored into any attempts at evaluation of the process.
> **[R316]**

It's clear that a false negative is a disaster. Discuss the consequences of a high false positive rate.

Exercise 13.7.7. [S][W][Section 13.2] [Goal 13.1][Goal 13.3] Breast cancer screening.

In his "Chances Are" column in *The New York Times* on April 25, 2010, Steven Strogatz wrote about a diagnostic puzzle presented to several doctors:

> The probability that [a woman in this cohort] has breast cancer is 0.8 percent. If a woman has breast cancer, the probability is 90 percent that she will have a positive mammogram. If a woman does not have breast cancer, the probability is 7 percent that she will still have a positive mammogram. Imagine a woman who has a positive mammogram. What is the probability that she actually has breast cancer?
> ...
> [When 24 doctors were asked this question], their estimates whipsawed from 1 percent to 90 percent. Eight of them thought the chances were 10 percent or less, 8 more said 90 percent, and the remaining 8

guessed somewhere between 50 and 80 percent. Imagine how upsetting it would be as a patient to hear such divergent opinions. [**R317**]

(a) What is the correct answer?

[See the back of the book for a hint.]

(b) What percentage of the 24 doctors got the correct answer?

Exercise 13.7.8. [U][S][Section 13.2][Goal 13.1] [Goal 13.3] Identity fraud.

On July 17, 2011, in an article in *The Boston Globe* headlined "Caught in a dragnet" you could read that in 2010 the Massachusetts Registry of Motor Vehicles used software that cost $1.5 million to send 1,500 suspension letters a day, leading to 100 arrests for fraudulent identity and 1,860 revoked licenses. [**R318**]

On July 24 Jane Allen wrote a letter to the editor in response, to say that the time and money hardly seemed worth it since only

about 390,000 people were questioned for the sake of finding fewer than 2,000 transgressors. [**R319**]

(a) Check Allen's arithmetic in the second paragraph.

(b) Identify the false positives and calculate the false positive rate. Explain the costs and benefits.

Exercise 13.7.9. [U][C][Section 13.5][Goal 13.1] Candy leads to crime.

An article headlined "Happy Halloween! Kids who eat candy every day grow up to be violent criminals" in the October 2, 2009, *Daily Finance*, begins

Quick, hide the candy jar! Feeding your child candy every day could help turn Junior into a violent criminal, according to a large study in Britain, which found that 69 percent of the participants who had committed violence by 34 had eaten sweets or chocolate nearly every day during childhood. [**R320**]

You can find the full text at www.aol.com/2009/10/02/happy-halloween-kids-who-eat-candy-every-day-grow-up-to-be-viol/.

(a) Read the rest of the article. Build the contingency table with columns for whether or not someone ate candy as a child, rows for whether or not they committed violence as an adult.

(b) Explain why this is an example of the prosecutor's fallacy.

(c) Some of the online comments on that article recognize the fallacy — for example

10-03-2009 @ 10:21PM
Bski said...
I bet you, 99% of criminals ate bread daily by the time they were 10 years old!!!!

Write your own blog entry, using your understanding of two way contingency tables to enlighten any readers. If you like what you've written you may still be able to post your comment on the article's blog.

Exercise 13.7.10. [U][Section 13.2][Goal 13.3] Domestic violence.

In Andrew Gelman's blog on "Statistical Modeling, Causal Inference, and Social Science" commenter Mike Spagat writes that

> Even within exceptionally violent environments most households will still not have a violent death. So a very small false positive rate in a household survey will cause substantial upward bias in violence estimates. [R321]

Write a paragraph or two explaining this to someone who is interested and smart enough to understand this but has not studied the material in this chapter. Consider making up some numbers to illustrate your argument.

Exercise 13.7.11. [U][Section 13.1][Goal 13.1] Surgery for prostate cancer?

An article in *The Boston Globe* headlined "Surgery offers no advantage for early prostate cancer, study finds" reported on a clinical trial involving 731 men diagnosed with prostate cancer. About half had surgery; the rest were monitored.

> After 12 years, nearly 6 percent of men who had immediate surgery died of the cancer, compared with slightly more than 8 percent of those patients who were observed, which was not a great enough difference to reach statistical significance. [R322]

(a) About how many men were in each category?

(b) About how many deaths were there in each category?

(c) Construct the contingency table for this study.

Exercise 13.7.12. [S][A][W][Section 13.2][Goal 13.1][Goal 13.3] Teenage drug use.

Here's a made-up story.

The dean at a fancy private high school is very worried. She suspects that about 20% of the 1,000 students on campus are using drugs. She has asked all the parents to administer a home drug test to their kids (since it's a private school she can actually require them to do it). She has read on the web that

> With home drug testing methods believed to produce reliable and accurate results, many of us overlook the cases of false positives and draw conclusions on the suspect before reconfirming the result. But, researchers from the Boston University have found out that drug tests may produce false positives in 5–10% of cases and false negatives in 10–15% of cases. [R323]

We found several blogs that seem to report on this same study. None gives a link or a precise reference. We haven't been able to locate the original.

Answer the following questions, assuming the worst cases (10% false positive rate, 15% false negative rate).

(a) Build the contingency table for this drug screening scenario. To do that you will have to figure out:

> How many students are drug users?
>
> How many of the drug users test positive? How many test negative?

How many students are drug free?

How many of the drug-free students test positive? How many test negative?

You may do the arithmetic by hand or with the spreadsheet at `ContingencyTable.xlsx`.

(b) What is the true positive rate?

(c) Student John Smith tested positive. What is the probability that he is really on drugs?

(d) Student Jane Doe tested negative. What is the probability that she is really drug free?

(e) Answer the previous two questions if you assume the best cases for reported false values in the Boston University study.

Exercise 13.7.13. [U][Section 13.6][Goal 13.1][Goal 13.3] The boy who cried "wolf". Use Table 13.7 to analyze the children's story with that title.

Exercise 13.7.14. [S][Section 13.6][Goal 13.1] Playing the lottery.
Table 13.8 illustrates the ultimate example of the error you can make reading a column instead of a row.

		bought a ticket		total
		yes	no	
won the lottery	yes	1	0	1
	no	many	very many	very many
	total	many	very many	very many

Table 13.8. Playing the lottery

(a) Suppose you won the lottery. What is the probability that you bought a ticket?

(b) Suppose you bought a ticket. What is the probability that you won the lottery?

Hints

Exercise 1.8.4. To answer the question you'll have to think about the size of the restaurant, the number of years it's been open, the number of customers served in a day and the fraction of those customers who order soup with matzoh balls (probably not a big fraction).

   You may need to do some research if you don't know what a matzoh ball is.

   Please don't contact the restaurant and ask them to answer the question for you.

Exercise 1.8.12. For the second question you have to think about whether people trying in one minute are the same as people trying the next minute, or trying an hour later.

Exercise 1.8.20. By now you shouldn't need this hint, but here it is: you will need to know the population of Peru to answer some of the questions.

Exercise 1.8.32. Make and justify estimates for

• the redemption value of a soda can or bottle,

• the 2006 cost of years at MIT and at California state schools.

   Then estimate how many cans or bottles you need to collect for one year's tuition. How many is that per day or per hour? Do your answers seem reasonable? Do you have to think about how many days per year or hours per day you could (reasonably) spend collecting cans?

Exercise 1.8.57. You may have to look up some baseball data to answer this question. How many pitches per game? Games per season?

Exercise 2.9.1. From: Laura M Keegan
To: eb@cs.umb.edu
Subject: Homework Help
Date: Mon, 16 Sep 2013 03:21:07 +0000
Hello Professor Bolker,
I have a questions about number 2 on the homework, I have not been
able to figure out exactly what I am supposed to do, I am not sure
of the type of article I am supposed to be searching for or just
how to get started looking for one? Sorry, this is kind of a silly
question but this part of the homework has taken me a really long
time and I still can't find/don't know exactly what to look for.
Thank you,
Laura Keegan

From: Ethan Bolker
To: Laura M Keegan
Subject: Homework Help
Date: Mon, 16 Sep 2013 09:21:15 -0400 (EDT)
Laura
There really is no "exactly what [you are] supposed to do". Just
read the newspaper or a magazine, find a story where numbers with
units appear in several versions (miles per gallon and gallons per
mile, dollars per hour and dollars per year, barrels of oil and liters
of gasoline, accidents per week and accidents per person, ...) and
show me that the arithmetic is right and the numbers make sense. You
don't need to discuss the whole article - just extract a paragraph or
a sentence two with the numbers.
The exercises in the book suggest the kind of thing I have in mind. If
you find a good one I can put it in the book.
There's no particular "type of article" to look for. If you find
one on a subject you're interested in rather than just thinking of
this as a homework problem you have to do you might even enjoy it.
Although I'd like to encourage you to read the newspaper, you can look
on the net for a story.
Ethan

Exercise 2.9.2. Warning: it's not 45 miles/hour.
The problem does not say how far it is from Here to There. You're free to pick some convenient distance to work with if you like.

Exercise 2.9.28. A search with the keywords *automobile crash statistics* leads to many links with data to confirm the numbers in these quotations.

Exercise 2.9.41. The metric system will help: a cubic centimeter of water weighs one gram. You can convert pints to cubic centimeters; pounds to kilograms and then to grams.

Exercise 2.9.52. You can find an inflation calculator on the web, or read ahead in Chapter 4.

Exercise 2.9.64. Remember that the volume is proportional to the cube of the height.

Exercise 3.8.11. The answers to some of these questions will be very small numbers. Keep careful track of the decimal points.
For the last question, which is more informative, the absolute or the relative comparison?

Exercise 3.8.24. You can do this problem with algebra, but you may find it easier by working with some definite numbers that aren't given in the problem. So pretend that fuel economy before the increase is, say, 20 miles per gallon and you will drive 100 miles. Then work out how much gas you save (in percentage terms) if that 20 miles per gallon increases 20% to 24 miles per gallon. For the second part of the problem, convert 20 miles per gallon to gallons per mile, or to gallons per hundred miles. Then a 20% improvement will decrease that number by 20%, since "improvement" means fewer gallons per mile.

Exercise 3.8.30. You can't really get started on this problem if you don't know what "wholesale price" and "markup" mean. If that's the case, look them up.

You may find it easier to work with a book whose wholesale price is $1 or $100.

Exercise 3.8.35. Read carefully. The $6.6 billion figure is what the banks will lose in revenue, not what they make now or will make after the cut takes effect.

Exercise 3.8.43. For the last part of the exercise you will need estimates for the student populations in each of those years. That number was growing, but not nearly as fast as the total debt.

Exercise 4.8.3. You won't find any help visiting the State House website.

The Bureau of Labor Statistics data don't go back far enough. There's a hint in the text about what to do.

Exercise 4.8.5. The inflation calculator can tell you the year when a nickel would buy what you need a dime for in 2011.

Exercise 4.8.9. Be careful when you move the decimal point to find the percent, which will be negative and pretty small.

Exercise 4.8.12. One of the raises in this 2010 article hadn't taken place when the article was written, so the inflation rate for that year wasn't known then. But it is known now.

When you use the inflation calculator, remember that the base pay for these raises was the 2005 value, not the one in 2006. That was after the first raise.

Exercise 4.8.13. Consider computing with a particular salary and inflation rate as examples. Choose numbers that are easy to work with.

Exercise 4.8.14. You may want to use the internet for part of your answer to the last question. You should not need it before that.

Exercise 4.8.23. (a) Careful. Why isn't the answer 30%?

(b) Use trial-and-error. Guess an answer, test it using the argument from the first part of the exercise, adjust your answer until it's right.

Exercise 5.7.13. If you find an "average" statistic for the last question you will need to think about which kind of average it is.

Exercise 6.12.5. You may draw your bar chart neatly by hand or use Excel.

Exercise 6.12.6. (c) If you want the absolute change in column E then put formula

$$=D7-C7$$

and copy the formula to the rest of the rows.

For the relative increase, divide instead of subtracting. Then subtract 1 to get the percentage increase.

Exercise 6.12.14. Start with a copy of `WingAeroHistogram.xlsx`. Create new data in columns D and E below the existing data there for the two new histograms. Fill in column F appropriately. Then you can copy and paste from what's there to fill in columns G and H and the sums and the mean.

If Excel wants to treat your salary ranges as dates, try formatting the cells as text.

To make the charts, copy the histogram that's there; then find the place in Excel where you can change the source data to the new rows in columns D and E.

Exercise 6.12.16. When you enter the data in two columns put the categories (usability scores) on the left since they are the labels for the *x*-axis. Put the numbers of websites on the right since they are the values that go with the categories, and should plot vertically, on the *y*-axis.

If your spreadsheet is anything like ours, it may well think that you want to display both columns on the *y*-axis, since both columns are numbers. If it does that it will label the *x*-axis with the numbers 1 to 10.

If that happens, delete the data series corresponding to the bars you don't want (the percentages). Then right click on the chart and explore until you find the place that allows you to enter the fields you want to use as *x*-axis category labels.

Exercise 6.12.20. The mean amount spent for Fire/EMS services per person is not the Excel `AVERAGE` of the amounts spent per person by each city. It's wrong to average those numbers since they are already averages. You must weight them by the city populations in order to compute the total amount spent by all the people in all the cities. Then divide by the total population.

You can't compute the median with Excel's `MEDIAN` function for the same reason. Further hint: Almost everyone lives in New York.

You can't answer the question "What do firefighters earn?" by finding the mean of the twelve numbers in column C. Compute the mean correctly as a weighted average. You will probably want to start by creating a column labeled

`total Fire/EMS expenses`

and fill in the value for each city.

Exercise 6.12.27. Warning: 2:30 is 2 hours and 30 minutes. That's 2.5 hours, not 2.3 hours.

Exercise 6.12.32. Your spreadsheet can do the arithmetic using the formula
$$\frac{\text{new value} - \text{original value}}{\text{original value}}.$$

Exercise 6.12.32. For part (d), use guess-and-check in Excel. Set up the computation so that when you change the percentage of meat the percentage of cheese and the total change automatically.

Exercise 7.8.9. Finding the places where the lines in your Excel chart cross is the key to the last part of the problem.

Exercise 7.8.13. For part (a) you will have to look up the conversion factors among various forms of energy in order to convert watt-hours to gallons of gasoline. Part (d) is about psychology, not quantitative reasoning. Your answer might begin "The display is valuable because .... On the other hand ...."

Exercise 7.8.19. (b) It should not be hard to find a website that helps with the second question.

Exercise 7.8.22. What did electricity cost in the spring of 2011?

Exercise 7.8.30. The answer depends on the relative lengths of the lines, not on the absolute difference in the lengths.

Exercise 8.5.3. Look at the data starting in about 1980.

Exercise 8.5.5. Try a Google search for

Per capita consumption of selected beverages in gallons .

Exercise 8.5.16. For the second question, all you can really look for is the order of magnitude. If that doesn't match, try to explain why.

Exercise 9.7.2. (a) Inflation is usually reported as a percent increase.

(b) No hint needed.

(c) No hint needed.

(d) Read this one carefully to think about what depends on what. Don't just jump at the word "percent".

(e) The washing happens over and over again on the same day — that's how dirty they were.

(f) No hint needed.

(g) Interest on unpaid balances accumulates.

(h) What units would the weatherman use to report the rate at which snow was accumulating?

(i) Think about how the number of people exposed to germs depends on the number of people sick.

(j) (Electronic) word of mouth generates new subscribers from old ones.

Exercise 9.7.15. Use the ExponentialGrowth.xlsx spreadsheet or find the answer by trying different values for T (# of years) in the formula until you find one that gets you close to $3,000.)

Exercise 9.7.16. Your computations should suggest that it won't be infinite, which might have been your first guess.

Exercise 9.7.17. You can answer the first question using guess-and-check until you're close enough. You'll need a web search for the second and a very good web search for the third, which is open-ended.

Exercise 9.7.19. Guess a number of hours, try your guess in the two exponential equations, then adjust your guess up or down until the answers match.

Exercise 9.7.22. Compare the relative change in the number of email accounts for the two time periods (early 1900's to end of 1999, end of 1999 to the time of the blog post).

Exercise 10.6.8. Start with the calculation $(1 + 0.0143)^{365}$. The answer is hard to believe.

Exercise 10.6.9. What annual interest rate would you need to turn $4,000 into $1.5 million in 94 years? Compare that rate to the increase due just to inflation.

Exercise 11.9.4. You might want to do this exercise in Excel. Then you can see the odds for all the horses and see how the payoffs change when you change the track's take.

Exercise 11.9.11. You might find it easiest to answer this question by imagining that you played the lottery 100 days in a row.

Exercise 12.7.3. There are 36 equally likely outcomes. One possibility is that you roll a 3 with die A (5 ways) and your friend rolls a 2 with die B (3 ways). So there are 15 ways that can happen, for a probability of 15/36 that you win that way. Now figure out the probabilities for the other possibilities.

Exercise 12.7.7. The answer is very small, but it's not zero. Use the Google calculator to find it.

Exercise 12.7.12. In `cancerclusters.xlsm` you can get the state populations from the data or find it on the web and download it. We might do it for you in a later edition of this book.

Exercise 12.7.21. Think about the number of births in that small village, and the number of small villages.

Exercise 13.7.1. The first quote tells you the false positive and false negative rates. The second tells you the incidence.

Exercise 13.7.7. Build the contingency table, based on a population of 1,000 women tested. You may do this by hand or with the spreadsheet `ContingencyTable.xlsx`.

References

The references here identify the sources for data and quotations in the text and exercises. Citing sources is a necessary part of good academic work. That does not mean you need to follow these links: you should be able to read the text and work the exercises without having to consult the original sources.

[R1] Lewis Carroll, *A Tangled Tale*, Answers to Knot 4, `www.gutenberg.org/files/29042/29042-8.txt` (last visited July 14, 2015).

[R2] Henry Wadsworth Longfellow, *Kavanagh*, 1849, `www.archive.org/stream/talekavanagh00longrich/talekavanagh00longrich_djvu.txt` (last visited July 14, 2015).

Chapter 1

[R3] A. Vaccaro, There were nearly 100,000 Uber and Lyft rides per day in Boston last year, *The Boston Globe*, May 1, 2018, `www.bostonglobe.com/business/2018/05/01/there-were-nearly-uber-and-lyft-rides-day-boston-streets-last-year/yzOWJ9PdVg8KKQMQSKSF2K/story.html` (last visited July 31, 2019).

[R4] N. Boroyan, Here's How Many Uber Drivers There Are in Boston, BostINNO, January 22, 2015, `www.americaninno.com/boston/uber-driver-data-boston-has-10000-uber-drivers/` (last visited January 12, 2019).

[R5] J. Beltrane, Artificial Mini-Hearts Developed, Historic Canada Blog, *The Canadian Encyclopedia*, `thecanadianencyclopedia.ca/en/article/artificial-mini-hearts-developed/` (updated December 16, 2013) (last visited August 16, 2015).

[R6] O. Judson, Darwin Got It Going On, *The New York Times* (May 4, 2010), `opinionator.blogs.nytimes.com/2010/05/04/darwin-got-it-going-on/` (last visited December 6, 2015).

[R7] Baba Brinkman bio, `bababrinkman.com/wp-content/uploads/2019/03/baba-bio-longform.pdf` , `www.bababrinkman.com/bio/` (last visited February 15, 2020).

[R8] Redrawn from data accompanying D. C. Denison, Taking a different measure, *The Boston Globe* (October 14, 2010), `www.boston.com/news/science/articles/2010/10/14/akamai_keeping_an_eye_on_its_greenhouse_emissions/` (last visited August 1, 2015). The original graphic is no longer available.

[R9] E. Horowitz, Want to join Forbes 400? It's $1.7b minimum, *The Boston Globe* (September 29, 2015), `www.bostonglobe.com/business/2015/09/29/how-rich-are-forbes-really/Zz8CaFo6iYMyamIKf6MFsJ/story.html` (last visited September 30, 2015).

[R10] Donnelly/Colt, search for "every minute" at `donnellycolt.com` (last visited October 11, 2015).

[R11] H. Bray, Smartphone apps may help retail scanning catch on, *The Boston Globe* (March 12, 2012), bostonglobe.com/business/2012/03/11/modiv-and-aislebuyer-apps-turn-smartphones-into-retail-scanners/DYgfE8JmimVWDjblpbKDzI/story.html (last visited February 3, 2019) (quoted with permission).

[R12] N. Trent, Ending Take Out Waste, Whole Foods Magazine, January 24, 2011, wholefoodsmagazine.com/blog/ending-take-out-waste/ (last visited July 25, 2019).

[R13] C. Shea, Our blog vocabulary, our selves, *The Boston Globe* (July 25, 2010), www.boston.com/bostonglobe/ideas/articles/2010/07/25/our_blog_vocabulary_our_selves/ (last visited July 12, 2015) (quoted with permission).

[R14] F. Cook, *To the Top of the Continent*, 1908. Reissued by The Mountaineers Books, Seattle, 2001, pp. 107–108.

[R15] Millions jam street-level crime map website, BBC News (February 1, 2011), www.bbc.co.uk/news/uk-12336381 (last visited July 20, 2015).

[R16] Comment on J. Calmes, Obama Draws New Hard Line on Long-Term Debt Reduction, *The New York Times* (September 19, 2011), www.nytimes.com/2011/09/20/us/politics/obama-vows-veto-if-deficit-plan-has-no-tax-increases.html (last visited February 21, 2016).

[R17] A. Waters and K. Heron, No Lunch Left Behind, *The New York Times* (February 19, 2009), www.nytimes.com/2009/02/20/opinion/20waters.html (last visited July 20, 2015).

[R18] Earth Month Tip: Turn off the tap, Environmental Protection Agency, blog.epa.gov/2014/04/23/earth-month-tip-turn-off-the-tap/ (last visited February 24, 2019).

[R19] Associated Press, Lady Liberty's crown to reopen July 4, *The Boston Globe* (May 9, 2009), www.boston.com/news/nation/articles/2009/05/09/lady_libertys_crown_to_reopen_july_4/ (last visited July 20, 2015).

[R20] L. Collins, Just a Minute, *The New Yorker* (December 7, 2009), www.newyorker.com/magazine/2009/12/07/just-a-minute (last visited July 26, 2015).

[R21] R. Vecchio, Global Psyche: On Their Own Time, *Psychology Today* (July 1, 2007), www.psychologytoday.com/articles/200707/global-psyche-their-own-time (last visited July 31, 2019).

[R22] John McPhee, "Season on the Chalk", *The New Yorker* (March 12 2008), reprinted in *Silk Parachute*, Farrar Straus and Giroux, New York, 2010, page 23.

[R23] A. Singer, The Future Buzz, Social Media, Web 2.0 and Internet Stats (2009), thefuturebuzz.com/2009/01/12/social-media-web-20-internet-numbers-stats/ (last visited February 15, 2020). Quoted with permission.

[R24] D. Abel, State panel OKs expansion of nickel deposit to bottled water, *The Boston Globe* (July 15, 2010), www.boston.com/news/local/massachusetts/articles/2010/07/15/state_panel_oks_expansion_of_nickel_deposit_to_bottled_water/ (last visited March 25, 2020).

[R25] Graphic on a T-shirt from the Homemade Cafe, www.homemade-cafe.com (last visited February 24, 2019). They're delighted by the publicity.

[R26] A. Doerr, So many books, so little time, *The Boston Globe* (March 4, 2012), www.bostonglobe.com/arts/books/2012/03/04/many-books-little-time/znyJ5MhKbMweJNNMk3yIYM/story.html (last visited August 17, 2015).

[R27] S. J. Dubner, Should the U.S. Really Try to Host Another World Cup?, *The New York Times* (July 19, 2010), `freakonomics.com/2010/07/19/should-the-u-s-really-try-to-host-another-world-cup/` (last visited July 31, 2019).

[R28] Putting Three Kids Through College by Redeeming Cans and Bottles, ABC News (March 30, 2006), `abcnews.go.com/2020/story?id=1787254` (last visited October 1, 2015).

[R29] E. Kolbert, Flesh of Your Flesh, *The New Yorker* (November 29, 2009), `www.newyorker.com/arts/critics/books/2009/11/09/091109crbo_books_kolbert` (last visited July 20, 2015).

[R30] Giga Quotes, `www.giga-usa.com/` (last visited July 20, 2015).

[R31] The 2014 population pyramid is no longer easily found on the internet. The 2016 pyramid is available at `www.indexmundi.com/united_states/age_structure.html` (last visited February 16, 2020).

[R32] S. Mayerowitz, United lures top fliers with promise of a hot meal, Associated Press, reported in *The Boston Globe* (August 22, 2014), `www.bostonglobe.com/business/2014/08/21/united-lures-top-fliers-with-promise-hot-meal/uBUbYBhinOZvUptp37MbYM/story.html` (last visited July 20, 2015).

[R33] K. Porzecanski, G. Tan and S. Basak, Billionaires Under Fire Confront Wealth Gap at Milken Conference, *Bloomberg*, May 1, 2019, `www.bloomberg.com/news/articles/2019-05-01/milken-conference-billionaires` (last visited May 2, 2019).

[R34] S. Stolzoff, Jeff Bezos will still make the annual salary of his lowest-paid employees every 11.5 seconds, October 2, 2018, `qz.com/work/1410621/jeff-bezos-makes-more-than-his-least-amazon-paid-worker-in-11-5-seconds/` (last visited April 5, 2019).

[R35] S. Grossfeld, Thinking inside the (batter's) box at Fenway, *The Boston Globe*, June 16, 2017, `www.bostonglobe.com/sports/redsox/2017/06/16/thinking-inside-batter-box-fenway/PMDhKueeIESuf0IMPf1zvM/story.html` (last visited March 25, 2020).

[R36] 11 Facts About High School Dropout Rates, `www.dosomething.org/us/facts/11-facts-about-high-school-dropout-rates` (last visited November 9, 2016).

[R37] S. Dreyer, Who's protecting the Internet? Five guys at a nonprofit, *The Boston Globe*, March 25, 2020, `www.bostonglobe.com/ideas/2019/01/10/who-protecting-internet-five-guys-nonprofit/XR3wg7FHfYYJcPTyrIdzHI/story.html` (last visited January 13, 2019).

[R38] How many galaxies are in the universe? Scientists' guess just got 10 times larger, Associated Press, October 13, 2016, `www.washingtonpost.com/lifestyle/kidspost/how-many-galaxies-are-in-the-universe-scientists-guess-just-got-10-times-larger/2016/10/13/268178bc-918b-11e6-9c85-ac42097b8cc0_story.html` (last visited February 16, 2020).

[R39] P. Krugman, Bubble, Bubble, Fraud and Trouble, *The New York Times*, January 29, 2018, `www.nytimes.com/2018/01/29/opinion/bitcoin-bubble-fraud.html` (last visited January 31, 2018).

[R40] T. Bernard, Phone Companies Are Testing Tech to Catch Spam Calls. Let's Hope It Works, *The New York Times*, April 25, 2019, `www.nytimes.com/2019/04/26/your-money/robocalls-spam-calls.html` (last visited April 26, 2019).

[R41] D. Durbin (Associated Press), Beverage companies investing millions in push to recycle bottles, Portland, Maine, *PressHerald*, October 29, 2019, `www.pressherald.com/2019/10/29/beverage-companies-investing-millions-in-push-to-recycle-bottles/` (last visited October 30, 2019).

Chapter 2

[R42] Distracted Driving, U. S. Centers for Disease Control and Prevention, `www.cdc.gov/motorvehiclesafety/distracted_driving/index.html` (last visited February 24, 2019).

[R43] R. P. Larrick and J. B. Soll, The MPG Illusion, *Science* (June 20, 2008), Vol. 320, No. 5883, pp. 1593–1594.

[R44] Gasoline Vehicle Label, U. S. Environmental Protection Agency, `www.epa.gov/carlabel/gaslabel.htm` (last visited July 31, 2015).

[R45] Federal Bureau of Investigation, Uniform Crime Reports, Crime in the United States 2011, `www.fbi.gov/about-us/cjis/ucr/crime-in-the-u.s/2011/crime-in-the-u.s.-2011/tables/table8statecuts/table_8_offenses_known_to_law_enforcement_california_by_city_2011.xls` (last visited July 20, 2015).

[R46] NASA's metric confusion caused Mars orbiter loss, CNN (September 30, 1999), `www.cnn.com/TECH/space/9909/30/mars.metric/` (last visited July 20, 2015).

[R47] E. Moskowitz, Urgent fixes will disrupt rail lines, *The Boston Globe* (April 28, 2010), `www.boston.com/news/local/massachusetts/articles/2010/04/28/urgent_fixes_will_disrupt_rail_lines/` (last visited March 26, 2020).

[R48] Screen captures from `www.online-calculator.com/` (last visited September 12, 2015). That website says, "Want to use Online-Calculator commercially or for business, time your employees (nasty!), use it in power-point documents, in presentations. — Go for it. Anything else — I'm sure it will be Fine." An email from them makes permission even clearer: "Yes - please feel free to use the images shown, or any other images/text/calculators in the textbook."

[R49] Christopher Marlowe, `allpoetry.com/The-Face-That-Launch'd-A-Thousand-Ships` (last visited August 23, 2015).

[R50] FAA shutdown would cost gov't $200 million a week, Associated Press story reported by *USA Today* (2011), `usatoday30.usatoday.com/news/washington/2011-07-22-faa-aviation-shutdown_n.htm` (last visited July 20, 2015).

[R51] L. Landes, 10 Examples of How You Can Be Penny Wise, Pound Foolish, Consumerism Commentary, October 29, 2020, `www.consumerismcommentary.com/10-examples-of-how-you-can-be-penny-wise-pound-foolish/` (last visited November 19, 2020).

[R52] Data from U.S. Mint annual reports, `www.usmint.gov/about_the_mint/?action=annual_report` (last visited July 17, 2015).

[R53] B. Baker, How hot was it? It was so hot that ..., *The Boston Globe* (June 22, 2012), `www.bostonglobe.com/2012/06/21/howhot/hwZQeTGjrRLU8v1XYYOPyJ/story.html` (last visited March 26, 2020).

[R54] M. Gromov, Crystals, Proteins, Stability and Isoperimetry, *Bulletin (New Series) of the American Mathematical Society*, Volume 48, Number 2, April 2011, p. 240, © 2011, American Mathematical Society. MR2774091

[R55] Sticker Shock, Editorial, *The New York Times* (June 5, 2011), `www.nytimes.com/2011/06/05/opinion/05sun1.html` (last visited July 17, 2015).

[R56] M. L. Wald, Hybrids vs. Nonhybrids: The 5-Year Equation, *The New York Times* Green Blog (February 23, 2011), `green.blogs.nytimes.com/2011/02/23/hybrids-vs-nonhybrids-the-5-year-equation/` (last visited August 22, 2015).

[R57] Clean Energy, calculations and references, Environmental Protection Agency, `www.epa.gov/cleanenergy/energy-resources/refs.html` (last visited July 16, 2015).

[R58] Comments on M. D. Shear, Rising Gas Prices Give G.O.P. Issue to Attack Obama, *The New York Times* (February 19, 2012), `www.nytimes.com/2012/02/19/us/politics/high-gas-prices-give-gop-issue-to-attack-obama.html` (last visited July 17, 2015).

[R59] Press release from Scholastic, Harry Potter Make Publishing History With 8.3 Million Copies Sold In First 24 Hours, *What they think*, July 23, 2007, `whattheythink.com/news/1894-harry-potter-make-publishing-history-with-83-million/` (last visited July 24, 2019).

[R60] Senior-citizen volunteers fight Medicare fraud, Associated Press, reported in *The Star Democrat* (Easton, MD) (2010), `www.stardem.com/article_c3eb003d-d240-59ba-99bd-ac0f458c5cfe.html` (last visited July 17, 2015).

[R61] T. Haupert, Carrots 'N' Cake, Grocery Shopping 101: Unit Price (2011), `carrotsncake.com/2011/01/grocery-shopping-101-unit-price.html` (last visited February 17, 2010). Quoted with permission.

[R62] K. Chang, With shuttle's end, X Prize race to the moon begins, *The Seattle Times* (2011). `seattletimes.com/text/2015689643.html` (last visited July 18, 2015).

[R63] Astrobiotic Technology, `www.astrobotic.com/` (last visited August 22, 2015).

[R64] J. Buzby and H. Farah, Chicken Consumption Continues Longrun Rise, USDA Economic Research Service, Web Archive, `www.ers.usda.gov/amber-waves/2006/april/chicken-consumption-continues-longrun-rise/` (last visited July 25, 2019).

[R65] Comment on G. Collins, Send in the Clowns, and Cheese, *The New York Times* (April 4, 2012), `www.nytimes.com/2012/04/05/opinion/collins-send-in-the-clowns-and-cheese.html` (last visited July 27, 2015).

[R66] TPC at Sawgrass, wikipedia, `en.wikipedia.org/wiki/TPC_at_Sawgrass` (last visited July 18, 2015) (Creative Commons Attribution-ShareAlike License, `en.wikipedia.org/wiki/Wikipedia:Text_of_Creative_Commons_Attribution-ShareAlike_3.0_Unported_License`).

[R67] E. Hazen, USA Coverage, How Many Driving Accidents Occur Each Year?, `www.usacoverage.com/auto-insurance/how-many-driving-accidents-occur-each-year.html` (last visited July 16, 2015). Quoted with permission.

[R68] Car-Accidents.com, Car Accident Statistics, `www.car-accidents.com/pages/stats.html` (last visited July 18, 2015).

[R69] B. P. Phillips, An Impartial Price Survey of Various Household Liquids as Compared to a Gallon of Gasoline, Annals of Improbable Research (14.4), `www.improbable.com/2008/09/17/an-impartial-price-survey-of-various-household-liquids-as-compared-to-a-gallon-of-gasoline/` (last visited July 18, 2015). Reproduced with permission.

[R70] Wikipedia, from Federal Highway Administration - MUTCD, `mutcd.fhwa.dot.gov/shsm_interim/index.htm` [Public domain], `commons.wikimedia.org/wiki/File:Speed_Limit_55_sign.svg` (last visited July 25, 2019) (Creative Commons Attribution-ShareAlike License, `en.wikipedia.org/wiki/Wikipedia:Text_of_Creative_Commons_Attribution-ShareAlike_3.0_Unported_License`).

[R71] A. L. Myers and J. Billeaud, Arizona using donations for border fence, *The Washington Times* (November 24, 2011), `www.washingtontimes.com/news/2011/nov/24/arizona-using-donations-for-border-fence` (last visited July 18, 2015).

[R72] K. Chang, Rosetta Spacecraft Set for Unprecedented Close Study of a Comet, *The New York Times* (August 5, 2014), `www.nytimes.com/2014/08/06/science/space/rosetta-spacecraft-set-for-unprecedented-close-study-of-a-comet.html` (last visited July 18, 2015).

[R73] D. Streitfeldmarch, In a Flood Tide of Digital Data, an Ark Full of Books, *The New York Times* (March 3, 2012), `www.nytimes.com/2012/03/04/technology/internet-archives-repository-collects-thousands-of-books.html` (last visited July 18, 2015).

[R74] C. Li and W. Qian, Floating trash threatens Three Gorges Dam, *China Daily* (August 2, 2010), `www.chinadaily.com.cn/m/hubei/2010-08/02/content_11083052.htm` (last visited July 17, 2015).

[R75] Smoot, Wikipedia, `en.wikipedia.org/wiki/Smoot`, Creative Commons Attribution-ShareAlike License, `en.wikipedia.org/wiki/Wikipedia:Text_of_Creative_Commons_Attribution-ShareAlike_3.0_Unported_License` (last visited February 17, 2020).

[R76] S. Nordenstam and B. Hirschler, Nobel Prize for seeing how life works at molecular level, Reuters (October 8, 2014), `uk.reuters.com/article/2014/10/08/us-nobel-prize-chemistry-idUKKCN0HX0V220141008` (last visited July 27, 2015).

[R77] C. Lyons, A Sticky Tragedy: The Boston Molasses Disaster, *History Today* (Volume 59, Issue 1, January 2009), `www.historytoday.com/chuck-lyons/sticky-tragedy-boston-molasses-disaster` (last visited October 21, 2015).

[R78] Great Molasses Flood, Wikipedia, `en.wikipedia.org/wiki/Great_Molasses_Flood` (last visited October 21, 2015).

[R79] Americans' Data Usage More Than Doubled in 2015, CTIA, `www.prnewswire.com/news-releases/americans-data-usage-more-than-doubled-in-2015-300272913.html` (last visited March 3, 2019).

[R80] U.S. 2015 data use hit nearly 1T GB, *Baltimore Sun*, `http://digitaledition.baltimoresun.com/tribune/article_popover.aspx?guid=c16cbb9e-acbf-4c99-b305-7d998acb0b02` (last visited March 2, 2019).

[R81] Walmart Launches Project Gigaton to Reduce Emissions in Company's Supply Chain, Walmart press release, `news.walmart.com/2017/04/19/walmart-launches-project-gigaton-to-reduce-emissions-in-companys-supply-chain` (last visited February 28, 2019).

[R82] Walmart Announces 20 MMT of Supplier Emission Reductions through Project Gigaton; Unveils Plans for Expanding Electric Vehicle Charging Stations and Doubling U.S. Wind and Solar Energy Use. Walmart press release, *Market Watch*, April 18, 2018, `www.marketwatch.com/press-release/walmart-announces-20-mmt-of-supplier-emission-reductions-through-project-gigaton-unveils-plans-for-expanding-electric-vehicle-charging-stations-and-doubling-us-wind-and-solar-energy-use-2018-04-18` (last visited July 25, 2019).

[R83] K. Chang, A Metropolis of 200 Million Termite Mounds Was Hidden in Plain Sight, *The New York Times*, November 20, 2018, `www.nytimes.com/2018/11/20/science/termite-mounds-brazil.html` (last visited November 22, 2018).

[R84] P. Annin, Tough Times Along the Colorado River, *The New York Times*, January 30, 2019, `www.nytimes.com/2019/01/30/opinion/tough-times-along-the-colorado-river.html` (last visited January 30, 2019).

[R85] Arizona Water Use by Sector, `www.arizonawaterfacts.com/water-your-facts` (last visited January 31, 2019).

[R86] S. Kohn, Why does it take so long to transmit an image from New Horizons to Earth?, `astronomy.stackexchange.com/questions/28969/why-does-it-take-so-long-to-transmit-an-image-from-new-horizons-to-earth` (last visited January 2, 2019) (Creative Commons Share Alike License: `creativecommons.org/licenses/by-sa/4.0/legalcode`).

[R87] I. Austen, Ron Joyce, Force Behind Tim Hortons Doughnut Shops, Dies at 88, *The New York Times*, February 6, 2019, `www.nytimes.com/2019/02/06/obituaries/roy-joyce-dead.html` (last visited February 6, 2019).

[R88] S. Buell, Equal pay for Massachusetts women in the spotlight this week, Metro Boston, April 14, 2016, `www.metro.us/boston/equal-pay-for-massachusetts-women-in-the-spotlight-this-week/zsJpdm---adHOzsodOHwWI` (last visited July 25, 2019).

Chapter 3

[R89] Congressional Budget Office, `www.cbo.gov/sites/default/files/2018-05/53624-fy17budget.png` (last visited March 23, 2019).

[R90] Ann Hulsing, using data from Policy Basics: Non-Defense Discretionary Programs, Center on Budget and Policy Priorities, `www.cbpp.org/research/federal-budget/policy-basics-non-defense-discretionary-programs` (last visited March 24, 2019).

[R91] Graphic redrawn from data in *The Boston Globe* (September 9, 2008).

[R92] N. Grimley, DeWine proposes 95-percent increase to children services investment, WKBN, Youngstown, Ohio, March 08, 2019, `https://www.wkbn.com/news/local-news/dewine-proposes-95-percent-increase-to-children-services-investment/1836391305` (last visited March 26, 2019).

[R93] George Pólya, *How to Solve It*, Princeton University Press (Reprint edition, 2014). MR3289212

[R94] After Portman-Carper PSI Investigative Report Exposed Price Hikes of More Than 600 Percent, Drug Manufacturer Kaléo Reduces Price of Naloxone Drug, `www.carper.senate.gov/public/index.cfm/2018/12/after-portman-carper-psi-investigative-report-exposed-price-hikes-of-more-than-600-percent-drug-manufacturer-kal-o-reduces-price-of-naloxone-drug` (last visited July 28, 2019).

[R95] E. Kolbert, Unnatural Selection, *The New Yorker*, April 18, 2016, p. 28, `www.newyorker.com/magazine/2016/04/18/a-radical-attempt-to-save-the-reefs-and-forests` (last visited July 31, 2019).

[R96] R. Munroe, `xkcd.com/985/` (last visited February 17, 2019). This work is licensed under a Creative Commons Attribution-NonCommercial 2.5 License, `creativecommons.org/licenses/by-nc/2.5/` .

[R97] Baby Infant Growth Chart Calculator, `www.infantchart.com/` (last visited February 19, 2019).

[R98] J. Miller, State's rainy day fund has dwindled over past decade, *The Boston Globe* (October 4, 2015), `www.bostonglobe.com/metro/2015/10/04/state-rainy-day-fund-depleted/gX3v8T9erHrlyBvGFpOefL/story.html` (last visited March 26, 2020).

[R99] J. Surowicki, The Financial Page, *The New Yorker* (March 29, 2010). p. 45, `archives.newyorker.com/?i=2010-03-29\#folio=044` (last visited July 27, 2015).

[R100] B. Reddall and E. Scheyder, Obama seeks new drill ban as oil still spews, Reuters (June 23, 2010), `in.reuters.com/article/2010/06/23/idINIndia-49558320100623` (last visited July 19, 2015).

[R101] J. McGuire, Data on Brain Injury in Massachusetts: A Snapshot, Massachusetts Executive Office of Health and Human Services, Presentation to the Brain Injury Commission (February 7 2011), `www.mass.gov/eohhs/docs/eohhs/braininjury/201102-presentation-mcguire.rtf` (last visited July 19, 2015).

[R102] Federal Funding for Public Broadcasting: Q&A, Western Reserve Public Media (2015), `westernreservepublicmedia.org/federal-funding.htm` (broken link) (last visited October 1, 2015), no longer visible.

[R103] D. Gonzalez, Births by U.S. visitors: A real issue?, *The Arizona Republic* (August 17, 2011), `www.azcentral.com/news/articles/2011/08/17/20110817births-by-us-visitors-smaller-issue.html` (last visited July 19, 2015).

[R104] Mercury in the Fog, *City on a Hill Press* (April 21, 2012), `www.cityonahillpress.com/2012/04/21/mercury-in-the-fog/` (last visited July 19, 2015).

[R105] J. Garthwaite, Coastal California Fog Carries Toxic Mercury, Study Finds, *The New York Times* (March 28, 2012), `green.blogs.nytimes.com/2012/03/28/coastal-california-fog-carries-toxic-mercury-study-finds/` (last visited July 19, 2015).

[R106] Harkin, Hagan Re-Introduce Bill to Promote Responsible Use of Taxpayer Dollars in Higher Ed, U.S. Senate Committee on Health, Education, Labor and Pensions (March 12, 2013), `www.help.senate.gov/ranking/newsroom/press/harkin-hagan-re-introduce-bill-to-promote-responsible-use-of-taxpayer-dollars-in-higher-ed` (last visited July 19, 2015).

[R107] Affiliated Managers, *The Boston Globe* (October 6, 2010), `www.boston.com/yourtown/beverly/articles/2010/10/06/affiliated_managers/` (broken link) (last visited August 12, 2015).

[R108] A. Kingswell, Five nines: chasing the dream?, Continuity Central (2010), `www.continuitycentral.com/feature0267.htm` (last visited July 20, 2015). Quoted with permission.

[R109] Digest of Educational Statistics, National Center for Educational Statistics, `nces.ed.gov/programs/digest/d10/tables/dt10_285.asp` (last visited July 19, 2015).

[R110] J. Eilperin, New Nature Conservancy atlas aims to show the state of the world's ecosystems, *The Washington Post* (April 12, 2010), `www.washingtonpost.com/wp-dyn/content/article/2010/04/11/AR2010041103556.html` (last visited July 19, 2015).

[R111] M. Iwata, Gulf oil spill could lead to drop in global output, Dow Jones Newswires reported in *The Denver Post*, `www.denverpost.com/nacchio/ci_15330002` (last visited September 21, 2015).

[R112] Goldman Sachs limits pay, earns $4.79 billion in fourth quarter, Associated Press, reported in the *Tampa Bay Times* (January 21, 2010), www.tampabay.com/news/business/banking/goldman-sachs-limits-pay-earns-479-billion-in-fourth-quarter/1067187 (broken link) (last visited September 12, 2015).

[R113] A. Bjerga, Food Stamps Went to Record 41.8 Million in July, Bloomberg (October 5, 2010), www.bloomberg.com/news/articles/2010-10-05/food-stamp-recipients-at-record-41-8-million-americans-in-july-u-s-says (last visited August 23, 2015).

[R114] W. Parry, Gamblers spending less time, money in AC casinos, Associated Press, reported in on NBC news (December 7, 2010), www.nbcnews.com/id/40547628/ns/business-us_business/t/gamblers-spending-less-atlantic-city-casinos (last visited September 12, 2015).

[R115] S. Syre, Harvard reports big gains on investments, *The Boston Globe* (September 23, 2011), www.boston.com/business/markets/articles/2011/09/23/harvard_endowment_posts_big_investment_gain/ (last visited August 1, 2015). Graphic showing the historical data no longer available.

[R116] T. S. Bernard and B. Protess, Banks to Make Customers Pay Fee for Using Debit Cards, *The New York Times* (September 29, 2011), www.nytimes.com/2011/09/30/business/banks-to-make-customers-pay-debit-card-fee.html (last visited July 19, 2015).

[R117] W. Yardley, Oregon town weighs future with fossil fuel, New York Times News Service, reported in *The Bulletin* (Bend, Oregon) (April 22, 2012), www.bendbulletin.com/article/20120422/NEWS0107/204220325/ (last visited July 19, 2015).

[R118] J. Abelson, Seeking savings, some ditch brand loyalty, *The Boston Globe* (January 29, 2010), www.boston.com/business/articles/2010/01/29/shoppers_are_ditching_name_brands_for_store_brands/ (last visited March 26, 2020).

[R119] D. C. Denison, UMass research funding reaches record, *The Boston Globe* (February 10, 2011), www.boston.com/business/articles/2011/02/10/umass_research_funding_reaches_record/ (last visited March 27, 2020).

[R120] A. Beam, Long gone?, *The Boston Globe* (March 4, 2011), www.boston.com/lifestyle/articles/2011/03/04/from_landlines_to_e_mail_to_movies_to_print_their_demise_has_been_greatly_exaggerated/ (last visited March 27, 2020).

[R121] J. Sununu, Creature double feature, *The Boston Globe* (March 21, 2011), www.boston.com/bostonglobe/editorial_opinion/oped/articles/2011/03/21/creature_double_feature/ (last visited March 26, 2020).

[R122] H. Bray, Facebook vs. Google+, *The Boston Globe* (September 29, 2011), www.bostonglobe.com/business/2011/09/28/facebook-google/yWa8anwKHgelcklsasoDkP/story.html (last visited August 27, 2015).

[R123] G. Washburn, Game time, *The Boston Globe* (June 30, 2010), www.boston.com/sports/basketball/celtics/articles/2010/06/30/game_time_when_james_makes_decision_dominoes_to_fall/ (last visited March 26, 2020).

[R124] S. Ohlemacher, 13 million get unexpected tax bill from Obama tax credit, Associated Press reported in Wilmington, NC, *StarNewsOnline* (December 17, 2010), www.starnewsonline.com/article/20101217/ARTICLES/101219738 (last visited July 20, 2015).

[R125] Loan Repayment/Debt Management, Red Rock Community College, `www.rrcc.edu/financial-aid/loans` (last visited July 27, 2019).

[R126] M. Viser, Benefits take hit in Patrick budget, *The Boston Globe* (January 13, 2008), `www.boston.com/news/local/articles/2008/01/13/benefits_take_hit_in_patrick_budget/` (last visited March 29, 2020).

[R127] Driving More Efficiently, U. S. Department of Energy, `www.fueleconomy.gov/feg/driveHabits.jsp\#speed-limit` (last visited July 20, 2015).

[R128] Labels redrawn from `www.fueleconomy.gov/feg/images/speed_vs_mpg_2012_sm.jpg` (last visited August 2. 2015).

[R129] G. Edgers, Holiday shows bringing box office joy, *The Boston Globe* (December 27, 2012), `www.bostonglobe.com/arts/2012/12/27/boston-ballet-holiday-pops-leading-wave-holiday-arts-show-boom/FELXeihGF2IT2SiooJ7JPI/story.html` (last visited March 29, 2020). Graphic redrawn from data in the article.

[R130] T. Luna, Amazon to begin collecting Mass. sales tax Friday, *The Boston Globe* (October 28, 2013), `www.bostonglobe.com/business/2013/10/28/amazon-begin-collecting-massachusetts-sales-tax-friday/acyUMexp4ChUk2I6JtdJyI/story.html` (last visited March 29, 2020).

[R131] M. Yglesias, Last year, 25 hedge fund managers earned more than double every kindergarten teacher combined, Vox.com (May 6, 2014), `www.vox.com/2014/5/6/5687788/last-year-25-hedge-fund-managers-earned-more-than-double-every` (last visited July 20, 2015).

[R132] S. Carrell and J. Zinman, In Harm's Way? Payday Loan Access and Military Personnel Performance, January 2013, *Rev. Financ. Stud.* (2014) 27 (9): 2805–2840, first published online May 28, 2014, doi:10.1093/rfs/hhu034, *The Review of Financial Studies*, `rfs.oxfordjournals.org/`.

[R133] J. Boak, School spending by affluent is widening wealth gap, Associated Press (September 30, 2014), `www.csmonitor.com/Business/Latest-News-Wires/2014/0930/How-school-spending-by-the-rich-is-widening-the-wealth-gap` (last visited July 27, 2019).

[R134] R. Burns and L. C. Baldor, U.S. nuclear woes: Pentagon chief orders a shakeup, Associated Press report in the *Vancouver Columbian* (November 14, 2014), `www.columbian.com/news/2014/nov/14/us-nuclear-woes-pentagon-chief-orders-a-shakeup/` (last visited August 28, 2015).

[R135] M. O'Brien, The top 400 households got 16 percent of all capital gains in 2010, *The Washington Post* (November 25, 2014), `www.washingtonpost.com/blogs/wonkblog/wp/2014/11/25/the-top-400-households-got-16-percent-of-all-capital-gains-in-2010/` (last visited July 31, 2010).

[R136] Ted Williams, Wikipedia, `en.wikipedia.org/wiki/Ted_Williams` (last visited July 21, 2015), Creative Commons Attribution-ShareAlike License, `en.wikipedia.org/wiki/Wikipedia:Text_of_Creative_Commons_Attribution-ShareAlike_3.0_Unported_License`.

[R137] A. Ryan and M. E. Irons, Boston city councilors vote for $20,000-a-year raise, *The Boston Globe* (October 8, 2014), `www.bostonglobe.com/metro/2014/10/08/boston-city-councilors-set-consider-raise/FqmrkRlm5wNDkvTBTEv8uI/story.html` (last visited March 29, 2020).

[R138] P. Whittle, Dogfish remain abundant off of Maine, East Coast, Associated Press reported in *The Washington Post* (September 28, 2014), www.washingtontimes.com/news/2014/sep/28/dogfish-remain-abundant-off-of-maine-east-coast/ (last visited July 21, 2015).

[R139] New Zealand Permanent Warning, Wikipedia, commons.wikimedia.org/wiki/File:New_Zealand_Permanent_Warning_-_Steep_Up_Grade.svg (last visited August 13, 2015). The copyright holder of this work allows anyone to use it for any purpose including unrestricted redistribution, commercial use, and modification.

[R140] E. Wagner, Democrats Propose 3.6 Percent Raise for Feds in 2020, *Government Executive*, www.govexec.com/pay-benefits/2019/02/dems-propose-36-percent-raise-feds-2020/154746/ , February 8, 2019 (last visited March 26, 2019).

[R141] K. McWilliams, Newington town manager proposes 4.7 percent budget increase, *Hartford Courant*, March 6, 2019, www.courant.com/community/newington/hc-news-newington-budget-increase--20190306-my34pf7gx5hrhihagjcilonv3y-story.html (last visited March 26, 2019).

[R142] User kmbunday, Two examples of innumeracy in books for parents about gifted children, Davidson Institute (April 8, 2013), giftedissues.davidsongifted.org/BB/ubbthreads.php/topics/152941/Re_Innumeracy_in_Gifted_Educat.html (last visited July 29, 2015).

[R143] T. Hsu, Bigger, Saltier, Heavier: Fast Food Since 1986 in 3 Simple Charts, *The New York Times*, March 4, 2019, www.nytimes.com/2019/03/03/business/fast-food-health-salt-calories-portions.html (last visited March 4, 2019).

[R144] T. Barrabi, After Bryce Harper deal, Phillies ticket prices surge, *FOXBusiness*, March 1, 2019, www.foxbusiness.com/retail/after-bryce-harper-deal-phillies-ticket-prices-surge .

[R145] M. Grunwald, An Unexpected Current That's Remaking American Politics, *Politico*, April 29, 2019, www.politico.com/magazine/story/2019/04/29/trump-wrong-about-wind-power-electricity-battery-storage-226755 (last visited April 29, 2019).

[R146] L. Laro, U.S. roadways more lethal during pandemic, safety group says, *The Washington Post*, May 20, 2020, www.washingtonpost.com/transportation/2020/05/20/us-roadways-more-lethal-during-pandemic-lockdown-national-safety-council-data-shows/ (last visited May 21, 2020).

Chapter 4

[R147] Screenshot from www.bls.gov/data/inflation_calculator.htm .

[R148] Health Care Cost Institute, Changes in Health Care Spending in 2011 (2012). www.healthcostinstitute.org/research/publications/hcci-research/entry/changes-in-health-care-spending-in-2011 (last visited July 27, 2019).

[R149] Screenshot from www.usinflationcalculator.com/ (last visited September 27, 2019).

[R150] Historical Consumer Price Index (CPI-U) Data, inflationdata.com/inflation/consumer_price_index/historicalcpi.aspx (last visited July 29, 2019).

[R151] M. M. Crow, Growing a better NIH, *The Boston Globe* (June 19, 2011), www.boston.com/bostonglobe/ideas/articles/2011/06/19/growing_a_better_nih (last visited March 29, 2020).

[R152] History of Federal Minimum Wage Rates Under the Fair Labor Standards Act, 1938–2009, Wage and Hour Division (WHD), United States Department of Labor, `www.dol.gov/whd/minwage/chart.htm` (last visited February 18, 2019).

[R153] Chart from data at `www.dol.gov/whd/minwage/chart.htm` (last visited February 18, 2019).

[R154] J. Schuesslerfer, In Search of the Slave Who Defied George Washington, *The New York Times*, February 6, 2017, `static01.nyt.com/images/2017/02/07/arts/07MTVERNONJP4/07MTVERNONJP4-blog427.jpg` (last visited February 25, 2017).

[R155] U. S. Historical Inflation, `www.in2013dollars.com/1796-dollars-in-2016` (last visited February 25, 2017).

[R156] P. Gillespie, It's official: America has deflation, CNN Money, February 26, 2015, `money.cnn.com/2015/02/26/news/economy/inflation-january-negative/` (last visited February 18, 2019).

[R157] AppleInsider Staff, Rare working Apple I goes for record-smashing $905,000 at auction, *appleinsider*, October 22, 2014, `appleinsider.com/articles/14/10/22/rare-working-apple-i-goes-for-record-smashing-905000-at-auction` (last visited February 20, 2020).

[R158] The Massachusetts State House Model, Office of the Secretary of the Commonwealth of Massachusetts, William Francis Galvin, `www.sec.state.ma.us/trs/trsbok/mod.htm` (last visited May 6, 2020).

[R159] C. Lohmann, Raise bottle deposit to 10 cents, Letter to the Editor, *The Boston Globe* (December 29, 2011), `www.bostonglobe.com/opinion/letters/2011/12/29/raise-bottle-deposit-cents/0xOnLO7DKi10wE69Ru5GiN/story.html` (last visited March 29, 2020).

[R160] M. Cieply, Hollywood Math: Bad to Worse. *The New York Times* (January 10, 2011), `query.nytimes.com/gst/fullpage.html?res=9E0CE1DE173CF933A25752C0A9679D8B63` (last visited October 2, 2015).

[R161] A. J. Liebling, The Jollity Building, *The New Yorker* (April 26, 1941), `www.newyorker.com/magazine/1941/04/26/the-jollity-building` (last visited July 27, 2015).

[R162] A. J. Liebling; Introduction by David Remnick, *Just Enough Liebling*, North Point Press, `us.macmillan.com/books/9780865477278` (last visited July 27, 2015).

[R163] S. Bauer, Committee approves 1 percent pay raise for state, University of Wisconsin workers, Associated Press reported in the Minneapolis-St. Paul *Star Tribune* (June 26, 2013), `www.startribune.com/nation/213097941.html` (last visited July 22, 2015).

[R164] Private colleges vastly outspent public peers, Bloomberg News reported in *The Boston Globe* (July 10, 2010), `www.boston.com/news/education/higher/articles/2010/07/10/private_colleges_vastly_outspent_public_peers/` (last visited July 22, 2015).

[R165] N. Mitchell, DPS, teachers' union reach accord, Chalkbeat Colorado (June 19, 2012), `co.chalkbeat.org/2012/06/19/dps-teachers-union-reach-accord/` (last visited July 22, 2015).

[R166] S. E. Harger, County wages dropped nearly 14 percent in last decade, *The Portland Tribune* (June 6, 2012), `portlandtribune.com/scs/83-news/110775-report-county-wages-dropped-nearly-14-percent-in-last-decade` (last visited July 22, 2015).

[R167] Newspaper sales slid to 1984 level in 2011, Reflections of a Newsosaur, newsosaur.blogspot.com/2012/03/newspaper-sales-slid-to-1984-level-in.html (last visited July 22, 2015).

[R168] M. J. Perry, Free-fall: Adjusted for Inflation, Print Newspaper Advertising Will be Lower This Year Than in 1950, *Carpe Diem* (September 6, 2012), mjperry.blogspot.com/2012/09/freefall-adjusted-for-inflation-print.html (last visited July 22, 2015).

[R169] D. Owen, Penny Dreadful, *The New Yorker* (March 31, 2008), www.newyorker.com/magazine/2008/03/31/penny-dreadful (last visited July 28, 2015).

[R170] S. Greenhouse, The West Virginia Teacher Strike Was Just the Start, *The New York Times*, March 7, 2018, https://www.nytimes.com/2018/03/07/opinion/teachers-west-virginia-strike.html (last visited March 8, 2018).

Chapter 5

[R171] Grading System, Office of the Registrar, UMass Boston, www.umb.edu/registrar/grades_transcripts/grading_system (last visited September 30, 2019).

[R172] Grading System, Office of the Registrar, UMass Boston, www.umb.edu/registrar/grades_transcripts/grading_system (last visited July 28, 2015).

[R173] United States Department of Labor, Bureau of Labor Statistics, Consumer Price Index Frequently Asked Questions, www.bls.gov/cpi/questions-and-answers.htm\#Question_1 (last visited February 14, 2019).

[R174] Components in the Consumer Price Index, Pittsburgh, PA, www.bls.gov/regions/mid-atlantic/news-release/consumerpriceindex_pittsburgh.htm, www.bls.gov/cpi/tables/relative-importance/2018.pdf (last visited February 15, 2019).

[R175] D. Strumpf, New-vehicle prices plunge, report says, Associated Press reported in *The Columbus Dispatch* (September 5, 2008), www.dispatch.com/content/stories/business/2008/09/05/new_vehicle_prices_0905.ART_ART_09-05-08_C10_AQB7T2V.html (last visited July 22, 2015).

[R176] H. Dondis and P. Wolff, Chess Notes, *The Boston Globe* (March 10, 2008), secure.pqarchiver.com/boston-sub/doc/405109540.html (broken link) (last visited July 22, 2015), www.newspapers.com/newspage/444326494/ (last visited May 6, 2020).

[R177] A. Ryan, Month may become dimmest on record, *The Boston Globe* (June 23, 2009), www.boston.com/news/local/massachusetts/articles/2009/06/23/so_far_june_sunlight_in_boston_is_lowest_in_past_century/ (last visited March 29, 2020).

[R178] *The Hightower Lowdown*, May 2010, Volume 12, Number 5. www.hightowerlowdown.org/node/2330 (last visited July 28, 2015).

[R179] Data from: Rising food prices mean a more costly Thanksgiving, *The Boston Globe* (November 19, 2011), www.bostonglobe.com/business/2011/11/19/rising-food-prices-mean-more-costly-thanksgiving/nlGoSN1DsnVRfn1MOhTXdI/igraphic.html (last visited March 29, 2020).

[R180] From wire reports, No fill-ups at Kyrgyzstan base for U.S., *USA Today* (June 2, 2010), usatoday30.usatoday.com/printedition/news/20100602/capcol02_st.art.htm (last visited July 22, 2015).

[R181] D. Hemenway, Why your classes are larger than "average", *Mathematics Magazine*, Vol. 55, No. 3 (May 1982), pp. 162–164, Mathematical Association of America, Article DOI: 10.2307/2690083, `www.jstor.org/stable/2690083` (last visited August 1, 2019). MR1572422

[R182] M. Unser, Penny Costs 1.82 Cents to Make in 2017, Nickel Costs 6.6 Cents; US Mint Realizes \$391.5M in Seigniorage, CoinNews.net, February 26, 2018, `www.coinnews.net/2018/02/26/penny-costs-1-82-cents-to-make-in-2017/` (last visited February 21, 2019).

[R183] Figure redrawn with data from E. Oster and G. Kocks, After a Debacle, How California Became a Role Model on Measles, *The New York Times*, January 16, 2018, `www.nytimes.com/2018/01/16/upshot/measles-vaccination-california-students.html` (last visited January 17, 2018).

Chapter 6

[R184] `en.wikipedia.org/wiki/File:Standard_deviation_diagram.svg` (last visited August 3, 2015). Licensed under the Creative Commons Attribution 2.5 Generic License.

[R185] Data source: Surveillance, Epidemiology, and End Results (SEER) Program (`www.seer.cancer.gov`) SEER*Stat Database: Incidence - SEER 9 Regs Limited-Use, Nov 2008 Sub (1973–2006), National Cancer Institute, DCCPS, Surveillance Research Program, Cancer Statistics Branch, released April 2009, based on the November 2008 submission. Graphic drawn by Ben Bolker.

[R186] Raising Taxes on Rich Seen as Good for Economy, Fairness, Pew Research Center (July 16, 2012), `www.people-press.org/2012/07/16/raising-taxes-on-rich-seen-as-good-for-economy-fairness` (last visited July 28, 2015). Quoted with permission.

[R187] E. Moskowitz, Cash-strapped T proposes 23 percent fare increase, *The Boston Globe* (March 28, 2012), `bostonglobe.com/metro/2012/03/28/mbta-unveils-percent-fare-hike-limited-service-cuts-also-proposed/moC142rwrONf5xyx20ZQGP/story.html` (last visited March 29, 2020).

[R188] M. C. Fisk and J. Lawrence, Walmart to Settle Massachusetts Suit for \$40 Million (Update2), *Bloomberg News* (December 2, 2009), `www.bloomberg.com/apps/news?pid=newsarchive&sid=a2AC1c9J8WwE` (last visited October 2, 2015).

[R189] Graphic redrawn from data from data J. P. Kahn, Missed connections in our digital lives, *The Boston Globe* (April 15, 2012), `www.bostonglobe.com/metro/2012/04/14/missed-connections-our-digital-lives/bPHauWdvU15XAd1ol7SOQL/igraphic.html` (last visited July 31, 2019).

[R190] Jakob Nielsen, Aspects of Design Quality Nielsen Norman Group (November 2, 2008), retrieved (with permission) from `www.nngroup.com/articles/aspects-of-design-quality/` (last visited February 28, 2020).

[R191] K. Geldis, The Richest Counties in America, *TheStreet* (February 13, 2012), `www.thestreet.com/story/11415107/3/the-richest-counties-in-america.html` (last visited July 22, 2015).

[R192] Wikipedia, `https://upload.wikimedia.org/wikipedia/commons/3/3d/Distribution_of_Annual_Household_Income_in_the_United_States_2011.png` (last visited September 22, 2019) (Creative Commons Attribution-ShareAlike License, `en.wikipedia.org/wiki/Wikipedia:Text_of_Creative_Commons_Attribution-ShareAlike_3.0_Unported_License`).

[R193] Graphic redrawn from data scraped from a Nate Silver *New York Times* graphic published October 31, 2012. The original seems not to be available.

[R194] Data from D. Slack, Boston spends most on firefighters in US, *The Boston Globe* (March 30, 2009), www.boston.com/news/local/massachusetts/articles/2009/03/30/boston_spends_most_on_firefighters_in_us/ , data for graphic at www.boston.com/news/local/massachusetts/articles/2009/03/30/fire_spending/ (last visited March 29, 2020).

[R195] D. Slack and J. C. Drake, Error made in fire dept. report, *The Boston Globe* (March 31, 2009), www.boston.com/news/local/massachusetts/articles/2009/03/31/error_made_in_fire_dept_report/ (last visited July 22, 2015).

[R196] Data from J. Stripling and A. Fuller, Presidents Defend Their Pay as Public Colleges Slash Budgets, *The Chronicle of Higher Education* (April 3, 2011), chronicle.com/article/Presidents-Defend-Their/126971 (last visited August 11, 2015).

[R197] Paul Erdős, en.wikipedia.org/wiki/Paul_Erdos (last visited July 22, 2015) (Creative Commons Attribution-ShareAlike License, en.wikipedia.org/wiki/Wikipedia:Text_of_Creative_Commons_Attribution-ShareAlike_3.0_Unported_License).

[R198] The distribution of Erdős numbers, The Erdos Number Project, wwwp.oakland.edu/enp/trivia/ (last visited October 11, 2015), reproduced with permission. MR1459693

[R199] M. Schlueb and D. Damron, Activists press officials to put sick-leave proposal to voters, *Orlando Sentinel* (August 6, 2012), www.orlandosentinel.com/news/os-xpm-2012-08-06os-sick-leave-ballot-race-20120806-story.html (last visited July 28, 2019).

[R200] M. Woolhouse, A Boston taco tells the tale of far-reaching food cost woe, *The Boston Globe* (February 06, 2015), www.bostonglobe.com/business/2015/02/05/food-prices-spike-increasing-cost-taco/vU3c42L99X9fBt25opSKkO/story.html (last visited December 16, 2015).

Chapter 7

[R201] Based on sample electricity bill at www.squashedfrogs.co.uk/ (broken link) (last visited August 31, 2015).

[R202] The Cook Nuclear Plant, www.cookinfo.com/cookplant.htm (last visited July 17, 2015).

[R203] K. Phillips Erb, IRS Announces 2019 Tax Rates, Standard Deduction Amounts and More, Forbes, November 15, 2018, www.forbes.com/sites/kellyphillipserb/2018/11/15/irs-announces-2019-tax-rates-standard-deduction-amounts-and-more/#44c55bc12081 (last visited October 21, 2020).

[R204] Data from G. Anrig, 10 Reasons to Eliminate the Tax Break for Capital Gains, The Century Foundation (October 20, 2011), tcf.org/blog/detail/10-reasons-to-eliminate-the-tax-break-for-capital-gains (last visited August 4, 2015).

[R205] United States of America 1789 (rev. 1992), www.constituteproject.org/constitution/United_States_of_America_1992 (last visited September 25, 2019).

[R206] Bradford Tax Institute, History of Tax Rates: 1913–2019, bradfordtaxinstitute.com/Free_Resources/Federal-Income-Tax-Rates.aspx (last visited February 2, 2019).

[R207] K. Phillips Erb, New: IRS Announces 2018 Tax Rates, Standard Deductions, Exemption Amounts And More, Forbes, March 7, 2018, `www.forbes.com/sites/kellyphillipserb/2018/03/07/new-irs-announces-2018-tax-rates-standard-deductions-exemption-amounts-and-more/\#16f3fe9f3133` (last visited February 7, 2019).

[R208] M. Kanellos, From Edison's Trunk, Direct Current Gets Another Look, *The New York Times* (November 17, 2011), `www.nytimes.com/2011/11/18/business/energy-environment/direct-current-technology-gets-another-look.html` (last visited July 22, 2015).

[R209] From "Public Street Trees — A Choice" by J. DiMiceli in Newton Conservators Newsletter (April 2012), `newtonconservators.org/public-street-trees-a-choice/` (last visited July 28, 2019). Quoted with permission.

[R210] Data from H. Bray, Pay full price for iPhone, avoid contract, *The Boston Globe* (June 14, 2012), `bostonglobe.com/business/2012/06/14/bgcom-techlab/AskcWIPBv1qccvmqIx7DmK/story.html` (broken link) (last visited July 22, 2015).

[R211] H. McGee, How Much Water Does Pasta Really Need?, *The New York Times* (February 24, 2009), `www.nytimes.com/2009/02/25/dining/25curi.html` (last visited July 22, 2015).

[R212] R. Randazzo, New Arlington Valley solar site packs power, *The Arizona Republic* (May 1, 2013). `www.azcentral.com/business/arizonaeconomy/articles/20130501new-arlington-valley-solar-site-packs-power.html` (last visited July 22, 2015).

[R213] M. Dickerson, Wind-power industry seeks trained workforce, *Los Angeles Times* (March 1, 2009), `www.latimes.com/archives/la-xpm-2009-mar-01-fi-wind-bootcamp1-story.html` (last visited July 28, 2019).

[R214] J. M. Roney, World Solar Power Topped 100,000 Megawatts in 2012, Earth Policy Institute (July 31, 2013), `www.earth-policy.org/indicators/C47/solar_power_2013` (last visited July 23, 2015).

[R215] Average electricity consumption per electrified household, `www.wec-indicators.enerdata.eu/household-electricity-use.html` (last visited November 14, 2015).

[R216] E. Ailworth, Chilling out by the quarry, *The Boston Globe* (August 16, 2010), `www.boston.com/business/technology/articles/2010/08/16/chilling_out_by_the_quarry/` (last visited March 29, 2020).

[R217] J. Coifman, An energy program too efficient for its own good, *The Boston Globe* (March 26, 2011), `www.boston.com/bostonglobe/editorial_opinion/oped/articles/2011/03/26/an_energy_program_too_efficient_for_its_own_good/` (last visited July 23, 2015).

[R218] Attorney General: Connecticut electricity tax could cost Mass. $26M, *The Norwich Bulletin* (June 7, 2011), `www.norwichbulletin.com/x832282338/Attorney-General-Connecticut-electricity-tax-could-cost-Mass-26M` (last visited July 23, 2015).

[R219] J. Carney, President Obama and Vice President Biden's 2013 Tax Returns, The White House Blog (April 11, 2014), `www.whitehouse.gov/blog/2014/04/11/president-obama-and-vice-president-biden-s-2013-tax-returns` (last visited July 23, 2015).

[R220] Adjusted gross income, Wikipedia, `en.wikipedia.org/wiki/Adjusted_gross_income` (last visited November 14, 2015) (Creative Commons Attribution-ShareAlike License, `en.wikipedia.org/wiki/Wikipedia:Text_of_Creative_Commons_Attribution-ShareAlike_3.0_Unported_License`).

[R221] B. Rooney, Pandora raises IPO target to $200 million, CNNMoneyTech (June 10, 2011), `money.cnn.com/2011/06/10/technology/pandora_ipo/index.htm` (last visited July 23, 2015).

[R222] Sotomayor will help usher out 2013 in NYC, Associated Press report in *The Boston Globe* (December 30, 2013), `www.bostonglobe.com/news/nation/2013/12/30/justice-sotomayor-lead-times-square-ball-drop/NacOVg3INmXYcLcLIYSz4J/story.html` (last visited July 29, 2015).

[R223] sifxtreme, Explain travel times and distances on flight, Travel Stack Exchange (April 15, 2014), `travel.stackexchange.com/questions/26083/explain-travel-times-and-distances-on-flight` (last visited July 23, 2015) (Creative Commons Share Alike License: `creativecommons.org/licenses/by-sa/4.0/legalcode`).

[R224] B. Teitell, 17 holiday blunders (and how to avoid them), *The Boston Globe*, November 27, 2015, `www.bostonglobe.com/lifestyle/2015/11/27/the-mistakes-you-make-this-holiday-season-and-how-avoid-them-maybe/cvAPjtOkhi8QoHcZDTig1L/story.html` (last visited March 29, 2020).

[R225] Internal Revenue Service Form 1040 (2018), p. 79, `www.irs.gov/pub/irs-pdf/i1040gi.pdf` (last visited February 15, 2019).

Chapter 8

[R226] Scientific Consensus: Earth's Climate is Warming, Earth Science Communications Team, NASA Jet Propulsion Laboratory, `climate.nasa.gov/scientific-consensus/` (last visited September 18, 2019).

[R227] J. Keohane, Imaginary fiends, *The Boston Globe* (2010), `www.boston.com/bostonglobe/ideas/articles/2010/02/14/imaginary_fiends/` (last visited July 17, 2015).

[R228] Data from `www2.fbi.gov/ucr/cius2008/data/table_01.html` (last visited August 21, 2015) and `www.gallup.com/poll/123644/Americans-Perceive-Increased-Crime.aspx` (last visited August 21, 2015).

[R229] P. E. Meehl, Theory-Testing in Psychology and Physics: A Methodological Paradox, *Philosophy of Science*, Vol. 34, No. 2 (June 1967), pp. 103–115, `www.fisme.science.uu.nl/staff/christianb/downloads/meehl1967.pdf` (last visited April 6, 2019).

[R230] Wikipedia, Anscombe's quartet, `en.wikipedia.org/wiki/Anscombe's_quartet` (last visited July 23, 2015) (Creative Commons Attribution-ShareAlike License, `en.wikipedia.org/wiki/Wikipedia:Text_of_Creative_Commons_Attribution-ShareAlike_3.0_Unported_License`).

[R231] Photo: "The Leaning Tower of Pisa SB" by Saffron Blaze — Own work. Licensed under CC BY-SA 3.0 via Wikimedia Commons `commons.wikimedia.org/wiki/File:The_Leaning_Tower_of_Pisa_SB.jpeg` (last visited July 28, 2019). The data can be found in D. S. Moore and G. McCabe, Introduction to the Practice of Statistics (last visited August 31, 2015).

[R232] United States Postal Service, Postage Rates and Historical Statistics, `about.usps.com/who-we-are/postal-history/rates-historical-statistics.htm` (last visited June 23, 2019).

[R233] D. Dineen, Despite its many benefits, corporate use of aircraft still vilified, *The Boston Globe* (May 26, 2012), `www.bostonglobe.com/opinion/letters/2012/05/25/despite-its-many-benefits-corporate-use-aircraft-still-vilified/mbQ6mINMQXbAayzWvFn6NI/story.html` (last visited July 23, 2015).

[R234] Mark Twain, Life on the Mississippi, `www.gutenberg.org/files/245/245-0.txt` (last visited July 28, 2019).

[R235] R. Munroe, xkcd, `xkcd.com/552/` (last visited August 14, 2015). This work is licensed under a Creative Commons Attribution-NonCommercial 2.5 License, `creativecommons.org/licenses/by-nc/2.5/`.

Chapter 9

[R236] High-Level Radioactive Waste, Nuclear Information and Resource Service, `http://archives.nirs.us/factsheets/hlwfcst.htm` (last visited March 2, 2020). Quoted with permission.

[R237] *Growth rates of different Burkholderia cenocepacia mutants* (reported in `doi.org/10.1073/pnas.1207025110`) in minimal galactose medium provided by the laboratory of Vaughn Cooper.

[R238] S. Thompson, These Ads Think They Know You, *The New York Times*, April 30, 2019, `www.nytimes.com/interactive/2019/04/30/opinion/privacy-targeted-advertising.html?action=click&module=Opinion&pgtype=Homepage` (last visited April 30, 2019).

[R239] D. Genis, Blurred Lines at NY Sketchbook Museum, *Daily Beast*, November 1, 2014, `www.thedailybeast.com/blurred-lines-at-ny-sketchbook-museum` (last visited April 22, 2019).

[R240] J, Silva, On Transhumanism and Why Technology Is Our Silicon Nervous System, *Daily Beast*, April 26, 2014, `www.thedailybeast.com/on-transhumanism-and-why-technology-is-our-silicon-nervous-system` (last visited April 22, 2019).

[R241] V. Markovitz, Sizing Up Wind Energy: Bigger Means Greener, Study Says, National Geographic, July 20, 2012, `news.nationalgeographic.com/news/energy/2012/07/120720-bigger-wind-turbines-greener-study-says/` (last visited May 1, 2019).

[R242] T. R. Malthus, An Essay on the Principle of Population, `www.gutenberg.org/etext/4239` (last visited July 17, 2015).

[R243] S. Clifford, Other Retailers Find Ex-Blockbuster Stores Just Right, *The New York Times* (April 8, 2011), `www.nytimes.com/2011/04/09/business/09blockbuster.html` (last visited July 23, 2015).

[R244] E. Osnos, Green Giant, *The New Yorker* (December 21, 2009), `www.newyorker.com/magazine/2009/12/21` (last visited July 29, 2015).

[R245] S. Graham and M. Hebert, Writing to Read, Carnegie Corporation (2010), `all4ed.org/wp-content/uploads/2010/04/WritingToRead.pdf` (last visited July 23, 2015).

[R246] Educating mothers saves lives, study says, Associated Press reported in *The Boston Globe* (September 17, 2010), `www.boston.com/news/world/europe/articles/2010/09/17/educating_mothers_saves_lives_study_says/` (last visited March 30, 2020).

[R247] V. Heffernan, The Trouble With E-Mail, *The New York Times* (May 29, 2011), `opinionator.blogs.nytimes.com/2011/05/29/the-trouble-with-e-mail/` (last visited July 23, 2015).

[R248] M. Rosenberg, India's Population, About.com (April 1, 2011), `geography.about.com/od/obtainpopulationdata/a/indiapopulation.htm` (last visited July 23, 2015).

[R249] M. B. Farrell, MIT grad led team that built faster YouTube player, *The Boston Globe* (September 24, 2012), `www.bostonglobe.com/business/2012/09/23/building-faster-youtube/JqbVsEFUJfa5tpQmgbujkL/story.html` (last visited March 30, 2020).

[R250] N. Silver, *The Signal and the Noise*, page 32, Penguin Press (September 27, 2012).

[R251] L. Neyfakh, Cuba, you owe us $7 billion, *The Boston Globe* (April 18, 2014), `www.bostonglobe.com/ideas/2014/04/18/cuba-you-owe-billion/jHAufRfQJ9Bx24TuzQyBNO/story.html` (last visited March 30, 2020).

[R252] J. Barron, As Time Goes By, What's This Piano Worth?, *The New York Times* (December 13, 2012), `cityroom.blogs.nytimes.com/2012/12/13/as-time-goes-by-whats-this-piano-worth/` (last visited July 23, 2015).

[R253] Lewis Carroll, Sylvie and Bruno, Project Gutenberg, `www.gutenberg.org/files/620/620-h/620-h.htm` (last visited July 27, 2019).

[R254] K. Johnson, Is the Evening Sky Doomed?, *The New York Times*, August 18, 2019, `www.nytimes.com/2019/08/17/opinion/sunday/light-pollution.html` (last visited August 18, 2019).

[R255] J. Hecht, Awash in Artificial Light, the World Gets 2 Percent Brighter Each Year, *IEEE Spectrum*, November 22, 2017, `https://spectrum.ieee.org/energywise/energy/environment/awash-in-artificial-light-the-world-gets-2-percent-brighter-each-year` (last visited August 18, 2019).

[R256] George V. Higgins, *Swan Boats at Four*, Henry Holt and Company, 1995, pp. 198–199.

[R257] Disrupting the cow, T. Soba and C. Tubb, *The Boston Globe*, November 29, 2019, `www.bostonglobe.com/2019/11/29/opinion/disrupting-cow/` (last visited December 3, 2019).

Chapter 10

[R258] P. Parker, Debit Card Processing Fees Explained, July 2, 2018, `www.cardpaymentoptions.com/fee-sweep/debit-fees-explained/` (last visited June 8, 2019).

[R259] B. Luthi, The History of Overdraft Fees, *Chime*, June 29, 2018, `www.chimebank.com/2018/06/29/the-history-of-overdraft-fees/` (last visited May 6, 2019).

[R260] S. Tierney, How to Avoid Overdraft Fees, *Nerdwallet*, November 2, 2018, `www.nerdwallet.com/blog/banking/avoid-overdraft-fees/` (last visited June 5, 2019).

[R261] Adapted from a sample statement no longer available at PracticalMoneySkills.com, `www.practicalmoneyskills.com` .

[R262] United States Federal Trade Commission, Consumer Information, Free Credit Reports (2013), `www.consumer.ftc.gov/articles/0155-free-credit-reports` (last visited July 17, 2015).

[R263] Credit Reports and Scores, USA.gov, `www.usa.gov/topics/money/credit/credit-reports/bureaus-scoring.shtml` (last visited July 23, 2015).

[R264] J. B. McKim, Rates for big loans tumble, *The Boston Globe* (November 20, 2010), `archive.boston.com/business/personalfinance/articles/2010/11/20/rates_for_big_loans_tumble/` (last visited March 31, 2020).

[R265] B. Applebaum, Without Loan Giants, 30-Year Mortgage May Fade Away, *The New York Times* (March 3, 2011), `www.nytimes.com/2011/03/04/business/04housing.html` (last visited July 23, 2015).

[R266] By P. McMorrow, The end of 30-year fixed-rate mortgage?, *The Boston Globe* (March 4, 2011), `www.boston.com/bostonglobe/editorial_opinion/oped/articles/2011/03/04/the_end_of_30_year_fixed__rate_mortgage/` (last visited March 30, 2020). MR2493146

[R267] N.H. caps rates on payday loans, Associated Press reported in *The Boston Globe* (January 1, 2009), `www.boston.com/business/articles/2009/01/01/nh_caps_rates_on_payday_loans/` (last visited March 30, 2020).

[R268] J. Saltzman, Charity sues R.I. hospital over donation in 1912, *The Boston Globe* (February 23, 2008), `www.boston.com/news/local/articles/2008/02/23/charity_sues_ri_hospital_over_donation_in_1912/` (last visited March 30, 2020).

[R269] Data from The Nilson Report (May 19, 2014), `www.nilsonreport.com`.

[R270] G. Stohr, American Express Fee Accusations Get U.S. High Court Hearing, *Bloomberg News*, October 16, 2017, `www.bloomberg.com/news/articles/2017-10-16/american-express-fee-accusations-get-u-s-supreme-court-hearing` (last visited October 18, 2017).

[R271] A. Liptak, Supreme Court Sides With American Express on Merchant Fees, *The New York Times*, June 25, 2018, `www.nytimes.com/2018/06/25/us/politics/supreme-court-american-express-fees.html` (last visited April 28, 2019).

Chapter 11

[R272] J. Jacoby, Lottery games online? Scratch that idea, *The Boston Globe*, July 10, 2016, `www.bostonglobe.com/opinion/2016/07/10/lottery-games-online-scratch-that-idea/H0j1VLsYDJa5RZqbbPdfzO/story.html` (last visited March 30, 2020).

[R273] `www.census.gov/programs-surveys/state.html` (last visited March 17. 2019).

[R274] M. Brown, Did We Get Lucky? LendEDU's Lottery Study & Report, lendedu.com, August 31, 2018, `lendedu.com/blog/lottery-study-report/` (last visited March 13, 2019).

[R275] `upload.wikimedia.org/wikipedia/commons/5/5d/13-02-27-spielbank-wiesbaden-by-RalfR-094.jpg` (last visited March 5, 2020) (Creative Commons Attribution-ShareAlike License, `en.wikipedia.org/wiki/Wikipedia:Text_of_Creative_Commons_Attribution-ShareAlike_3.0_Unported_License`, Permission is granted to copy, distribute and/or modify this document under the terms of the GNU Free Documentation License, Version 1.2 or any later version published by the Free Software Foundation; with no Invariant Sections, no Front-Cover Texts, and no Back-Cover Texts. A copy of the license is at `commons.wikimedia.org/wiki/Commons:GNU_Free_Documentation_License,_version_1.2`).

[R276] L. P. Weston, Dump the Insurance on your Clunker, *MSN Money* (March 2007), reposted at `www.insurancemommy.com/Images/dumpyourclunker.pdf` (last visited October 4, 2015).

[R277] G. B. Shaw, "The Vice of Gambling and the Virtue of Insurance", Chapter 3 of T. W. Korner, *Naive Decision Making: Mathematics Applied to the Social World*, Cambridge University Press, 2008, and in Volume 3 of J. R. Newman, *The World of Mathematics*, Dover, 2003, `www.unz.org/Pub/NewmanJames-1957v03-01524` (last visited October 10, 2016). MR2482423

[R278] M. Novak, 9 Albert Einstein Quotes That Are Totally Fake,
paleofuture.gizmodo.com/9-albert-einstein-quotes-that-are-totally-
fake-1543806477 (last visited March 13, 2014).

[R279] D. Fears, Possible cut to beach testing a health threat, critics say, Washington Post report
in *The Boston Globe* (March 4, 2012), www.bostonglobe.com/news/nation/2012/
03/04/elimination-funding-for-beach-contamination-monitoring-could-
health-hazard-environmentalists-say/9sGB4SzlU2m3CvM6jaINhN/story.html
(last visited July 24, 2015).

[R280] M. Torres, What Is a TV Extended Warranty?, About.com, tv.about.com/od/
warranties/a/buyexwarranty.htm (last visited July 29, 2015).

[R281] www.masslottery.com/winners/pf_faqs.html (last visited July 30, 2015).

[R282] S. Bishop, Lottery suspense builds in Mass. for $355m prize, *The Boston Globe* (Janu-
ary 5, 2011), www.boston.com/news/local/massachusetts/articles/2011/01/05/
lottery_suspense_builds_in_mass_for_355m_prize/ (last visited March 30, 2020).

[R283] M. Levenson, Megabucks plan rankles lottery players, *The Boston Globe* (March
21, 2009), www.boston.com/news/local/massachusetts/articles/2009/03/21/
megabucks_plan_rankles_lottery_players/ (last visited March 30, 2020).

[R284] C. Bialik, Lottery Math 101, *The Wall Street Journal* (September 22, 2009),
blogs.wsj.com/numbersguy/lottery-math-101-801/ (last visited July 30, 2015).

[R285] CrzRsn, Roulette strategy …, March 27, 2008, forums.finalgear.com/off-topic/
roulette-strategy-26295/ (broken link) (last visited September 3, 2015).

[R286] Roll two dice. What is the probability that one die shows exactly two more than the
other die?, math.stackexchange.com/questions/1716651/roll-two-dice-what-
is-the-probability-that-one-die-shows-exactly-two-more-than (last visited
March 28, 2016) (Creative Commons Share Alike License: creativecommons.org/
licenses/by-sa/4.0/legalcode).

[R287] P. Abraham, Red Sox Notebook, *The Boston Globe*, April 4, 2017, www.bostonglobe.
com/sports/redsox/2017/04/03/snotes/ehRUEIoZ7rVl86N11w46sK/story.html (last visited March 30, 2020).

[R288] R. Oppel Jr., How Bettors in the Know Cashed In on 'Game of Thrones', *The
New York Times*, May 21, 2019, www.nytimes.com/2019/05/21/us/game-of-thrones-
predictions-betting.html (last visited May 22, 2019).

[R289] B. Schoenfeld, How Data (and Some Breathtaking Soccer) Brought Liverpool to the Cusp
of Glory, *The New York Times*, May 22, 2019, www.nytimes.com/2019/05/22/magazine/
soccer-data-liverpool.html (last visited May 24, 2019).

Chapter 12

[R290] S. Hazzard, *The Transit of Venus*, Viking Press, New York, 1980, p. 62.

[R291] D. Robinson, Cancer Clusters: Findings Vs Feelings, Medscape, www.medscape.com/
viewarticle/442554_5 (last visited July 15, 2015).

[R292] When "1-In-100-Year" Floods Happen Often, What Should You Call Them?, R. Her-
sher, *All Things Considered*, May 8, 2019, www.npr.org/2019/05/08/720737285/when-
1-in-100-year-floods-happen-often-what-should-you-call-them .

[R293] C. Ingraham, Houston is experiencing its third '500-year' flood in 3 years. How is that possible?, *The Washington Post*, August 29, 2017, `www.washingtonpost.com/news/wonk/wp/2017/08/29/houston-is-experiencing-its-third-500-year-flood-in-3-years-how-is-that-possible` (last visited July 20, 2019),

[R294] G. Trudeau, Doonesbury (March 21, 2009), `www.gocomics.com/doonesbury/2009/03/21` (last visited August 11, 2015).

[R295] C. Bialik, Odds Are, Stunning Coincidences Can Be Expected, *The Wall Street Journal* (September 24, 2009), `www.wsj.com/articles/SB125366023562432131` (last visited October 27, 2020).

[R296] A. Gelman, Lottery probability update, Statistical Modeling, Causal Inference, and Social Science (May 26 2011), `andrewgelman.com/2011/05/26/lottery_probabi/` (last visited July 24, 2015). Quoted with permission.

[R297] Sicherman dice, Wikipedia, `en.wikipedia.org/wiki/Sicherman_dice` (last visited August 11, 2015) (Creative Commons Attribution-ShareAlike License, `en.wikipedia.org/wiki/Wikipedia:Text_of_Creative_Commons_Attribution-ShareAlike_3.0_Unported_License`).

[R298] J. Pope, Colleges bewildered by anonymous major gifts, Associated Press report in *The Boston Globe* (April 24, 2009), `www.boston.com/news/nation/articles/2009/04/24/colleges_bewildered_by_anonymous_major_gifts/` (last visited March 30, 2020).

[R299] Ever wonder if the lottery numbers you play everyday have actually already hit, before you started playing them?, *Numbers Planet*, `www.numbersplanet.com/` (broken link) (last visited July 24, 2015).

[R300] Odd vs. Even statistics, *Numbers Planet*, `www.numbersplanet.com/` (broken link) (last visited July 24, 2015).

[R301] R. Highfield, Olympic Games drug testing means 'cheaters escape and innocents tarnished', *The Telegraph*, August 6, 2008, `www.telegraph.co.uk/news/science/science-news/3348936/Olympic-Games-drug-testing-means-cheaters-escape-and-innocents-tarnished.html` (last visited July 29, 2019).

[R302] From an email from Old Friends Farm, `www.oldfriendsfarm.com/` (last visited August 3, 2015). "Go ahead and use the quote — thanks for asking!"

[R303] Alexander McCall Smith, *The Double Comfort Safari Club*, Pantheon Books, 2010, page 116.

[R304] D. Grinberg, My tax refund is exactly $3000. Did I hit a cap?, `money.stackexchange.com/questions/90936/my-tax-refund-is-exactly-3000-did-i-hit-a-cap`, February 16, 2018 (last visited February 19, 2018) (Creative Commons Share Alike License: `creativecommons.org/licenses/by-sa/4.0/legalcode`).

[R305] J. Doe, Have only girls been born for a long time in this village?, skeptics.stackexchange.com, August 19, 2019, `skeptics.stackexchange.com/questions/44732/have-only-girls-been-born-for-a-long-time-in-this-village` (last visited August 20, 2019) (Creative Commons Share Alike License: `creativecommons.org/licenses/by-sa/4.0/legalcode`).

[R306] M. Lampert, J. Levin, M. Vounatsos and J. Maraganore, Cautiously optimistic about discovering drugs to treat the coronavirus, *The Boston Globe*, April 27, 2020, `www.bostonglobe.com/2020/04/27/opinion/cautiously-optimistic-about-discovering-drugs-treat-coronavirus/` (last visited April 28, 2020).

Chapter 13

[R307] M. S. Klinkman, J. C. Coyne, S. Gallo and T. L. Schwenk, False Positives, False Negatives, and the Validity of the Diagnosis of Major Depression in Primary Care, *Arch. Fam. Med.* 1998;7(5):451–461, www.ncbi.nlm.nih.gov/pubmed/9755738 (last visited October 4, 2015). Licensed under a Creative Commons Attribution-Noncommercial-No Derivative Works 3.0 United States License (creativecommons.org/licenses/by-nc-nd/3.0/.

[R308] T. Davies, Prenatal screening for trisomy 18: Should not be contemplated, *British Medical Journal*, February 12, 1994, doi: doi.org/10.1136/bmj.308.6926.471, www.bmj.com/content/308/6926/471.1 (last visited July 28, 2019).

[R309] Amniocentesis, American Pregnancy Association, americanpregnancy.org/prenataltesting/amniocentesis.html (last visited July 15, 2015).

[R310] M. Lynn and C. Maier, McDaniel v. Brown (08-559) Legal Information Institute, LII Supreme Court Bulletin, Cornell University Law School, www.law.cornell.edu/supct/cert/08-559 (last visited September 18, 2015). Michelle Jessica Lynn and Chris Maier are the authors of that Supreme Court Preview. Legal Information Institute at Cornell Law School.

[R311] D. Badertscher, U.S. Supreme Court Update: McDaniel v. Brown, Criminal Law Library Blog (January 26, 2010), www.criminallawlibraryblog.com/2010/01/us_supreme_court_update_mcdani.html (last visited July 25, 2015).

[R312] R. Stein, Researchers link chronic fatigue syndrome to class of virus, Washington Post report in *The Boston Globe* (August 24, 2010), www.boston.com/news/nation/articles/2010/08/24/researchers_link_chronic_fatigue_syndrome_to_class_of_virus (last visited March 30, 2020).

[R313] M. Reyes et al., Prevalence and incidence of chronic fatigue syndrome in Wichita, Kansas. *Arch. Intern. Med.* 2003 Jul 14;163(13):1530–6, www.ncbi.nlm.nih.gov/pubmed/12860574 (last visited December 15, 2015).

[R314] When should I test with a pregnancy test?, Yourdays (free information for women), www.yourdays.com/when-pregnancy-test.htm (last visited March 12, 2020).

[R315] M. Specter, Damn Spam, Annals of Technology, *The New Yorker* (August 6, 2007), www.newyorker.com/reporting/2007/08/06/070806fa_fact_specter (last visited July 31, 2019).

[R316] G. Karthikeyan, The cost of a "negative test", response to Screening programme evaluation applied to airport security, *British Medical Journal* (December 27, 2007), www.bmj.com/rapid-response/2011/11/01/cost-negative-test (last visited September 4, 2015). Quoted with permission.

[R317] S. Strogatz, Chances Are, *The New York Times* (April 25, 2010), opinionator.blogs.nytimes.com/2010/04/25/chances-are/ (last visited March 2, 2016).

[R318] M. E. Irons, Caught in a dragnet, *The Boston Globe*, July 17, 2011, archive.boston.com/news/local/massachusetts/articles/2011/07/17/man_sues_registry_after_license_mistakenly_revoked/ (last visited March 12, 2020).

[R319] J. Allen, Identity fraud dragnet hardly seems worth the expense or trouble, *The Boston Globe* (July 24, 2011), www.boston.com/bostonglobe/editorial_opinion/letters/articles/2011/07/24/identity_fraud_dragnet_hardly_seems_worth_the_expense_or_trouble/ (last visited March 30, 2020).

[R320] E. Wahlgren, Happy Halloween! Kids who eat candy every day grow up to be violent criminals, www.aol.com/2009/10/02/happy-halloween-kids-who-eat-candy-every-day-grow-up-to-be-viol/, originally published on DailyFinance.com (October 2, 2009) (last visited March 12, 2020). Quoted with permission.

[R321] A. Gelman, The Reliability of Cluster Surveys of Conflict Mortality: Violent Deaths and Non-Violent Deaths, Statistical Modeling, Causal Inference, and Social Science (August 11, 2011), andrewgelman.com/2011/08/the_reliability/ (last visited July 25, 2015). Quoted with permission.

[R322] D. Kotz, Surgery offers no advantage for early prostate cancer, study finds, *The Boston Globe* (July 18, 2012), bostonglobe.com/lifestyle/health-wellness/2012/07/18/surgery-offers-survival-advantage-for-older-men-with-early-stage-prostate-cancer-study-finds/T5XM7APIuoZuav6PbJzYuI/story.html (last visited July 25, 2015).

[R323] How to Avoid False Positives While Conducting a Home Drug Test, lapoliticaesotracosa.blogspot.com/2012/05/how-to-avoid-false-positives-while.html (last visited July 25, 2015).

Index

Intercultural Competence

Interpersonal Communication Across Cultures

Seventh Edition

Myron W. Lustig
San Diego State University

Jolene Koester
California State University, Northridge

PEARSON

Boston   Columbus   Indianapolis   New York   San Francisco   Upper Saddle River
Amsterdam   Cape Town   Dubai   London   Madrid   Milan   Munich   Paris   Montréal   Toronto
Delhi   Mexico City   São Paulo   Sydney   Hong Kong   Seoul   Singapore   Taipei   Tokyo

Editor-in-Chief, Communication: Karon Bowers
Assistant Editor: Stephanie Chaisson
Editorial Associate: Megan Sweeney
Associate Managing Editor: Bayani Mendoza de Leon
Senior Marketing Manager: Blair Zoe Tuckman
Executive Digital Producer: Stefanie Snajder
Senior Digital Editor: Paul DeLuca
Digital Editor: Lisa Dotson
Project Coordination, Text Design, and Electronic
    Page Makeup: Integra Software Services, Inc.
Manufacturing Buyer: Mary Ann Gloriande
Senior Cover Design Manager: Nancy Danahy
Image Permission Coordinators: Lee Scher/Annette Linder
Photo Researchers: Melody English/Sarah E. Peavey
Text Rights Clearance Editor: Danielle Simon
Permissions Project Manager/Specialist: Joseph Croscup/
    Dana Weightman
Cover Image: ©Nancy Danahy

Credits and acknowledgments borrowed from other sources and reproduced, with permission, in this textbook appear on pages 366–370.

Library of Congress Cataloging-in-Publication Data
Lustig, Myron W.
    Intercultural competence: interpersonal communication across cultures / Myron W. Lustig, & Jolene Koester.—7th ed.
        p. cm.
    Includes bibliographical references and index.
    ISBN 978-0-205-21124-1
    1. Intercultural communication.   2. Communicative competence—United States.   3. Interpersonal communication—United States.   I. Koester, Jolene.
    HM1211.L87 2012
    303.48'20973—dc23   2012008753

1 2 3 4 5 6 7 8 9 10—RRD-Crawfordsville—15 14 13 12

PEARSON

ISBN-13: 978-0-205-21124-1
ISBN-10:   0-205-21124-0

Why Do You Need this New Edition?

If you're wondering why you should buy this new edition of *Intercultural Competence*, here are 6 good reasons!

1. The visual elements of the seventh edition have been updated to full color and improved significantly to "catch" your eye and direct attention to the various concepts and ideas throughout the text.

2. Chapter 11 has been revised and updated to reflect current ideas related to health care, education, business, and other contexts outside of the classroom.

3. The cultural taxonomies chapter (Chapter 5) includes major updates and additions to the ideas of Geert Hofstede and the GLOBE researchers, an examination of Shalom Schwartz's ideas on the dimensions of culture, and a synthesis of the various cultural taxonomies.

Taken together, these changes provide you with greater accessibility to the ideas about cultural differences and similarities.

4. New "Culture Connections" boxes are carefully placed throughout the seventh edition to clearly underscore the conceptual issues being discussed.

5. The section "Intercultural Contact" in Chapter 12 has been substantially changed with an additional, elaborated discussion of tourism as it relates to intercultural communication.

6. Updated and new examples throughout the text place emphasis on the use of current technologies in your everyday life and that affect intercultural communication.

Let us not be blind to our differences, but let us also direct our attention to our common interests and the means by which those differences can be resolved. And if we cannot end now our differences, at least we can help make the world safe for [cultural] diversity. For in the final analysis, our most basic common link is that we all inhabit this small planet. We all breathe the same air. We all cherish our children's future. And we are all mortal.
— John F. Kennedy June 10, 1963

In this second decade of the twenty-first century, the world is vastly different from what it was in 1963, or in 1983, or even in 2003. Innovations in communications, transportation, and information technologies—on a global and unprecedented scale—have created vast economic interdependencies, demographic shifts both within and across nations, and the greatest mingling of cultures the world has ever seen. But as the opening quote from U.S. President John F. Kennedy suggests, and as the French proverb wryly states, "The more things change, the more they stay the same," at least in one important way: competence in intercultural communication is vital if you want to function well in your private and public life. This creates a very strong imperative for you to learn to communicate with people whose cultural heritage makes them very different from you. Our goal in this book is to give you the knowledge, motivation, and skills to accomplish that objective.

NEW TO THIS EDITION

Considerable progress has been made by scholars and practitioners of intercultural communication and related disciplines, and this edition reflects those changes. Many of the substantial changes may not be obvious to the casual reader, nor should they be. For instance, there is an extensive update of the research citations that undergird the presentation of information and ideas. These changes help the book remain contemporary. They appear at the back in the Notes section, where they are available to the interested reader without intruding on the flow of the text. Similar changes occur in the end-of-chapter materials, where the "For Discussion" questions and the "For Further Reading" suggestions have been updated substantially.

Among the major changes are the following:

■ Many examples have been updated or added, and many new ideas are explicated in detail. Similarly, we have heightened our emphasis on the use of current technologies that affect intercultural communication.
■ In Chapter 1, the sections on "Imperatives for Intercultural Competence" and "The Challenge of Communicating in an Intercultural World" have been updated and improved.
■ The "Metaphors of U.S. Cultural Diversity" in Chapter 3 have been revised, and the ideas are now more tightly integrated with the discussion of acculturation in Chapter 12.
■ Chapter 5, on cultural taxonomies, has been extensively revised. The new material includes major updates and additions to the ideas of Geert Hofstede and

the GLOBE researchers, an examination of Shalom Schwartz's ideas on the dimensions of culture, and a synthesis of the various cultural taxonomies. Taken together, these changes provide greater student accessibility to the ideas about cultural differences and similarities.

■ Chapter 8 has been substantially revised to reflect current ideas about nonverbal communication in intercultural interactions. A discussion of the functions of nonverbal communication has been added, and the analysis of the types of nonverbal messages has been reorganized and updated.

■ Chapter 11 has been revised and updated to reflect current ideas related to the health care, education, and business contexts.

■ In Chapter 12, the section on "Intercultural Contact" has been substantially changed, and an elaborated discussion of tourism has been added.

■ The back-of-book "Resources" section has been repurposed. Now included is detailed information about the cultural patterns of specific groups on the Hofstede, GLOBE, and Schwartz cultural taxonomies; these taxonomies are discussed conceptually in Chapter 5. No longer included in the "Resources" section are the suggestions for online sites that students might use to investigate a specific culture; various comprehensive Internet search engines have eliminated the need for this information.

■ About half of the "Culture Connections" boxes are new, as are essentially all of the photographs. We have selected and placed these elements very carefully to underscore more clearly the conceptual issues being discussed.

■ The book's visual elements have been improved significantly to support reader interest and involvement; new to this edition is the full use of color to "catch the eye" and direct attention to the various ideas that we include.

Additional changes to this addition are too numerous to enumerate completely. Changes, updates, and many small improvements have been made throughout the book.

UNCHANGED IN THIS EDITION

Some things have not changed, nor should they. Our students and colleagues have helped to guide the creation of this seventh edition of *Intercultural Competence*. They have affirmed for us the critical features in this book that provide the reader with a satisfying experience and are useful for learning and teaching about intercultural communication. These features include:

■ **An easy-to-read, conversational style.** Students have repeatedly praised the clear and readable qualities of the text. We have tried, in this and previous editions, to ensure that students have an "easy read" as they access the book's ideas.

■ **A healthy blend of the practical and the theoretical, of the concrete and the abstract.** We believe strongly that a textbook on intercultural communication needs to include both a thorough grounding in the conceptual ideas and an applied orientation that makes those ideas tangible.

■ **Culture Connections boxes that provide emotional connections.** The Culture Connections boxes exemplify and integrate important concepts while providing access to the affective dimension of intercultural competence. These boxes also illustrate the lived experiences of intercultural communicators. About half of the Culture Connections boxes are new to this edition, and we chose each selection carefully to provide the opportunity for students to "feel" some aspect of intercultural competence.

■ **A strong grounding in theory and research.** Intercultural communication theories and their supporting research provide powerful ways of viewing and understanding intercultural communication phenomena. We also link the presentation of theories to numerous illustrative examples. These conceptual underpinnings to intercultural communication have been updated, and we have incorporated ideas from literally hundreds of new sources across a wide spectrum of inquiry. These sources form a solid bibliography for those interested in pursuing specific topics in greater depth. As we have done in the past, however, we have chosen to maintain the text's readability by placing the citations at the end of the book, where they appear in detailed endnotes that are unobtrusive but available to interested readers.

■ **A focus on the significance and importance of cultural patterns.** Cultural patterns provide the underlying set of assumptions for cultural and intercultural communication. The focus on cultural patterns as the lens through which all interactions are interpreted is thoroughly explored in Chapters 4 and 5, and the themes of these two chapters permeate the concepts developed in all subsequent chapters.

■ **Attention to the impact of technology on intercultural communication.** From Chapter 1, where we describe the technological imperative for intercultural communication that challenges us to be interculturally competent, to Chapter 12, where we analyze the perils and possibilities for living in an intercultural world, and throughout each of the intervening chapters, this edition is focused on the new information technologies and their effects on intercultural communication.

■ **A consideration of topics not normally emphasized in intercultural communication textbooks.** Although it is standard fare for most books to consider verbal and nonverbal code systems, we provide a careful elaboration of the nature of differing logical systems, or preferred reasoning patterns, as well as a discussion of the consequences for intercultural communication when the expectations for the language-in-use are not widely shared. Similarly, drawing heavily on the available information about interpersonal communication, we explore the dynamic processes of establishing and developing relationships between culturally different individuals, including an elaboration of issues related to "face" in interpersonal relationships.

■ **Pedagogical features that enhance student retention and involvement.** Concluding each chapter are For Discussion questions; they can be used to guide in-class conversations, or they may serve as the basis for short, focused assignments. Similarly, the For Further Reading suggestions can be readily understood by the beginning student and provide additional entry into that chapter's ideas.

ACKNOWLEDGMENT OF CULTURAL ANCESTRY

At various points in our writing, we were amazed at how subtly but thoroughly our own cultural experiences had permeated the text. Lest anyone believe that our presentation of relevant theories, examples, and practical suggestions is without the distortion of culture, we would like to describe our own cultural heritage. That heritage shapes our understanding of intercultural communication, and it affects what we know, how we feel, and what we do when we communicate with others.

Our cultural ancestry is European, and our own cultural experiences are predominantly those that we refer to in this book as European American. Both of our family backgrounds and the communities in which we were raised have influenced and reinforced our cultural perspectives. The European American cultural experience is the one we know best, simply because it is who we are. Many of our ideas and examples about intercultural communication, therefore, draw on our own cultural experiences.

We have tried, however, to increase the number and range of other cultural voices through the ideas and examples that we provide. These voices and the lessons and illustrations they offer represent our colleagues, our friends, and, most important, our students.

IMPORTANCE OF VOICES FROM OTHER CULTURES

Although we have attempted to include a wide range of domestic and international cultural groups, inevitably we have shortchanged some simply because we do not have sufficient knowledge, either through direct experience or through secondary accounts, of all cultures. Our errors and omissions are not meant to exclude or discount. Rather, they represent the limits of our own intercultural communication experiences. We hope that you, as a reader with a cultural voice of your own, will participate with us in a dialogue that allows us to improve this text over a period of time. Readers of previous editions were generous with their suggestions for improvement, and we are very grateful to them for these comments. We ask that you continue this dialogue by providing us with your feedback and responses. Send us examples that illustrate the principles discussed in the text. Be willing to provide a cultural perspective that differs from our own and from those of our colleagues, friends, and students. Our commitment now and in future editions of this book is to describe a variety of cultural voices with accuracy and sensitivity. We ask for your help in accomplishing that objective.

ISSUES IN THE USE OF CULTURAL EXAMPLES

Some of the examples in the following pages may include references to a culture to which you belong or with which you have had substantial experiences, and our examples may not match your personal knowledge. As you will discover in the opening chapters of this book, both your own experiences and the examples we recount could be accurate. One of the tensions we felt in writing this book was in making statements that are broad enough to provide reasonably accurate generalizations but specific and tentative enough to avoid false claims of universal applicability to all individuals in a given culture.

We have struggled as well with issues of fairness, sensitivity, representativeness, and inclusiveness. Indeed, we have had innumerable discussions with our colleagues across the country—colleagues who, like ourselves, are committed to making the United States and its colleges and universities into truly multicultural institutions—and we have sought their advice about appropriate ways to reflect the value of cultural diversity in our writing. We have responded to their suggestions, and we appreciate the added measure of quality that these cultural voices supply.

TEXT ORGANIZATION

Our goal in this book is to provide ideas and information that can help you achieve competence in intercultural communication. Part One, Communication and Intercultural Competence, orients you to the central ideas that underlie this book. Chapter 1 begins with a discussion of five imperatives for attaining intercultural competence. We also define and discuss the nature of communication generally and interpersonal communication specifically. In Chapter 2, we introduce the notion of culture and explain why cultures differ. Our focus then turns to intercultural communication, and we distinguish that form of communication from others. As our concern in this book is with interpersonal communication among people from different cultures, an understanding of these key concepts is critical. Chapter 3 begins with a focus on the United States as an intercultural community, as we address the delicate but important issue of how to characterize its cultural mix and the members of its cultural groups. We then lay the groundwork for our continuing discussion of intercultural competence by explaining what competence is, what its components are, and how people can achieve it when they communicate with others. The chapter also focuses on two communication tools that could help people to improve their intercultural competence.

Part Two, Cultural Differences in Communication, is devoted to an analysis of the fundamental ways that cultures vary. Chapter 4 provides a general overview of the ways in which cultures differ, and it emphasizes the importance of cultural patterns in differentiating among communication styles. This chapter also examines the structural features that are similar across all cultures. Chapter 5 offers several taxonomies that can be used to understand systematic differences in the ways in which people from various cultures think and communicate. Chapter 6 underscores the importance of cultural identity and the consequences of biases within intercultural communication.

In Part Three, Coding Intercultural Communication, we turn our attention to verbal and nonverbal messages, which are central to the communication process. Chapter 7 examines the coding of verbal languages and the influences of linguistic and cultural differences on attempts to communicate interculturally. Chapter 8 discusses the effects of cultural differences on nonverbal codes, as the accurate coding and decoding of nonverbal symbols is vital in intercultural communication. Chapter 9 investigates the effects or consequences of cultural differences in coding systems on face-to-face intercultural interactions. Of particular interest are those experiences involving participants who were taught to use different languages and organizational schemes.

Part Four, Communication in Intercultural Relationships, emphasizes the associations that form among people as a result of their shared communication experiences. Chapter 10 looks at the all-important issues related to the development and maintenance of interpersonal relationships among people from different cultures. Chapter 11 highlights the processes by which communication events are grouped into episodes and interpreted within such contexts as health care, education, and business. Finally, Chapter 12 emphasizes intercultural contacts and highlights the ethical choices individuals must face when engaged in interpersonal communication across cultures. The chapter concludes with some remarks about the problems, possibilities, and opportunities for life in our contemporary intercultural world.

A NOTE TO INSTRUCTORS

Accompanying the text is an Instructor's Manual, Test Bank, and PowerPoint presentation which are available to instructors who adopt the text for their courses. They provide pedagogical suggestions and instructional activities to enhance students' learning of course materials. Please contact your Pearson representative for these materials. These resources are available for download at www.pearsonhighered.com/irc (instructor login required).

Teaching a course in intercultural communication is one of the most exciting assignments available. It is difficult to convey in writing the level of involvement, commitment, and interest displayed by typical students in such courses. These students are the reason that teaching intercultural communication is, quite simply, so exhilarating and rewarding.

ACKNOWLEDGMENTS

Many people have assisted us, and we would like to thank them for their help. Literally thousands of students and faculty have now reviewed this text and graciously shared their ideas for improvements. A substantial portion of those ideas and insightful criticisms has been incorporated into the current edition, and we continue to be grateful for the helpful comments and suggestions that have spurred vital improvements. The following reviewers contributed detailed comments for this edition: Kumi Ishii, Western Kentucky University; Robert Hertzog, Northern Indiana University; Craig Nelson, Eastern Carolina University; Joann Brown, Florida International University; and Tasha Szousa, Humboldt State University. We are indebted to the students and faculty at our respective institutions, to our colleagues in the communication discipline, and to many people throughout higher education who have willingly shared their ideas and cultural voices with us.

We continue to be very grateful that the study of intercultural communication has become an increasingly vital and essential component of many universities' curricula. While we harbor no illusions that our influence was anything but minor, it is nevertheless gratifying to have been a "strong voice in the chorus" for these positive changes. Finally, we would like to acknowledge each other's encouragement and support throughout the writing of this book. It has truly been a collaborative effort.

Myron W. Lustig
Jolene Koester

Give your students choices

In addition to the traditional printed text, *Intercultural Competence,* 7th Edition, is available in the following formats to give you and your students more choices—and more ways to save.

MySearchLab® with Pearson eText

MySearchLab is an interactive website that features an eText, access to the EBSCO ContentSelect database and multimedia, and step-by-step tutorials that offer complete overviews of the entire writing and research process. MySearchLab is designed to amplify a traditional course in numerous ways or to administer a course online. Additionally, MySearchLab offers course-specific tools to enrich learning and help students succeed.

- **eText:** Identical in content and design to the printed text, the Pearson eText provides access to the book wherever and whenever it is needed. Students can take notes and highlight, just like a traditional book. The Pearson eText is also available on the iPad for all registered users of MySearchLab.
- **Flashcards:** Review important terms and concepts from each chapter online. Students can search by chapters or within a glossary and also access drills to help them prepare for quizzes and exams. Flashcards can be printed or exported to your mobile device.
- **Chapter-Specific Content:** Each chapter contains Learning Objectives, Quizzes, Media, and Flashcards. These can be used to enhance comprehension, help students review key terms, prepare for tests, and retain what they've learned. To order this book with MySearchLab access at no extra charge use ISBN 0-205-91204-4.

Learn more at www.mysearchlab.com

The **CourseSmart eTextbook** offers the same content as the printed text in a convenient online format—with highlighting, online search, and printing capabilities. www.coursesmart.com

Introduction to Intercultural Competence

KEY TERMS

In this, the second decade of the twenty-first century, culture, cultural differences, and intercultural communication are among the central ingredients of your life. As inhabitants of this post-millennium world, you no longer have a choice about whether to live and communicate with people from many cultures. Your only choice is whether you will learn to do it well.

These U.S. American tourists plot a day's sightseeing in Amsterdam. Tourism is a major international industry, bringing people from many cultures into contact with one another.

The world has changed dramatically from what it was even a generation ago. Across the globe and throughout the United States, there is now a heightened emphasis on culture. Similarly, there is a corresponding interplay of forces that both encourage and discourage accommodation and understanding among people who differ from one another. This emphasis on culture is accompanied by numerous opportunities for experiences with people who come from vastly different cultural backgrounds. Intercultural encounters are now ubiquitous; they occur within neighborhoods, across national borders, in face-to-face interactions, through mediated channels, in business, in personal relationships, in tourist travel, and in politics. In virtually every facet of life—in work, play, entertainment, school, family, community, and even in the media that you encounter daily—your experiences necessarily involve intercultural communication.

What does this great cultural mixing mean as you strive for success, satisfaction, well-being, and feelings of involvement and attachment to families, communities, organizations, and nations? It means that the forces that bring people from other cultures into your life are dynamic, potent, and ever present. It also means that competent intercultural communication has become essential.

Our purpose in writing this book is to provide you with the conceptual tools for understanding how cultural differences can affect your interpersonal communication. We also offer some practical suggestions concerning the adjustments necessary to achieve competence when dealing with these cultural differences. We begin by examining the forces that create the need for increased attention to intercultural communication competence.

IMPERATIVES FOR INTERCULTURAL COMPETENCE

The need to understand the role of culture in interpersonal communication is growing. Because of demographic, technological, economic, peace, and interpersonal concerns, intercultural competence is now more vital than ever.

The Demographic Imperative for Intercultural Competence

The United States—and the world as a whole—is currently in the midst of what is perhaps the largest and most extensive wave of cultural mixing in recorded history. Recent census figures provide a glimpse into the shape of the changing demographics of the U.S. population.

The U.S. population is now more than 308 million, of which 62.3 percent are European American, 16.3 percent are Latino, 12.6 percent are African American, 4.8 percent are Asian American, and 1.1 percent are Native American. About 2.9 percent of people identify themselves as belonging to two or more cultural groups.[1] Although all U.S. cultures increased in size over the past decade, the average 0.9 percent annual rate of U.S. population growth, while modest and declining, was not uniform. Since 2000, over half of the overall U.S. population growth has been Latino.[2] Similarly, the increase in Asian Americans has also been substantial at 43 percent.[3] If current trends continue, by 2050 the U.S. population of about 439 million will be about 46 percent European American, 30 percent Latino, 14 percent African American, 9 percent Asian American, and 1 percent Native American.[4] As William A. Henry has said of these changes, "The browning of America will alter everything in society, from politics and education to industry, values, and culture."[5]

People are also becoming more comfortable with multiple racial and cultural identities. In 2010, about 9 million people identified with more than one racial group, an increase of 32 percent from the previous decade. While very small in terms of sheer numbers, by 2050 the multiracial population is expected to triple, and since most of those people will be children, the multiracial children's demographic is among the fastest-growing.[6] Demographers attribute this population growth to more social acceptance of multiracial individuals; more than half of those who identified themselves as multiracial were younger than twenty years old.[7]

Census figures indicate that cultural diversity is a nationwide phenomenon. Over half of the states in the United States have at least 50,000 Native American residents, over half have at least 100,000 Asian American residents, and over 40 percent of the states exceed these numbers for both cultural groups. Latinos make up over a third of the populations of California, New Mexico, and Texas, and they constitute at least 20 percent of the people in Arizona, Colorado, Florida, and Nevada.[8] Over the past decade, the Latino population more than doubled in Kentucky, Alabama, Arkansas, Mississippi, North Carolina, and South Carolina. Eighteen states have African American populations that exceed a million;[9] African Americans constitute a majority of the District of Columbia's population, and they comprise at least a fifth of the population in nine states including Alabama, Delaware, Maryland, Louisiana, and Virginia.[10] There are already "minority-majorities"—populations of African Americans, Native Americans, Pacific Islanders, Latinos, and Asian Americans that, when combined, outnumber the European American population—in California, Hawaii, New Mexico, and Texas,[11] and Arizona, Florida, Georgia, Maryland, and Nevada will also soon reach this "minority-majority" status. Cities such as Atlanta, Chicago, Dallas, Detroit, Fresno, Houston, Las Vegas, Los Angeles, Memphis, Miami, New York, San Antonio, San Diego, and Washington already have "minority-majorities," and more than 10 percent of all counties within the United States have "minority-majorities" as well.

While population growth over the past decade among Latinos has been fueled more by natural increases (births minus deaths) than by new immigrants,[12] much of

The United States is a nation comprised of many cultural groups. These immigrants are becoming new citizens of the United States.

the overall U.S. population shift can be attributed to immigration. In 2009, about 38.5 million people—or about 12.5 percent of the U.S. population—were immigrants, and there are fifteen states with foreign-born populations greater than 10 percent.[13] While numerically this is the largest quantity of immigrants, in percentage terms it is substantially lower than the peak immigration years of 1880 through 1920 and about the same as it was in 1850, the first year the Census Bureau asked people for their place of birth. What distinguishes the current wave of immigrants from those of the early 1900s, however, is the country of origin. In 1900, the proportion of European immigrants to the United States was 86 percent; by 1960, Europeans still comprised 75 percent of the immigrant population. By 2009, however, only 12.7 percent of immigrants to the United States were European. Conversely, in 1970 only 19 percent of the foreign-born U.S. population was from Latin America, and 9 percent was from Asia. In 2009, more than half of the immigrants to the United States came from Latin America, and more than a quarter came from Asia.[14]

Recent data clearly show that the United States is now a multicultural society. Over 20 percent of the people in the United States—or one in five—speak a language other than English at home,[15] and that ratio holds for college students as well.[16] There are more Muslims, Hindus, and Buddhists in the United States than there are Lutherans or Episcopalians. However, the "typical" foreign-born resident in the United States is actually quite different from what many people suppose. She or he has lived in the United States for more than twenty years. Of those over age twenty-five, more than two-thirds have a high school diploma, and more than a quarter are college graduates. This latter figure is essentially the same as the college graduation rate of the native-born U.S. population. Foreign-born adults in the United States are likely to be employed, married,

CULTURE CONNECTIONS

If the world was a village of 100 people,

There would be:

61	Asians
	20 Chinese
	17 Asian Indians
14	Africans
11	Europeans
9	Latin Americans (Central Americans and South Americans)
5	North Americans
0	Australians/Oceanians

There would be:

33	Christians
20	Muslims
13	Hindus
6	Buddhists
14	People practicing other religions
14	Atheists or nonreligious ■

and living with their spouse and with one or two children.[17] As Antonia Pantoja and Wilhelmina Perry note about the U.S. demographics,

> The complete picture is one of change where large numbers of non-European immigrants from Africa, Asia, South and Central America, and the Caribbean will constitute majorities in many major cities. These immigrants will contribute to existing social movements. Many of these new immigrants are skilled workers and professionals, and these qualities will be highly valued in a changing United States economy. They come from countries with a history of democratic civil struggles and political revolutions. They arrive with a strong sense of cultural and ethnic identity within their intact family and social networks and strong ties to their home countries. At the same time they have a strong determination to achieve their goals, and they do not intend to abandon or relinquish their culture as the price for their success.[18]

The consequences of this "browning of America" can be seen in every major cultural and social institution. In Los Angeles public schools, for example, a third of the students speak a language other than English at home, and ninety-two different first languages are spoken.[19] Institutions of higher education are certainly not exempt from the forces that have transformed the United States into a multicultural society.[20] The enrollment of "minority-group" college students is increasing annually and now comprises more than a third of all students.[21] Additionally, more than 723,000 international students attend U.S. universities; this figure represents 3.5 percent of the total enrollment, a percentage that has remained relatively stable for more than a decade.[22] Recent projections suggest that by 2019, minority students will comprise 38 percent and international students another 4 percent of college students.[23] Similarly, there were more than 270,000 U.S. students studying abroad in 2009, which is double what it was a decade ago and four times what it was two decades ago.[24] Estimates are that, worldwide, the number of students who study outside their home countries will grow from 3 million in 2009 to 8 million in 2025.[25]

The United States is not alone in the worldwide transformation into multicultural societies. Throughout Europe, Asia, Africa, South America, and the Middle East, there is an increasing pattern of cross-border movements that is both changing the distribution

U.S. Americans are as varied as the landscape. Here, a group of colleagues, who represent several different cultures, have a friendly conversation.

of people around the globe and intensifying the political and social tensions that accompany such population shifts. This demographic imperative requires a heightened emphasis on intercultural competence.

The Technological Imperative for Intercultural Competence

Marshall McLuhan coined the term global village to describe the consequences of the mass media's ability to bring events from the far reaches of the globe into people's homes, thus shrinking the world.[26] Today, the "global village" is an image that is used to describe the worldwide web of interconnections that modern technologies have created. Communications media such as the Internet, global positioning satellites, and cell phones now make it possible to establish virtually instantaneous links to people who are thousands of miles away. In the past fifteen years, international telephone traffic has more than tripled. Simultaneously, the number of cell phone users has grown from virtually zero to nearly 4 billion—nearly 58 percent of the world's population—and Internet users now approach 2 billion, a quarter of whom have broadband access.[27]

Modern transportation systems contribute to the creation of the global village. A visit to major cities such as New York, Los Angeles, Mexico City, London, Nairobi, Istanbul, Hong Kong, or Tokyo, with their multicultural populations, demonstrates that the movement of people from one country and culture to another has become commonplace. "There I was," said Richard W. Fisher, president of the Federal Reserve Bank of Dallas, "in the middle of a South American jungle, thumbing out an e-mail [on my mobile phone] so work could get done thousands of miles away.... Technology, capital, labor, and ideas, now able to move at unprecedented speed across national boundaries, have integrated the world to an unprecedented degree."[28]

Modern information technologies allow people in the United States and throughout the world to participate in the events and lives of people in other places. Many world events are experienced almost instantaneously and are no longer separated from us in time and space. Scenes of a flood in China, of an earthquake in California or Virginia, of a drought in Texas, or of a tsunami in Japan are viewed worldwide on local television stations; immigrants and expatriates maintain their cultural ties by participating in Internet chat groups; the Travel Channel and similar fare provide insights into distant cultures; and grandmothers in India use webcams to interact with their granddaughters in New York and London. As blogger and entrepreneur Vinnie Mirchandani concludes,

> When I travel around the world, I see grandmas with headsets speaking on Skype to their loved ones at the other end of world. I see people crowding in Internet cafes. I see teenagers furiously texting each other. I hear unusual phrases like "I left you a missed call." I see people using their mobile phones to buy from vending machines.[29]

CULTURE CONNECTIONS

Flushing [New York] is a sea. A baptismal sea that churns out New Americans. It admits a constant influx of new people, not so much from other parts of America as from the rest of the world, people who come from other continents across seas and deserts and rivers and over mountains. You see them everywhere in Flushing. On the subway. On the street. At stores. The new people. You can always tell them right away from the way they dress or wear their hair; or from the language they speak or the subtle scents they carry; or from other such myriads of small things. Some carry their villages in their walk, and others wear the terrain they come from on their faces. As unmistakable as their hard-to-erase accents.

It never ceases to amaze me that they all find their way and manage to build a new life here. It seems a miracle that they all somehow survive. Some of them come here with nothing. Nothing but memories and a dream and a will. Some smuggled in as stowaways on a ship. So awfully unprepared. But even they manage. Most of them, anyway. They find places to live. They find work. They put food on the table for their families. They buy their first TV set. Their first dining-room table. Their first car. Their first apartment or house. And their children start school, and are on their way to becoming Americans. It's nothing special. Really. As they say, people do it every day. And so many people have done it before them. And so many will do it, long, long after them. And after all, we did that. There's no mystery at all. Remember? Once we were that new people on the street, shopping for our first whatever, and once we were the kids on the street in our fresh-off-the-boat clothes. But I don't remember how we did it. It was our parents' responsibility to put food on the table, to buy that first TV set and the first house.

—Mia Yun ■

These increased contacts, which are facilitated by recent technological developments, underscore the significant interdependencies that now link people to those from other cultures. Intercultural links are reinforced by the ease with which people can now travel to other places. Nearly 64 million U.S. residents travel abroad annually.[30] Likewise, citizens of other countries are also visiting the United States in record-setting numbers.

Technology allows and facilitates human interactions across the globe and in real time. Such instantaneous communication has the potential to increase the amount of

This Vietnamese man, who is checking his e-mail, demonstrates the technological imperative for intercultural communication.

communication that occurs among people from different cultures, and this expansion will necessarily add to the need for greater intercultural competence. "The world is flat," as Thomas Friedman so aptly suggested, because the convergence of technologies is creating an unprecedented degree of global competitiveness based on equal opportunities and access to the marketplace.[31] Similarly, consider YouTube, which has encouraged the widespread dissemination of visual and auditory ideas by anyone with access to an inexpensive digital video camera. Unlike more restrictive and more expensive television stations, which require access to sophisticated equipment and distribution networks, Internet-based social networking sites such as Facebook, Baidu, LinkedIn, and Google+, as well as such technological innovations as Skype, Twitter, Chat, and Yahoo Messenger, are used by an extraordinarily large number of people to connect with others whom they may have never met—and perhaps will never meet—in face-to-face interactions.

The technological imperative has increased the urgency for intercultural competence. Because of the widespread availability of technologies and long-distance transportation systems, intercultural competence is now as important as it has ever been.

The Economic Imperative for Intercultural Competence

Globalization—the integration of capital, technology, and information across national borders—is creating a global marketplace. Consequently, the economic success of the United States in the global arena increasingly depends on individual and collective abilities to communicate competently with people from other cultures. Clearly, U.S. economic relationships require global interdependence and intercultural competence, since economic growth or retrenchment in one part of the world now reverberates and affects many others. The European banking and debt crisis, for instance, is having worldwide consequences, since about 90 percent of Greece's bonds, and over half of Portugal's and Italy's bonds, are held by investors outside of those countries.[32] Similarly, the integration into the global workforce of workers from the so-called BRIC countries—Brazil, Russia, India, and China—is driving down labor costs worldwide.[33] Consequently, the economic health of the United States is inextricably linked

The economic imperative for intercultural competence is exemplified by these Western and Middle Eastern businesspeople, who are developing their interpersonal relationships.

to world business partners. U.S. international trade has more than doubled every decade since 1960, and it now exceeds $3.5 trillion annually.[34]

International tourism is one of the "growth industries" that is burgeoning worldwide, and the United States benefits substantially from it. In 2011, a record 64 million international tourists to the United States spent over $150 billion during their stays, and this figure is projected to grow by more than 5 percent annually for the foreseeable future.[35] By 2020, as many as 100 million Chinese tourists are likely to descend upon popular destinations throughout the world, and the United States is poised to garner a substantial share of this emerging market.[36] Roger Dow, president of the U.S. Travel Association, recently testified about the economic impact of international travel. Said Dow:

> When visitors travel to the United States from abroad, they inject new money into the U.S. economy by staying in U.S. hotels, spending in U.S. stores, visiting U.S. attractions, and eating at U.S. restaurants—purchases that are all chalked up as U.S. exports that contribute positively to America's trade balance....
>
> International travel [to the U.S.] is already America's largest export, representing 8.7 percent of U.S. exports of goods and services in 2010 and nearly one-fourth of services exports alone. The travel industry's $134.4 billion in exports contributed more than any other industry to America's $1.8 trillion worth of total goods and services exports. And in a time of yawning national trade deficits, the travel sector enjoys an overall trade surplus: $31.7 billion in 2010.[37]

U.S. higher education is also a major service-sector export. Each year, international students spend more than $21 billion on their living expenses, and about 70 percent of this total comes from sources outside the United States.[38]

Corporations can also move people from one country to another, so within the workforce of most nations, there are representatives from cultures throughout the world. However, even if one's work is within the national boundaries of the United States, intercultural competence is imperative. The citizenry of the United States includes many individuals who are strongly identified with a particular culture. Thus, it is no longer safe to assume that clients, customers, business partners, and coworkers will have similar cultural views about what is important and appropriate.

The U.S. workplace reflects the increasing cultural diversity that comprises the nation as a whole. For example, in recent years the number of businesses opened by Asian Americans grew at twice the national average. These businesses generate more than $326 billion in revenues annually.[39] The number of African American businesses also continues to grow; more than a million small businesses are now owned by African Americans.[40] Similarly, Latinos and Latinas are opening businesses at a faster pace than ever before. In 2012, there were 4.3 million Latino-owned businesses that generated more than $539 billion in annual gross receipts.[41]

In sum, the economic imperative for intercultural competence is powerful, pervasive, and likely to increase in the coming years. As Andrew Romano says:

> For more than two centuries, Americans have gotten away with not knowing much about the world around them. But times have changed.... While isolationism is fine in an isolated society, we can no longer afford to mind our own business. What happens in China and in India (or at a Japanese nuclear plant) affects the autoworkers in Detroit; what happens in the statehouse and the White House affects the competition in China and India. Before the Internet, brawn was enough; now the information economy demands brains instead.[42]

The peace inperative for intercultural communication is exemplified by the goals and experiences of the athletes and fans of the Olympics.

The Peace Imperative for Intercultural Competence

The vision of interdependence among cultural groups throughout the world and in the United States has led Robert Shuter to declare that "culture is the single most important global communication issue" that humans face.[43] The need to understand and appreciate those who differ from ourselves has never been more important. As the President's Commission on Foreign Language and International Studies has said,

> Nothing less is at stake than the nation's security. At a time when the resurgent forces of nationalism and of ethnic and linguistic consciousness so directly affect global realities, the United States requires far more reliable capabilities to communicate with its allies, analyze the behaviors of potential adversaries, and earn the trust and sympathies of the uncommitted. Yet, there is a widening gap between these needs and the American competence to understand and deal successfully in a world in flux.[44]

Witness the problems in Darfur, Chechnya, Zimbabwe, Central Asia, Indonesia, the Middle East, Afghanistan, Venezuela, and Iraq; in each instance, cultures have clashed over the right to control resources and ideologies. But such animosities do not occur just outside the borders of the United States. Consider the proliferation of such groups as the neo-Nazis, white nationalists, racist skinheads, and those with links to the Ku Klux Klan. The Southern Poverty Law Center has estimated that, as of 2010, the number of hate groups is increasing; there are currently more than a thousand hate groups in the United States, and multiple hate groups flourish in every U.S. state.[45] The most recent data on hate crimes reveal that, in 2009, there were more than six thousand hate crimes committed against individuals because of their race, culture, religion, or social group membership. About 35 percent of these hate crimes were directed against African Americans, another 14 percent targeted Jews, an additional 7 percent were aimed at Latinos, and 1.6 percent of the U.S. hate crimes were directed against Muslims.[46] As Catharine R. Stimpson has said of these "culture clashes,"

> the refusal to live peaceably in pluralistic societies [has been] one of the bloodiest problems—nationally and internationally—of the 20th century. No wizard, no fairy godmother is going to make this problem disappear. And I retain a pluralist's stubborn, utopian hope that people can talk about, through, across, and around their differences and that these exchanges will help us live together justly.[47]

The Interpersonal Imperative for Intercultural Competence

The demographic, technological, economic, and peace imperatives all combine to create a world in which human interactions are dominated by culture, cultural differences, and the ability of humans to understand and interact within multiple cultural frameworks. In short, the quality of your daily life—from work to play to family to community interactions—will increasingly depend upon your ability to communicate competently with people from other cultures. Your neighbors may speak different first languages, have different values, and celebrate different customs. Your family members may include individuals from cultural backgrounds other than yours. Your colleagues

CULTURE CONNECTIONS

I saw America in a mall the other night, and it was a good feeling.

I saw a man in a turban pushing a baby carriage. I joked with a group of Japanese teenagers standing in line for pizza. I saw a woman chasing her infant son, calling to him in Spanish.

I heard languages I didn't even recognize in Macy's at one end of the mall and in the women's department of Nordstrom at the other.

I communicated my weariness to a young Chinese man sitting next to me on a bench whose response was interpreted by a Brit sitting next to him who had been raised in Hong Kong, and for a moment we were the world.

—Al Martinez ■

at work may belong to various cultures, and you may be expected to participate in intercultural and cross-national teams with them; a team leader in Chicago might have a supervisor in Switzerland and team members located in New York, San Francisco, Buenos Aires, Melbourne, Johannesburg, and Shanghai.

There are some obvious consequences to maintaining competent interpersonal relationships in an intercultural world. Such relationships will inevitably introduce doubt about others' expectations and will reduce the certainty that specific behaviors, routines, and rituals mean the same things to everyone. Cultural mixing implies that people will not always feel completely comfortable as they attempt to communicate in another language or as they try to talk with individuals who are not proficient in theirs. Their sense of "rights" and "wrongs" will be threatened when challenged by the actions of those with an alternative cultural framework. Many people will need to live in two or more cultures concurrently, shifting from one to another as they go from home to school, from work to play, and from the neighborhood to the shopping mall. The tensions inherent in creating successful intercultural communities are obvious as well. Examples abound that underscore how difficult it is for groups of culturally different

The challenge of communicating in an intercultural world occurs in our families, homes, work settings, schools, and neighborhoods.

individuals to live, work, play, and communicate harmoniously. The consequences of failing to create a harmonious intercultural society are also obvious—human suffering, hatred passed on from one generation to another, disruptions in people's lives, and unnecessary conflicts that sap people's creative talents and energies and that siphon off scarce resources from other important societal needs.

The challenge of the twenty-first century—and our challenge to you in this book—is to understand and to appreciate cultural differences and to translate that understanding into competent interpersonal communication.

COMMUNICATION

To understand intercultural communication events, you must first study the more general processes involved in all human communication transactions. All communication events, including intercultural ones, are made up of a set of basic characteristics. Once these characteristics are known, they can be applied to intercultural interactions in order to analyze the unique ways in which intercultural communication differs from other forms of communication.

Defining Communication

The term communication has been used in many ways for varied, and often inconsistent, purposes.[48] For example, Frank Dance identified 15 different conceptual components for the term,[49] and Dance and Carl Larson listed 126 different definitions for *communication*.[50] Like all terms or ideas, we chose our specific definition because of its usefulness in explaining the thoughts and ideas we wish to convey. Consequently, our definition is not the "right" one, nor is it somehow "more correct" than the others. Indeed, as you might expect, our definition is actually very similar to many others with which you may be familiar. However, the definition we have selected is most useful for our purpose of helping you to achieve interpersonal competence when communicating in the intercultural setting.

> Communication is a symbolic, interpretive, transactional, contextual process in which people create shared meanings.

To understand what this definition means, we will explore its implications for the study of intercultural communication.

Characteristics of Communication

Six characteristics of our definition of communication require further elaboration. Our definition asserts that communication is symbolic, interpretive, transactional, contextual, and a process, and it involves shared meanings. Let's examine each of these characteristics more closely.

COMMUNICATION IS SYMBOLIC Symbols are central to the communication process because they represent the shared meanings that are communicated. A symbol is a word, action, or object that stands for or represents a unit of meaning. Meaning, in turn, is a perception, thought, or feeling that a person experiences and might want to communicate to others. These meaning-full experiences could include sensations resulting from a room's temperature, thoughts about a teacher in a particular course, or feelings of happiness or anger because of what someone said. However, the private meanings within a person cannot be shared directly with others. They can become shared and understood only when they are interpreted as a message. A message, then, refers to the "package" of symbols used to create shared meanings. For example, the words in this book are symbols that, taken together, form the message that we, the authors, want to communicate to you.

People's behaviors are frequently interpreted symbolically, as an external representation of feelings, emotions, and internal states. To many people in the United States, for example, raising an arm with the hand extended and moving the hand and arm up and down symbolizes saying good-bye. Flags can symbolize a country, and most of the world's religions have symbols that are associated with their beliefs.

There is an important characteristic of symbols that might not be obvious to you but that nevertheless affects your ability to be a competent participant in intercultural communication: Symbols vary in their degree of arbitrariness. That is, the relationships between symbols and their referents can vary in the extent to which they are fixed or arbitrary.

Some symbol systems, such as verbal languages and a special class of nonverbal symbols called *emblems,* are completely unrelated to their referents except by common agreement among a group of people to refer to things in a particular way. (Emblems are discussed more fully in Chapter 8.) There is nothing peacelike, for instance, in the peace symbol, which is a nonverbal emblem that can be displayed by extending the index and middle fingers upward from a clenched fist. The same symbol was used by Winston Churchill to indicate victory, but to many people in South American countries it is regarded as an obscene gesture. Similarly, there is nothing booklike in the object you are holding as you read these words. We call the object a book not because there is anything inherent in the object that suggests "book" but simply because, by common agreement among users of English, we have agreed to do so. Those who speak other languages have other symbols, which are equally arbitrary, to refer to the same referents.

It is quite possible for a community of language users to agree to refer to some objects by using symbols that differ from the common ones. For example, the people in a class (perhaps even your class) could decide to change the symbols and refer to the teacher as a *door,* the students as *cows,* the blackboard as a *pancake,* the classroom as a *bar,* and the desks as *pineapples.* A description of the classroom with these new and arbitrarily assigned symbols might read as follows: "When the cows entered the bar, they sat down at their pineapples, and the door began to write on the pancake." Although the sentence sounds strange (and perhaps quite humorous), if everyone consistently referred to the objects in the same way, the meaning that would be created in using these symbols would soon become widely shared.

For many symbol systems, such as most nonverbal and visual ones, the relationship between the symbols and their referents is much less arbitrary than that of verbal languages. Such symbols as a growling stomach when hungry, a child's tears when sad, or a portrait that details a person's facial features are all so intrinsically associated with their referents that the range of expected meanings is very restricted. However, these types of symbols are useful precisely because much less knowledge of a specific language and culture is required to understand them. Thus, international traffic symbols, which consist of easily understood pictures, are frequently used in place of words to instruct drivers who might otherwise be a major hazard. Yet even with these traffic symbols, which are designed specifically to be understood easily, it cannot be assumed that everyone will automatically interpret the symbols in an identical fashion.

COMMUNICATION IS INTERPRETIVE   Messages do not have to be consciously or purpose-fully created with the specific intention of communicating a certain set of meanings for others to be able to make sense of the symbols forming the message. Rather, commu-nication is always an interpretive process. Whenever people communicate, they must interpret the symbolic behaviors of others and assign significance to some of those behaviors in order to create a meaningful account of the others' actions. This idea sug-gests that each person in a communication transaction may not necessarily interpret the messages in exactly the same way. Indeed, during episodes involving intercultural com-munication, the likelihood is high that people will interpret the meaning of messages differently.

Many people incorrectly use the word *communication* to represent an acceptable level of similarity or agreement in their conversations. They might use the phrase "I really could communicate with her" when they have had a very pleasant conversation in which the other person expressed a similar point of view, or they might say, "I just can't communicate with him anymore" when disagreements exist. These errors confuse two very different outcomes of the communication process.

The first outcome of communication is understanding what the others are trying to communicate. Understanding means that the participants have imposed similar or shared interpretations about what the messages actually mean. Indeed, without some degree of understanding between the participants, it would be inaccurate to claim that communication has even occurred. Thus, failed attempts at communication, such as when an accident victim calls for help and no one is nearby to hear, are not actually communication.

The second outcome is reaching agreement on the particular issues that have been discussed. Agreement means that each participant not only understands the other's interpretations but also holds a view that is similar. However, although understanding is a necessary ingredient to say that communication has occurred, agreement is not a requirement of communication. It is possible, and often quite likely, that people will understand one another's position or ideas yet not agree with them. For instance, two people who differ in their basic beliefs about religion or politics can still communicate about their personal preferences in a meaningful and fulfilling way without necessarily expecting the other person to agree.

CULTURE CONNECTIONS

I love traveling. Being an outsider in so many places makes me very aware of my own beliefs, values, and assumptions about the world. Traveling in places so different from my own experience vividly engages my mind and body, and everything seems fascinating. One very interesting thing I have noticed as a Latina visiting other countries is that I have a tendency to identify myself as a Texan, a Tejana, rather than an American. When I talked about this to a girlfriend, I realized why. I say I am *Tejana* because, in Texas, there are many women like me—minorities who are sandwiched between the Mexican and American cultures. To me, Texas repre-sents that netherworld of "not American/not Mexican" citizenry. It's a place where there are a lot of people like me, living in the bor-derlands between two cultures, not fully one or the other.

—Laurie L. Lopez Charlés ∎

It should be obvious that complete accuracy in interpreting the meanings that are shared by people is rare, if not impossible. Such a level of accuracy would require symbols to be understood by the participants in *exactly* the same way. Further, even if complete understanding was possible in a given instance, it would be impossible to verify that the meanings that were created for the symbols were identical in the minds of all participants.

Our stipulation that communication requires understanding does not imply that because completely accurate interpretations are impossible, communication is also impossible. Rather, we need to recognize that there are different levels or degrees of understanding. Communication requires a degree of understanding sufficient to accomplish the purposes of the participants, which can vary from one experience to another. For example, it may or may not be communication if a man who is dressed in unfamiliar clothes and who is obviously from another culture walks up to you and, after bowing, utters some sounds that seem like they could be language but whose meaning is unknown to you. If his purpose is merely to provide you with a ritualistic greeting and, recognizing this, you return his bow, then, relative to the purposes of the participants, we would say that the two of you have created shared meanings for your behaviors, and, consequently, communication has occurred. However, if he is asking you for directions and you merely return his bow without even recognizing his intended goal, then shared meanings do not exist, and communication has not occurred relative to the task at hand.

COMMUNICATION IS TRANSACTIONAL   To suggest that communication is transactional implies that all participants in the communication process work together to create and sustain the meanings that develop. A transactional view holds that communicators are simultaneously sending and receiving messages at every instant that they are involved in conversations.

The earliest views of the communication process were *actional*. An actional view held that communication was a linear, one-way flow of ideas and information and that the focus of this view was primarily on information transmission, or what the sender should do to structure a message that would achieve a desired result. As Figure 1.1 indicates, the earliest actional models did not even include the receivers of the messages. Later actional models added a receiver at the end of the message arrow, but those who held this view were still not very concerned with the receiver's characteristics.[51]

Actional views of the communication process are not very useful in the study of intercultural communication for two very important reasons. First, the underlying assumption of the actional view is that the sender's goal is to persuade the receiver. The sender is not really interested in understanding others, being sensitive to cultural differences, or developing better interpersonal relationships; rather, the focus is on telling and selling. Second, actional views of communication assume that the receivers

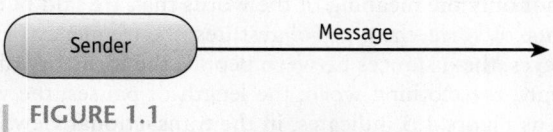

FIGURE 1.1

An actional view of communication.

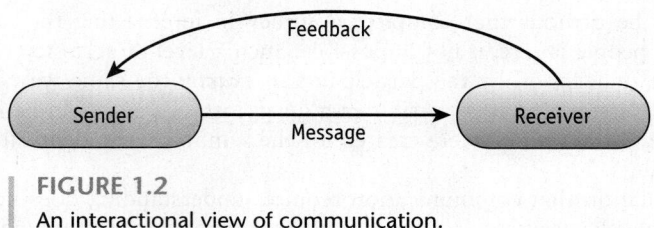

FIGURE 1.2

An interactional view of communication.

of messages are somehow inferior to the senders, with little ability to become involved in or to influence the communication process. In this view, those who create the messages should merely manipulate the receivers.

The limitations of the actional view led to the development of the *interactional* view of the communication process. Whereas the former emphasizes transmission of the message, the latter emphasizes interpretation. The interactional view explicitly includes the receiver in the communication process, and it recognizes that the receivers provide the senders with ongoing responses, called feedback, about how the messages are received.[52] As Figure 1.2 indicates, the focus of this view is still primarily sender-oriented. The model merely recognizes that senders must continually adapt their messages to the changing perceptions of the receivers to be most effective in influencing them. The implied goal of the interactional view of communication is to influence and control the receiver.

Like actional views, interactional views of the communication process are not very useful in the study of intercultural communication. The goal of the sender is still one of influencing others rather than being culturally sensitive and thereby improving intercultural relationships. The receivers, in the interactional view, need to be understood only insofar as that understanding is necessary to manipulate them more effectively. A final criticism of the interactional view is that it is not really a model of true interaction. Rather, it suggests a sequence of action–reaction behaviors in which messages are exchanged between a sender and a receiver, who perhaps alternate in these roles. Absent from the interactional view is any sense that the participants co-produce and co-interpret the messages that are communicated.

The limitations of the actional and interactional views have led to the development of the *transactional* view, which emphasizes the construction or shared creation of messages and meanings. The transactional view differs from the earlier views in two ways. First, it recognizes that the goal of communication is not merely to influence and persuade others but also to improve one's knowledge, to seek understanding, to develop agreements, and to negotiate shared meanings. Second, it recognizes that, at any given instant, no one is just sending or just receiving messages; therefore, there are no such entities as pure senders or pure receivers. Nor does it make sense to describe a single message as being the exclusive one at any selected moment. Rather, all participants are simultaneously interpreting multiple messages at all moments. These messages include not only the meaning of the words that are said but also the meaning conveyed by the tone of voice, the types of gestures, the frequency of body movements, the motion of the eyes, the distances between people, the formality of the language, the seating arrangements, the clothing worn, the length of pauses, the words unsaid, and much more. Thus, as Figure 1.3 indicates, in the transactional view, it is impossible to describe one person as exclusively the sender and the other as exclusively the receiver.

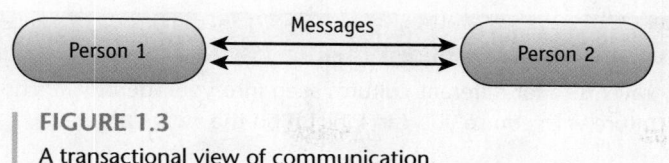

FIGURE 1.3

A transactional view of communication.

COMMUNICATION IS CONTEXTUAL All communication takes place within a setting or situation called a context. By context, we mean the place where people meet, the social purpose for being together, and the nature of the relationship. Thus, the context includes the physical, social, and interpersonal settings within which messages are exchanged.

The Physical Context The physical context includes the actual location of the interactants: indoors or outdoors, crowded or quiet, public or private, close together or far apart, warm or cold, bright or dark. The physical context influences the communication process in many obvious ways. Dean Barnlund captures its importance: "The streets of Calcutta, the avenues of Brazilia, the Left Bank of Paris, the gardens of Kyoto, the slums of Chicago, and the canyons of lower Manhattan provide dramatically different backgrounds for human interaction."[53] An afternoon conversation at a crowded sidewalk café and an evening of candlelight dining in a private salon will differ in the kinds of topics that are covered and in the interpretations that are made about the meanings of certain phrases or glances. As Donald Klopf so poignantly illustrates, knowledge of the physical context often provides important information about the meanings that are intended and the kinds of communication that are possible:

> I wanted to see her one more time before leaving Hong Kong. So I called her at work and she agreed to lunch. Near her office was a traditional *dim sum* restaurant. Sounded good to me; *dim sum* literally means "to touch the heart" and she had done that to me. What a mistake! Noisy?! The place was bedlam. The waitresses shouted out their wares—some sixty to seventy *dim sum* choices. We shared a table with a couple of tourists who griped about the food, and everything else. Crowded, every steno around must have decided on a *yam cha* meal today. Words of endearment didn't seem appropriate there. "Let's go next door to the Lau Ling Bar," I suggested, "for our black tea." Quiet and refined, it was the proper site to touch the heart, and I think I did.[54]

The Social Context The social context refers to the widely shared expectations people have about the kinds of interactions that normally should occur given different kinds of social events. Of course we realize that communication at funerals differs from that at a party; the social context of a classroom makes us expect certain forms of communication that differ from those at a soccer game. However, there is often a great deal of difficulty in understanding the social contexts for communication events that involve other cultures, as the common expectations about what behaviors are preferred or prohibited may be very different. For instance, before a funeral in Ireland, an all-night celebration called a wake is sometimes held. Such festive behaviors, which are appropriate to the social context of an Irish funeral, would be completely inappropriate where the social context dictates that alternative behaviors are more fitting.

CULTURE CONNECTIONS

The joy of travel is to let different cultures seep into your identity. It's not to bring your own culture with you so you can inflict it on the native populace.

—Seth Stevenson ∎

The Interpersonal Context   The interpersonal context refers to the expectations people have about the behaviors of others as a result of differences in the relationships between them. Communication between teachers and students, even outside the classroom context, differs from communication between close friends. Communication among friends differs from communication among acquaintances, coworkers, or family members. As people get to know each other and develop shared experiences, the nature of their interpersonal relationships is altered. This change in the interpersonal context is accompanied by alterations in the kinds of messages created and in the interpretations made about the meanings of the messages exchanged. As we suggested about physical and social contexts, people behave differently from one interpersonal context to another. The meanings assigned to particular behaviors can differ dramatically as different definitions of the context are imposed. As John Condon has said about differences in male–female relationships in the United States and Mexico,

> There are meanings to be read into settings and situations which must be learned if one is to avoid misunderstandings and even unpleasant experiences. A boss who invites his secretary out for a drink after work may or may not have ulterior motives in either country. However, the assumption that this was a romantic overture would be far more common in Mexico City than in New York or Los Angeles.[55]

COMMUNICATION IS A PROCESS   People, relationships, activities, objects, and experiences can be described either in static terms or as part of a dynamic process. Viewing communication in static terms suggests that it is fixed and unchanging, whereas viewing it as a process implies that things are changing, moving, developing, and evolving. A process is a sequence of many distinct but interrelated steps. To understand communication as a process, it is necessary to know how it can change over time.

Like the adage "You can't stand in the same stream twice," communication events are unique, as seemingly identical experiences can take on vastly different meanings at different stages of the process. This stream of events, which involves both past experiences and future expectations, is always moving and changing. Thus, the very same message may be interpreted very differently when said at different stages of the communication process.

COMMUNICATION INVOLVES SHARED MEANINGS   The interpretive and transactional nature of communication suggests that correct meanings are not just "out there" to be discovered. Rather, meanings are created and shared by groups of people as they participate in the ordinary and everyday activities that form the context for common interpretations. The focus, therefore, must be on the ways that people attempt to "make sense" of their common experiences in the world.

Interpersonal Communication

Definitions, as we have said, are chosen because they are useful for conveying the thoughts, ideas, and distinctions that one wishes to explain.

Interpersonal communication is a form of communication that involves a small number of individuals who are interacting exclusively with one another and who therefore have the ability both to adapt their messages specifically for those others and to obtain immediate interpretations from them.

Each of the four characteristics of this definition will now be discussed.

A SMALL NUMBER OF PEOPLE   To a certain degree, all communication could be called interpersonal, as it occurs between (*inter*) two or more people. However, we think it useful and practical to differentiate those relationships that involve a relatively small number of people—such as couples, families, friends, work groups, and even classroom groups—from those involving much larger numbers of people—such as public rallies or massive television audiences. Unlike other forms of communication, interpersonal communication involves person-to-person interactions. In addition, the perception that a social bond has developed between the interactants, however tenuous and temporary it may seem, is also much more likely.

PEOPLE INTERACTING EXCLUSIVELY WITH ONE ANOTHER   Unlike public speaking or mass media communication events, in which messages are sent to large, undifferentiated, and heterogeneous audiences, interpersonal communication typically involves clearly identified participants who are able to select those with whom they interact. In addition, when people interact directly with one another, they may use many sensory channels to convey information. Such details as looks, grunts, touches, postures, nods, smells, voice changes, and other specific behaviors are all available for observation and interpretation. Interestingly, given the ease with which people worldwide can use telephones and the Internet—for email, list serves, chatrooms, and instant messaging—interpersonal communication can now occur across global distances, and interpersonal relationships now can be built and sustained with these non–face-to-face channels.

ADAPTED TO SPECIFIC OTHERS   Because interpersonal communication involves a small number of people who can speak exclusively to one another, it is possible for the participants to assess what is being understood and how the messages are being interpreted. Because many of the messages are designed to evoke a particular effect in other people, the messages can be adapted to fit the specific people for whom they are intended.

CULTURE CONNECTIONS

I fear that Mexican writer Carlos Fuentes is right when he says: "What the U.S. does best is understand itself. What it does worst is understand others." As Senator Paul Simon has put it, "We Americans simply cannot afford cultural isolation." We need to understand our co-inhabitants on this earth so that we can compete effectively in an increasingly global economy. But we also need to better understand the diverse peoples of our world in the interest of peaceful coexistence.

—Johnnetta R. Cole ■

CULTURE CONNECTIONS

I knew we differed culturally, but different isn't wrong; different is merely different. Basically we're all human.

—Deon Meyer ■

IMMEDIATE INTERPRETATIONS   In interpersonal communication, in contrast to books or newspapers, the interpretation of messages can occur essentially simultaneously with their creation. The swift and instantaneous adaptations that people can make as a consequence of these immediate interpretations can permit a subtle and ongoing adjustment to the setting and the other participants.

THE CHALLENGE OF COMMUNICATING IN AN INTERCULTURAL WORLD

Throughout the remaining chapters of this book, we have much to say about *culture* and about becoming a competent intercultural communicator. By way of previewing what these subsequent chapters contain, we now want to discuss the Iceberg Analogy of Culture.

Edward T. Hall provided an apt analogy of *culture* as an iceberg.[56] An iceberg, you may recall, has a visible portion that is above the water, and a hidden portion—comprising nearly 90 percent of it—that lies below the water's surface. Similarly, all cultures have a small visible component and a larger (and arguably more important) part that is hidden from view.

The visible component is what is observable and can be perceived directly with our senses. It includes everyday features such as the culture's food, music, tools, and other preferences. The visible component also includes behavioral differences, which we regard as more central to our purposes in this book about intercultural communication: the ways people actually communicate—verbally, nonverbally, and interpersonally—and thereby maintain an ongoing sense of community and connection.

The hidden component of culture, which is perhaps more central and important for understanding the core aspects of culture and intercultural communication, is comprised of a culture's deeply held beliefs, values, and norms about the "correct" ways to behave and the "right" ways to interpret what is happening in the world. These below-the-water cultural patterns are largely unspoken and are typically out of awareness, but they constitute the unchallenged assumptions that can determine—often to a very great extent—how and why members of a given culture behave and communicate as they do. In sum, the Iceberg Analogy of Culture suggests that none of the visible elements of a culture can ever make sense without an understanding of the deeper and more important components that remain hidden from view.

To begin our exploration of intercultural competence, Chapters 2 and 3 discuss some important foundational and definitional issues about the nature of culture and intercultural communication, and they develop a vocabulary for talking about cultural diversity and intercultural competence. Then, because of their centrality in understanding intercultural competence, Chapters 4 and 5 provide an extended treatment of cultural patterns, which are those below-the-water features that differentiate one culture from another. Chapter 6 extends these unseen aspects of culture to emphasize cultural identity—who, culturally and fundamentally, we are—and the hidden biases that often hinder our attempts to understand, appreciate, and communicate with people from cultures that differ substantially from our own.

The remainder of the book deals more fully with above-the-water features of the intercultural competence iceberg, but throughout there is close attention paid to the hidden qualities that drive those observable behaviors. Chapters 7 through 9 focus on intercultural communication; they describe, in substantial detail, the ways that messages are structured and are coded so that they are culturally and interculturally appropriate and effective. Chapters 10 and 11 emphasize many important issues for understanding and thriving in intercultural relationships, and they draw attention to specific settings where intercultural communication frequently occurs. Finally, Chapter 12 highlights the ongoing relational and ethical issues that affect the potential for competent intercultural communication, and it concludes with encouragement to use your increased intercultural competence regularly.

There are no simple prescriptions or pat answers that can guarantee competent interpersonal communication among people from different cultures, nor has anyone discovered how to eliminate the destructive consequences of prejudice and racism. The importance of maintaining one's cultural identity—and therefore the need to preserve, protect, and defend one's culturally shared values—often creates a rising tide of emotion that promotes fear and distrust while encouraging cultural autonomy and independence. This emotional tide, whose beneficial elements increase people's sense of pride and help to anchor a people in time and place, can also be a furious and unbridled force of destruction.

Nevertheless, the joys and benefits of embracing an intercultural world are many. As the world is transformed into a place where cultural boundaries cease to be impenetrable barriers, differences among people become reasons to celebrate and share rather than to fear and harm. As Richard Rodriguez puts it,

We are at one of the great moments of civilization. Nothing as audacious as what we are trying to achieve has been attempted in human history, where you have the Iranian living next door to the Pakistani, living next door to the Cambodian, living next door to the Irishman, living next door to the Mexican. There is just no country that has ever tried this at the level at which we're trying it.[57]

CULTURE CONNECTIONS

What sets worlds in motion is the interplay of differences, their attractions and repulsions. Life is plurality, death is uniformity. By suppressing differences and peculiarities, by eliminating different civilizations and cultures, progress weakens life and favors death. The ideal of a single civilization for everyone, implicit in the cult of progress and technique, impoverishes and mutilates us. Every view of the world that becomes extinct, every culture that disappears, diminishes a possibility of life.

—Octavio Paz ■

In sum, the opportunities to understand, experience, and benefit from unfamiliar ways are unprecedented. You are—we are all—twenty-first-century pioneers on an incredible voyage, a new kind of pilgrim on a new frontier.

SUMMARY

The chapter began with descriptions of five imperatives for achieving intercultural communication competence: the economic, technological, demographic, peace, and interpersonal imperatives. Intercultural competence is now more important than it has ever been.

Next, the chapter provided a general analysis of the human communication process and discussed the topics of communication and interpersonal communication. These topics are of central importance to an understanding of intercultural communication, as they form the foundation for all of our subsequent ideas about the nature of intercultural transactions.

We defined communication as a symbolic, interpretive, transactional, contextual process in which people create shared meanings. Each of the characteristics of communication included in the definition was considered in turn. The role of symbols, which are the words, actions, or objects that stand for or represent units of meaning, was discussed. The consequences of the interpretation of symbols by both senders and receivers of messages on the outcomes of communication were explored. We also described the transactional nature of communication as involving the mutual influence of all individuals on communicative outcomes, so that every person simultaneously creates and interprets meanings and messages. The physical, social, and interpersonal contexts that bind each message were explained. Finally, we described communication as a process, an always-changing flow of interpretations.

We then narrowed our focus to the study of interpersonal communication, which we defined as a form of communication that involves a small number of people who can interact exclusively with one another and who therefore have the ability both to adapt their messages specifically for those others and to obtain immediate interpretations from them. Again, each of the important characteristics of interpersonal communication contained in this definition was considered.

Culture, we suggested, is like an iceberg; many cultural differences are hidden "below the surface" and are the assumptions that people make about how to communicate "correctly." Consequently, living in an intercultural world provides numerous challenges and opportunities, as your success and well-being increasingly depend on your ability to behave competently in intercultural encounters. In Chapter 2, we consider two concepts that are central to this book: culture and intercultural communication.

FOR DISCUSSION

1. What are some of the implications for a United States in which, within your lifetime, European Americans will no longer comprise a majority of the population?
2. Identify some of the ways in which your life is influenced by the presence of individuals from cultures that differ from your own.

3. Which of the imperatives for intercultural competence—demographic, technological, economic, peace, or interpersonal—is the most powerful motivator for you to improve your intercultural competence?

4. Communication has been defined as "a symbolic, interpretive, transactional, contextual process in which people create shared meanings." What do each of the elements in this definition mean? In your view, which is/are the most important?

FOR FURTHER READING

Joseph A. DeVito, *Human Communication: The Basic Course*, 12th ed. (Boston: Allyn & Bacon, 2012). Useful background for students who have not previously studied the human communication process.

Shelley D. Lane, *Interpersonal Communication: Competence and Contexts*, 2nd ed. (Boston: Allyn & Bacon, 2010). Provides insights into the special characteristics that influence interpersonal communication. This approach to interpersonal communication focuses on competence and fits well with the general approach of this book.

Vincent N. Parrillo, *Strangers to These Shores: Race and Ethnic Relations in the United States*, 10th ed. (Boston: Allyn & Bacon, 2011). Redefines the United States from a nation with only a European American history to one that, from its very beginning, has been a racially and culturally diverse country.

Sarah Trenholm, *Thinking through Communication: An Introduction to the Study of Human Communication*, 6th ed. (Boston: Allyn & Bacon, 2011). Another excellent introduction to the study of human communication that will be particularly useful to the beginning student of communication.

U.S. Census Bureau, www.census.gov. This Web site is a veritable treasure trove of statistical information about the multicultural character of the population of the United States of America.

CHAPTER

2

Culture and Intercultural Communication

KEY TERMS

This book is about interpersonal communication among people from different cultures. Our goal is to explain how you can achieve interpersonal competence in interactions that involve intercultural communication. This chapter provides a general understanding of culture and intercultural communication. It also includes a discussion about why one culture differs from another. Chapter 3 will continue the discussion by exploring the nature of intercultural communication competence.

CULTURE

Definitions of *culture* are numerous. In 1952, Alfred L. Kroeber and Clyde Kluckhohn published a list of 164 definitions of the term.[1] More recently, John R. Baldwin and his colleagues have listed over three hundred meanings for *culture*,[2] and other scholars have offered additional definitions and approaches.

Our concern in this book is with the link between culture and communication. Consequently, our definition of *culture* is one that allows us to investigate how culture contributes to human symbolic processes.

Defining Culture for the Study of Communication

Our goal in presenting a particular definition of culture is to explain the important link between culture and communication. However, we emphasize that the way we define culture is not the "right" or "best" way. Rather, it is a definition that is useful for our purpose of helping you to understand the crucial link between culture and communication as you set out to improve your intercultural competence.

Culture is a learned set of shared interpretations about beliefs, values, norms, and social practices, which affect the behaviors of a relatively large group of people.

CULTURE IS LEARNED Humans are not born with the genetic imprint of a particular culture. Instead, people learn about their culture through interactions with parents, other family members, friends, and even strangers who are part of the culture. Later in this chapter we explain why some cultures are so different from others. For now, we want to describe the general process by which people learn their culture.

Culture is learned from the people you interact with as you are socialized. Watching how adults react and talk to new babies is an excellent way to see the actual symbolic transmission of culture among people. Two babies born at exactly the same time in two parts of the globe may be taught to respond to physical and social stimuli in very different ways. For example, some babies are taught to smile at strangers, whereas others are taught to smile only in very specific circumstances. In the United States, most children are asked from a very early age to make decisions about what they want to do and what they prefer; in many other cultures, a parent would never ask a child what she or he wants to do but would simply tell the child what to do.

Culture is also taught by the explanations people receive for the natural and human events around them. Parents tell children that a certain person is a good boy because____. People from different cultures would complete the blank in contrasting ways. The people with whom the children interact will praise and encourage particular kinds of behaviors (such as crying or not crying, being quiet or being talkative). Certainly there are variations in what a child is taught from family to family in any given culture. However, our interest is not in these variations but in the similarities across most or all families that form the basis of a culture. Because our specific interest is in the relationship between culture and interpersonal communication, we focus on how cultures provide their members with a set of interpretations that they then use as filters to make sense of messages and experiences.

CULTURE IS A SET OF SHARED INTERPRETATIONS Shared interpretations establish the very important link between communication and culture. Cultures exist in the minds

CULTURE CONNECTIONS

Culture's Core

I recall now, so very clearly,
as evening clings like tapestry,
a distant time when I was small
and loved to creep along the wall
toward the circle cast by light
where elders talked, among themselves,
  into the night.

They filled the room with stirring tales, as I—
in my pajamas with the little feet—would lie
behind the outsized chair, hair pressed to rug,
listening invisibly, 'til wakened by the hug
of arms that cradled me 'round knee and head
and placed me back in sagging bed.

They told their stories,
one by one, of hardships suffered,
  and of glories—
times endured, evils feared,
stunning triumphs engineered
by luck, effort, patience, cunning,
and those who saved themselves by running.

I remember, too, the sagas told
about the turning points in growing old
amidst the tempests once withstood,
and tender details of first kisses, which were good
for waves of jokes and laughter
that I scarcely understood 'til after
I had aged, and learned of love affairs,
and private things that people do in pairs.

So now, like sages who have been and done,
who've told their tales of favors lost and won,
I primp my heirs with stories from my youth
of the vainglorious pursuits of truth,
justice, and the 'Merican way,
'til a still small voice can guide, I pray,
the journey forth where only they may go,
toward a promised land, which I will never know.

Thus repeats the simple lore
of passion, pleasure, pain, and pride
that marks us all, deep down inside,
as humans with a common core.

—Myron W. Lustig ■

of people, not just in external or tangible objects or behaviors. Integral to our discussion of communication is an emphasis on symbols as the means by which all communication takes place. The meanings of symbols exist in the minds of the individual communicators; when those symbolic ideas are shared with others, they form the basis for culture. Not all of an individual's symbolic ideas are necessarily shared with other people, and some symbols will be shared only with a few. A culture can form only if symbolic ideas are shared with a relatively large group of people.

CULTURE INVOLVES BELIEFS, VALUES, NORMS, AND SOCIAL PRACTICES The shared symbol systems that form the basis of culture represent ideas about beliefs, values, norms, and social practices. Because of their importance in understanding the ways in which cultures vary and their role in improving intercultural communication competence, the first section of Chapter 4 is devoted to their detailed explanation. For now, it is enough to know that beliefs refer to the basic understanding of a group of people about what the world is like or what is true or false. Values refer to what a group of people defines as good and bad or what it regards as important. Norms refer to rules for appropriate behavior, which provide the expectations people have of one another and of themselves. Social practices are the predictable behavior patterns that members of a culture typically follow. Taken together, the shared beliefs, values, norms, and social practices provide a "way of life" for the members of a culture.

CULTURE AFFECTS BEHAVIOR If culture were located solely in the minds of people, we could only speculate about what a culture is, since it is impossible for one person to see into the mind of another. However, these shared interpretations about beliefs, values, and norms affect the behaviors of large groups of people. In other words, the social practices that characterize a culture give people guidelines about what things mean, what is important, and what should or should not be done. Thus, culture establishes predictability in human interactions. Cultural differences are evident in the varying ways in which people conduct their everyday activities, as people "perform" their culture in their behavioral routines.

Within a given geographical area, people who interact with one another will, over time, form social bonds that help to stabilize their interactions and patterns of behavior. These social practices become the basis for making predictions and forming expectations about others. However, no one is entirely "typical" of the culture to which she or he belongs; each person differs, in unique ways, from the general cultural tendency to think and to behave in a particular way. Nor is "culture" the complete explanation for why people behave as they do: differences in age, gender, social status, and many other factors also affect the likelihood that people will enact specific behaviors. Thus, "culture" is an important, but not the only, explanation for people's conduct.

The beliefs, values, norms, and social practices of one's culture are learned. This birthday celebration provides an important setting in which cultural patterns are acquired.

CULTURE INVOLVES LARGE GROUPS OF PEOPLE We differentiate between smaller groups of individuals, who may engage in interpersonal communication, and larger groups of people more traditionally associated with cultures. For example, if you work every day with the same group of people and you regularly see and talk to them, you will undoubtedly begin to develop shared perceptions and experiences that will affect the way

CULTURE CONNECTIONS

"Electrons don't have culture!" an engineer protested recently during a corporate training program. "Scientists think the same all over the world; it's just logic and reason."

Ah, I thought, but *whose* logic and reason?

And therein lies one of the questions at the core of globalization. Whose values, whose thinking patterns, whose communication styles,

whose negotiation model will shape our interactions?

The answers are no longer quite so simple as they may have been in past decades, when we might possibly have survived assuming that culture-free electrons truly eradicated the complexity of human interaction.

—Janet M. Bennett ■

By teaching and explaining to their children, parents help them develop a common set of meanings and expectations.

you communicate. Although some people might want to use the term *culture* to refer to the bonds that develop among the people in a small group, we prefer to distinguish between the broad-based, culturally shared beliefs, values, norms, and social practices that people bring to their interactions and the unique expectations and experiences that arise as a result of particular interpersonal relationships that develop. Consequently, we will restrict the use of the term *culture* to much larger, societal levels of organization.

Culture is also often used to refer to other types of large groups of people. Mary Jane Collier and Milt Thomas, for example, assert that the term "can refer to ethnicity, gender, profession, or any other symbol system that is bounded and salient to individuals."[3] Our definition does not exclude groups such as women, the deaf, gays and lesbians, and others identified by Collier and Thomas. However, our emphasis is primarily on culture in its more traditional forms, which Collier and Thomas refer to as *ethnicity*.

Culture and Related Terms

Terms such as *nation, race,* and *ethnic group* are often used synonymously with the term *culture. Subculture* and *co-culture* are other terms that are sometimes used in talking about groups of people. There are important distinctions, however, between these terms and the groups of people to which they might refer.

NATION   In everyday language, people commonly treat *culture* and *nation* as equivalent terms. They are not. Nation is a political term referring to a government and a set of formal and legal mechanisms that regulate the political behavior of its people. These regulations often encompass such aspects of a people as how leaders are chosen, by what rules the leaders must govern, the laws of banking and currency, the means to establish military groups, and the rules by which a legal system is conducted. Foreign policies, for instance, are determined by a nation and not by a culture. The culture, or cultures, that exist within the boundaries of a nation-state certainly influence the regulations that a nation develops, but the term *culture* is not synonymous with *nation.* Although one cultural group predominates in some nations, most nations contain multiple cultures within their boundaries.

The United States is an excellent example of a nation that has several major cultural groups living within its geographical boundaries; European Americans, African

Americans, Native Americans, Latinos, and various Asian American cultures are all represented in the United States. All the members of these different cultural groups are citizens of the nation of the United States.

Even the nation of Japan, often regarded as so homogeneous that the word *Japanese* is commonly used to refer both to the nation and to the culture, is actually multicultural. Though the Yamato Japanese culture overwhelmingly predominates within the nation of Japan, there are other cultures living there. These groups include the Ainu, an indigenous group with their own culture, religion, and language; other cultures that have lived in Japan for many generations and originate mainly from Okinawa, Korea, and China; and more recent immigrants also living there.[4]

RACE   Race commonly refers to certain physical similarities, such as skin color or eye shape, that are shared by a group of people and are used to mark or separate them from others. Contrary to popular notions, however, race is not primarily a biological term; it is a political and societal one that was invented to justify economic and social distinctions. In the United States, for example, various non–Anglo-Saxon and non-Nordic cultural groups that would now be regarded as predominantly "white"— European Jews and people from such places as Ireland, Italy, Poland, and other eastern and southern European locales—were initially derided as being racial "mongrels" and therefore nonwhite. Conversely, Latinos who were classified as "white" through the 1960 census are now regarded by the United States as "not-quite whites, as in Hispanic whites."[5] Similarly, the U.S. Census Bureau has changed the racial classification of various Asian American groups from white to Asian. Thus, one's "race" is best understood as a social and legal construction.[6]

Although racial categories are inexact as a classification system, it is generally agreed that race is a more all-encompassing term than either *culture* or *nation*. Whereas many western European countries principally include people from the Caucasian race, not all Caucasian people are part of the same culture or nation. Consider the cultural differences among the primarily Caucasian countries of Great Britain, Norway, Germany, and Italy to understand the distinction between culture and race.

Sometimes race and culture do seem to work hand in hand to create visible and important distinctions among groups within a larger society; and sometimes race plays a part in establishing separate cultural groups. An excellent example of the interplay of culture and race is in the history of African American people in the United States. Although race may have been used initially to set African Americans apart from

CULTURE CONNECTIONS

I took a drink from every bottle held out to me. The man leaned over and slapped my knee—I was a good chap, I did not get drunk and fall on the floor. They laughed. I smiled to show I was cheerful and friendly, but I couldn't laugh because I didn't know what they were laughing at.

This, I thought, is what being a foreigner is really all about. It is not wandering through a strange country seeing unfamiliar people: it's when all the unfamiliar people stare at *you*, and find *you* strange: when you can't fit anonymously into a crowd: when your passing is an uncommon event. It's when you don't understand the joke, and the joke may very well be you.

—Liza Cody ■

Caucasian U.S. Americans, African American culture provides a strong and unique source of identity to members of the black race in the United States. Scholars now acknowledge that African American culture, with its roots in traditional African cultures, is separate and unique and has developed its own set of cultural patterns. Although a person from Nigeria and an African American are both from the same race, they are from distinct cultures. Similarly, not all black U.S. Americans are part of the African American culture, since many have a primary cultural identification with cultures in the Caribbean, South America, or Africa.

Race can, however, form the basis for prejudicial communication that can be a major obstacle to intercultural communication. Categorization of people by race in the United States, for example, has been the basis of systematic discrimination and oppression of people of color. We will explore the impact of racism more fully in Chapter 6.

ETHNICITY   *Ethnic group* is another term often used interchangeably with culture. Ethnicity is actually a term that is used to refer to a wide variety of groups who might share a language, historical origins, religion, nation-state, or cultural system. The nature of the relationship of a group's ethnicity to its culture will vary greatly depending on a number of other important characteristics. For example, many people in the United States still maintain an allegiance to the ethnic group of their ancestors who emigrated from other nations and cultures. It is quite common for people to say they are German or Greek or Armenian when the ethnicity indicated by the label refers to ancestry and perhaps some customs and practices that originated with the named ethnic group. Realistically, many of these individuals now are typical members of the European American culture. In other cases, the identification of ethnicity may coincide more completely with culture. In the former Yugoslavia, for example, there are at least three major ethnic groups—Slovenians, Croatians, and Serbians—each with its own language and distinct culture, who were forced into one nation-state following World War II. It is also possible for members of an ethnic group to be part of many different cultures and/or nations. For instance, Jewish people share a common ethnic identification even though they belong to widely varying cultures and are citizens of many different nations.

SUBCULTURE AND CO-CULTURE   Subculture is also a term sometimes used to refer to racial and ethnic minority groups that share both a common nation-state with other cultures and some aspects of the larger culture. Often, for example, African Americans, Arab Americans, Asian Americans, Native Americans, Latinos, and other groups are referred to as subcultures within the United States. The term, however, has connotations that we find problematic, because it suggests subordination to the larger European American culture. Similarly, the term co-culture has become more commonly used in

CULTURE CONNECTIONS

What I understand by manners is a culture's hum and buzz of implication. I mean the whole evanescent context in which its explicit statements are made. It is that part of a culture which is made up of half-uttered or unuttered or unutterable expressions of value. They are hinted at by small actions, sometimes by... tone, gesture, emphasis, or rhythm, sometimes by the words that are used with a special frequency or a special meaning.

—Lionel Trilling ■

CULTURE CONNECTIONS

"Once I asked her what was the most difficult thing about living in Iceland or moving to Iceland from Thailand and she talked about how Icelanders were a bit reserved compared to the Thais. She said personal contact was more open over there. Everyone talks to everyone else, complete strangers will discuss anything quite happily. If you're sitting out on the pavement having a meal you're not shy about inviting passers-by to join you."

"And the weather's not quite the same," Elínborg said.

"No. People stay outside in all that good weather, of course. We spend most of the year indoors and everyone here lives in his own private world. You run into closed doors everywhere. Just look at this corridor. I'm not saying it's better or worse, but it's different. It's two different worlds. When you get to know Sunee you have the feeling that life in Thailand is much calmer and more relaxed. Do you think it would be all right for me to drop in on her?"

"Perhaps you should wait a day or two, she's under a lot of strain."

"The poor woman," Fanney said. "It's not *sanuk sanuk* any more."

"What do you mean?"

"She's tried to teach me a few words of Thai. Like *sanuk sanuk*. She said that's typical of all Thais. It means simply enjoying life, doing something nice and fun. Enjoy life! And she taught me *pay nay*. That's the usual greeting in Thailand, like we say hello. But it means something completely different. *Pay nay* means 'where are you going?' It's a friendly question and a greeting at the same time. It conveys respect. Thais have great respect for the individual."

—Arnaldur Indridason ∎

an effort to avoid the implication of a hierarchical relationship between the European American culture and these other important cultural groups that form the mosaic of the United States. This term, too, is problematic and should be avoided. *Co-culture* suggests, for instance, that there is a single overarching culture in the United States, thus giving undue prominence to the European American cultural group and implicitly suggesting, as Thierry Devos and Mahzarin Banaji note, that "American equals White."[7] We view the United States as a nation within which there are many cultures, and we regard African Americans, Arab Americans, Chinese Americans, Native Americans, Latinos, and similar groups of people as cultures in their own right.[8] When used to refer to cultural groups within a nation, therefore, the term *co-culture* strikes us as redundant. When used to refer to one's identity as a member of various groups based on occupation, hobbies, interests, and the like, *co-culture* seems less precise than such alternative terms as *lifestyle* or *social group*. Chapter 6 elaborates on this distinction between one's cultural and social identities.

WHY CULTURES DIFFER

Cultures look, think, and communicate as they do for very practical reasons: to have a common frame of reference that provides a widely shared understanding of the world and of their identities within it; to organize and coordinate their actions, activities, and social relationships; and to accommodate and adapt to the pressures and forces that influence the culture as a whole.

Members of a culture seldom notice these motives because they usually exert a steady and continuous effect on everyone. Few people pay attention to the subtleties of commonplace events and circumstances. Instead, they remain oblivious to the powerful forces that create and maintain cultural differences. This tendency has led Gustav Ichheiser to declare that "nothing evades our attention as persistently as that which is taken for granted."[9]

In this section, we ask you to explore with us the taken-for-granted forces that create and maintain cultural differences. Our goal is to explain *why* one culture differs from another. As you read, consider your own culture and compare it to one that is very different or foreign to you. Why are they different? Why aren't all cultures alike? Why do cultures develop certain characteristics? Why do cultures communicate as they do? Why are they changing?

Forces That Maintain Cultural Differences

Cultural differences are created and sustained by a complex set of forces that are deeply embedded within the culture's members. We have selected six forces that help to generate cultural differences, including a culture's history, ecology, technology, biology, institutional networks, and interpersonal communication patterns. Of course, this list is by no means exhaustive. Consider these forces as representing factors with the potential to influence the ways in which cultures develop and maintain their differences yet change over time.

HISTORY The unique experiences that have become part of a culture's collective wisdom constitute its history. Wars, inheritance rules, religious practices, economic consequences, prior events, legislative acts, and the allocation of power to specific individuals are all historical developments that contribute to cultural differences.

As one of literally thousands of possible examples of the effects of historical forces on the development and maintenance of a culture, let us briefly consider a set of events that occurred in Europe during the late fourteenth century. The experience of bubonic plague, commonly known as the Black Death, was widely shared throughout most of Europe as well as in portions of Africa and Asia. It affected subsequent beliefs and behaviors for many generations.

In 1347, a trading ship traveling to Europe from the Black Sea carried an inadvertent cargo—a horrible disease known as the Black Death. It was spread rapidly by infected fleas carried by rats, and within two years it had traveled from the southern tip of Italy across the entire European continent, killing between one-third and one-half of the European population. There were recurring outbreaks about every decade until the early eighteenth century. Unlike famine, the

The Great Wall of China symbolizes the importance and centrality of historical events on modern Chinese cultures.

Black Death attacked every level and social class. When the initial wave of the epidemic was over, the survivors began a reckless spending spree, fueled in large measure by the newly acquired wealth left by the dead and by a sense of anarchy. The diminished availability of workers meant that labor, now a scarce commodity, was in demand. Workers throughout Europe organized to bargain for economic and political parity, and revolts against religious and political institutions were commonplace over the next several centuries. Often, however, workers' demands for retribution were either unrealistic or unrelated to the actual causes of the unrest, and the targets of the revolts were frequently foreigners or other helpless victims who were used as scapegoats for political purposes.

Although the Black Death was not the only historical force behind European cultural change, and indeed is insufficient by itself as an explanation for the changes in modern Europe, it was certainly a crucial experience that was recounted across the generations and influenced the development of European cultures. For instance, although the Black Death can now be controlled with modern-day antibiotics, a recent outbreak of bubonic plague in India caused mass panic and widespread evacuations to other cities, thus inadvertently spreading the disease to uninfected areas before it was contained. One important consequence of the Black Death was the unchallenged expectation that all population increases were desirable, as new births would replace those who had died and would thereby lead to increased standards of living. This belief predominated for more than four hundred years, leading people to ignore the evidence that overcrowding in some cities was the cause of disease and famine. In 1798, Thomas Malthus challenged this belief, arguing that human populations might be limited by their available food supply.

Recent examples of the influence of historical events are abundant. In the United States, for instance, consider the economic depression of 1929 and the fear of hyperinflation in 1979; the lessons learned in the "Cold War" with the Soviet Union and in "hot" wars with Germany, Japan, Korea, Vietnam, Somalia, Afghanistan, and Iraq; bread lines and gas lines; the proliferation of AIDS, cancer, and drugs; and the deaths of John F. Kennedy, Martin Luther King, Jr., John Lennon, Ronald Reagan, the astronauts aboard the *Challenger,* and the firefighters at the World Trade Center on September 11, 2001. All of these events have had profound effects on the ways in which U.S. Americans view themselves and their country. You have undoubtedly heard parents or other elders describe historical events as significantly influencing them and the lives of everyone in their generation. Descriptions of these events are transmitted across generations and form the shared knowledge that guides a culture's collective actions. As David

CULTURE CONNECTIONS

He looked at her now, the smile lighting her narrow eyes, eyes sometimes hazel, sometimes a light brown, sometimes verging on a mossy green. He'd never been close enough for long enough to figure out which was the one true color. Her hair was thick and black and as shiny as a raven's wing, and had once hung to her belt in a neat French braid. Now it was cropped short, brushed straight back from a broad brow, falling into a natural part over her right temple, the ends apt to curl into inky commas around her ears. Her cheekbones were high and flat and just beginning to take on that bronze tint he had noticed during previous summers, all gifts of her Aleut heritage, although the high bridge of her nose was all Anglo and the jut of her chin as Athabascan as it got.

—Dana Stabenow ■

McCullough says of such events and experiences, "You have to know what people have been through to understand what people want and what they don't want. That's the nub of it. And what people have been through is what we call history."[10]

ECOLOGY   The external environment in which the culture lives is the culture's ecology. It includes such physical forces as the overall climate, the changing weather patterns, the prevailing land and water formations, and the availability or unavailability of certain foods and other raw materials.

There is a considerable amount of evidence to demonstrate that ecological conditions affect a culture's formation and functioning in many important and often subtle ways. Often, the effects of the culture's ecology remain hidden to the members of a culture because the climate and environment are a pervasive and constant force. For example, the development and survival of cultures living in cold-weather climates demand an adaptation that often takes the form of an increased need for technology, industry, urbanization, tolerance for ambiguity, and social mobility.[11] High levels of involvement and closer physical distances in communication characterize cultures that develop in warm climates. High-contact cultures tend to be located in such warm-weather climates as the Middle East, the Mediterranean region, Indonesia, and Latin America, whereas low-contact cultures are found in cooler climates such as Scandinavia, northern Europe, England, portions of North America, and Japan.[12]

In the United States, differences in climate are related to variations in self-perceptions and interaction patterns. Compared with residents in the warmer areas of the South, for instance, those living in the colder areas of the northern United States tend to be less verbally dramatic, less socially isolated, less authoritarian in their communication style, more tolerant of ambiguity, more likely to avoid touching others in social situations, and lower in feelings of self-importance or self-worth.[13] Surviving a harsh cold-weather climate apparently requires that people act in a more constrained and organized fashion, maintain flexibility to deal with an ambiguous and unpredictable environment, cooperate with others to stave off the wind and the weather, and recognize how puny humans are when compared to such powerful forces as ice storms and snow drifts.

Another important aspect of the ecological environment is the predominant geographical and geological features. For instance, an abundant water supply shapes the economy of a region and certainly influences the day-to-day lifestyles of people. If water is a scarce commodity, a culture must give a major portion of its efforts to locating and providing an item that is essential to human life. Energy expended to maintain a water supply is not available for other forms of accomplishment. Likewise, the shape and contour of the land, along with the strategic location of a culture in relation to other people and places, can alter the mobility, outlook, and frequency of contact

Global warming and its corresponding climate changes, including melting of the Perito Moreno glacier field in southern Argentina, require cultures throughout the world to adapt in order to survive.

with others. Natural resources such as coal, tin, wood, ivory, silver, gold, spices, precious stones, agricultural products, and domesticated animals all contribute to the ecological forces that help to create differences among cultures.

TECHNOLOGY   The inventions that a culture has created or borrowed are the culture's technology, which includes such items as tools, microchips, hydraulic techniques, navigational aids, paper clips, barbed wire, stirrups, and weapons. Changes in the available technology can radically alter the balance of forces that maintain a culture. For instance, the invention of barbed wire allowed the U.S. American West to be fenced in, causing range wars and, ultimately, the end of free-roaming herds of cattle.[14] Similarly, stirrups permitted the Mongols to sweep across Asia, because they allowed riders to control their horses while fighting with their hands.

You have undoubtedly experienced the relationship of technology to culture. Most likely, you have always lived in a home with a microwave oven, though this technology was less common a generation ago. Two generations before microwave ovens became common, most homes also did not have refrigerators and freezers, relying instead on daily trips to the butcher and the baker and on regular visits from the milkman and the iceman to keep foods from spoiling. Think about how a family's food preparation has changed in the United States. Grocery stores now stock very different food products because of the prevalence of refrigerators and microwave ovens; entirely new industries have developed as well (as shown by the many freezer-to-microwave dishes).

Other examples of technological changes with even greater consequences are microprocessors and nanotechnologies, which have encouraged artificial intelligence, stronger and lighter materials, wireless communications, Internet search engines, iPods and other mp3 players, and "smart" machines that are capable of adapting to changing circumstances. The corresponding revolution in the storage, processing, production, and transmission of printed words (such as this textbook), spoken words and other sounds, and visual images is leading to a world in which there is access to an abundance of information.

One special form of technology that has had a major influence on cultures around the world is the media. Media are any technologies that extend the ability to communicate beyond the limits of face-to-face encounters. The media allow humans to extend their sensory capabilities, to communicate across time and long distances, and to duplicate messages with ease. Traditional media, such as books, newspapers, magazines, telegraph, telephone, photography, radio, phonograph, and television, have had a major influence in shaping cultures. New media technologies, such as satellites, DVDs, high-definition TVs, iPod players, "smart" mobile phones with 3G wireless capabilities, computers, streaming videos, and the Internet with its chatrooms, instant messaging, blogs, and email, further extend the capabilities of the traditional media to influence cultures.

The Internet alone has radically transformed the ways in which people are able to interact with one another. Email, in the form of text-based messages, was the first widely used Internet tool; it allowed family

Harnessing the energy of animals has been a major labor-saving technology throughout the world. These Amish men use horses to harvest their corn crop.

Technology is now an everyday part of people's lives. Here a new father balances work and family responsibilities.

members, friends, and coworkers who were widely dispersed to maintain contact easily. Now one's photos, videos, and other visual images can also move almost instantaneously across cyberspace, allowing people to share their key experiences with others. Similarly, Internet-based phone services allow people to talk just as if they were on a landline telephone and for little or no extra cost beyond that of the Internet connection. MySpace, Facebook, and other rapidly developing social networking sites create other Internet-based mechanisms for people—regardless of geographic location—to stay in touch. Thus the Internet helps people to sustain their culture's beliefs, values, norms, and social practices. For example, cyber-worshippers who are unable to take part in a *puja* (Hindu prayer ceremony) in India's holy city of Varanasi can now participate via cyberspace.[15] A counterpoint to these beneficial consequences of the Internet is its availability for introducing new ideas and images into cultures, which may speed up and change the nature of the culture itself.

Media can be responsible for minimizing the effects of geographic distance by increasing the speed, volume, and opportunities with which ideas can be introduced from one culture to another in a matter of seconds. The latest designs from a Paris fashion show can be captured digitally and instantly transmitted to Hong Kong manufacturers within minutes of their display in France, and accurate copies of the clothing can be ready for sale in the United States within a very short time.

Especially relevant is the way in which media technologies influence people's perceptions about other cultures. How do *Seinfeld* reruns, beamed by satellite to Rio de Janeiro, influence the way Brazilians try to communicate with someone from the United States?[16] In what ways do U.S. action-adventure films, in which many of the characters commit acts of violence to resolve interpersonal disagreements, affect the expectations of people from Egypt when they visit the United States? To what extent do media programs accurately reflect a culture and its members? With the ready availability of television programs, music, and movies from multiple cultures, what are the consequences to specific cultures that these different ways of behaving and living have? As George A. Barnett and Meihua Lee have suggested:

> Throughout the world, many cultures depend heavily on imported television programs, primarily from the United States, Western Europe, or Japan. Most are entertainment and sports programs. Since the early 1980's, along with the emergence of a genuinely global commercial media market, the newly developing global media system has been dominated by three or four dozen large, transnational corporations, with fewer than 10 mostly U.S.-based conglomerates towering over the global market.[17]

So what are people learning from these pervasive media messages? In a comprehensive study of U.S. prime-time television shows, James W. Chesebro concludes that the messages are clear and consistent: the values portrayed include individuality and authority.[18]

> Marine major J. D. "Mac" MacGillis on *Major Dad,* Jack Taylor of *The Family Man,* Murphy Brown, Nash Bridges, Commander Harmon "Harm" Rabb, Jr. on *Jag,* and Detective John Munch and Lieutenant Al Giardello on *Homicide: Life on the Streets,*

all are individuals, but they also wield authority, authority directed toward a liberal end. For the last twenty-five years on American television, such leader-centered figures have accounted for 35 percent of all primetime television series....

The vast majority of [U.S.] primetime television entertainment—some 70 percent of all series—has promoted the same two values—individuality and authority—during the last twenty-five years. The repetition of these same values from season to season and from series to series—particularly when placed in the elegantly produced settings and as enacted by actors and actresses who have already captured the public's imagination—is likely to leave its influence, if not overtly change the attitudes, beliefs, and actions of viewers.[19]

Thus, media-generated stereotypes, and technology in general, have important consequences for the processes and outcomes of intercultural communication.

BIOLOGY   The inherited characteristics that cultural members share are the result of biology, as people with a common ancestry have similar genetic compositions. These hereditary differences often arise as an adaptation to environmental forces, and they are evident in the biological attributes often referred to as *race*. Depending on how finely you wish to make distinctions, there are anywhere from three to hundreds of human races. Biologists are quick to point out, however, that there is far more genetic diversity within each race than there is among races, as humans have had both the means and the motives to mate with others across the entire spectrum of human genetic differences. This makes race an arbitrary but sometimes useful term.[20]

Although it is undeniable that genetic variations among humans exist, it is equally clear that biology cannot explain all or even most of the differences among cultures. For example, the evidence from studies that have been conducted in the United States on differences in intelligence suggests that most of the variation in intelligence quotient (IQ) scores is unrelated to cultural differences. Studies of interracial adoption, for instance, reveal that educational and economic advantages, along with the prebirth intrauterine environment, are the critical factors in determining children's IQ scores.[21] The data therefore suggest that, although hereditary differences certainly exist, most of the distinctions among human groups result from cultural learning or environmental causes rather than from genetic or biological forces. As Michael Winkelman suggests, "Biology provides the basis for acquiring capacities, while culture provides those specific skills as related to specific tasks and behaviors."[22]

Readily observable biological differences among groups of people have been amply documented, particularly for external features such as body shape, skin color, and other physical attributes. These visible differences among cultures are often used to define racial boundaries, although they can be affected by climate and other external constraints and are therefore not reliable measures of racial makeup. Better indicants of genetic group distinctions, according to scientists who study their origins and changes, are the inherited single-gene characteristics such as differences in blood types, earwax, and the prevalence of wisdom teeth. Type B blood, for instance, is common among Asian and African races, whereas Rh-negative blood is relatively common among Europeans but rare among other races; Africans and Europeans have soft, sticky earwax, whereas Japanese and many Native American groups have dry, crumbly earwax; and many Asians lack the third molars, or wisdom teeth, whereas about 15 percent of Europeans and almost all West Africans have them.[23] Of course, racial distinctions such as these are not what is intended by those who differentiate among individuals based on their physical or "racial" characteristics.

CULTURE CONNECTIONS

A sense of tribe is deeply embedded in the human soul. All of us are appropriately dependent upon and interdependent with networks of belonging. Grounded in the quest for survival and security, millennia of human existence in bands and tribes have fundamentally shaped our attitudes and behaviors as a species.... So we use the word "tribe" with caution, as a metaphor to acknowledge an aspect of human evolution at the core of our social identity. Our prosperity as social beings is woven by patterns of kinship, mutual assistance, and affection. Despite individualism's triumph, the industrial revolution, the power of nation states and international corporations, nothing has finally erased the human need to belong, to share an identity with others whom we recognize as "like me" and "one of us."

—Laurent A. Parks Daloz, Cheryl H. Keen, James P. Keen, and Sharon Daloz Parks ■

A complicating factor in making racial distinctions is that virtually all human populations have the same genetic origins. One theory about human biological differences holds that all humans descended from common ancestors who lived in northeast Africa, just south of what is now the Red Sea, more than 100,000 years ago.[24] By 50,000 years ago, they were living in various tribes and spoke a common language. Migrations of small groups of these ancestral humans began and continued incrementally in a slow expansion by successive generations, through the Middle East, India, Asia, Europe, and across Ice Age land-bridges to Australia, the Americas, and the rest of the world.[25]

In 1950 the United Nations declared that *race* is not a biological term, and scholars have generally agreed that there is no scientific basis for *race*.[26] However, research on the study of human genomes (the hereditary information in one's DNA) has reintroduced the notion that *race* may be biologically based. These studies have been propelled by pharmaceutical companies that want to target race-based health differences and by crime scene investigators who want to narrow their search for suspects.[27] Luigi Cavalli-Sforza agrees that clustering individuals based on biological similarities may be helpful for many types of medical investigations, but he argues that basing those groupings on current notions of *race* are not very useful.[28] Dismissing the "one-per-continent" view of common conceptions of *race*, Cavalli-Sforza offers the following illustration:

> The remarkable differences in genetic pathologies found among Ashkenazim and Sephardim (Jews of northern-European and Mediterranean origin) populations indicate that genetic clusters of individuals of medical interest may have to be small to be really useful; even a relatively small group like Jews has to be split into subclusters for medical genetics research. If the size of these groups should be an example of the size of useful genetic clusters from a medical point of view, then one would need on the order of a thousand genetic clusters for the whole species.[29]

Such a shift in viewpoint would eliminate, from a biological perspective, the need for racial groupings and would instead encourage alignment by human genome variability.[30]

The lack of a substantial biological basis for racial distinctions does not mean that race is unimportant. Rather, *race* should be understood as a social, political, and personal term that is used to refer to those who are believed by themselves or by others to constitute a group of people who share common physical attributes.[31]

Racial differences are often used politically to define those features that are used to include some individuals in a particular group while excluding others. Thus race can form the basis for prejudicial communication that can be a major obstacle to competent intercultural communication. Categorization of people by race in the United States, for example, has been the source of systematic discrimination and oppression of many people of color. Consequently, race-based distinctions have left a substantial legacy of harmful social and political consequences. As Audrey Smedley and Brian D. Smedley aptly suggest about this legacy, "race as biology is fiction, racism as a social problem is real."[32] We will explore the impact of racism more fully in Chapter 6.

INSTITUTIONAL NETWORKS   Institutional networks are the formal organizations in societies that structure activities for large numbers of people. These include government, education, religion, work, professional associations, and even social organizations. With the new media that have recently become available, institutional networks can be created and sustained more readily through the power provided by information technologies.

The importance of government as an organizing force is acknowledged by the emphasis placed in secondary schools on the different types of government systems around the globe. Because their form of government influences how people think about the world, this institutional network plays an important role in shaping culture.

The importance of institutional networks is also illustrated by the variability in the ways that people have developed to display spirituality, practice religion, and confront their common mortality. Indeed, religious practices are probably as old as humankind. Even 50,000 years ago, Neanderthal tribes in western Asia buried their dead with food, weapons, and fire charcoal, which were to be used in the next life.

Religion is an important institutional network that binds people to one another and helps to maintain cultural bonds. However, the manner in which various religions organize and connect people differs widely. In countries that practice Christianity or Judaism, people who are deeply involved in the practice of a religion usually belong to a church or synagogue. The congregation is the primary means of affiliation, and religious services are attended at the same place each time. As people become more involved in religious practices, they meet others and join organizations in the congregation, such as men's and women's clubs, Bible study, youth organizations, and Sunday school.

Buddhists pray at a temple. The practice of religion provides important institutional networks in most cultures.

Through the institutional network of the church or synagogue, religious beliefs connect people to one another and reinforce the ideas that initially led them to join.

Religious organizations in non-Christian cultures are defined very differently, and the ways they organize and connect people to one another are also very different. In India, for example, Hindu temples are seemingly everywhere. Some are very small and simple, whereas others are grand and elaborate. The idea of a stable congregation holding regularly scheduled services, as is done in the religious practices of Christianity and Judaism, is unknown. People may develop a level of comfort and affiliation with a particular temple, but they don't "join the congregation" and attend prayer meetings. They simply worship in whatever temple and at whatever time they deem appropriate.

INTERPERSONAL COMMUNICATION PATTERNS The face-to-face verbal and nonverbal coding systems that cultures develop to convey meanings and intentions are called inter-personal communication patterns. These patterns include links among parents, siblings, peers, teachers, relatives, neighbors, employers, authority figures, and other social contacts.

Differences in interpersonal communication patterns both cause and result from cultural differences. Verbal communication systems, or languages, give each culture a common set of categories and distinctions with which to organize perceptions. These common categories are used to sort objects and ideas and to give meaning to shared experiences. Nonverbal communication systems provide information about the meanings associated with the use of space, time, touch, and gestures. They help to define the boundaries between members and nonmembers of a culture.

Interpersonal communication patterns are also important in maintaining the structure of a culture because they are the means through which a culture transmits its beliefs and practices from one generation to another. The primary agents for conveying these basic tenets are usually parents, but the entire network of interpersonal relationships provides unrelenting messages about the preferred ways of thinking, feeling, perceiving, and acting in relation to problems with which the culture must cope. For instance, when a major storm causes death and the destruction of valuable property, the explanations given can shape the future of the culture. An explanation that says, "God is punishing the people because they have disobeyed" shapes a different perception of the relationship among humans, nature, and spirituality than an explanation that says, "Disasters such as this one happen because of tornadoes and storms that are unrelated to human actions."

Cultures organize and assign a level of importance to their interpersonal communication patterns in various ways, and the level of importance assigned in turn influences other aspects of the culture. Ideas concerning such basic interpersonal relationships as *family* and *friend* often differ because of unique cultural expectations about the obligations and privileges that should be granted to a particular network

CULTURE CONNECTIONS

Being a good host, being a good guest, proving one's generosity, and putting kinship above all—these are the central preoccupations of an Afghan....

We developed a culture that said, No one is ever on their own. Everyone belongs to a big group. The prosperity and survival of the group comes first. And no, everyone is not equal. Some are patriachs, and some are poor relations; that's life. But generosity is the value that makes it all work.

—Tamim Ansaray ∎

of people. In the United States, for instance, college students consider it appropriate to live hundreds of miles from home if doing so will allow them to pursue the best education. Many Mexican college students, however, have refused similar educational opportunities because, in Mexico, one's family relationships are often more important than individual achievement. In the Republic of Korea, family members are so closely tied to one another in a hierarchy based on age and gender that the oldest male relative often has the final say on such important matters as where to attend school, what profession to pursue, and whom to marry.

Because an understanding of cultural differences in interpersonal communication patterns is so crucial to becoming interculturally competent, it is a central feature of this book. Subsequent chapters will focus specifically on the importance of interpersonal communication patterns and will consider more general issues about the nature of interpersonal communication among cultures.

The Interrelatedness of Cultural Forces

Although we have discussed the forces that influence the creation and development of cultural patterns as if each operated independently of all the others, we do wish to emphasize that they are all interrelated. Each force affects and is affected by all of the others. Each works in conjunction with the others by pushing and pulling on the members of a culture to create a series of constraints that alter the cultural patterns.

As an example of the interrelationship among these powerful forces, consider the effects of population, religion, resource availability, and life expectancy on the formation of certain cultural values and practices in Ireland and India during the late nineteenth century.[33] In Ireland, the population was large relative to the available food, and severe food shortages were common. Therefore, there was a pressing need to reduce the size of the population. Because the Irish were predominantly Catholic, artificial methods of birth control were unacceptable. Given the negative cultural value associated with birth control and the problems of overpopulation and lack of food, a cultural practice evolved that women did not marry before the age of about thirty. The population was reduced, of course, by the delay in marriages. India, at about that same time, also had harsh economic conditions, but the average life expectancy was about twenty-eight years, and nearly half of the children died before age five. Given that reality, a cultural value evolved that the preferred age for an Indian woman to marry was around twelve or thirteen. That way, all childbearing years were available for procreation, thus increasing the chances for the survival of the culture.

Cultural adaptations and accommodations, however extreme, are rarely made consciously. Rather, cultures attempt to adjust to their unique configuration of forces by altering the shared and often unquestioned cultural assumptions that guide their thoughts and actions. Thus, changes in a culture's institutions or traditions cause its members to alter their behaviors in some important ways. These alterations, in turn, foster additional adjustments to the institutions or traditions in a continual process of adaptation and accommodation.

Jared Diamond has suggested that it is the interrelationship of these cultural forces that explains the European conquest of the Americas.[34] Why, Diamond asks, were western Europeans able to conquer the indigenous cultures that were living in the Americas, rather than the other way around? Why didn't the native cultures of North and South America conquer Europe? Why, in other words, did the European cultures become disproportionately powerful? In a thorough and well-reasoned argument,

Diamond concludes very convincingly that the answers to his questions are unrelated to any biological differences that might have existed; cultural differences in intellect, initiative, ingenuity, and cognitive adaptability have been minor, and the variability of these mental attributes within each culture has been so much greater than their variations across cultures. Rather, the explanation for who-conquered-whom begins with two important environmental or ecological advantages that western Europeans had over the native peoples of the Americas, which led to institutional, technological, and biological advantages.

The first ecological advantage of the Europeans was the availability of a large number of wild species that had the potential to be domesticated. The variety of these domesticated plants and large animals provided food, transportation, mechanical power, carrying ability, and military advantages. By happenstance, an enormous range of possibilities for domestication was available to the Europeans but not to the cultures living in the Americas. Both massive varieties of plants (grains, fruits, vegetables) and the ready availability of many types of large animals (sheep, goats, cows, pigs, and horses) provided the Europeans with many opportunities for domestication that their counterparts in the Americas simply did not have.

The second ecological advantage of the Europeans was the shape and topography of Europe. Unlike the Americas, Europe has an east-west axis. Axis orientation affects the likelihood that domesticated crops will be able to spread, since locales that are east and west of each other, and therefore at about the same latitude, share the same day length and its seasonable variations. Seeds that are genetically programmed to germinate and grow in specific climatic conditions will likely be able to produce food to the east or the west of their initial locations, but typically they will not grow if planted far to the north or south.

These ecological advantages led to the production and storage of large quantities of food, which in turn encouraged the growth of larger communities. This is so because an acre of land can feed more herders and farmers—often, up to a hundred times more—than it can hunter-gatherers. Domesticated plants and animals provided not just food but also the raw materials for making many other useful items: clothing, blankets, tools, weapons, machinery, and much more.

As populations grew, so too did two additional forces: an increased complexity in the institutional networks, and biological changes in the form of resistance to infectious diseases. First, large population densities must pay much more attention to issues of social control. Rulers, bureaucracies, complex political units, hierarchical organizational structures, and the concentration of wealth are all required to accomplish large public projects and to sustain armies for defense and conquest. Domestication of plants and animals also meant that some people who were not needed for food production could become the specialists who managed the bureaucracies or who developed and manufactured useful products.

Second, there is a powerful biological relationship among population density, the availability of domesticated animals, and the spread of infectious diseases. Most of the major epidemics—smallpox, tuberculosis, malaria, plague, measles, influenza, cholera, typhus, diphtheria, mumps, pertussis, yellow fever, syphilis, gonorrhea, AIDS, and many more—originally jumped to humans from diseases that were carried by animals. Epidemics end when those who are available to be infected either die or become immune. Because many generations of Europeans had been exposed to the infectious diseases that initially came from their domesticated animals, they were able to survive the epidemics that they subsequently spread to the Americas (inadvertently or otherwise) with devastating consequences. (The number of native peoples in the Americas,

CULTURE CONNECTIONS

Sometimes the worst things are not what people say to your face or what they say at all, it is the things that are assumed. I am in line at the grocery store, studying at a café, on a plane flying somewhere.

"Her English is excellent, she must have grown up here," I hear a lady whisper. "But why on earth does she wear that thing on her head?"

"Oh that's not her fault," someone replies, "her father probably forces her to wear that."

I am still searching for a profound thirty-second sound byte to use when I hear comments like that. The trouble is that things like that never take thirty seconds to say. So I say nothing, but silence does not belong there. I want to grasp their hands and usher them home with me. Come, meet my father. Don't look at the wrinkles, don't look at the scars, don't mind the hearing aid, or the thick accent.

Don't look at the world's effect on him; look at his effect on the world. Come to my childhood and hear the lullabies, the warm hand on your shoulder on the worst of days, the silly jokes on the mundane afternoons. Come meet the woman he has loved and respected his whole life, witness the confidence he has nurtured in his three daughters. Stay the night; hear his footsteps come in at midnight after a long day's work. That thumping is his head bowing in prayer although he is exhausted. Granted, the wealth is gone and the legacy unknown, but look at what the bombs did not destroy. Now tell me, am I really oppressed? The question alone makes me laugh. Now tell me, is he really the oppressor? The question alone makes me cry.

—Waheeda Samady ■

estimated at about 20 million when Columbus arrived, was down to about 1 million two centuries later.) In sum, Diamond asserts,

> Plant and animal domestication meant much more food and hence much denser human populations. The resulting food surpluses, and (in some areas) the animal-based means of transporting those surpluses, were a prerequisite for the development of settled, politically centralized, socially stratified, economically complex, technologically innovative societies. Hence the availability of domestic plants and animals ultimately explains why empires, literacy, and steel weapons developed earliest in Eurasia and later, or not at all, on other continents. The military uses of horses and camels, and the killing-power of animal derived germs, complete the list of major links between food production and conquest.[35]

INTERCULTURAL COMMUNICATION

A simple way to define the term *intercultural communication* is to use the definition of *communication* that was provided in the previous chapter and insert the phrase "from different cultures." This addition would yield the following definition:

> *Intercultural communication is a symbolic, interpretive, transactional, contextual process in which people from different cultures create shared meanings.*

This definition, although accurate, is difficult to apply. In the following examples, we describe several situations and ask you to analyze them with this definition in mind. Our intention in the discussion that follows is to give you a more sophisticated understanding of the term *intercultural communication* by exploring more fully the meaning of the phrase "people from different cultures."

Examples of Intercultural Interactions

Read the description of each interaction and think carefully about the questions that follow. Decide whether you think the communication between the people involved is or is not intercultural. Our answers to these questions are provided in the subsequent discussion.

Example 1

Dele is from Nigeria, and Anibal is from Argentina. Both young men completed secondary education in their own countries and then came to the United States to study. They studied at the same university, lived in the same dormitory their first year on campus, and chose agriculture as their major. Eventually, they became roommates, participated in many of the same activities for international students, and had many classes together. After completing their bachelor's degrees, they enrolled in the same graduate program. After four more years in the United States, each returned to his home country and took a position in the country's Agricultural Ministry. In emails, phone calls, and the occasional visit with each other, both comment on the difficulties that they are experiencing in working with farmers and the larger agribusiness interests within their own country.

Questions for Example 1

- When they first begin their studies in the United States, is the communication between Dele and Anibal intercultural communication?
- When they complete their studies in the United States, is the communication between Dele and Anibal intercultural communication?
- After they return to their home countries, is the communication between each man and the agricultural business managers with whom they work intercultural communication?

Example 2

Janet grew up in a small town of about 3,500 people in western Massachusetts. She is surrounded by her immediate family, many other relatives, and lots of friends. Her parents grew up in this same town, but Janet is determined to have experiences away from her family and away from the small portion of New England that has formed the boundaries of her existence. Despite parental concerns, Janet goes to one of Colorado's major public universities, and she begins her life in the West. Janet is at first excited and thrilled to be living in Colorado, but within a very short period of time, she begins to feel very isolated. She is assigned to live in a coeducational dormitory, and she finds it disconcerting to be meeting male students as she walks down the hallway in her bathrobe. Although her fellow students seem friendly, her overtures for coffee or movies or even studying together are usually met with a smile and a statement that "It would be great, but...." The superficial friendliness of most of the people she meets starts to annoy her, and Janet becomes bad-tempered and irritable.

Questions for Example 2

- Is the culture of Massachusetts sufficiently different from that of Colorado to characterize Janet's communication with her fellow students as intercultural?

- Would Janet have had the same kinds of feelings and reactions if she had moved into a coeducational dormitory at a public university in Massachusetts?

Example 3

Even though Hamid's parents immigrated to the United States from Iran (Persia) before he was born, they speak Persian at home and expect Hamid to behave according to Persian family values and norms. Because Hamid is the eldest child, his parents have additional expectations for him. Hamid loves his parents very much, but he finds their expectations difficult to fulfill. He thinks he speaks respectfully to his mother and father when he tells them that he is going out with friends rather than staying for a celebration to which his extended family has been invited, but his parents tell him that he is being disrespectful. The family reaches a major crisis when Hamid announces that he is going to go to a college that has a good studio arts program, rather than pursue the business degree that his parents want him to earn in preparation for taking over the family business.

Question for Example 3

- Is Hamid's communication with his parents intercultural, either because Hamid is very U.S. American and his parents are Persian or because parents and children have different cultures?

Example 4

Jane Martin works for a U.S. company that has a major branch in South Korea. Although Jane is fairly young, her boss has asked her to travel to Seoul to teach her Korean counterparts a new internal auditing system. Despite Jane's lack of linguistic skill in Korean (she speaks no Korean) and little experience in another country (she has spent a week in London and a week in Paris on holiday), she is confident that she will be successful in teaching the Korean employees the new system. She has won high praise for her training skills in the United States, and the company promises to provide her with a good interpreter. "After all," Jane thinks, "we're all part of the same company—we do the same kinds of work with the same kinds of corporate regulations and expectations. Besides, Koreans are probably familiar with U.S. Americans."

Questions for Example 4

- Is Jane's communication with South Koreans intercultural, or does working for the same corporation mean that Jane and her South Korean counterparts share a common culture?

- Is Jane's age and gender a factor in communication with her Korean counterparts?

- Would you answer the previous questions any differently if Jane's company were sending her to the branch office in England rather than to the one in South Korea?

Example 5

Jody has been fascinated with South Asian cultures since she was a child, when her family lived next door to someone from India and her best friend was a girl named Priya. As Jody got older, she began to read about India. More importantly, she became "hooked" on movies from or about India. Because she lived in a large urban area, "Bollywood" films were regularly available, and she often went to a theater to see them. Jody has also rented many DVDs that are directed by and populated with people from India, and she has watched them carefully as well. Though many of these films are in Hindi or another Indian language that Jody doesn't speak, she is confident that she understands them. Jody is now a third-year student at her state university, and recently she was offered an opportunity to study in India for a semester. She is very excited about this chance to live with an Indian family and take classes at an Indian university. She is very familiar with Indian food, dress, and films, and she really regards herself as very knowledgeable about and comfortable with Indian culture. Jody believes that she will have no trouble adjusting to life in India.

Questions for Example 5

■ How accurate is Jody's assessment that she understands Indian culture from her reading and extensive exposure to Indian films?

■ Can intercultural communication take place even when people do not share a common language?

■ Will Jody's communication with her Indian host family be less intercultural because of her familiarity with Indian films?

Example 6

John has worked for the same company, based in Minneapolis, Minnesota, for the six years since his graduation from college. A recent promotion means that John has to move to his company's branch office in Milwaukee, Wisconsin. John faces difficulties almost immediately after beginning work in Milwaukee. His boss has a much different management style than the one with which John is familiar. His new job responsibilities require some knowledge and sophistication in areas in which John is not an expert. After several months on the job, John is feeling fairly beleaguered and is beginning to lose confidence in his abilities.

Question for Example 6

■ Is John's communication with his boss intercultural communication?

Example 7

When he was eight years old, Jorge's family immigrated to Texas from Mexico. At first, Jorge did not know English, but, as he progressed through school, he gradually became a fluent English speaker. The language spoken at home, however, was always Spanish. Jorge felt proud to have a family that celebrated its Mexican cultural heritage; his parents had made great sacrifices to move to the United States, but their hopes for a better life for their children motivated them to immigrate. Jorge was an excellent student in college, ultimately received an MBA, and landed a challenging job in a well-known company. One day Jorge learned that his company was going to expand its operations to Chile and, because Jorge was one of the few Spanish-speaking managers in the company, he was going to be sent to Chile to manage the business there. His bosses presumed that because Jorge knows Spanish, he will be the ideal person for this important work. Jorge, on the other hand, is nervous and tries to explain that, while knowing the Spanish language is important, the culture in Chile is a very different from that of Mexico. Even the spoken and written forms of Spanish are quite different in the two cultures.

Questions for Example 7

■ Is Jorge correct that Mexican and Chilean cultures are sufficiently different to make his communication with Chileans intercultural?

■ How important is it to know how to speak a language in intercultural communication?

Each of these examples represents a likely communication event in today's world. It is very probable that two people from different countries will spend an extended period of time in a third country, as Dele and Anibal have. It is also very likely that these two people will, over time, form relationships that create a shared set of experiences. Moving from one part of a country to another is a commonplace occurrence, whether the goal is to attend a university, as Janet did, or to advance professionally, as John did. Immigration of people from one country to another also occurs frequently, producing communication problems typical of those experienced by U.S.-born-and-raised Hamid with his Persian-born-and-identified parents. The significance of the global marketplace means that work often takes people to countries around the world, as companies like Jane's and Jorge's become increasingly multinational. With the advent of modern communication technologies, many more people will be able to select television programs, films, music, radio shows, and computerized messages that are arranged in verbal and nonverbal codes different from their own. Jody's experience with the Indian films will be repeatable almost everywhere. But are these examples, all of which involve communication, also examples of intercultural communication? Do any of them clearly *not* involve intercultural communication? In the next sections, we attempt to provide answers to these questions.

Similarities and Differences Between Communicators

By applying the definition of intercultural communication given at the beginning of the chapter, it would be relatively simple to categorize each example. You would go through the examples and make a bipolar choice—either yes or no—based on whether the people in the examples were from different cultures. Thus, you would probably decide that the communication between Anibal and Dele was intercultural when they arrived in the United States. It would be much more difficult to judge their communication after they completed their studies. Perhaps you would decide that their communication with people from their own country following their return home was not intercultural, or perhaps you would say that it was. Similarly, you might be convinced that Colorado is indeed a different culture from Massachusetts,

**Most
intercultural**   **Least
intercultural**

FIGURE 2.1
A continuum of interculturalness.

or you might argue vehemently that it is not. Most likely you would decide that Jane's communication with her Korean counterparts was intercultural, even though they undoubtedly did share some common expectations about work performance because the same company employed them. Had her company decided to send Jane to England instead of to Korea, her communication with her English coworkers would have been similarly intercultural. Yet you might feel a bit uncomfortable, as we are, with the idea of putting U.S.–Korean communication into the same category as U.S.–English communication. Similarly, Jorge has language skills and some cultural knowledge, which makes him a better choice for the Chilean assignment than his European American colleagues who speak only English and know little about the cultures in Latin America, but his communication in Chile will certainly be more challenged than it would be if his assignment had been to Mexico City.

The difficulties encountered in a simple yes-or-no decision lead us to suggest an alternative way of thinking about intercultural communication. What is missing is an answer to three questions that emerge from the preceding examples:

1. What differences among groups of people constitute cultural differences?
2. How extensive are those differences?
3. How does extended communication change the effects of cultural differences?

This last question suggests the possibility that initially one's interactions could be very intercultural, but subsequent communication events could make the relationship far less intercultural.

To demonstrate the importance of these questions, we would like you to take the examples presented earlier and arrange them in order from most intercultural to least intercultural.[36] Use a continuum like the one shown in Figure 2.1.

Thus, you will be identifying the degree of interculturalness in each interaction and, in effect, you will be creating an "interculturalness" scale. It should even be possible to make distinctions among those communication situations that are placed in the middle, with some closer and some farther from the most intercultural end. When you place the examples on a continuum, they might look something like Figure 2.2.

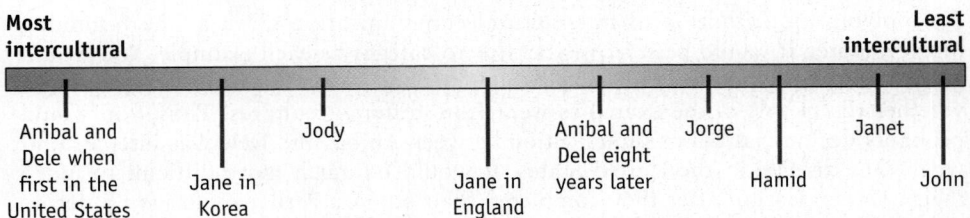

FIGURE 2.2
A continuum of interculturalness, with examples.

We suspect that the continuum you have created is very similar to ours. Where we might disagree is on how we ordered the examples placed near the middle.

The next important issue for understanding the definition of intercultural communication concerns the characteristics present in the encounters. What is it about the people, the communication, the situation, or some combination of those factors that increases the likelihood that the communication will be intercultural?

What varies and changes across the examples is the degree of similarity or the amount of difference between the interactants. For instance, Anibal (from Argentina) and Dele (from Nigeria) are very different when they first come to study in the United States. Each speaks English but as a second language to Spanish and Yoruba, respectively; their facility with English is initially weak, and they are uncomfortable with it. In addition, their values, social customs, gestures, perceptions of attractiveness, and expectations about personal space and how friendships are established differ. Initially, Anibal and Dele are culturally very different, or heterogeneous, and their communication should certainly be placed near the "most intercultural" end of the continuum. However, after eight years in the United States, having studied the same academic subjects, shared many of the same friends, and participated in many common experiences, their communication with each other does not have the same degree of interculturalness as it did initially. Certainly, each still retains part of his own cultural heritage and point of view, but the two men have also created an important set of common understandings between themselves that is not grounded in their respective cultural frameworks.

Janet, in contrast, was placed near the "least intercultural" end of the continuum because of the degree of similarity, or homogeneity, she shares with Coloradans. They speak the same language, and their values, gestures, social perceptions, and expectations about relationships are all similar. Certainly, Coloradans use slang and jargon with which Janet is not familiar, but they speak, read, and study in English. And certainly, Coloradans, particularly urban Coloradans, seem to place importance on different things than Janet does. She also thinks it unusual and a bit uncomfortable to be sharing a living space with men she does not even know. Nevertheless, the magnitude of these differences is relatively small.

There are learned differences among groups of people that are associated with their culture, such as cultural patterns, verbal and nonverbal codes, relationship rules and roles, and social perceptions. When such important differences are relatively large, they lead to dissimilar interpretations about the meanings of the messages that are created, and they therefore indicate that people are from different cultures. Thus,

People are from different cultures whenever the degree of difference between them is sufficiently large and important that it creates dissimilar interpretations and expectations about what are regarded as competent communication behaviors.

Definition of Intercultural Communication

Previous definitions have described the central terms *communication* and *culture*. By combining the meanings of these terms with the ideas suggested in our discussion about the degrees of difference that can occur among people from dissimilar cultures, we offer the following definition of intercultural communication:

Intercultural communication occurs when large and important cultural differences create dissimilar interpretations and expectations about how to communicate competently.

CULTURE CONNECTIONS

I was looking very hard at these people, because this is how it was with them: the boy's father had dark skin, darker even than my own, and the boy's mother was a white woman. They were holding hands and smiling at their boy, whose skin was light brown. It was the color of the man and the woman joined in happiness. It was such a good color that tears came into my eyes. I would not even try to explain this to the girls from back home because they would not believe it. If I told them that there were in this city children that were born of black and white parents, holding hands in the street and smiling with pride, they would only shake their heads and say, *Little miss been-to is making up her tales again.*

But I saw it with my eyes.... and I saw his mother and father lift him up, and I saw the three of them hugging one another tight and laughing while the crowd looked and laughed with them. This I saw with my own eyes, and when I looked around the crowd I saw that there was more of it. There were people in that crowd, and strolling along the walkway, from all of the different colors and nationalities of the earth. There were more races even than I recognized from the detention center. I stood with my back against the railings and my mouth open and I watched them walking past, more and more of them. And then I realized it. I said to myself, Little Bee, there is no *them.* This endless procession of people, walking along beside this great river, these people are *you.*

—Chris Cleave ■

The degree to which individuals differ is the degree to which there is interculturalness in a given instance of communication. Situations in which the individuals are very different from one another are most intercultural, whereas those in which the individuals are very similar to one another are least intercultural.

Intercultural Communication and Related Terms

The relationship between culture and communication is important to many disciplines. Consequently, many terms have been used to describe the various ways in which the study of culture and communication intersect: *cross-cultural communication, international communication, intracultural communication, interethnic communication,* and *interracial communication.* The differences among these terms can be confusing, so we would like to relate them to the focus of study in this book.[37]

INTRACULTURAL COMMUNICATION   The term intercultural, used to describe one endpoint of the continuum, denotes the presence of at least two individuals who are culturally different from each other on such important attributes as their value orientations, preferred communication codes, role expectations, and perceived rules of social relationships. We would now like to relabel the "least intercultural" end of the continuum, which is used to refer to communication between culturally similar individuals, as intracultural. John's communication with his new boss in Milwaukee is intracultural. Janet's communication with her fellow students in Colorado is more intracultural than intercultural. Both *intercultural* and *intracultural* are comparative terms. That is, each refers to differences in the magnitude and importance of expectations that people have about what constitutes competent communication behaviors.

INTERETHNIC AND INTERRACIAL COMMUNICATION   Just as *race* and *ethnic group* are terms commonly used to refer to cultures, *interethnic* and *interracial communication* are two labels commonly used as substitutes for *intercultural communication*. Usually, these terms are used to explain differences in communication between members of racial and ethnic groups who are all members of the same nation-state. For example, communication between African Americans and European Americans is often referred to as interracial communication. The large numbers of people of Latino origin who work and live with people of European ancestry produce communication characterized as interethnic. Sometimes the terms are also used to refer to communication between people from various ethnic or racial groups who are not part of the same nation but live in specific geographic areas. Although it may be useful in some circumstances to use the terms *interethnic* and *interracial,* we believe these types of communication are most usefully categorized as subsets of intercultural communication.

Both ethnicity and race contribute to the perceived effects of cultural differences on communication, which moves that communication toward the "most intercultural" end of the continuum. We will therefore rely on the broader term of *intercultural communication* when discussing, explaining, and offering suggestions for increasing your degree of competence in interactions that involve people from other races and ethnic groups. In Chapter 6, however, when considering particular cultural biases, we will give special attention to the painful and negative consequences of racism.

CROSS-CULTURAL COMMUNICATION   The term *cross-cultural* is typically used to refer to the study of a particular idea or concept within many cultures. The goal of such investigations is to conduct a series of intracultural analyses in order to compare one culture with another on the attributes of interest. For example, someone interested in studying the marriage rituals in many cultures would be considered a cross-cultural researcher. Scholars who study self-disclosure patterns, child-rearing practices, or educational methods as they

Intercultural communication occurs when large and important cultural differences create dissimilar interpretations and expectations about how to communicate competently.

exist in many different cultures are doing cross-cultural comparisons. Whereas intercultural communication involves interactions among people from different cultures, cross-cultural communication involves a comparison of interactions among people from the same culture to those from another culture. Although cross-cultural comparisons are very useful for understanding cultural differences, our principal interest is in using these cross-cultural comparisons to understand intercultural communication competence.

INTERNATIONAL COMMUNICATION International communication refers to interactions among people from different nations. Scholars who compare and analyze nations' media usage also use this term. Certainly, communication among people from different countries is likely to be intercultural communication, but that is not always true, as illustrated by the example of Anibal and Dele after eight years together in the United States. As we suggested with the terms *interracial* and *interethnic communication,* we prefer to focus on *intercultural communication.*

SUMMARY

Our goal in this chapter has been to provide an understanding of some of the key concepts underlying the study of intercultural competence. We began with a discussion of the concept of culture. From the many available approaches to defining culture, we selected one that emphasizes the close relationship between culture and communication. We defined *culture* as a learned set of shared interpretations about beliefs, values, norms, and social practices, which affect the behaviors of a relatively large group of people. We emphasized that people are not born with a culture but learn it through their interactions with others. Our definition located culture in the minds of people, and in the shared ideas that can be understood by their effects on behavior. We distinguished between culture and other groups to which people belong by suggesting that culture occurs only when beliefs, values, norms, and social practices affect large groups of people. We next made some important distinctions among terms such as *culture, nation, race, ethnic group,* and *subculture.*

We also explored some of the reasons that cultures differ. The shared experiences remembered by cultural members, or a culture's history, were considered first. In the United States, for instance, the lesson of the country's historical experiences affects U.S. Americans' views of their government's relationships with other countries. The ways in which a culture's unique ecology profoundly alters the collective actions of its people were then illustrated. Next, we discussed the biological or genetic forces affecting cultures. Genetic variations among people are only a small source of cultural differences. We also explained the role of the formal organizations of a culture, the institutional networks such as government, religion, work organizations, and other social organizations. These institutional networks organize groups of individuals and provide the regulations by which the culture functions as a collective. The undisputed effects of technology on a culture were explored next. Technological differences promote vast changes in the ways cultures choose to function. Finally, interpersonal communication patterns, the means by which cultural patterns are transmitted from one generation to another, were considered. These interpersonal communication patterns include the links a culture emphasizes among parents, siblings, peers, teachers, relatives, neighbors, authority figures, and other social contacts. The reciprocal relationship among these forces suggests the inevitability and constancy of accommodations and changes that characterize all cultures.

The chapter concluded with a discussion of a topic that is central to this book: intercultural communication. We began with several examples, which were followed by an exploration of issues related to similarities and differences among communicators that produce intercultural communication. Finally, after providing our definition of intercultural communication, we differentiated between that term and related terms, including *intracultural, interethnic, interracial, cross-cultural,* and *international* communication. In Chapter 3, we consider an additional concept that is the focal point of this book—intercultural communication competence.

FOR DISCUSSION

1. What differences are there between the view that "people are born into a culture" versus the opinion that "one becomes a member of a culture through a process of learning"?
2. In the United States, how are the terms *nation, race, culture,* and *co-culture* used inaccurately?
3. How do you think the ever-present cell phone, as a medium to communicate with others, is changing interpersonal relationships within the cultures of the United States?
4. What is implied by the statement that "race as biology is fiction, racism as a social problem is real"? Do you agree or disagree with this statement?
5. What links are there between intercultural communication and interpersonal communication?

FOR FURTHER READING

William B. Gudykunst, *Theorizing about Intercultural Communication* (Thousand Oaks, CA: Sage, 2005). A lucid and insightful guide to theories and theorizing about intercultural communication phenomena.

Myron W. Lustig and Jolene Koester (eds.), *AmongUS: Essays on Identity, Belonging, and Intercultural Competence*, 2nd ed. (Boston: Allyn & Bacon, 2006). This collection includes many first-person essays that document the emotions and experiences of people living in an intercultural world.

Richard E. Nisbett, *Intelligence and How to Get It: Why Schools and Cultures Count* (New York: W. W. Norton & Co., 2009). Argues convincingly that social and cultural experiences—that is, what is external to the individual and is controllable, rather than what is innate and genetic—are the biggest determinants of intelligence.

Joseph Shaules, *A Beginner's Guide to the Deep Culture Experience: Beneath the Surface* (Boston: Intercultural Press, 2010). A practical book that emphasizes becoming aware of one's cultural programming in order to engage people from other cultures in a substantial and deep (rather than superficial) way.

Ronald Takaki, *A Different Mirror: A History of Multicultural America*, 1st rev. ed. (New York: Little, Brown, 2008). Recounts the history of the United States from the perspective of various cultural groups that comprise today's U.S. Americans.

CHAPTER

3

Intercultural Communication Competence

KEY TERMS

When does communication become intercultural communication? What distinguishes intercultural communication from communication that is not intercultural? What does it mean to be a competent intercultural communicator? In Chapters 1 and 2, we defined the terms *communication, culture,* and *intercultural communication*. In this chapter, we first discuss the multicultural nature of the United States, where intercultural competence is essential. Then we focus our attention on the components and characteristics of intercultural communication competence. Our purpose is to establish boundaries and common understandings about this central idea.

THE UNITED STATES AS AN INTERCULTURAL COMMUNITY

A set of complicated issues underlies our discussion in Chapter 1 about the imperatives for intercultural communication. Stated most simply, these issues focus on what it means to be an American and on decisions about how to refer to the various cultural groups that reside within the borders of the United States. In the following sections, we first examine the implications of five metaphors that have been used to describe U.S. cultural diversity, and we suggest others that might also be used. Next, we analyze the question of what to call someone from the United States. Finally, we describe the difficult choices we faced in selecting labels to refer to the domestic cultures within the United States.

Metaphors of U.S. Cultural Diversity

Many cultural groups live within the borders of the United States. When people talk about the blend of U.S. cultural groups, their ideas are often condensed into a few key words or phrases. These summary images, called metaphors, imply both descriptions of what is and, less obviously, prescriptions of what should be. Although we will have much more to say in subsequent chapters about the effects of language and labeling on the intercultural communication process, we would like to focus now on five metaphors that have been used to describe the cultural mix within the United States: a melting pot, a set of tributaries, a rainbow, a tapestry, and a garden salad.

THE MELTING POT METAPHOR   Perhaps the oldest metaphor for describing multiple cultures in the United States is the melting pot.[1] America, according to this image, is like a huge crucible, a container that can withstand extremely high temperatures and can therefore be used to melt, mix, and ultimately fuse together metals or other substances. This image was the dominant way to represent the ideal blending of cultural groups at a time when the hardened steel that was forged in the great blast furnaces of Pittsburgh helped to make the United States into an industrial power. According to this view, immigrants from many cultures came to the United States to work, live, mix, and blend together into one great assimilated culture that is stronger and better than the unique individual cultures of which it is composed.

CULTURE CONNECTIONS

The fact is that you can never tell who someone is or where their ancestors came from just by looking at them. Identity is complex. Its roots lie beneath the surface. It's a product of events that we don't know about ourselves.

—Henry Louis Gates, Jr. ∎

Dynamic as the melting pot metaphor has been in the United States, it has never been an accurate description. The tendency for diverse cultures to melt together and assimilate their unique heritages into a single cultural entity has never really existed. Rather, the many cultural groups within the United States have continuously adapted to one another as they have accommodated and perhaps adopted some of the practices and preferences of other groups while maintaining their own unique and distinctive heritages.

THE TRIBUTARIES METAPHOR   A currently popular metaphor for describing the mix of cultures in the United States is that of tributaries or tributary streams. America, according to this image, is like a huge cultural watershed, providing numerous paths in which the many tributary cultures can flow. The tributaries maintain their unique identities as they surge toward their common destination. This view is useful and compelling. Unlike the melting pot metaphor, which implies that all cultures in the United States ought to be blended to overcome their individual weaknesses, the tributary image seems to suggest that it is acceptable and desirable for cultural groups to maintain their unique identities. However, when the metaphor of tributaries is examined closely, there are objections to some of its implications.

Tributary streams are small, secondary creeks that ultimately flow into a common stream, where they combine to form a major river. Our difficulty with this notion rests in the hidden assumption that the cultural groups will ultimately and inevitably blend together into a single, common current. Indeed, there are far fewer examples of cultures that have totally assimilated into mainstream U.S. culture than there are instances of cultures that have remained unique. Further, the idea of tributaries blending together to form one main stream suggests that the tributaries are somehow subordinate to or less important than the mighty river into which they flow.

THE RAINBOW METAPHOR   A rainbow is a multicolored arc of light that can be seen, often after a rainstorm, when the sun's rays are redirected by drops of water. The rainbow's colors represent the wide variety of cultures that comprise the United States.

Intercultural communication occurs when there are significant differences among the communicators. What do you suppose are the consequences of these differences for the people shown here?

These cultures coexist amicably, without clash or conflict; together they form an attractive national pattern that is pleasing, likeable, and beautiful. Thus a rainbow is often invoked as a metaphor of people from disparate cultures who enjoy living together in peaceful harmony.

The rainbow metaphor is not without its supporters. For example, the term "rainbow nation" was used by Nelson Mandela when he became the president of a post-apartheid South Africa; it was intended to summarize the idea of unity among people who were once separated by a strict division of racial and cultural groups. When scrutinized thoroughly, however, we find the rainbow metaphor to be problematic. To us, it suggests that U.S. cultures should be separate, unequal, and hierarchically ordered. The colors in a rainbow, after all, have fixed and unchanging positions. Red will always be separated from yellow; green will forever be above purple. Thus, though the rainbow metaphor has desirable qualities, it also could suggest an undesirable separation of U.S. cultures.

THE TAPESTRY METAPHOR A tapestry is a decorative cloth made up of many strands of thread. The threads are woven together into an artistic design that may be pleasing to some but not to others. Each thread is akin to a person, and groups of similar threads are analogous to a culture. Of course, the types of threads differ in many ways; their thickness, smoothness, color, texture, and strength may vary. The threads can range from gossamer strands to inch-thick yarn, from soft silk to coarse burlap, from pastel hues to fluorescent radiance, and from fragile spider webs to steel cables. The weaving process itself can vary from one location to another within the overall tapestry. Here, a wide swatch of a single type of thread may be used; there, many threads might be interwoven with many others, so no single thread is distinguished; and elsewhere, the threads may have been grouped together into small but distinguishable clumps.

Although the metaphor of a tapestry has much to commend it, like all metaphors the image is not flawless. After all, a tapestry is rather static and unchangeable. One does not typically unstring a bolt of cloth, for instance, only to reassemble the threads elsewhere in a different configuration. Cultural groups in the United States are more fluid than the tapestry metaphor might imply; migrations, immigrations, and mortality patterns all alter the cultural landscape. Despite its limitations, however, we find this metaphor preferable to the previous three.

THE GARDEN SALAD METAPHOR Like a garden salad made up of many distinct ingredients that are being tossed continuously, some see the United States as made up of a complex array of distinct cultures that are blended into a unique, and one hopes tasteful, mixture. Substitute one ingredient for another, or even change how much of each ingredient is present, and the entire flavor of the salad may be changed. Mix the salad differently and the look and feel will also differ. A salad contains a blend of ingredients, and it provides a unique combination of tints, textures, and tastes that tempt the palate.

Like the other metaphors, the garden salad is not without its flaws. In contrast to the tapestry image, which implies that the United States is too fixed and unchanging, a garden salad suggests an absence of firmness and stability. A typical garden salad has no fixed arrangement; it is always in a state of flux. Cultural groups in the United States, however, are not always moving, mixing, and mingling with the speed and alacrity that the metaphor would suggest. Nevertheless, we recommend this metaphor, and that of the tapestry, as two images that are likely to be useful in characterizing the diversity of cultural groups in the United States.

OTHER METAPHORS   Our list of common metaphors for cultural diversity in the United States is certainly not complete, nor can it ever be. Other metaphors that have been proposed include those of a symphony, a stew, a kaleidoscope, a jazz band, and a mosaic, among others. In addition to thinking about the implications of each of these metaphors, we encourage you to develop your own metaphor possibilities; they can help you to understand the inherently intercultural nature of today's United States. Of course, each metaphor should help to grasp the desirable characteristics of the intercultural mixing of the United States, both now and in the future. But each metaphor will also have limits or adverse qualities that restrict its usefulness. No metaphor can capture completely the cultural dynamics of today's United States. Instead, a set of metaphors is needed to help us to envision, and subsequently to act upon, what it means to live in an intercultural United States.[2]

What Do You Call Someone from the United States of America?

Many people who live in the United States of America prefer to call themselves American. However, people from Brazil, Argentina, Guatemala, Mexico, and many other Central and South American countries also consider themselves American, as they are all part of the continents known collectively as the Americas. Indeed, people from these countries consider the choice of *American* for those from the United States to be imperialistic and insulting. They resent the implication that they are less central or less important.

An alternative choice for a name, which is frequently selected by those who are trying to be more sensitive to cultural differences, is *North American*. *North American* is the English translation of the Spanish label that is commonly used by people from many Central and South American countries to refer to people from the United States, and the name is widely regarded as far less insulting and imperialistic. However, this label still has the potential for creating friction and causing misunderstanding. *North American* refers to an entire continent, and people from Mexico and Canada are, strictly speaking, also North Americans. Indeed, conversations with Canadians and Mexicans have confirmed for us that *North American* is not the ideal term.

One possibility that is often overlooked is to refer to people from the United States as *United Statians* or *United Staters*. These labels have the obvious advantage of being unambiguous, as they specifically identify people from a single country. Realistically, however, these are not labels that citizens of the United States would regard as comfortable and appropriate, and we agree that they are artificial and unlikely to be widely used.

CULTURE CONNECTIONS

I only have one-eighth Shoshone blood, but it shows in my hair color and high cheek bones. I seldom think of myself in terms of either my Indian heritage or the Scotch-Irish blood that makes up the remainder of my genetic composition. My attitude is a symptom of what's happened to ethnic groups in America, and I suppose in some ways the blurring of differences is a good thing. But on the other hand, there's an inherent sadness in the loss of consciousness of our roots, the loss of touch with the history and traditions that make us who we are.

—Marcia Muller ∎

Our preference is the label U.S. Americans. This referent retains the word *American* but narrows its scope to refer only to those from the United States. The term retains the advantages of a name that is specific enough to be accurate, yet it does not resort to a form of address that people would be unlikely to use and would regard as odd.

Cultural Groups in the United States

It is also important to select terms that adequately and sensitively identify the variety of cultural groups that make up the U.S. citizenry. As the population of the United States becomes increasingly more varied culturally, it is extremely urgent that we find ways to refer to these cultures with terms that accurately express their differences but avoid negative connotations and evaluations.

Some of the terms used in the past have negative associations, and we, as authors, have struggled to find more appropriate alternatives. For example, earlier writings about intercultural communication often referred to the culture associated with white U.S. Americans as either the "dominant" culture or the "majority" culture. The term *dominant* usually suggested the economic and political power of white U.S. America, referring to the control of important sources of institutional and economic power. The term often conveyed a negative meaning to members of other cultural groups, as it suggested that white U.S. Americans were somehow better or superior. It also implied that people from nondominant cultures were somehow subordinate or inferior to the dominant group. As more and more cultural groups have gained political and economic power in the United States, *dominant* no longer accurately reflects the current reality.

An alternative label for white U.S. Americans was *majority culture*. This term was intended to reflect a numerical statement that the majority of U.S. Americans are from a particular cultural group. *Majority* was often coupled with the term *minority,* which also had negative connotations for many members of other cultural groups: it suggested to some people that they were not regarded as important or significant as members of the majority. In addition, as previously suggested, nonwhite cultural groups now make up a sufficient proportion of the total population, so white U.S. Americans no longer constitute an absolute majority in many places. Thus, we prefer to avoid such emotionally charged words as *majority, minority, dominant, nondominant,* and *subordinate* when we discuss the cultural groups residing in the United States.

We have also elected not to use the term *white* or *Caucasian* in all subsequent discussions about a specific cultural group of U.S. Americans. *White* and *Caucasian* refer to a particular race. As suggested in Chapter 2, a racial category does not necessarily identify and distinguish a particular culture. Although many members of this group may prefer to use the term *white,*[3] we think it is less useful in this book on inter*cultural* communication to refer to a cultural group in the United States by a term that denotes race. Consequently, because their common cultural heritage is predominantly European, we have chosen to describe white U.S. cultural members as *European Americans*.

Many black Americans prefer to be identified by a term that distinguishes them by their common cultural characteristics rather than by their racial attributes. *African American* recognizes the effects of traditional African cultural patterns on U.S. Americans of African heritage, and it acknowledges that African American cultural patterns are distinct from those of European Americans. Because it denotes a cultural rather than a racial distinction, we will use the term *African American* in this book.

Another set of terms is usually applied to those residents of the United States whose surname is Spanish. *Hispanic, Chicano, Mexican American,* and *Latino* are often used interchangeably, but the distinctions between the terms can be quite important.[4] *Hispanic* derives from the dominant influences of Spain and the Spanish language, but many shy away from this term because it tends to homogenize all groups of people who have Spanish surnames and who use the Spanish language. *Chicano* (or *Chicana*) refers to the "multiple-heritage experience of Mexicans in the United States" and speaks to a political and social consciousness of the Mexican American.[5] Specific terms such as *Mexican American* or *Cuban American* are preferred by those who wish to acknowledge their cultural roots in a particular national heritage while simultaneously emphasizing their pride in being U.S. Americans.[6] Finally, *Latino* (or *Latina*) is a cultural and linguistic term that includes "all groups in the Americas that share the Spanish language, culture, and traditions."[7] As Earl Shorris notes, "Language defines the group, provides it with history and home; language should also determine its name—Latino."[8] Because *Latino* and *Latina* suggest cultural distinctions, we will use them in this book.

Terms routinely used to describe members of other cultural groups include *Native American, Arab American, Asian American,* and *Pacific Islander.* Each of these labels, as well as those previously described, obscures the rich variety of cultures that the single term represents. For instance, many tribal nations can be included under the term *Native American,* and members of those groups prefer a specific reference to their culture (e.g., Chippewa, Sioux, Navajo, Choctaw, Cherokee, and Inuit). Similarly, *Asian American* is a global term that can refer to Japanese Americans, Chinese Americans, Malaysian Americans, Korean Americans, and people from many other cultures that geographically originated in the part of the world loosely referred to as Asia. Even *European American* obscures differences among those whose heritage may be English, French, Italian, or German. Our use of these overly broad terms is not meant to deny the importance of cultural distinctions but to allow for an economy of words. We will use the broader, more inclusive, and less precise terms when making a generalization that describes a commonality among these cultures. When using examples that are limited to a particular culture, we will use the more specific nomenclature.

Notice that there are some inherent difficulties in our choices of cultural terms to refer to U.S. Americans. If precision were our only criterion, we would want to make

CULTURE CONNECTIONS

In a world composed of a few hundred nations, thousands of groups speaking thousands of languages, and more than 6 billion inhabitants, what is a reasonable goal? Clearly, we can no longer simply draw a curtain or build a wall that isolates groups from one another indefinitely. We homo sapiens must somehow learn how to inhabit neighboring places—and the same planet—without hating one another, without lusting to injure or kill one another, without acting on xenophobic inclinations even if our own group might emerge triumphant in the short run. Often the desideratum *tolerance* is invoked,

and it may be the case that it is all that we can aspire to. Wordsmiths of a more optimistic temperament opt for romantic language; on the eve of World War II, poet W. H. Auden declared, "We must love one another or die."

I prefer the concept of respect. Rather than ignoring differences, being inflamed by them, or seeking to annihilate them through love or hate, I call on human beings to accept the differences, learn to live with them, and value those who belong to other cohorts.

—Howard Gardner ∎

many further distinctions. But we are also aware of the need for economy and the force of common usage. Although it is not our intent to advocate terms that ignore or harm particular cultural groups, we do prefer a vocabulary that is easily understood, commonly used, and positively regarded. Please remember, however, that the term preferred by specific individuals is an important reflection of the way they perceive themselves and others. A "Chicana" defines herself differently from someone who labels herself as a "Mexican American" or a "Latina." Similarly, the expressed identities of those who call themselves "black" or "black American" or "African American" all differ.[9] Likewise, individuals who describe themselves as "European American" rather than as "white" express less racial prejudice and are more likely to support multiculturalism.[10]

COMPETENCE AND INTERCULTURAL COMMUNICATION

Competent interpersonal communication is a worthy and often elusive goal. Interpersonal competence in intercultural interactions is an even more difficult objective to achieve, because cultural differences create dissimilar meanings and expectations that require even greater levels of communication skill. We base our understanding of intercultural competence on the work of scholars who have studied communicative competence from a primarily intracultural perspective and on the conclusions of other scholars who have studied intercultural competence.

The study of intercultural competence has been motivated primarily by practical concerns. Businesses, government agencies, and educational institutions want to select people for intercultural assignments who will be successful. Lack of intercultural competence means failed business ventures, government projects that have not achieved their objectives, and unsuccessful learning experiences for students.

Intercultural Communication Competence

Although there is still some disagreement among scholars about how best to conceptualize and measure communication competence, there is increasing agreement about certain of its fundamental characteristics.[11] In our discussion, we draw heavily on the work of Brian Spitzberg and his colleagues. The following definition of communication competence illustrates the key components of their approach:

> Competent communication is interaction that is perceived as effective in fulfilling certain rewarding objectives in a way that is also appropriate to the context in which the interaction occurs.[12]

This definition provides guidance for understanding communicative and intercultural competence in several ways. A key word is *perceived* because it means that competence is best determined by the people who are interacting with each other. In other words, communicative competence is a social judgment about how well a person interacts with others. That competence involves a social perception suggests that it will always be specific to the context and interpersonal relationship within which it occurs. Therefore, whereas judgments of competence are influenced by an assessment of an individual's personal characteristics, they cannot be wholly determined by them, because competence involves an interaction between people.

Competent interpersonal communication results in behaviors that are regarded as *appropriate*. That is, the actions of the communicators fit the expectations and demands

In this Cinco de Mayo celebration, Latinas celebrate their cultural traditions.

of the situation. Appropriate communication means that people use the symbols they are expected to use in a given context.

Competent interpersonal communication also results in behaviors that are *effective* in achieving desired personal outcomes. Satisfaction in a relationship or the accomplishment of a specific task-related goal is an example of an outcome people might want to achieve through their communication with others.

Thus, communication competence is a social judgment that people make about others. The judgment depends on the context, the relationship between the interactants, the goals or objectives that the interactants want to achieve, and the specific verbal and nonverbal messages that are used to accomplish those goals.

The Components of Intercultural Competence

Our central concern in this book is improving your intercultural competence, and the ideas presented here are the key to doing so. In the remaining chapters, we will return to the concepts that follow to suggest ways to improve your ability.

A word of caution is necessary before we begin, however. We cannot write a prescription guaranteed to ensure competence in intercultural communication. The complexity of human communication in general, and intercultural communication in particular, denies the possibility of a quick fix. There is not necessarily only one way to be competent in your intercultural interactions. Even within the context of a specific person and specific setting, there may be several paths to competent interaction. The goal here is to understand the many ways that a person can behave in an interculturally competent manner.

The remaining portion of this chapter provides a description of the characteristics of people, what they bring to the intercultural communication situation, and the nature of the communication itself, all of which increase the possibility of competence in intercultural communication. Subsequent chapters build on this discussion by offering guidelines for achieving competence. The summary of previous research suggests that competent intercultural communication is contextual; it produces behaviors that are both appropriate and effective; and it requires sufficient knowledge, suitable motivations, and the right skills. Let's examine each of these components.

CONTEXT   Intercultural competence is contextual. An impression or judgment that a person is interculturally competent is made with respect to both a specific relational context and a particular situational context. Competence is not independent of the relationships and situations within which communication occurs.

Thus, competence is not an individual attribute; rather, it is a characteristic of the association between individuals.[13] It is possible, therefore, for someone to be perceived as highly competent in one set of intercultural interactions and only moderately competent in another. For example, a Canadian woman living with a family in India might establish competent relationships with the female family members but be unable to relate well to the male members.

Judgments of intercultural competence also depend on cultural expectations about the permitted behaviors that characterize the settings or situations within which people communicate. The settings help to define and limit the range of behaviors that are regarded as acceptable. Consequently, the same set of behaviors may be perceived as very competent in one cultural setting and much less competent in another. As an obvious example that competence is situationally determined, consider what might happen when two people who come from very different cultural backgrounds are involved in a close business relationship. Whereas one person might want to use highly personalized nicknames and touching behaviors in public, the other person might regard such visible displays as unwarranted and therefore incompetent.

Many previous attempts to describe intercultural competence have erroneously focused on the traits or individual characteristics that make a person competent. Thus, in the past, individuals have been selected for particular intercultural assignments based solely on such personality attributes as authoritarianism, empathy, self-esteem, and world-mindedness. Because intercultural competence is contextual, these trait approaches have been unsuccessful in identifying competent intercultural communicators. Although specific personality traits might allow a person to be more or less competent on particular occasions, there is no prescriptive set of characteristics that inevitably guarantees competence in all intercultural relationships and situations.

APPROPRIATENESS AND EFFECTIVENESS   Both interpersonal competence and intercultural competence require behaviors that are appropriate and effective. By appropriate we mean those behaviors that are regarded as proper and suitable given the expectations generated by a given culture, the constraints of the specific situation, and the nature of the relationship between the interactants. By effective we mean those behaviors that lead to the achievement of desired outcomes. The following example illustrates this important distinction between appropriateness and effectiveness.

Brian Holtz is a U.S. businessperson assigned by his company to manage its office in Thailand. Mr. Thani, a valued assistant manager in the Bangkok office, has recently been arriving late for work. Holtz has to decide what to do about this problem. After carefully thinking about his options, he decides there are four possible strategies:

1. Go privately to Mr. Thani, ask him why he has been arriving late, and tell him that he needs to come to work on time.
2. Ignore the problem.

3. Publicly reprimand Mr. Thani the next time he is late.
4. In a private discussion, suggest that he is seeking Mr. Thani's assistance in dealing with employees in the company who regularly arrive late for work, and solicit his suggestions about what should be done.

Holtz's first strategy would be effective, as it would probably accomplish his objective of getting Mr. Thani to arrive at work more promptly. However, given the expectations of the Thai culture, which are that one person never directly criticizes another, such behavior would be very inappropriate. Conversely, Holtz's second strategy would be appropriate but not effective, as there would probably be no change in Mr. Thani's behavior. The third option would be neither appropriate nor effective because public humiliation might force Mr. Thani, a valuable employee, to resign. The fourth option, which is the best choice, is both appropriate and effective. By using an indirect means to communicate his concerns, Mr. Thani will be able to "save face" while Holtz accomplishes his strategic goals.

KNOWLEDGE, MOTIVATIONS, AND SKILLS   Intercultural competence requires sufficient knowledge, suitable motivations, and the right skills. Each of these components alone is insufficient to achieve intercultural competence.

Knowledge   Knowledge refers to the cognitive information you need to have about the people, the context, and the norms of appropriateness that operate in a specific culture. Without such knowledge, it is unlikely that you will interpret correctly the meanings of other people's messages, nor will you be able to select behaviors that are appropriate and that allow you to achieve your objectives. Consequently, you will not be able to determine what the appropriate and effective behaviors are in a particular context.

The kinds of knowledge that are important include culture-general and culture-specific information. Culture-general information provides insights into the intercultural communication process abstractly and can therefore be a very powerful tool in making sense of cultural practices, regardless of the cultures involved. For example, the

CULTURE CONNECTIONS

Opening to the World

Human beings tend to regard the conventions of their own societies as natural, often as sacred. One of the great steps forward in history was learning to regard those who spoke odd-sounding languages and had different smells and habits as fully human, as similar to oneself. The next step from this realization, the step which we have still not fully made, is the willingness to question and purposefully alter one's own conditions and habits, to learn by observing others.

If a particular arrangement is not necessary, it might be possible to choose to change it. Still aristocratic Chinese ladies of the old regime, crippled for life by the binding of their feet, looked down on peasant women with unbound feet. Exposure to other ways of doing things is insufficient if it is not combined with empathy and respect.

—Mary Catherine Bateson ■

knowledge that cultures differ widely in their preferred patterns (or rules) of interaction should help to sensitize you to the need to be aware of these important differences. This book is an excellent example of a source for culture-general knowledge. Knowledge about interpersonal communication and the many ways in which culture influences the communication process is very useful in understanding actual intercultural interactions.

Intercultural competence also depends on culture-specific information, which is used to understand a particular culture. Such knowledge should include information about the forces that maintain the culture's uniqueness (see Chapter 2) and facts about the cultural patterns that predominate (see Chapters 4 and 5). The type of intercultural encounter will also suggest other kinds of culture-specific information that might be useful. Exchange students might want to seek out information about the educational system in the host country. Businesspeople may need essential information about the cultural dynamics of doing business in a specific country or with people from their own country who are members of different cultural groups. Tourists would benefit from guidebooks that provide information about lodging, transportation, food, shopping, and entertainment.

At an official ceremony in New Zealand, Prince William receives a *hongi* from a Maori tribal elder, which is a traditional greeting in which the participants press their noses together. Intercultural competence requires an understanding of the appropriate and effective communication behaviors that are expected in a given setting.

An additional—and crucial—form of culture-specific knowledge involves information about the specific customs that govern interpersonal communication in the culture. For example, before traveling to Southeast Asia, it would be very useful to know that many Southeast Asian cultures regard a display of the soles of the feet as very offensive. This small bit of information can be filed away for later recall when travelers visit temples and attempt to remove their shoes. The imperative to learn about other cultures is equally strong for those cultures with which you interact on a daily basis. Culture-specific knowledge about the rules and customs of the multiple cultures that make up the cultural landscape of the United States is essential information if you are to be interculturally competent.

Often overlooked is knowledge of one's own cultural system. Yet the ability to attain intercultural competence may be very closely linked to this kind of knowledge. Knowledge about your own culture will help you to understand another culture. Fathi Yousef has even suggested that the best way to train businesspeople who must deal with cultural differences might be to teach them about the characteristics of their own culture rather than those of others.[14] The idea behind this admonition is that, if people are able to understand how and why they interpret events and experiences, it is more likely that they would be able to select alternative interpretations and behaviors that are more appropriate and effective when interacting in another culture.

Motivations Motivations include people's overall set of emotional associations as they anticipate and actually communicate interculturally. Some scholars have suggested that people's motivations—particularly those that make one more willing to participate in intercultural communication encounters—are often the starting point for becoming interculturally competent.[15]

As with knowledge, different aspects of the emotional terrain contribute to the achievement of intercultural competence. Human emotional reactions include both feelings and intentions.

Feelings refer to the emotional or affective states that you experience when communicating with someone from a different culture. Feelings are not thoughts, though people often confuse the two; rather, feelings are your emotional and physiological reactions to thoughts and experiences. Feelings of happiness, sadness, eagerness, anger, tension, surprise, confusion, relaxation, and joy are among the many emotions that can accompany the intercultural communication experience. Feelings involve your general sensitivity to other cultures and your attitudes toward the specific culture and individuals with whom you must interact. How would you characterize your general motivation toward other cultures? Are you excited by the thought of talking with someone from a culture that is different from yours? Or are you anxious at the prospect? Do you think your culture is superior to other cultures? Are you even willing to entertain the idea that another culture's ways of doing various life activities might be as good as, or even better than, your culture's ways? Some people simply do not want to be confronted with things that differ from what they are used to. The different sights, sounds, and smells of another culture are often enough to send them running back to the safety of a hotel room. Eagerness and a willingness to experience some uncertainty is a necessary part of your motivation to achieve intercultural competence.

Intentions are what guide your choices in a particular intercultural interaction. Your intentions are the goals, plans, objectives, and desires that focus and direct your behavior. Intentions are often affected by the stereotypes you have of people from other cultures because stereotypes reduce the number of choices and interpretations you are willing to consider. For instance, if you begin an intercultural interaction having already formed a

CULTURE CONNECTIONS

"Everyone's staring at us." It was her mother's stage whisper again.

Margaret glanced down the carriage and saw that nearly everyone was indeed watching them, in silent but unabashed curiosity. It was something Margaret had long since ceased to notice. But even today the sight of a westerner still drew stares of astonishment. Sometimes people would ask to touch Margaret's hair, and they would gaze, unblinking, into her eyes, amazed at their clear, blue colour. "That's because we look so strange," she said.

"*We* look strange?" Mrs. Campbell said indignantly.

"Yes," Margaret said. "We're a curiosity. A couple of bizarre-looking, round-eyed foreign devils."

"Foreign devils!"

"*Yangguizi.* That's the word they have for us when they're not being too polite. Literally, foreign devils. And then there's *da bidze.* Big noses. You see, *you* might think the Chinese have got flat faces and slanted eyes. *They* think we've got prominent brows and gross features, and have more in common with Neanderthal Man. That's because they consider themselves to be a more highly evolved strain of the species."

"Ridiculous," Mrs. Campbell said, glaring at the Chinese faces turned in her direction.

"No more ridiculous than those white, Anglo-Saxon Americans who think they're somehow better than, say, the blacks or the Hispanics."

—Peter May   ■

negative judgment of the other person's culture, it will be very difficult for you to develop accurate interpretations of the behaviors that you observe. Intentions toward the specific interaction partner also must be positive. If your intentions are positive, accurate, and reciprocated by the people with whom you are interacting, your intercultural competence will likely be enhanced.

Skills Finally, skills refer to the actual performance of those behaviors that are regarded as appropriate and effective. Thus, you can have the necessary information, be motivated by the appropriate feelings and intentions, and still lack the behavioral skills necessary to achieve competence. For example, students from other cultures who enroll in basic public speaking classes often have an excellent understanding of the theory of speech construction. In addition, they have a positive attitude toward learning U.S. speaking skills; they want to do well and are willing to work hard in preparation. Unfortunately, their speaking skills sometimes make it difficult for them to execute the delivery of a speech with the level of skill and precision that they would like.[16]

BASIC TOOLS FOR IMPROVING INTERCULTURAL COMPETENCE

In the preceding section, we suggested that intercultural competence means using your knowledge, motivation, and skills to deal appropriately and effectively with cultural differences. We now offer two tools to assist you in becoming more interculturally competent. These tools can help you improve your interpersonal interactions and will facilitate the development of intercultural relationships.

The BASICs of Intercultural Competence

The Behavioral Assessment Scale for Intercultural Competence (BASIC), developed by Jolene Koester and Margaret Olebe,[17] is based on work done originally by Brent Ruben and his colleagues.[18] A very simple idea provides the key to understanding how to use these BASIC skills: what you actually do, rather than your internalized attitudes or your projections of what you might do, is what others use to determine whether you are interculturally competent. The BASIC skills are a tool for examining people's communication behaviors—yourself included—and in so doing provides a guide to the very basics of intercultural competence.

Eight categories of communication behavior are described in the BASIC instrument, each of which contributes to the achievement of intercultural competence. As each of the categories is described, mentally assess your own ability to communicate. Do you display the behaviors necessary to achieve intercultural competence? From what you now know about intercultural communication, what kinds of changes might make your behavior more appropriate and effective?

Before we describe each of the BASIC skills, we would like to emphasize that the BASIC descriptions of behaviors are culture-general. That is, most cultures use the types of behaviors that are described to make judgments of competence about themselves and others. But within each culture there may be, and in all likelihood will be, different ways of exhibiting these behaviors. For example, actions that show respect for others, and the ability to maintain conversations and manage communicative interactions, are necessary in all cultures for someone to be judged as competent. However, the way each culture teaches its members to exhibit these actions is culture-specific. Even among the various cultural groups that live in the United States, the rules for taking turns in a conversation vary widely. The eight types of communication behaviors are each discussed and are summarized in Figure 3.1.

DISPLAY OF RESPECT   Although the need to display respect for others is a culture-general concept, within every culture there are specific ways to show respect and specific expectations about those to whom respect should be shown. What constitutes respect in one culture, then, will not necessarily be so regarded in another culture.

Respect is shown through both verbal and nonverbal symbols. Language that can be interpreted as expressing concern, interest, and an understanding of others will often convey respect, as will formality in language, including the use of titles, the absence of jargon, and an increased attention to politeness rituals. Nonverbal displays of respect include showing attentiveness through the position of the body, facial expressions, and the use of eye contact in prescribed ways. A tone of voice that conveys interest in the other person is another vehicle by which respect is shown. The action of displaying respect increases the likelihood of a judgment of competence.

ORIENTATION TO KNOWLEDGE   Orientation to knowledge refers to the terms people use to explain themselves and the world around them. A competent orientation to knowledge occurs when people's actions demonstrate that all experiences and interpretations are individual and personal rather than universally shared by others.

Many actions exhibit people's orientation to knowledge, including the specific words that are used. Among European Americans, for instance, declarative statements that

Display of Respect	The ability to show respect and positive regard for other people and their cultures
Orientation to Knowledge	The recognition that individuals' experiences shape what they know
Empathy	The capacity to behave as though you understand the world as others do
Interaction Management	Skill in regulating conversations and taking turns
Task Role Behavior	Behaviors that involve the initiation of ideas related to group problem-solving activities
Relational Role Behavior	Behaviors associated with interpersonal harmony and mediation
Tolerance for Ambiguity	The ability to react to new and ambiguous situations with little visible discomfort
Interaction Posture	The ability to respond to others in descriptive, nonevaluative, and nonjudgmental ways

FIGURE 3.1

BASIC dimensions of intercultural competence.

express personal attitudes or opinions as if they were facts and an absence of qualifiers or modifiers would show an ineffective orientation to knowledge:

- "New Yorkers must be crazy to live in that city."
- "Parisians are rude and unfriendly."
- "The custom of arranged marriages is barbaric."
- "Every person wants to succeed—it's human nature."

In contrast, a competent intercultural communicator acknowledges a personal orientation to knowledge, as illustrated in the following examples:

- "I find New York a very difficult place to visit and would not want to live there."
- "Many of the people I interacted with when visiting Paris were not friendly or courteous to me."
- "I would not want my parents to arrange my marriage for me."
- "I want to succeed at what I do, and I think most people do."

At least some of the time, all people have an orientation to knowledge that is not conducive to intercultural competence. In learning a culture, people develop beliefs about the "rightness" of a particular way of seeing events, behaviors, and people. It is actually very natural to think, and then to behave, as if your personal knowledge and experiences are universal. Intercultural competence, however, requires an ability to move beyond the perspective of your cultural framework.

EMPATHY   Empathy is the ability of individuals to communicate an awareness of another person's thoughts, feelings, and experiences, and such individuals are regarded as more competent in intercultural interactions. Alternatively, those who lack empathy, and who therefore indicate little or no awareness of even the most obvious feelings and thoughts of others, will not be perceived as competent. Empathetic behaviors include verbal statements that identify the experiences of others and nonverbal codes that are complementary to the moods and thoughts of others.

It is necessary to make an important distinction here. Empathy does not mean "putting yourself in the shoes of another." It is both physically and psychologically impossible to do so. However, it is possible for people to be sufficiently interested and aware of others that they appear to be putting themselves in others' shoes. The skill we are describing here is the capacity to *behave as if one understands the world as others do*. Of course, empathy is not just responding to the tears and smiles of others, which may, in fact, mean something very different than your cultural interpretations would suggest. Although empathy does involve responding to the emotional context of another person's experiences, tears and smiles are often poor indicators of emotional states.

INTERACTION MANAGEMENT   Some individuals are skilled at starting and ending interactions among participants and at taking turns and maintaining a discussion. These interaction management skills are important because, through them, all participants in an interaction are able to speak and contribute appropriately. In contrast, dominating a conversation or being nonresponsive to the interaction is detrimental to competence. Continuing to engage people in conversation long after they have begun to display signs of disinterest and boredom or ending conversations abruptly may also pose problems. Interaction management skills require knowing how to indicate turn taking both verbally and nonverbally.

TASK ROLE BEHAVIOR   Because intercultural communication often takes place where individuals are focused on work-related purposes, appropriate task-related role behaviors are very important. Task role behaviors are those that contribute to the group's problem-solving activities—for example, initiating new ideas, requesting further information or facts, seeking clarification of group tasks, evaluating the suggestions of others, and keeping a group on task. The difficulty in this important category is the display of culturally appropriate behaviors. The key is to recognize the strong link to a culture's underlying patterns and to be willing to acknowledge that tasks are accomplished by cultures in multiple ways. Task behaviors are so intimately entwined with cultural expectations about activity and work that it is often difficult to respond appropriately to task expectations that differ from one's own. What one culture defines as a social activity, another may define as a task. For example, socializing at a restaurant or a bar may be seen as a necessary prelude to conducting a business negotiation. Sometimes that socializing is expected to occur over many hours or days, which surprises and dismays many European Americans, who believe that "doing business" is separate from socializing.

CULTURE CONNECTIONS

The interpretation of historical data from a strictly Eurocentric perspective can lead to serious intercultural conflict, based on wrong premises.

Let me give an example of how cultural misunderstandings can be propagated on the basis of different views. In the nineteenth century, Cecil John Rhodes sought to gain control of a large territory of southern Africa, controlled by the Ndebele King Lobengula, and sent emissaries to the powerful king in an effort to secure his consent. After many days of discussion with Lobengula, the white emissaries returned to Rhodes with the signature of Lobengula on a piece of paper. They told Rhodes that the king had given Rhodes all of his territory. Rhodes sent a column of soldiers into the area with instructions to shoot any black on sight. Thus began the country of Rhodesia.

Rhodes may have believed that King Lobengula gave him title to the land, but Lobengula never believed that he had. Two cultural views of the world clashed, and the Europeans automatically assumed the correctness of their view. An Afrocentric analysis points out that Lobengula could never have sold or given the land away, since it did not belong to him

but to the ancestors and the community. He could grant Rhodes permission to hunt, to farm, and even to build a house, but not to own land. Only in this manner could the king follow the discourse of his ancestors. It took nearly one hundred years, two revolts, and a seven-year war to correct the situation. A rigid Eurocentrism made Rhodes believe that Lobengula had signed his country over to him.

Similarly, I am certain that the Indians did not believe they had sold Manhattan Island for twenty-three dollars worth of trinkets, no matter what the Dutch thought. Native Americans revered the land in much the same way as Africans. No king or clan leader could sell what did not belong to him. On the basis of European contractual custom, the Dutch may have actually thought they were purchasing the island from the Indians; but this was obviously a view based on their own commercial traditions.

Ascertaining the view of the other is important in understanding human phenomena. African responses and actions, however, have too often been examined from Eurocentric perspectives.

—Molefi K. Asante ∎

RELATIONAL ROLE BEHAVIOR Relational role behaviors concern efforts to build or maintain personal relationships with group members. These behaviors may include verbal and nonverbal messages that demonstrate support for others and that help to solidify feelings of participation. Examples of competent relational role behaviors include harmonizing and mediating conflicts between group members, encouraging participation from others, general displays of interest, and a willingness to compromise one's position for the sake of others.

TOLERANCE FOR AMBIGUITY Tolerance for ambiguity concerns a person's responses to new, uncertain, and unpredictable intercultural encounters. Some people react to new situations with greater comfort than do others. Some are extremely nervous, highly frustrated, or even hostile toward new situations and those who may be present in them. Those who do not tolerate ambiguity well may respond to new and unpredictable situations with hostility, anger, shouting, sarcasm, withdrawal, or abruptness.

Others view new situations as a challenge; they seem to do well whenever the unexpected or unpredictable occurs, and they quickly adapt to the demands of changing environments. Competent intercultural communicators are able to cope with the nervousness and frustrations that accompany new or unclear situations, and they are able to adapt quickly to changing demands.

Learning about other cultures is a necessary prerequisite to achieving intercultural competence. Sometimes a member of the culture, such as this tour guide, can help.

INTERACTION POSTURE Interaction posture refers to the ability to respond to others in a way that is descriptive, nonevaluative, and nonjudgmental. Although the specific verbal and nonverbal messages that express judgments and evaluations can vary from culture to culture, the importance of selecting messages that do not convey evaluative judgments is paramount. Statements based on clear judgments of rights and wrongs indicate a closed or predetermined framework of attitudes, beliefs, and values, and they are used by the evaluative, and less competent, intercultural communicator. Nonevaluative and nonjudgmental actions are characterized by verbal and nonverbal messages based on descriptions rather than on interpretations or evaluations.

Description, Interpretation, and Evaluation

We have approached the study of intercultural competence by looking at the elements of culture that affect interpersonal communication. There is, however, a tool that allows people to control the meanings they attribute to the verbal and nonverbal symbols used by others. The tool is based on the differences in how people think about, and then verbally speak about, the people with whom they interact and the events in which they participate.

The interaction tool is called description, interpretation, and evaluation (D-I-E). It starts with the assumption that, when most people process the information around them, they use a kind of mental shorthand. Because people are taught what symbols mean, they are not very aware of the information they use to form their interpretations. In other words, when people see, hear, and in other ways receive information from the world around them, they generally form interpretations and evaluations of it without being aware of the specific sensory information they have perceived. For example, students and teachers alike often comment about the sterile, institutional character of many of the classrooms at universities. Rarely do these conversations detail the specific perceptual information on which that interpretation is based. Rarely does someone say, for instance, "This room is about twenty by forty feet in size, the walls are painted a cream color, there is no artwork on the walls, it is lit by eight fluorescent bulbs, and the floors are cream-colored tiles with multiple pieces of dirt." Yet when students and professors say that their classroom is "sterile, institutional-looking, and unattractive," most people who have spent a great deal of time in such rooms have a fairly accurate image of the classroom. Similarly, if a friend is walking toward you, you might say, "Hi! What's wrong? You look really tired and upset." That kind of comment is considered normal, but if you said instead, "Hi! Your shoulders are drooping, you're not standing up straight, and you are walking much slower than usual," it would be considered strange. In both examples, the statements considered to be normal are really interpretations and evaluations of sensory information the individual has processed.

The skill we are introducing trains you to distinguish among statements of description, interpretation, and evaluation. These statements can be made about all characteristics, events, persons, and objects. A statement of description details the specific perceptual cues and information a person has received, without judgments or interpretations—in other words, without being distorted by opinion. A statement of interpretation provides a conjecture or hypothesis about what the perceptual information might mean. A statement of evaluation indicates an emotional or affective judgment about the information.

Often, the interpretations people make of perceptual information are very closely linked to their personal evaluation of that information. Any description can have many different interpretations, but because most people think in a mental shorthand, they are generally aware of only the interpretation that immediately comes to mind, which they use to explain the event. For example, teachers occasionally have students who arrive late to class. A statement of description about a particular student engaging in this behavior might be as follows:

■ Kathryn arrived ten minutes after the start of the class.
■ Kathryn also arrived late each of the previous times the class has met.

Statements of interpretation, which are designed to explain Kathryn's behavior, might include some of the following:

- Kathryn doesn't care very much about this particular class.
- Kathryn is always late for everything.
- Kathryn has a job on the other side of campus and is scheduled to work until ten minutes before this class. The person who should relieve her has been late, thus not allowing Kathryn to leave to be on time for class.
- Kathryn is new on campus this semester and is misinformed about the starting time for the class.

For each interpretation, the evaluation can vary. If the interpretation is "Kathryn doesn't care very much about this class," different professors will have differing evaluations:

- I am really offended by that attitude.
- I like a student who chooses to be enthusiastic only about classes she really likes.

The interpretation a person selects to explain something like Kathryn's behavior influences the evaluation that is made of that behavior. In people's everyday interactions, distinctions are rarely made among description, interpretation, and evaluation. Consequently, people deal with their interpretations and evaluations as if these were actually what they saw, heard, and experienced.

The purpose of making descriptive statements when you are communicating interculturally is that they allow you to identify the sensory information that forms the basis of your interpretations and evaluations. Descriptive statements also allow you to consider alternative hypotheses or interpretations. Interpretations, although highly personal, are very much affected by underlying cultural patterns. Sometimes when you engage in intercultural communication with specific persons or groups of people for an extended period of time, you will be able to test the various interpretations of behavior that you are considering. By testing the alternative interpretations, it is also possible to forestall the evaluations that can negatively affect your interactions. Consider the

CULTURE CONNECTIONS

Not far from the giraffes' trees we saw two distinct tracks, like footpaths, and I was told it was a road. It didn't make sense to me, because it looked like a footpath and I couldn't imagine what a car was, never having seen one before. In my neighborhood at night I had seen lights that people said belonged to cars, but I had never actually seen one. I asked the boys how big a car is, and they told me there were different sizes. "There are small ones and real big ones." "Do they trot, gallop, or walk?" I asked. One of the boys said they ran, and not having seen the continuous motion of a wheel before, I had trouble understanding what he meant. I straddled the two trails, started to hop like a frog, and asked them if the car ran that way, but it was too much for them to explain, and they dismissed me as a fool.

—Tepilit Ole Saitoti ■

following situation, and notice how differences among description, interpretation, and evaluation affect John's intercultural competence:

> John Richardson has been sent by his U.S.-based insurance company to discuss, and possibly to sell, his company's products with an Argentinean company that has expressed great interest in them. His secretary has set up four appointments with key company officials. John arrives promptly at his first appointment, identifies himself to the receptionist, and is asked to be seated. Some thirty minutes later he is ushered into the office of the company official, who has one of his employees in the office with whom he is discussing another issue. John is brought into the office of his second appointment within a shorter period of time, but the conversation is constantly disrupted by telephone calls and drop-in visits from others. At the end of the day, John is very discouraged; he calls the home office and says, "This is a waste of time; these guys aren't interested in our products at all! I was left cooling my heels in their waiting rooms. They couldn't even give me their attention when I got in to see them. There were constant interruptions. I really tried to control myself, but I've had it. I'm getting on a plane and coming back tomorrow.

John would be better off if he approached this culturally puzzling behavior by separating his descriptions, interpretations, and evaluations. By doing so, he might choose very different actions for himself. Descriptive statements might include the following:

- My appointments started anywhere from fifteen to thirty minutes later than the time I scheduled them.
- The people with whom I had appointments also talked to other company employees when I was in their offices.
- The people with whom I had appointments accepted telephone calls when I was in their offices.

Interpretations of this sensory information might include the following:

- Company officials were not interested in talking with me or in buying my company's products.
- Company officials had rescheduled my appointments for a different time, but they neglected to tell my secretary about the change.
- In Argentina, attitudes toward time are very different from those in the United States; although appointments are scheduled for particular times, no one expects that people will be available at precisely that time.
- In Argentina, it is an accepted norm of interaction between people who have appointments with each other to allow others to come into the room, either in person or by telephone, to ask their questions or to make their comments.

These interpretations suggest very different evaluations of John's experiences. His frustration with the lack of punctuality and the lack of exclusive focus on him and his ideas may still be a problem even if he selects the correct cultural interpretation, which is that in Argentina, time is structured and valued very differently than it is in the United States. But by considering other interpretations, John's evaluations and his actions will be more functional, as he might say the following:

- I don't like waiting around and not meeting according to the schedule I had set, but maybe I can still make this important sale.
- Some of the people here are sure interesting and I am enjoying meeting so many more people than just the four with whom I had scheduled appointments.

Contrast the difference between your own culture and the culture of those in this picture. Use the skills of Description, Interpretation, and Evaluation to understand these Indian women eating a traditional Indian meal. How difficult would it be for you to communicate competently in this setting?

The tool of description, interpretation, and evaluation increases your choices for understanding, responding positively to, and behaving appropriately with people from different cultures. The simplicity of the tool makes it available in any set of circumstances and may allow the intercultural communicator to suspend judgment long enough to understand the symbols used by the culture involved.

SUMMARY

The United States is an intercultural community, and five metaphors—melting pots, tributaries, rainbows, tapestries, and garden salads—were introduced to describe its diversity. We suggested that the term *U.S. American* should be used to characterize someone from the United States but noted that a variety of terms are used to refer to the nation's cultural groups. The goal is to find ways to refer to cultural groups that reflect their differences accurately while avoiding negative connotations and evaluations.

This chapter next focused on intercultural communication competence. We began by explaining *intra*cultural communication competence, which was followed by an examination of *inter*cultural competence. Three components of intercultural competence were discussed, including the interpersonal and situational contexts within which the communication occurs; the degree of appropriateness and effectiveness in the interaction; and the importance of knowledge, motivations, and skills.

Two tools were provided to improve intercultural competence. The first is the culture-general Behavioral Assessment Scale for Intercultural Competence (BASIC), which includes the ability to display respect, a recognition that knowledge is personal rather than universal, an empathetic sense about the experiences of others that results in behaviors appropriate to those experiences, the ability to manage interactions with others, skills in enacting appropriate task and relational role behaviors, the capacity to tolerate uncertainty without anxiety, and a nonevaluative posture toward the beliefs and actions of others. Within each culture there will be culturally specific ways of behaving that are used to demonstrate these competencies. The second tool is the ability to distinguish among the techniques of description, interpretation, and evaluation. This tool encourages communicators to describe the sensory information they receive and then to construct alternative evaluations about their perceptions by making correspondingly different interpretations.

FOR DISCUSSION

1. What do you think about using the terms *United Statians, United Staters, North Americans,* or *U.S. Americans* to refer to people in the United States? What alternative phrases might accurately and sensitively be used to refer to people from the United States?
2. What do we lose, and what do we gain, by using general terms such as *Asian American* and *Native American* when referring to cultural groups in the United States?
3. What does it mean to say that communication competence is a social judgment that people make about others? Would people from different cultures likely judge the same kinds of communication behaviors as competent? Why or why not?
4. Why is it impossible to "put yourself in someone else's shoes"?
5. What three BASIC skills would you argue are most important for developing intercultural communication competence?
6. How would you describe your own interaction posture?

FOR FURTHER READING

Gary Althen and Janet Bennett, *American Ways: A Cultural Guide to the United States,* 3rd ed. (Boston: Intercultural Press, 2011). Written as a guide for foreign visitors to the United States., this book also provides insights that U.S. Americans can use to understand their own cultural and communication patterns.

Darla K. Deardorff (ed.), *The Sage Handbook of Intercultural Competence* (Thousand Oaks, CA: Sage, 2009). A compendium of recent theorizing and research on topics related to intercultural communication competence.

Alberto González, Marsha Houston, and Victoria Chen (eds.), *Our Voices: Original Essays in Culture, Ethnicity, and Communication,* 5th ed. (New York: Oxford University Press, 2012). A collection of first-person essays that explore intercultural communication concepts through the personal experiences of people who are living multicultural lives.

Stephen W. Littlejohn and Karen A. Foss (eds.), *Encyclopedia of Communication Theory* (Thousand Oaks, CA: Sage, 2009). A comprehensive state-of-the-art compendium of information about various theories that pertain to intercultural communication.

Sherwyn P. Morreale, Brian H. Spitzberg, and J. Kevin Barge, *Human Communication: Motivation, Knowledge, and Skills,* 2nd ed. (Belmont, CA: Wadsworth, 2007). A basic communication textbook that also provides an intellectual foundation for understanding interpersonal communication competence.

4

Cultural Patterns and Communication: Foundations

KEY TERMS

If you have had even limited contact with people from other cultures, you know that they differ in both obvious and subtle ways. An obvious cultural difference is in the food people eat, such as the ubiquitous hamburger, the U.S. offering to the world's palate. We identify pasta with Italy, stuffed grape leaves with Greece and Turkey, sushi with Japan, curry with India and Southeast Asia, and kimchee with Korea.

Another obvious difference between cultures is the clothing people wear. Walk down the streets near United Nations Plaza in New York City or in diplomatic areas of Washington, D.C., and you will see men wearing colorful African dashikis, women in graceful and flowing Indian saris, and men from Middle Eastern cultures with long robes and

People from all cultures teach their children the norms for proper dress and behavior.

headdresses. Most television cable services now provide an array of shows that are set in various cultures. Watching those shows for even a short period of time can help to create an immediate awareness of some obvious differences between and among cultures.

Other cultural differences are more subtle and become apparent only after more extensive exposure. This chapter and the next are about those subtle, less visible differences that are taken for granted within a culture. In defining culture, we called the effects of these subtle differences *shared interpretations*. Shared interpretations lead to actions that are regarded as appropriate and effective behaviors within a culture. They are therefore very important, and they result from the culture's collective assumptions about what the world is, shared judgments about what it should be, widely held expectations about how people should behave, and predictable behavior patterns that are commonly shared. We are going to call these unseen but shared expectations *cultural patterns*.

It is extremely important that you understand differences in cultural patterns if you wish to develop competence in intercultural communication. Cultural patterns are the basis for interpreting the symbols used in communication. If the cultural patterns between people are sufficiently different, the symbols used in communicating will be interpreted differently and may be misunderstood—unless people are aware that no common set of behaviors is universally interpreted in the same way nor regarded with the same degree of favorability.

DEFINING CULTURAL PATTERNS

Shared beliefs, values, norms, and social practices that are stable over time and that lead to roughly similar behaviors across similar situations are known as cultural patterns. These cultural patterns affect perceptions of competence. Despite their importance in the development and maintenance of cultures, they cannot be seen, heard, or experienced directly. However, the consequences of cultural patterns—shared interpretations that are evident in what people say and do—are readily observable. Cultural patterns are primarily inside people, in their minds. They provide a way of thinking about the world, of orienting oneself to it. Therefore, cultural patterns are shared mental programs that govern specific behavior choices.

Cultural patterns provide the basic set of standards that guide thought and action. Some aspects of this mental programming are, of course, unique to each individual. Even within a culture, no two people are programmed identically, and these distinctive personality differences separate the members of a culture. In comparisons across cultures, some mental programs are essentially universal. A mother's concern for her newborn infant, for example, reflects a biological program that exists across all known cultures and is part of our common human experience.

In addition to those portions of our mental programs that are unique or universally held, there are those that are widely shared only by members of a particular group or culture. These collective programs can be understood only in the context of a particular culture, and they include such areas as the preferred degree of social equality, the importance of group harmony, the degree to which emotional displays are permitted, the value ascribed to assertiveness, and the like.

Cultural patterns are not so much consciously taught as unconsciously experienced as a by-product of day-to-day activities. Most core assumptions are programmed at a very early age and are reinforced continuously. Saudi Arabians, for example, are taught to admire courage, patience, honor, and group harmony. European Americans are trained to admire achievement, practicality, material comfort, freedom, and individuality.

Because of their importance in shaping judgments about intercultural competence, we will discuss cultural patterns in great detail through several approaches. We emphasize both what is similar about all cultural patterns and what is different among them. We begin by describing the basic components of all cultural patterns: beliefs, values, norms, and social practices. We then turn to characteristics of cultural patterns. Chapter 5 presents several scholarly approaches, or taxonomies, to describe the ways in which cultures differ, followed by our synthesis of these cultural taxonomies.

COMPONENTS OF CULTURAL PATTERNS

In Chapter 2, we offered a definition of culture as a learned set of shared interpretations about beliefs, values, norms, and social practices. At that point, however, we left these four key terms undefined. We now explain in some detail the nature of beliefs, values, norms, and social practices, which together constitute the components of cultural patterns.

Beliefs

A belief is an idea that people assume to be true about the world. Beliefs, therefore, are a set of learned interpretations that form the basis for cultural members to decide what is and what is not logical and correct.

Beliefs can range from ideas that are central to a person's sense of self to those that are more peripheral. Central beliefs include the culture's fundamental teachings about what reality is and expectations about how the world works. Less central, but also important, are beliefs based on or derived from the teachings of those regarded as authorities. Parents, teachers, and other important elders transmit the culture's assumptions about the nature of the physical and interpersonal world. Peripheral beliefs refer to matters of personal taste. They contribute to each person's unique configuration of ideas and expectations within the larger cultural matrix.[1]

Discussing culturally shared beliefs is difficult because people are usually not conscious of them. Culturally shared beliefs are so fundamental to assumptions about what the world is like and how the world operates that they are typically unnoticed. We hope you will come to realize through this discussion of cultural beliefs that much of what you consider to be reality may, in fact, not be reality to people from other cultures. What you consider to be the important "givens" about the world, such as the nature of people and their relationships with one another, are based on your culturally shared beliefs, which have been transmitted to and learned by you and are not a description of some invariant, unchanging characteristic of the world.

A well-known example of a widely shared belief dates back to the time when Europeans believed that the earth was flat. That is, people "knew" that the earth was flat. Most people now "know" (believe) that the earth is basically round and would scoff at any suggestion that it is flat. Yet we still talk about Asia as "the East" and about Europe and the United States as "the West," even though California is due east of the major population centers in China.

Another example of a belief for many European Americans is that in "reality" there is a separation between the physical and spiritual worlds. If a teacher one day started kicking the doorsill at the front of the room, the students might begin to worry about the teacher's mental health. The students would probably not be concerned about the doorsill itself, nor would they be alarmed about the spirits who might reside there. Of course, you and they "know" that there are no spirits in doorsills. But people from Thailand and elsewhere "know" that spirits do indeed reside in inanimate objects such as doorsills, which is why doorsills should always be stepped over rather than on. In addition to their concern about the teacher, therefore, people from other cultures might conceivably worry about upsetting the spirits who dwell in the doorsill.

These Peruvian villagers have gathered to honor and pray for those who have died. Prayer is an activity that reflects aspects of one's culture.

Members of the European American culture see humans as separate from nature. Based on this set of beliefs about the world, European Americans have set out to control nature. From the viewpoint of the typical European American, a person who believes, as the typical Indian woman does, that she "catches colds and fevers from evil spirits that lurk in trees"[2] would be seen as strange. European Americans "know" that people do not become ill from spirits that live in trees. Yet, in the Indian culture, people "know" that human illness is caused by such spirits.

Values

Cultures differ not only in their beliefs but also in what they value. Values involve what a culture regards as good or bad, right or wrong, fair or unfair, just or unjust, beautiful or ugly, clean or dirty, valuable or worthless, appropriate or inappropriate, and kind or cruel.[3] Because values are the *desired* characteristics or goals of a culture, a culture's values do not necessarily describe its *actual* behaviors and characteristics. However, values are often offered as the explanation for the way in which people communicate. Thus, as Shalom Schwartz suggests, values serve as guiding principles in people's lives.[4]

From culture to culture, values differ in their valence and intensity. Valence refers to whether the value is seen as positive or negative. Intensity indicates the strength or importance of the value, or the degree to which the culture identifies the value as significant. For example, in some U.S. American cultures, the value of respect for elders is negatively valenced and held with a modest degree of intensity. Many U.S. Americans value youth rather than old age. In Korea, Japan, and Mexico, however, respect for elders is a positively valenced value, and it is very intensely held. It would be possible after studying any particular culture to determine its most important values and each value's valence and intensity.

Norms

Norms are the socially shared expectations of appropriate behaviors. When a person's behaviors violate the culture's norms, social sanctions are usually imposed. Like values, norms can vary within a culture in terms of their importance and intensity. Unlike values, however, norms may change over a period of time, whereas beliefs and values tend to be much more enduring.

Norms exist for a wide variety of behaviors. For example, the greeting behaviors of people within a culture are governed by norms. Similarly, good manners in a variety of situations are based on norms. Norms also exist to guide people's interactions and

CULTURE CONNECTIONS

None of my Afghan relatives was ever alone or ever wanted to be. And that's so different from my life today, here in the West.... Most of the people I know are like this. We need solitude, because when we're alone, we're free from obligations, we don't need to put on a show, and we can hear our own thoughts.

My Afghan relatives achieved this same state by being with one another. Being at home with the group gave them the satisfactions we associate with solitude—ease, comfort, and the freedom to let down one's guard.

—Tamim Ansary ∎

to indicate how to engage in conversation, what to talk about, and how to disengage from conversations. Because people are expected to behave according to their culture's norms, they therefore come to see their own norms as constituting the "right" way of communicating. Norms, then, are linked to the beliefs and values of a culture. Because they are evident through behaviors, norms can be readily inferred.

Social Practices

Social practices are the predictable behavior patterns that members of a culture typically follow. Thus, social practices are the outward manifestations of beliefs, values, and norms. In the United States, lunch is usually over by 1:30 p.m., gifts brought by dinner guests are usually opened in the presence of the guests, television watching dramatically increases during the annual Super Bowl, and children sleep alone or with other children. In Italy, lunch hasn't even begun by 1:30 p.m., and soccer is more popular than American football. In Malaysia, gifts are never opened in front of the giver; doing so is considered bad manners. In many Middle Eastern, Latin American, and Asian families, children routinely share beds with adult relatives.[5]

One type of social practice is informal and includes everyday tasks such as eating, sleeping, dressing, working, playing, and talking to others. Such behaviors are so predictable and commonplace within a culture that the subtle details about how they are accomplished may pass nearly unnoticed. For instance, cultures have social practices about eating with "good manners." Slurping one's food in Saudi Arabia and in many Asian cultures is the usual practice, and it is regarded favorably as an expression of satisfaction and appreciation for the quality of the cooking. But good manners in one culture may be bad manners in another; European Americans typically consider such sounds to be inappropriate.[6]

Another type of social practice is more formal and prescriptive. These include the rituals, ceremonies, and structured routines that are typically performed publicly and collectively: saluting the flag, praying in church, honoring the dead at funerals, getting married, and many other social practices. Of course, all members of a culture do not necessarily follow that culture's "typical" social practices; each person differs, in unique and significant ways, from the general cultural tendency to think and behave in particular ways. As William B. Gudykunst and Carmin M. Lee suggest, "Individuals in a culture generally are socialized in ways consistent with the cultural-level tendencies, but some individuals in every culture learn different tendencies."[7]

CHARACTERISTICS OF CULTURAL PATTERNS

In this section, we describe a set of similarities underlying all cultural patterns. In so doing, we draw heavily on the work of Kluckhohn and Strodtbeck and their theory of value orientations. Next, we elaborate on those ideas to provide a general overview of cultural patterns.

The Functions of Cultural Patterns

Florence Kluckhohn and Fred Strodtbeck wanted to make sense of the work of cultural anthropologists who, for many years, had described systematic variations both between and within cultures. That is, cultures clearly differed from one another, but within every culture there were individuals who varied from the cultural patterns most often associated with it.[8] To explain both these cultural-level and individual-level differences, Kluckhohn and Strodtbeck offered four conclusions about the functions of cultural patterns that apply to all cultures:

1. People in all cultures face common human problems for which they must find solutions.
2. The range of alternative solutions to a culture's problems is limited.
3. Within a given culture, there will be preferred solutions, which most people within the culture will select, but there will also be people who will choose other solutions.
4. Over time, the preferred solutions shape the culture's basic assumptions about beliefs, values, norms, and social practices—the cultural patterns.

The first conclusion, that all cultures face similar problems, is not just about everyday concerns such as "Do I have enough money to get through the month?" or "Will my parent overcome a serious illness?" Rather, the cultural problems involve three types of difficulties that affect the entire culture and therefore require widely shared solutions if the culture is to thrive. These problems are those of internal integration (how to form and maintain human social relationships), external adaptation (how to cope with the environment and survive), and issues that arise from the passage of time.[9] Kluckhohn and Strodtbeck describe five problems or orientations that each culture must address:

1. What is the human orientation to activity?
2. What is the relationship of humans to each other?
3. What is the nature of human beings?
4. What is the relationship of humans to the natural world?
5. What is the orientation of humans to time?

CULTURE CONNECTIONS

We know that the white man does not understand our ways. One portion of the land is the same to him as the next, for he is a stranger who comes in the night and takes whatever he needs from the land. The earth is not his brother, but his enemy, and when he has conquered it, he moves on. He leaves his fathers' graves behind him, and he does not care. He kidnaps the earth from his children. He does not care. His father's graves and his children's birthright are forgotten. He treats his mother, the earth, and his brother, the sky, as things to be bought, plundered, sold like sheep or bright beads. His appetite will devour the earth and leave behind only a desert.

You must teach your children that the ground beneath their feet is the ashes of our grandfathers. So that they will respect the land, tell your children that the earth is rich with the lives of our kin. Teach your children what we have taught our children, that the earth is our mother. Whatever befalls the earth, befalls the sons of the earth. If men spit upon the ground they spit upon themselves.

This we know. The earth does not belong to man; man belongs to the earth. This we know. All things are connected like the blood which unites one family. All things are connected.

Whatever befalls the earth befalls the sons of the earth. Man did not weave the web of life; he is merely a strand in it. Whatever he does to the web, he does to himself.

—Chief Seattle, 1854 ■

Each culture, in its own unique way, must provide answers to these questions in order to develop a coherent and consistent interpretation of the world. We will return to these questions, in modified form, in our discussion of cultural patterns.

Kluckhohn and Strodtbeck's second conclusion is that a culture's possible responses to these universal human problems are limited, as cultures must select their solutions from a range of available alternatives. Thus, a culture's orientation to the importance and value of activity can range from passive acceptance of the world (a "being" orientation), a preference for a gradual transformation of the human condition (a "being-in-becoming" orientation), or more direct intervention (a "doing" orientation). A culture's solution to how it should organize itself to deal with interpersonal relationships can vary along a continuum from hierarchical social organization ("linearity") to group identification ("collaterality" or collectivism) to individual autonomy ("individualism"). The available alternatives to the problem "What is the nature of human beings?" can range from "Humans are evil" to "Humans are a mixture of good and evil" to "Humans are good." A culture's response to the preferred relationship of humans to the natural world can range from a belief that "People are subjugated by nature" to "People live in harmony with nature" to "People master nature." Finally, the culture's preferred time orientation can emphasize events and experiences from the past, the present, or the future. Table 4.1 summarizes the Kluckhohn and Strodtbeck value orientation theory.

Kluckhohn and Strodtbeck's third conclusion is their answer to an apparent contradiction that scholars found when studying cultures. They argued that, within any culture, a preferred set of solutions will be chosen by most people. However, not all people from a culture will make exactly the same set of choices, and, in fact, some people from each culture will select other alternatives. For example, most people who are part of European American culture have a "doing" orientation, a veneration for the future, a belief in control over nature, a preference for individualism, and a belief that people are basically good and changeable. But clearly not everyone identified with the European American culture shares all of these beliefs.

The fourth conclusion by Kluckhohn and Strodtbeck explains how cultural patterns develop and are sustained. A problem that is regularly solved in a similar way creates an underlying premise or expectation about the preferred or appropriate way to accomplish a specific goal. Such preferences, chosen unconsciously, implicitly define the shared meanings of the culture. Over time, certain behaviors to solve particular problems become preferred, others permitted, and still others prohibited.

TABLE 4.1

KLUCKHOHN AND STRODTBECK'S VALUE ORIENTATIONS

Orientation	Postulated Range of Variations		
Activity	Being	Being-in-becoming	Doing
Relationships	Linearity	Collaterality	Individualism
Human nature	Evil	Mixture of good and evil	Good
People-nature	Subjugation to nature	Harmony with nature	Mastery over nature
Time	Past	Present	Future

Source: Adapted from Florence R. Kluckhohn and Fred Strodtbeck, *Variations in Value Orientations* (Evanston, IL: Row, Peterson, 1960).

Kluckhohn and Strodtbeck's ideas have been very influential among intercultural communication scholars, and they form the foundation for our understanding of cultural patterns. In the following section, we extend their work to explain, in a general way, the variations in beliefs, values, norms, and social practices that are typically associated with cultural patterns. Chapter 5 extends this overview to focus on specific conceptual taxonomies that can be used to understand cultural differences.

An Overview of Cultural Patterns

Members of a culture generally have a preferred set of responses to the world. Imagine that, for each experience, there is a range of possible responses from which a culture selects its preferred response. In this section we extend the thoughts of Kluckhohn and Strodtbeck as we describe these alternative responses.[10] In so doing, we will compare and contrast the cultural patterns of different cultural groups and suggest their implications for the process of interpersonal communication. Comparing the patterns of different cultures can sometimes be tricky because a feature of one culture, when compared with another culture, may appear very different than it would when compared with a third culture. Kluckhohn and Strodtbeck's cultural orientations are especially useful because they describe a broad range of cultural patterns against which a particular culture can be understood.

The five major elements in Kluckhohn and Strodtbeck's description of cultural patterns address the manner in which a culture orients itself to activities, social relations, the self, the world, and the passage of time. Note that there are strong linkages among the various elements. As you read the descriptions in the sections that follow, try to recognize the preferred patterns of your culture. Also, focus on your own beliefs, values, norms, and social practices, as they may differ in certain respects from your culture's predominant pattern.

ACTIVITY ORIENTATION   An activity orientation defines how the people of a culture view human actions and the expression of self through activities. This orientation provides answers to questions such as the following:

- Is it important to be engaged in activities in order to be a "good" member of one's culture?
- Can and should people change the circumstances of their lives?
- Is work very different from play?
- Which is more important, work or play?
- Is life a series of problems to be solved or simply a collection of events to be experienced?

To define their activity orientation, cultures usually choose a point on the being–becoming–doing continuum. "Being" is an activity orientation that values inaction and an acceptance of the status quo. African American and Greek cultures are usually regarded as "being" cultures. Another characterization of this orientation is a belief that all events are determined by fate and are therefore inevitable. Hindus from India often espouse this view.

A "becoming" orientation sees humans as evolving and changing; people with this orientation, including Native Americans and most South Americans, are predisposed to think of ways to change themselves as a means of changing the world.

"Doing" is the dominant characteristic of European Americans, who rarely question the assumption that it is important to get things done. Thus, European Americans often ask, "What do you do?" When they first meet someone, a common greeting is "Hi! How are you doing?" and Monday morning conversations between coworkers often center on what each person "did" over the weekend. Similarly, young children are asked what they want to be when they grow up (though what is actually meant is "What do you want to do when you grow up?"), and cultural heroes are those who do things. The "doing" culture is often the striving culture, in which people seek to change and control what is happening to them. The common adage "Where there's a will there's a way" captures the essence of this cultural pattern. When faced with adversity, for example, European Americans encourage one another to fight on, to work hard, and not to give up.

How a person measures success is also related to the activity orientation. In cultures with a "doing" orientation, activity is evaluated by scrutinizing a tangible product or by evaluating some observable action directed at others. In other words, activity should have a purpose or a goal. In the "being" and "becoming" cultures, activity is not necessarily connected to external products or actions; the contemplative monk or the great thinker is most valued. Thus the process of striving toward the goal is sometimes far more important than accomplishing it.

In "doing" cultures, work is seen as a separate activity from play and an end in itself. In the "being" and "becoming" cultures, work is a means to an end, and there is no clear-cut separation between work and play. For these individuals, social life spills over into their work life. When members of a "being" culture work in the environment of a "doing" culture, their behavior is often misinterpreted. A Latina employee described her conversation with a European American coworker who expressed anger that she spent so much "work" time on the telephone with family and friends. For the Latina, it was important to keep in contact with her friends and family; for the European American, only work was done at work, and one's social and personal relationships were totally separated from the working environment. In a "doing" culture, employees who spend too much time chatting with their fellow employees may be reprimanded by a supervisor. In the "being" and "becoming" cultures, those in charge fully expect their employees to mix working and socializing. Along with the activity orientation of "doing" comes a problem–solution orientation. The preferred way of dealing with a difficulty is to see it as a challenge to be met or a problem to be solved. The world is viewed as something that ought to be changed in order to solve problems, rather than as something that ought to be accepted as it is, with whatever characteristics it has.

CULTURE CONNECTIONS

"Do you see those trees there?" he asks, gesturing across the pond toward a gathering of large, tall deciduous trees. They look like ficus trees, with large green leaves and knobby, wise-looking trunks. "Pipal trees," he says. "It is believed that the god Brahma resides in the roots, the god Vishnu in the trunk, and the god Shiva in the crown. People believe that the god Krishna breathed his last underneath a pipal, and that the tree is invested with a sacred thread. In the Konkan, where your Jews originally resided, I have read that people who are wishing for a particular outcome used to worship the tree, and walk around its base several times a day. There is also a belief there that the spirits of the dead are reborn in their descendants, especially if they have wishes that remain unfulfilled. Sometimes if a child resembles a relative who has died it is believed that the ancestor has returned to the family in the form of the child. You are seeking a ghost, aren't you? Perhaps that's why you like to come here."

—Sadia Shepard ∎

Cultures with a "being" orientation value contemplative behaviors. In Thailand and Cambodia, all men are encouraged to serve as Buddhist monks, at least for a short period of time. These Cambodian monks are talking to a local woman.

In every culture, these preferences for particular orientations to activities shape the interpersonal communication patterns that will occur. In "doing" cultures, interpersonal communication is characterized by concerns about what people do and how they solve problems. There are expectations that people should be involved in activities, that work comes before play, and that people should sacrifice in other parts of their lives in order to meet their work responsibilities. In "being" cultures, interpersonal communication is characterized by being together rather than by accomplishing specific tasks, and there is generally greater balance between work and play. Figure 4.1 summarizes the alternative cultural orientations to activities.

1. How do people define activity?

    doing ———————————— becoming ———————————— being

    striving ——————————————————————————————— fatalistic

    compulsive ——————————————————————————— easygoing

2. How do people evaluate activity?

    techniques ——————————————————————————————— goals

    procedures ——————————————————————————————— ideals

3. How do people regard and handle work?

    an end in itself ——————————————————— a means to other ends

    separate from play ——————————————————— integrated with play

    a challenge ———————————————————————————— a burden

    problem solving ——————————————————— coping with situations

FIGURE 4.1

Activity orientations.

SOCIAL RELATIONS ORIENTATION   The social relations orientation describes how the people in a culture organize themselves and relate to one another. This orientation provides answers to questions such as the following:

- To what extent are some people in the culture considered better or superior to others?
- Can social superiority be obtained through birth, age, good deeds, or material achievement and success?
- Are formal, ritualized interaction sequences expected?
- In what ways does the culture's language require people to make social distinctions?
- What responsibilities and obligations do people have to their extended families, their neighbors, their employers or employees, and others?

A social relations orientation can range from one that emphasizes differences and social hierarchy to one that strives for equality and the absence of hierarchy. Many European Americans, for example, emphasize equality and evenness in their interpersonal relationships, even though certain groups have been treated in discriminatory and unequal ways. Equality as a value and belief is frequently expressed and is called on to justify people's actions. The phrase "We are all human, aren't we?" captures the essence of this cultural tenet. From within this cultural framework, distinctions based on age, gender, role, or occupation are discouraged. Conversely, other cultures, such as the Korean, emphasize status differences between individuals. Mexican American culture, drawing on its cultural roots in traditional Mexican values, also celebrates status differences and formalizes different ways of communicating with people depending on who they are and what their social characteristics happen to be.

One noticeable difference in social relations orientations is in the degree of importance a culture places on formality. In cultures that emphasize formality, people address others by appropriate titles, and highly prescriptive rules govern the interaction. Conversely, in cultures that stress equality, people believe that human relationships develop best when those involved can be informal with one another. Students from other cultures who study in the United States are usually taken aback by the seeming informality that exists between teachers and students. Many professors allow, even ask, students to call them by their first names, and students disagree with and challenge their teachers in front of the class. The quickness with which interpersonal relationships in the United States move to a first-name basis is mystifying to those from cultures where the personal form of address is used only for selected, special individuals. Many U.S. Americans who share aspects of both European American culture and another culture also express difficulty with this aspect of cultural behavior.

In cultures such as those of Japan, Korea, and China, individuals identify with only a few distinct groups, and the ties that bind people to these groups are so strong that group membership may endure for a lifetime. Examples of these relationships include nuclear and extended families, friends, neighbors, work groups, and social organizations. In contrast, European Americans typically belong to many groups throughout their lifetimes. Although the groups may be very important for a period of time, they are often discarded when they are no longer needed. That is, voluntary and informal groups are meant to be important for brief periods of time, but they typically serve a transitory purpose. In addition, it is accepted and even expected that European Americans often change jobs and companies. "Best friends" may only be best friends for brief periods.

Another important way in which social relations orientations can vary is how people define their social roles or their place in a culture. In some cultures, the family and the position into which a person is born determine a person's place. At the other extreme are cultures in which all people, regardless of family position, can achieve success and high status. Among African Americans and European Americans, for instance, there is a widespread belief that social and economic class should not predetermine a person's opportunities and choices. For example, consider the tale of Abraham Lincoln, a poor boy who went from a log cabin to the White House; or the books of Horatio Alger, the nineteenth-century author who wrote numerous rags-to-riches stories of success and happiness that were achieved through hard work and perseverance; or the Sylvester Stallone hero in the movie *Rocky*, who went from journeyman boxer to heavyweight champion of the world; or *Working Girl* Tess McGill, the Melanie Griffith character who went from her entry-level job in the typing pool to a senior executive; or the saga of washed-up boxer Jim Braddock, played by Russell Crow in the movie *Cinderella Man*, who became a champion in the 1930s; or the heartwarming and true story of Chris Gardner, played by Will Smith in *The Pursuit of Happyness*, who overcame poverty and hardships to become a successful stockbroker and raise his son. In each of these examples, there is a common belief that people should not be restricted by the circumstances of their birth.

Some cultures believe that humans both can and should control the powerful forces of nature. Hoover Dam in Nevada represents one example of this worldview.

Cultural patterns can also prescribe appropriate behaviors for men and women. In some cultures, very specific behaviors are expected; other cultures allow more ambiguity in the expected roles of women and men.

A culture's social relations orientation affects the style of interpersonal communication that is most preferred. Cultures may emphasize indirectness, obliqueness, and ambiguity, which is the typical pattern for most Eastern European cultures and Mexican Americans, or they may emphasize directness and confrontation, which is the typical European American pattern.

The European American preference for "putting your cards on the table" and "telling it like it is" presupposes a world in which it is desirable to be explicit, direct, and specific about personal reactions and ideas, even at the expense of social discomfort on the part of the person with whom one is interacting. For European Americans, good interpersonal communication skills include stating directly one's personal needs and reactions to the behaviors of others. Thus, if European Americans hear that others have complained about them, they would probably ask, "Why didn't they tell me directly if they have a problem with something that I have done?"

Contrast this approach to that of Asian cultures such as those in Japan, Korea, Thailand, and China, where saving face and maintaining interpersonal harmony are so highly valued that it would be catastrophic to confront another person directly and verbally express anger. The same values are usually preferred in India[11] and in many Eastern European cultures, where saying "no" might be regarded as offensive. A U.S. American scholar working in Hungary tells of asking his Hungarian colleague if he could borrow a particular book. Instead of saying "no," the Hungarian repeatedly provided other reasons why he couldn't loan the book at that specific moment: he was using it just now; his mother was sick; he needed the book to complete an essay before his mother died. The Hungarian's strong sense of connectedness to family and friends meant that it would not be polite to say "no" directly.[12]

The tendency to be verbally explicit in face-to-face interactions is related to a preference for direct interaction rather than interaction through intermediaries. Among European Americans there is a belief that, ideally, people should depend only on themselves to accomplish what needs to be done. Therefore, the notion of using intermediaries to accomplish either personal or professional business goals is not widely accepted.

Although African Americans prefer indirectness and ambiguity in conversations with fellow cultural members, they do not choose to use intermediaries in these conversations. In many cultures, however, the use of intermediaries is the preferred method of conducting business or passing on information.[13] Marriages are arranged, business deals are made, homes are purchased, and other major negotiations are all conducted through third parties. These third parties soften and interpret the messages of both sides, thereby shielding the parties from direct, and therefore risky and potentially embarrassing, transactions with each other. Similarly, among many cultures from southern Africa, such as Swaziland, there is a distinct preference for the use of intermediaries to deal with negotiations and conflict situations. Consider the experience of the director of an English program in Tunisia, a culture that depends on intermediaries. One of the Tunisian teachers had been consistently late to his morning classes. Rather than calling the teacher in and directly explaining the problem, the director asked the teacher's friend about the teacher's health and happiness. The director indicated that the teacher's late arrival for class might have been a sign that something was wrong. The friend then simply indirectly conveyed the director's concern to the late teacher, who was late no more.

A culture's social relations orientation also affects the sense of social reciprocity—that is, the underlying sense of obligation and responsibility between people. Some cultures prefer independence and a minimum number of obligations and responsibilities; alternatively, other cultures accept obligations and encourage dependence. The nature of the dependence is often related to the types of status and the degree of formality that exist between the individuals. Cultures that depend on hierarchy and formality to guide their social interactions are also likely to have both a formal means for fulfilling social obligations and clearly defined norms for expressing them. Figure 4.2 summarizes the alternative cultural orientations to social relations.

SELF-ORIENTATION Self-orientation describes how people's identities are formed, whether the culture views the self as changeable, what motivates individual actions, and the kinds of people who are valued and respected. A culture's self-orientation provides answers to questions such as the following:

- Do people believe they have their own unique identities that separate them from others?
- Does the self reside in the individual or in the groups to which the individual belongs?
- What responsibilities does the individual have to others?
- What motivates people to behave as they do?
- Is it possible to respect a person who is judged "bad" in one part of life but is successful in another part of life?

For most European Americans, the emphasis on the individual self is so strong and so pervasive that it is almost impossible for them to comprehend a different point of view. Thus, many European Americans believe that the self is located solely within the individual, and the individual is definitely separate from others. From a very young

1. How do people relate to others?

as equals ——————————————————————————————————— hierarchical

informal ——————————————————————————————————————— formal

member of many groups ————————————————————— member of few groups

weak group identification ————————————————— strong group identification

2. How are roles defined and allocated?

achieved ————————————————————————————————————— ascribed

gender roles similar ————————————————————— gender roles distinct

3. How do people communicate with others?

directly ——————————————————————————————————— indirectly

no intermediaries ————————————————————————— intermediaries

4. What is the basis of social reciprocity?

independence ——————————— interdependence ——————————— dependence

autonomy ——————————————————————————————————— obligation

FIGURE 4.2
Social relations orientations.

age, children are encouraged to make their own decisions. Alternatively, cultures may define who people are only through their associations with others because an individual's self-definition may not be separate from that of the larger group. Consequently, there is a heightened sense of interdependence, and what happens to the group (family, work group, or social group) happens to the person. For example, Mary Jane Collier, Sidney Ribeau, and Michael Hecht found that Mexican Americans "place a great deal of emphasis on affiliation and relational solidarity."[14] The sense of being bonded or connected to others is very important to members of this cultural group. Vietnamese Americans have a similarly strong affiliation with their families.

The significance to intercultural communication of a culture's preferences for defining the self is evident in the statement of a Latina student describing her friendship with a second-generation Italian American woman, whose family has also maintained "traditional values."

> I think we are able to communicate so well because our cultural backgrounds are very similar. I have always been family-oriented and so has she. This not only allowed us to get along, but it allowed us to bring our families into our friendship. [For instance] a rule that the two of us had to live by up to this point has been that no matter how old we may get, as long as we are living at home we must ask our parents for permission to go out.

Related to self-orientation is the culture's view of whether people are changeable. Naturally, if a culture believes that people can change, it is likely to expect that human beings will strive to be "better," as the culture defines and describes what "better" means.

The source of motivation for human behavior is also part of a culture's self-orientation. Among African Americans and European Americans, individuals are motivated

CULTURE CONNECTIONS

"Some men in safari suits came one day and grabbed the chest. Nobody was sure who they were or where they were planning to take the puppets. The abbot charged with their safe-keeping was shown a government directive that the chest was to be moved for security reasons. When the abbot asked for details, they told him it was all confidential. There wasn't much he could do about it.

"And that's how the chest ended up in the archive department of the Ministry and why all hell broke loose. You see, the chest can't be opened by just anyone whenever they feel like it. The spirits of the puppets are incredibly powerful and amazingly temperamental. They were already—"

"How can puppets have spirits?" Civilai interrupted.

"What?"

"Puppets aren't people, and they aren't dead. So how—?"

"Ah, but the puppets are made of balsa, and before the wood to carve them is cut from the tree, the puppet-maker has to get permission from the tree spirits. The balsa is a gentle wood and spirits are plentiful in it. When they learn that the wood is going to be made into the image of a person, it's awfully tempting for the more nostalgic spirits to jump ship and settle in the form of the puppet. It's as if they've returned to their lost host.

—Colin Cotterill ■

to achieve external success in the form of possessions, positions, and power. Self-orientation combines with the "doing" orientation to create a set of beliefs and values that place individuals in total control of their own fate. Individuals must set their own goals and identify the means necessary to achieve them. Consequently, failure is viewed as a lack of willpower and a disinclination to give the fullest individual effort. In this cultural framework, individuals regard it as necessary to rely on themselves rather than on others.

Another distinguishing feature of the cultural definition of self is whether the members of the culture believe that people are inherently bad, good, or some combination of these two. The Chinese, for example, believe people are inherently good, and they must therefore be protected from exposure to corrupting influences. Conversely, other cultures are influenced by religious tenets that regard humans as intrinsically bad. A related issue is whether the culture emphasizes duties or rights. One culture that expects its members to act because it is their duty to do so is the Japanese. In contrast, for European Americans, the concept of duties and obligations to others is not as powerful a motivator.

An additional part of self-orientation is the set of characteristics of those individuals who are valued and cherished. Cultures vary in their allegiance to the old or to the young, for example. Many cultures venerate their elders and view them as a source of wisdom and valuable life experience. Individuals in these cultures base decisions on the preferences and desires of their elders. Many Asian and Asian American cultures illustrate this preference. The value on youth typifies the European American culture, in which innovation and new ideas, rather than the wisdom of the past, are regarded as important. European Americans venerate the upstart, the innovator, and the person who tries something new. Figure 4.3 summarizes the alternative cultural orientations to the self.

1. How should people form their identities?
   by themselves ———————————————————————————— with others
2. How changeable is the self?
   changeable ———————————————————————————— unchangeable
   self-realization stressed ———————————————— self-realization not stressed
3. What is the source of motivation for the self?
   reliance on self ———————————————————————— reliance on others
   rights ———————————————————————————————— duties
4. What kind of person is valued and respected?
   young ———————————————————————————————— aged
   vigorous ——————————————————————————————— wise
   innovative —————————————————————————————— prominent
   material attributes ———————————————————— spiritual attributes

FIGURE 4.3
Self-orientations.

WORLD ORIENTATION   Cultural patterns also tell people how to locate themselves in relation to the spiritual world, nature, and other living things. A world orientation provides answers to questions such as the following:

- Are human beings intrinsically good or evil?
- Are humans different from other animals and plants?
- Are people in control of, subjugated by, or living in harmony with the forces of nature?
- Do spirits of the dead inhabit and affect the human world?

In the African and African American worldview, human beings live in an interactive state with the natural and spiritual worlds. Daniel and Smitherman describe a fundamental tenet of the traditional African worldview as that of "a dynamic, hierarchical unity between God, man, and nature, with God serving as the head of the hierarchy."[15] In this view of the relationship between the spiritual and material worlds, humans are an integral part of nature. Thus, in the African and African American worldview, "One becomes a 'living witness' when he aligns himself with the forces of nature and instead of being a proselytized 'true believer' strives to live in harmony with the universe."[16] Native American groups, as well, clearly have a view of humans as living in harmony with nature.[17] Latino culture places a great value on spirituality but views humans as being subjugated to nature, with little power to control circumstances that influence their lives.[18] Asian Indians also have a worldview that humans are subjugated to nature.[19]

Most European Americans view humans as separate and distinct from nature and other forms of life. Because of the supremacy of the individual and the presumed uniqueness of each person, most European Americans regard nature as something to be manipulated and controlled in order to make human life better. Excellent examples of this cultural belief can be found in news reports whenever a natural disaster occurs in the United States. For instance, when Hurricane Katrina hit New Orleans and elsewhere along the Gulf Coast in August 2005 and nearly two thousand individuals died in the subsequent flooding, people were outraged that the flood protection and levee systems could be so unsafe. The assumption in these pronouncements was that the consequences of natural forces such as hurricanes could have been prevented simply by using better technology and by reinforcing the levees and other structures to withstand the forces of nature. Similarly, in August 2007, within a day of the collapse of a bridge that spanned the Mississippi River in Minneapolis, syndicated television news stations were broadcasting a headline that asked, "Deadly Bridge Collapse: Who's to Blame?"

The position that humans are separate and distinct from nature is also associated with a belief that disease, poverty, and adversity can be overcome to achieve health and wealth. In this cultural framework, the "natural" part of the human experience— illness, loss, even death—can be overcome, or at least postponed, by selecting the right courses of action and having the right kinds of attitudes.

The spiritual and physical worlds can be viewed as distinct or as one. Among European Americans there is generally a clear understanding that the physical world, of which humans are a part, is separate from the spiritual world. If people believe in a spiritual world, it exists apart from the everyday places where people live, work, and play. Individuals who say they are psychic or who are mind readers are viewed with suspicion and curiosity. Those who have seen ghosts are questioned in an effort to find a more "logical" and "rational" explanation. In other cultural frameworks, however, it is "logical" and "rational" for spirits to live in both animate and inanimate objects.[20] Alternatives in cultural orientations to the world are summarized in Figure 4.4.

TIME ORIENTATION   The final aspect of cultural patterns concerns how people conceptualize time. Time orientation provides answers to questions such as the following:

- How should time be valued and understood?
- Is time a scarce resource, or is it unlimited?
- Is the desirable pace of life fast or slow?
- Is time linear or cyclical?

1. What is the nature of humans in relation to the world?

   separate from nature ———————————————— integral part of nature

   humans modify nature ———————————————— humans adapt to nature

   health natural ———————————————— disease natural

   wealth expected ———————————————— poverty expected

2. What is the world like?

   spiritual-physical dichotomy ———————————————— spiritual-physical unity

   empirically understood ———————————————— magically understood

   technically controlled ———————————————— spiritually controlled

FIGURE 4.4
World orientations.

Some cultures choose to describe the future as most important, others emphasize the present, and still others emphasize the past. In Japanese and Chinese cultures, the anniversary of the death of a loved one is celebrated, illustrating the value these cultures place on the past. In contrast, Native Americans and Latinos are present-oriented. European Americans, of course, are future-oriented.

Most European Americans view time as a scarce and valuable commodity akin to money or other economic investments. They strive to "save time," "make time," "spend time," and "gain time." Events during a day are dictated by a schedule of activities, precisely defined and differentiated. Most cultures in Latin America bring an entirely different orientation to time, responding to individuals and circumstances rather than following a scheduled plan for the day. Similarly, Romanians also do not define punctuality as precisely as European Americans do. Thus, time is viewed within these cultural frames as endless and ongoing.

A culture's time orientation also suggests the pace of life. The fast, hectic pace of European Americans, governed by clocks, appointments, and schedules, has become so commonly accepted that it is almost a cliché. The pace of life in cultures such as India, Kenya, and Argentina and among African Americans is less hectic, more relaxed, and more comfortably paced. In African American culture, for example, orientations to time are driven less by a need to "get things done" and conform to external demands than by a sense of participation in events that create their own rhythm. As Jack Daniel and Geneva Smitherman suggest about time in African American culture,

> Being on time has to do with participating in the fulfillment of an activity that is vital to the sustenance of a basic rhythm, rather than with appearing on the scene at, say, "twelve o'clock sharp." The key is not to be "on time" but "in time."[21]

Alternatives in cultural orientations to time are summarized in Figure 4.5. In Chapter 8, the discussion of nonverbal communication codes also considers the influence of a culture's orientation toward time on aspects of communication.

A culture's underlying patterns consist of orientations to activity, social relations, the self, the world, and time. The interdependence among these aspects of culture is obvious from the preceding discussion. Kluckhohn and Strodtbeck provide a way to understand, rather than to judge, different cultural predispositions, and it demonstrates that there are different ways of defining the "real," "good," and "correct" ways to behave.

---

1. How do people define time?

   future ——————————— present ——————————— past

   precisely measurable ————————————————— undifferentiated

   linear ——————————————————————— cyclical

2. How do people value time?

   scarce resource ——————————————— unlimited

   fast pace ——————————————————— slow pace

FIGURE 4.5

Time orientations.

CULTURAL PATTERNS AND INTERCULTURAL COMPETENCE

There is a strong relationship between the foundations of cultural patterns and intercultural competence. Remember that intercultural competence depends on knowledge, motivation, and skills, which occur in specific contexts with messages that are both appropriate and effective.

The patterns of a culture create the filter through which all verbal and nonverbal symbols are interpreted. Because all cultures have distinct beliefs, values, norms, and social practices, symbols do not have universal interpretations, nor will the interpretations have the same degree of favorability. Judgments of competence are strongly influenced by the underlying patterns of a person's cultural background. In every intercultural interaction, a cultural pattern that is different from one's own may be used to interpret one's messages. Every intercultural interaction, then, can be viewed as a puzzle or a mystery that needs to be solved.

How individuals define the relational context is always related to the mental programming that cultures provide. One person's definition of the relationship (e.g., friend) may not match that of the person with whom he or she interacts (e.g., fellow student), causing radically different expectations and interpretations of behaviors.

In solving the intercultural puzzle, it is critical to remember another valuable insight from Kluckhohn and Strodtbeck's foundational work. Although a culture (the collectivity of people) will make preferred choices about beliefs, values, norms, and social practices, not all cultural members will necessarily share all of those preferred choices, nor will they share them with the same degree of intensity. The immediate consequence of this conclusion for the development of intercultural competence is that every person represents the cultural group with which he or she identifies, but to a greater or lesser degree. A cultural pattern may be the preferred

In this wedding ceremony, Christians express their relationship to the spiritual world.

choice of most cultural members, but what can accurately be described for the culture in general cannot necessarily be assumed to be true for a specific individual. In simple terms, this principle translates into an important guideline for the development of intercultural competence: even though you may have culture-specific information, you can never assume that every person from that culture matches the profile of the typical cultural member.

Because cultural patterns describe what people perceive as their reality, what they view as desirable, how they should behave, and what they typically do, there are repercussions for the motivational component of intercultural competence. Recall that communicators' feelings in intercultural interactions affect their ability to be open to alternative interpretations. Yet if cultural patterns predispose people to a particular definition of what is real, good, and right, reactions to others as unreal, bad, or wrong may create psychological distance between interactants. When confronting a set of beliefs, values, norms, and social practices that are inconsistent with their own, many people will evaluate them negatively. Strong emotional reactions are often predictable; after all, these variations from other cultures challenge the basic view people have of their world. Nevertheless, other cultures' ways of believing and their preferred values are not crazy or wrong, just different.

Cultural patterns form the basis for what is considered to be communicatively appropriate and effective. Examples that illustrate how beliefs, values, norms, and social

CULTURE CONNECTIONS

"Meanwhile, a Korean ambulance arrives but at the same time the girl's father erupts on the scene. He shoves Jill out of the way, sees his daughter on the ground and then, realizing she's still breathing—barely—he lifts her up and starts to carry her home. The Korean paramedics stand by and do nothing. Jill's shocked. She's sure the girl's suffering from internal bleeding, and she knows enough about first aid to know that in order to save her life the girl has to be taken to an emergency room immediately, if not sooner."

"When nobody acts, she does. Jill grabs the father and holds him, screaming and pointing to the ambulance. The father won't hear of it. Why, no one knows."

I did. Or at least I thought I did. I explained it to Sergeant Bernewright. And to Ernie. In Korean tradition, it is believed that if someone dies away from home their spirit, when it rises and leaves the body, will become disoriented. It will become lost and then, being away from home, away from the shrine set up by its family, the spirit will become a wandering ghost. Without the proper ceremonies, without offerings of incense and food, without the prayers of the people who loved the spirit in life, it will never be able to make the transition from wandering ghost to revered ancestor. So Chon Un-suk's father's reaction was rational from his point of view. He didn't want his daughter to be hauled away by strangers to die alone in some emergency room. He wanted to take care of her. He wanted to make sure she died at home, not on the street where she'd be lost and would wander alone for eternity—with no one to burn incense at her shrine, no one to pray at her gravesite, and no one to make offerings of food and drink to ease her sojourn through the underworld.

A hungry ghost, the Koreans call such a creature. A spirit whom no one remembers. A spirit who can't find its way home.

—Martin Limón ■

practices set the boundaries of appropriateness and effectiveness include such instances as speaking your mind, in contrast to being quiet; defending yourself against a criticism, in contrast to accepting it; confronting another person about a problem rather than indirectly letting her or him know of the concern; and emphasizing the differences in status in relationships, in contrast to emphasizing commonality. Simply knowing what is appropriate and what has worked to accomplish your personal objectives in your own culture may not, and in all likelihood will not, have similar results when you interact with culturally different others. Intercultural competence usually requires alternative choices for actions. Consequently, we recommend that, before acting, you should contemplate, draw on your culture-specific knowledge, and make behavioral choices that are appropriate for interacting with members of the other culture.

The patterns of a culture shape, but do not determine, the mental programming of its people. Because cultural patterns define how people see and define reality, they are a powerful emotional force in competent intercultural interaction.

SUMMARY

This chapter began the discussion about cultural patterns, which are invisible differences that characterize cultures. Beliefs, values, norms, and social practices are the ingredients of cultural patterns. Beliefs are ideas that people assume to be true about the world. Values are the desired characteristics of a culture. Norms are socially shared expectations of appropriate behaviors. Social practices, the final component of cultural patterns, are the predictable behavior patterns that people typically follow.

Cultural patterns are shared among a group of people, and they form the foundation for the maintenance of cultures. They are stable over relatively long periods of time, and they lead most members of a culture to behave in roughly similar ways when they encounter similar situations.

Florence Kluckhohn and Fred Strodtbeck suggest that each culture selects a preferred set of choices to address common human issues. While not all people in the culture make exactly the same choices, the preferred solutions define the shared meanings of the culture.

Cultural patterns focus on the way cultures orient themselves to activities, social relations, the self, the world, and time. The activity orientation defines how people express themselves through activities and locate themselves on the being–becoming–doing continuum. The social relations orientation describes the preferred forms of interpersonal relationships within a culture. The self-orientation indicates the culture's conception of how people understand who they are in relation to others. The world orientation locates a culture in the physical and spiritual worlds. The time orientation directs a culture to value the past, present, or future.

FOR DISCUSSION

1. How might individuals from *doing, being,* and *becoming* cultures engage in conflict in the workplace, in school, or in interpersonal relationships?
2. One person comes from a culture that believes "We're all humans, aren't we?" Another person comes from a culture that says, "Status is everything." What might occur as these two individuals try to communicate with each other?
3. *Truth* or *lie, just* or *unjust, right* or *wrong,* and *good* or *bad* are all common human judgments of the actions of others. How does your awareness of cultural patterns affect your understanding of each of these sets of terms?
4. Using the five dimensions of cultural patterns described in this chapter, describe how you think each is displayed in your own culture.

FOR FURTHER READING

Martin J. Gannon and Rajnandini Pillai, *Understanding Global Cultures: Metaphorical Journeys through 29 Nations, Clusters of Nations, Continents, and Diversity*, 4th ed. (Los Angeles: Sage, 2010). An excellent guide to understanding the worldviews and perspectives of many cultures throughout the world.

Helga Kotthoff and Helen Spencer-Oatey (eds.), *Handbook of Intercultural Communication* (New York: Mouton de Gruyter, 2009). Scholarly essays on intercultural communication from the perspective of applied linguistics.

Samuel Roll and Marc Irwin, *The Invisible Border: Latino Culture in America* (Boston: Intercultural Press, 2008). An example of a culture-specific investigation of one of the cultures comprising the United States. For those interested in understanding and bridging the cultural differences between Latinos and other U.S. Americans.

Craig Storti, *Americans at Work: A Guide to Can-Do People* (Yarmouth, ME: Intercultural Press, 2004). With a focus on understanding and explaining the cultural patterns of European Americans, this book provides an excellent overview of European American cultural patterns.

Cultural Patterns and Communication: Taxonomies

KEY TERMS

In the previous chapter, we provided an overview of the patterns that underlie all cultures. We described the nature of cultural patterns and the importance of beliefs, values, norms, and social practices in helping cultures to cope with problems. We now focus on specific conceptual taxonomies that are useful for understanding cultural differences.

We have chosen four different but related taxonomies to describe variations in cultural patterns. The first was developed by Edward Hall, who noted that cultures differ in the extent to which their primary message patterns are high context or low context. The second describes the ideas of Geert Hofstede, who identifies six dimensions along which cultures vary. The third taxonomy explains the ideas of Shalom Schwartz, who reasoned that there are three problems that all cultural groups must solve, which results in seven dimensions of culture. The fourth taxonomy, by a group of researchers collectively known as the GLOBE team, incorporates many of the previously described ideas and identifies nine dimensions of culture. Finally, we provide a synthesis of these four taxonomies and propose seven key features or dimensions that differ across cultures. These cultural dimensions are individualism versus collectivism, power distance, gender expectations, task versus relationship focus, uncertainty avoidance, harmony versus mastery, and time orientation.

As you read the descriptions of cultural patterns by Hall, Hofstede, Schwartz, and the GLOBE researchers, we caution you to remember three points. First, there is nothing sacred about these approaches and the internal categories they employ. Each approach takes the whole of cultural patterns (beliefs, values, norms, and social practices) and divides them in different ways.

Second, the parts of each of the systems are interrelated. We begin the description of each system at an arbitrarily chosen point, presupposing other parts of the system that have not yet been described. Cultural patterns are best understood as a unique whole rather than as an isolated dimension or characteristic, even if a given attribute is distinctive or predominates within a specific culture.

Finally, individual members of a culture may vary greatly from the pattern that is typical of that culture. Therefore, as you study these approaches to cultural patterns, we encourage you to make some judgments about how your own culture fits into the pattern. Then, as you place it within the pattern, also try to discern how you, as an individual, fit into the patterns described. Similarly, as you learn about other cultural patterns, please remember that a specific person may or may not be a typical representative of that culture. As you study your own cultural patterns and those of other cultures, you improve the knowledge component of intercultural competence.

CULTURE CONNECTIONS

Iranian culture explicitly accepts certain types of deception and dissimulation. To begin with, the values of kindness, courtesy, and hospitality stand higher in many contexts than the values of frankness and honesty. Why tell the truth when feelings will be hurt? Why, indeed, linger on painful truths? Even deaths in the family may be concealed when circumstances are not conducive to informing the bereaved tactfully and supportively. It is important to note that this is not a case where honesty is not valued but where, given a certain type of dilemma in which honesty will cause pain, a kind deception is preferred—which would not be regarded as a lie, for real lies are very much condemned.

—Mary Catherine Bateson   ■

HALL'S HIGH- AND LOW-CONTEXT CULTURAL TAXONOMY

Edward T. Hall, whose writings about the relationship between culture and communication are well known, organizes cultures by the amount of information implied by the setting or context of the communication itself, regardless of the specific words that are spoken.[1] Hall argues that every human being is faced with so many perceptual stimuli— sights, sounds, smells, tastes, and bodily sensations—that it is impossible to pay attention to them all. Therefore, one of the functions of culture is to provide a screen between the person and all of those stimuli to indicate what perceptions to notice and how to interpret them. Hall's approach is compatible with the other approaches discussed in this chapter. Where it differs is in the importance it places on the role of context.

According to Hall, cultures differ on a continuum that ranges from high to low context. High-context cultures prefer to use high-context messages, in which most of the meaning is either implied by the physical setting or is presumed to be part of the individual's internalized beliefs, values, norms, and social practices; very little is provided

A Japanese tea ceremony is an example of a high-context message. Nearly every movement, gesture, and action has significance to those who understand the "code" being used.

in the coded, explicit, transmitted part of the message. Examples of high-context cultures include Japanese, African American, Mexican, and Latino. Low-context cultures prefer to use low-context messages, in which the majority of the information is vested in the explicit code. Low-context cultures include German, Swedish, European American, and English.

A simple example of high-context communication is the interactions that take place in a long-term relationship between two people who are often able to interpret even the slightest gesture or the briefest comment. The message does not need to be stated explicitly because it is carried in the shared understandings about the relationship.

A simple example of low-context communication is now experienced by more and more people as they interact with computers. For computers to "understand" a message, every statement must be precise. Many computers will not accept or respond to instructions that do not have every space, period, letter, and number in precisely the right location. The message must be overt and very explicit.

Hall's description of high- and low-context cultures is based on the idea that some cultures have a preponderance of messages that are high context, others have messages that are mostly low context, and yet others have a mixture of both. Hall also describes other characteristics of high- and low-context cultures that reveal the beliefs, values, norms, and social practices of the cultural system. These characteristics include the use of indirect or direct messages, the importance of ingroups and outgroups, and the culture's orientation to time.

Use of Indirect and Direct Messages

In a high-context culture such as that of Japan, meanings are internalized and there is a large emphasis on nonverbal codes. Hall describes messages in high-context cultures as almost preprogrammed, in which very little of the interpretation of the message is left to chance because people already know that, in the context of the current situation, the communicative behaviors will have a specific and particular message. In low-context cultures, people look for the meaning of others' behaviors in the messages that are plainly and explicitly coded. The details of the message are expressed precisely and specifically in the words that people use as they try to communicate with others.

Another way to think about the difference between high- and low-context cultures is to imagine something with which you are very familiar, such as repairing a car, cooking, sewing, or playing a particular sport. When you talk about that activity with someone else who is very familiar with it, you will probably be less explicit and instead use a more succinct set of verbal and nonverbal messages. You will talk in a verbal shorthand that does not require you to be specific and precise about every aspect of the ideas that you are expressing, because the others will know what you mean without the ideas' specific presentation. However, if you talk to someone who does not know very much about the activity, you will have to explain more, be more precise and specific, and provide more background information.

In a high-context culture, much more is taken for granted and assumed to be shared, and consequently the overwhelming preponderance of messages are coded in such a way that they do not need to be explicitly and verbally transmitted. Instead, the demands of the situation and the shared meanings among the interactants mean that the preferred interpretation of the messages is already known.

Consider, as an example of high-context messages, an event that occurred in Indonesia. A young couple met, fell in love, and wanted to marry. She was from a wealthy and well-connected family, whereas he was from a family of more modest means, but the young couple did not regard this difference as a problem. So they shared their happy news with their respective families; shortly thereafter, the young man's parents were invited to the woman's home to socialize and to meet her parents. The social occasion was very cordial; the conversation was pleasant, and the two sets of parents were very gracious toward one another. At the appropriate time, the woman's parents served *nasi goreng* (fried rice) and star fruit, two foods that are very common in Indonesia. Finally, after an appropriate interval, the young man's parents thanked their hosts and left. Throughout the entire episode, the topic of the wedding was never broached. However, everyone knew that the wedding would never occur. After all, *nasi goreng* doesn't go with star fruit; the high-context and face-saving message that the woman's parents communicated, and that the man's parents clearly understood, was that they disapproved of the marriage.

Reactions in high-context cultures are likely to be reserved, whereas reactions in low-context cultures are frequently very explicit and readily observable. It is easy to understand why this is so. In high-context cultures, an important purpose in communicating is to promote and sustain harmony among the interactants. Unconstrained reactions could threaten the face or social esteem of others. In low-context cultures, however, an important purpose in communicating is to convey exact meaning. Explicit messages help to achieve this goal. If messages need to be explicit, so will people's reactions. Even when the message is understood, a person cannot assume that the meanings are clear in the absence of verbal messages coded specifically to provide feedback.

Importance of Ingroups and Outgroups

In high-context cultures, it is very easy to determine who is a member of the group and who is not. Because so much of the meaning of messages is embedded in the rules and rituals of situations, it is easy to tell who is acting according to those norms. As there are fixed and specific expectations for behaviors, deviations are easy to detect.

Another distinction concerns the emphasis placed on the individual in contrast to the group as a source of self-identity. In a high-context culture, the commitment between people is very strong and deep, and responsibility to others takes precedence over responsibility to oneself. Loyalties to families and the members of one's social and work groups are long-lasting and unchanging. This degree of loyalty differs from that found in a low-context culture, in which the bonds between people are very fragile and the extent of involvement and commitment to long-term relationships is lower.

Orientation to Time

The final distinguishable characteristic of high- and low-context cultures is their orientation to time. In the former, time is viewed as more open, less structured, more responsive to the immediate needs of people, and less subject to external goals and constraints. In low-context cultures, time is highly organized, in part because of the additional energy required to understand the messages of others. Low-context cultures are almost forced to pay more attention to time in order to complete the work of living with others.

As Table 5.1 indicates, Edward Hall's placement of cultures onto a continuum that is anchored by preferences for high-context messages and low-context messages offers a way to understand other variations in cultural patterns. A high-context culture chooses to use indirect and implicit messages that rely heavily on nonverbal code systems. In a high-context culture, the group is very important, as are traditions, and members of the

TABLE 5.1

CHARACTERISTICS OF LOW- AND HIGH-CONTEXT CULTURES

High-Context Cultures	Low-Context Cultures
Indirect and implicit	Direct and explicit
Messages internalized	Messages plainly coded
Much nonverbal coding	Details verbalized
Reactions reserved	Reactions on the surface
Distinct ingroups and outgroups	Flexible ingroups and outgroups
Strong interpersonal bonds	Fragile interpersonal bonds
Commitment high	Commitment low
Time open and flexible	Time highly organized

ingroup are easily recognized. Time is less structured and more responsive to people's needs. Low-context cultures are characterized by the opposite attributes: messages are explicit and dependent on verbal codes, group memberships change rapidly, innovation is valued, and time is highly structured.

HOFSTEDE'S CULTURAL TAXONOMY

Geert Hofstede's impressive studies of cultural differences in value orientations offer another approach to understanding the range of cultural differences.[2] Hofstede's approach is based on the assertion that people carry mental programs, or "software of the mind," that are developed during childhood and are reinforced by their culture. These mental programs contain the ideas of a culture and are expressed through its dominant values. To identify the principal values of different cultures, Hofstede surveyed more than 100,000 IBM employees in seventy-one countries, and he has subsequently broadened his analysis to include many others.

Through theoretical reasoning and statistical analyses, Hofstede identified five dimensions along which dominant patterns of a culture can be ordered: power distance, uncertainty avoidance, individualism versus collectivism, masculinity versus femininity, and time orientation. Recently an additional dimension has been added: indulgence versus restraint. Hofstede's work provides an excellent synthesis of the relationships between cultural values and social behaviors.[3]

Power Distance

One of the basic concerns of all cultures is the issue of human inequality. Contrary to the claim in the U.S. Declaration of Independence that "all men are created equal," all people in a culture do not have equal levels of status or social power. Depending on the culture, some people might be regarded as superior to others because of their wealth, age, gender, education, physical strength, birth order, personal achievements, family background, occupation, or a wide variety of other characteristics.

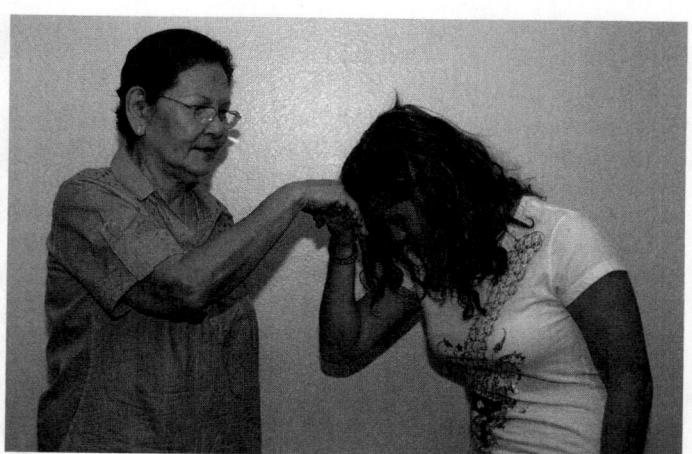

In the Philippines, large power distance is shown in this traditional gesture of greeting by a granddaughter to her grandmother.

Cultures also differ in the extent to which they view such status inequalities as good or bad, right or wrong, just or unjust, and fair or unfair. That is, all cultures have particular value orientations about the appropriateness or importance of status differences and social hierarchies. Thus power distance refers to the degree to which the culture believes that institutional and organizational power should be distributed unequally and the decisions of the power holders should be challenged or accepted.

Cultures that prefer small power distances—such as Austria, Denmark, Israel, and New Zealand—believe in the importance of minimizing social or class inequalities, questioning or challenging authority figures, reducing hierarchical organizational structures, and using power only for legitimate purposes. Conversely, cultures that prefer large power

CULTURE CONNECTIONS

The value systems of Australians and Americans combine competitive and cooperative strands, but in different ways. The Australian harmonizes them while the American sees them as mutually exclusive and is torn between them. Americans are always ready to put themselves in competition with the group or groups to which they belong; it is often "either the group or me." For the Australian it is "the group *and* me, with a great deal of personal privacy as well." Australians search for ways to collaborate with the competition while Americans seek ways to "beat" it. The American position seems to be that too much cooperation weakens one's advantage. This may stem, in part,

from the different ways such values are inculcated. For example, much is made of mandatory participation in team sports in Australian schools. Americans place more emphasis on the outstanding individual and early on learn "spectatorism," with its powerful identification with the few superior performers. Social welfare legislation is much more comprehensive and more readily accepted in Australia than in the United States. The degree to which social welfare is a continuing social, political, economic, and ideological battleground in the United States surprises Australians.

—George W. Renwick ■

distances—such as those in Arab countries, Guatemala, Malaysia, and the Philippines—believe that each person has a rightful and protected place in the social order, that the actions of authorities should not be challenged or questioned, that hierarchy and inequality are appropriate and beneficial, and that those with social status have a right to use their power for whatever purposes and in whatever ways they deem desirable.

The consequences of the degree of power distance that a culture prefers are evident in family customs, the relationships between students and teachers, organizational practices, and in other areas of social life. Even the language systems in high power-distance cultures emphasize distinctions based on a social hierarchy.

Children raised in high power-distance cultures are expected to obey their parents without challenging or questioning them, while children raised in low power-distance cultures put less value on obedience and are taught to seek reasons or justifications for their parents' actions. Even the language of high power-distance cultures is more sensitive to hierarchical distinctions; the Chinese language, for instance, has separate terms for older brother, oldest brother, younger sister, youngest sister, and so on.

Students in high power-distance cultures are expected to comply with the wishes and requests of their teachers, and conformity is regarded very favorably. As a consequence, the curriculum in these cultures is likely to involve a great deal of rote learning, and students are discouraged from asking questions because questions might pose a threat to the teacher's authority. In low power-distance cultures, students regard their independence as very important, and they are less likely to conform to the expectations of teachers or other authorities. The educational system itself reinforces the low power-distance values by teaching students to ask questions, to solve problems creatively and uniquely, and to challenge the evidence leading to conclusions.

In the business world, managers in high power-distance cultures are likely to prefer an autocratic or centralized decision-making style, whereas subordinates in these cultures expect and want to be closely supervised. Alternatively, managers in low power-distance cultures prefer a consultative or participative decision-making style, and their subordinates expect a great deal of autonomy and independence as they do their work.

European Americans tend to have a relatively low power distance, though it is by no means exceptionally low. However, when European Americans communicate with people from cultures that value a relatively large power distance, problems related to differences in expectations are likely. For example, European American exchange students in a South American or Asian culture sometimes have difficulty adapting to a world in which people are expected to do as they are told without questioning the reasons for the requests. Conversely, exchange students visiting the United States from high power-distance cultures sometimes feel uneasy because they expect their teachers to direct and supervise their work closely, and they may also have been taught that it would be rude and impolite to ask for the kinds of information that might allow them to be more successful.

Uncertainty Avoidance

Another concern of all cultures is how they will adapt to changes and cope with uncertainties. The future will always be unknown in some respects. This unpredictability and the resultant anxiety that inevitably occurs are basic in human experience.

Cultures differ in the extent to which they prefer and can tolerate ambiguity and, therefore, in the means they select for coping with change. Thus, all cultures differ in their perceived need to be changeable and adaptable. Hofstede refers to these variations as the uncertainty avoidance dimension, the extent to which the culture feels threatened by ambiguous, uncertain situations and tries to avoid them by establishing more structure.

At one extreme on this dimension are cultures such as those of Denmark, Jamaica, India, and Ireland, which are all low in uncertainty avoidance and therefore have a high tolerance for uncertainty and ambiguity. They believe in minimizing the number of rules and rituals that govern social conduct and human behavior, in accepting and encouraging dissent among cultural members, in tolerating people who behave in ways that are considered socially deviant, and in taking risks and trying new things. Conversely, the cultures of Greece, Guatemala, Portugal, and Uruguay are among those that prefer to avoid uncertainty as a cultural value. They desire or even demand consensus about societal goals, and they do not like to tolerate dissent or allow deviation in the behaviors of cultural members. They try to ensure certainty and security through an extensive set of rules, regulations, and rituals.

Cultures must cope with the need to create a world that is more certain and predictable, and they do so by inventing rules and rituals to constrain human behaviors. Because members of high uncertainty avoidance cultures tend to be worried about the future, they have high levels of anxiety and are highly resistant to change. They regard the uncertainties of life as a continuous threat that must be overcome. Consequently, these cultures develop many rules to control social behaviors, and they often adopt elaborate rituals and religious practices that have a precise form or sequence.

Members of low uncertainty avoidance cultures tend to live day to day, and they are more willing to accept change and take risks. Conflict and competition are natural, dissent is acceptable, deviance is not threatening, and individual achievement is regarded as beneficial. Consequently, these cultures need few rules to control social behaviors, and they are unlikely to adopt religious rituals that require precise patterns of enactment.

Differences in level of uncertainty avoidance can result in unexpected problems in intercultural communication. For instance, European Americans tend to have a moderately low level of uncertainty avoidance. When these U.S. Americans communicate with someone from a high uncertainty avoidance culture, such as those in Japan or France, they are likely to be seen as too nonconforming and unconventional, and they may view their Japanese or French counterparts as rigid and overly controlled. Conversely, when these U.S. Americans communicate with someone from an extremely low uncertainty avoidance culture, such as the Irish or Swedes, they are likely to be viewed as too structured and uncompromising, whereas they may perceive their Irish or Swedish counterparts as too willing to accept dissent.

Individualism Versus Collectivism

Another concern of all cultures, and a problem for which they must all find a solution, involves people's relationships to the larger social groups of which they are a part. People must live and interact together for the culture to survive. In doing so, they must develop a way of relating that strikes a balance between showing concern for themselves and concern for others.

Cultures differ in the extent to which individual autonomy is regarded favorably or unfavorably. Thus, cultures vary in their tendency to encourage people to be unique and independent or conforming and interdependent. Hofstede refers to these variations as the individualism–collectivism dimension, the degree to which a culture relies on and has allegiance to the self or the group.

CULTURE CONNECTIONS

My job was managing the household. I told Joseph I didn't want any servants. Joseph wasn't listening.

"We can't afford servants," I protested....

"You'll like Kamau," he said. "He is a good man."

"It's not a question of liking," I said. "I don't want a cook. It's ridiculous for the two of us to have a cook and a yard man."

"A man in my position must have servants. It's expected. In America, I kept silent and learned your ways. Now you must learn."

"But servants, Joseph? It's so un-American."

He laughed and took my hand. "Kamau is of my age group—we were initiated together. We will help him. It is not a matter of choice." As it turned out, many things were not a matter of choice—my husband's monthly salary contributed to the school fees of several brothers or cousins and our garden was freely harvested by his family.

—Geraldine Kennedy ■

Highly individualistic cultures, such as the dominant cultures in Belgium, Hungary, the Netherlands, and the United States, believe that people are only supposed to take care of themselves and perhaps their immediate families. In individualist cultures, the autonomy of the individual is paramount. Key words used to invoke this cultural pattern include *independence*, *privacy*, *self*, and the all-important *I*. Decisions are based primarily on what is good for the individual, not for the group, because the person is the primary source of motivation. Similarly, a judgment about what is right or wrong can be made only from the point of view of each individual.

Highly collectivist cultures such as those in Guatemala, Indonesia, Pakistan, and West Africa value a collectivist orientation. They require an absolute loyalty to the group, though the relevant group might be as varied as the nuclear family, the extended family, a work group, a social organization, a caste, or a jati (a subgrouping of a caste). In collectivist cultures, decisions that juxtapose the benefits to the individual and the benefits to the group are always based on what is best for the group, and the groups to which a person belongs are the most important social units. In turn, the group is expected to look out for and take care of its individual members. Consequently, collectivist cultures believe in obligations to the group, dependence of the individual on organizations and institutions, a "we" consciousness, and an emphasis on belonging.

Huge cultural differences can be explained by differences on the individualism–collectivism dimension. We have already noted that collectivistic cultures tend to be group-oriented. A related characteristic is that they typically impose a very large psychological distance between those who are members of their group (the ingroup) and those who are not (the outgroup). Ingroup members are required to have unquestioning loyalty, whereas outgroup members are regarded as almost

The importance of preserving one's cultural patterns is reflected in this protest against Canada's First Nations Governance Act.

inconsequential. Conversely, members of individualistic cultures do not perceive a large chasm between ingroup and outgroup members; ingroup members are not extremely close, but outgroup members are not as distant.

Individualist cultures train their members to speak out as a means of resolving difficulties. In classrooms, students from individualistic cultures are likely to ask questions of the teacher; students from collectivistic cultures are not. Similarly, people from individualistic cultures are more likely than those from collectivistic cultures to use confrontational strategies when dealing with interpersonal problems; those with a collectivistic orientation are likely to use avoidance, third-party intermediaries, or other face-saving techniques. Indeed, a common maxim among European Americans, who are highly individualistic, is that "the squeaky wheel gets the grease" (suggesting that one should make noise in order to be rewarded); the corresponding maxim among the Japanese, who are somewhat collectivistic, is "the nail that sticks up gets pounded" (so one should always try to blend in).

Masculinity Versus Femininity

A fourth concern of all cultures, and for which they must all find solutions, pertains to gender expectations and the extent to which people prefer achievement and assertiveness or nurturance and social support. Hofstede refers to these variations as the masculinity–femininity dimension. This dimension indicates the degree to which a culture values "masculine" behaviors, such as assertiveness and the acquisition of wealth, or "feminine" behaviors, such as caring for others and the quality of life.

At one extreme are masculine cultures such as those in Austria, Italy, Japan, and Mexico, which believe in achievement and ambition. In this view, people should be judged on their performance, and those who achieve have the right to display the material goods that they acquired. The people in masculine cultures also believe in ostentatious manliness, and very specific behaviors and products are associated with appropriate male behavior.

At the other extreme are feminine cultures such as those of Chile, Portugal, Sweden, and Thailand, which believe less in external achievements and shows of manliness and more in the importance of life choices that improve intrinsic aspects of the quality of life, such as service to others and sympathy for the unfortunate. People in these feminine cultures are also likely to prefer equality between the sexes, less prescriptive role behaviors associated with each gender, and an acceptance of nurturing roles for both women and men.

Members of highly masculine cultures believe that men should be assertive and women should be nurturing. Sex roles are clearly differentiated, and sexual inequality is regarded as beneficial. The reverse is true for members of highly feminine cultures: men are far less interested in achievement, sex roles are far more fluid, and equality between the sexes is the norm.

Teachers in masculine cultures praise their best students because academic performance is rewarded highly. Similarly, male students in these masculine cultures strive to be competitive, visible, successful, and vocationally oriented. In feminine cultures, teachers rarely praise individual achievements and academic performance because social accommodation is more highly regarded. Male students try to cooperate with one another and develop a sense of solidarity, they try to behave modestly and properly, they select subjects because they are intrinsically interesting rather than vocationally rewarding, and friendliness is much more important than brilliance.

Time Orientation

A fifth concern of all cultures relates to its orientation to time. Hofstede has acknowledged that the four previously described dimensions have a Western bias, as they were developed by scholars from Europe or the United States who necessarily brought to their work an implicit set of assumptions and categories about the types of cultural values they would likely find. His time-orientation dimension is based on the work of Michael H. Bond, a Canadian who has lived in Asia for many years and who assembled a large team of researchers from Hong Kong and Taiwan to develop and administer a Chinese Value Survey to university students around the world.[4]

The time-orientation dimension refers to a person's point of reference about life and work. It ranges from long-term to short-term. Cultures with a long-term time orientation toward life include those of Germany, Japan, Russia, and South Korea. They all admire persistence, thriftiness, and humility. Linguistic and social distinctions between elder and younger siblings are common, and deferred gratification of needs is widely accepted. Conversely, cultures with a short-term orientation toward changing events include those from Australia, Colombia, Iran, and Morocco. These cultures have an expectation of quick results following one's actions.[5] The Chinese, for example, typically have a long-term time orientation—note the tendency to mark time in year-long increments, as in the Year of the Snake or the Year of the Horse—whereas Europeans typically have a short-term time orientation and aggregate time in month-long intervals (such as Aries, Gemini, Pisces, or Aquarius).

Indulgence Versus Restraint

Recently Hofstede has included an additional dimension to those previously described. Based on recent research, including ideas from Middle Eastern, Nordic, and Eastern European perspectives,[6] Hofstede has added the dimension of indulgence versus restraint to his taxonomy.

The *hejab*, or head scarf, is worn by many Muslim women as a statement of their cultural values.

The indulgence versus restraint dimension juxtaposes hedonism with self-discipline. Indulgence—the view that pleasure and the enjoyment of life are very desirable—puts the focus on happiness as a way of life. Having fun, fulfilling one's appetites for delectable foods and drinks, indulging in social and sexual pleasures, and generally enjoying life by having pleasant and pleasurable experiences are characteristic. Cultures high on indulgence include those of El Salvador, Mexico, New Zealand, and Sweden. They all tend to encourage pleasure, enjoyment, spending, consumption, sexual gratification, and general merriment.

At the other extreme on this dimension are cultures that emphasize restraint. These cultures focus on self-discipline and believe that individuals should curb their urges and desires for unrestrained fun. Self-control, characterized as willpower, modesty, moderation, and self-discipline, is typical. Cultures high on restraint include those of Bulgaria, Italy, Morocco, and Pakistan. These cultures value the control of personal indulgences, they prefer to restrict "worldly" pleasures, and they discourage the seeking of enjoyments associated with leisure activities.

Comparing Hofstede's Dimensions

Hofstede's foundational work has been widely cited and appropriately praised for its importance, clarity, straightforwardness, simplicity, and excellence. Each of Hofstede's dimensions provides insights into the influence of culture on the communication process. Every culture, of course, forms an intricate and interrelated pattern; no one cultural dimension is sufficient to describe or understand this complexity.

Hofstede's dimensions describe cultural expectations for a range of social behaviors: *power distance* refers to relationships with people higher or lower in rank, *uncertainty avoidance* to people's search for truth and certainty, *individualism-collectivism* to expected behaviors toward the group, *masculinity–femininity* to the expectations surrounding achievement and gender differences, *time orientation* to people's search for virtue and lasting ideals, and *indulgence–restraint* to psychological impulse control.

For additional numerical information on more than seventy cultures on Hofstede's dimensions, please turn to the Resources section at the back of this book. There you will find information that is grouped by geographic region.

To guide you in understanding and using Hofstede's numerical data in the Resources section, let's use the United States as an example. As we noted in Chapter 3, there are important differences between *nations* and *cultures*. Though Hofstede's data focus on national characteristics, the information is best understood as representing the dominant culture within a nation or group. When Hofstede did his research, the dominant culture in the United States was European American.

A look at Hofstede's U.S. data reveals that European Americans tend to be at the extremes: low on power distance (–92), uncertainty avoidance (–93), and time orientation (–84) and high on individualism (195), masculinity (68), and indulgence (104). Translating Hofstede's data into specific cultural characteristics suggests a cultural orientation in which European Americans prefer to minimize status differences (power distance), encourage risk-taking (uncertainty avoidance), prefer short-term goals (time orientation), emphasize individual rights (individualism), value achievement (masculinity), and desire pleasurable consumption (indulgence).

Similar analyses can be done with data from other cultures. If two cultures have similar configurations on Hofstede's dimensions, they would likely have similar

Time management, productivity, and communication all depend on the patterns of one's culture to define their importance.

communication patterns; conversely, cultures that are very different from one another would probably behave dissimilarly. Note, however, that even cultures that are located very near others are not entirely similar; these differences underscore the importance of being cautious when making generalizations about cultures, even when they are within the same regions of the world (e.g., Latin America or the Middle East).

SCHWARTZ'S CULTURAL TAXONOMY

Another set of ongoing studies on differences in cultural patterns was conducted by Shalom Schwartz. He also began with Kluckhohn and Strodtbeck's premise (which is discussed more extensively in the previous chapter) that all cultures face common problems for which they must find a solution. Schwartz reasoned that there are three problems or issues that all groups must resolve.[7]

Schwartz's first problem is concerned with a cultural preference for the kinds of relationships and boundaries that ought to exist between individuals and the larger group. Schwartz calls this dimension autonomy versus embeddedness. At one extreme are cultures that value autonomy. In autonomy cultures, people are regarded as independent, and they are encouraged to express their unique preferences, tendencies, abilities, and feelings. Expressions of autonomy, Schwartz further reasoned, can occur in two ways, leading to two types of cultural autonomy: intellectual autonomy and affective autonomy. Cultures that value intellectual autonomy, such as those in France and Japan, promote and support people's independent pursuit of thoughts, ideas, and knowledge; curiosity, creativity, and a broadminded view of the world are all encouraged. Cultures that value affective autonomy, such as those in Denmark and England, encourage and reinforce each individual's pursuit of pleasurable emotional states, enjoyable feelings, varied experiences, and an exciting life.

In contrast to autonomy cultures, cultures that value embeddedness view people as nested within a collective social network. Identification with the group is a central concern, and maintenance of harmony in social relationships is paramount. Meaning in one's life is found primarily through identification with the group and through one's interpersonal relationships within that group. In embeddedness cultures, the preference is for one's routines, activities, goals—indeed, one's entire life—to be shared communally. Anything that might threaten or disrupt the sense of ingroup solidarity could be viewed as a threat. Consequently, in cultures such as those in Nigeria and Pakistan, the ideals of predictability, obedience to traditional authorities, maintenance of social order, and respect for elders' wisdom are usually central concerns.

The second cultural problem that Schwartz addresses is that people must organize and coordinate their activities in a way that preserves and fulfills the needs and goals of the social group. That is, within every culture, each person's survival requires that people must work together productively, must consider and adapt to the needs and wants of others, must coordinate and manage their actions with those of others, and must do their fair share of helping with the activities that are communally required. Schwartz calls this dimension egalitarianism versus hierarchy. At one extreme are egalitarian cultures, which encourage people to view others as social and moral equals who voluntarily choose to work together as peers to fulfill shared interests. People in egalitarian cultures, such as those in Spain and Belgium, are taught to be concerned about the welfare of others, to cooperate, and to be responsible and honest about helping others. At the other extreme are hierarchy cultures, which see the unequal distribution of social, political, and economic power as legitimate and desirable. People in hierarchy cultures, such as those in Thailand and Turkey, are taught to defer to those with higher status and to value authority, humility, and social power.

Schwartz's third cultural problem, which he calls harmony versus mastery, deals with people's orientations to social and natural resources. Harmony cultures encourage acceptance and blending into the natural and social worlds, as humans are seen as an integral part of nature. The view of this cultural orientation, which is held by

CULTURE CONNECTIONS

A powwow arena is a place for celebration by Native Indian people. It is an opportunity for Native Indian people from all parts of North America and Canada to share their music and their communal beliefs in the nature of life. As one powwow host stated before an initiation ceremony for a young girl, "This circle (the powwow arena) is the Creator's circle. It's a sacred place." For many Native Indian people, attending a powwow has the same characteristics as attending church. However, most Anglos usually cannot see the analogy. A "religious" service has different qualities for Anglos, and the celebratory atmosphere of powwows, as well as the presence of contests and vendors and grandstands, makes it difficult for many Anglos to recognize the sacred nature of what is occurring in front of them. Further, when Anglos behave inappropriately at powwows by being scantily clad or by walking into the arena to take pictures, few Native people will overtly criticize their actions. This is especially true when visitors are perceived as "guests." Numerous times I was encouraged to move ahead of Navajos when waiting in line for activities, told as they moved aside, "You're our guest." There may often be disapproving glances toward the Anglos, especially from the elderly Indians, but no direct confrontations. Except for children and some teenagers, most Native peoples at the powwows wear long pants or long skirts and do not expose their bodies unnecessarily.

—Charles A. Braithwaite ■

the dominant cultures in Italy and Mexico, is to accept rather than to change, to fit in rather than to exploit, and to limit rather than to control. Mastery cultures encourage their members to direct and control the natural and social worlds. One's goals can be achieved most effectively by changing and adapting the social and natural environments. The view of this cultural orientation, which is held by the dominant cultures in China and India, is that one should be self-sufficient, self-assertive, daring, ambitious, and, ultimately, successful.

In sum, Schwartz maintains that there are three primary cultural dimensions: autonomy versus embeddedness, egalitarianism versus hierarchy, and harmony versus mastery. Each cultural dimension identifies alternative solutions to a central problem that every culture must resolve. Whereas each dimension represents a continuum of possible cultural responses, a culture's tendency to prefer one pole of a given dimension means that the opposite pole is less emphasized and therefore less important to that culture. Figure 5.1 depicts the relationships among Schwartz's value orientations.

For additional numerical information that locates eighty cultures on Schwartz's dimensions, please turn to the Resources section at the back of this book. There you will find information about many cultures, grouped by geographic region. As we suggested about Hofstede's information, Schwartz's data can also be used to understand the tendencies of cultures that are of interest to you. Our earlier caution still applies: no one cultural dimension is sufficient to describe or understand the complexity of cultural differences.

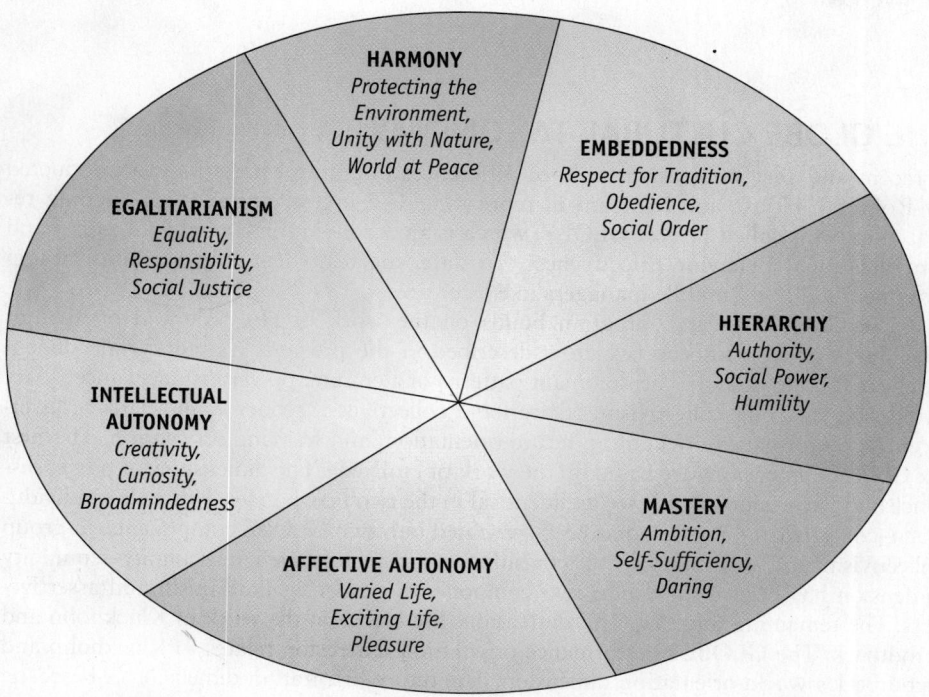

FIGURE 5.1

Schwartz's cultural value orientations.[8]

CULTURE CONNECTIONS

Ghote wanted to say that his name was Ghote, and that it was spelt with the H as the second letter. But he knew at least something about Americans. They believed in informality.

"I am Ganesh," he said. "Ganesh."

"Well, this is how it is, Gan," Hoskins said. "I'm the guy who picked up the trail of the Shahaneye kid and I'm the guy who found the ashram. So I'm in a position to inform you that I know as much about that little piece of ass as anyone. And you can take my word for it, she's not going to leave that place any time soon. She's gone off on a religion kick, and that's the way she's gonna stay."

Ghote, his head still thickly muzzy from his long flight, felt as if a hammer was being repeatedly banged down on the top of his skull. But he had to make some sort of a reply.

"Yes, Mr Hoskins," he began, "I very well understand what is the position, but—"

"Listen, if we're gonna work together on this case we're gonna have to work as a team. So you're gonna have to call me Fred. In this United States we don't stand on ceremony. You're just gonna have to learn that."

"Yes," Ghote said.

He wished with all his might that this yammering giant could simply vanish into thin air. But he was dependent on the fellow. Without him he would have the greatest difficulty getting to the ashram at all. He did not even know its address, just that it was not in Los Angeles but somewhere outside. He could make inquiries if he had to, and in the end he would find it. But if he was to act at all quickly Fred Hoskins stood, giant-like, squarely in his path.

"Fred," he said. "Yes, I will call you Fred."

—H. R. F. Keating ■

THE GLOBE CULTURAL TAXONOMY

A recent and very impressive study of differences in cultural patterns was conducted by Robert J. House and his team of more than 170 investigators.[9] This ongoing research effort is called Project GLOBE, which is an acronym for Global Leadership and Organizational Behavior Effectiveness. To date, the team has collected information from nearly 20,000 middle managers in 61 cultures.

The GLOBE research program builds on the work of Hofstede and on that of Kluckhohn and Strodtbeck (which is described in the previous chapter). Nine dimensions are used to describe the dominant patterns of a culture: power distance, uncertainty avoidance, in-group collectivism, institutional collectivism, gender egalitarianism, assertiveness, performance orientation, future orientation, and humane orientation. The first six GLOBE dimensions are based on the work of Hofstede. The dimensions of power distance and uncertainty avoidance are identical in the two taxonomies. Hofstede's individualism–collectivism dimension has been separated into two GLOBE components: in-group collectivism and institutional collectivism. Similarly, Hofstede's masculinity–femininity dimension has been divided into two components: gender egalitarianism and assertiveness. The remaining three GLOBE dimensions are based on the work of Kluckhohn and Strodtbeck. The GLOBE's performance orientation dimension relates to Kluckhohn and Strodtbeck's world-orientation dimension. The future orientation dimension is based on Kluckhohn and Strodtbeck's concept of time and the distinctions among past-, present-, and future-oriented cultures. The GLOBE's humane orientation dimension is anchored in Kluckhohn and Strodtbeck's view of human nature, especially their distinction that

TABLE 5.2

GLOBE DIMENSIONS AND CULTURAL CHARACTERISTICS[10]

Dimension	Cultural Characteristics	Sample Items
Power Distance	The degree to which people believe that power should be stratified, unequally shared, and concentrated at higher levels of an organization or government.	Followers are expected to obey their leaders without question.
Uncertainty Avoidance	The extent to which people strive to avoid uncertainty by relying on social norms, rules, rituals, and bureaucratic practices to alleviate the unpredictability of future events.	Most people lead highly structured lives with few unexpected events.
In-Group Collectivism	The degree to which people express pride, loyalty, and cohesiveness in their families.	Parents take great pride in the accomplishments of their children.
Institutional Collectivism	The degree to which a culture's institutional practices encourage collective actions and the collective distribution of resources.	Leaders encourage group loyalty even if individual goals suffer.
Gender Egalitarianism	The extent to which people minimize gender-role differences and gender discrimination while promoting gender equality.	Boys are encouraged more than girls to attain a higher education. (scored inversely)
Assertiveness	The degree to which people are assertive, confrontational, and aggressive in social relationships.	People are generally dominant in their relationships with each other.
Performance Orientation	The extent to which people encourage others to improve their task-oriented performance and excel.	Students are encouraged to strive for continuously improved performance.
Future Orientation	The degree to which people engage in future-oriented behaviors such as planning, investing in the future, and delaying gratification.	Most people live in the present rather than for the future. (scored inversely)
Humane Orientation	The degree to which people encourage others to be fair, altruistic, friendly, generous, caring, and kind.	Most people are generally very tolerant of mistakes.

cultures may regard humans on a continuum ranging from inherently "good" to inherently "bad." Table 5.2 provides the nine cultural dimensions studied in the GLOBE research, their cultural characteristics, and sample items. The information in this table provides a useful reference guide to help you understand the GLOBE ideas more easily.

Power Distance

As Hofstede suggested, one of the basic concerns of all cultures is the issue of human inequality. Cultures differ in the extent to which they view status inequalities as desirable or undesirable. Thus power distance refers to the degree to which cultures believe that social and political power should be distributed disproportionately, shared unequally, and concentrated among a few top decision makers.

High power-distance cultures, such as those in France, Argentina, and Nigeria, believe it is very appropriate to have differences among social classes. Upward mobility ought to be limited, because people already occupy their correct places in the social hierarchy. The decisions of the powerful authorities should be met with unchallenged acceptance.

Conversely, low power-distance cultures like those in Australia, Denmark, and Albania believe it is important to minimize or even eliminate social class differences. Upward mobility is high, because an equal opportunity for each person is an overriding goal. Questioning and challenging the decisions of authorities is regarded as each person's duty and responsibility, as only through such challenges will social and political power be used well.

Uncertainty Avoidance

All cultures need to have some degree of predictability in their social worlds. While complete certainty can never be achieved, humans could not survive in a world of total and chaotic uncertainty. Thus cultures vary in the degree of predictability they prefer. These variations constitute the uncertainty avoidance dimension, which is the extent to which cultures feel threatened by the unpredictability of the future and therefore try to establish more structure in the form of rules, regulations, rituals, and mandatory practices.

Cultures such as those in Sweden, Switzerland, and China are relatively high on uncertainty avoidance. Therefore, they prefer to avoid uncertainty as a cultural value, desire or even demand consensus about societal goals, and do not tolerate dissent or allow deviation in the behaviors of cultural members. They try to ensure certainty and security through an extensive set of instructions about how one ought to behave. As a result, cultures that are high on uncertainty avoidance prefer to develop many ways to control people's social behaviors. These controls exist as formal regulations and as informal rules about acceptable conduct, and they also include elaborate rituals and religious practices that have a precise form or sequence.

Cultures such as those in Russia, Bolivia, and South Korea are relatively low on uncertainty avoidance. Therefore, they have a higher tolerance for uncertainty and ambiguity and are much more comfortable with the unpredictability of life. Consequently, rules and regulations are kept to a minimum, dissent is tolerated, and deviance is more likely to be regarded as peculiar or eccentric rather than as threatening.

In-Group Collectivism

The in-group collectivism dimension is similar to what Hofstede calls individualism–collectivism. Individualistic cultures have low in-group collectivism, whereas collectivistic cultures rate high on this dimension.

In-group collectivism reflects the degree to which people express pride, loyalty, and solidarity with their family or similar group. In cultures with high in-group collectivism, individuals take pride in and define their sense of self—quite literally, their sense of who they are—in terms of their family or similar group. That is, people's identities within collectivistic cultures are closely tied to their ingroups, and strong group memberships are both required and desired. As the African saying suggests, in collectivist cultures "I am because we are." Representative cultures that are high on in-group collectivism include those in Georgia, Morocco, and the Philippines.

In individualistic cultures—those that are low on in-group collectivism—the independence and autonomy of the individual is an overriding feature. People's identities within individualistic cultures are separate from, and perhaps very distant from, those of the group. Group membership is often regarded as voluntary, and allegiance with one's ingroup—even with one's family—is not expected to be overly strong. Included in this category are such cultures as those in New Zealand, Finland, and the Netherlands.

Institutional Collectivism

Another aspect of the dimension that Hofstede called individualism–collectivism is concerned with the basis upon which decisions are made and the group's resources are allocated. The dimension of institutional collectivism represents the degree to which cultures support, value, and prefer to distribute rewards based on group versus individual interests.

In cultures that are high on institutional collectivism, decisions that juxtapose the benefits to the group with the benefits to the individual nearly always base the decision on what is best for the group. Thus, in cultures like those in Qatar and Japan, group activities are typically preferred to individual actions.

In cultures that are low on institutional collectivism, decisions are based on what is good for the individual, with little regard for the group. Because the person is the primary source of motivation, individual autonomy and actions tend to dominate. Thus, in Italy and Greece, decisions are based on individual merit rather than on collective involvement.

Gender Egalitarianism

If you carefully read the description of Hofstede's masculinity–femininity dimension, you will note that it combines two related attributes that, in the GLOBE project, have been separated into separate dimensions: a belief in equality between women and men and a preference for forceful assertiveness. The first of these attributes is called

Within Massai culture, which is high on institutional collectivism, there is a strong identification with the tribe. Here, a group of Massai women live in a communal village.

Gender egalitarianism, which minimizes differences among men and woman, is evident in this conversation between two business people.

gender egalitarianism and is the extent to which a culture minimizes differences in gender expectations for men versus women.

Cultures such as those in Hungary and Poland, which are near the midpoint of the gender egalitarianism dimension, believe that gender equality is preferred, that men and women should be treated in the same way, and that unequal treatment solely because of one's biological sex or gender constitutes discrimination and should not occur. Conversely, cultures like those in Austria and Egypt, which are low in gender egalitarianism, engage in unequal treatment of men and women. In this view, there are inherent differences between men and women, and these differences require dissimilar expectations and treatments. Rather than regarding these fundamental differences negatively, cultures that are low on gender egalitarianism view the divergence in gender roles and expectations as normal and natural.

Assertiveness

Another concern of all cultures, which also requires every culture to find a solution, pertains to the cultural preference for dominance and forcefulness or nurturance and social support. This assertiveness dimension describes the extent to which people value and prefer tough aggressiveness or tender nonaggressiveness.

Cultures high on the assertiveness dimension value strength, success, and taking the initiative. Competition is good, winning is desirable, and rewards should go to those who are victorious. People are encouraged to be competitive, visible, and successful. Representative cultures include those in Germany and Hong Kong.

Conversely, cultures low on the assertiveness dimension value modesty, tenderness, warm relationships, and cooperation. Competition is bad, a win–lose orientation is unacceptable, and rewards should be shared among all. Nurturance and social support are important, as are modesty, cooperation with others, and a sense of solidarity. Friendliness is much more important than brilliance. Typical of this orientation are the cultures in Kuwait and Thailand.

Performance Orientation

The degree to which a culture encourages and rewards people for their accomplishments is called the performance orientation dimension. Depending on the culture, some people might be regarded as superior to others because of who they are—the "correct" family background, age, gender, birth order, or school—whereas others may acquire status based on personal achievements such as the amount of education, success in business, physical strength, occupation, or a wide variety of other characteristics.

In high performance-oriented cultures, such as those in Canada and Singapore, status is based on what a person has accomplished. Schooling and education are critical to one's success, people are expected to demonstrate some initiative in work-related tasks, and expectations are high. Conversely, in low performance-oriented cultures, like those in Colombia and Guatemala, status is based on who you are. Attending the "right" school is important, as are family connections, seniority, loyalty, and tradition.

An important component of performance orientation is people's preferred relationship to the natural and spiritual worlds. As Kluckhohn and Strodtbeck suggested in the previous chapter, some cultures view nature as something to be conquered and controlled, others see themselves as living in harmony with nature, and still others view themselves as subjugated to nature.

High performance-oriented cultures assert their dominance over nature, and they try to shape the world to fit their needs. Getting the job done is far more important than maintaining effective relationships, for what really matters is the task-related results that show what someone has accomplished. People in high performance-oriented cultures value competitiveness, assertiveness, and achievement. In contrast, people in low performance-oriented cultures feel more controlled by nature and want to live in harmony with the natural and spiritual environments. Maintaining effective relationships is more important in such cultures than is getting the job done; what matters most are cooperation, integrity, and loyalty.

Another important distinction related to performance orientation is Edward Hall's concept of low-context versus high-context messages, which we discussed earlier. High performance-oriented cultures tend to be low context; they prefer to use messages that are clear, explicit, and direct. They also have a monochronic approach to time; time is valuable and limited, events are sequential, and punctuality is preferred. Conversely, low performance-oriented cultures use high-context messages more often; their intent is to avoid direct confrontations and maintain harmony in their relationships.

Future Orientation

Locating one's world in time—and thereby giving structure, coherence, and significance to events—creates order and meaning in people's lives. The extent to which a culture plans for forthcoming events is the future orientation dimension. Related slightly to Hofstede's long-term dimension and very directly to Kluckhohn and Strodtbeck's ideas on time orientation (see Chapter 4), the future orientation dimension describes the degree to which cultures advocate long-term planning and deferred gratification or the deeply felt satisfaction that comes from experiencing the simple pleasures of the present moment.

Cultures differ, of course, in the extent to which they prefer to focus on the future rather than on the spontaneity of the present. Those high in future orientation, such as Iran and Hong Kong, believe that current pleasures are less important than future benefits, so they believe in planning, self-control, and activities that have a delayed impact. Cultures like those in Portugal and Venezuela are low in future orientation and thus prefer to enjoy fully the experiences currently under way; they like to live "in the moment" and are less constrained by doubts about the past or concerns about the future.

People from cultures that are high in future orientation want to save money and other resources. They believe in strategic planning, and they value economic success. People from cultures that are low in future orientation are more likely to spend now rather than save for later. They view material and spiritual achievements as opposing goals, and they prefer the latter.

CULTURE CONNECTIONS

At the other end of the line is Vitalie, a blogger, one of the few in Moldova. He seemed like someone who could shed some light on Moldovan misery. We agree to meet the next day. Vitalie asks me if I know any good restaurants. This strikes me as odd, since he's lived here all of his life and I've been in Moldova for about one hour. I take this as a discouraging sign about the culinary prospects in Moldova. Later, someone explained that in Moldova the relationship between host and guest is reversed. It is the guest's obligation to make the host feel at ease. Reverse hospitality. One of the many peculiar customs in this country.

—Eric Weiner ■

Humane Orientation

The final GLOBE dimension, humane orientation, refers to the extent to which cultures encourage and reward their members for being benevolent and compassionate toward others or are concerned with self-interest and self-gratification.

Cultures high in humane orientation value expressions of kindness, genero-sity, caring, and compassion, and people who express social support for others are admired. Members of humane-oriented cultures are expected to help others financially and emotionally, to share information that others may need, to spend time with others, and to offer empathy and love. Representative cultures include those in Zambia and Indonesia.

Cultures low in humane orientation value comfort, pleasure, satisfaction, and personal enjoyment. People from low humane orientation cultures are expected to confront personal problems by themselves, and they are concerned primarily with individual gratification. Typical of this orientation are the cultures of Spain and white South Africa.

Comparing the GLOBE Dimensions

As we suggested in Chapter 4, cultural patterns represent a universal social choice that must be made by each culture and that is learned from the family and throughout the social institutions of a culture: in the degree to which children are encouraged to have their own desires and motivations, in the solidarity and unity expected in the family, in the role models that are presented, and throughout the range of messages that are conveyed.

Additional numerical information about sixty-one cultures on the GLOBE dimensions can be found in the Resources section at the back of this book. The regional groupings organize the cultures in the GLOBE studies by geographic areas. As we suggested previously, the data can be used to understand the tendencies of cultures that are of interest to you.

The GLOBE research expands our understanding of cultural patterns. By providing updated information on a wide range of cultures, and by revising and expanding the cultural dimensions that are relevant, this effort substantially increases our understanding of cultures and of intercultural communication. To provide just one example of the usefulness of the expanded GLOBE dimensions, consider the information (in the Resources section at the back of this book) about the Japanese culture on the dimensions of institutional collectivism and in-group collectivism. Whereas the Japanese are extremely high in institutional collectivism (+222), they are below the average for in-group collectivism (–68). This information clarifies why decisions in Japan are most often made, and resources among the Japanese are typically distributed, in a very collectivist fashion, but the collective group for the Japanese—that is, the group with which people identify most closely—is not necessarily the family but rather the organization, the nation, or some other social unit. New Zealanders, Swedes, and Danes, among others, have patterns on these two dimensions that are similar to the Japanese; Greeks, Guatemalans, Colombians, and others have the opposite pattern.

A SYNTHESIS OF CULTURAL TAXONOMIES

We began this chapter by noting that each attempt to identify the fundamental ways that cultures can differ—that is, the core ideas of Kluckhohn and Strodtbeck (in the previous chapter), Hall, Hofstede, Schwartz, and the GLOBE team—takes the whole of cultural patterns and divides them in different ways. Yet we are sure you noticed that there are many commonalities in these approaches.

One idea that is central to each of these approaches was first discussed in the previous chapter. Kluckhohn and Strodtbeck noted that all cultures face a common set of problems for which they must find solutions. Based on the ideas presented in this chapter, we now have a more refined basis for suggesting what those fundamental cultural problems might be, and therefore what aspects of culture would likely make one cultural group similar to or different from another.

Based on the extensive scholarly research that is discussed in this chapter and the previous one, we suggest that there are seven universal problems, and therefore seven cultural dimensions, that are fundamental to understanding a culture. Each dimension can be viewed as a continuum of choices that a culture must make. To avoid confusion, we label these dimensions by their commonly used names, when such a name exists. Each label describes one or both of the end points—the extremes—of the continuum. The cultural dimensions are: individualism versus collectivism, power distance, gender expectations, task versus relationship, uncertainty avoidance, harmony versus mastery, and time orientation.

Perhaps the most essential issue that all cultures must confront involves the requirement for *a balance between the needs of individuals versus the priorities of the group.* The individualism–collectivism dimension highlights this important issue. Scholars such as Harry Triandis have suggested that the individualism–collectivism dimension is, by far, the most important attribute that distinguishes one culture from another[11]; thus it is not surprising that it appears in each of the previously discussed taxonomies.

Another universal issue faced by all cultures involves *expectations about the behaviors of people with higher or lower status.* This power-distance dimension emphasizes the cultural choices related to equality versus hierarchy. Both the *value* of status differences (how important they are) and the *basis* of status differences (the characteristics that give someone increased status) are culturally based.

A third issue that all cultures must confront involves their *beliefs about appropriate behaviors for men and women.* This gender expectations dimension includes both expectations about suitable role behaviors—to what extent, for example, are men and

CULTURE CONNECTIONS

Balinese culture is one of the most methodical systems of social and religious organization on earth, a magnificent beehive of tasks and role and ceremonies. The Balinese are *lodged*, completely held, within an elaborate lattice of customs. A combination of several factors created this network but basically we can say that Bali is what happens when the lavish rituals of traditional Hinduism are superimposed over a vast rice-growing agricultural society that operates, by necessity, with elaborate communal cooperation. Rice terraces require an unbelievable amount of shared labor, maintenance and engineering in order to prosper, so each Balinese village has a *banjar*—a united organization of citizens who administer, through consensus, the village's political and economic and religious and agricultural decisions. In Bali, the collective is absolutely more important than the individual, or nobody eats.

—Elizabeth Gilbert ∎

women encouraged to be assertive or nurturing—and expectations about the preferred similarities or differences in the behaviors of men and women.

A fourth cultural issue requires a *balance between task-related and relationship-building activities*. This task-relationship dimension involves a cultural choice that emphasizes the relative importance of task concerns (getting the job done) and relational concerns (maintaining good social relationships) among the cultural members.

A fifth concern of all cultures arises through the culture's efforts at *coping with the unknown*. This uncertainty avoidance dimension highlights cultural choices that involve a preference for impulsiveness versus predictability, and it therefore emphasizes the culture's desire for risk versus caution.

A sixth cultural issue entails *locating the culture in space and place*. The harmony–mastery dimension is concerned with a culture's approach to its physical and material environments. Cultural preferences on this dimension can range from complete acceptance of and "fitting into" the natural world to a preference for absolute control and conquest of the physical environment.

The final cultural issue, which is general and overarching, is concerned with *locating the culture in time*. This time-orientation dimension includes four very different types of problems that arise from the passage of time: the culture's preferred *goals*, which can range from short term to long term; its *emphasis*, which can focus on past, present, or future moments; its *use*, which involves a penchant for organizing activities one-at-a-time or many-things-at-once; and the culture's preferred *rhythm* for pacing activities, which can range from a view that time is open and flexible to one in which time is highly structured and organized.

Table 5.3 highlights the correspondence between the seven universal problems we have identified, the cultural dimensions that address them, and the cultural taxonomies that we discussed previously. As you can see, there is substantial commonality across the taxonomies. But they also differ in what they emphasize, in the distinctions they regard as significant, and in the terminology they use to feature these universal cultural issues.

Figure 5.2 summarizes the range of cultural orientations that can occur on these seven dimensions. Taken together, these seven dimensions provide a culture-general "map" or framework that can be filled in with culture-specific information. This gives you an accessible way to understand specific cultural patterns.

TABLE 5.3

A SYNTHESIS OF CULTURAL TAXONOMIES

Synthesis of Taxonomies	Kluckhohn & Strodtbeck	Hall	Hofstede	Schwartz	GLOBE
Individualism–Collectivism	Social Relations Orientation	Context	Individualism–Collectivism	Autonomy vs. Embeddedness	Ingroup Collectivism; Institutional Collectivism
Power Distance	Social Relations Orientation	Context	Power Distance	Egalitarianism vs. Hierarchy	Power Distance
Gender Expectations	Social Relations; Self-Orientation	Context	Masculinity–Femininity	—	Gender Egalitarianism; Assertiveness
Task Relationship	Activity Orientation	—	Indulgence–Restraint	Intellectual Autonomy; Affective Autonomy	Performance Orientation; Humane Orientation
Uncertainty Avoidance	—	Context	Uncertainty Avoidance	—	Uncertainty Avoidance
Harmony–Mastery	World Orientation	—	—	Harmony vs. Mastery	—
Time Orientation	Time Orientation	Context	Time Orientation	—	Future Orientation

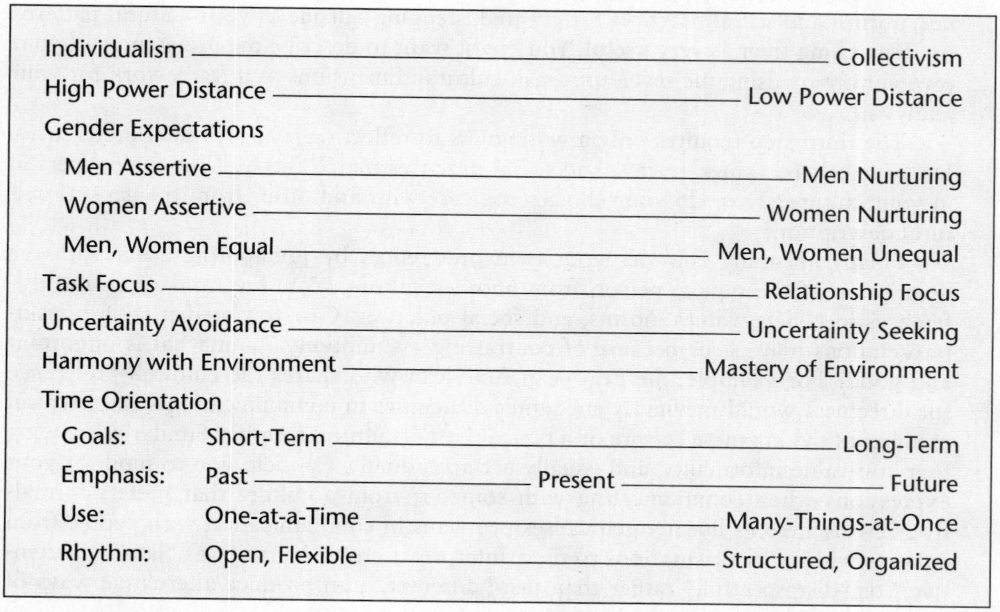

```
Individualism ————————————————————————— Collectivism
High Power Distance ——————————————— Low Power Distance
Gender Expectations
    Men Assertive ————————————————————— Men Nurturing
    Women Assertive ————————————————— Women Nurturing
    Men, Women Equal ————————————— Men, Women Unequal
Task Focus ————————————————————————— Relationship Focus
Uncertainty Avoidance ——————————— Uncertainty Seeking
Harmony with Environment ————— Mastery of Environment
Time Orientation
    Goals:      Short-Term ——————————————————— Long-Term
    Emphasis:   Past ——————— Present ——————————— Future
    Use:        One-at-a-Time ————————— Many-Things-at-Once
    Rhythm:     Open, Flexible ————————— Structured, Organized
```

FIGURE 5.2

Cultural orientations.

CULTURAL TAXONOMIES AND INTERCULTURAL COMPETENCE

The major lesson in this chapter is that cultures vary systematically in their choices about solutions to basic human problems. The taxonomies offer lenses through which cultural variations can be understood and appreciated, rather than negatively evaluated and disregarded. The categories in these taxonomies can help you to describe the fundamental aspects of cultures. As frames of reference, they provide mechanisms to understand many intercultural communication events. In any intercultural encounter, people may be communicating from very different perceptions of what is "real," what is "good," and what is "correct" behavior. The competent intercultural communicator must recognize that there will be cultural differences in addressing the seven universal cultural issues, and these differences will always be a factor in intercultural communication.

The taxonomies allow you to use culture-specific knowledge to improve intercultural competence. First, begin by seeking out information about the cultural patterns of those individuals with whom you engage in intercultural communication. To assist your analysis and understanding of the culture, use the seven cultural dimensions as an organizing framework for the information you gather, and create a profile of the culture's preferred choices. Libraries and the Internet are natural starting places for this kind of knowledge. So, too, are representatives of the culture. Engage them in conversation as you try to understand their culture. Most people welcome questions from a genuinely curious person. Be systematic in your search for information by using the categories thoroughly. Think about the interrelatedness of the various aspects of the culture's patterns.

Second, study the patterns of your own culture. Because you take your beliefs, values, norms, and social practices for granted, stepping outside of your cultural patterns by researching them is very useful. You might want to describe the preferences of your own culture by using the seven universal cultural dimensions as a framework for your analysis.

The third step requires only a willingness to reflect on your personal preferences. Do your beliefs, values, norms, and social practices match those of the typical person in your culture? How do your choices coincide with and differ from the general cultural description?

Finally, mentally consider your own preferences by juxtaposing them with the description of the typical person from another culture. Note the similarities and differences in beliefs, values, norms, and social practices. Can you predict where misinterpretations may occur because of contrasting assumptions about what is important and good? For example, the European American who shares the culture's preference for directness would inevitably encounter difficulties in communication with a typical member of the Japanese culture or a typical Latino cultural member. Similarly, knowing that you value informality, and usually act accordingly, can help you to monitor your expressions when communicating with someone from a culture that prefers formality. Viewing time as linear often causes problems in communication with people from cultures with other orientations to time. Interpretations of behavior as "late," "inattentive," or "disrespectful," rather than just "different," can produce alternative ways of viewing the ticking of the clock.

SUMMARY

This chapter first discussed four important taxonomies that can be used to describe cultural variations. Edward Hall placed cultures on a continuum from high context to low context. High-context cultures prefer messages in which most of the meaning is either implied by the physical setting or is presumed to be part of the individual's internalized beliefs, values, norms, and social practices; low-context cultures prefer messages in which the information is contained within the explicit code.

Geert Hofstede described six dimensions along which dominant patterns of a culture can be ordered: power distance, uncertainty avoidance, individualism–collectivism, masculinity–femininity, time orientation, and indulgence–restraint. The power-distance dimension assesses the degree to which the culture believes that institutional power should be distributed equally or unequally. The uncertainty avoidance dimension describes the extent to which cultures prefer and can tolerate ambiguity and change. The individualism–collectivism dimension describes the degree to which a culture relies on and has allegiance to the self or the group. The masculinity–femininity dimension indicates the degree to which a culture values assertiveness and "manliness" or caring for others and the quality of life. The time-orientation dimension refers to a long-term versus short-term orientation toward life and work. The indulgence–restraint dimension contrasts pleasure-seeking with self-restraint.

Shalom Schwartz maintained that there are three aspects of culture that are primary: autonomy versus embeddedness, egalitarianism versus hierarchy, and harmony versus mastery. Each of these cultural continua identifies alternative solutions to a central problem that every culture must resolve. The autonomy–embeddedness dimension is concerned with preferences for an individual versus collective orientation; the egalitarianism–hierarchy dimension is concerned with preferences for power and control; and the harmony–mastery dimension emphasizes the culture's approach to the social and natural worlds.

The GLOBE researchers identified nine dimensions of culture. The power-distance and uncertainty avoidance dimensions are similar to those that Hofstede described. In-group collectivism and institutional collectivism refine Hofstede's individualism–collectivism dimension; in-group collectivism is concerned with family loyalty, whereas societal collectivism refers to group-oriented actions. Gender egalitarianism and assertiveness refine Hofstede's masculinity–femininity dimension; gender egalitarianism is about equality between men and women, while assertiveness is about social dominance. Performance orientation refers to task- or work-related accomplishments, future orientation is about preferences for delayed versus immediate gratifications, and humane orientation is concerned with fairness and generosity.

Our synthesis of the various cultural taxonomies proposes seven cultural dimensions: individualism versus collectivism, power distance, gender expectations, task versus relationship, uncertainty avoidance, harmony versus mastery, and time orientation. Each of these dimensions is based on—and is therefore similar to—the analogous ideas of Kluckhohn and Strodtbeck, Hall, Hofstede, Schwartz, and the GLOBE researchers.

The ideas presented in this chapter and in the previous one offer alternative lenses through which cultures can be understood and appreciated. Taken together, these two chapters provide multiple frames of reference that can enhance your knowledge, motivations, and skills in intercultural communication.

FOR DISCUSSION

1. What does Edward Hall mean when he refers to culture as a "screen" for its members?
2. Describe how each of Hofstede's dimensions of cultural patterns is displayed within your culture.
3. Does Schwartz's taxonomy coincide with your own intercultural experiences? Explain.
4. Consider the following two philosophical statements: "I think; therefore, I am" and "I am because we are." What do these two statements reveal about the underlying cultural values of those who use them?
5. Compare your own values with the GLOBE's cultural values for the "typical" person from your culture. In what ways are they the same? Different? What might this suggest about intercultural communication?

FOR FURTHER READING

Jagdeep S. Chhokar, Felix C. Brodbek, and Robert J. House (eds.), *Culture and Leadership across the World: The GLOBE Book of In-Depth Studies of 25 Societies* (Mahwah, NJ: Erlbaum, 2007). A companion to the earlier GLOBE book, this volume provides in-depth qualitative information about the dimensions of culture.

Edward T. Hall, *Beyond Culture* (New York: Anchor Books, 1989). Describes, in great detail, the cultural variations among high- and low-context cultures.

Geert Hofstede, Gert Jan Hofstede, and Michael Minkov, *Cultures and Organizations: Software of the Mind: Intercultural Cooperation and Its Importance for Survival*, 3rd ed. (New York: McGraw-Hill, 2010). Hofstede's recent and most comprehensive book, which extensively describes his ideas about the dimensions on which cultures can vary.

Robert J. House, Paul J. Hanges, Mansour Javidan, Peter W. Dorfman, and Vipin Gupta (eds.), *Culture, Leadership, and Organizations: The GLOBE Study of 62 Societies* (Thousand Oaks, CA: Sage, 2004). A momentous work that presents ground-breaking research on the current practices and value dimensions that differ among cultures. Provides extensive quantitative evidence for cultural variations on the nine GLOBE dimensions.

Shalom H. Schwartz, "A Theory of Cultural Value Orientations: Explication and Applications," *Measuring and Mapping Cultures: 25 Years of Comparative Value Surveys*, ed. Yilmaz Esmer and Thorleif Pettersson (Boston: Brill, 2007), 33–78. A concise and useful summary of Schwartz's research on cultural values.

For detailed numerical information about the classification of many cultures on the Hofstede, Schwartz, and GLOBE cultural taxonomies, please turn to the Resources section at the back of this book.

Cultural Identity and Cultural Biases

KEY TERMS

I n the previous two chapters, we emphasized the critical importance of cultural patterns in shaping the preferred ways to think, feel, and act in a variety of situations. An equally interesting and important question in the development of intercultural communication competence concerns how people come to identify themselves as belonging to a particular cultural group. For example, how and when does a child begin to think of herself as a Latina, Japanese American, or Japanese? When do adults who are born into one culture and living in another begin to think of their cultural identity as embracing parts of both their original culture and the later culture? Similarly, how are some people defined as "not members" of our cultural group? How and why do groups of people from one culture develop negative attitudes and actions toward other cultural groups?

CULTURE CONNECTIONS

Sixteen hours later, we arrived in Pakistan in the hazy light of dawn. The crew burst open the doors of our 747 to receive those heady scents of arrival: jasmine flowers, gasoline, burning trash, and cow dung. The heat, even at this early hour, seeped into the cabin almost instantly. On subsequent trips I would come to think of these first sensations of touching down in Pakistan as one united feeling, the pungent smell of difference, the awareness that I had entered a new world, one with a separate set of laws. Nana handed us kurtas to change into, tucking our Western clothes into her carry-on luggage, and our transformation from half-Pakistani American children to half-American Pakistani children was complete.

On the plane ride back to Boston, I tried to write my report for school in my blue notebook, drifting in and out of sleep. Half Pakistani, half American. Half Muslim, half Christian. Half-half.

—Sadia Shepard ■

The present chapter discusses some aspects of cultural identity that can have a very large effect on intercultural communication. We begin with a discussion of cultural identity and the powerful ways in which our self-concept as a member of a particular cultural group filters our interpretations of the world. Then we explore the nature of cultural biases, rooted in cultural identity, as we examine the effects of ethnocentrism, stereotyping, prejudice, discrimination, and racism on intercultural interactions.

CULTURAL IDENTITY

As part of the socialization process, children learn to view themselves as members of particular groups. Children in all cultures, for example, are taught to identify with their families (even though, as Chapter 10 indicates, whom to include as part of one's "family" differs across cultures). As a child becomes a teenager and then an adult, the development of vocational and avocational interests creates new groups with which to identify. "Baseball player," "ballet dancer," or "scientist" may become important labels to describe the self.

Another feature of socialization is that people are taught about groups to which they do not belong, and they often learn that certain groups should be avoided. This tendency to identify as a member of some groups, called *ingroups,* and to distinguish these ingroups from *outgroups* is so prevalent in human thinking that it has been described as a universal human tendency.[1] Recent scholarship is investigating the role of new media in supporting or diminishing this human tendency to define others as either part of our own ingroup or as part of our outgroup.[2]

The Nature of Identity

Related to the distinction between ingroup and outgroup membership is the concept of one's *identity* or self-concept. An individual's self-concept is built on cultural, social, and personal identities.[3]

Cultural identity refers to one's sense of belonging to a particular culture or ethnic group. It is formed in a process that results from membership in a particular culture, and it involves learning about and accepting the traditions, heritage, language, religion, ancestry, aesthetics, thinking patterns, and social structures of a culture. That is, people internalize the beliefs, values, norms, and social practices of their culture and identify with that culture as part of their self-concept.

Social identity develops as a consequence of memberships in particular groups within one's culture.[4] The characteristics and concerns common to most members of such social groups shape the way individuals view their characteristics. The types of groups with which people identify can vary widely and might include perceived similarities due to age, gender, work, religion, ideology, social class, place (neighborhood, region, and nation), and common interests. For instance, those baseball players, ballet dancers, and scientists who strongly identify with their particular professions likely view themselves as "belonging" to "their" group of professionals, with whom they have similar traits and share similar concerns.

Finally, personal identity is based on people's unique characteristics, which may differ from those of others in their cultural and social groups. You may like cooking or chemistry, singing or sewing; you may play tennis or trombones, soccer or stereos; you may view yourself as studious or sociable, goofy or gracious; and most assuredly you have abilities, talents, quirks, and preferences that differ from those of others.

For ease and clarity, we have chosen to present aspects of a person's identity as separate categories. There is a great deal of interdependence, however, among these three aspects of identity. Characteristics of people's social identities will inevitably be linked to the preferences shaped by their cultural identities. Similarly, how people enact their unique interests will also be heavily influenced by their cultural identities. Thus, for example, a teenage girl's identity will likely be strongly linked to her culture's preferences for gendered role behaviors as well as to her social class and her personal characteristics and traits.[5]

The Formation of Cultural Identity

Cultural identities often develop through a process involving three stages: unexamined cultural identity, cultural identity search, and cultural identity achievement.[6] During the unexamined cultural identity stage, one's cultural characteristics are taken for granted, and consequently there is little interest in exploring cultural issues. Young children, for instance, typically lack an awareness of cultural differences and the distinguishing characteristics that differentiate one culture from another. Teenagers and adults may not want to categorize themselves as belonging to any particular culture. Some people may not have explored the meanings and consequences of their cultural membership but may simply have accepted preconceived ideas about it that were obtained from parents, the community, the mass media, and others. Consequently, some individuals may unquestioningly accept the prevailing stereotypes held by others and may internalize common stereotypes of their own culture and of themselves. Scholars have suggested that the cultural identities of many European Americans, in particular, have remained largely unexamined, a consequence of the power, centrality, and privilege that the European American cultural group has had in the United States.[7] As Judith Martin, Robert Krizek, Thomas Nakayama, and Lisa Bradford suggest,

In this intercultural family, the child will draw upon the cultures of both parents in forming her own cultural identity.

CULTURE CONNECTIONS

"I'm Kate Shugak," she said. "I met your wife at church this morning."

"Kate Shugak?" She nodded. "Any relation to Ekaterina Shugak?" She nodded again. He took in the color of her skin and the epicanthic folds of her eyes, she the slant of his cheekbones and the thick, straight black hair.

He didn't say, "Aleut?" and she didn't say, "Athabaskan?" but they both relaxed a little, the way people of color always do when the door closes after the last white person has left the room.

—Dana Stabenow ∎

This lack of attention to white identity and self-labeling reflects the historical power held by Whites in the United States. That is, Whites as the privileged group take their identity as the norm or standard by which other groups are measured, and this identity is therefore invisible, even to the extent that many Whites do not consciously think about the profound effect being white has on their everyday lives.[8]

Cultural identity search involves a process of exploration and questioning about one's culture in order to learn more about it and to understand the implications of membership in that culture. By exploring the culture, individuals can learn about its strengths and may come to a point of acceptance both of their culture and of themselves. For some individuals, a turning point or crucial event precipitates this stage, whereas for others it just begins with a growing awareness and reinterpretation of everyday experiences. Common to this stage is an increased social and political awareness along with an increased desire to learn more about one's culture. Such learning may be characterized by an increased degree of talking with family and friends about cultural issues, independent reading of relevant sources, enrolling in appropriate courses, or increased attendance at cultural events such as festivals and museums. There may also be an emotional component to this stage, of varying intensity, that involves tension, anger, and perhaps even outrage directed toward other groups. These emotions may intensify as people become aware of and wrestle with the effects of discrimination on their present and future lives and the potential difficulties in attaining educational, career, and personal objectives.

Cultural identity achievement is characterized by a clear, confident acceptance of oneself and an internalization of one's cultural identity. Such acceptance can calmly and securely be used to guide one's future actions. People in this stage have developed ways of dealing with stereotypes and discrimination so that they do not internalize others' negative perceptions and are clear about the personal meanings of

The use of the Menorah and related artifacts in the celebration of Chanukah helps this Jewish American family to strengthen their cultural identity.

TABLE 6.1

STAGES IN THE DEVELOPMENT OF CULTURAL IDENTITY

Stage	Sample Comments	Source of Comments
Unexamined cultural identity	"My parents tell me about where they lived, but what do I care? I've never lived there."	Mexican American male
	"Why do I have to learn who was the first black woman to do this or that? I'm just not too interested."	African American female
	"I don't have a culture. I'm just an American."	European American male
Cultural identity search	"I think people should know what black people had to go through to get to where we are now."	African American female
	"There are a lot of non-Japanese people around me, and it gets pretty confusing to try and decide who I am."	Japanese American male
	"I want to know what we do and how our culture is different from others."	Mexican American female
Cultural identity achievement	"My culture is important, and I am proud of what I am. Japanese people have so much to offer."	Japanese American male
	"It used to be confusing to me, but it's clear now. I'm happy being black."	African American female

Source: Adapted from Jean S. Phinney, "A Three-Stage Model of Ethnic Identity Development in Adolescence," *Ethnic Identity: Formation and Transmission among Hispanics and Other Minorities,* ed. Martha E. Bernal and George P. Knight (Albany: State University of New York Press, 1993), 61–79.

their culture. This outcome contributes to increased self-confidence and positive psychological adjustment. Table 6.1 provides sample comments from individuals in each of the three stages of cultural identity development.

Characteristics of Cultural Identity

Once formed, cultural identities provide an essential framework, organizing and interpreting our experiences of others. This is because cultural identities are central, dynamic, and multifaceted components of one's self-concept.

Cultural identities are central to a person's sense of self. Like gender and race, your culture is more "basic" because it is broadly influential and is linked to a great number of other aspects of your self-concept. These core aspects of your identity are likely to be important in most of your interactions with others. Most components of your identity, however, become important only when they are activated by specific circumstances. For many people, the experience of living in another culture or interacting with a person from a different culture triggers an awareness of their own cultural identities that they did not have before. When a component of your identity becomes conscious and important to you, or "activated," your experiences get filtered through that portion of your identity. Aspects of one's cultural identity can be activated not only by direct experiences with others but also by the media reports, by artistic portrayals that have particular cultural themes, by musical performances (such as rap music) that are identified with specific cultural groups, and by a range of other personal and mass-mediated experiences.[9] Thus, if individuals from one's culture are frequently portrayed in popular films and television programs, this can provide a sense of legitimacy for the culture and can help to establish that the culture's members are attractive, desirable, and good. Conversely, the absence of such role models in the media can dampen one's identification with the culture and the individual's perceptions that the culture is vital and vibrant. Because your cultural identity is likely to be central to your sense of self, most of your experiences are interpreted or "framed" by your cultural membership.

Because cultural identities are dynamic, your cultural identity—your sense of the culture to which you belong and who you are in light of this cultural

CULTURE CONNECTIONS

When people ask me my nationality, I find it difficult to give a sufficient answer.

Who am I really? Am I an Asian-Indian? American? Both? Neither? These are questions that many "bi-cultured" teenagers can't help but ask themselves.

I was born in India, and when I was 2 years old my parents emigrated to the United States. They didn't leave India because of any hardships, but simply because they were curious—curious about what America might hold for them. Their curiosity has led to a unique lifestyle for me, as an Indian-American teenager.

But what kind of lifestyle? Do I really fit in here? Do I fit in India? No. I don't. One adventurous move by my parents has turned me, like many others in my situation, into a misfit. I am a person who doesn't have one nationality.

The biggest problem is finding out who I really am. Our parents want us to be like teens in our original countries; however, we want to stay and fit in here. We want to be like everybody around us, and to most of our parents that's a threat.

So is there a middle ground? Or will our teenage years simply end up being a power struggle—in my case a power struggle between India, America and me?

One of the biggest disadvantages and one of the biggest advantages to being "bi-cultured" is that we live in two cultures at once. You could say that we lead double lives. In school, we act, dress and eat like everyone else, but when we get home life changes.

In my home, my parents usually speak in Hindi, our native language. Dad will say something to me in Hindi, and I'll answer back in a mixture of Hindi and English. When Mom says dinner's ready, I go down and eat a delicious Indian meal that most people never get the opportunity to eat. Mom will say "Get ready for temple," and I'll automatically don a salwar-kameez (an Indian two-piece outfit). How many actually lead a life like mine?

My parents emigrated from India, and they are Indians. My children will be born here, they will be Americans. I was born in India and brought here, I am neither Indian nor American, yet I am both.

My life is and will continue to be a struggle between nationalities, but I'm sure I can face this challenge head on. This determination I get from both of my countries.

—Smriti Aggarwal ■

membership—exists within a changing social context. Consequently, your identity is not static, fixed, and enduring; rather, it is dynamic and changes with your ongoing life experiences. In even the briefest encounter with people whose cultural backgrounds differ from your own, your sense of who you are *at that instant* may well be altered, at least in some small ways. Over time, as you adapt to various intercultural challenges, your cultural identity may be transformed into one that is substantially different from what it used to be.[10] The inaccurate belief that cultural identities are permanent, that "Once a Swedish American, always a Swedish American," ignores the possibility of profound changes that people may experience as a result of their intercultural contacts. Indeed, recent communication technologies have made it easier, and therefore more common, for those living within a "foreign" culture to maintain connections to their culture-of-origin—both those "back home" and others who, like themselves, are experiencing the changes and disconnections of living in a new culture.[11]

Cultural identities are also multifaceted. At any given moment, you have many "components" that make up your identity. For instance, a specific person may simultaneously view herself as a student, an employee, a friend, a woman, a Southerner, a daughter, a Methodist, a Millennial, and more. Similarly, there are typically many facets or components to your cultural identity.

Many people incorrectly assume that an individual could, or perhaps should, identify with only one cultural group. However, as Young Yun Kim suggests,

> If someone sees himself or herself, or is seen by others, as a Mexican-American, then this person's identity is [commonly] viewed to exclude all other identities. This tendency to see cultural identity in an "all-or-none" and "either-or" manner glosses over the fact that many people's identities are not locked into a single, uncompromising category, but incorporate other identities as well.[12]

Given our increasingly multicultural world, in which people from many cultures coexist and in which the United States has become a country where individuals from many cultures live and interact, the multifaceted characteristic of cultural identity becomes even more important.[13]

CULTURAL BIASES

In Chapter 2, we defined culture as a learned set of shared interpretations about beliefs, values, norms, and social practices that affect the behaviors of a relatively large group of people. We also pointed out that culture really exists in people's minds, but that the consequences of culture—the shared interpretations—can be seen in people's communication behaviors. Shared interpretations, which we have called cultural patterns, provide guidelines about how people should behave, and they indicate what to expect in interactions with others. In other words, a culture's shared interpretations create predictability and stability in people's lives. Cultural similarity allows people to reduce uncertainty and to know what to expect when interacting with others.

Interaction only within one's own culture produces a number of obvious benefits. Because the culture provides predictability, it reduces the threat of the unknown. When something or someone that is unknown or unpredictable enters a culture, the culture's beliefs, values, norms, and social practices tell people how to interpret and respond appropriately, thus reducing the perceived threat of the intrusion. Cultural patterns also allow for automatic responses to stimuli; in essence, cultural patterns save people time and energy.

Intercultural communication, by definition, means that people are interacting with at least one culturally different person. Consequently, the sense of security, comfort, and predictability that characterizes communication with culturally similar people is lost. The greater the degree of interculturalness, the greater the loss of predictability and certainty. Assurances about the accuracy of interpretations of verbal and nonverbal messages are lost.

Terms that are often used when communicating with culturally different people include *unknown, unpredictable, ambiguous, weird, mysterious, unexplained, exotic, unusual, unfamiliar, curious, novel, odd, outlandish,* and *strange.* As you read this list, consider how the choice of a particular word might also reflect a particular value. What characteristics, values, and knowledge allow individuals to respond more competently to the threat of dealing with cultural differences? What situations heighten the perception of threat among members of different cultural groups? To answer questions such as these, we need to explore how people make sense of information about others as they categorize or classify others in their social world.

Social Categorizing

Three features in the way all humans process information about others are important to your understanding of intercultural competence. First, as cognitive psychologists have repeatedly demonstrated, people impose a pattern on their world by organizing the stimuli that bombard their senses into conceptual categories. Every waking moment, people are presented with literally hundreds of different perceptual stimuli. Therefore, it becomes necessary to simplify the information by selecting, organizing, and reducing it to less complex forms. That is, to comprehend stimuli, people organize them into categories, groupings, and patterns. As a child, you might have completed a drawing by connecting numbered dots. Emerging from the lines was the figure of an animal or a familiar toy. Even though its complete form was not drawn, it was relatively easy to identify. This kind of recognition occurs simply because human beings have a tendency to organize perceptual cues to impose meaning, usually by using familiar, previous experiences.

Second, most people tend to think that other people perceive, evaluate, and reason about the world in the same way that they do. In other words, humans assume that other people with whom they interact are like themselves. Indeed, it is quite common for people to draw on their personal experiences to understand and evaluate the motivations of others. This common human tendency is sometimes called ethnocentrism.

Third, humans simplify the processing and organizing of information from the environment by identifying certain characteristics as belonging to certain categories of persons and events. For example, a child's experiences with several dogs that growled and snapped are likely to result in a future reaction to other dogs as if they will also growl and snap. The characteristics of particular events, persons, or objects, once experienced, are often assumed to be typical of similar events, persons, or objects. Though these assumptions are sometimes accurate, often they are not. Not all dogs necessarily growl and snap at young children. Nevertheless, information processing results in a simplification of the world, so that prior experiences are used as the basis for determining both the categories and the attributes of the events. This process is called stereotyping.

Please note that we are describing these human tendencies nonevaluatively. Their obvious advantage is that they allow people to respond efficiently to a variety of perceptual stimuli. Nevertheless, this organization and simplification can create some genuine obstacles to intercultural competence because they may lead to prejudice, discrimination, and racism.

Ethnocentrism

Twenty-five hundred years ago, the Greek historian Herodotus, whom Cicero called "The Father of History," related a story about Darius, the first monarch of the great Persian empire. Darius became king of Persia (now Iran) in 521 B.C., and he ruled a vast empire that, for a time, included most of the "known" world, including southeastern Europe, northern Africa, India, southern Russia, and the Middle East. Darius, so the story goes,

> sent for the Greeks at his court to ask them their price for devouring the corpses of their ancestors. They replied that no price would be high enough. Thereupon the Persian king summoned the representatives of an Indian tribe which habitually practiced the custom from which the Greeks shrank, and asked them through the interpreter, in the presence of the Greeks, at what price they would burn the corpses of their ancestors. The Indians cried aloud and besought the king not even to mention such a horror. From these circumstances the historian drew the following notable moral for human guidance: If all existing customs could somewhere be set before all men in order that they might select the most beautiful for themselves, every nation would choose out, after the most searching scrutiny, the customs they had already practiced.[14]

In the preceding passage, Herodotus described what is now called *ethnocentrism*, which is the notion that the beliefs, values, norms, and practices of one's own culture are superior to those of others.

All cultures teach their members the "preferred" ways to respond to the world, which are often labeled as "natural" or "appropriate." Thus, people generally perceive their own experiences, which are shaped by their own cultural forces, as natural, human, and universal.

Cultures also train their members to use the categories of their own cultural experiences when judging the experiences of people from other cultures. Our culture tells us that the way we were taught to behave is "right" or "correct," and those who do things differently are wrong. William G. Sumner, who first introduced the concept of

CULTURE CONNECTIONS

I do still encounter incidents and people who don't understand Koreans or Korean culture. At work I was hanging out in the kitchen tossing French fries back and forth with a cook who is African American. He said, "Hey, don't be doing any of that Kung-fu stuff on me! I know you Chinese people!" I said, "Um, I'm not Chinese. I'm Korean." He said, "What's the difference?" like it was unimportant. I said, "It's a big difference. What's the difference between a Jamaican and a Nigerian?" He said, "I don't know." "The difference is culture! Different languages, food, values, beliefs, traditions, music, ways of life!"

—Mei Lin Kroll ■

ethnocentrism, defined it as "the view of things in which one's own group is the center of everything, and all others are scaled and rated with reference to it."[15] Sumner illustrates how ethnocentrism works in the following example:

> When Caribs were asked whence they came, they replied, "We alone are people." "Kiowa" means real or principal people. A Laplander is a "man" or "human being." The highest praise a Greenlander has for a European visiting the island is that the European by studying virtue and good manners from the Greenlanders soon will be as good as a Greenlander. Nature peoples call themselves "men" as a rule. All others are something else, but not men. The Jews divide all mankind into themselves and Gentiles—they being the "chosen people." The Greeks and Romans called outsiders "barbarians." Arabs considered themselves as the noblest nation and all others as barbarians. Russian books and newspapers talk about its civilizing mission, and so do the books and journals of France, Germany, and the United States. Each nation now regards itself as the leader of civilization, the best, the freest, and the wisest. All others are inferior.[16]

Ethnocentrism is a learned belief in cultural superiority. Because cultures teach people what the world is "really like" and what is "good," people consequently believe that the values of their culture are natural and correct. Thus, people from other cultures who do things differently are wrong. When combined with the natural human tendency to prefer what is typically experienced, ethnocentrism produces emotional reactions to cultural differences that reduce people's willingness to understand disparate cultural messages.

Ethnocentrism tends to highlight and exaggerate cultural differences. As an interesting instance of ethnocentrism, consider beliefs about body odor. Most U.S. Americans spend large sums of money each year to rid themselves of natural body odor. They then replace their natural odors with artificial ones as they apply deodorants, bath powders, shaving lotions, perfumes, hair sprays, shampoos, mousse, gels, toothpaste, mouthwash, and breath mints. Many U.S. Americans probably believe that they do not have an odor—even after they have routinely applied most, if not all, of the artificial ones in the preceding list. Yet the same individuals will react negatively to culturally different others who do not remove natural body odors and who refuse to apply artificial ones.

Another example of ethnocentrism concerns the way in which cultures teach people to discharge mucus from the nose. Most U.S. Americans purchase boxes of tissues and strategically place them at various points in their homes, offices, and cars so that they will be available for use when blowing their noses. In countries where paper products have historically been scarce and very expensive, people blow their noses onto the ground or the street. Pay attention to your reaction as you read this last statement. Most U.S. Americans, when learning about this behavior, react with a certain amount of disgust. But think about the U.S. practice of blowing one's nose into a tissue or handkerchief, which is then placed on the desk or into a pocket or purse. Now ask yourself which is really more disgusting—carrying around tissues with dried mucus in them or blowing the mucus onto the street? Described in this way, both practices have a certain element of repugnance, but because one's culture teaches that there is one preferred way, that custom is familiar and comfortable and the practices of other cultures are seen as wrong or distasteful.

Ethnocentrism can occur along all of the dimensions of cultural patterns discussed in the previous two chapters. People from individualistic cultures, for instance, find the idea that a person's self-concept is tied to a group to be unfathomable. To most U.S. Americans, the idea of an arranged marriage seems strange at best and a confining and reprehensible limitation on personal freedom at worst.

One area of behavior that quickly reveals ethnocentrism is personal hygiene. For example, U.S. Americans like to see themselves as the cleanest people on Earth. In the United States, bathrooms contain sinks, showers or bathtubs, and toilets, thus allowing the efficient use of water pipes. Given this arrangement, people bathe themselves in close proximity to the toilet, where they urinate and defecate. Described in this way, the cultural practices of the United States may seem unclean, peculiar, or even absurd. Why would people in a so-called modern society place two such contradictory functions next to each other? People from many other cultures, who consider the U.S. arrangement to be unclean and unhealthy, share that sentiment. Our point here is that what is familiar and comfortable inevitably seems the best, right, and natural way of doing things. Judgments about what is "right" or "natural" create emotional responses to cultural differences that may interfere with our ability to understand the symbols used by other cultures. For example, European Americans think it is "human nature" to orient oneself to the future and to want to improve one's material status in life. Individuals whose cultures have been influenced by alternative forces, resulting in contrary views, are often judged negatively and treated with derision.

To be a competent intercultural communicator, you must realize that you typically use the categories of your own culture to judge and interpret the behaviors of those who are culturally different from you. You must also be aware of your own emotional reactions to the sights, sounds, smells, and variations in message systems that you encounter when communicating with people from other cultures. The competent intercultural communicator does not necessarily suppress negative feelings but acknowledges their existence and seeks to minimize their effect on her or his communication. If you are reacting strongly to some aspect of another culture, seek out an explanation in the ethnocentric preferences that your culture has taught you.

The U.S. American preoccupation with body odors can be seen on the shelves of many stores.

CULTURE CONNECTIONS

He gave us some food called rice, which looked as revolting as tapeworms, so I refused to eat it. The other boys insisted that it tasted good. I still would not eat it. There were pieces of meat in the rice and I picked at them so the teacher would not say I was ungrateful. Until I went to school my staple foods had been milk, meat, and in dry seasons, maize. Now for the first time I would taste sweets and biscuits, European and African fruits such as bananas and oranges, and even grains like beans. The weird smell of soap, which stunk at first, would eventually become acceptable.

—Tepilit Ole Saitoti ∎

Stereotyping

Journalist Walter Lippmann introduced the term *stereotyping* in 1922 to refer to a selection process that is used to organize and simplify perceptions of others.[17] Stereotypes are a form of generalization about some group of people. When people stereotype others, they take a category of people and make assertions about the characteristics of all people who belong to that category. The consequence of stereotyping is that the vast degree of differences that exists among the members of any one group may not be taken into account in the interpretation of messages.

To illustrate how stereotyping works, read the following list: college professors, surfers, Marxists, Democrats, bankers, New Yorkers, Californians. Probably, as you read each of these categories, it was relatively easy for you to associate particular characteristics and traits with each group. Now imagine that a person from one of these groups walked into the room and began a conversation with you. In all likelihood, you would associate the group's characteristics with that specific individual.

Your responses to this simple example illustrate what typically occurs when people are stereotyped. First, someone identifies an outgroup category—"they"—whose characteristics differ from those in one's own social ingroup. Next, the perceived dissimilarities between the groups are enlarged and accentuated, thereby creating differences that are clearer and more distinct. By making sharper and more pronounced boundaries between the groups, it becomes more difficult for individuals to move from one group to another. Concurrently, an evaluative component is introduced, whereby the characteristics of the outgroup are negatively judged; that is, the outgroup is regarded as wrong, inferior, or stigmatized as a result of given characteristics. Finally, the group's characteristics are attributed to all people who belong to the group, so that a specific person is not treated as a unique individual but as a typical member of a category.

Categories that are used to form stereotypes about groups of people can vary widely, and they might include the following:

- Regions of the world (Asians, Arabs, South Americans, Africans)
- Countries (Kenya, Japan, China, France, Great Britain)
- Regions within countries (Northern Indians, Southern Indians, U.S. Midwesterners, U.S. Southerners)
- Cities (New Yorkers, Parisians, Londoners)
- Cultures (English, French, Latino, Russian, Serbian, Yoruba, Mestizo, Thai, Navajo)
- Race (African, Caucasian)
- Religion (Muslim, Hindu, Buddhist, Jewish, Christian)
- Age (young, old, middle-aged, children, adults)
- Occupations (teacher, farmer, doctor, housekeeper, mechanic, architect, musician)
- Relational roles (mother, friend, father, sister, brother)
- Physical characteristics (short, tall, fat, skinny)
- Social class (wealthy, poor, middle class)

All cultures are ethnocentric and consider outsiders to be foreigners.

This list is by no means exhaustive. What it should illustrate is the enormous range of possibilities for classification

CULTURE CONNECTIONS

A great many people think they are thinking when they are merely rearranging their prejudices.

—Attributed to William James ■

and simplification. Consider your own stereotypes of people in these groups. Many may have been created by direct experience with only one or two people from a particular group. Others are probably based on secondhand information and opinions, output from the mass media, and general habits of thinking; they may even have been formed without any direct experience with individuals from the group. Yet many people are prepared to assume that their stereotypes are accurate representations of all members of specific groups.[18] Interestingly, stereotypes that are based on secondhand opinions—that is, stereotypes that are derived from the opinions of others or from the media—tend to be more extreme, less variable from one person to another, more uniformly applied to others, and more resistant to change than are stereotypes based on direct personal experiences and interactions.[19]

Stereotypes can be inaccurate in three ways.[20] First, as we have suggested, stereotypes often are assumed to apply to all or most of the members of a particular group or category, resulting in a tendency to ignore differences among the individual members of the group. This type of stereotyping error is called the *outgroup homogeneity effect* and results in a tendency to regard all members of a particular group as much more similar to one another than they actually are.[21] Arab Americans, for instance, complain that other U.S. Americans often hold undifferentiated stereotypes about members of their culture. Albert Mokhiber laments that,

> if there's problem in Libya we're all Libyans. If the problem is in Lebanon we're all Lebanese. If it happens to be Iran, which is not an Arab country, we're all Iranians. Conversely, Iranians were picked on during the Gulf War as being Arabs. Including one fellow who called in who was a Polynesian Jew. But he looked like what an Arab should look like, and he felt the wrath of anti-Arab discrimination. Nobody's really free from this. The old civil rights adage says that as long as the rights of one are in danger, we're all in danger. I think we need to break out of our ethnic ghetto mentality, all of us, from various backgrounds, and realize that we're in this stew together.[22]

A second form of stereotype inaccuracy occurs when the group average, as suggested by the stereotype, is simply wrong or inappropriately exaggerated. This type of inaccuracy occurs, for instance, when Germans are stereotypically regarded as being very efficient, or perhaps very rigid, when they may actually be less efficient or less rigid than the exaggerated perception of them would warrant.

A third form of stereotype inaccuracy occurs when the degree of error and exaggeration differs for positive and negative attributes. For instance, imagine that you have stereotyped a culture as being very efficient (a positive attribute) but also very rigid and inflexible in its business relationships (a negative attribute). If you tend to overestimate the prevalence and importance of the culture's positive characteristics, such as its degree of efficiency, while simultaneously ignoring or underestimating its rigidity and other negative characteristics, you would have a "positive valence inaccuracy." Conversely, a "negative valence inaccuracy" occurs if you exaggerate the negative attributes while ignoring or devaluing its positive ones. This latter condition, often called *prejudice,* will be discussed in greater detail later.

The problems associated with using stereotyping as a means of understanding individuals is best illustrated by identifying the groups to which you belong. Think about the characteristics that might be stereotypically assigned to those groups. Determine whether the characteristics apply to you or to others in your group. Some of them may be accurate descriptions; many, however, will be totally inaccurate, and you would resent being thought of in that way. Stereotypes distort or hide the individual. Ultimately, people may become blind to the actual characteristics of the group because not all stereotypes are accurate. Most are based on relatively minimal experiences with particular individuals.

Stereotype inaccuracy can lead to errors in interpretations and expectations about the behaviors of others. Interpretation errors occur because stereotypes are used not only to categorize specific individuals and events but also to judge them. That is, one potentially harmful consequence of stereotypes is that they provide inaccurate labels for a group of people, which are then used to interpret subsequent ambiguous events and experiences involving members of those groups. As Ziva Kunda and Bonnie Sherman-Williams note,

> Consider, for example, the unambiguous act of failing a test. Ethnic stereotypes may lead perceivers to attribute such failure to laziness if the actor is Asian but to low ability if the actor is Black. Thus stereotypes will affect judgments of the targets' ability even if subjects base these judgments only on the act, because the stereotypes will determine the meaning of the act.[23]

Because stereotypes are sometimes applied indiscriminately to members of a particular culture or social group, they can also lead to errors in one's expectations about the future behaviors of others. Stereotypes provide the bases for estimating, often inaccurately, what members of the stereotyped group are likely to do. Most disturbingly, stereotypes will likely persist even when members of the stereotyped group repeatedly behave in ways that disconfirm them. Once a stereotype has taken hold, members of the stereotyped group who behave in nonstereotypical ways will be expected to compensate in their future actions in order to "make up for" their atypical behavior. Even when some individuals from a stereotyped group repeatedly deviate from expectations, they may be regarded as exceptions or as atypical members of their group. Indeed, stereotypes may remain intact, or may even be strengthened, in the face of disconfirming experiences; those who hold the stereotypes often expect that the other members of the stereotyped social group will be even *more* likely to behave as the stereotype predicts, in order to "balance out" or compensate for the "unusual" instances that they experienced. That is, stereotypes encourage people to expect future behaviors that compensate for perceived inconsistencies and thus allow people to anticipate future events in a way that makes it unnecessary to revise their deeply held beliefs and values.[24]

The process underlying stereotyping is absolutely essential for human beings to function. Some categorization is necessary and normal. Indeed, there is survival value in the ability to make accurate generalizations about others, and stereotypes function as mental "energy-saving devices" to help make those generalizations efficiently.[25] However, stereotypes may also promote prejudice and discrimination directed toward members of cultures other than one's own. Intercultural competence requires an ability to move beyond stereotypes and to respond to the individual. Previous experiences should be used only as guidelines or suggested interpretations rather than as hard-and-fast categories. Judee Burgoon, Charles Berger, and Vincent Waldron suggest that mindfulness—that is, paying conscious attention to the nature and basis of one's stereotypes—can help to reduce stereotype inaccuracies and thereby decrease intercultural misunderstandings.[26]

Prejudice

Prejudice refers to negative attitudes toward other people that are based on faulty and inflexible stereotypes. Prejudiced attitudes include irrational feelings of dislike and even hatred for certain groups, biased perceptions and beliefs about the group members that are not based on direct experiences and firsthand knowledge, and a readiness to behave in negative and unjust ways toward members of the group. Gordon Allport, who first focused scholarly attention on prejudice, argued that prejudiced people ignore evidence that is inconsistent with their biased viewpoint, or they distort the evidence to fit their prejudices.[27]

The strong link between prejudice and stereotypes should be obvious. Prejudiced thinking is dependent on stereotypes and is a fairly normal phenomenon.[28] To be prejudiced toward a group of people sometimes makes it easier to respond to them. We are not condoning prejudice or the hostile and violent actions that may occur as a result of prejudice. We are suggesting that prejudice is a universal psychological process; all people have a propensity for prejudice toward others who are unlike themselves. For individuals to move beyond prejudicial attitudes and for societies to avoid basing social structures on their prejudices about groups of people, it is critical to recognize the prevalence of prejudicial thinking.

What functions does prejudice serve? We have already suggested that the thought process underlying prejudice includes the need to organize and simplify the world. Richard Brislin describes four additional benefits, or what he calls functions, of prejudice.[29] First, he suggests that prejudice satisfies a *utilitarian* or adjustment function. Displaying certain kinds of prejudice means that people receive rewards and avoid punishments. For example, if you express prejudicial statements about certain people, other people may like you more. It is also easier to simply dislike and be prejudiced toward members of other groups because they can then be dismissed without going through the effort necessary to adjust to them. Another function that prejudice serves is an *ego-defensive* one; it protects self-esteem.[30] If others say or do things that are inconsistent with the images we hold of ourselves, our sense of self may be deeply threatened, and we may try to maintain our self-esteem by scorning the sources of the message. So, for example, people who are unsuccessful in business may feel threatened by groups whose members are successful. Prejudice may function to protect one's self-image by denigrating or devaluing those who might make us feel less worthy.[31] Still another

This Italian train, spray-painted in German with "Jews have money. And you???" illustrates the prejudice, discrimination, and racism (called "anti-Semitism" when referring to Jews) experienced by many cultural groups.

CULTURE CONNECTIONS

I had left Nigeria with the impression that African Americans were confined to a life of silliness and crime in America, and that the experience of slavery had wrought considerable havoc on the psyches of blacks. My experiences in Nigeria with some African Americans living and working there (my boss was African American, as was one of my good high school friends) did little to alter the very negative perceptions reflected in media images that had become the prism through which I saw life in America. My initial interactions with African Americans upon my arrival in the United States were therefore filled with curiosity, because I wanted to get to know them, and caution—one could even say fear—because of the negative media portrayals. Such fear was manifested in my polite refusal of an offer from a young African American male staffer to assist me with my luggage at Dulles; yet I felt safe when a similar offer came, moments later, from a white male attendant! For most new-comers into a foreign land, their knowledge of a place is typically informed by media images.

—Peter O. Nwosu ∎

advantage of prejudicial attitudes is the *value-expressive* function. If people believe that their group has certain qualities that are unique, valuable, good, or in some way special, their prejudicial attitudes toward others is a way of expressing those values. Finally, Brislin describes the *knowledge function* as prejudicial attitudes that people hold because of their need to have the world neatly organized and boxed into categories. This function takes the normal human proclivity to organize the world to an extreme. The rigid application of categories and the prejudicial attitudes assigned to certain behaviors and beliefs provide security and increase predictability. Obviously, these functions cannot be neatly applied to all instances of prejudice. Nor are people usually aware of the specific reasons for their prejudices. For each person, prejudicial attitudes may serve several functions.

Discrimination

Whereas *prejudice* refers to people's attitudes or mental representations, the term discrimination refers to the behavioral manifestations of that prejudice. Thus discrimination can be thought of as prejudice "in action."

Discrimination can occur in many forms. From the extremes of segregation and apartheid to biases in the availability of housing, employment, education, economic resources, personal safety, and legal protections, discrimination represents unequal treatment of certain individuals solely because of their membership in a particular group.

Teun van Dijk has conducted a series of studies of people's everyday conversations as they discussed different racial and cultural groups. Van Dijk concludes that, when individuals make prejudicial comments, tell jokes that belittle and dehumanize others, and share negative stereotypes about others, they are establishing and legitimizing the existence of their prejudices and are laying the "communication groundwork" that will make it acceptable for people to perform discriminatory acts.[32]

Often, biases and displays of discrimination are motivated not by direct hostility toward some other group but merely by a strong preference for, and loyalty to, one's own culture.[33] Thus, the formation of one's cultural identity, which we discussed earlier in this chapter, can sometimes lead to hostility, hate, and discrimination directed against nonmembers of that culture.

Racism

One obstacle to intercultural competence to which we want to give special attention is racism. Because racism often plays such a major role in the communication that occurs between people of different races or ethnic groups, it is particularly important to understand how and why it occurs.

The word racism itself can evoke very powerful emotional reactions, especially for those who have felt the oppression and exploitation that stems from racist attitudes and behaviors. For members of the African American, Asian American, Native American, and Latino cultures, racism has created a social history shaped by prejudice and discrimination.[34] For individual members of these groups, racism has resulted in the pain of oppression. To those who are members of cultural groups that have had the power to oppress and exploit others, the term *racism* often evokes equally powerful thoughts and emotional reactions that deny responsibility for and participation in racist acts and thinking. In this section, we want to introduce some ideas about racism that illuminate the reactions of both those who have received racist communication and those who are seen as exhibiting it.

Robert Blauner has described racism as a tendency to categorize people who are culturally different in terms of their physical traits, such as skin color, hair color and texture, facial structure, and eye shape.[35] Dalmas Taylor offers a related approach that focuses on the behavioral components of racism. Taylor defines racism as the cumulative effects of individuals, institutions, and cultures that result in the oppression of ethnic minorities.[36] Taylor's approach is useful in that it recognizes that racism can occur at three distinct levels: individual, institutional, and cultural.

Racism is a force with which both individuals and social systems must grapple. This monument, in California's Manzanar War Relocation Center, stands as a reminder that about 120,000 Japanese American citizens were rounded up and placed in one of ten internment camps in the United States during World War II.

At the individual level, racism is conceptually very similar to prejudice. Individual racism involves beliefs, attitudes, and behaviors of a given person toward people of a different racial group.[37] Specific European Americans, for example, who believe that African Americans are somehow inferior, exemplify individual racism. Positive contact and interaction between members of the two groups can sometimes change these attitudes. Yet, as the preceding discussion of prejudice suggests, people with prejudicial beliefs about others often distort new information to fit their original prejudices.

At the institutional level, racism is the exclusion of certain people from equal participation in the society's institutions solely because of their race.[38] Institutional racism is built into such social structures as the government, schools, the media, and industry practices. It leads to certain patterns of behaviors and responses to specific racial or cultural groups that allow those groups to be systematically exploited and oppressed. For example, institutional racism has precluded both Jews and African Americans from attending certain public schools and universities, and at times it has restricted their participation in particular professions.[39] Repeated instances of institutional racism, which commonly appear in the popular media, can be especially difficult to overcome. By focusing on some topics or characteristics and not on others, the media often "prime" people's attention and thereby influence the interpretations and evaluations one makes of others. Such biased portrayals can be particularly salient when the media provide people's primary or only knowledge of particular cultures and their members. Consider, for example, Elizabeth Bird's insightful analysis of the ways in which Native Americans are marginalized by the popular media's portrayal of their sexuality:

> The representations we see are structured in predictable, gendered ways. Women are faceless, rather sexless squaws in minor roles, or sexy exotic princesses or maidens who desire White men. Men are either handsome young warriors, who desire White women, or safe sexless wise elders, who dispense ancient wisdom. Nowhere, in this iconography, do the male and female images meet. The world where American Indian men and women love, laugh, and couple *together* lurks far away in the shadows. These days, representations of American Indians are more accurate, in terms of costume, cultural detail, and the like than in the 1950s, when White actors darkened their skins to play American Indians. As far as suggesting an authentic, subjective American Indian experience, though, there has been little progress.[40]

As Bird concludes, current portrayals of Native Americans "may be more benign images than the squaw or the crazed savage, but they are equally unreal and, ultimately, equally dehumanizing."[41]

At the cultural level, racism denies the existence of the culture of a particular group[42]—for example, the denial that African Americans represent a unique and distinct culture that is separate from both European American culture and all African cultures. Cultural racism also involves the rejection by one group of the beliefs and values of another, such as the "negative evaluations by whites of black cultural values."[43]

Although *racism* is often used synonymously with *prejudice* and *discrimination,* the social attributes that distinguish it from these other terms are oppression and power. Oppression refers to "the systematic, institutionalized mistreatment of one group of people by another."[44] Thus, racism is the tendency by groups in control of institutional and cultural power to use it to keep members of groups who do not have access to the same kinds of power at a disadvantage. Racism oppresses entire groups of people, making it very difficult, and sometimes virtually impossible, for their members to have access to political, economic, and social power.[45]

Forms of racism vary in intensity and degree of expression, with some forms far more dangerous and detrimental to society than others. The most extreme form of racism is *old-fashioned racism.* Here, members of one group openly display obviously bigoted views about those from another group. Judgments of superiority and inferiority are commonplace in this kind of racism, and there is a dehumanizing quality to it. African Americans and other cultural groups in the United States have often experienced this form of racism from other U.S. Americans.

Symbolic racism, which is sometimes called *modern racism,* is currently prevalent in the United States. In symbolic racism, members of a group with political and economic power believe that members of some other group threaten their traditional values, such as individualism and self-reliance. Fears that the outgroup will achieve economic or social success, with a simultaneous loss of economic or social status by the ingroup, typify this form of racism. In many parts of the United States, for instance, this type of racism has been directed toward Asians and Asian Americans, who are accused of being too "pushy" because they have achieved economic success. Similarly, symbolic racism includes the expression of feelings that members of cultures such as African Americans and Mexican Americans are moving too fast in seeking social change, are too demanding of equality and social justice, are not playing by the "rules" established in previous generations, and simply do not deserve all that they have recently gained. Paradoxically, symbolic racists typically do not feel personally threatened by the successes of other cultures, but they fear for their core values and the continued maintenance of their political and economic power.[46]

Tokenism as a form of racism occurs when individuals do not perceive themselves as prejudiced because they make small concessions to, while holding basically negative attitudes toward, members of the other group. Tokenism is the practice of reverse discrimination, in which people go out of their way to favor a few members of another group in order to maintain their own self-concepts as individuals who believe in equality for all. Such behaviors may increase a person's esteem, but they may also decrease the possibilities for more meaningful contributions to intercultural unity and progress.

Aversive racism, like tokenism, occurs when individuals who highly value fairness and equality among all racial and cultural groups nevertheless have negative beliefs and feelings about members of a particular race, often as a result of childhood socialization experiences. Individuals with such conflicting feelings may restrain their overt racist behaviors, but they may also avoid close contact with members of the other group and may express their underlying negative attitudes subtly, in ways that appear rational and that can be justified on the basis of some factor other than race or culture. Thus, the negativity of aversive racists "is more likely to be manifested in discomfort, uneasiness, fear, or avoidance of minorities rather than overt hostility."[47] An individual at work, for instance, may be polite but distant to a coworker from another culture but may avoid that person at a party they both happen to attend.

Genuine likes and dislikes may also operate as a form of racism. The cultural practices of some groups of people can form the basis for a prejudicial attitude simply because the group displays behaviors that another group does not like. For example, individuals from cultures that are predominantly vegetarian may develop negative attitudes toward those who belong to cultures that eat meat.

Finally, the least alarming form of racism, and certainly one that everyone has experienced, is based on the degree of unfamiliarity with members of other groups. Simply responding to unfamiliar people may create negative attitudes because of a lack of experience with the characteristics of their group. The others may look, smell, talk, or act differently, all of which can be a source of discomfort and can form the basis for prejudicial attitudes or racist actions.

IDENTITY, BIASES, AND INTERCULTURAL COMPETENCE

In Chapters 4 and 5, we suggested that learning about the preferences that describe your own culture's patterns, in order to understand better your own beliefs, values, norms, and social practices, is an important step toward improving intercultural competence. The discussion of cultural identity in this chapter should serve to reinforce this guideline. A good place to begin is by describing your own cultural identity. Is this relatively easy for you to do? Have you always been aware of your cultural background, or have you experienced events that have caused you to search for an understanding

CULTURE CONNECTIONS

Niran was not an Icelander and had no interest in becoming one, but living up here in the Arctic meant that he could hardly call himself Thai either. He realised that he was neither. He belonged to neither country, belonged nowhere except in some invisible, intangible no man's land. Previously he had never had to think about where he came from. He was a Thai, born in Thailand. Now he drew strength from the company of other immigrant children with similar backgrounds and made his best friends among them. He became fascinated with his heritage, with the history of Thailand and the story of his ancestors. The feeling had only intensified when he got to know other, older immigrant children at his last school.

—Arnaldur Indridason   ■

of your cultural identity? Do you place your cultural identity primarily in one cultural group or in several cultural groups? How does your cultural identity shape your social and personal identities? Does your cultural identity result in a strong sense of others as either in or out of your cultural group? If so, were you taught to evaluate negatively those who are not part of your cultural group? Conversely, do you sometimes feel excluded from and evaluated negatively by people from cultures that differ from your own? The answers to these questions will help you to understand the possible consequences, both positive and negative, of your cultural identity as you communicate interculturally.

To improve your intercultural competence by building positive motivations, or emotional reactions, to intercultural interactions, take an honest inventory of the various ways in which you categorize other people. Can you identify your obvious ethnocentric attitudes about appearance, food, and social practices? Make a list of the stereotypes, both positive and negative, that you hold about the various cultural groups with which you regularly interact. Now identify those stereotypes that others might hold about your culture. By engaging in this kind of self-reflective process, you are becoming more aware of the ways in which your social categorizations detract from an ability to understand communication from culturally different others.

Ethnocentrism, stereotyping, prejudice, discrimination, and racism are so familiar and comfortable that overcoming them requires a commitment both to learning about other cultures and to understanding one's own. A willingness to explore various cultural experiences without prejudgment is necessary. An ability to behave appropriately and effectively with culturally different others, without invoking prejudiced and stereotyped assumptions, is required. Although no one can completely overcome the obstacles to intercultural competence that naturally exist, the requisite knowledge, motivation, and skill can certainly help to minimize the negative effects of prejudice and discrimination.

The intercultural challenge for all of us now living in a world where interactions with people from different cultures are common features of daily life is to be willing to grapple with the consequences of prejudice, discrimination, and racism at the individual, social, and institutional levels. Because *prejudice* and *racism* are such emotionally charged concepts, it is sometimes very difficult to comment on their occurrence in our interactions with others. Individuals who believe that they have perceived discriminatory remarks and prejudicial actions often recognize that there may be substantial social costs associated with speaking out, and consequently they may sometimes be unwilling to risk the negative evaluations from their coworkers, fellow students, teachers, or service providers that would likely occur should they directly confront such biases and demand interactions that do not display them.[48] Conversely, those who do not regard themselves as having prejudiced or racist attitudes and who believe they never behave in discriminatory ways are horrified to learn that others might interpret their attitudes as prejudiced and their actions as discriminatory. Although discussions about prejudice, discrimination, and racism can lead to a better understanding of the interpersonal dynamics that arise as individuals seek to establish mutually respectful relationships, they can just as easily lead to greater divisions and hostilities between people. The challenge for interculturally competent communicators is to contend with the pressing but potentially inflammatory issues of prejudice and discrimination in a manner that is both appropriate and effective.

We are also challenged to function competently in a world that, increasingly, is characterized by multiple cultures inhabiting adjacent and often-overlapping terrain. The ability to adapt to these intercultural settings—to maintain positive, healthy relationships with people from cultures other than your own—is the hallmark of the interculturally competent individual.

SUMMARY

This chapter began with a discussion of cultural identity. The cultures with which you identify affect your views about where you belong and whom you consider to be "us" and "them."

Next, we discussed the biases that impede the development of intercultural competence. Ethnocentrism, stereotyping, prejudice, discrimination, and racism occur because of the human need to organize and streamline the processing of information. When people assume that these "thinking shortcuts" are accurate representations, intercultural competence is impaired.

Cultural biases are based on normal human tendencies to view ourselves as members of a particular group and to view others as not belonging to that group. Status, power, and economic differences heavily influence all intercultural contacts. Cultural biases are a reminder that all relationships take place within a political, economic, social, and cultural context. The intercultural challenge for all of us, as we live in a world where interactions with people from different cultures are common features of daily life, is to be willing to grapple with the consequences of prejudice, discrimination, and racism at the individual, social, and institutional levels.

FOR DISCUSSION

1. If people are born into one culture but raised in another, to which culture(s) do they belong?
2. What are the advantages and disadvantages for U.S. Americans who grow up with multiple cultural heritages?
3. What do people lose, and what do they gain, from having an ethnocentric perspective?
4. Is it possible for European Americans to be the recipients of any form of racism in the United States?
5. Why might less obvious or less alarming forms of racism be just as dangerous as old-fashioned or symbolic racism?

FOR FURTHER READING

Gordon W. Allport, *The Nature of Prejudice* (Cambridge, MA: Addison-Wesley, 1954). A classic work that established our understanding of the hows, whys, and nature of prejudice in human interactions.

Yoshihisa Kashima, Klaus Fiedler, and Peter Freytag (eds.), *Stereotype Dynamics: Language-Based Approaches to the Formation, Maintenance, and Transformation of Stereotypes* (New York: Erlbaum, 2008). Details how our stereotypes of others develop, persist, and (sometimes) change.

Todd D. Nelson (ed.), *Handbook of Prejudice, Stereotyping, and Discrimination* (New York: Psychology Press, 2009). Provides a thorough and comprehensive understanding of the causes and consequences of stereotyping, prejudice, and discrimination, and offers some approaches for reducing prejudice.

Linda R. Tropp and Robyn K. Mallett (eds.), *Moving Beyond Prejudice Reduction: Pathways to Positive Intergroup Relations* (Washington, DC: American Psychological Association, 2011). Examines what the research on intercultural contacts suggests about the best ways to reduce prejudice and discrimination. Also explores ideas that can improve intercultural relationships.

Verbal Intercultural Communication

The Power of Language in Intercultural Communication

Definition of Verbal Codes

The Features of Language
Rule Systems in Verbal Codes
Interpretation and Intercultural
 Communication

Language, Thought, Culture, and Intercultural Communication

The Sapir–Whorf Hypothesis of Linguistic Relativity
Language and Intercultural Communication

Verbal Codes and Intercultural Competence

Summary

KEY TERMS

In this chapter, we consider the effects of language systems on people's ability to communicate interculturally. In so doing, we explore the accuracy of a statement by the world-famous linguistic philosopher Ludwig Wittgenstein, who asserted that "the limits of my language are the limits of my world."

THE POWER OF LANGUAGE IN INTERCULTURAL COMMUNICATION

Consider the following examples, each of which illustrates the pivotal role of language in human interaction:

Example 1

A U.S. business executive is selected by her company for an important assignment in Belgium, not only because she has been very successful but also because she speaks French. She prepares her materials and presentation and sets off for Belgium with high expectations for landing a new contract for her firm. Once in Belgium she learns that, although the individuals in the Belgian company certainly speak French, and there are even individuals who speak German or English, their first language and the preferred language for conducting their business is Flemish. Both the U.S. business executive and her company failed to consider that Belgium is a multicultural and multilingual country populated by Walloons who speak French and Flemings who speak Flemish.

Example 2

Vijay is a student from India who has just arrived in the United States to attend graduate school at a major university. Vijay began to learn English in primary school, and since his field of study is engineering, even his classes in the program leading to his bachelor's degree were conducted in English. Vijay considers himself to be proficient in the English language. Nevertheless, during his first week on campus, the language of those around him is bewildering. People seem to talk so fast that Vijay has difficulty differentiating one word from another. Even when he recognizes the words, he cannot quite understand what people mean by them. His dormitory roommate seemed to say, "I'll catch you later" when he left the room. The secretary in the departmental office tried to explain to him about his teaching assistantship and the students assigned to the classes he was helping to instruct. Her references to students who would attempt to "crash" the course were very puzzling to him. His new faculty advisor, sensing Vijay's anxiety about all of these new situations, told him to "hang loose" and "go with the flow." When Vijay inquired of another teaching assistant about the meaning of these words, the teaching assistant's only reaction was to shake his head and say, "Your advisor's from another time zone!" Needless to say, Vijay's bewilderment continued.

Language—whether it is English, French, Swahili, Flemish, Hindi, or one of the world's other numerous languages—is a taken-for-granted aspect of people's lives. Language is learned without conscious awareness. Children are capable of using their language competently before the age of formal schooling. Even during their school years, they learn the rules and words of the language and do not attend to how the language influences the way they think and perceive the world. It is usually only when people speak their language to those who do not understand it or when they struggle to become competent in another language that they recognize language's central role in the ability to function, to accomplish tasks, and, most important, to interact with others. It is only when the use of language no longer connects people to others or when individuals are denied the use of their language that they recognize its importance.

There is a set of circumstances involving communication with people from other cultural backgrounds in which awareness of language becomes paramount. Intercultural communication usually means interaction between people who speak different languages. Even when the individuals seem to be speaking the same language—a person from Spain interacting with someone from Venezuela, a French Canadian conversing with a French-speaking citizen of Belgium, or an Australian visiting the United States—the differences in the specific dialects of the language and the different cultural practices that govern language use can mystify those involved, and they can realistically be portrayed as two people who speak different languages.

In this chapter, we explore the nature of language and how verbal codes affect communication between people of different cultural backgrounds. Because this book is written in English and initially intended for publication and distribution in the United States, many of the examples and comparisons refer to characteristics of the English language as it is used in the United States. We begin with a discussion of the characteristics and rule systems that create verbal codes and the process of interpretation from one verbal code to another. We then turn to a discussion of the all-important topic of the relationship among language, culture, thought, and intercultural communication. As we consider this issue, we explore the Sapir–Whorf hypothesis of linguistic relativity and assess the scholarly evidence that has been amassed both in support of the hypothesis and in opposition to it. We also consider the importance of language in the identity of ethnic and cultural groups. The chapter concludes with a consideration of verbal codes and intercultural competence.

CULTURE CONNECTIONS

It was my first year of school, my first days away from the private realm of our house and tongue. I thought English would be simply a version of our Korean. Like another kind of coat you could wear. I didn't know what a difference in language meant then. Or how my tongue would tie in the initial attempts, stiffen so, struggle like an animal booby-trapped and dying inside my head. Native speakers may not fully know this, but English is a scabrous mouthful. In Korean, there are no separate sounds for L and R, the sound is singular and without a baroque Spanish trill or roll. There is no B and V for us, no P and F. I always thought someone must have invented certain words to torture us. *Frivolous. Barbarian....*

I will always make bad errors of speech. I remind myself of my mother and father, fumbling in front of strangers. Lelia says there are certain mental pathways of speaking that can never be unlearned. Sometimes I'll say *riddle* for *little*, or *bent* for *vent*, though without any accent and so whoever's present just thinks I've momentarily lost my train of thought. But I always hear myself displacing the two languages, conflating them—maybe conflagrating them—for there's so much rubbing and friction, a fire always threatens to blow up between the tongues. Friction, affliction. In kindergarten, kids would call me "Marble Mouth" because I spoke in a garbled voice, my bound tongue wrenching myself to move in the right ways.

—Chang-rae Lee ■

DEFINITION OF VERBAL CODES

Discussions about the uniqueness of human beings usually center on people's capabilities to manipulate and understand symbols that allow interaction with others. In a discussion of the importance of language, Charles F. Hockett noted that language allows people to understand messages about many different topics from literally thousands of people. Language allows a person to talk with others, to understand or disagree with them, to make plans, to remember the past, to imagine future events, and to describe and evaluate objects and experiences that exist in some other location. Hockett also pointed out that language is taught to individuals by others and, thus, is transmitted from generation to generation in much the same way as culture. In other words, language is learned.[1]

Popular references to language often include not only spoken and written language but also "body language." However, we will discuss the latter topic in the next chapter on nonverbal codes. Here, we will concentrate on understanding the relationship of spoken and written language, or verbal codes, to intercultural communication competence.

The Features of Language

Verbal means "consisting of words." Therefore, a verbal code is a set of rules about the use of words in the creation of messages. Words can obviously be either spoken or written. Verbal codes, then, include both oral (spoken) language and non-oral (written) language.

Children first learn the oral form of a language. Parents do not expect two-year-olds to read the words on the pages of books. Instead, as parents speak aloud to a child, they identify or name objects in order to teach the child the relationship between the language and the objects or ideas the language represents. In contrast, learning a second language as an adolescent or adult often proceeds more formally, with a combination of oral and non-oral approaches. Students in a foreign language class are usually required to buy a textbook that contains written forms of the language, which then guide students in understanding both the oral and the written use of the words and phrases.

The concept of a written language is familiar to all students enrolled in U.S. college and university classes, as they all require at least reasonable proficiency in the non-oral form of the English language. Fewer and fewer languages exist only in oral form. When anthropologists and linguists discover a culture that has a unique oral language, they usually attempt to develop a written form of it in order to preserve it. Indeed, many Hmong who immigrated to the United States from their hill tribes in Southeast Asia have had to learn not only the new language of English but also, in many instances, the basic fact that verbal codes can be expressed in written form. Imagine the enormous task it must be not only to learn a second language but also first to understand that language can be written.

Our concern in this chapter is principally with the spoken verbal codes that are used in face-to-face intercultural communication. Nevertheless, because the written language also influences the way the language is used orally, written verbal codes play a supporting role in our discussion, and some of our examples and illustrations draw on written expressions of verbal codes in intercultural communication.

An essential ingredient of both verbal and nonverbal codes is symbols. As you recall from Chapter 1, symbols are words, actions, or objects that stand for or represent units of meaning. The relationship between symbols and what they stand for is often highly arbitrary, particularly for verbal symbols.

Another critical ingredient of verbal codes is the system of rules that governs the composition and ordering of the symbols. Everyone has had to learn the rules of a language—how to spell, use correct grammar, and make appropriate vocabulary choices—and thereby gain enough mastery of the language to tell jokes, to poke fun, and to be sarcastic. Even more than differences in the symbols themselves, the variations in rules for ordering and using symbols produce the different languages people use.

Rule Systems in Verbal Codes

Five different but interrelated sets of rules combine to create a verbal code, or language. These parts, or components, of language are called phonology, morphology, semantics, syntactics, and pragmatics.

PHONOLOGY When you listen to someone who speaks a language other than your own, you will often hear different (some might even say "strange") sounds. The basic sound units of a language are called phonemes, and the rules for combining phonemes constitute the phonology of a language. Examples of phonemes in English include the sounds you make when speaking, such as [k], [t], or [a].

The phonological rules of a language tell speakers which sounds to use and how to order them. For instance, the word *cat* has three phonemes: a hard [k] sound, the short [a] vowel, and the [t] sound. These same three sounds, or phonemes, can be rearranged to form other combinations: *act, tack,* or even *tka*. Of course, as someone who speaks and writes English, your knowledge of the rules for creating appropriate combinations of phonemes undoubtedly suggests to you that *tka* is improper. Interestingly, you know that *tka* is incorrect even though you probably cannot describe the rules that make it so.[2]

Languages have different numbers of phonemes. English, for example, depends on about forty-five phonemes. The number of phonemes in other languages ranges from as few as fifteen to as many as eighty-five.[3]

Although English is spoken in many parts of the world, its use varies greatly. This sign of caution in a Shanghai shopping mall differs from the words you would find on an escalator in the United States.

CULTURE CONNECTIONS

An increasing number of my students who come from homes in which a language other than English is spoken, and in which at least two cultures coexist, may not, in fact, experience a disruption in their sense of themselves. Code-switching (the alternate use of two or more languages in the same utterance) is, for many of them, a normal state of discourse. That is the case for not only my international students, but also for many of my students who are members of various ethnic groups. Hispanic-American students, in particular, now proudly assert their once-precarious position on the borderlands, refusing to choose among identities or languages. Indeed, despite the politics that swirl around the issue, we are increasingly coming to realize that code-switching is a sophisticated communicative strategy.

—Isabelle de Courtivron ■

Mastery of another language requires practice in reproducing its sounds accurately. Sometimes it is difficult to hear the distinctions in the sounds made by those proficient in the language. Native U.S. English speakers often have difficulty in hearing phonemic distinctions in tonal languages, such as Chinese, that use different pitches for many sounds, which then represent different meanings. Even when the differences can be heard, the mouths and tongues of those learning another language are sometimes unable to produce these sounds. In intercultural communication, imperfect rendering of the phonology of a language—in other words, not speaking the sounds as native speakers do—can make it difficult to be understood accurately. Accents of second-language speakers, which we discuss in more detail later in this chapter, can sometimes provoke negative reactions in native speakers.

MORPHOLOGY   Phonemes combine to form morphemes, which are the smallest units of meaning in a language. The forty-five English phonemes can be used to generate more than 50 million morphemes! For instance, the word *comfort,* whose meaning refers to a state of ease and contentment, contains one morpheme. But the word *comforted* contains two morphemes: *comfort* and *-ed.* The latter is a suffix that means that the comforting action or activity happened in the past. Indeed, although all words contain at least one morpheme, some words (such as *uncomfortable,* which has three morphemes) can contain two or more. Note that morphemes refer only to meaning units. Though the word *comfort* contains smaller words such as *or* and *fort,* these other words are coincidental to the basic meaning of *comfort.* Morphemes, or meaning units in language, can also differ depending on the way they are pronounced. In Chinese, for instance, the word pronounced as "ma" can have four different meanings—mother, toad, horse, or scold—depending on the tone with which it is uttered.[4] Pronunciation errors can have very unintended meanings!

SEMANTICS   As noted earlier, morphemes—either singly or in combination—are used to form words. The study of the meaning of words is called semantics. The most convenient and thorough source of information about the semantics of a language is the dictionary, which defines what a word means in a particular language. A more formal way of describing the study of semantics is to say that it is the study of the relationship between words and what they stand for or represent. You can see the semantics of a language in action when a baby is being taught to name the parts of the body. Someone skilled in the language points to and touches the baby's nose and simultaneously vocalizes the word *nose.* Essentially, the baby is being taught the vocabulary of a language. Competent communication in any language requires knowledge of the words needed to express ideas. You have probably experienced the frustration of trying to describe an event but not being able to think of words that accurately convey the intended meaning. Part of what we are trying to accomplish with this book is to give you a vocabulary that can be used to understand and explain the nature of intercultural communication competence.

Can you order from this menu in Italy?

Communicating interculturally necessitates learning a new set of semantic rules. The baby who grows up where people speak Swahili does not learn to say *nose* when the protruding portion of the face is touched; instead, she or he is taught to say *pua*. For an English speaker to talk with a Swahili speaker about his or her nose, at least one of them must learn the word for nose in the other's language. When learning a second language, much time is devoted to learning the appropriate associations between the words and the specific objects, events, or feelings that the language system assigns to them. Even those whose intercultural communication occurs with people who speak the "same" language must learn at least some new vocabulary. The U.S. American visiting Great Britain will confront new meanings for words. For example, *boot* refers to the storage place in a car, or what the U.S.-English-speaking person would call the *trunk*. *Chips* to the British are *French fries* to the U.S. American. A *Band-Aid* in the United States is called a *plug* in Great Britain. As Winston Churchill so wryly suggested, the two countries are indeed "divided by a common language."

The discussion of semantics is incomplete without noting one other important distinction: the difference between the denotative and connotative meanings of words. Denotative meanings are the public, objective, and legal meanings of a word. Denotative meanings are those found in the dictionary or law books. In contrast, connotative meanings are personal, emotionally charged, private, and specific to a particular person.

As an illustration, consider a common classroom event known as a *test*. When used by a college professor who is speaking to a group of undergraduate students, *test* is a relatively easy word to define denotatively. It is a formal examination that is used to assess a person's degree of knowledge or skill. But the connotative meanings of *test* probably vary greatly from student to student; some react to the idea with panic, and others are blasé and casual. Whereas denotative meanings tell, in an abstract sense, what the words mean objectively, our interest in intercultural communication suggests that an understanding of the connotative meanings—the feelings and thoughts evoked in others as a result of the words used in the conversation—is critical to achieving intercultural competence.

As an example of the importance of connotative meanings, consider the experience reported by a Nigerian student who was attending a university in the United States. When working with a fellow male student who was African American, the Nigerian called to him by saying, "Hey, boy, come over here." To the Nigerian student, the term *boy* connotes a friendly and familiar relationship, is a common form of address in Nigeria, and is often used to convey a perception of a strong interpersonal bond. To the African American student, however, the term *boy* evokes images of racism, oppression, and an attempt to place him in an inferior social status. Fortunately, the two students were friends and were able to talk to each other to clarify how they each interpreted the Nigerian student's semantic choices; further misunderstandings were avoided. Often, however, such opportunities for clarification do not occur.

Another example is seen in the casual conversation of a U.S. American student and an Arab student. The former had heard a radio news story about the intelligence of pigs and was recounting the story as "fact" when the Arab student forcefully declared, "Pigs are dirty animals, and they are very dumb." The U.S. American student describes her reaction: "In my ignorance, I argued with him by telling him that it was true and had been scientifically proven." It was only later that she learned that as part of the religious beliefs of devout Muslims, pigs are believed to be unclean. Learning the connotative meanings of language is essential in achieving competence in another culture's verbal code.

SYNTACTICS   The fourth component of language is syntactics, the relationship of words to one another. When children are first learning how to combine words into phrases, they are being introduced to the syntactics of their language. Each language stipulates the correct way to arrange words. In English it is not acceptable to create a sentence such as the following: "On by the book desk door is the the." It is incorrect to place the preposition *by* immediately following the preposition *on*. Instead, each preposition must have an object, which results in phrases such as "on the desk" and "by the door." Similarly, articles such as *the* in a sentence are not to be presented one right after the other. Instead, the article is placed near the noun, which produces a sentence that includes "the book," "the door," and "the desk." The syntactics of English grammar suggest that the words in the preceding nonsense sentence might be rearranged to form the grammatically correct sentence "The book is on the desk by the door." The order of the words helps establish the meaning of the utterance.

Each language has a set of rules that govern the sequence of the words. To learn another language, you must learn those rules. The sentence "John has, to the store to buy some eggs, gone" is an incorrect example of English syntax but an accurate representation of German syntax.

PRAGMATICS   The final component of all verbal codes is pragmatics, the effect of language on human perceptions and behaviors. The study of pragmatics focuses on how language is actually used. A pragmatic analysis of language goes beyond phonology, morphology, semantics, and syntactics. Instead, it considers how users of a particular language are able to understand the meanings of specific utterances in particular contexts. For example, some people regard the U.S. American greeting ritual that asks *"How are you?"* as insincere and perhaps even hypocritical. As an Israeli woman observed,

> No matter if your kids are on drugs, your spouse is leaving you, and you just declared bankruptcy, you are expected to smile, and say, "Everything is great!" Why do Americans ask if they don't really want to know?[5]

Of course, to U.S. Americans the frequently asked question *"How are you?"* is simply intended as a pleasant and polite greeting ritual and is not expected to be an inquiry into one's well-being.

By learning the pragmatics of language use, you understand how to participate in a conversation and you know how to sequence the sentences you speak as part of the conversation. Thus, when you are eating a meal with a group of people and somebody says, "Is there any salt?" you know that you should give the person the salt shaker rather than simply answer "yes."

CULTURE CONNECTIONS

"Were you enjoying a moment of privacy?" Vicky asked.

"Well, 'privacy' is a word that is difficult to translate into Chinese."

He had stumbled over it several times. There was not a single-word equivalent to "privacy" in his language. Instead, he had to use a phrase or sentence to convey its meaning.

—Qiu Xiaolong ∎

To illustrate how the pragmatics of language use can affect intercultural communication, imagine yourself as a dinner guest in a Pakistani household. You have just eaten a delicious meal. You are relatively full but not so full that it would be impossible for you to eat more if it was considered socially appropriate to do so. Consider the following dialogue:

Hostess: I see that your plate is empty. Would you like some more curry?

You: No, thank you. It was delicious, but I'm quite full.

Hostess: Please, you must have some more to eat.

You: No, no thank you. I've really had enough. It was just great, but I can't eat another bite.

Hostess: Are you sure that you won't have any more? You really seemed to enjoy the brinjals. Let me put just a little bit more on your plate.

What is your next response? What is the socially appropriate answer? Is it considered socially inappropriate for a dinner guest not to accept a second helping of food? Or is the hostess pressing you to have another helping because, in her culture, your reply is not interpreted as a true negative response? Even if you knew Urdu, the language spoken in Pakistan, you would have to understand the pragmatics of language use to respond appropriately—in this instance, to say "no" at least three times.

The rules governing the pragmatics of a language are firmly embedded in the larger rules of the culture and are intimately associated with the cultural patterns discussed in Chapters 4 and 5. For example, cultures vary in the degree to which they encourage people to ask direct questions and to make direct statements. Imagine a student from the United States who speaks some Japanese and who subsequently goes to Japan as an exchange student. The U.S. American's culturally learned tendency is to deal with problems directly, and she may therefore confront her Japanese roommate about the latter's habits in order to "clear the air" and establish an "open" relationship. Given the Japanese cultural preference for indirectness and face-saving behaviors, the U.S. American student's skill in Japanese does not extend to the pragmatics of language use. As Wen Shu Lee suggests, these differences in the pragmatic rule systems of languages also make it very difficult to tell a joke—or even to understand a joke—in a second language.[6] Humor requires a subtle knowledge of both the expected meanings of the words (semantics) and their intended effects (pragmatics).

CULTURE CONNECTIONS

Even though I've learned to say knickers when I mean underpants, people still remark, 'Oh, you're *American*' every time I open my mouth. As if American is a synonym for no culture, no history, the wrong vocabulary. I pronounce flower with an er in it, God with an awe. Although inside me there's a different voice, refined and full of ironic insight, it always comes out *awe, er, um,* and instead of 'Absolutely!' I say 'You bet!'.... In England, a less extreme landscape, people seem more sure of their footing. They're always saying 'Absolutely!'...

—Leslie Forbes ∎

Interpretation and Intercultural Communication

Translation can be defined as the use of verbal signs to understand the verbal signs of another language.[7] Translation usually refers to the transfer of written verbal codes between languages. Interpretation refers to the oral process of moving from one code to another. When heads of state meet, an interpreter accompanies them. The translator, in contrast to the interpreter, usually has more time to consider how she or he wants to phrase a particular passage in a text. Interpreters must make virtually immediate decisions about which words or phrases would best represent the meanings of the speaker.

THE ROLE OF INTERPRETATION IN TODAY'S WORLD   Issues surrounding the interpretation of verbal codes from one language to another are becoming more and more important for all of us. Such issues include whether the words or the ideas of the original should be conveyed, whether the translation should reflect the style of the original or that of the translator, and whether an interpreter should correct cultural mistakes.

In today's global marketplace, health care workers, teachers, government workers, and businesspeople of all types find that they are increasingly required to use professional interpreters to communicate verbally with their clients and, thus, fulfill their professional obligations.[8] Similarly, instructions for assembling consumer products that are sold in the United States but manufactured in another country often demonstrate the difficulty in moving from one language to another. Even though the words on the printed instruction sheet are in English, the instructions may not be correct or accurately interpreted.

Issues in interpretation, then, are very important. People involved in intercultural transactions must often depend on the services of multilingual individuals who can help to bridge the intercultural communication gap.

TYPES OF EQUIVALENCE   If the goal in interpreting from one language to another is to represent the source language as closely as possible, a simpler way of describing the goal is with the term equivalence. Those concerned about developing a science of translation have described a number of different types of equivalence. Dynamic equivalence has been offered as one goal of good translation and interpretation.[9] Five kinds of equivalence must be considered in moving from one language to another: vocabulary, idiomatic, grammatical–syntactical, experiential, and conceptual equivalence.[10]

This sign reflects the many languages used by U.S. Americans.

Vocabulary Equivalence   To establish vocabulary equivalence, the interpreter seeks a word in the target language that has the same meaning in the source language. This is sometimes very difficult to do. Perhaps the words spoken in the source language have no direct equivalents in the target language. For instance, in Igbo, a language spoken in Nigeria, there is no word for *window*. The word in Igbo that is used to represent a window, *mpio*, actually means "opening." Likewise, there is no word for *efficiency* in the Russian language, and the English phrase "A house is not a home" has no genuine vocabulary equivalent in some languages. Alternatively, there may be several words in the target language that have similar meanings to the word in the source language, so the interpreter must select the word that best fits the intended ideas. An interpreter will sometimes use a combination of words in the target language to approximate the original word, or the interpreter may offer several different words to help the listener understand the meaning of the original message.

Idiomatic Equivalence   An idiom is an expression that has a meaning contrary to the usual meaning of the words. Phrases such as "Eat your heart out," "It's raining cats and dogs," and "Eat humble pie" are all examples of idioms. Idioms are so much a part of language that people are rarely aware of using them. Think of the literal meaning of the following idiom: "I was so upset I could have died." Or consider the plight of a Malaysian student who described his befuddlement when his fellow students in the United States initiated conversations by asking, "What's up?" His instinctive reaction was to look up, but after doing so several times he realized that the question was an opening to conversation rather than a literal reference to something happening above him. Another example is the request a supervisor in a university media center made to a student assistant from India, who tended to take conversations and instructions literally. The supervisor instructed the assistant to "put this DVD on the television." The supervisor was later surprised to learn that the DVD was literally placed on top of the television, instead of being played for the class. The challenge for interpreters is to understand the intended meanings of idiomatic expressions and to translate them into the other language.

Grammatical–Syntactical Equivalence   The discussion later in this chapter about some of the variations among grammars highlights the problems in establishing equivalence in grammatical or syntactical rule systems. Quite simply, some languages make grammatical distinctions that others do not. For instance, when translating from the Hopi language into English, the interpreter has to make adjustments for the lack of verb tenses in Hopi because tense is a necessary characteristic of every English utterance.

Experiential Equivalence   Differing life experiences are another hurdle the interpreter must overcome. The words presented must have some meaning within the experiential framework of the person to whom the message is directed. If people have never seen a television, for instance, a translation of the phrase "I am going to stay home tonight and watch television" would have virtually no meaning to them. Similarly, although clocks are a common device for telling time and they govern the behaviors of most U.S. Americans, many people live in cultures in which there are no clocks and no words for this concept. Some Hmong people, upon moving to the United States, initially had difficulty with the everyday experience of telling time with a clock.

CULTURE CONNECTIONS

Arabic is a language fond of formal indirectness, and, during the first planning sessions of the Arab-language edition of "The Apprentice," the producers decided to replace "You're fired!"—Donald Trump's catchphrase of blunt humiliation—with a line that translates into English as "May God be kind to you."

—Ian Parker ■

Conceptual Equivalence   Conceptual equivalence takes us back to the discussions in Chapters 4 and 5 about cultural patterns being part of a person's definition of reality. Conversation with people with radically different cultural patterns requires making sense of the variety of concepts that each culture defines as real and good.

LANGUAGE, THOUGHT, CULTURE, AND INTERCULTURAL COMMUNICATION

Every language has its unique features and ways of allowing those who speak it to identify specific objects and experiences. These linguistic features, which distinguish each language from all others, affect how the speakers of the language perceive and experience the world. To understand the effects of language on intercultural communication, questions such as the following must be explored:

- How do initial experiences with language shape or influence the way in which a person thinks?
- Do the categories of a language—its words, grammar, and usage—influence how people think and behave?

More specifically, consider the following question:

- Does a person growing up in Saudi Arabia, who learns to speak and write Arabic, "see" and "experience" the world differently than does a person who grows up speaking and writing Tagalog in the Philippines?

Although many scholars have advanced ideas and theories about the relationships among language, thought, culture, and intercultural communication, the names most often associated with these issues are Benjamin Lee Whorf and Edward Sapir. Their theory is called linguistic relativity.

The Sapir–Whorf Hypothesis of Linguistic Relativity

Until the early part of the twentieth century, in western Europe and the United States, language was generally assumed to be a neutral medium that did not influence the way people experienced the world.[11] During that time, the answer to the preceding question would have been that, regardless of whether people grew up learning and speaking Arabic or Tagalog, they would experience the world similarly. The varying qualities of language would not have been expected to affect the people who spoke those languages. Language, from this point of view, was merely a vehicle by which ideas were presented, rather than a shaper of the very substance of those ideas.

In 1921, anthropologist Edward Sapir began to articulate an alternative view of language, asserting that language influenced or even determined the ways in which people thought.[12] Sapir's student, Benjamin Whorf, continued to develop Sapir's ideas through the 1940s. Together, their ideas became subsumed under several labels, including the theory of linguistic determinism, the theory of linguistic relativity, the Sapir–Whorf hypothesis, and the Whorfian hypothesis. The following quotation from Sapir is typical of their statements:

> Human beings do not live in the objective world alone, nor alone in the world of social activity as ordinarily understood, but are very much at the mercy of the particular language which has become the medium of expression for their society.

It is quite an illusion to imagine that one adjusts to reality essentially without the use of language and that language is merely an incidental means of solving specific problems of communication or reflection. The fact of the matter is that the "real world" is to a large extent unconsciously built up on the language habits of the group.... The worlds in which different societies [cultures] live are distinct worlds, not merely the same world with different labels attached.... We see and hear and otherwise experience very largely as we do because the language habits of our community predispose certain choices of interpretation.[13]

Our discussion of the Sapir–Whorf hypothesis is not intended to provide a precise rendering as articulated by Sapir and Whorf, which is virtually impossible to do. During the twenty years in which they formally presented their ideas to the scholarly community, their views shifted somewhat, and their writings include both "firmer," or more deterministic views of the relationship between language and thought, and "softer" views that describe language as merely influencing or shaping thought.

In the "firm," or deterministic, version of the hypothesis, language functions like a prison—once people learn a language, they are irrevocably affected by its particulars. Furthermore, it is never possible to translate effectively and successfully between languages, which makes competent intercultural communication an elusive goal.

The "softer" position is a less causal view of the nature of the relationship between language and thought. In this version, language shapes how people think and experience their world, but this influence is not unceasing. Instead, it is possible for people from different initial language systems to learn words and categories sufficiently similar to their own so that communication can be accurate.

If substantial evidence had been found to support the firmer version of the Sapir–Whorf hypothesis, it would represent a dismal prognosis for competent intercultural communication. Because so few people grow up bilingually, it would be impossible to transcend the boundaries of their linguistic experiences. Fortunately, the weight of the scholarly evidence, which we summarize in the following section, debunks the notion that people's first language traps them inescapably in a particular pattern of thinking. Instead, evidence suggests that language plays a powerful role in *shaping* how people think and

These schoolchildren in Kenya are practicing their lessons. The Sapir–Whorf hypothesis underscores the relationship between their language and their experiences in the world.

experience the world. Although the shaping properties of language are significant, linguistic equivalence can be established between people from different language systems.[14]

Sapir and Whorf's major contribution to the study of intercultural communication is that they called attention to the integral relationship among thought, culture, and language. In the following section, we discuss some of the differences in the vocabulary and grammar of languages and consider the extent to which these differences can be used as evidence to support the two positions of the Sapir–Whorf hypothesis. As you consider the following ideas, examine the properties of the languages you know. Are there specialized vocabularies or grammatical characteristics that shape how you think and experience the world as you use these languages?

VARIATIONS IN VOCABULARY   The best known example of vocabulary differences associated with the Sapir–Whorf hypothesis is the large number of words for snow in the Eskimo language. (The language is variously called Inuktitut in Canada, Inupit in Alaska, and Kalaallisut in Greenland.) Depending on whom you ask, there are from seven to fifty different words for snow in the Inuktitut language.[15] For example, there are words that differentiate falling snow (*gana*) and fully fallen snow (*akilukak*). The English language has fewer words for snow and no terms for many of the distinctions made by Eskimos. The issue raised by the Sapir–Whorf hypothesis is whether the person who grows up speaking Inuktitut actually perceives snow differently than does someone who grew up in Southern California and may only know snow by secondhand descriptions. More important, could the Southern Californian who lives with the Inupit in Alaska learn to differentiate all of the variations of snow and to use the specific Eskimo words appropriately? The firmer version of the Sapir–Whorf hypothesis suggests that linguistic differences are accompanied by perceptual differences, so that the English speaker looks at snow differently than does the Eskimo speaker.

Numerous other examples of languages have highly specialized vocabularies for particular features of the environment. For instance, in the South Sea islands, there are numerous words for coconut, which not only refer to the object of a coconut but also indicate how the coconut is being used or to a specific part of the coconut.[16] Similarly, in classical Arabic, thousands of words are used to refer to a camel.[17]

Another variation in vocabulary concerns the terms a language uses to identify and divide colors in the spectrum. For example, the Kamayura Indians of Brazil have a single word that refers to the colors that English speakers would call blue and green. The best translation of the word the Kamayuras use is "parakeet colored."[18] The Dani of West New Guinea divide all colors into only two words, which are roughly equivalent in English to "dark" and "light."[19] The important issue, however, is whether speakers of these languages are able to distinguish among the different colors when they see them or can experience only the colors suggested by the words available for them to use. Do the Kamayura Indians actually see blue and green as the same color because they use the same word to identify both? Or does their language simply identify colors differently than does English?

CULTURE CONNECTIONS

In the Hopi language there are no words for war or aggression.

—Thom Akeman   ∎

Do you think that you could learn to distinguish all of the variations of the object "snow" that are important to the Eskimos? Could you be taught to see all of the important characteristics of a camel or a coconut? Such questions are very important in accepting or rejecting the ideas presented in the firm and soft versions of the Sapir–Whorf hypothesis.

Researchers looking at the vocabulary variations in the color spectrum have generally found that, although a language may restrict how a color can be labeled verbally, people can still see and differentiate among particular colors. In other words, the Kamayura Indians can, in fact, see both blue and green, even though they use the same linguistic referent for both colors.[20] The evidence on color perception and vocabulary, then, does not support the deterministic version of the Sapir–Whorf hypothesis.

What about all those variations for snow, camels, and coconuts? Are they evidence to support the firm version of the Sapir–Whorf hypothesis? A starting point for addressing this issue is to consider how English speakers use other words along with essentially the one word English has for "particles of water vapor that, when frozen in the upper air, fall to earth as soft, white, crystalline flakes." English speakers are able to describe verbally many variations of snow by adding modifiers to the root word. People who live in areas with a lot of snow are quite familiar with *dry snow, heavy snow, slush,* and *dirty snow.* Skiers have a rich vocabulary to describe variations in snow on the slopes. It is possible, therefore, for a person who has facility in one language to approximate the categories of another language. The deterministic position of Sapir–Whorf, then, is difficult to support. Even Sapir and Whorf's own work can be used to argue against the deterministic interpretation of their position because, in presenting all of the Eskimo words for snow, Whorf provided their approximate English equivalents.

A better explanation for linguistic differences is that variations in the complexity and richness of a language's vocabulary reflect what is important to the people who speak that language. To an Eskimo, differentiating among varieties of snow is much more critical to survival and adaptation than it is to the Southern Californian, who may never see snow. Conversely, Southern Californians have numerous words to refer to four-wheeled motorized vehicles, which are very important objects in their environment. However, we are certain that differences in the words and concepts of a language do affect the ease with which a person can change from one language to another because there is a dynamic interrelationship among language, thought, and culture.

VARIATIONS IN LINGUISTIC GRAMMARS A rich illustration of the reciprocal relationship among language, thought, and culture can be found in the grammatical rules of different languages. In the following discussion, you will once again see how the patterns of

CULTURE CONNECTIONS

It looked like I was going to be here for a while, so I sat next to the young Maasai and said hello. He spoke English. The oldest man was his father, he explained.

"He's very old to be your father," I said.

"Yes, the oldest of my father's twelve brothers."

"Ah." And I'd thought father only meant one thing.

—Karin McQuillan ∎

a culture's beliefs, values, norms, and social practices, as discussed in Chapters 4 and 5, permeate all aspects of the culture. Because language shapes how its users organize the world, the patterns of a culture will be reflected in its language and vice versa.

Cultural Conceptions of Time   Whorf himself provided detailed descriptions of the Hopi language that illustrate how the grammar of a language is related to the perceptions of its users. Hopi do not linguistically refer to time as a fixed point or place but rather as a movement in the stream of life. The English language, in contrast, refers to time as a specific point that exists on a linear plane divided into past, present, and future. Hopi time is more like an ongoing process; the here and now (the present) will never actually arrive, but it will always be approaching. The Hopi language also has no tenses, so the people do not place events into the neat categories of past, present, and future that native speakers of English have come to expect. As Stephen Littlejohn has suggested, the consequences of these linguistic differences is that

> Hopi and SAE [Standard Average European] cultures will think about, perceive, and behave toward time differently. For example, the Hopi tend to engage in lengthy preparing activities. Experiences (getting prepared) tend to accumulate as time gets later. The emphasis is on the accumulated experience during the course of time, not on time as a point or location. In SAE cultures, with their spatial treatment of time, experiences are not accumulated in the same sense. Elaborate and lengthy preparations are not often found. The custom in SAE cultures is to record events (space-time analogy) such that what happened in the past is objectified in space (recorded).[21]

Because a culture's linguistic grammar shapes its experiences, the speakers of Hopi and English will experience time differently, and each may find it difficult to understand the view of time held by the other. Judgments about what is "natural," "right," or "common sense" will obviously vary and will be reinforced by the linguistic habits of each group.

Showing Respect and Social Hierarchy   Languages allow, and to a certain extent force, speakers to display respect for others. For instance, it is much easier to show respect in Spanish than it is in English. Consider the following sentences:

> ¿Sabe usted dónde está la profesora?
> Know you where is the professor? [Do you know where the professor is?]
> ¿Sabes dónde está la profesora?
> Know you where is the professor? [Do you know where the professor is?]

These distinctly different Spanish sentences are identical when translated into English. The sentences in Spanish reflect the differences in the level of respect that must be shown between the person speaking and the person being addressed. The pronoun *usted* is used in the first example to mark the speaker's question as particularly formal or polite. The *s* in *Sabes* in the second example marks the relationship between the speaker and the person being addressed as familiar or informal. In the actual practice of Spanish, a younger person would not use the informal grammatical construction to address an older person, just as an older person would not use the formal *usted* with a person who is much younger.

This example illustrates once again that the grammar of a language can at least encourage its users to construct their interactions with others in particular ways. When a language directs a speaker to make distinctions among the people with whom the speaker interacts, in this instance by showing linguistically a greater respect for some and not others, the language helps to remind its users of social distinctions and the behaviors that are appropriate to them. Thus, language professors who teach Spanish to English-speaking students often note that the English-speaker is not behaving respectfully.

The degree to which a language demands specific words and grammatical structures to show the nature of the relationship between the communicators suggests how much a culture values differences between people. In the frameworks of the ideas presented in Chapters 4 and 5, Spanish-speaking cultures would be more likely to value a hierarchical social organization and a large power distance. Chinese, Japanese, and Korean languages also reflect the relative social status between the addresser and addressee. In Hindi, Korean, and other languages, there are specific words for older brother, older sister, younger brother, and younger sister, which remind all siblings of their relative order in the family and the norms or expectations appropriate to specific familial roles. Languages with grammatical and semantic features that make the speakers decide whether to show respect and social status to others are constant reminders of those characteristics of social interaction. In contrast, a language with few terms to show status and respect tends to minimize those status distinctions in the minds of the language's users.

Pronouns and Cultural Characteristics   English is the only language that capitalizes the pronoun *I* in writing. English does not, however, capitalize the written form of the pronoun *you*. Is there a relationship between the individualism that characterizes most of the English-speaking countries and this feature of the English language? In contrast, consider that there are more than twelve words for *I* in Vietnamese, more than ten in Chinese, and more than one hundred in Japanese.[22] Does a language that demands a speaker to differentiate the self (the "I") from other features of the context (for example, other people or the type of event) shape the way speakers of that language think about themselves? If "I" exist, but "I" am able to identify myself linguistically only through reference to someone else, will "I" not have a different sense of myself than the English-speaking people who see themselves as entities existing apart from all others?[23]

As an example of the extreme contrasts that exist in the use and meanings of pronouns, consider the experiences of Michael Dorris, who lived in Tyonek, Alaska, an Athabaskan-speaking Native American community:

> Much of my time was spent in the study of the local language, linguistically related to Navajo and Apache but distinctly adapted to the subarctic environment. One of its most difficult features for an outsider to grasp was the practice of almost always speaking and thinking in a collective plural voice. The word for people, "dene," was used as a kind of "we"—the subject for virtually every predicate requiring a personal pronoun—and therefore any act became, at least in conception, a group experience.[24]

Imagine having been trained in the language that Dorris describes. Would speaking such a language result in people who think of themselves as part of a group rather than as individuals?[25] Alternatively, if you are from a culture that values individualism, would you have difficulty communicating in a language that requires you always to say *we* instead of *I*? If your cultural background is more group-oriented, would it be relatively easy for you to speak in a language that places you as part of a group?

CULTURE CONNECTIONS

After five weeks of intense study, I panicked. I suddenly realized that I had not learned the Woleaian equivalent for "to have." How could I have overlooked something so basic? In English, you learn "to have" shortly after "to be." "Have" is the 11th most commonly used word in English; the concept is essential.

I had previously studied German and Spanish and knew that *haben* and *tener* held equally important places in those languages, respectively. I must have been doing a terribly inept job of learning Woleaian. What else had I missed?

Well, whatever else I'd missed, I hadn't missed "to have." It wasn't there to be missed. There are Woleaian equivalents for some of its uses, but not for the term itself. You cannot, in Woleaian, say "I have food," or "I have a car." You cannot "have" a wife, a good time, or a seat on an airplane. You cannot even "have" the flu. (The equivalent is "the flu saw me.")

Why do Woleaians lack a term so central to the other languages I knew? I have a theory.

Language reveals culture, and Woleaian life is generally based on sharing, rather than owning. "To have," which is basically an ownership term, is simply not that important there.

In a society where food and many other things are automatically shared, "Is there breadfruit?" is a more reasonable question than "May I have some breadfruit?" The first is a standard question in Woleaian, the second cannot be said. Woleaians do have ways to denote ownership, of course. But sharing plays a larger role in their lives, and ownership plays a smaller role in their language.

Whatever the reason for the lack of "to have," there's no mystery behind the most common form of "hello" among Woleaians. *Butog mwongo*! (Come and eat!) expresses their hospitality, their love of food, and their love of sharing. It is not just a greeting, it is a genuine invitation—at any time of the day or night.

—Jerry Miller ■

LINGUISTIC RELATIVITY AND INTERCULTURAL COMMUNICATION   The semantic and syntactic features of language are powerful shapers of the way people experience the physical and social worlds. Sapir and Whorf's assertions that language *determines* our reality have proved to be false. Language does not determine our ability to sense the physical world, nor does the language first learned create modes of thinking from which there is no escape. However, language shapes and influences our thoughts and behaviors. The vocabulary of a language reflects what you need to know to cope with the environment and the patterns of your culture. The semantics and syntactics of language gently nudge you to notice particular kinds of things in your world and to label them in particular ways. All of these components of language create habitual response patterns to the people, events, and messages that surround you. Your language intermingles with other aspects of your culture to reinforce the cultural patterns you are taught.

The influence of a particular language is something you can escape; it is possible to translate to or interact in a second language. But as the categories for coding or sorting the world are provided primarily by your language, you are predisposed to perceive the world in a particular way, and the reality you create is different from the reality created by those who use other languages with other categories.

When the categories of languages are vastly different, people will have trouble communicating with one another. Differences in language affect what is relatively easy to say and what seems virtually impossible to say. As Wilma M. Roger has suggested, "Language and the cultural values, reactions, and expectations of speakers of that language are subtly melded."[26]

We offer one final caution. For purposes of discussion, we have artificially separated vocabulary and grammar, as if language is simply an adding together of these two elements. In use, language is a dynamic and interrelated system that has a powerful effect on people's thoughts and actions. The living, breathing qualities of language as spoken and used, with all the attendant feelings, emotions, and experiences, are difficult to convey adequately in an introductory discussion such as this one.

Language and Intercultural Communication

The earlier sections of this chapter may have given the impression that language is stable and used consistently by all who speak it. However, even in a country that has predominantly only one language, there are great variations in the way the language is spoken (accents), and there are wide deviations in how words are used and what they mean. Among U.S. Americans who speak English, it is quite common to hear many different accents. It is also quite common to hear words, phrases, and colloquial expressions that are common to only one region of the country. Think of the many voices associated with the speaking of English in the United States. Do you have an auditory image of the way someone sounds who grew up in New York City? How about someone who grew up in Georgia? Wisconsin? Oregon? The regional variations in the ways English is spoken reflect differences in accents and dialects.

Increasingly, U.S. Americans speak many first languages other than English. As noted in Chapter 1, multiple language systems are represented in U.S. schools. Employers in businesses must now be conscious of the different languages of their workers. In addition, specialized linguistic structures develop for other functions within the context of a larger language. Because language differences are powerful factors that influence the relationships between ethnic and cultural groups who live next to and with each other in communities and countries, we will examine the variations among languages of groups of people who essentially share a common political union.[27] We begin by considering the role of language in maintaining the identity of a cultural group and in the relationship between cultural groups who share a common social system. We then talk about nonstandard versions of a language, including accents, dialects, and argot, and we explore their effects on communication with others.

LANGUAGE, ETHNIC GROUP IDENTITY, AND DOMINANCE   Each person commonly identifies with many different social groups. For example, you probably think of yourself as part of a certain age grouping, as male or female, as married or unmarried, and as a college student or someone who is simply interested in learning about intercultural communication. You may also think of yourself as African American, German American, Vietnamese American, Latino, Navajo, or one of the many other cultural groups comprising the population of the United States. You may also identify with a culture from outside of the United States.

Henri Tajfel argues that humans categorize themselves and others into different groups to simplify their understanding of people. When you think of someone as part of a particular social group, you associate that person with the values of that group.[28] In this section we are particularly concerned with the ways in which language is used to identify people in a group, either by the group members themselves or by outsiders from other groups. Some of the questions we are concerned with include the following: How important is language to the members of a culture? What is the role of language in the maintenance of a culture? Why do some languages survive over time while others do not? What role does language play in the relationship of one culture to another?

Can you guess what this business in Vietnam is selling?

The importance that cultures attribute to language has been well established.[29] In fact, some would argue that the very heart of a culture is its language and that a culture dies if its language dies.[30] However, it is difficult to determine the exact degree of importance that language has for someone who identifies with a particular group because there are so many factors that affect the strength of that identification. For example, people are more likely to have a strong sense of ethnic and linguistic identity if members of other important cultural groups acknowledge their language in some way. In several states within the United States, for example, there have been heated legal battles to allow election ballots to be printed in languages other than English. Those advocating this option are actually fighting to gain official status and support for their languages.

A language will remain vital and strong if groups of people who live near one another use the language regularly. The sheer number of people who identify with a particular language and their distribution within a particular country or region have a definite effect on the vigor of the language. For people who are rarely able to speak the language of their culture, the centrality of the language and the cultural or ethnic identity that goes with it are certainly diminished. Their inability to use the language results in lost opportunities to express their identification with the culture that it symbolizes.

The extent to which a culture maintains a powerful sense of identification with a particular language is called perceived ethnolinguistic vitality, which refers to "the individual's subjective perception of the status, demographic characteristics, and institutional support of the language community."[31] Very high levels of perceived ethnolinguistic vitality mean that members of a culture will be unwilling to assimilate their linguistic behavior with other cultures that surround them.[32] Howard Giles, one of the foremost researchers in how languages are used in multilingual societies, concludes that there are likely to be intense pressures on cultural members to adopt the language of the larger social group and to discontinue the use of their own language when

1. the members of a culture lack a strong political, social, and economic status;
2. there are few members of the culture compared to the number of people in other groups in the community; and
3. institutional support to maintain their unique cultural heritage is weak.[33]

When multiple languages are spoken within one political boundary, there are inevitably political and social consequences. In the United States, for example, English has maintained itself as the primary language over a long period of time. Immigrants to the United States have historically been required to learn English in order to participate in the wider political and commercial aspects of the society. Schools offered classes only in English, television and radio programs were almost exclusively in English, and the work of government and business also required English. The English-only requirement has not been imposed without social consequences, however. In Micronesia, for example, where there are nine major languages and many dialects, people are demonstrably apprehensive about communicating with others when they must use English instead of their primary language.[34]

In recent years in the United States, there has been a change in the English-only pattern. Now in many areas of the country there are large numbers of people for whom English is not the primary language. As a consequence, teaching staffs are multilingual; government offices provide services to non-English speakers; and cable television has an extensive array of entertainment and news programming in Spanish, Chinese, Japanese, Arabic, and so on.

In some countries, formal political agreements acknowledge the role of multiple languages in the government and educational systems. Canada has two official languages: English and French. Belgium uses three: French, German, and Flemish. In Singapore, English, Mandarin, Malay, and Tamil are all official languages, and India has more than a dozen.

When India was established in 1948, one of the major problems concerned a national language. Although Hindi was the language spoken by the largest number of people, the overwhelming majority of the people did not speak it. India's solution to this problem was to identify sixteen national languages, thus formalizing in the constitution the right for government, schools, and commerce to operate in any of them. Even that solution has not quelled the fears of non-Hindi speakers that Hindi will predominate. In the mid-1950s, there was political agitation to redraw the internal state boundaries based on the languages spoken in particular regions. Even now, major political upheavals periodically occur in India over language issues.

Because language is such an integral part of most people's identities, a great deal of emotion is attached to political choices about language preferences. However, what is most central to intercultural competence is the way in which linguistic identification influences the interaction that occurs between members of different cultural groups. In interpersonal communication, language is used to discern ingroup and outgroup members. That is, language provides an obvious and highly accurate cue about whether people share each other's cultural background. If others speak as you do, you are likely to assume that they are similar to you in other important ways.

Howard Giles has developed communication accommodation theory to explain why people in intercultural conversations may choose to *converge* or *diverge* their communication behaviors to that of others.[35] At times, interactants will converge their language use to that of their conversational partners by adapting their speech patterns to the behaviors of others. They do so when they desire to identify with others, appear similar to them, gain their approval, and facilitate the development of smooth and harmonious relationships. At other times, interactants' language use will diverge from their conversational partners and will thus accentuate their own cultural memberships, maintain their individuality, and underscore the differences between themselves and others. Giles suggests that

CULTURE CONNECTIONS

The in-law situation was overwhelming for me, an only child. There is no Kikuyu word for uncle or aunt so Joseph had six fathers and the mamas were even more numerous: small, wiry women who smelled like butter when they hugged me.

—Kathlean Coskran ■

the likelihood that people will adapt and accommodate to others depends on such factors as their knowledge of others' communication patterns, their motivations to converge or diverge, and their skills in altering their preferred repertoire of communication behaviors.

People also make a positive or negative evaluation about the language that others use. Generally speaking, there is a pecking order among languages that is usually buttressed and supported by the prevailing political order. Thus,

> In every society the differential power of particular social groups is reflected in language variation and in attitudes toward those variations. Typically, the dominant group promotes its patterns of language use as dialect or accents by minority group members reduce their opportunities for success in the society as a whole. Minority group members are often faced with difficult decisions regarding whether to gain social mobility by adopting the language patterns of the dominant group or to maintain their group identity by retaining their native speech style.[36]

In the United States, there has been a clear preference for English over the multiple other languages that people speak, and those who speak English are evaluated according to their various accents and dialects. African Americans, for instance, have often been judged negatively for their use of Black Standard English, which has grammatical forms that differ from those used in Standard American English.[37] In the next section, we discuss the consequences of these evaluations and the effects of alternative forms of language use on intercultural communication competence.

A big challenge in learning a new language is to be understood clearly when speaking.

ALTERNATIVE VERSIONS OF A LANGUAGE   No language is spoken precisely the same way by all who use it. The sounds made when speaking English by someone from England, Australia, or Jamaica differ from the speech of English-speaking U.S. Americans. Even among those who share a similar language and reside in the same country, there are important variations in the way the language is spoken. These differences in language use include the way the words are pronounced, the meanings of particular words or phrases, and the patterns for arranging the words (grammar). Terms often associated with these alternative forms of a language include *dialect, accent, argot* (pronounced "are go"), and *jargon*.

DIALECTS   Dialects are versions of a language with distinctive vocabulary, grammar, and pronunciation that are spoken by particular groups of people or within particular regions. Dialects can play an important role in intercultural communication because they often trigger a judgment and evaluation of the speaker. Dialects are measured against a "standard" spoken version of the language. The term *standard* does not describe inherent or naturally occurring characteristics but, rather, historical circumstances. For example, among many U.S. Americans, Standard American English is often the preferred dialect and conveys power and dominance. But as John R. Edwards has suggested, "As a dialect, there is nothing intrinsic, either linguistically or esthetically, which gives Standard English special status."[38]

Occasionally, use of a nonstandard dialect may lead to more favorable evaluations of the speaker. Thus, a U.S. American may regard someone speaking English with a British accent as more "cultured" or "refined." However, most nonstandard dialects of English are frequently accorded less status and are often considered inappropriate or unacceptable in education, business, and government. For example, speakers of Spanish- or Appalachian-accented English, as well as those who speak Black Standard English, are sometimes unfairly assumed to be less reliable, less intelligent, and of lower status than those who speak Standard American English.[39]

One dialect frequently used in the United States has been variously called Black Standard English, Black English, African American Vernacular English, and Ebonics. Linguists have estimated that about 90 percent of the African American community uses Ebonics at least some of the time. Geneva Smitherman explains some of the linguistic forces that underlie Ebonics by providing an example of some African American women at a beauty shop, one of whom exclaims, "The Brotha be looking good; that's what got the Sista nose open!" According to Smitherman:

> In this statement, *Brotha* refers to an African American man, *looking good* refers to his style (not necessarily the same thing as physical beauty in Ebonics), *Sista* is an African American woman, and her passionate love for the Brotha is conveyed by the phrase *nose open* (the kind of passionate love that makes you vulnerable to exploitation). *Sista nose* is standard Ebonics grammar for denoting possession, indicated by adjacency/context (rather than the /'s, s'/). The use of *be* means that the quality of *looking good* is not limited to the present moment but reflects the Brotha's past, present, and future essence. As in the case of Efik and other West African languages, aspect is important in the verb system of US Ebonics, conveyed by the use of the English verb *be* to denote a recurring, habitual state of affairs. (Contrast *He be looking good* with *He looking good,* which refers to the present moment only—certainly not the kind of *looking good* that opens the nose!)[40]

Like all dialects, Ebonics is not slang, sloppy speech, incorrect grammar, or broken English. Rather, it reflects an intersection of West African languages and European American English that initially developed during the European slave trade and the enslavement of African peoples throughout the Americas and elsewhere.

Accents Distinguishable marks of pronunciation are called *accents*. Accents are closely related to dialects. Research studies repeatedly demonstrate that speakers' accents are used as a cue to form impressions of them.[41] Those of you who speak English with an accent or in a nonstandard version may have experienced the negative reactions of others, and you know the harmful effects such judgments can have on intercultural communication. Studies repeatedly find that accented speech and dialects provoke stereotyped reactions in listeners, so that the speakers are usually perceived as having less status, prestige, and overall competence. Interestingly, these negative perceptions and stereotyped responses sometimes occur even when the listeners themselves use a nonstandard dialect.[42]

If you are a speaker of Standard American English, you speak English with an "acceptable" accent. Can you recall conversations with others whose dialect and accent did not match yours? In those conversations, did you make negative assessments of their character, intelligence, or goodwill? Such a response is fairly common. Negative judgments that are made about others simply on the basis of how

CULTURE CONNECTIONS

"What makes you a Cherokee if you don't have Cherokee thoughts?" asked Rita Bunch, superintendent of the tribe's Sequoyah Schools.

Tribal officials thus decided to develop the language immersion school, in which students would be taught multiple subjects in a Cherokee-only environment.

The Oklahoma school began in 2001 and now has 105 students in kindergarten through fifth grade. They work on Apple laptops already loaded with the Cherokee language—the Macintosh operating system has supported Cherokee since 2003—and featuring a unique keypad overlay with Cherokee's 85 characters, each of which represents a different syllable.

—Murray Evans ∎

they speak are obviously a formidable barrier to competence in intercultural communication. For example, an Iranian American woman describes the frustration and anger experienced by her father, a physician, and her mother, a nurse, when they attempted to communicate with others by telephone. Although both of her parents had immigrated to the United States many years before, they spoke English with a heavy accent. These educated people were consistently responded to as if they lacked intelligence simply because of their accent. Out of sheer frustration, they usually had their daughter, who spoke English with a U.S. accent, conduct whatever business needed to be accomplished on the telephone.

Jargon and Argot Both jargon and argot are specialized forms of vocabulary. Jargon refers to a set of words or terms that are shared by those with a common profession or experience. For example, students at a particular university share a jargon related to general education requirements, registration techniques, add or drop procedures, activity fees, and so on. Members of a particular profession depend on a unique set of meanings for words that are understood only by other members of that profession. The shorthand code used by law-enforcement officers, lawyers, those in the medical profession, and even professors at colleges and universities are all instances of jargon.

Argot refers to a specialized language that is used by a large group within a culture to define the boundaries of their group from others who are in a more powerful position in society. As you might expect, argot is an important feature in the study of intercultural communication. Unlike jargon, argot is typically used to keep those who are not part of the group from understanding what members say to one another. The specialized language is used to keep those from the outside, usually seen as hostile, at bay.

CULTURE CONNECTIONS

We were weaving between Spanish and English, all of us fluent in both. It was something I had to get more accustomed to now, living on the West Coast, where people wove their conversations between the two tongues as commonly as if they were one.

—Marcos M. Villatoro ∎

CODE SWITCHING   Because of the many languages spoken in the United States, you will likely have many opportunities to hear and perhaps to participate in a form of language use called code switching. Code switching refers to the selection of the language to be used in a particular interaction by individuals who can speak multiple languages. The decision to use one language over another is often related to the setting in which the interaction occurs—a social, public, and formal setting versus a personal, private, and informal one. In his poignant exploration about speaking Spanish in an English-speaking world, Richard Rodriguez describes his attachment to the language associated with this latter setting.

> When I was a boy, things were different. The accent of *los gringos* was never pleasing nor was it hard to hear. Crowds at Safeway or at bus stops would be noisy with sound. And I would be forced to edge away from the chirping chatter above me....
>
> But then there was Spanish. *Español:* my family's language. *Español:* the language that seemed to me to be a private language. I'd hear strangers on the radio and in the Mexican Catholic church across town speaking in Spanish, but I couldn't really believe that Spanish was a public language, like English. Spanish speakers, rather, seemed related to me, for I sensed that we shared—through our language—the experience of feeling apart from *los gringos*.... Spanish seemed to me the language of home....
>
> A family member would say something to me and I would feel myself specially recognized. My parents would say something to me and I would feel embraced by the sounds of their words. Those sounds said: *I am speaking with ease in Spanish. I am addressing you in words I never use with* los gringos. *I recognize you as someone special, close, like no one outside. You belong with us. In the family.*[43]

A person's conversational partner is another important factor in code-switching decisions. Many African Americans, for instance, switch their linguistic codes based on the culture and gender of their conversational partners.[44]

The topic of conversation is another important influence on the choice of a linguistic code. One study found that Moroccans, for instance, would typically use French when discussing scientific or technological topics and Arabic when discussing cultural or religious ones. Interestingly, people's attitudes toward a particular topic were found to be consistent with the underlying beliefs, values, norms, and social practices of the culture whose language they choose to speak.[45]

VERBAL CODES AND INTERCULTURAL COMPETENCE

The link between knowledge of other verbal codes and intercultural competence is obvious. To speak another language proficiently requires an enormous amount of effort, energy, and time. The opportunity to study another language in your college curriculum is a choice we highly recommend to prepare you for a multicultural and multilingual world. Those world citizens with facility in a second or third language will be needed in every facet of society.

Many English speakers have a false sense of security because English is studied and spoken by so many people around the world. There is arrogance in this position that should be obvious because it places all of the responsibility for learning another language on the non-English speaker. Furthermore, even if two people from different cultures are using the verbal code system of one of the interactants, significant influences on their communication arise from their initial languages.

CULTURE CONNECTIONS

We can't really know what a pleasure it is to run in our own language until we're forced to stumble in someone else's. It was a great relief when Khaderbhai spoke in English.

—Gregory David Roberts ■

The multicultural nature of the United States and the interdependence of world cultures means that multiple cultures and multiple languages will be a standard feature of people's lives. Despite our strong recommendation that you learn and be tolerant of other languages, it is virtually impossible for anyone to be proficient in all of the verbal codes that might be encountered in intercultural communication. However, there are important ways to improve competence in adjusting to differences in verbal codes when communicating interculturally.

First, the study of at least one other language is extraordinarily useful in understanding the role of differences in verbal codes in intercultural communication. Genuine fluency in a second language demonstrates experientially all of the ways in which language embodies another culture. It also reveals the ways in which languages vary and how the nuances of language use influence the meanings of symbols. Even if you never become genuinely proficient in it, the study of another language teaches much about the culture of those who use it and the categories of experience the language can create. Furthermore, such study demonstrates, better than words written on a page or spoken in a lecture, the difficulty in gaining proficiency in another language and may lead to an appreciation of those who are struggling to communicate in second or third languages.

Short of becoming proficient in another language, learning about its gram-matical features can help you understand the messages of the other person. Study the connections between the features of a verbal code and the cultural patterns of those who use it. Even if you are going to communicate with people from another culture in your own first language, there is much that you can learn about the other person's language and the corresponding cultural patterns that can help you to behave appropriately and effectively.

Knowledge of another language is one component of the link between competence and verbal codes. Motivation, in the form of your emotional reactions and your intentions toward the culturally different others with whom you are communicating, is another critical component. Trying to get along in another language can be an exhilarating and very positive experience, but it can also be fatiguing and frustrating. The attempt to speak and understand a new verbal code requires energy and perseverance. Most second-language learners, when immersed in its cultural setting, report a substantial toll on their energy.

Functioning in a culture that speaks a language different from your own can be equally tiring and exasperating. Making yourself understood, getting around, obtaining food, and making purchases all require a great deal of effort. Recognizing the possibility of irritability and fatigue when functioning in an unfamiliar linguistic environment is an important prerequisite to intercultural competence. Without such knowledge, the communicator may well blame his or her personal feelings of discomfort on the cultures that are being experienced.

The motivation dimension also concerns your reactions to those who are attempting to speak your language. In the United States, for example, those who speak English often lack sympathy for and patience with those who do not. If English is your first language, notice those learning it and provide whatever help you can. Respond patiently. If you do not understand, ask questions and clarify. Try making your verbal point in alternative ways by using different sets of words with approximately equivalent meanings. Speak slowly, but do not yell. Lack of skill in a new language is not caused by a hearing impairment.

Learning other languages is an important feature of intercultural competence.

Be aware of the jargon in your speech, and provide a definition of it. Above all, to the best of your ability, withhold judgments and negative evaluations; instead, show respect for the enormous difficulties associated with learning a new language.

An additional emotional factor to monitor in promoting intercultural competence is your reaction to nonstandard versions of a language. The negative evaluations that nonstandard speech often triggers are a serious impediment to competence.

Competence in intercultural communication can be assisted by behaviors that indicate interest in the other person's verbal code. Even if you have never studied the language of those with whom you regularly interact, do attempt to learn and use appropriate words and phrases. Get a phrase book and a dictionary to learn standard comments or queries. Learn how to greet people and to acknowledge thanks. At the same time, recognize your own limitations and depend on a skilled interpreter when needed.

Intercultural competence requires knowledge, motivation, and actions that recognize the critical role of verbal codes in human interaction. Although learning another language is a very important goal, it is inevitable that you will need to communicate with others with whom you do not share a common verbal code.

SUMMARY

In this chapter, we have explored the vital role of verbal codes in intercultural communication. The features of language and the five rule systems were discussed. Phonology, the rules for creating the sounds of language, and morphology, the rules for creating the meaning units in a language, were described briefly. The study of the meaning of words (semantics), the rules for ordering the words (syntactics), and the effects of language on human perceptions and behaviors (pragmatics) were also described. We then discussed the difficulties in establishing equivalence in the process of interpretation from one language to another.

The important relationships among language, thought, culture, and behavior were explored. The Sapir–Whorf hypothesis of linguistic relativity, which concerns the effects of language on people's thoughts and perceptions, was discussed. We noted that the firmer version of the hypothesis portrays language as the determiner of thought, and the softer version portrays language as a shaper of thought; that variations in words and grammatical structures from one language to another provide important evidence in the debate on the Sapir–Whorf hypothesis; and that each language, with its own unique features, serves as a shaper rather than determiner of human thought, culture, and behavior.

Finally, variations in language use within a nation were considered. Language plays a central role in establishing and maintaining the identity of a particular culture. Language variations also foster a political hierarchy among cultures within a nation; nonstandard versions of a language, including accents, dialects, jargon, and argot, are often regarded less favorably than the standard version. The concept of code switching, and some factors that affect the selection of one language over another, were also discussed. The chapter concluded with a discussion of intercultural competence and verbal communication.

FOR DISCUSSION

1. Based on the examples at the beginning of this chapter, what do you think Ludwig Wittgenstein meant when he said that "the limits of my language are the limits of my world"?
2. Is accurate translation and interpretation from one language to another possible? Explain.
3. What is the difference between a dialect and an accent? Between jargon and argot? Give an example of each of these terms.
4. If you speak more than one language (or language dialect), when is each of them used? That is, in what places, relationships, or settings do you use each of them?
5. If you could construct an ideal society, would it be one in which everyone spoke the same language? Or does a society in which people speak different languages offer greater advantages? Explain.

FOR FURTHER READING

Mark Abley, *Spoken Here: Travels Among Threatened Languages* (Boston: Houghton Mifflin, 2003). An examination of the fascinating subject of languages that have only a few native speakers remaining and the efforts that are being made to preserve these languages. It also looks at what is lost when a language dies, as well as the forces, from pop culture to global politics, that threaten to wipe out 90 percent of all languages in the next twenty years.

John J. Gumperz and Stephen C. Levinson (eds.), *Rethinking Linguistic Relativity* (Cambridge: Cambridge University Press, 1996). A good source for additional information on the Sapir–Whorf hypothesis. A modern classic.

Steven Pinker, *The Stuff of Thought: Language as a Window into Human Nature* (New York: Viking, 2007). A very readable explanation of how our use of words in everyday life reveals our human nature.

Geneva Smitherman, *Word from the Mother: Language and African Americans* (New York: Routledge, 2006). Offers insights into the language use that helps to shape the culture and experiences of African Americans.

Eva Alcón Soler and Maria Pilar Safont Jordà (eds.), *Intercultural Language Use and Language Learning* (Dordrecht, Netherlands: Springer, 2007). Offers examples of the semantic, syntactic, and pragmatic issues that arise in teaching and using a foreign language.

Maryanne Wolf, *Proust and the Squid: The Story and Science of the Reading Brain* (New York: HarperCollins, 2007). An intriguing exploration of the ways the human brain adapts to be able to learn to read, with varying adaptations linked to alphabets and languages. A scholarly and engaging "good read" that will both inform and entertain with its prose.

CHAPTER

8

Nonverbal Intercultural Communication

KEY TERMS

Learning to communicate as a native member of a culture involves knowing both the verbal and the nonverbal code systems that are used. The verbal code system, which was considered in the last chapter, constitutes only a portion of the messages that people exchange when they communicate. In this chapter, we explain the types of messages that are often regarded as more foundational or more elemental to human communication. Taken together, these messages constitute the nonverbal communication system.

CHARACTERISTICS OF NONVERBAL CODES

Whereas verbal codes are language-based, nonverbal codes are not. Nonverbal codes encompass the ways that people communicate without words, and they include all forms of communication other than linguistic ones.

The importance of nonverbal codes in communication has been well established. Nonverbal behaviors can become part of the communication process when someone intentionally tries to convey a message or when someone attributes meaning to the nonverbal behaviors of another, whether or not the person intended to communicate a particular meaning.

An important caution related to the distinction between nonverbal and verbal communication must be made as you learn about nonverbal code systems. Though we describe the communication of verbal and nonverbal messages in separate chapters for explanatory convenience, it would be a mistake to assume that they are actually separate and independent communication systems.[1] In fact, they are inseparably linked together to form the code systems through which the members of a culture convey their beliefs, values, thoughts, feelings, and intentions to one another. As Sheila Ramsey suggests:

> Verbal and nonverbal behaviors are inextricably intertwined; speaking of one without the other is, as Birdwhistell says, like trying to study "noncardiac physiology." Whether in opposition or complementary to each other, both modes work to create the meaning of an interpersonal event. According to culturally prescribed codes, we use eye movement and contact to manage conversations and to regulate interactions; we follow rigid rules governing intra-and interpersonal touch, our bodies synchronously join in the rhythm of others in a group, and gestures modulate our speech. We must internalize all of this in order to become and remain fully functioning and socially appropriate members of any culture.[2]

Thus, our distinction between verbal and nonverbal messages is a convenient, but perhaps misleading, way to sensitize you to the communication exchanges within and between cultures.

Nonverbal codes have several characteristics that make them different from verbal codes. Unlike verbal communication, nonverbal communication is multichanneled; this means that nonverbal messages can occur in a variety of ways simultaneously. Nonverbal codes are also multifunctional. As we will elaborate shortly, nonverbal codes can fulfill several goals or communicative functions simultaneously. Moreover, nonverbal codes are typically enacted spontaneously and subconsciously, and oftentimes they convey their meanings in subtle and covert ways.[3] People process nonverbal messages, both the sending and receiving of them, with less awareness than they process verbal messages. Contributing to the silent character of nonverbal messages is the fact that most of them are continuous and

CULTURE CONNECTIONS

A patter of footsteps announced my first customer—a skinny little girl, maybe four years old, with long black hair and a runny nose. She regarded the strange *naluaqmiu* before her with alarm. When I smiled, she steadied herself and solemnly laid a grubby handful of change on the counter, still eyeing me warily. In my best store-keeper's voice, I asked her what she needed today.

Silence.

"Candy?" I prompted.

She didn't answer, but her eyes widened at the array behind the counter—cases of Milky Ways, Twizzlers, Drax Snax, LifeSavers, Garbage Candy—at least twenty varieties.

"Which one?"

More wide-eyed silence.

"This one?"...

"What about this one?"

Finally in exasperation I laid a Drax Snax and some Twizzlers on the counter and sorted out her change. With an expression of complete ecstasy the pretty little girl opened her mouth....

It took me a couple weeks to figure out that she'd been talking to me all along. The Inupiat are subtle, quiet people, and much of their communication hinges on nonverbal cues. Raising the eyebrows or widening the eyes means yes; a wrinkled nose is a negative. The poor girl had been shouting at me, "Yes! Yes! YES!" All these years later, I still recall that first simple failure to understand; it reminds me of all my failures since then, and of the distance that remains.

—Nick Jans ■

natural, and they tend to blur into one another. For example, raising one's hand to wave good-bye is a gesture made up of multiple muscular movements, yet it is interpreted as one continuous movement.

Unlike verbal communication systems, however, there are no dictionaries or formal sets of rules to provide a systematic list of the meanings of a culture's nonverbal code systems. The meanings of nonverbal messages are usually less precise than are those of verbal codes. It is difficult, for example, to define precisely the meaning of a raised eyebrow in a particular culture. Skill in the use of nonverbal message systems has only recently begun to receive formal attention in the educational process, a reflection of the out-of-awareness character of nonverbal codes.

CULTURAL UNIVERSALS IN NONVERBAL COMMUNICATION

Charles Darwin believed that certain nonverbal displays are universal.[4] The shoulder shrug, for example, is used to convey such messages as "I can't do it," "I can't stop it from happening," "It wasn't my fault," "Be patient," and "I do not intend to resist." Many hand gestures (such as pointing) and vocal characteristics (such as the intonation patterns of sentences) are cultural universals as well.[5] Even children who are blind from birth use many of the same gestures and emotional expressions as sighted children do, which suggests their universality.[6]

Another universal aspect of nonverbal communication is the need to be territorial. Territoriality is an innate, evolutionary characteristic that occurs in both animals and humans.[7] Humans from all cultures mark and claim certain spaces as their own.

Paul Ekman's research on facial expressions demonstrates the universality of many nonverbal emotional displays.[8] Ekman described three separate sets of facial muscles that operate independently and can be manipulated to form a variety of emotional expressions. These muscle sets include the forehead and brow; the eyes, eyelids, and base of the nose; and the cheeks, mouth, chin, and rest of the nose. The muscles in each of these facial regions are combined in a variety of unique patterns to display emotional states. For example, fear is indicated by a furrowed brow, raised eyebrows, wide-open eyes, creased or pinched base of the nose, taut cheeks, partially open mouth, and upturned upper lip. Because the ability to produce such emotional displays is consistent across cultures, there is probably a biological or genetic basis that allows these behaviors to be produced in all humans in a particular way.[9]

More recently, David Matsumoto's research has found that, independent of culture, people use the same facial cues when judging emotional expressions.[10] Recent evidence suggests that basic emotions—happiness, sadness, anger, and the like—can be universally recognized not just from the face but from the voice as well.[11] These studies strongly indicate that the expression and recognition of emotions are universally shared by all humans.

Michael Argyle has summarized a number of characteristics of nonverbal communication that are universal across all cultures: (1) the same body parts are used for nonverbal expressions; (2) nonverbal channels are used to convey similar information, emotions, values, norms, and self-disclosing messages; (3) nonverbal messages accompany verbal communication and are used in art and ritual; (4) the motives for using the nonverbal channel, such as when speech is impossible, are similar across cultures; and (5) nonverbal messages are used to coordinate and control a range of contexts and relationships that are similar across cultures.[12]

Although some aspects of nonverbal code systems are universal, it is also clear that cultures choose to express emotions and territoriality in differing ways. These variations are of particular interest in intercultural communication.

CULTURAL VARIATIONS IN NONVERBAL COMMUNICATION

Many instances of nonverbal communication can be interpreted only within the framework of the culture in which they occur. Cultures vary in their nonverbal behaviors in three ways. First, cultures differ in the specific repertoire of behaviors that are enacted. Certain movements, body positions, postures, vocal intonations, gestures, spatial requirements, and even dances and ritualized actions are specific to a particular culture.

Second, all cultures have display rules that govern when and under what circumstances various nonverbal expressions are required, preferred, permitted, or prohibited. Thus, children learn both how to communicate nonverbally and the appropriate display rules that govern their nonverbal expressions. Display rules indicate such things as how far apart people should stand while talking, whom to touch and where, the speed and timing of movements and gestures, when to look directly at others in a conversation and when to look away, whether loud talking and expansive gestures or quietness and controlled movements should be used, when to smile and when to frown, and the overall pacing of communication.

The norms for display rules vary greatly across cultures.[13] Latinos and European Americans, for instance, differ substantially in their use of eye contact during family conversations. Compared to European Americans, fathers and children in Latino families gaze at one another less, and Latino mothers and sons (but not mothers and daughters) are less likely to look at one another as well.[14] These Latino display rules are based on a cultural value of *respeto*, or respect, which is important in that culture for maintaining cordial relationships.[15]

Cultural differences in nonverbal display rules are certainly not confined to U.S. cultures. The use of eye contact during conversations, for example, is more common among Italians and the British than it is among Japanese and Chinese.[16] Similarly, U.S. pedestrians are more likely than their Japanese counterparts to make eye contact, smile, and nonverbally greet other pedestrians.[17] Even photographs that people attach online to their instant messaging (IM) sites differ across cultures: IM users from Eastern Europe are less likely to smile than IM users from Western Europe.[18]

Such differences in display rules can cause discomfort and misinterpretations. To illustrate, consider a Vietnamese woman named Hoa who is visiting her cousin in the United States. As Hoa arrives,

> her cousin Phuong and some of his American friends are waiting at the airport to greet her. Hoa and Phuong are both excited about their meeting because they have been separated for seven years. As soon as Hoa enters the passenger terminal, Phuong introduces her to his friends, Tom, Don, and Charles. Tom steps forward and hugs and kisses Hoa. She pushes him away and bursts into tears.[19]

The difference in when, where, and whom it is acceptable to kiss was the source of the discomfort for Hoa; in her culture's display rules, it is an insult for a boy to hug and kiss a girl in public.

Display rules also indicate the intensity of the behavior that is acceptable. In showing grief or intense sadness, for instance, people from southern Mediterranean cultures may tend to exaggerate or amplify their displays, European Americans may try to remain calm and somewhat neutral, the British may understate their emotional displays by showing only a little of their inner feelings, and the Japanese and Thai may attempt to mask their sorrow completely by covering it with smiling and laughter.[20]

The third way that cultures differ in their nonverbal behaviors is in the interpretations, or meanings, that are attributed to particular nonverbal behaviors. Three possible interpretations could be imposed on a given instance of nonverbal behavior: it is random, it is idiosyncratic, or it is shared.[21] An interpretation that the behavior is random means that it has no particular meaning to anyone. An idiosyncratic interpretation suggests that the behaviors are unique to special individuals or relationships, and they therefore have particular meanings only to these people. For example, family members often recognize that certain unique behaviors of a person signify a specific emotional state. Thus, a family member who tugs on her ear may indicate, to other family members, that she is about to explode in anger. The third interpretation is that the behaviors have shared meaning and significance, as when a group of people jointly attribute the same meaning to a particular nonverbal act.

Behaviors that are insignificant in one culture may be very meaningful in another. This woman relaxes and shows the soles of her feet to others. In many cultures, this would be regarded as an insult.

However, cultures differ in what they regard as random, idiosyncratic, and shared. Thus, behaviors that are regarded as random in one culture may have shared significance in another. For example, when a British professor in Cairo inadvertently showed the soles of his shoes to his class while leaning back in his chair, the Egyptian students were very insulted.[22] The professor's random behavior of leaning back and allowing the soles of his shoes to be seen was a nonverbal behavior with the shared meaning of insult in Egyptian culture. Similarly, a female graduate student from Myanmar was aghast at the disrespectful behavior of her U.S. American classmates, some of whom put their feet on empty desks, thus showing the soles of their shoes to the professor.[23] Such differences in how cultures define behaviors as *random*, *idiosyncratic*, or *shared* can lead to problems in intercultural communication; if one culture defines a particular behavior as random, that behavior will probably be ignored when someone from a different culture uses it to communicate what is assumed to be *shared* meanings.

Even nonverbal behaviors that have shared significance in each of two cultures may mean something very different to their members. As Ray Birdwhistell suggested, "A smile in one society portrays friendliness, in another embarrassment, and in still another [it] may contain a warning that unless tension is reduced, hostility and attack will follow."[24] U.S. Americans typically associate the smile with happiness or friendliness. To the Japanese, the smile can convey a much wider range of emotions, including happiness, agreement, sadness, embarrassment, and disagreement. And though the Thais smile a lot, Koreans rarely smile and regard those who do as superficial.

Nonverbal repertoires, their corresponding display rules, and their preferred interpretations are not taught verbally. Rather, they are learned directly through observation and personal experience in a culture. Because they are frequently acquired outside of conscious awareness, they are rarely questioned or challenged by their users and are often noticed only when they are violated. In intercultural communication, therefore, misunderstandings often occur in the interpretations of nonverbal behaviors because different display rules create very different meanings about the appropriateness and effectiveness of particular interaction sequences. Consider, for instance, the following example:

> An American college student, while having a dinner party with a group of foreigners, learns that her favorite cousin has just died. She bites her lip, pulls herself up, and politely excuses herself from the group. The interpretation given to this behavior will vary with the culture of the observer. The Italian student thinks, "How insincere; she doesn't even cry." The Russian student thinks, "How unfriendly; she didn't care enough to share her grief with her friends." The fellow American student thinks, "How brave; she wanted to bear her burden by herself."[25]

As you can see, cultural variations in nonverbal communication alter the behaviors that are displayed, the meanings that are imposed on those behaviors, and the interpretations of the messages.

CULTURE CONNECTIONS

Smiling, I shook my head and offered Arthur a virulently pink bun that someone had forced on me. He took it in the Batswana manner, placing his left hand on his right forearm as a sign of politeness and murmured, 'Thank you, Rra'.

—Will Randall ∎

NONVERBAL FUNCTIONS IN INTERCULTURAL COMMUNICATION

Before we discuss specific nonverbal codes, let us first consider the following question: What does nonverbal communication *do*? That is, what are the functions of nonverbal communication? Functions are the purposes, meanings, motives, reasons, or goals of communication. As Miles Patterson suggests, "nonverbal communication is best understood as a coordinated system that facilitates adaptive, interpersonal goals."[26]

Nonverbal codes can be used to fulfill five functions: to provide information, manage impressions, express emotions, regulate interactions, and convey relationship messages.[27] These five functions are interrelated; multiple goals are usually being accomplished during every interaction. For ease of explanation, we discuss these functions as though they are separate and independent. Please remember, however, that multiple functions—indeed, often all of them—can be fulfilled simultaneously.

One function served by nonverbal communication is that of providing information. This function is similar to a primary function of verbal communication. The difference, of course, is in the kinds of information that can be conveyed by nonverbal and verbal codes. Nonverbal codes are most useful to convey global meanings and emotional information; verbal codes are most useful to convey logical and factual information. The old saying that "A picture is worth a thousand words" underscores this holistic emphasis that nonverbal codes can provide.

A second function of nonverbal codes is that of managing impressions of oneself and others. What we wear, where we move, how we stand, where we look, and many other nonverbal behaviors all help to convey messages that say, in an all-inclusive way, "This is the perception or image I want you to have of me. And this is how I view you."

Perhaps the most obvious function of nonverbal communication is that of expressing emotions. The primary way that emotions are expressed is nonverbally. Nonverbal facial expressions that convey feelings often occur spontaneously, without conscious or intentional control: a smile of happiness, a frown of sadness, and other facial expressions that display emotions such as pride, surprise, warmth, fear, anger, disgust, shame, guilt, or jealousy.[28] Facial displays are just one way that emotions can be shown: body cues and actions such as crying or agitated movements also contribute to the expression of emotions.[29] Even one's tone of voice can accurately convey a person's emotional state.[30] People with whom we interact use our emotional displays as social cues to infer what we are feeling and what we might say and do. As we suggested previously, however, culturally determined display rules can alter both how emotions are expressed and what sense to make of them. But emotional displays don't just function as a communication cue for others to interpret. A growing body of evidence suggests that our emotions are also a fundamental aid in cognitive reasoning and decision making. Throughout even a single day, we make literally thousands of culturally programmed choices about what to do and how to respond to events that are unfolding. To make

CULTURE CONNECTIONS

Nayir grinned and embraced his friend. Mutlaq greeted him with the traditional kiss on the nose, the only Bedouin gesture Nayir had never dared to imitate.

—Zoë Ferraris ∎

all of these choices thoughtfully, or consciously, would render us incapable of functioning at all.[31] But even though we often can't say, and perhaps consciously and logically may not even know, why we chose one action over another, many of these decisions are accompanied by positive emotional feelings that are powered by tiny surges of dopamine and other chemicals within the brain. As Jonah Lehrer suggests, "If it weren't for our emotions, reason wouldn't exist at all.... The process of thinking requires feeling, for feelings are what let us understand all the information that we can't directly comprehend."[32]

Another function of nonverbal codes is that of regulating interactions. Conversations are highly structured, with people typically taking turns at talking in a smooth and highly organized sequence. Nonverbal codes help to maintain the back-and-forth sequencing of conversations. Speakers use nonverbal means to convey that they want the other person to talk or that they do not wish to be interrupted, just as listeners indicate when they wish to talk and when they prefer to continue listening. Looking behaviors, vocal inflections, gestures, and general cues of readiness or relaxation all help to signal a person's conversational intentions.

A final function of nonverbal codes is that of conveying relationship messages. Interpersonal relationships develop, and they are sustained, primarily through the exchange of nonverbal communication. Messages of equality or inequality, love or hate,

CULTURE CONNECTIONS

No discovery pleased me more, on that first excursion from the city, than the full translation of the famous Indian head-wiggle. The weeks I'd spent in Bombay with Prabaker had taught me that the shaking or wiggling of the head from side to side—that most characteristic of Indian expressive gestures—was the equivalent of a forward nod of the head, meaning *Yes*. I'd also discerned the subtler senses of *I agree with you*, and *Yes, I would like that*. What I learned, on the train, was that a universal message attached to the gesture, when it was used as a greeting, which made it uniquely useful.

Most of those who entered the open carriage greeted the other seated or standing men with a little wiggle of the head. The gesture always drew a reciprocal wag of the head from at least one, and sometimes several of the passengers. I watched it happen at station after station, knowing that the newcomers couldn't be indicating *Yes*, or *I agree with you* with the head-wiggle because nothing had been said, and there was no exchange other than the gesture itself. Gradually, I realised that the wiggle of

the head was a signal to others that carried an amiable and disarming message: *I'm a peaceful man. I don't mean any harm.*

Moved by admiration and no small envy for the marvellous gesture, I resolved to try it myself. The train stopped at a small rural station. A stranger joined our group in the carriage. When our eyes met for the first time, I gave the little wiggle of my head, and a smile. The result was astounding. The man beamed a smile at me so huge that it was half the brilliance of Prabaker's own, and set to such energetic head waggling in return that I was, at first, a little alarmed. By journey's end, however, I'd had enough practice to perform the movement as casually as others in the carriage did, and to convey the gentle message of the gesture. It was the first truly Indian expression my body learned, and it was the beginning of a transformation that has ruled my life, in all the long years since that journey of crowded hearts.

—Gregory David Roberts ∎

and trust or distrust are among the relationship messages that nonverbal communication can convey. Standing above or below, and near or far; looking intently at another, or looking away; using the body to control, to invite, or to display respect; speaking loudly, or softly, or not at all: these and many other nonverbal behaviors all function to express, at every instant of a communicative encounter, each person's view of their shared relationship.

NONVERBAL MESSAGES IN INTERCULTURAL COMMUNICATION

Messages are transmitted between people over some sort of channel. Unlike written or spoken words, however, nonverbal communication can occur in multiple channels simultaneously. Thus, several types of nonverbal messages can be generated by a single speaker or listener at any given instant. When you "read" or observe the nonverbal behaviors of others, you might notice how they appear, what they wear, where they look, how they move, the characteristics of their voice, and how they orient themselves in space and time. All of these nonverbal codes use particular channels or means of communicating messages, which are usually interpreted in a similar fashion by members of a given culture.

We will discuss several types of nonverbal codes to demonstrate their importance in understanding how members of a culture attempt to understand, organize, and interpret the behaviors of others. We will first consider nonverbal codes that are relatively static and unchanging within communication interactions: the physical attributes of people's bodies and the environment and artifacts—the setting—within which communication occurs. Then we focus on nonverbal codes that are dynamic and can change during interactions: body movements, personal space, touching, and the characteristics of the voice. We conclude this section with an examination of cultural differences in time usage, which includes both static and dynamic features.

Physical Appearance

When you first see someone, even before you interact with that person, what do you notice? A likely answer is that you observe, at least in part, the person's physical attributes or physical appearance. Some aspects of a person's physical appearance, such as the actual body characteristics, are relatively permanent; they are stable features that are regarded as an inherent part of that person. Included in this list are one's body shape, body size, body type, facial features, height, weight, skin color, eye color, and various qualities that denote age and gender. Other aspects of one's physical

---

CULTURE CONNECTIONS

"You're Iranian in a superficial way," he said one day, after I rescheduled one of our meetings for the eighth time. "You come across warm, but your affective nature is really Western. Eastern affection involves generosity with time. You drench people with warmth and charm, to distract them from how miserly you are with your time. You handle minutes like an accountant."

—Azadeh Moaveni ∎

appearance involve body modifications such as piercings, tattoos, and cosmetic procedures that are also relatively permanent. Finally, some aspects of one's physical appearance can and usually do change from one situation to another, but they usually don't change within a specific interaction. These body adornments may include one's clothing, makeup, jewelry, glasses, hair characteristics, and body scents both natural (such as from sweat) and artificial (such as from perfumes and colognes).

Physical appearance becomes a nonverbal code when it creates shared meanings among individuals. For example, clothing and body decorations can be, and often are, intentionally used to shape others' understanding of one's personal identity, cultural identity, social affiliations, preferences, moods, status, attractiveness, relationships, and other aspects of oneself. Clothing styles may also fulfill the culture's needs for modesty, self-expression, or privacy.

Sometimes there are large and obvious differences in physical appearance across cultural, ethnic, and racial groups. The information in Chapter 6 suggests that both cultural identities and cultural biases are often based on these differences. Conversely, similarities in these physical attributes can be used to identify fellow ingroup members.

Cultural standards for beauty and attractiveness vary greatly, as do expectations about how people should look and smell. People have distinct scents that can be affected by their way of living, food preferences, habits, and environment. These differences are often used to make judgments or interpretations about members of a culture. For instance, most meat-eating Westerners have a distinct body odor that may be unpleasant to cultures that do not consume red meat. Similarly, many hotels in Malaysia have posted signs that say "No Durians" to discourage their guests from eating the pungent, sweet-tasting fruit that many consider to be a delicacy. Among many Arabic-speaking cultures, attempts to mask body odors with perfumes is sometimes considered an insult; for both Arabs and Filipinos, smelling another person's breath may be so favorably regarded that close spatial distances in conversations are used to obtain that smell.

Environment

Another nonverbal code that does not change during a specific interaction is the environment, which encompasses the physical features or characteristics of our surroundings. The environment might be a home, a classroom, a store, or a specific outdoor location. Within that environment are the objects, tools, furniture, adornments, lighting, colors, decorations, sounds, and creations that people have put there to influence potential interactions and to send distinctive messages about themselves.

People organize their perceptions of their environments in similar ways. Environments differ in their formality, warmth, privacy, familiarity, constraint, and distance.[33]

Formality refers to the heightened sense of decorum and politeness that some environments seem to require. Informal environments allow you to have a more relaxed and casual demeanor. Like many nonverbal codes, perceptions of formality are specific to a culture and its interpretations. U.S. students taking Spanish lessons in Costa Rica, for example, are sometimes regarded as rude and impolite because they act too casually and informally toward their teacher. Conversely, a Nigerian exchange student at a U.S. university who refers to his teacher as "Madam Professor" is displaying a misunderstanding of U.S. classroom norms for appropriate (in)formality.

Warmth refers not to the physical temperature of the setting but to the emotional tone conveyed by the environment. A warm environment feels comfortable and seems to invite you in; it is appealing and welcoming. The colors in the room, the type of the furniture and how it is arranged, and other aspects of the environment all help to communicate the degree to which the physical setting is warm and intimate.

Privacy refers to the degree to which the environment allows you to be surrounded by others or isolated from those who might learn what you are saying and doing. Differences in privacy needs, for example, are often indicated by such features as closed doors in the United States, soundproof doors in Germany, tree-lined barriers at property lines in England, and paper-thin walls in Japan.

The dimension of familiarity describes the degree to which the environment is well known and therefore predictable to you, or strange and unpredictable to you. In familiar environments within your own culture, you are more likely to be relaxed and to feel at ease. Conversely, the sights, sounds, smells, and objects in "foreign" environments can sometimes seem strange and unsettling.

Constraint refers to your perception of the extent to which you feel "stuck" in a particular environment or free to leave it. If you have ever been on a lengthy airplane flight while sitting next to an unwelcome stranger—perhaps a stranger who talks too much and who "invades" your very limited space while you are in your seat with your seat belt securely fastened—then you have experienced environmental constraint.

Distance refers to the spatial arrangements of the environment. How far away, for example, is the door? Does the space seem to "fit" the number of people in it, or does it feel too large or small? Perceptions of spaciousness or crowding are often related to these spatial arrangements, and cultures differ widely in what they regard as typical or unusual.

Body Movements

In addition to nonverbal codes that don't change in an interaction, there are many nonverbal messages that do. Perhaps the most researched of these is body movements. The study of body movements, often inaccurately called body language, is known as kinesics.[34]

Kinesic behaviors include gestures, head movements, facial expressions, eye behaviors, and other physical displays that can be used to communicate. Of course, like all other forms of communication, no single type of behavior exists in isolation. Specific body movements can be understood only by taking the person's total behavior into account.

Paul Ekman and Wallace Friesen have suggested that there are five categories of kinesic behaviors: emblems, illustrators, affect displays, regulators, and adaptors.[35] We will consider each type of kinesic behavior in turn.

CULTURE CONNECTIONS

A classic example, for instance, is the Asian cultural belief that too much eye contact is disrespectful and even confrontational. However, not making eye contact can be misconstrued within the American system as an indication of insincerity or discomfort. Similarly, deference to authority may be likened to timidity or a lack of opinions. Self-effacement suggests indifference or a lack of ambition; avoidance of shame prevents one from publicly acknowledging his or her aspirations for fear of failing; self-control and modesty inhibit one from social interaction and public speaking; and so on.

—Renu Khator ∎

The *wai* gesture is an emblem that is used throughout Thailand both as a greeting and to say goodbye.

EMBLEMS Emblems are nonverbal behaviors that have a direct verbal counterpart. Emblems that are familiar to most U.S. Americans include such gestures as the two-fingered peace symbol and arm waving to indicate hello or good-bye. Emblems are typically used as a substitute for the verbal channel, either by choice or when the verbal channel is blocked for some reason. Underwater divers, for example, have a rich vocabulary of kinesic behaviors that they use to communicate with their fellow divers. Similarly, a baseball coach uses kinesic signals to indicate a particular pitch or type of play, which is usually conveyed by an elaborate pattern of hand motions that involve touching the cap, chest, wrist, and other areas in a pattern known to the players.

Emblems, like all verbal languages, are symbols that have been arbitrarily selected by the members of a culture to convey their intended meanings. For example, there is nothing peacelike in the peace symbol, which is a nonverbal emblem that can be displayed by extending the index and middle fingers upward from a clenched fist. Indeed, in other cultures the peace symbol has other meanings: Winston Churchill used the same symbol to indicate victory, but to many people in South American countries, it is regarded as an obscene gesture. The meanings of emblems are learned within a culture and, like verbal codes, are used consciously by the culture's members when they wish to convey specific ideas to others. Because emblems have to be learned to be understood, they are culture-specific.

Emblems can be a great source of misunderstanding in intercultural communication because the shared meanings for an emblem in one culture may be different in another. In many South Pacific islands, for instance, people raise their eyebrows to indicate "yes." Albanians and Bulgarians signal "yes" by shaking their heads from side to side, and they signal "no" by moving their heads up and down.[36] Similarly, in Turkey, to say "no" nonverbally, just

> nod your head up and back, raising your eyebrows at the same time. Or just raise your eyebrows; that's "no."...
>
> By contrast, wagging your head from side to side doesn't mean "no" in Turkish; it means "I don't understand." So if a Turk asks you, "Are you looking for the bus to Ankara?" and you shake your head, he'll assume you don't understand English, and will probably ask you the same question again, this time in German.[37]

Sometimes these mix-ups might be seen as humorous, such as the time a German student tried, unsuccessfully, to order a beer in a Canadian bar by gesturing with a raised thumb—the common German hand gesture for the number one—only to be met with indifference.[38] At other times these misunderstandings can have serious consequences. A U.S. engineer, for example, unintentionally offended his German counterpart by giving the common U.S. gesture for "OK": hand up, thumb and forefinger held in a circle, to indicate that he had done a good job. The German interpreted the gesture's meaning as a crude reference to a body orifice and walked off the job.[39]

This speaker gestures as he provides his comments at a health conference. These illustrators accompany his talk and underscore the words he is using.

ILLUSTRATORS   Illustrators are nonverbal behaviors that are directly tied to, or accompany, the verbal message. They are used to emphasize, explain, and support a word or phrase. They literally illustrate and provide a visual representation of the verbal message. In saying "the huge mountain," for example, you may simultaneously lift your arms and move them in a large half-circle. Similarly, you may point your index finger to emphasize an important idea or use hand motions to convey directions to a particular address. Unlike emblems, however, none of these gestures has meaning in itself. Rather, the meaning depends on the verbal message it underscores.

Illustrators are less arbitrary than emblems, which makes them more likely to be universally understood. But differences in both the rules for displaying illustrators and in the interpretations of them can be sources of intercultural misunderstanding. For example, calling for a person or a taxi while pointing an index finger is very inappropriate in many Asian cultures, akin to calling a dog.

During his presidency, Bill Clinton travelled to China and, at one stop during the trip, spoke to university students in Beijing. President Clinton's remarks were generally well received and were followed by a lively question and answer session.

When interviewed for the American press, one student remarked, "During the question and answer period, I did not understand why the president pointed his finger at us to select a person. We would not use such a rude gesture." Puzzled, the American reporter asked the student what gesture the president should have used. The student answered using a sweep of the open hand—palm upward.[40]

Likewise, beckoning someone with the palm facing the body, fingers turned upward, can be offensive; Filipinos, Vietnamese, Mexicans, Pakistanis, and many others regard this gesture as disrespectful, because it is used to call those who are inferior in status.[41] Instead, the whole right hand is used, palm down, with the fingers together in a scooping motion toward the body. Similarly, punching the fist into the open palm as a display of strength may be misinterpreted as an obscene gesture whose meaning is similar to a Westerner's use of the middle finger extended from a closed fist.

AFFECT DISPLAYS   Affect displays are facial and body movements that show feelings and emotions. Expressions of happiness or surprise, for instance, are displayed by the face and convey a person's inner feelings. Though affect displays are shown primarily through the face and eyes, postures and other body displays can also convey an emotional state.

Many affect displays may be universally recognized. The research of Paul Ekman and his colleagues indicates that, regardless of culture, the primary emotional states include happiness, sadness, anger, fear, surprise, disgust, contempt, and interest.[42] In addition to these primary affect displays, there are about thirty affect blends, combinations of the primary emotions.

Affect displays may be conscious and intentional, as when we purposely smile and look at another person to convey warmth and affection. Or affect displays may be unconscious and unintentional, such as a startled look of surprise, a blush of embarrassment, or dilated pupils due to pleasure or interest.

Cultural norms often govern both the kind and amount of affect displays shown. The Chinese, for instance, typically have lower frequency, intensity, and duration of affect displays than their European counterparts.[43] Particularly in high-context cultures, subtle changes in skin tonalities due to blushing, blanching, goose flesh, and related experiences may be carefully observed to learn what the other person may be feeling.

REGULATORS   Regulators are nonverbal behaviors that help to synchronize the back-and-forth nature of conversations. This class of kinesic behaviors helps to control the flow and sequencing of communication and may include head nods, eye contact, postural shifts, back-channel signals (such as "Uh-huhm" or "Mmm-mmm"), and other turn-taking cues.

Regulators are used by speakers to indicate whether others should take a turn and by listeners to indicate whether they wish to speak or would prefer to continue listening. They also convey information about the preferred speed or pacing of conversations and the degree to which the other person is understood and believed.

Regardless of culture, taking turns is required in all conversations. Thus, for interpersonal communication to occur, talk sequences must be highly coordinated. Regulators are those subtle cues that allow people to maintain this high degree of coordination.

Regulators are culture-specific. For instance, people from high-context cultures such as Korea and Japan are especially concerned with meanings conveyed by the eyes. In an interesting study comparing the looking behaviors of African Americans and European Americans in a conversation, Marianne LaFrance and Clara Mayo found that there were many differences in the interpretations of turn-taking cues. European Americans tend to look directly into the eyes of the other person when they are the listeners, whereas African Americans prefer to look away. Unfortunately, to African Americans, such behaviors by European Americans may be regarded as invasive or confrontational, when interest and involvement are intended instead. Conversely, the behaviors of African Americans could be regarded by European Americans as a sign of indifference or inattention, when respect is intended. LaFrance and Mayo also found that, when African American speakers pause while simultaneously looking directly at their European American listeners, the listeners often interpret this as a signal to speak, only to find that the African American person is also speaking.[44]

ADAPTORS   Adaptors are personal body movements that occur as a reaction to an individual's physical or psychological state. Scratching an itch, fidgeting, tapping a pencil, and smoothing one's hair are all behaviors that fulfill some individualized need.

Adaptors are usually performed unintentionally, mindlessly, and without conscious awareness. They seem to be more frequent under conditions of stress, impatience, enthusiasm, excitement, or nervousness, and they are often interpreted by others as a sign of discomfort, uneasiness, agitation, irritation, or other negative feelings.

Personal Space

The use of space functions as an important communication system in all cultures. Cultures are organized in some spatial pattern, and that pattern can reveal the character of the people in that culture. Two important features of the way cultures use the space around them are the different needs for personal space and the messages that are used to indicate territoriality.

CULTURAL DIFFERENCES IN THE USE OF PERSONAL SPACE   Wherever you go, whatever you do, you are surrounded at all moments by a personal space "bubble." Edward Hall coined the term proxemics to refer to the study of how people differ in their use of personal space.

Personal space distances are culture-specific.[45] People from colder climates, for instance, typically use large physical distances when they communicate, whereas those from warm-weather climates prefer close distances. The personal space bubbles for northern Europeans are therefore large, and people expect others to keep their distance. The personal space bubbles for Europeans get smaller and smaller, however, as one travels south toward the Mediterranean. Indeed, the distance that is regarded as intimate in Germany, Scandinavia, and England overlaps with what is regarded as a normal conversational distance in France and the Mediterranean countries of Italy, Greece, and Spain. Thus, when an Italian and a Norwegian attempt to have a simple informal conversation, for example, the Italian tries to move closer—into his comfort zone for such conversations—whereas the Norwegian continually tries to move backward in an attempt to maintain the "correct" conversational distance. The resulting interaction, which might look like a slow-motion dance across the room, could be comical except that it results in negative evaluations on both sides; the Norwegian thinks his southern European counterpart is "too close for comfort," whereas the Italian regards his northern European neighbor as "too distant and aloof."

The habitual use of the culturally proper spacing distance is accompanied by a predictable level and kind of sensory information. For example, if the standard cultural

CULTURE CONNECTIONS

"Do you find Chinese ways strange?" I looked at her closely.

Cynthia scrunched her nose up and twisted her lips. "Sometimes." She smiled. "I'm not sure why exactly. At first, it's just the language, I think, I couldn't understand the accents—it's not like school."...

"And the way people move, squatting, and always bumping into each other on the sidewalk, the buses. Chinese people touch each other more. Women and women. Men and men. When you would come up to me and touch my arm, that surprised me at first."

"We're always bumping each other but we aren't like Americans. We touch the outside but never show our inside."

—May-lee Choi   ■

spacing distance in a personal conversation with an acquaintance is about three feet, people will become accustomed to the sights, sounds, and smells of others that are usually acquired at that distance. For someone who is accustomed to a larger spacing distance, at three feet the voices will sound too loud, it might be possible to smell the other person's breath, the other person will seem too close and perhaps out of the "normal" focal range, and the habitual ways of holding the body may no longer work. Then, the culturally learned cues that are so helpful within one's culture can become a hindrance. One European American student, for instance, in commenting on a party that was attended by many Italians and Spaniards, exclaimed, "They would stand close enough that I could almost feel the air coming from their mouths." Similar reactions to intercultural encounters are very common. As Edward and Mildred Hall have suggested:

> Since most people don't think about personal distance as something that is culturally patterned, foreign spatial cues are almost inevitably misinterpreted. This can lead to bad feelings which are then projected onto the people from the other culture in a most personal way. When a foreigner appears aggressive and pushy, or remote and cold, it may mean only that her or his personal distance is different from yours.[46]

CULTURAL DIFFERENCES IN TERRITORIALITY   Do you have a favorite chair or classroom seat that you think "belongs" to you? Or do you have a room, or perhaps just a portion of a room, that you consider to be off limits to others? The need to protect and defend a particular spatial area is known as territoriality, a set of behaviors that people display to show that they "own" or have the right to control the use of a particular geographic area.

People mark their territories in a variety of ways. It can be done formally using actual barriers such as fences and signs that say, "No Trespassing" or "Keep Off the Grass." Territories can also be marked informally by nonverbal markers such as clothing, books, and other personal items that indicate a person's intent to control or occupy a given area.

Cultural differences in territoriality can be exhibited in three ways. First, cultures can differ in the general degree of territoriality that its members tend to exhibit. Some cultures are far more territorial than others. For instance, as Hall and Hall point out in their comparison of Germans and French:

> People like the Germans are highly territorial; they barricade themselves behind heavy doors and soundproof walls to try to seal themselves from others in order to concentrate on their work. The French have a close personal distance and are not as territorial. They are tied to people and thrive on constant interaction and high-information flow to provide them the context they need.[47]

Second, cultures can differ in the range of possible places or spaces about which they are territorial. A comparison of European Americans with Germans, for example, reveals that both groups are highly territorial. Both have a strong tendency to establish areas that they consider to be their own. In Germany, however, this feeling of territoriality extends to "all possessions, including the automobile. If a German's car is touched, it is as though the individual himself has been touched."[48]

Finally, cultures can differ in the typical reactions exhibited in response to invasions or contaminations of their territory. Members of some cultures prefer to react by withdrawing or avoiding confrontations whenever possible. Others respond by insulating themselves from the possibility of territorial invasion, using barriers or other boundary markers. Still others react forcefully and vigorously in an attempt to defend their "turf" and their honor.

Touch

Touch is probably the most basic component of human communication. It is experienced long before we are able to see and speak, and it is a fundamental part of the human experience.

THE MEANINGS OF TOUCH   Stanley E. Jones and A. Elaine Yarbrough have identified five meanings of touch that are important in understanding the nature of intercultural communication.[49] Touch is often used to indicate affect, the expression of positive and negative feelings and emotions. Protection, reassurance, support, hatred, dislike, and disapproval are all conveyed through touch; hugging, stroking, kissing, slapping, hitting, and kicking are all ways in which these messages can be conveyed. Touch is also used as a sign of playfulness. Whether affectionately or aggressively, touch can be used to signal that the other's behavior should not be taken seriously. Touch is frequently used as a means of control. "Stay here," "Move over," and similar messages are communicated through touch. Touching for control may also indicate social dominance. High-status individuals in most Western countries, for instance, are more likely to touch than to be touched, whereas low-status individuals are likely to receive touching behaviors from their superiors.[50] Touching for ritual purposes occurs mainly on occasions involving introductions or departures. Shaking hands, clasping shoulders, hugging, and kissing the cheeks or lips are all forms of greeting rituals. Touching is also used in task-related activities. These touches may be as casual as a brief contact of hands when passing an object, or they may be as formal and prolonged as a physician taking a pulse at the wrist or neck.

CULTURAL DIFFERENCES IN TOUCH   Cultures differ in the overall amount of touching they prefer. People from high-contact cultures such as those in the Middle East, Latin America, and southern Europe touch each other in social conversations much more than do people from noncontact cultures such as Asia and northern Europe. These cultural differences can lead to difficulties in intercultural communication. Germans, Scandinavians, and Japanese, for example, may be perceived as cold and aloof by Brazilians and Italians, who in turn may be regarded as aggressive, pushy, and overly familiar by northern Europeans. As Edward and Mildred Hall have noted, "In northern Europe one does not touch others. Even the brushing of the overcoat sleeve used to elicit an apology."[51] A comparable difference was observed by Dean Barnlund, who found that U.S. American students reported being touched twice as much as did Japanese students.[52]

Cultures also differ in where people can be touched. In Thailand and Malaysia, for instance, the head should not be touched because it is considered to be sacred and the locus of a person's spiritual and intellectual powers. In the United States, the head is far more likely to be touched.[53]

CULTURE CONNECTIONS

Irene put down the tray and held her hand out to Thatcher. "How do you do," she said. She spoke in Navajo to Leaphorn, using the traditional words, naming her mother's clan, the Towering House People, and her father's, the Paiute Dineh. She didn't hold out her hand. He wouldn't expect it. This touching of strangers was a white man's custom that some traditional Navajos found difficult to adopt.

—Tony Hillerman ■

Cultures differ in the use of touching and space. These men from Syria hold hands, which in many cultures is a commonly accepted behavior among male friends.

Cultures vary in their expectations about who touches whom. In Japan, for instance, there are deeply held feelings against the touch of a stranger. These expectations are culture-specific, and even cultures that exist near one another can have very different norms. Among the Chinese, for instance, shaking hands among people of the opposite sex is perfectly acceptable; among many Malay, it is not. Indeed, for those who practice the Muslim religion, casual touching between members of the opposite sex is strictly forbidden. Both men and women have to cleanse themselves ritually before praying if they happen to make physical contact with someone of the opposite sex. Holding hands, for example, or walking with an arm across someone's shoulder or around the waist, or even grabbing an elbow to help another cross the street, are all considered socially inappropriate behaviors between men and women. In some places there are legal restrictions against public displays of hugging and kissing, even among married couples. However, this social taboo refers only to opposite-sex touching; it is perfectly acceptable for two women to hold hands or for men to walk arm in arm. Many European Americans, of course, have the opposite reaction; they react negatively to same-sex touching (particularly among men) but usually do not mind opposite-sex touching.

Finally, cultures differ in the settings or occasions in which touch is acceptable. Business meetings, street conversations, and household settings all evoke different norms for what is considered appropriate. Cultures make distinctions between those settings that they regard as public and those considered private. Although some cultures regard touching between men and women as perfectly acceptable in public conversations, others think that such activities should occur only in the privacy of the home; to them, touch is a highly personal and sensitive activity that should not occur where others might see it.

Voice

Nonverbal messages are often used to accent or underscore the verbal message by adding emphasis to particular words or phrases. Indeed, the many qualities of the voice itself, in addition to the actual meaning of the words, form the vocalic nonverbal communication system. Vocalics also include many nonspeech sounds, such as belching, laughing, and crying, and vocal "filler" sounds such as *uh, er, um,* and *uh-huh.*

VOCAL VERSUS VERBAL COMMUNICATION   Vocalic qualities include pitch (high to low), rate of talking (fast to slow), conversational rhythm (smooth to staccato), and volume (loud to soft). Because spoken (i.e., verbal) language always has some vocal elements, it is difficult to separate the meaning conveyed by the language from that conveyed by the vocalic components. However, if you can imagine that these words you are now reading are a transcript of a lecture we have given, you will be able to understand clearly the distinctions we are describing. Although our words—the language spoken— are here on the printed page, the vocalics are not. Are we speaking rapidly or slowly? How does our inflection change to emphasize a point or to signal a question? Are we yelling, whispering, drawling, or speaking with an accent? Do our voices indicate that we are tense, relaxed, strained, calm, bored, or excited? The answers to these types of questions are conveyed by the speaker's voice.

CULTURAL DIFFERENCES IN VOCAL COMMUNICATION   There are vast cultural differences in vocalic behaviors. For example, unlike English, many Asian languages are tonal. The same Chinese words when said with a different vocalic tone or pitch can have vastly different meanings. In addition to differences in tone or pitch, there are large cultural differences in the loudness and frequency of speaking. Latinos, for instance, perceive themselves as talking more loudly and more frequently than European Americans.[54]

The emotional meanings conveyed by the voice are usually taken for granted by native language users, but they can be the cause of considerable problems when they fail to conform to preconceived expectations. For instance, when a Saudi Arabian man is speaking in English, he will usually transfer his native intonation patterns without necessarily being aware that he has done so. In Arabic, the intonation pattern is such that many of the individual words in the sentence are stressed. Although a flat intonation pattern is used in declarative sentences, the intonation pattern for exclamatory sentences is much stronger and more emotional than that in English. The higher pitch of Arabic speakers also conveys a more emotional tone than that of English speakers. Consequently, differences in vocal characteristics may result in unwarranted negative impressions. The U.S. American may incorrectly perceive that the Saudi Arabian is excited or angry when in fact he is not. Questions by the Saudi that merely seek information may sound accusing. The monotonous tone of declarative sentences may be perceived as demonstrating apathy or a lack of interest. Vocal stress and intonation differences may be perceived as aggressive or abrasive, when only polite conversation is intended. Conversely, the Saudi Arabian may incorrectly interpret certain behaviors of the U.S. American speaker as an expression of calmness and pleasantness when anger or annoyance is being conveyed. Similarly, a statement that seems to be a firm assertion to the U.S. American speaker may sound weak and doubtful to the Saudi Arabian.[55]

Time

The study of time—how people use it, structure it, interpret it, and understand its passage—is called chronemics. We consider chronemics from two perspectives: time orientations and time systems.

TIME ORIENTATIONS   Time orientation refers to the value or importance the members of a culture place on the passage of time. In Chapter 1, we indicated that communication is a process, which means that people's behaviors must be understood as part of an ongoing stream of events that changes over a period of time. Chapters 4 and 5 suggested that members of a culture share a similar worldview about the nature of time. We also indicated that different cultures can have very different conceptions and values about the appropriate ways to comprehend events and experiences. Specifically, some cultures are predominantly past-oriented, others are present-oriented, and still others prefer a future-oriented worldview. As we briefly review these cultural orientations about time, take note of the amazing degree of interrelationship—in this case the link between a culture's nonverbal code system and its cultural patterns—that characterizes the various aspects of a culture.

Past-oriented cultures regard previous experiences and events as most important. These cultures place a primary emphasis on tradition and the wisdom passed down from older generations. Consequently, they show a great deal of deference and respect for parents and other elders, who are the links to these past sources of knowledge. Events are circular, as important patterns perpetually recur in the present; therefore, tried-and-true methods for overcoming obstacles and dealing with problems can be applied to current difficulties. Many aspects of the British, Chinese, and Native American experiences, for instance, can be understood only by reference to their reverence for traditions, past family experiences, or tribal customs. Consider this example of a past-oriented culture, the Samburu, a nomadic tribe from northern Kenya that reveres its elders:

> The elders are an invaluable source of essential knowledge, and in an environment that by its very nature allows only a narrow margin for error, the oldest survivors must possess the most valuable knowledge of all. The elders know their environment intimately—every lie and twist of it. The land, the water, the vegetation; trees, shrubs, herbs—nutritious, medicinal, poisonous. They know each cow, and have a host of specific names for the distinctive shape and skin patterns of each animal in just the same way that Europeans distinguish within the general term flower, or tree.[56]

Present-oriented cultures regard current experiences as most important. These cultures place a major emphasis on spontaneity and immediacy and on experiencing each moment as fully as possible. Consequently, people do not participate in particular events or experiences because of some potential future gain; rather, they participate because of the immediate pleasure the activity provides. Present-oriented cultures typically believe that unseen and even unknown outside forces, such as fate or luck, control their lives. Cultures such as those in the Philippines and many Central and South American countries are usually present-oriented, and they have found ways to encourage a rich appreciation for the simple pleasures that arise in daily activities.

Future-oriented cultures believe that tomorrow—or some other moment in the future—is most important. Current activities are not accomplished and appreciated for their own sake but for the potential future benefits that might be obtained. For example, you go to school, study for your examinations, work hard, and delay or deny present rewards for the potential future gain that a rewarding career might provide. People from future-oriented cultures, which include many European Americans, believe that their fate is at least partially in their own hands and that they can control the consequences of their actions.

TIME SYSTEMS   Time systems are the implicit cultural rules that are used to arrange sets of experiences in some meaningful way. There are three types of time systems: technical, formal, and informal.

Technical time systems are the precise, scientific measurements of time that are calculated in such units as nanoseconds. Typically, members of a culture do not use technical time systems because they are most applicable to specialized settings

CULTURE CONNECTIONS

Smelling another person's cheeks as a form of greeting is also used by the Arabs. To the Arab, to be able to smell a friend is reassuring. Smelling is a way of being involved with another, and to deny a friend his breath would be to act ashamed. In some rural Middle Eastern areas, when Arab intermediaries call to inspect a prospective bride for a relative, they sometimes ask to smell her. Their purpose is not to make sure that she is freshly scrubbed: apparently what they look for is any lingering odor of anger or discontent.

—A. J. Almany and A. J. Alwan   ■

such as the research laboratory. Consequently, technical time systems are of little relevance to the common experiences that members of a culture share.

Formal time systems refer to the ways in which the members of a culture describe and comprehend units of time. Time units can vary greatly from culture to culture. Among many Native American cultures, for instance, time is segmented by the phases of the moon, the changing seasons, the rise and fall of the tides, or the movements of the sun. Similarly, when a Peruvian woman was asked for the distance to certain Inca ruins, she indicated their location by referring nonverbally to a position in the sky that represented the distance the sun would travel toward the horizon before the journey would be complete.[57] Among European Americans, the passage of time is segmented into seconds, minutes, hours, days, weeks, months, and years.

Time's passage may likewise be indicated by reference to significant events such as the birth of a royal son or an important victory in battle. Time intervals for particular events or activities may also be based on significant external events, such as the length of a day or the phases of the moon. Alternatively, time intervals may be more arbitrary, as in the length of a soccer game or the number of days in a week. These ways of representing the passage of time, however arbitrary, are the culture's formal time system. Sequences such as the months in the year are formally named and are explicitly taught to children and newcomers as an important part of the acculturation process.

The formal time system includes agreements among the members of a culture on such important issues as the extent to which time is regarded as valuable and tangible. European Americans, of course, typically regard time as a valuable, tangible commodity that is used or consumed to a greater or lesser degree.

Informal time systems refer to the assumptions cultures make about how time should be used or experienced. How long should you wait for someone who will be ready soon, in a minute, in a while, or shortly? When is the proper time to arrive for a 9:00 a.m. appointment or an 8:00 p.m. party? As a dinner guest, how long after your arrival would you expect the meal to be served? How long should you stay after the meal has been concluded? Cultures have unstated expectations about the timing and duration of such events. In this regard, Edward Hall has reported:

> The time that it takes to reach an agreement or for someone to make up his mind operates within culturally defined limits. In the U.S. one has about four minutes in the business world to sell an idea. In Japan the well-known process of "nema-washi"—consensus building, without which nothing can happen—can take weeks or months. None of this four-minute sell.[58]

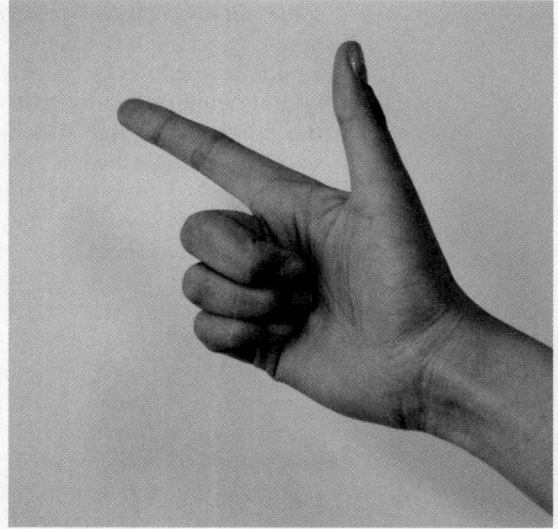

Although these expectations differ, depending on such factors as the occasion and the relative importance of those being met or visited, they are widely held and consistently imposed as the proper or appropriate way to conduct oneself as a competent member of the culture. For example, in business appointments among Koreans, those with lower status are expected to arrive promptly while those with higher status are allowed, and sometimes expected, to arrive late.[59]

Can you guess what this hand gesture means? In China, it represents the number "eight" and is commonly used.

CULTURE CONNECTIONS

Dtui had been sitting for an hour in front of the office of the politburo member. She hadn't made an appointment with Civilai. That wasn't a particularly Lao thing to do. Appointments were rarely kept. She knew he had to come to his office eventually, and much sooner than she'd expected she was proven right.

—Colin Cotterill ∎

Perhaps the most important aspect of the culture's informal time system is the degree to which it is monochronic or polychronic.[60] A monochronic time system means that things should be done one at a time, and time is segmented into precise, small units. In a monochronic time system, time is viewed as a commodity; it is scheduled, managed, and arranged. European Americans, like members of other monochronic cultures, are very time-driven. Similarly, within Swiss-German culture, people will often interpret tardiness as a personal insult.[61] The ubiquitous calendar or scheduler that many people carry, which tells them when, where, and with whom to engage in activities, is an apt symbol of a monochronic culture. An event is regarded as separate and distinct from all others and should receive the exclusive focus of attention it deserves. These events also have limits or boundaries, so that there are expected beginning and ending points that have been scheduled in advance. Thus people from monochronic cultures

> find it disconcerting to enter an office overseas with an appointment only to discover that other matters require the attention of the man we are to meet. Our ideal is to center the attention first on one thing and then move on to something else.[62]

A polychronic time system means that several things are being done at the same time. In Spain and among many Spanish-speaking cultures in Central and South America, for instance, relationships are far more important than schedules. Appointments will be quickly broken, schedules readily set aside, and deadlines unmet without guilt or apology when friends or family members require attention. Those who use polychronic time systems often schedule multiple appointments simultaneously, so keeping "on schedule" is an impossibility that was never really a goal. European Americans, of course, are upset when they are kept waiting for a scheduled appointment, particularly when they discover that they are the third of three appointments that have been scheduled at exactly the same hour.

CULTURAL DIFFERENCES IN PERCEPTIONS AND USE OF TIME   Cultures differ in their time orientations and in the time systems they use to give order to experiences. Misunderstandings can occur between people who have different time orientations. For instance, someone from a present-oriented culture might view people from past-oriented cultures as too tied to tradition and people from future-oriented cultures as passionless slaves to efficiency and materialism. Alternatively, someone from a future-oriented culture might view those from present-oriented cultures as self-centered, hedonistic, inefficient, and foolish.[63] This natural tendency to view one's own practices as superior to all others is a common source of problems in intercultural communication.

Cultures also differ in the formal and informal time systems they use to determine how long an event should take, and even how long "long" is. Misinterpretations often occur when individuals from monochronic and polychronic cultures attempt to inter-act. Each usually views the other's responses to time "commitments" as disrespectful and unfriendly. Interculturally competent individuals, however, are typically aware of the time systems they are using to regulate their behaviors, and they are able to adapt their time orientations to the prevailing social and situational constraints. For example, participants at a board meeting of a Puerto Rican Community Center used European American references to time when they were focused on their work but employed "Puerto Rican time" when the goal was socializing.[64] Similarly, in some intercultural situations, the name of a culture will be added after a meeting time to designate if the time is to be regarded as fixed or flexible.[65]

SYNCHRONY OF NONVERBAL COMMUNICATION CODES

Cultures train their members to synchronize the various nonverbal behaviors to form a response pattern that typifies the expected behaviors in that culture. Subtle variations in the response patterns are clearly noticed, even when they differ by only a few thou-sandths of a second. William Condon, who describes himself as "a white, middle-class male," suggests that interactional synchrony is learned from birth and occurs within a fraction of a second. Condon compares the differences in the speech and gestures of African Americans and European Americans:

> If I say the word "because" both my hands may extend exactly together. In Black behavior, however, the right hand may begin to extend with the "be" portion slightly ahead of the left hand and the left hand will extend rapidly across the "cause" portion. This creates the syncopation, mentioned before, which can ap-pear anywhere in the body. A person moves in the rhythm and timing of his or her culture, and this rhythm is in the whole body. ... It may be that those having different cultural rhythms are unable to really "synch-in" fully with each other. ... I think that infants from the first moments of life and even in the womb are getting the rhythm and structure and style of sound, the rhythms of their culture, so that they imprint to them and the rhythms become part of their very being.[66]

Behavioral synchrony in the use of nonverbal codes can be found in virtually all cultures. Not only must an individual's many behaviors be coordinated appropriately, they must also mesh properly with the words and movements of the other interactants. Coordination in Japanese bowing behaviors, for example, requires an adaptation to the status relationships of the participants; the inferior must begin the bow, and the superior decides when the bow is complete. If the participants are of equal status, they must begin and end their bows simultaneously. This is not as easy as it seems. As one Japanese man relates:

> Perfect synchrony is absolutely essential to bowing. Whenever an American tries to bow to me, I often feel extremely awkward and uncomfortable because I simply cannot synchronize bowing with him or her. ... bowing occurs in a flash of a sec-ond, before you have time to think. And both parties must know precisely when to start bowing, how deep, how long to stay in the bowed position, and when to bring their heads up.[67]

Similar degrees of coordination and synchrony can be found in most everyday activities. Sensitivity to these different nonverbal codes can help you to become more interculturally competent.

NONVERBAL COMMUNICATION AND INTERCULTURAL COMPETENCE

Nonverbal codes are important to an understanding of intercultural communication because virtually everything we say, do, create, and wear can communicate messages about our culture and ourselves. Indeed, as Peter A. Andersen suggests, "One of the most basic and obvious functions of nonverbal communication is to communicate one's culture."[68]

The rules and norms that govern most nonverbal communication behaviors are both culture-specific and outside of conscious awareness. That is, although members of a culture know and follow their culture's expectations, they probably learned the norms for proper nonverbal expressiveness very early in childhood, and these norms may never have been articulated verbally.[69] Sometimes, therefore, the only way you will know that a cultural norm exists is when you break it!

An important consequence of this out-of-awareness aspect is that members of a culture use their norms to determine appropriate nonverbal behaviors and then make negative judgments about others' feelings, motives, intentions, and even their attractiveness if these norms are violated.[70] Often the violations will be inaccurately attributed to aspects of personality, attitudes, or intelligence rather than to a mismatch between learned nonverbal codes. U.S. Americans, for instance, highly value positive nonverbal displays and typically regard someone who smiles as more intelligent than someone who does not; the Japanese, however, whose cultural norms value constraint in nonverbal expressiveness, do not equate expressiveness with intelligence.[71] The very nature of nonverbal behaviors makes inaccurate judgments difficult to recognize and correct.

The following suggestions will help you use your knowledge of nonverbal communication to improve your intercultural competence. These suggestions are designed to help you notice, interpret, and use nonverbal communication behaviors to function more appropriately and more effectively in intercultural encounters.

Researchers have been known to take weeks or even months to analyze the delicate interaction rhythms involved in a single conversation. Of course, most people do not have the luxury of a month to analyze someone's comments before responding. However, the knowledge that the patterns of behavior will probably be very complex will help sensitize you to them and may encourage you to notice more details.

No set of behaviors is universally correct, so the "right" behaviors can never be described in a catalog or list. Rather, the proper behaviors are those that are appropriate and effective in the context of the culture, setting, and occasion. What is right in one set of circumstances may be totally wrong in another. Although it is useful to gather culture-specific information about appropriate nonverbal behaviors, even this knowledge should be approached as relative because prescriptions of "right" behavior rarely identify all of the situational characteristics that cultural natives "know."

CULTURE CONNECTIONS

By then, I was used to the averted gaze of devout Muslim men, and it seemed normal to me to be conversing with someone whose eyes were focused on a floor tile an inch in front of my shoe. He was considering whether to let me meet his wife.

—Geraldine Brooles ∎

By monitoring your emotional reactions to differences in nonverbal behaviors, you can be alert to the interpretations you are making and therefore to the possibility of alternative meanings. Strong visceral responses to differences in smell, body movement, and personal spacing are quite common in intercultural communication. Knowledge that these might occur, followed by care in the interpretation of meanings, is critical.

Skillful interpretation includes observation of general tendencies. Focus on what members of the other culture prefer and the ways in which they typically behave. How, when, and with whom do they gesture, move, look, and touch? How are time and space used to define and maintain social relationships? It is much harder to pay attention to these general tendencies than you might think because, in all likelihood, you have not had much practice in consciously looking for patterns in the commonplace, taken-for-granted activities through which cultural effects are displayed. Nevertheless, it is possible, with practice, to improve your observation skills.

These business people illustrate several kinds of nonverbal messages. Notice the meanings that you attribute to their gestures, facial expressions, body postures, clothing, and their distance from one another.

Even after making observations, be tentative in your interpretations and generalizations. You could be wrong. You will be far more successful in making sense of others' behaviors if you avoid the premature closure that comes with assuming you know for certain what something means. Think of your explanations as tentative working hypotheses rather than as unchanging facts. Next, look for exceptions to your generalizations. These exceptions are very important because they help you recognize that no one individual, regardless of the thoroughness and accuracy with which you have come to understand a culture, will exactly fit the useful generalizations you have formed. The exceptions that you note can help you limit the scope of your generalizations and recognize the boundaries beyond which your judgments may simply not apply. Maybe your interpretations apply only to men, or students, or government officials, or strangers, or the elderly, or potential customers. Maybe your evaluations of the way time and space are structured apply only to business settings, or among those whose status is equal, or with particular people like yourself. Though it is necessary to make useful generalizations to get along in another culture, it is equally necessary to recognize the limits of these generalizations.

Finally, practice to improve your ability in observing, evaluating, and behaving in appropriate and effective ways. Practice increases your skills in recognizing specific patterns to people's behaviors, in correctly interpreting the meanings and likely consequences of those behaviors, and in selecting responses that are both appropriate and effective. Like all skills, your level of intercultural competence will improve with practice. Of course, the best form of practice is one that closely approximates the situations in which you will have to use the skills you are trying to acquire. Therefore, we encourage you to seek out and willingly engage in intercultural communication experiences.

SUMMARY

Although there is some evidence that certain nonverbal communication tendencies are common to all humans, cultures vary greatly in the repertoire of behaviors and circumstances in which nonverbal exchanges occur. A smile, a head nod, and eye contact may all have different meanings in different cultures.

This chapter considered the importance of nonverbal code systems in intercultural communication. Nonverbal code systems are the "silent language" of communication. They are less precise and less consciously used and interpreted than verbal code systems, but they can have powerful effects on perceptions of and interpretations about others. To explore what nonverbal communication does, five functions of nonverbal code systems were discussed. Next, the nonverbal code systems relating to physical appearance, the environment, body movements, personal space, touch, the voice, and the use of time were each described. Finally, the interrelationship of these nonverbal code systems with one another and with the verbal code system was explored.

FOR DISCUSSION

1. What are some examples of cultural universals? Can you think of examples from your personal experiences that either confirm or contradict the idea of cultural universals?
2. It is widely believed by many that "a smile is universally understood." Do you agree with this statement? Why or why not?
3. Touch is one of the most fundamental parts of the human experience. But cultural differences in the norms for touching can cause problems in intercultural interactions. Provide examples of your touching norms that you believe differ for people from cultures other than your own.
4. Each culture socializes its members to speak at its "preferred" rate and volume. Can you think of instances when you have made judgments about others because they spoke louder or softer, or faster or slower, than you wanted? If so, what were the evaluations you made? Were these judgments connected to cultural differences in vocal communication?
5. We know that cultures use and value time differently. What kinds of judgments might be made of those who use time differently from the ways that your culture does?
6. What are some of the ways that U.S. Americans have been taught (or have unconsciously learned) to synchronize their nonverbal behaviors?

FOR FURTHER READING

Peter A. Andersen, *Nonverbal Communication: Forms and Functions*, 2nd ed. (Long Grove, IL: Waveland Press, 2008). Thoroughly researched and enjoyable to read. A scholarly summary of nonverbal communication that highlights cultural differences.

Judee K. Burgoon, Laura K. Guerrero, and Cory Floyd, *Nonverbal Communication* (Boston: Allyn & Bacon, 2010). A very readable introduction to the topic of nonverbal communication. This is a useful summary because it is current and links well to other key concepts presented in this textbook.

Edward T. Hall, *The Hidden Dimension* (New York: Anchor Books, 1990). An exploration of the variations in the use of space across cultures and how that use reflects cultural values and establishes rules for interactions. A classic book that was influential in the study of nonverbal communication across cultures.

Mark L. Knapp and Judith A. Hall, *Nonverbal Communication in Human Interaction*, 7th ed. (Belmont, CA: Thomson/Wadsworth, 2010). Offers the basic perspectives and literature on nonverbal communication. Focuses on the people, behaviors, environments, and messages that affect communication.

The Effects of Code Usage in Intercultural Communication

KEY TERMS

Practical, everyday communication experiences—greeting a friend, buying something from a shopkeeper, asking directions, or describing a common experience—require messages to be organized in a meaningful way. Cultures differ, however, in the patterns that are preferred for organizing ideas and communicating them to others. These differences affect what people regard as logical, rational, and a basis for sound reasoning and conclusions.

This chapter focuses on the consequences for intercultural communication of differences in the way cultures use verbal and nonverbal communication. Do people in particular cultures have distinctive preferences for what, where, when, and with whom they speak? Are there differences in what are regarded as the ideal ways to organize ideas and present them to others? What constitutes appropriate forms of reasoning, evidence, and proof in a discussion or argument? Is proof accomplished with a statistic, an experience, an expert's testimony, or a link between some aspect of the problem and the emotions of the listener? What is considered "rational" and "logical"? In short, how do conversations differ because of the differences in culture, language, and nonverbal codes?

Competent intercultural communication requires more than just an accurate rendition of the verbal and nonverbal codes that others use. The "logic" of how those codes are organized and used must also be understood.

This chapter begins by considering alternative preferences for the organization of messages. Next we discuss cultural variations in persuasive communication. Finally, differences in the structure of conversations are presented as another way in which code systems influence intercultural competence.

PREFERENCES IN THE ORGANIZATION OF MESSAGES

Cultures have distinct preferences for organizing ideas and presenting them in writing and in public speeches. Consider what you have been taught in English composition courses as the "correct" way to structure an essay, or recall the organizational patterns you have used to structure the content of a speech. The premise underlying our discussion is that cultures provide preferred ways for people to organize and convey thoughts and feelings. These preferences influence the ways people communicate and the choices they make to arrange ideas in specific patterns.

In this section, we first describe the organizational features of the English language as it is used in the United States. We then explore the organizational features associated with other languages used in particular cultures.

Organizational Preferences in the Use of U.S. English

For most cultures, the correct use of language is most easily observed when the language is formally taught in the school system. English is a standard feature of the U.S. high school curriculum, and English composition is a requirement for virtually all U.S. college students. The development of oral communication skills, which usually includes training in public speaking, is also a common requirement for many college students. In both written and oral communication courses, users of U.S. English explicitly learn rules that govern how ideas are to be presented. Indeed, the features that characterize a well-organized essay in U.S. English are very similar to the features of a well-organized public speech.

The structure of a good essay or speech in U.S. English requires the development of a specific theme. A thesis statement, which is the central organizing idea of the speech or essay, is the foundation on which speakers or writers develop their speech or essay. Ideally, thesis statements are clear and specific; speakers and writers must present their ideas in a straightforward and unambiguous manner. Often, the thesis statement is provided in the opening paragraph of an essay or in the beginning of a speech.

The paragraph is the fundamental organizational unit of written English. Paragraphs are composed of sentences and should express a single idea. A straightforward presentation of the main idea typically appears as the topic sentence of the paragraph, and it is often located at or near the beginning of the paragraph.[1]

There are other rules that guide how paragraphs are combined into an essay or main points into a speech. Generally, correct organization in U.S. English means that writers or speakers clearly state their thesis at the beginning and provide the audience with an overview of their main points. As the key to good organization, students are taught to outline the main points of the essay or speech by subordinating supporting ideas to the main ideas. In fact, most teachers give students explicit instructions to help them learn to organize properly.

In U.S. English there is also a preferred way to develop the main points. If a speaker is talking about scuba diving, with main points on equipment and safety tips, he or she is expected to develop the point on equipment by talking only about equipment. Safety tips should not be mentioned in the midst of the discussion about equipment. If the speaker gave examples related to safety tips in the middle of the discussion on equipment, listeners (or readers) trained in the preferences embedded in U.S. English would become confused and would think the speaker was disorganized. A teacher would probably comment on the organizational deficiencies and might lower the student's grade because the speech does not match the expected form of a well-organized speech.

The organizational pattern preferred in the formal use of U.S. English can best be described as linear. This pattern can be visualized as a series of steps or progressions that move in a straight line toward a particular goal or idea. Thus, the preferred organizational pattern forms a series of "bridges," where each idea is linked to the next.

Organizational Preferences in Other Languages and Cultures

Some years ago, Robert Kaplan systematically began to study the preferred organizational patterns of nonnative speakers of English. In launching a specialization that is now called intercultural rhetoric, Kaplan characterized the preferences for the organization of paragraphs among people from different language and cultural groups.[2] Scholars such as Ulla Connor have extended Kaplan's work to look at additional cultural differences in language use.[3]

CULTURE CONNECTIONS

A Puerto Rican manager, Juan Marin, was asked to give a brown-bag luncheon talk at the mortgage company where he worked in Houston. The topic for the series of discussions was cross-cultural communication. As he spoke, Juan drew on the white board to illustrate the difference in the preferred reasoning style of his American co-workers. "You talk from point A to point B." Juan drew a straight line connecting the two letters.

"In my culture, it is different. We do it like this." At this point, Juan drew circles that overlapped eventually forming the pattern of a flower. His artwork drew lots of laughs and comments and was a revelation for those from low-context cultures who sometimes were impatient with Juan's tendency to talk "around" a subject. Most participants did not realize that preference for circular or indirect reasoning is culturally influenced.

—Sana Reynolds, Deborah Valentine, Mary Munter ■

The practical consequences of language use on the organization of ideas are often obvious to teachers of English as a second language (ESL). Even after nonnative English speakers have mastered the vocabulary and grammar of the English language, they are unlikely to write an essay in what is considered "correct" English form. In fact, because of the particular style for the organization and presentation of ideas, ESL teachers can often identify the native language of a writer even when the essay is written in English.

One important difference in organizational structure concerns languages that are speaker-responsible versus those that are listener-responsible. In English, which is a speaker-responsible language, the speaker is expected to provide the structure and, therefore, much of the specific meaning of the statements.[4] Because the speaker tells the listener exactly what is going to be talked about and what the speaker wants the listener to know, prior knowledge of the speaker's intent is not necessary. In Japanese, which is a listener-responsible language, speakers need to indicate only indirectly what they are discussing and what they want the listener to know when the conversation is over. The listener is forced to construct the meaning, and usually does so, based on shared knowledge between the speaker and the listener.

The U.S. English concepts of thesis statements and paragraph topic sentences have no real equivalents in many languages. Studies of Japanese, Korean, Thai, and Chinese language use indicate that, in these languages, the thesis statement is often buried in the passage.[5] Thus, for example, the preferred structure of a Japanese paragraph is often called a "gyre" or a series of "stepping stones" that depend on indirection and implication to connect ideas and provide the main points.[6] The rules for language use in Japan mean that speakers may not tell the listener the specific point being conveyed; rather, the topic is circled delicately to imply its domain.[7]

U.S. Secretary of State Hillary Rodham Clinton uses her persuasive skills to convince her audience. Cultures differ in the evidence and arguments they regard as persuasive.

Imagine the consequences of an intercultural interaction between a Japanese person and a U.S. American. What might happen if one of them is able to speak in the other's language and is sufficiently skilled to convey meaning linguistically but is not adept at the logic of the language? The Japanese person is likely to think that the U.S. American is rude and aggressive. Conversely, the U.S. American is likely to think that the Japanese person is confusing and imprecise. Both people in this intercultural interaction are likely to feel dissatisfied, confused, and uncomfortable.

The nonlinear structure of Japanese language use also characterizes Hindi, one of India's national languages. In Hindi, one does not typically develop just one unified thought or idea; rather, the preferred style contains digressions and includes related material.[8] When using English, speakers of Hindi exhibit the characteristics of the Hindi organizational style and provide many minor contextual points before advancing the thesis.[9]

How do you think U.S. teachers of English would grade an assignment that was written in English by a native Hindi speaker? We can easily imagine the comments about the lack of organization and the poor development of the ideas. Because the Indian writing and speaking conventions have an obvious preference for nonlinearity, they are likely to be perceived, from a U.S. perspective, as illogical.

Chinese discourse styles are similar to those of Hindi English. Rather than relying on a preview statement to orient the listener to the discourse's overall direction, the Chinese rely heavily on contextual cues. Chinese speech also tends to use single words such as *because, as,* and *so* to replace whole clause connectives, such as "in view of the fact that," "to begin with," or "in conclusion," that are commonly used in English.[10]

Cultural patterns interact with code systems to create expectations about what is considered the proper or the logical way to organize the presentation of ideas. What is considered the right way to organize ideas within one culture may be regarded in another as some combination of illogical, disorganized, unclear, confusing, imprecise, rude, discourteous, aggressive, and ineffective.[11] In intercultural communication, people make judgments about the appropriateness and effectiveness of others' thoughts, and these assumptions about the "right" way to communicate may vary greatly and may lead to misunderstandings.

CULTURAL VARIATIONS IN PERSUASION

Persuasion involves the use of symbols to influence others. Persuasion may occur in formal, public settings, such as when a candidate for political office tries to win votes through speeches and advertisements. On many occasions, persuasive messages are mediated through television, video, film, photographs, and even music and art. More commonly, persuasion occurs in everyday interactions between people. Our daily conversations include attempts to influence others to accept our ideas, agree with our preferences, or engage in behaviors that we want. In other words, we all take part in persuasion on a regular basis. For example, you might try to convince your roommate to clean the apartment. Or perhaps you want your coworkers to increase their involvement in your work group's project. You might even attempt to persuade your professor to give you an extension on the due date for your term paper.

Persuasion in Intercultural Encounters

In today's multicultural world, many of our persuasive encounters will likely involve culturally heterogeneous individuals. An African American salesperson may have her territory expanded to include Australia and find that her primary contact there is Chinese. A European American college student may need to negotiate a desired absence from class with a Latino professor. Japanese tourists in South Africa may want room service even though the service is shut down for the night. An Indian manager for a major international bank may have employees reporting to him who are Bengali, New Zealanders, Swiss, British, and Chinese. All of these communicative situations require knowledge and skill in using the appropriate means of persuasion; whereas members of some cultures genuinely enjoy the persuasive or argumentative encounter, many others shun such confrontations.[12]

The effective use of verbal and nonverbal codes to persuade another varies greatly from culture to culture. For instance, there are differences in *what* cultures consider to be acceptable evidence, in *who* can be regarded as an authority, in *how* evidence is used to create persuasive arguments, and in *when* ideas are accepted as reasonable. These preferred ways to persuade others are called the culture's persuasive style. When people from diverse cultures communicate, the differences in their persuasive styles are often very evident. Even the persuasive value of written versus spoken communication varies across cultures. For example, whereas written messages in the United States are commonly used to convey organizational policies and procedures, among Ecuadorians and people from other South American cultures—with their cultural patterns of collectivism and hierarchy (power distance)—spoken messages are preferred for providing this feedback.[13]

The word *logical* is often used to describe the preferred persuasive style of a culture. Logic and rationality seem to be invoked as though there were some firm "truth" somewhere that simply has to be discovered and used in order to be convincing. We agree with James F. Hamill, however, that "because logic has cultural aspects, an understanding of social life requires an understanding of how people think in their own cultural context."[14] In fact, Stephen Toulmin, a leading philosopher who studies human reasoning, claims that what people call "rationality" varies from culture to culture and from time to time.[15] A phrase that sums up these variations is ethno-logics or alternative logics.

Persuasion involves an interaction between a speaker and his or her audience in which the speaker intends to have the audience accept a point of view or a conclusion. Persuasion usually involves evidence, establishing "logical" connections between the pieces of evidence, and ordering the evidence into a meaningful arrangement—all of which are used to persuade the audience to accept the speaker's point of view or conclusion. In each of these elements of persuasion, there are substantial variations from culture to culture. In the following sections, we describe some of these variations.[16]

Cultural Differences in What Is Acceptable as Evidence

Evidence is what a persuader offers to those she or he is trying to persuade. In any given persuasive situation, we have available to us a myriad of sensory information or ideas. For example, suppose that students are trying to persuade a teacher to give an extension on a paper's due date. There have probably been numerous events that students could select as evidence. Maybe many students had been ill for a day, or perhaps the teacher had been sick. Or perhaps during one critical lecture, the noise of construction workers just outside the classroom made it difficult to pay attention to the discussion. Any of these events might be used as evidence to support the conclusion that the paper's due date ought to be postponed. An idea or experience does not become evidence, however, until it is selected for use in a persuasive interaction. What we choose from among all of the available cues is highly influenced by our culture.

There are no universally accepted standards about what constitutes evidence or about how evidence should be used in support of claims or conclusions. In many cultures, people use parables or stories as a form of evidence. But the contents of those stories, and their use in support of claims, differ widely. Devout Muslims and Christians, for example, may use stories from the Koran or the Bible as a powerful form of evidence; the story is offered, the lesson from the story is summarized, and the evidence is regarded as conclusive. In other cultures, the story itself must be scrutinized to determine how illustrative it is compared with other possibilities. Native American stories, for instance, rarely provide a deductive conclusion; consistent with a cultural value of indirectness, the story may simply end with "That's how it was" and expect the listeners to infer the relationship between the story and the current circumstances.[17] For Kenyans, persuasive messages depend on narratives and personal stories as supporting evidence in speeches.[18] Similarly, cultures influenced by Confucianism often rely on metaphors and analogies as persuasive evidence.[19]

The European American culture prefers physical evidence and eyewitness testimony, and members of that culture see "facts" as the supreme kind of evidence. Popular mysteries on television or in best-selling books weave their tales by giving clues through the appearance of physical evidence or facts—a button that is torn off a sleeve, a record of calls made from one person's telephone to another's, or a bankbook that shows regular deposits or withdrawals. From all of these pieces of evidence, human behavior and motivation are regarded as apparent. In some cultures, however, physical evidence is discounted because no connection is seen between those pieces of the physical world and human actions. People from cultures that view the physical world as indicative of human motivation have difficulty understanding this point of view.

The use of expert testimony in the persuasive process also varies greatly from one culture to another. In certain African cultures, the words of a witness would be discounted or even totally disregarded because people believe that, if you speak up about seeing something, you must have a particular agenda in mind; in other words, no one is regarded as objective. In the Chinese legal system, the primary purpose of testimony is not to gather information but to persuade the court while shaming the defendant into confessing.[20] The U.S. legal system, however, depends on the testimony of others; witnesses to traffic accidents, for example, are called to give testimony concerning the behavior of the drivers involved in the accident. Teachers of adults who are learning English indicate that these students may not understand the relative weight or authority to give to a scholarly presentation in a journal or academic work versus opinions found in an editorial column of a newspaper or magazine.[21] Such differences are not exclusive

To persuade potential customers, this Latina is arranging the information in her presentation. Evidence, logic, and reasoning are all shaped by one's cultural background.

to new-language learners. Professional historians from German and Russian cultures, for instance, also diverge in their preferences and expectations for the kind and amount of evidence needed to support an argument or persuasive message.[22]

Cultural Differences in Styles of Persuasion

Perhaps the best way to understand what a culture might regard as logical or reasonable is to refer to the culture's dominant patterns, which we discussed in Chapters 4 and 5. Cultural patterns supply the underlying assumptions that people within a culture use to determine what is "correct" and reasonable, and they therefore provide the persuader's justification for linking the evidence to the conclusions desired from the audience.[23] These differences in the ways people prefer to arrange the evidence, assumptions, and claims constitute the culture's persuasive style.[24]

Much of the tradition of persuasion and rhetoric among European Americans is influenced by the rhetoric of Aristotle, who emphasized the separation of logic and reason from emotion. This Aristotelian perspective is also common to the rhetorical traditions of many European cultures, but it is antithetical to good rhetorical practices in numerous other cultures.[25]

Thai, Arab, and Chinese discourse all have rhetorical traditions that emphasize the importance of emotion in assessing the truthfulness of a situation. Thus an examination of persuasive letters written by native Thai speakers and native English speakers found that the Thai used a combination of logical arguments, emotional pleas, and requests based on the speaker's *ethos* (credibility, character, and appeal to others), while the English speakers depended almost solely on logical arguments.[26] Similarly, a comparison of newspaper editorials from the United States (in English) to those from Saudi Arabia (in Arabic) found substantial differences in the persuasive techniques employed; though the editorials from both cultures made strong assertions, the persuasive appeals in the U.S. editorials were direct and explicit, whereas the Saudi editorials used implied arguments and indirect language that provided an opportunity to disagree without requiring the reader to concede the disagreement.[27] Likewise U.S. Americans tend to vary their phrasing and utilize alternate ways of saying the same thing during business negotiations, while the Chinese typically just restate their position repeatedly without changing it publicly before first consulting privately among themselves.[28]

Lakota speakers may offer stories that are related to their persuasive point, but the persuader may not make explicit the link to the conclusion.[29] Likewise, Latin American speakers may use passive sentence constructions and descriptions to lead others to a conclusion that is not stated specifically.[30] Mexicans, among others, are sometimes very emotional and dramatic, and they may subordinate the goal of accuracy to one involving pleasantness. Thus, when a U.S. American asks a Mexican shopkeeper a question, the shopkeeper may be concerned less with the correctness of details than with the maintenance of a harmonious relationship.[31]

Even seemingly "objective" reporting of news may convey a subtle persuasive message. Japanese newspapers and television news shows, for example, routinely refer to Japanese adults by their family name plus *san*, the latter word being an address term denoting respect or honor. As Daniel Dolan suggests, however, "This respect is *conditional*, because in most instances of reporting about a person associated with criminal activity, mass media reporters will publicly divest an individual of the personal address term *san* and in its place use *yogisha* [suspect], *hikoku* [accused], or family name alone. The effect is to banish the person, at least temporarily, from functioning citizenry."[32]

We would like to elaborate on the idea of persuasive style because, like many of the other characteristics of a culture, it is an important cultural attribute that is taken for granted within a culture but affects communication between cultures. As we have cautioned elsewhere, not every person in a culture will select the culture's preferred style. Rather, we are describing a cultural tendency, a choice or preference that most people in the culture will select most of the time.

Barbara Johnstone describes three general strategies of persuasion that can form a culture's preferred style: the quasilogical, presentational, and analogical.[33] Each of these styles depends on different kinds of evidence, organizational patterns, and conclusions. As you read the descriptions of these styles, try to imagine what a persuasive encounter would be like and what might happen if others preferred a different persuasive style.

QUASILOGICAL STYLE   The preferred style for members of many Western cultures is the quasilogical style. In this style, the preference is to use objective statistics and testimony from expert witnesses as evidence. The evidence is then connected to the conclusion in a way that resembles formal logic. In formal logic, once the listener accepts or believes the individual pieces of evidence, the conclusions follow "logically" and must also be

CULTURE CONNECTIONS

About "English only": Let me tell you a little story that happened in Houston, Texas, just before I became a national best-selling author. I met this young woman at the University of Houston. She looked like she was part Black, part White, and part American Indian. She was stunningly beautiful, with huge greenish eyes. She spoke Spanish. I asked her where she was from. She said Panama. I asked how she liked the United States. She said she didn't, and that as soon as she graduated she wanted to return to Panama. I asked why. She told me that she'd had a boyfriend for four years. "And the other day he said 'I think I love you,' so I dropped him as fast as I could. My God," she added, "after four years he was still thinking about our love. I can't stand to be around people who are always thinking so much."

I laughed. I could see her point completely, because in Spanish you'd never say, "I think I love you," especially after four years. That would be an insult. You'd say, "I feel love for you so deeply that when I just think of you, I start to tremble and feel my heart flutter." Why? Because Spanish is a feeling-based language that comes first from the heart, just as English is a thinking-based language that comes first from the head. And Yaqui, Navajo, and the fifty-seven dialects of Oaxacan are ever-changing languages that come first from the soul, then go to the heart, and lastly to the brain.

—Victor Villaseñor ■

accepted. In the quasilogical style, the speaker or persuader will connect the evidence to the persuasive conclusion by using such words as *thus*, *hence*, and *therefore*. The form or arrangement of the ideas is very important.

The dominance of the quasilogical style for English speakers is underscored by advice given to students of English as a second language: "English-speaking readers are convinced by facts, statistics, and illustrations in arguments; they move from generalizations to specific examples and expect explicit links between main topics and subtopics; and they value originality."[34] The underlying assumption of this style is that, if the idea is "true," it simply needs to be presented in a logical way so that its truthfulness becomes apparent to all. Those who prefer the quasilogical style assume that it is possible to discover what is true or false and right or wrong about a particular experience. In other words, they believe that events can be objectively established and verified.

PRESENTATIONAL STYLE   The presentational style emphasizes and appeals to the emotional aspects of persuasion. In this style, it is understood that people, rather than the idea itself, are what make an idea persuasive. That is, ideas themselves are not inherently persuasive; what makes them compelling is how they are presented to others. Thus, an immutable truth does not exist, and there are no clear rights or wrongs to be discovered.

In the presentational style, the persuader uses language to create an emotional response. The rhythmic qualities of words and the ability of words to move the hearer visually and auditorily are fundamental to this style of persuasion. You have probably read poetry or literature that stimulated a strong emotional reaction. Those who use a presentational style persuade in the same way. By the use of words, the ideas of the speaker become so vivid and real that the persuasive idea almost becomes embedded in the consciousness of the listener. The language of this style of persuasion is filled with sensory words that induce the listener to *look*, *see*, *hear*, *feel*, and ultimately *believe*.

ANALOGICAL STYLE   The analogical style seeks to establish an idea (a conclusion) and to persuade the listener by providing an analogy, a story, or a parable in which there is either an implicit or explicit lesson to be learned. The storybook pattern that begins "Once upon a time" is one example of this style, as are the sermons of many ministers and preachers. An assumption underlying the analogical style is that the collective

CULTURE CONNECTIONS

"We're not looking for a motive, Detective." Tao cut him off sharply, and the earlier tension immediately returned to the room. "We're looking for evidence. As much as we can accumulate. No matter how painful, or how slow. Only then will we see the bigger picture. There are no shortcuts."

It was the old argument, the traditional Chinese approach to criminal investigation.

Accumulate enough evidence and you will solve the crime. Unlike the approach of criminal investigators in the West, motive was regarded as being of secondary importance, something which would become self-apparent when enough evidence had been gathered.

—Peter May ■

experience of groups of people—the culture—is persuasive, rather than the ideas themselves or the characteristics of a dynamic individual. Historical precedent takes on great importance because what convinces is a persuader's ability to choose the right historical story to demonstrate the point. In the analogical style, skill in persuasion is associated with the discovery and narration of the appropriate story—a story that captures the essence of what the persuader wants the listeners to know.

Persuasive encounters involving people with different stylistic preferences may result in neither person being persuaded by the other. To a person with a cultural preference for a quasilogical style, the presentational style will appear emotional and intuitive, and the analogical style will appear irrelevant. To those using the presentational style, the quasilogical style will appear dull, insignificant, and unrelated to the real issues. To those using the analogical style, the quasilogical style will seem blunt and unappealing.

An interesting example of the clash in preferences for persuasive styles is described by Donal Carbaugh:

> I have heard several Russian speakers in public who were asked questions of fact yet responded with impassioned, even artful expressions of an image of the good, presenting a moral tale of an ideal world as it should be. Russians likewise heard many Americans stating—sometimes in great detail—troubling truths, rather than expressing common virtues or the shared fiber of a strong moral life. In fact, as a result in part of conversing in these distinctive cultural ways, Russians are often led to portray Americans as soul-less or immoral, too willing to spill the discreditable trust and unable to state any shared morality; Americans in turn are often led to portray Russians as not fully reasonable, as unable to answer basic questions of fact, too willing in public to be passionate, too righteous, to the point of being illogical.[35]

Both this Peruvian woman (left) and Dutch woman (right) use a sales pitch to influence shoppers. People the world over depend on persuasion to transform interested customers into purchasing customers.

CULTURE CONNECTIONS

"Have you eaten?" she asked in *putonghua* Chinese. It was the traditional Beijing greeting, born of a time when food was scarce and hunger a way of life. "Yes, I have eaten," Margaret replied. Also in *putonghua*.

—Peter May ■

Behind the misunderstandings that Carbaugh describes is a Russian preference for the presentational style and a European American preference for the quasilogical style.[36]

CULTURAL VARIATIONS IN THE STRUCTURE OF CONVERSATIONS

All conversations differ on a number of important dimensions: how long one talks; the nature of the relationship between the conversants; the kinds of topics discussed; the way information is presented; how signals are given to indicate interest and involvement; and even whether conversation is regarded as a useful, important, and necessary means of communicating. In this section, we explore some of the differences in the way cultures shape the use of codes to create conversations. Our usual caution applies: when cultural tendencies are described, remember that not all members of the culture will necessarily reflect these characteristics.

Value of Talk and Silence

The importance given to words varies greatly from one culture to the next. Among African Americans and European Americans, for example, words are considered very important. In informal conversations between friends, individuals often "give my word" to ensure the truth of their statement. In the legal setting, people swear that their words constitute "the truth, the whole truth, and nothing but the truth." Legal obligations are contracted with formal documents to which people affix their signatures—another set of words. The spoken word is seen as a reflection of a person's inner thoughts. Even the theories of communication that are presented in most books about communication—including this one—are highly influenced by underlying assumptions that give words the ability to represent thoughts. In this characteristically Western approach to communication, people need words to communicate accurately and completely with one another. Conversely, silence is often taken by many Western Europeans and European Americans to convey a range of negative experiences—awkwardness, embarrassment, hostility, uninterest, disapproval, shyness, an unwillingness to communicate, a lack of verbal skills, or an expression of interpersonal incompatibility.[37]

Some cultures are very hesitant about the value of words. Asian cultures, such as those of Japan, Korea, and China, as well as southern African cultures, such as those of Swaziland, Zambia, and Lesotho, have quite a different evaluation of words and talking. They all place much more emphasis on the value of silence, on the unspoken meaning or intentions, and on saying as little as is necessary.[38] Because of a combination of historical and cultural forces, spoken words in these cultures are sometimes viewed with some suspicion and disregard. Taoist sayings such as "One who speaks does not know" and "To be always talking is against nature" convey this distrust of talking and wordiness.[39]

Cultures influenced by Confucian and Buddhist values frequently disparage spoken communication. In Japan, the term *haragei* (wordless communication) describes the cultural preference to communicate without using language. Donald Klopf elaborates on this Japanese penchant:

> The desire not to speak is the most significant aspect or feature of Japanese language life. The Japanese hate to hear someone make excuses for his or her mistakes or failures. They do not like long and complicated explanations. Consequently, the less talkative person is preferred and is more popular than the talkative one, other conditions being equal. If one has to say something normally, it is said in as few words as possible.[40]

In Korea, the strong religious and cultural influence of Confucian values has devalued oral communication and made written communication highly regarded. June Ock Yum, in an interesting exploration of the relationship between Korean philosophy and communication, says, "Where the written communication was dominant, spoken words were underrated as being apt to run on and on, to be mean and low. To read was the profession of scholars, to speak the act of menials."[41] Buddhism, also a major influence on Korean thought, teaches, "True communication is believed to occur only when one speaks without the mouth and when one hears without the ears."[42]

Such cultural preferences for silence over talkativeness are not confined to those who have been influenced by Confucian and Buddhist values. In Swaziland, for example, people are suspicious of those who talk excessively. As Peter Nwosu has observed:

> The Swazis are quick to attribute motives when a person during negotiations is very pushy, engages in too much self-praise, or acts like he or she knows everything. "People who talk a lot are not welcome; be calm, but not too calm that they suspect you are up to some mischief," remarks an official of the Swazi Embassy in Washington, D.C. Indeed, there is such a thin line between talkativeness and calmness that it is difficult for a foreigner to understand when one is being "too talkative" or "too calm."[43]

People from Finland are also less willing than European Americans to talk, even among close friends.[44] Donal Carbaugh describes a visit that he and his wife had with some Finnish friends, who asked,

> "When you are with your friends in the United States, do you talk most all of the time?" My wife and I looked at each other, nodded, and smiled, while I responded: "Well, uhm, yes, pretty much." Liisa (Finnish friend) said: "How do you do that? That must be exhausting!" We all laughed as my wife and I admitted, "Yeah, at times, it can be."[45]

The consequences of these differences in preferences for talking are illustrated by a Japanese American student and an African American student who became roommates. Over a period of a few weeks, the African American student sought social interaction and conversation with his Japanese American roommate, who seemed to become less and less willing to converse. The African American student interpreted this reticence as an indication of dislike and disinterest rather than as an indication of cultural differences in conversational preferences. Finally, he decided to move to a different room, because he felt too uncomfortable with the silence to remain.

CULTURE CONNECTIONS

Afternoons are reserved for the "errands" of our life, and they generally involve walking somewhere in the village to do something vital and being waylaid by myriad friends and acquaintances and fresh gossip. There is no such thing as a straight line from home to destination. We are learning the protocol of greetings as we go. With gringos, it's "Hi." With the Mexicans, we are beginning to recognize, it can be "*Hola*" but often is more elaborate. There is a three-part greeting ritual to be followed: "*Buenos días,*" which must, in fact, be spoken at the correct time of day; if it is a few minutes past noon or dusk, it may be gently rectified by the recipient; followed by *señor, señora,* or, most trickily, *señorita,* which is often a judgment call and may also be corrected; and "*¿Cómo está usted?*" which in Mexico may suggest a courteous pause to listen to the answer, which will often involve matters of health or family. A couple of times a day we pass Gaby's parents, an old, dignified, and somewhat worn-out couple who spend most of their afternoons on chairs by their front stoop, but the routine is invariably the same.

"*Buenos días, Señor y Señora Ponce. ¿Cómo están ustedes?*"

"*Buenas tardes,*" Señor Ponce corrects, pointing up at the sun for emphasis. "*Estoy bien, pero mi pinche pierna ...*" "I'm fine, but that god damn leg of mine. ...."

—Barry Golson ∎

There are also different cultural preferences for silence and the place of silence in conversations. Keith Basso describes a number of interpersonal communication experiences in which members of the Apache tribe prefer silence, whereas non–Native Americans might prefer to talk a lot: meetings between strangers, the initial stages of a courtship, an individual returning home to relatives and friends after a long absence, a person verbally expressing anger, someone being sad, and during a curing ceremony.[46] Basso gives this assessment of the value placed on introductions:

> The Western Apache do not feel compelled to "introduce" persons who are unknown to each other. Eventually, it is assumed, strangers will begin to speak. However, this is a decision that is properly left to the individuals involved, and no attempt is made to hasten it. Outside help in the form of introductions or other verbal routines is viewed as presumptuous and unnecessary.
>
> Strangers who are quick to launch into conversations are frequently eyed with undisguised suspicion. A typical reaction to such individuals is that they "want something," that is, their willingness to violate convention is attributed to some urgent need which is likely to result in requests for money, labor, or transportation.[47]

In sum, the fundamental value and role of talk as a tool for conversation vary from culture to culture.

Rules for Conversations

Cultures provide an implicit set of rules to govern interaction. Verbal and nonverbal codes come with a set of cultural prescriptions that determine how they should be used. In this section, we explore some of the ways in which conversational structures can vary from one culture to another. In Chapter 11, we also consider some aspects of conversational structures that are particularly relevant to the development of intercultural relationships.

Some of the ways in which conversational rules can vary are illustrated in the following questions:

- How do you know when it is your turn to talk in a conversation?
- When you talk to a person you have never met before, how do you know what topics are acceptable for you to discuss?
- In a conversation, must your comments be directly related to those that come before?
- When you are upset about a grade, how do you determine the approach to take in a conversation with the teacher or even *if* you should have a conversation with the teacher?
- When you approach your employer to ask for a raise, how do you decide what to say?
- If you want someone to do something for you, do you ask for it directly, or do you mention it to others and hope that they will tell the first person what it is that you want?
- If you decide to ask for something directly, do you go straight to the point and say, "This is what I need from you," or do you hint at what you want and expect the other person to understand?
- When you speak, do you use grand language filled with images, metaphors, and stories, or do you simply and succinctly present the relevant information?

Cultural preferences would produce many different answers to these questions. For example, European Americans signal a desire to speak in a conversation by leaning forward a small degree, slightly opening their mouths, and establishing eye contact. In another culture, this same set of symbols could be totally disregarded because it has no meaning, or it could mean something totally different (for example, respectful listening). Acceptable topics of conversations for two U.S. American students meeting in a class might include their majors, current interests, and where they work. Those same topics in some other cultures—particularly if there happens to be a high unemployment rate—might be regarded as too personal for casual conversations, but discussions about religious beliefs and family history might seem perfectly acceptable. Though European Americans expect comments in a conversation to be related to previous ones, Japanese express their views without necessarily responding to what the other has said.[48]

William Gudykunst and Stella Ting-Toomey describe cultural variations in conversational style along four dimensions: direct–indirect, elaborate–succinct, personal–contextual, and instrumental–affective.[49] Cultures that prefer a direct style, such as European Americans, use verbal messages that are explicit in revealing the speaker's true intentions and desires. In contrast, those that prefer an indirect style will veil the speaker's true wants and needs with ambiguous statements. African Americans and Koreans, for example, prefer an indirect style, as do the Japanese, for whom interactions are governed in part by a desire to avoid saying "no" directly.

Cultural conversational styles also differ on a dimension of elaborate to succinct. The elaborate style, which is found in most Arab and Latino cultures, results in the frequent use of metaphors, proverbs, and other figurative language. The expressiveness of this style contrasts with the succinct style, in which people give precisely the amount of information necessary. In the succinct style, there is a preference for understatement and long pauses, as in Japanese American, Native American, and Chinese American cultures.

Like all exchanges, this conversation is governed by a complex set of rules about who talks to whom, for how long, and on what topics. Yet the participants are unlikely to be consciously aware of these rules until someone breaks them.

In cultures that prefer a personal style, in contrast to those that prefer a contextual style, there is an emphasis on conversations in which the individual, as a unique human being, is the center of action. This style is also characterized by more informality and less status-oriented talk. In the contextual style, the emphasis is on the social roles that people have in relationships with others. Japanese, Chinese, and Indian cultures all emphasize the social role or the interpersonal community in which a particular person is embedded. The style is very formal and heightens awareness of status differences by accentuating them.

In the instrumental style, communication is goal-oriented and depends on explicit verbal messages. Affective styles are more emotional and require sensitivity to the underlying meanings in both the verbal and nonverbal code systems. Min-Sun Kim suggests that the goal-oriented style, which is characterized by a heightened concern for "getting the job done," is preferred by people from individualistic cultures, whereas the affective style, which is concerned with the feelings and emotions of others, typifies people from collectivistic cultures.[50] Thomas Kochman has articulated some of the differences in conversational styles between European Americans and African Americans:

> The differing potencies of black and white public presentations are a regular cause of communicative conflict. Black presentations are emotionally intense, dynamic, and demonstrative; white presentations are more modest and emotionally restrained.[51]

Melanie Booth-Butterfield and Felecia Jordan similarly found that African American females were more expressive, more involved with one another, more animated, and more at ease than were their European American counterparts, who appeared more formal and restrained.[52] Because of these differences in conversational style, an African American and a European American may judge each other negatively.

Sometimes a cultural style that differs substantially from one's own can be so unfamiliar that it can seem chaotic—except, of course, that it works. A U.S. American working in Budapest described a meeting of his Hungarian colleagues as a group of people "sitting around a table shouting at each other, interrupting, not seeming to listen when others were speaking, but then suddenly reaching a decision which seemed to satisfy everyone."[53]

Ronald Scollon and Suzanne Wong-Scollon, who have studied the Athabaskan, a cultural group of native peoples in Alaska and northern Canada, describe similar problems in intercultural communication:

> When an Athabaskan and a speaker of English talk to each other, it is very likely that the English speaker will speak first.... The Athabaskan will feel it is important to know the relationship between the two speakers before speaking. The English speaker will feel talking is the best way to establish a relationship. While the Athabaskan is waiting to see what will happen between them, the English speaker will begin speaking, usually asking questions of fact, to find out what will happen. Only where there is a longstanding relationship and a deep understanding between the two speakers is it likely that the Athabaskan will initiate the conversation.[54]

Regulating conversations is also problematic for the English speaker and an Athabaskan because the latter uses a longer pause—about a half second longer—between turns. The effects of this slightly longer pause would be comical if the consequences for intercultural communication were not so serious.

> When an English speaker pauses, he waits for the regular length of time (around one second or less), that is, *his* regular length of time, and if the Athabaskan does not say anything, the English speaker feels he is free to go on and say anything else he likes. At the same time the Athabaskan has been waiting his regular length of time before coming in. He does not want to interrupt the English speaker. This length of time we think is around one and one-half seconds. It is just enough longer that by the time the Athabaskan is ready to speak the English speaker is already speaking again. So the Athabaskan waits again for the next pause. Again, the English speaker begins just enough before the Athabaskan was going to speak. The net result is that the Athabaskan can never get a word in edgewise (an apt metaphor in this case), while the English speaker goes on and on.[55]

These very real differences in the nature of conversations play a critical role in intercultural communication. The ultimate result is often a negative judgment of other people rather than a recognition that the variability in cultural preferences is creating the difficulties.

CULTURE CONNECTIONS

For a person from the Vietnamese culture, it is not typical to display emotions or tensions externally through gestures or facial expressions. If a boss becomes angry with an employee, the boss will respond by leaving the work area for a brief period of time. If a husband becomes angry with his wife, his likely response is to leave silently and stay with friends for a couple of days. Teachers, when they become angry with students, are very quiet. The influence of Confucianism is offered as explanation for this lack of the display of tension or emotion, because Confucianism suggests following a path of moderation, avoiding exaggeration, and cultivating equanimity. Westerners often mistake the lack of visible emotional expressiveness as impassiveness, placidity, or even hypocrisy. ■

EFFECTS OF CODE USAGE ON INTERCULTURAL COMPETENCE

Developing competence in the practical, everyday use of verbal and nonverbal codes is undoubtedly a major challenge for the intercultural communicator. But simply knowing the syntactic rules of other code systems is not sufficient to be able to use those code systems well.

The most important knowledge you can take away from this chapter is the realization that people from other cultures may organize their ideas, persuade others, and structure their conversations in a manner that differs from yours. You should attempt, to the greatest extent possible, to understand your own preferences for using verbal and nonverbal codes to accomplish practical goals. If you can, mentally set aside your beliefs and the accompanying evaluative labels. Instead, recognize that your belief system and the verbal and nonverbal symbols that are used to represent it were taught to you by your culture and constitute only one among many ways of understanding the world and accomplishing one's personal objectives.

Differences in the ways people prefer to communicate can affect their ability to interact competently in an intercultural encounter. Look for differences in the ways that people from other cultures choose to accomplish their interpersonal objectives. Look for alternative logics. Approach the unfamiliar as a puzzle to be solved rather than as something to be feared or dismissed as illogical, irrational, or wrong. Much can be learned about the effects of code usage by observing others. If your approach is not successful, notice how members of the culture accomplish their objectives.

SUMMARY

The chapter described the effects on intercultural communication of cultural differences in the way verbal and nonverbal codes are used. These differences affect how people attempt to understand messages, organize ideas, persuade others, and engage in discussions and conversations.

We began with a discussion of differences in cultural preferences for organizing and arranging messages, and we contrasted the organizational preferences of U.S. English, which are typically linear, with those of other languages and cultures.

Cultural variations in persuasion and argumentation were considered next. We emphasized that appropriate forms of evidence, reasoning, and rationality are all culturally based and can affect intercultural communication. Indeed, there are major differences in persuasive styles that are taken for granted within a culture but that affect the communication between cultures.

Cultural variations also exist in the structure of conversations. The importance given to talk and silence, the social rules and interaction styles that are used in conversations, and even the cues used to regulate the back-and-forth sequencing of conversations can all create problems for intercultural communicators.

Finally, we noted that differences in the way people prefer to communicate can affect the ability to behave appropriately and effectively in intercultural encounters. These cultural preferences typically operate outside of awareness and may lead to judgments that others are "wrong" or "incorrect" when they are merely different.

FOR DISCUSSION

1. What does it mean to learn the "logic" of a language?
2. In what ways does the U.S. legal system reflect the European American view of argumentation and persuasion?
3. Does your culture value a particular style of persuasion? Do your own preferred ways of persuading others reflect your culture's style of persuasion?
4. What does silence communicate to you? How is your culture's use of silence connected to Hall's cultural patterns of low and high context?
5. Members of some cultures will invariably say "yes" even though, given the situation and their true feelings, the answer is most likely "no." How do you explain this phenomenon?

FOR FURTHER READING

Donal Carbaugh, *Cultures in Conversation* (Mahwah, NJ: Erlbaum, 2005). Illustrates, through several in-depth studies of cultures, how human conversation or "talk" is embedded in one's culture.

Ulla Connor, *Intercultural Rhetoric in Second Language Writing* (Ann Arbor: University of Michigan Press, 2011). Explores and explains the importance of written intercultural communication across a broad array of cultures and rhetorical texts.

Richard D. Rieke, Malcolm O. Sillars, and Tarla Rai Peterson, *Argumentation and Critical Decision Making*, 7th ed. (Boston: Allyn & Bacon, 2009). A text on argumentation and critical thinking. Use the framework we provide in this chapter to read and evaluate the cultural framework embedded within their recommended use of evidence, argument, and reasoning.

Andrea Rocci, "Pragmatic Inference and Argumentation in Intercultural Communication," *Intercultural Pragmatics* 3–4 (2006): 409–422. While highly theoretical, this article provides a useful examination of the elements of arguments and conclusions. It provides a conceptual framework to help learn about variations across cultural traditions.

Ron Scollon, Suzanne Wong-Scollon, and Rodney H. Jones, *Intercultural Communication: A Discourse Approach*, 3rd ed. (Malden, MA: Blackwell, 2012). Offers numerous examples of the relationships between a person's language, thinking, and logical actions. Explores how language and culture provide a preferred structure to conversations.

Intercultural Competence in Interpersonal Relationships

KEY TERMS

All relationships imply connections. When you are in an interpersonal relationship, you are connected—in a very important sense, you are bound together—with another person in some substantial way. Of course, the nature of these ties is rarely physical. Rather, in interpersonal relationships, you are connected to others by virtue of your shared experiences, interpretations, perceptions, and goals.

CULTURAL VARIATIONS IN INTERPERSONAL RELATIONSHIPS

In Chapter 2, we indicated that communication is interpersonal as long as it involves a small number of participants who can interact directly with one another and who therefore have the ability to adapt their messages specifically for one another. Of course, different patterns of interpersonal communication are likely to occur with different types of interpersonal relationships. We believe it is useful to characterize the various types of interpersonal relationships by the kinds of social connections the participants share.

Types of Interpersonal Relationships

Some interpersonal connections occur because of blood or marriage. Others exist because of overlapping or interdependent objectives and goals. Still others bind people together because of common experiences that help to create a perception of "we-ness." However, all interpersonal relationships have as their common characteristic a strong connection among the individuals.

The number of interpersonal relationships that you have throughout your life is probably very large. Some of these relationships are complex and involved, whereas others are simple and casual; some are brief and spontaneous, while others may last a lifetime. Some of these relationships, we hope, have involved people from different cultures.

Interpersonal relationships between people from different cultures can be difficult to understand and describe because of the contrasts in culturally based expectations about the nature of interpersonal communication. However, regardless of the cultures involved or the circumstances surrounding the relationship's formation, there is always some sort of bond or social connection that links or ties the people to one another. The participants may be strangers, acquaintances, friends, romantic partners, or family or kinship members. Each relationship carries with it certain expectations for appropriate behaviors that are anchored within specific cultures. People in an intercultural relationship, then, may define their experiences very differently and may have dissimilar expectations; for example, a stranger to someone from one culture may be called a friend by someone from another culture.

STRANGERS   You will undoubtedly talk to many thousands of people in your lifetime, and most of them will be strangers to you. But what exactly is a stranger? Certainly, a stranger is someone whom you do not know and who is therefore unfamiliar to you. But is someone always a stranger the first time you meet? How about the second time, or the third? What about the people you talked with several times, although the conversation was restricted to the task of seating you in a restaurant or pricing your groceries, so names were never actually exchanged? Are these people strangers to you? Your answers to these questions, like so many of the ideas described in this book, depend on what you have been taught by your culture.

In the United States, for instance, the social walls that are erected between strangers may not be as thick and impenetrable as they are in some collectivistic cultures. European Americans, who are often fiercely individualistic as a cultural group, may not have developed the strong ingroup bonds that would promote separation from outsiders. Among the Greeks, however, who hold collectivistic values, the word for "non-Greek" translates as "stranger."

Even in the United States, the distinction between stranger and nonstranger is an important one; young children are often taught to be afraid of people they do not know. Compare, however, a U.S. American's reaction toward a stranger with that of a Korean in a similar situation. In Korea, which is a family-dominated collectivist culture, a stranger is anyone to whom you have not been formally introduced. Strangers in Korea are "nonpersons" to whom the rules of politeness and social etiquette simply do not apply. Thus, Koreans may jostle you on the street without apologizing or, perhaps, even noticing. However, once you have been introduced to a Korean, or the Korean anticipates in other ways that he or she may have an ongoing interpersonal relationship with you, elaborate politeness rituals are required.

ACQUAINTANCES   An acquaintance is someone you know, but only casually. Therefore, interactions tend to be on a superficial level. The social bonds that link acquaintances are very slight. Acquaintances will typically engage in social politeness rituals, such as greeting one another when first meeting or exchanging small talk on topics generally viewed as more impersonal such as the weather, hobbies, fashions, and sports. But acquaintances do not typically confide in one another about personal problems or discuss private concerns. Of course, the topics appropriate for small talk, which do not include personal and private issues, will differ from one culture to another. Among European Americans, it is perfectly appropriate to ask a male acquaintance about his wife; in the United Arab Emirates, it would be a major breach of social etiquette to do so. In New Zealand, it is appropriate to

CULTURE CONNECTIONS

An equally perplexing experience for me was the reaction of most U.S. Americans to my family background. I come from a fairly large extended family with some history of polygyny. Polygyny is the union between a man and two or more wives. (Polygamy, a more general term, refers to marriage among several spouses, including a man who marries more than one wife or a woman who marries more than one husband.) Polygyny is an accepted and respected marriage form in traditional Igbo society. My father, Chief Clement Muoghalu Nwosu, had two wives. My paternal grandfather, Chief Ezekwesili Nwosu, was married to four. My great grandfather, Chief Odoji, who also married four wives, was the chief priest and custodian of traditional religion in my town, Umudioka town, a small rural community in Anambra State of the Federal Republic of Nigeria....

The traditional economic structure in Igbo society dictated this familial arrangement whereby a man would have more than one spouse and produce several children, who would then assist him with farm work, which is regarded as the fiber and glue of economic life in traditional Igbo society. Each wife and her own children live in a separate home built by the husband. Each wife is responsible for the upkeep of her immediate family, with support from her husband.

—Peter O. Nwosu   ■

talk about national and international politics; in Pakistan, these and similar topics should be avoided. In Austria, discussions about money and religion are typically sidestepped; elsewhere, acquaintances may well be asked "personal" questions about their income and family background.

FRIENDS   As with many of the other terms that describe interpersonal relationships, *friend* is a common expression that refers to many different types of relationships. "Good friends," "close friends," and "just friends" are all commonly used expressions among U.S. Americans. Generally speaking, a friend is someone you know well, someone you like, and someone with whom you feel a close personal bond. A friendship usually includes higher levels of intimacy, self-disclosure, involvement, and intensity than does acquaintanceship. In many ways, friends can be thought of as close acquaintances.

Unlike kinships, friendships are voluntary, even though many friendships start because the participants have been thrust together in some way. Because they are voluntary, friendships usually occur between people who see themselves as similar in some important ways and who belong to the same social class.

European American friendships tend to be very compartmentalized because they are based on a shared activity, event, or experience. The European American can study with one friend, play racquetball with another, and go to the movies with a third. As suggested in Chapter 4, this pattern occurs because European Americans typically classify people according to what they do or have achieved rather than who they are. Relations among European Americans are therefore fragmented, and they view themselves and others as a composite of distinct interests.

These Italian men share a friendly moment of conversation. In every culture, expectations about social behaviors are influenced by a set of rules.

The Thai are likely to react more to the other person as a whole and will avoid forming friendships with those whose values and behaviors are in some way deemed undesirable.[1] Unlike friendships in the United States, in Thailand a friend is accepted completely or not at all; a person cannot disapprove of some aspect of another's political beliefs or personal life and still consider her or him to be a friend. Similarly, the Chinese typically have fewer friends than European Americans do, but Chinese friends expect one another to be involved in all aspects of their lives and to spend much of their free time together. Friends are expected to anticipate others' needs and to provide unsolicited advice about what to do. These differing expectations can cause serious problems as a Chinese and a European American embark on the development of what each sees as a "friendship."[2]

John Condon has noted that the language people use to describe their interpersonal relationships often reflects the underlying cultural values about their relationships' meaning and importance. Thus, Condon says, friendships among European Americans are expressed by terms such as *friends, allies,* and *neighbors,* all of which reflect an individualistic cultural value. However, among African Americans and some Southern whites, closeness between friends is expressed by such terms as *brother, sister,* or *cousin,* suggesting a collectivist cultural value. Mexican terms for relationships, like the cultural values they represent, are similar to those of African Americans. Thus, when European Americans and Mexicans speak of close friendships, the former will probably use a word such as *partner*, which suggests a voluntary association, whereas Mexicans may use a word such as *brother* or *sister*, which suggests a lasting bond that is beyond the control of any one person.[3]

As interpersonal relationships move from initial acquaintance to close friendship, five types of changes in perceptions and behaviors will probably occur. First, friends interact more frequently; they talk to each other more often, for longer periods of time, and in more varied settings than acquaintances do. Second, the increased frequency of interactions means that friends will have more knowledge about and shared experiences with each other than will acquaintances, and this unique common ground will probably develop into a private communication code to refer to ideas, objects, and experiences that are exclusive to the relationship. Third, the increased knowledge of the other person's motives and typical behaviors means that there is an increased ability to predict a friend's reactions to common situations. The powerful need to reduce uncertainty in the initial stages of relationships, which we discuss in greater detail later in this chapter, suggests that acquaintanceships are unlikely to progress to friendships without the ability to predict the others' intentions and expectations. Fourth, the sense of "we-ness" increases among friends. Friends often feel that their increased investment of time and emotional commitment to the relationship creates a sense of interdependence, so that individual goals and interests are affected by and linked to each person's satisfaction with the relationship. Finally, close friendships are characterized by a heightened sense of caring, commitment, trust, and emotional attachment to the other person, so that the people in a friendship view it as something special and unique.[4]

Intercultural friendships can vary in a variety of ways: whom a person selects as a friend, how long a friendship lasts, the prerogatives and responsibilities of being a friend, the number of friends that a person prefers to have, and even how long a relationship must develop before it becomes a friendship. African American friends, for instance, expect to be able to confront and criticize one another, sometimes in a loud and argumentative manner.[5] Latinos, Asian Americans, and African Americans feel that it takes them, on the average, about a year for an acquaintanceship to develop into a close friendship, whereas European Americans feel that it takes only a few months.[6] For intercultural friendships to be successful, therefore, they may require an informal agreement between the friends about each of these aspects for the people involved to have shared expectations about appropriate behaviors.

CULTURE CONNECTIONS

"Ah, you're a dreamer, I can tell," Daniel says. "This is my trouble, too." He sighs and looks out the window. "But this is why Leah is a good match for me."

I feel suddenly as if I am a spy, sent from the other side, but I'm thrilled to find myself in such a position.

"Did you want to have an arranged marriage?" I ask, casually.

"Arranged marriage is the best thing, I think," he says, nodding. "Somebody my mummy selects for me. Because she knows me better than anybody else. So she knows my... with what sort of girl I can spend the rest of my life. She's the best judge. So I left it on her to decide."

—Sadia Shepard   ■

ROMANTIC PARTNERS The diversity of cultural norms that govern romantic relationships is an excellent example of the wide range of cultural expectations. Consider, for instance, the enormous differences in cultural beliefs, values, norms, and social practices about love, romance, dating, and marriage.

Among European Americans, dating usually occurs for romance and companionship. A dating relationship is not viewed as a serious commitment that will necessarily, or even probably, lead to an engagement. If they choose to do so, couples will marry because of love and affection for each other. Although family members may be consulted before a final decision is made, the choice to marry is made almost exclusively by the couples themselves.

In Argentina and Spain, dating is taken more seriously. Indeed, dating the same person more than twice may mean that the relationship will lead to an engagement and, ultimately, marriage. Yet engagements in these Spanish-speaking cultures typically last a long time and may extend over a period of years, as couples work, save money, and prepare themselves financially for marriage.

In contrast, casual dating relationships and similar opportunities for romantic expression among unmarried individuals are still quite rare in India; marriages there are usually arranged by parents, typically with the consent of the couple. So when a European American couple—friends of the bride's family—was invited to a wedding in India, they brought their fourteen-year-old daughter. The day before the wedding, at the bride's home, a group of girls was seated tightly on a large bed in the parents' bedroom. Except for the European American, all the girls were Indian. Their conversation was raucous and rambling. As it turned to the topic of marriage—as such conversations often do at a wedding—the Indian girls chattered away about whom they hoped their parents might pick to be their husbands. Taken aback by the notion of an arranged marriage, the European American girl asserted her individualism, declared that *she* would find her *own* husband, and announced that she would make these choices without *any* intervention from her parents. The Indian girls initially reacted quizzically to this strange pronouncement; then, as its implications slowly sank in, they displayed looks of puzzlement, astonishment, concern, and finally fear. One of the girls asked, "Aren't your parents even going to help you?" To the Indian girls, it was unfathomable that they would have to select their life partners without the help of parents and other elders.

Similar patterns of familial involvement can also be found in Muslim cultures, where marriage imposes great obligations and responsibilities on the families of the couple. In Algeria, for instance, a marriage is seen as an important link between families, not individuals; consequently, the selection of a spouse may require the approval of the entire extended family. In Indonesia, the opportunities for men and women to be together, particularly in unchaperoned settings, are much more restricted.

In both India and Algeria, romantic love is believed to be something that develops after marriage, not before. Even in Colombia, where, because of changes in customs and cultural practices, arranged marriages are no longer fashionable, the decision to

Friendships are often based on common interests and experiences, as these cyclists know. Sometimes, however, differences in cultural patterns may cause difficulties in managing interpersonal relationships.

get married requires family approval. Yet research by Stanley Gaines on the nature of intercultural romantic relationships found a great deal of similarity in the communication across cultures.[7]

FAMILY   Family or kinship relationships are also characterized by large cultural variations. Particularly important to the development of intercultural relationships are these factors: how the family is defined, or who is considered to be a member of the family; the formality of roles and behavioral expectations for particular family members; and the importance of the family in social relationships and personal decisions.

Among European Americans, and even among members of most European cultures, family life is primarily confined to interactions among the mother, father, and children. Households usually include just these family members, though the extended family unit also includes grandparents, aunts, uncles, and cousins. Though the amount and quality of interaction among extended family members will vary greatly from family to family, members of the extended family rarely live together in the same household or take an active part in the day-to-day lives of the nuclear family members.

Family relationships in other cultures can be quite different. Among Latinos, for instance, the extended family is very important.[8] Similarly, in India the extended family dominates; grandparents, aunts, uncles, and many other relatives may live together in one household. Families in India include people who would be called second or third cousins in the European American family, and the unmarried siblings of those who have become family members through marriage may also be included in the household. These "family members" would rarely be defined as such in the typical European American family. Among Native Americans, family refers to all members of the clan.[9] No particular pattern of family relationships can be said to typify the world's cultures. Many Arab families, for instance, include multiple generations of the male line. Often three generations—grandparents, married sons and their wives, and unmarried children—will live together under one roof. Among certain cultural groups in Ghana, however, just the opposite pattern can be found; families have a matrilineal organization, and the family inheritance is passed down through the wife's family rather than the husband's.

Expected role behaviors and responsibilities also vary among cultures. In Argentina, family roles are very clearly defined by social custom; the wife is expected to raise the children, manage the household, and show deference to the husband. In India, the oldest male son has specific family and religious obligations that are not requirements for other sons in the same family. Languages sometimes reflect these specialized roles. In China, for example,

> [a] sister-in-law is called by various names, depending on whether she is the older brother's wife, the younger brother's wife, or the wife's sister. Aunts, uncles, and cousins are named in the same way. Thus, a father's sister is "ku," a mother's sister is "yi," an uncle's wife is "shen," and so on.[10]

Families also differ in their influence over a person's social networks and decision making. In some cultures, the family is the primary means through which a person's social life is maintained. In others, such as among European Americans, families are almost peripheral to the social networks that are established. In the more collectivist cultures such as Japan, Korea, and China, families play a pivotal role in making decisions for children, including the choice of university, profession, and even marital partner. In contrast, in individualistic cultures, where children are taught from their earliest years to make their own decisions, a characteristic of "good parenting" is to allow children to "learn for themselves" the consequences of their own actions.

The increasing number of people creating intercultural families, in which husband and wife represent different cultural backgrounds, poses new challenges for family communication. Often, the children in these families are raised in an intercultural household that is characterized by some blending of the original cultures. Differences in the expectations of appropriate social roles—of wife and husband, son and daughter, older and younger child, or husband's parents and wife's parents—require a knowledge of and sensitivity to the varying influences of culture on family communication.

Dimensions of Interpersonal Relationships

People throughout the world use at least three primary dimensions to interpret interpersonal communication messages: control, affiliation, and activation.[11]

CONTROL Control involves status or social dominance. We have control to the extent that we have the power and prestige to influence the events around us. Depending on the culture, control can be communicated by a variety of behaviors, including touching, looking, talking, and the use of space. Supervisors, for instance, are more likely to touch their subordinates than vice versa. In many cultures, excessive looking behaviors are viewed as attempts to "stare down" the other person and are usually seen as an effort to exert interactional control. Similarly, high-power individuals seek and are usually given more personal space and a larger territory to control than their low-power counterparts. Of course, many of these same behaviors, when used in a different context, could also indicate other aspects of the interpersonal relationship. Excessive eye contact, for example, might not be an indication of power; it may merely mean that the two individuals are deeply in love. Usually, however, there are other situational cues that can be used to help interpret the behaviors correctly.

Control is often conveyed by the specific names or titles used to address another person. Do you address physicians, teachers, and friends by their first names, or do you say *Doctor*, *Professor*, or *Mr.* or *Ms.*? In Malaysia and many other places, personal names are rarely used among adults because such use might imply that the other person has little social status. Instead,

> a shortened form or a pet name is often used if a kin term is not appropriate. This is to avoid showing disrespect, since it is understood that the more familiar the form of address to a person, the more socially junior or unimportant he must be regarded.[12]

In cultures that are very attuned to status differences among people, such as Japan, Korea, and Indonesia, the language system requires distinctions based on people's degree of social dominance. In Indonesia, for instance,

> the Balinese speak a language which reflects their caste, a tiered system where (like the Javanese) at each level their choice of words is governed by the social relationship between the two people having a conversation.[13]

Because culture influences expectations about appropriate and effective behaviors, intercultural families may need to pay more attention to the negotiation of their relationship rules.

CULTURE CONNECTIONS

To exit a marriage in Bali leaves a person alone and unprotected in ways that are almost impossible for a Westerner to imagine. The Balinese family unit, enclosed within the walls of a family compound, is merely everything— four generations of siblings, cousins, parents, grandparents and children all living together in a series of small bungalows surrounding the family temple, taking care of each other from birth to death. The family compound is the source of strength, financial security, health care, daycare, education and—most important to the Balinese—spiritual connection.

The family compound is so vital that the Balinese think of it as a single, living person. The population of a Balinese village is traditionally counted not by the number of individuals, but by the number of compounds. The compound is a self-sustaining universe. So you don't leave it.

—Elizabeth Gilbert   ■

Intercultural communication is often characterized by an increased tendency to *mis*interpret nonverbal control and status cues. In both the United States and Germany, for instance, private offices on the top floors and at the corners of most major businesses are reserved for the highest-ranking officials and executives; in France, executives typically prefer an office that is centrally located, in the middle of their subordinates if possible, in order to stay informed and to control the flow of activities. Thus, the French may infer that the Germans are too isolated and the Germans that the French are too easily interrupted to manage their respective organizations well.

AFFILIATION   Members of a culture use affiliation to interpret the degree of friendliness, liking, social warmth, or immediacy that is being communicated. Affiliation is an evaluative component that indicates a person's willingness to approach or avoid others. Albert Mehrabian suggests that we approach those people and things we like and we avoid or move away from those we do not like.[14] Consequently, affiliative behaviors are those that convey a sense of closeness, communicate interpersonal warmth and accessibility, and encourage others to approach.

Affiliation can be expressed through eye contact, open body stances, leaning forward, close physical proximity, touching, smiling, a friendly tone of voice, and other communication behaviors. Edward Hall has called those cultures that display a high degree of affiliation "high-contact" cultures; those that display a low level are called "low-contact" cultures.[15]

Compared with low-contact cultures, members of high-contact cultures tend to stand closer, touch more, and have fewer barriers, such as desks and doors, to separate themselves from others. High-contact cultures, which are generally located in warmer climates, include many of the cultures in South America, Latin America, southern Europe, and the Mediterranean region; most Arab cultures; and Indonesia. Low-contact cultures, which tend to be located in colder climates, include the Japanese, Chinese, U.S. Americans, Canadians, and northern Europeans. One explanation for these climate-related differences is that the harshness of cold-weather climates forces people to live and work closely with one another in order to survive, and some cultures have compensated for this forced togetherness by developing norms that encourage greater distance and privacy.[16]

ACTIVATION   Activation refers to the ways people react to the world around them. Some people seem very quick, excitable, energetic, and lively; others value calmness, peacefulness, and a sense of inner control. Your perception of the degree of activity that another person exhibits is used to evaluate that person as fast or slow, active or inactive, swift or sluggish, relaxed or tense, and spirited or deliberate.

Cultures differ in what they consider acceptable and appropriate levels of activation in a conversation. For instance, among many of the black tribes of southern Africa, loud talking is considered inappropriate. Similarly, among Malaysians,

> too much talk and forcefulness on the part of an adult speaker is disapproved.... A terse, harmonious delivery is admired.... The same values—of evenness and restraint—hold for Malay interpersonal relations generally. Thus Malay village conversation makes little use of paralinguistic devices such as facial expression, body movement, and speech tone.... Malays are not highly emotive people.[17]

Thais, like Malays, often dampen or moderate their level of responsiveness. As John Feig suggests,

> Thais have a tendency to neutralize all emotions; even in a very happy moment, there is always the underlying feeling: I don't want to be too happy now or I might be correspondingly sad later; too much laughter today may lead to too many tears tomorrow.[18]

Iranians tend to have the opposite reaction, as they are often very emotionally expressive in their conversations. Particularly when angry, a man's conversation may consist of behaviors such as "turning red, invoking religious oaths, proclaiming his injustices for all to hear, and allowing himself to be held back."[19]

European Americans are probably near the midpoint of this dimension. Compared to the Japanese, for instance, European Americans tend to be fairly active and expressive in their conversations. As Harvey Taylor suggested:

> An American's forehead and eyebrows are constantly in motion as he speaks, and these motions express the inner feelings behind the words. The "blank," nearly motionless Japanese forehead reveals very little of the Japanese person's inner feelings to the American (but not necessarily to the Japanese). Therefore the American feels that the Japanese is not really interested in the conversation or (worse yet) that the Japanese is hiding the truth.[20]

Compared to Jordanians, Iranians, African Americans, and Latinos, however, European Americans are passive and reserved in conversational expressiveness.

It is useful once again to remind you that all beliefs, values, norms, and social practices lie on a continuum. How a particular characteristic is displayed or perceived in a specific culture is interpreted against the culture with which it is being compared. Thus it is possible for an African American to seem very active and emotionally expressive to the Japanese but quite calm and emotionally inexpressive to the Kuwaitis.

Dynamics of Interpersonal Relationships

Interpersonal relationships are dynamic. That is, they are continually changing as they are pushed and pulled by the ongoing tugs of past experiences, present circumstances, and future expectations.

One useful way to think about relational dynamics is to view people in interpersonal relationships as continually attempting to maintain their balance amidst changing circumstances. To illustrate, imagine that you and your partner are attempting to do a common dance routine such as a country line dance, a tango, or a waltz. Now imagine that you are dancing aboard a ship at sea: the floor rises and falls to the pulsing of the waves; uneven electrical power makes the music speed up and slow down; and your partner wants to add graceful variations to the typical sequence of steps. Your efforts to stay "in rhythm" and coordinate your movements with the music and with your partner are analogous to the adaptations that people must continually make to the ongoing dynamics of interpersonal relationships.

Leslie Baxter suggests that the changing dynamics in interpersonal relationships are due to people's attempts to maintain a sense of "balance" among opposing and seemingly contradictory needs. These basic contradictions in relationships, called "dialectics," create ongoing tensions that affect the way people connect to one another.[21] Three dialectics have been identified as important in interpersonal relationships: autonomy–connection, novelty–predictability, and openness–closedness. Each of the dialectics has corresponding cultural-level components.

The autonomy–connection dialectic is perhaps the most central source of tensions in interpersonal relationships. Individuals inevitably vary, at different moments of their interpersonal relationships, in the extent to which they want a sense of separation from others (autonomy) and a feeling of attachment to others (connection). Note the word *and* in the previous sentence; both types of interpersonal needs, though they may seem contradictory, occur simultaneously. As we implied in our discussions of individualism–collectivism in Chapter 5, a culture teaches its members both the "correct" range of autonomy and connection and how these should be expressed when communicating with others. Thus, while the general level of autonomy desired by someone from an individualistic culture may be relatively high, one's specific needs for autonomy and connection will vary across time and relationships.

The novelty–predictability dialectic relates to people's desire for change and stability in their interpersonal relationships. All relationships require moments of novelty and excitement, or they will be emotionally dead. They also require a sense of predictability, or they will be chaotic. The novelty–predictability dialectic refers to the dynamic tensions between these opposing needs. The cultural dimension of uncertainty avoidance provides a way of understanding the general range of novelty and predictability that people desire. At specific moments within each relationship, however, individuals can vary in their preferences for novelty and predictability.

CULTURE CONNECTIONS

Again, the goodness and innocence of women are preserved by concealing them. Women, while not intrinsically evil, are corruptible, and therefore must be protected. Seclusion and the veil are not regarded as concealing something ugly and offensive but as concealing something precious that is to be kept unsullied, even though both also secondarily protect men from temptations and distractions. This, incidentally, is a misunderstanding that lies behind many discussions of the veil between Muslims and non-Muslims, who make the assumption that to conceal something is a mark of disesteem. Houses, similarly, traditionally turned drab faces outward, with high concealing walls, while within, the welcome visitor found the green garden and warm hospitality.

—Mary Catherine Bateson　∎

The openness–closedness dialectic relates to people's desire to share or withhold personal information. To some extent, openness and self-disclosure are necessary to establish and maintain relational closeness and intimacy. However, privacy is an equally important need; the desire to establish and maintain boundaries is basic to the human condition. For instance, a person may be open to interpersonal contact at certain moments, or with specific individuals, or about certain topics. There will also be times when that person may want to shut the office door or find another way to lessen the degree of interpersonal contact. The openness–closedness dialectic operates not only within a relationship but also in decisions about the public presentation of the relationship to others. Individuals in interpersonal relationships must continually negotiate what kinds of information about their relationship they want to reveal or withhold from others. Several cultural dimensions may affect openness–closedness. Collectivist cultures, for instance, with their tightly knit ingroups and relatively large social distances from outgroups, typically encourage openness within the ingroup and closedness to outgroup members. Alternatively, cultures that value large power distances may expect openness within interpersonal relationships to be asymmetric, such that those relatively lower in social status are expected to share personal information with their superiors.

Each of these relational dialectics, and others as well, contributes to a dynamically changing set of circumstances that affect what people expect, want, and communicate in interpersonal relationships. As the following section explains, how people in interpersonal relationships maintain an appropriate balance among these dialectics relates to their maintenance of face.

THE MAINTENANCE OF FACE IN INTERPERSONAL RELATIONSHIPS

A very important concept for understanding interpersonal communication among people from different cultures is that of face, or the public expression of the inner self. Erving Goffman defined face as the favorable social impression that a person wants others to have of him or her.[22] Face therefore involves a claim for respect and dignity from others.

The definition of *face* suggests that it has three important characteristics. First, face is *social*. This means that face is not what an individual thinks of himself or herself but rather how that person wants others to regard his or her worth. Face therefore refers to the public or social image of an individual that is held by others. Face, then, always occurs in a relational setting. Because it is social, one can only gain or lose face through actions that are known to others. The most heroic deeds, or the most bestial ones, do not affect a person's face if they are done in complete anonymity. Nor can face be claimed independent of the social perceptions of others. For instance, the statement "No matter what my teachers think of me, I know I am a good student" is not a statement about face. Because face has a social component, a claim for face would occur only when the student conveys to others the idea that teachers should acknowledge her or his status as a good student. In this sense, the concept of *face* is only meaningful when considered in relation to others in the social network.[23] Consequently, it differs from such psychological concepts as self-esteem or pride, which can be claimed for oneself independently of others and can be increased or decreased either individually or socially.

Second, face is an *impression,* which may or may not be shared by all, that may differ from a person's self-image. People's claims for face, therefore, are not requests to know what others actually think about them; instead, they are solicitations from others of favorable expressions about them. To maintain face, people want others to act toward them with respect, regardless of their "real" thoughts and impressions. Thus, face maintenance involves an expectation that people will act as though others are appreciated and admired.

Third, face refers only to the *favorable* social attributes that people want others to acknowledge. Unfavorable attributes, of course, are not what others are expected to admire. However, cultures may differ in the behaviors that are highly valued, and they may have very different expectations, or norms, for what are considered to be desirable face behaviors.

Types of Face Needs

Penelope Brown and Stephen Levinson extended Goffman's ideas by proposing a universal model of social politeness.[24] They pointed out that, regardless of their culture, all people have face and a desire to maintain and even gain more of it. Face is maintained through the use of various politeness rituals in social interactions, as people try to balance the competing goals of task efficiency and relationship harmony.[25] Tae-Seop Lim suggests that there are three kinds of face needs: the need for control, the need for approval, and the need for admiration.[26] We now describe these three universal face needs.

THE NEED FOR CONTROL   Control face is concerned with individual requirements for freedom and personal authority. It is related to people's need for others to acknowledge their individual autonomy and self-sufficiency. As Lim suggests, it involves people's

> image that they are in control of their own fate, that is, they have the virtues of a full-fledged, mature, and responsible adult. This type of face includes such values as "independent," "in control of self," "initiative," "mature," "composed," "reliable," and "self-sufficient." When persons claim these values for themselves, they want to be self-governed and free from others' interference, control, or imposition.[27]

The claim for control face, in other words, is embodied in the desire to have freedom of action.

CULTURE CONNECTIONS

In her first meeting as team leader for a North American software company, Jessica Shultz counted on several people in her team to be supportive. She had carefully built relationships over the years so, that when she became a team leader, she could rely on others for allegiance and cooperation.

At this meeting, Jessica asked a very direct question of a Korean American, Linn Park, an expert in product development pricing. Linn, however, didn't have the numbers readily available. Jessica stated that the lack of numbers was not a problem and that Linn could get back to the team with the data later in the week.

Linn, humiliated by the directness of Jessica's question and her inability to properly answer the financial question, immediately requested a transfer to a different division. Jessica had failed to understand the concept of face in maintaining good relationships with those influenced by the Asian culture. She repeatedly met with Linn and over time was able to rebuild a strong relationship.

—Sana Reynolds, Deborah Valentine, and Mary Munter ■

THE NEED FOR APPROVAL   Approval face is concerned with individual requirements for affiliation and social contact. It is related to people's need for others to acknowledge their friendliness and honesty. This type of face is similar to what the Chinese call *lien,*[28] or the integrity of moral character, the loss of which makes it impossible for a person to function appropriately within a social group. As Hsien Chin Hu relates,

> A simple case of *lien*-losing is afforded by the experience of an American traveler in the interior of China. In a little village she had made a deal with a peasant to use his donkey for transportation. On the day agreed upon the owner appeared only to declare that his donkey was not available, the lady would have to wait one day. Yet he would not allow her to hire another animal, because she had consented to use his ass. They argued back and forth first in the inn, then in the courtyard; a crowd gathered around them, as each stated his point of view over and over again. No comment was made, but some of the older people shook their heads and muttered something, the peasant getting more and more excited all the time trying to prove his right. Finally he turned and left the place without any more arguments, and the American was free to hire another beast. The man had felt the disapproval of the group. The condemnation of his community of his attempt to take advantage of the plight of the traveler made him feel he had "lost *lien.*"[29]

Lien is maintained by acting with good *jen,* the Chinese term for "human." As Francis Hsu explains:

> When the Chinese say of so-and-so "*ta pu shih jen*" (he is not *jen*), they do not mean that this person is not a human animal; instead they mean that his behavior in relation to other human beings is not acceptable."[30]

Hsu regards the term *jen* as similar in meaning to the Yiddish term *mensh,* which refers to a good human being who is kind, generous, decent, and upright. Such an individual should therefore be admired for his or her noble character. As Leo Rosten says of this term,

> It is hard to convey the special sense of respect, dignity, approbation, that can be conveyed by calling someone "a real *mensh.*" ... The most withering comment one might make on someone's character or conduct is: "He is not (did not act like) a *mensh.*" ... The key to being "a real *mensh*" is nothing less than—character: rectitude, dignity, a sense of what is right, responsible, decorous. Many a poor man, many an ignorant man, is a *mensh.*[31]

Thus, approval face reflects the desire to be treated with respect and dignity.

THE NEED FOR ADMIRATION   Admiration face is concerned with individual needs for displays of respect from others. It is related to people's need for others to acknowledge their talents and accomplishments. This type of face is similar to what the Chinese call *mien-tzu,* or prestige acquired through success and social standing. One's *mien-tzu*

> is built up through high position, wealth, power, ability, through cleverly establishing social ties to a number of prominent people, as well as through avoidance of acts that would cause unfavorable comment.... All persons growing up in any community have the same claim to *lien,* an honest, decent "face"; but their *mien-tzu* will differ with the status of the family, personal ties, ego's ability to impress people, etc.[32]

Thus, admiration face involves the need for others to acknowledge a person's success, capabilities, reputation, and accomplishments.

Facework and Interpersonal Communication

The term facework refers to the actions people take to deal with their own and others' face needs. Everyday actions that impose on another, such as requests, warnings, compliments, criticisms, apologies, and even praise, may jeopardize the face of one or more participants in a communicative act. Ordinarily, say Brown and Levinson,

> people cooperate (and assume each other's cooperation) in maintaining face in interaction, such cooperation being based on the mutual vulnerability of face. That is, normally everyone's face depends on everyone else's being maintained, and since people can be expected to defend their faces if threatened, and in defending their own to threaten others' faces, it is in general in every participant's best interest to maintain each others' face.[33]

The degree to which a given set of actions may pose a potential threat to one or more aspects of people's face depends on three characteristics of the relationship.[34] First, the potential for face threats is associated with the control dimension of interpersonal communication. Relationships in which there are large power or status differences among the participants have a great potential for people's actions to be interpreted as face-threatening. Within a large organization, for instance, a verbal disagreement between a manager and her employees will have a greater potential to be perceived as face-threatening than will an identical disagreement among employees who are equal in seniority and status.

Second, face-threat potential is associated with the affiliation dimension of interpersonal communication. That is, relationships in which participants have a large social distance, and therefore less social familiarity, have a great potential for actions to be perceived as face-threatening. Thus, very close family members may say things to one another that they would not tolerate from more distant acquaintances. Relationships where strangers have no formal connection to one another but are, for example, simply waiting in line at the train station, the taxi stand, or the bank may sometimes be seen as an exception to this general principle.[35] As Ron Scollon and Suzie Wong-Scollon suggest, "Westerners often are struck with the contrast they see between the highly polite and deferential Asians they meet in their business, educational, and governmental contacts and the rude, pushy, and aggressive Asians they meet on the subways of Asia's major cities."[36] At many train stations in the People's Republic of China, for example,

> people are not in the midst of members of their own community, so the drive to preserve face and act with proper behavior is much lower. Passengers usually wait in waiting rooms until the attendant moves a barrier and they can cross the area between them and the train. The competition is quite fierce as passengers rush toward the train with their luggage, and they have little regard for the safety of other passengers. Often, fellow travelers are injured by luggage, knocked to the ground, or even pushed between the platform and the train, where they fall to the tracks.[37]

Third, face-threat potential is related to culture-specific evaluations that people make. That is, cultures may make unique assessments about the degree to which particular actions are inherently threatening to a person's face. Thus, certain actions within one culture may be regarded as face-threatening, whereas those same actions

CULTURE CONNECTIONS

She was jolted by the ringing of her phone. It was Chen. There was traffic noise in the background.

"Where are you, Chief Inspector Chen?"

"On my way home. I had a call from Party Secretary Li. He invites you to a Beijing Opera performance this evening."

"Does Mr. Li want to discuss the Wen case with me?"

"I'm not sure about that. The invitation is to demonstrate our bureau's attention to the case, and to you, our distinguished American guest."

"Isn't it enough to assign you to me?" she said.

"Well, in China, Li's invitation gives more face."

"Giving face—I've heard only about losing face."

"If you are a somebody, you give face by making a friendly gesture."

"I see, like your visit to Gu. So I have no choice?"

"Well, if you say no, Party Secretary Li will lose face. The bureau will, too—including me."

"Oh no! Yours is one face I have to save." She laughed. "What shall I wear to the Beijing Opera?"

"Beijing Opera is not like Western opera. You don't have to dress formally, but if you do—"

"Then I'm giving face, too."

"Exactly. Shall I pick you up at the hotel?"

—Qiu Xiaolong ■

in another culture may be regarded as perfectly acceptable. In certain cultures, for instance, passing someone a bowl of soup with only one hand, or with one particular hand, may be regarded as an insult and therefore a threat to face; in other cultures, however, those same actions are perfectly acceptable.

Stella Ting-Toomey[38] and Min-Sun Kim[39] both suggest that cultural differences in individualism–collectivism affect the facework behaviors that people are likely to use. In individualist cultures, concerns about message clarity and preserving one's own face are more important than maintaining the face of others, because tasks are more important than relationships, and individual autonomy must be preserved. Consequently, direct, dominating, and controlling face-negotiation strategies are common, and there is a low degree of sensitivity to the face-threatening capabilities of particular messages. Conversely, in collectivist cultures, the mutual preservation of face is extremely important, because it is vital that people be approved and admired by others. Therefore, indirect, obliging, and smoothing face-negotiation strategies are common, direct confrontations between people are avoided, concern for the feelings of others is heightened, and ordinary communication messages are seen as having a great face-threatening potential.

Facework and Intercultural Communication

Competent facework, which lessens the potential for specific actions to be regarded as face-threatening, encompasses a wide variety of communication behaviors. These behaviors may include apologies, excessive politeness, the narration of justifications or excuses, displays of deference and submission, the use of intermediaries or other avoidance strategies, claims of common ground or the intention to act cooperatively, or the use of implication or indirect speech. The specific facework strategies a person uses, however, are shaped and modified by his or her culture. For instance, the Japanese and U.S. Americans have very different reactions when they realize that they have committed a face-threatening act and would like to restore the other's face. The Japanese prefer to adapt their messages to the social status of their interaction partners and provide an appropriate apology. They want to repair the damage, if possible, but without providing reasons that explain or justify their original error. Conversely, U.S. Americans would prefer to adapt their messages to the nature of the provocation and provide verbal justifications for their initial actions. They may use humor or aggression to divert attention from their actions but do not apologize for their original error.[40]

As another example of culture-specific differences in facework behaviors, consider the comments that are commonly appended to the report cards of high school students in the United States and in China. In the United States, evaluations of high school students include specific statements about students' strengths and weaknesses. In China, however, the high school report cards that are issued at the end of each semester never criticize the students directly; rather, teachers use indirect language and say, "I wish that you would make more progress in such areas as … " in order to save face while conveying their evaluations.[41]

Facework is a central and enduring feature of all interpersonal relationships. Facework is concerned with the communication activities that help to create, maintain, and sustain the connections between people. As Robyn Penman says:

> Facework is not something we do some of the time, it is something that we unavoidably do all the time—it is the core of our social selves. That it is called face and facework is curious but not critical here. What is critical is that the mechanism the label stands [for] seems to be as enduring as human social existence. In the very act of communicating with others we are inevitably commenting on the other and our relationship with them. And in that commenting we are maintaining or changing the identity of the other in relationship to us.[42]

IMPROVING INTERCULTURAL RELATIONSHIPS

Competent interpersonal relationships among people from different cultures do not happen by accident. They occur as a result of the knowledge and perceptions people have about one another, their motivations to engage in meaningful interactions, and their ability to communicate in ways that are regarded as appropriate and effective. To improve these interpersonal relationships, then, it is necessary to learn about and thereby reduce anxiety and uncertainty about people from other cultures, to share oneself with those people, and to handle the inevitable differences in perceptions and expectations that will occur.

Learning About People from Other Cultures

The need to know, to understand, and to make sense of the world is a fundamental necessity of life. Without a world that is somewhat predictable and that can be interpreted in a sensible and meaningful way, humankind would not survive.

We have already suggested in Chapter 5 that both individuals and cultures can differ in their need to reduce uncertainty and in the extent to which they can tolerate ambiguity and, therefore, in the means they select to adapt to the world. The human need to learn about others, to make sense of their actions, and to understand their beliefs, values, and behaviors has typically been studied under the general label of uncertainty reduction theory.[43] This theory explains the likelihood that people will seek out additional information about one another, but it deals primarily with the knowledge component of

CULTURE CONNECTIONS

Like many other newcomers to the Anglo world, I was struck by the elasticity of the English concept of "friend", which could be applied to a wide range of relationships, from deep and close, to quite casual and superficial. This was in stark contrast to the Polish words *przyjaciel* (male) and *przyjaciółka* (female), which could only stand for exceptionally close and intimate relationships. What struck me even more was the importance of the concept embodied in the Polish word *koledzy* (female counterpart *koleżanki*) as a basic conceptual category defining human relations—quite unlike the relatively marginal concept encoded in the English word *colleague,* relevant only to professional elites. It became clear to me that concepts such as "*koledzy*" ("*koleżanki*") and "*przyjaciele*" (*przyjaciótki*) (plural) organised the social universe quite differently from concepts such as "friends".

—Anna Wierzbicka ■

communication competence. William B. Gudykunst revised uncertainty reduction theory and renamed it anxiety/uncertainty management theory.[44] It now focuses more clearly on intercultural communication, incorporates the emotional or motivational component of intercultural competence, and emphasizes ways to cope with or manage the inherent tensions and anxieties that inevitably occur in many intercultural encounters. In the sections that follow, we describe the components, causes, and consequences of uncertainty management behaviors and some strategies for reducing uncertainty in interpersonal relationships among people from different cultures.

COMPONENTS OF UNCERTAINTY AND ANXIETY MANAGEMENT   Some degree of unpredictability exists in all interpersonal relationships, but it is typically much higher in intercultural interactions. There are two broad components involved in the management of uncertainty behaviors: uncertainty and anxiety. Uncertainty refers to the extent to which a person lacks the knowledge, information, and ability to understand and predict the intentions and behaviors of another. Anxiety refers to an individual's degree of emotional tension and her or his inability to cope with change, to live with stress, and to contend with vague and imprecise information.

Uncertainty and anxiety are influenced by culture. In Chapter 5, when we discussed Hofstede's value dimensions, we suggested that cultures differ in the extent to which they prefer or can cope with uncertainty. It should now be obvious that Hofstede's uncertainty avoidance dimension is related to what is here being referred to as anxiety/uncertainty management.

CAUSES OF UNCERTAINTY AND ANXIETY   Three conditions are related to uncertainty and anxiety management behaviors. These are your expectations about future interactions with other people, the incentive value or potential rewards that relationships with other people may have for you, and the degree to which other people exhibit behaviors that deviate from or do not match your expectations.

The first condition is your expectations about future interactions with another person. If you believe that you are very likely to interact with some person on future occasions, the degree to which you can live with ambiguity and insufficient information about that person will be low, and your need for more knowledge about that person will be high. Conversely, if you do not expect to see and talk with someone again, you will be more willing to remain uncertain about her or his motives and intentions, your anxiety level will be relatively low, and you will therefore not attempt to seek out any additional information. This person will continue to be a stranger. Anxiety/uncertainty management theory suggests that sojourners and immigrants who know they will be interacting in a new culture for a long period of time will be more likely to try to reduce their uncertainty about how and why people behave than will a tourist or temporary visitor.

The second condition, incentive value, refers to the perceived likelihood that the other person can fulfill various needs that you have, give you some of the resources that you want, or provide you with certain rewards that you desire. If a person's incentive value is high—that is, if the other person has the potential to be very rewarding to you—your need to find out more about that person will be correspondingly high. As you might expect, a high incentive value also increases the degree to which a person will be preferred or viewed as interpersonally attractive. Of course, the needs or rewards that people might want vary widely; the incentive value of a given person is related to his or her ability to provide such benefits as status, affection, information, services, goods, money, or some combination of these resources.[45]

One form of incentive value that has been widely investigated is the perceived similarity of the other person. The similarity–attraction hypothesis suggests that we like and are attracted to those whom we regard as comparable to ourselves in ways that we regard as important. Conversely, we are unlikely to be attracted to those who are very different from us. This hypothesis implies that, at least in the initial stages of intercultural encounters, the dissimilarities created by cultural differences may inhibit the development of new interpersonal relationships.

The third condition is the degree of deviance that the other person exhibits. Deviant behaviors are those that are not typically expected because they are inconsistent with the common norms that govern particular social situations. When a person acts deviantly, both your level of anxiety and your degree of uncertainty about that person increase, because he or she is far less predictable to you. Conversely, when a person conforms to your expectations by behaving in a predictable way, your level of anxiety and your degree of uncertainty about that person decrease. A person who behaves in deviant and unexpected ways is often disliked and is regarded as interpersonally unattractive, whereas one who conforms to others' expectations and is therefore predictable is often most liked and preferred. In intercultural communication, it is extremely likely that the other person will behave "deviantly" or differently from what you might expect. Thus, uncertainty about people from other cultures will typically be high, as will the level of anxiety and tension that you experience.

CONSEQUENCES OF UNCERTAINTY AND ANXIETY MANAGEMENT Because intercultural communication involves people from dissimilar cultures, each person's behaviors are likely to violate the others' expectations and create

Competent intercultural communication often leads to a reduction in uncertainty and anxiety in intercultural interactions.

uncertainty and anxiety. Consequently, there is always the possibility that fear, distrust, and similar negative emotions may prevail. Often, but not always, the negative emotions can be overcome, and positive outcomes can result.

Judee Burgoon has developed expectancy violations theory to explain when deviations from expectations will be regarded as positive or as negative.[46] All behaviors that differ from expectations will increase the degree of uncertainty in an interaction. Burgoon suggests that how a person interprets and reacts to the deviations of another depends on how favorably that person is perceived. If the other person is perceived positively, violations of your expectation that increase interaction involvement will be seen as favorable, whereas violations of expectations that decrease interaction involvement will be viewed as unfavorable. To illustrate, imagine that you are having a conversation with someone who is standing closer to you than you would expect. This is clearly a violation of your expectations, but how would you likely react to this situation? Burgoon suggests that, if the person is positively valenced—because, for example, you regard the person as physically attractive—then you may view the violation and the other person favorably, whereas if the other person is negatively valenced, then you will regard the violation and the other person unfavorably and may attempt to back away or escape. Conversely, imagine that your conversation is with someone who is standing farther away than your typical or expected interaction distance. If the person is positively valenced, you may attempt to compensate for the violation by moving closer, whereas if the person is negatively valenced, you will likely attribute negative connotations—he or she is aloof, cold—to the person.

The positive consequences of anxiety and uncertainty management behaviors that are applicable to intercultural communication can be grouped under two general labels: informational consequences and emotional consequences. *Informational consequences* result from the additional knowledge that has been gained about other people, including facts or inferences about their culture; increased accuracy in the judgments made about their beliefs, values, norms, and social practices; and an increased degree of confidence that they are being perceived accurately.

Emotional consequences may include increased levels of self-disclosure, heightened interpersonal attraction, increases in intimacy behaviors, more frequent nonverbal displays of positive emotions, and an increased likelihood that future intercultural contacts will be regarded as favorable. Of course, these positive outcomes all presume that the reduction in anxiety and uncertainty about another person will result in an increase in positive communicator valence, which is not necessarily so. Unfortunately, as Gudykunst suggests, negative perceptions in intercultural encounters frequently occur because people are not *mindful*—focused, aware, open to new information, and tolerant of differences. This allows our cultural assumptions to remain unchallenged. As we have seen, the perception that a person is acting in a deviant way (as defined by one's own cultural expectations) will often lead to decreased satisfaction with the encounter.

STRATEGIES FOR REDUCING UNCERTAINTY AND ANXIETY   To behave both appropriately and effectively in an intercultural encounter, you must make an accurate assessment about many kinds of information: the individual characteristics of the person with whom you interact, the social episodes that are typical of the particular setting and occasion, the specific roles that are being played within the episode, the rules of interaction that govern what people can say and do, the setting or context within which the interaction occurs, and the cultural patterns that influence what is regarded as appropriate and effective. Thus, uncertainty is not reduced for its own sake, but occurs every day for strategic purposes. As Charles R. Berger suggested:

> To interact in a relatively smooth, coordinated, and understandable manner, one must be able both to predict how one's interaction partner is likely to behave, and, based on these predictions, to select from one's own repertoire those responses that will optimize outcomes in the encounter.[47]

There are three general types of strategies—passive, active, and interactive—that can be used to gain information about other people and thus reduce one's level of uncertainty and degree of anxiety. *Passive* strategies involve quiet and surreptitious observation of another person to learn how he or she behaves. *Active* strategies include efforts to obtain information about another person by asking others or structuring the environment to place the person in a situation that provides the needed information. *Interactive* strategies involve actually conversing with the other person in an attempt to gather the needed information. As you might expect, there are large cultural differences in the preferred strategies that are used to reduce uncertainty and manage anxiety in intercultural encounters. For example, European Americans are more likely than their Japanese counterparts to use active strategies such as asking questions and self-disclosing as a way to obtain information about another person, whereas the Japanese are more likely to use passive strategies.[48]

Sharing Oneself with People from Other Cultures

The human tendency to reveal personal information about oneself and to explain one's inner experiences and private thoughts is called self-disclosure. Self-disclosure occurs among people of all cultures, but there are tremendous cultural differences in the breadth, depth, valence, timing, and targets of self-disclosing events.

The *breadth* of self-disclosing information refers to the range of topics that are revealed, and European Americans tend to self-disclose about more topics than do members of most other cultures. For example, Tsukasa Nishida found that European Americans discussed a much wider range of topics that were related to the self (such as health and personality) with strangers than did Japanese; also, Japanese had far more self-related topics than did European Americans that they would never discuss with others.[49] Ghanaians tend to self-disclose about family and background matters, whereas U.S. Americans self-disclose about career concerns.[50] In contrast,

> Chinese culture takes a conservative stand on self-disclosure. For a Chinese, self-centered speech would be considered boastful and pretentious. Chinese tend to scorn those who often talk about themselves and doubt their motives when they do

so. Chinese seem to prefer talking about external matters, such as world events. For Americans, self-disclosure is a strategy to make various types of relationships work; for Chinese, it is a gift shared only with the most intimate relatives and friends.[51]

The *depth* of the self-disclosing information refers to the degree of "personalness" about oneself that is revealed. Self-disclosure can reveal superficial aspects ("I like broccoli") or very private thoughts and feelings ("I'm afraid of my father"). Of the many cultures that have been studied, European Americans are among the most revealing self-disclosers. European Americans disclose more than African Americans, who in turn disclose more than Mexican Americans.[52] European Americans also disclose more than the British,[53] French,[54] Germans, Japanese,[55] and Puerto Ricans.[56]

Valence refers to whether the self-disclosure is positive or negative, and thus favorable or unfavorable. Not only do European Americans disclose more about themselves than do members of many other cultures but they are also more likely to provide negatively valenced information. Compared to many Asian cultures, for example, European Americans are far less concerned with issues of "face" and are therefore more inclined to share information that may not portray them in the most favorable way.

Timing refers to when the self-disclosure occurs in the course of the relationship. For European Americans, self-disclosure in new relationships is generally high because the participants share information about themselves that the others do not know. A person's name, hometown, employment or educational affiliations, and personal interests are all likely to be shared in initial interactions. As the relationship progresses, the amount of self-disclosure diminishes because the participants have already learned what they need to know to interact appropriately and effectively. Only if the relationship becomes more personal and intimate will the amount of self-disclosure again begin to increase. But the timing of the self-disclosure process can be very different in other cultures. For example, Native Americans typically reveal very little about themselves initially because they believe that too much self-disclosure at that stage is inappropriate. A similar pattern may be found among members of Asian cultures.

Target refers to the person to whom self-disclosing information is given. Among European Americans, spouses are usually the targets of a great deal of self-disclosure, and mutual self-disclosure is widely regarded as contributing to an ideal and satisfactory marriage.[57] The breadth and depth of self-disclosure among other European American family members are of much lesser degree. Other cultures have different patterns. Among the Igbos of Nigeria, for instance, age is used to determine the appropriate degree of self-disclosure among interactants, younger interactants being expected to self-disclose far more than their older counterparts. As a cultural norm, when elder Igbos are in an initial encounter with someone who is younger, they have the right to inquire about the young person's background, parents, hometown, and similar information that may ultimately lead to contact with distant relatives or old friends.

Issues of disclosure, conflict resolution, and face maintenance affect all interpersonal relationships. These people confer informally about their work.

Handling Differences in Intercultural Relationships

Conflict in interpersonal relationships is a major nemesis for most people. Add the complications of different cultural backgrounds, and problems in managing conflict can become even more severe. Stella Ting-Toomey and John Oetzel's work provides some direction for managing intercultural conflict.[58]

Ting-Toomey and Oetzel use the distinction between collectivism and individualism, which is discussed more fully in Chapter 5. Briefly, in collectivist cultures, interpersonal bonds are relatively enduring, and there are distinct ingroups and outgroups. Collectivist cultures are often very traditional. In individualistic cultures, the bonds between people are more fragile, and because people belong to many different groups that often change, membership in ingroups and outgroups is very flexible. Individualistic cultures are therefore often characterized by rapid innovation and change.

Conflict may involve either task or instrumental issues. Task issues are concerned with how to do something or how to achieve a specific goal, whereas instrumental issues are concerned with personal or relationship problems, such as hostility toward another person. The distinction therefore focuses on conflict about ideas versus conflict about people.

Ting-Toomey and Oetzel believe that people in collectivistic and individualistic cultures typically define and respond to conflict differently. In collectivistic cultures, people are more likely to merge task and instrumental concerns, and conflict is therefore likely to be seen as personal. To shout and scream publicly, thus displaying the conflict to others, threatens everyone's face to such an extreme degree that such behavior is usually avoided at all costs. In contrast, people from individualistic cultures are more likely to separate the task and the instrumental dimensions. Thus, they are able to express their agitation and anger (perhaps including shouting and strong nonverbal actions) about an issue and then joke and socialize with the other person once the disagreement is over. It is almost as if once the conflict is resolved, it is completely forgotten.

Because there is a great deal of volatility and variability in people's behaviors in individualistic cultures, there is often considerable potential for conflict. Because people are encouraged to be unique, their behaviors are not as predictable as they would be in collectivistic cultures. Also, because expectations are individually based rather than group-centered, there is always the possibility that the behavior of any one person will violate the expectations of another, possibly producing conflict.

Cultures also shape attitudes toward conflict. In collectivistic cultures, which value indirectness and ambiguity, conflicts and confrontations are typically avoided. Thus, rather than trying to resolve the problem directly, people in collectivistic cultures will attempt to maintain the external smoothness of the relationship. In individualistic cultures, which are also more likely to be "doing" or activity-oriented cultures, people's approach to conflict will be action-oriented. That is, the conflict precipitates actions, and the conflict is explicitly revealed and named.[59]

A very important concept for understanding how people from different cultures handle conflict is that of face, which we discussed earlier in this chapter. In conflicts, in particular, face is very likely to be threatened, and all participants are vulnerable to the face-threatening acts that can occur.

The actions of people in conflict can include attempts to save face for themselves, others, or all participants. Members of collectivistic cultures are likely to deal with face threats such as conflicts by selecting strategies that smooth over their disagreements and allow them to maintain the face of both parties, that is, mutual face-saving. As Ringo Ma suggests, however, such strategies do not simply ignore the conflicting issues; after all, conflicts do get resolved in high-context cultures. Rather, nonconfrontational alternatives are used to resolve differences. Often, for instance, a friend of those involved in the conflict, or an elderly person respected by all, will function as an unofficial intermediary who attempts to preserve the face of each person and the relationship by preventing rejection and embarrassment.[60] Members of individualistic cultures, conversely, are likely to deal with face threats in a direct, controlling way. It is important to their sense of self to maintain their own face, to take charge, to direct the course of action, and in so doing to protect their own dignity and self-respect even at the expense of others.[61]

Imagine a scene involving two employees assigned to an important and high-tension project. Perhaps they are operating under serious time constraints, or perhaps the lives of many people depend on their success. Inevitably, disagreements about how to approach the assignment, as well as the specifics of the assignment itself, are likely to occur. Now assume that one employee is from a collectivistic culture such as Korea and the other is from an individualistic culture such as England. The difficulties inherent in completing their assignment will probably be increased by the great differences that will arise in their approaches to the problems. Each person's attempt to maintain face may induce the other to make negative judgments and evaluations. Each person's attempt to cope with the conflict and accomplish the task may produce even more conflict.[62] As Ting-Toomey and Oetzel suggest, these differences will need to be addressed before the work can be accomplished successfully.

CULTURE CONNECTIONS

Presented by the couples as an often unresolvable problem was the loss of place, culture and family that resulted from one partner's leaving his or her home of origin. The immigrant partner in the couple almost invariably expressed a longing for his or her own landscape and climate, and a deep sadness about the distance from extended family members. Other topics that frequently arose were loss of religion and language. While many couples were originally optimistic about importing religion and language from their native lands, they found that the United States exerted a pressure to assimilate that was very difficult to resist. As a result, these essential elements of culture were preserved only through great effort, if at all. For example, while Hueping Chin has managed to teach her son to speak Chinese fluently, as he grows older, their dialogue in Chinese sometimes falters in the face of new topics; "How does the space shuttle work? I find it hard to explain that to him in Chinese." The words of Jat Aluwalia capture the sense of regret that some felt: "I admit defeat; I guess the sense of being Indian ends with me."

—Jessie Carroll Grearson and Lauren B. Smith ■

INTERPERSONAL RELATIONSHIPS AND INTERCULTURAL COMPETENCE

Intercultural competence in interpersonal relationships requires knowledge, motivation, and skill in using verbal and nonverbal codes, as described in previous chapters. In addition, it requires behaviors that are appropriate and effective for the different types and dimensions of interpersonal relationships described in this chapter.

Competence in intercultural relationships requires that you understand the meanings attributed to particular types of interpersonal relationships. Whom should you consider to be a stranger, an acquaintance, a friend, or a family member? What expectations should you have for people in these categories? What clues do people from other cultures offer about their expectations for you? Your expectations about the nature of interpersonal relationships affect how you assign meaning to other people's behaviors.

Your willingness to understand the face needs of people from other cultures and to behave appropriately to preserve and enhance their sense of face is critical to your intercultural competence. Always consider a person's need to maintain a favorable face in her or his interactions with others. Perceptions of autonomy, approval, and respect by others are important, but you must meet these face needs with facework that is appropriate to the other's cultural beliefs.

Your expectations about self-disclosure, obtaining information about others, and handling disagreements will not, in all likelihood, be the same as those of people from other cultures. Competence in developing and maintaining intercultural relationships requires knowledge of differences, a willingness to consider and try alternatives, and the skill to enact alternative relational dynamics.

SUMMARY

People in an intercultural relationship may have very different expectations about the preferred nature of their social interactions. The types of interpersonal relationships, including those among strangers, acquaintances, friends, romantic partners, and family members, may also vary greatly across cultures.

Interpersonal relationships can be interpreted along the three dimensions of control, affiliation, and activation. The control dimension provides interpretations about status or social dominance. The affiliation dimension indicates the degree of friendliness, liking, and social warmth that is being communicated. The activation dimension is concerned with interpersonal responsiveness.

The concept of face refers to the positive social impressions that people want to have and would like others to acknowledge. Face includes the need for autonomy or individual freedom of action, approval or inclusion in social groups, and admiration or respect from others because of one's accomplishments. The need for facework depends on the control and affiliation dimensions of interpersonal communication and on culture-specific judgments about the extent to which certain actions inherently threaten one's face.

To improve intercultural relationships, you must learn about people from other cultures and thereby reduce the degree of uncertainty. Sharing yourself in appropriate ways with people from other cultures and learning to use culturally sensitive ways to handle the differences and disagreements that may arise are additional ways to improve intercultural relationships.

FOR DISCUSSION

1. What is a friend to you? What do you expect of your friends?
2. What is meant by the concept of "face"? Have you ever experienced a loss of face?
3. Describe the relationship among the following terms: *face, face maintenance, facework, embarrassment, truthfulness, dishonesty, fear,* and *withdrawal.*
4. Why do anxiety and uncertainty management play a particularly powerful role in intercultural communication?
5. Do differences in what we categorize as "public" and "private" information hold any consequences for the development of a relationship?
6. How do you think email, text messaging, and other forms of Internet communication have affected the development of intercultural relationships?

FOR FURTHER READING

Francesca Bargiela-Chiappini and Michael Haugh (eds.), *Face, Communication, and Social Interaction* (Oakville, CT: Equinox, 2009). Both scholarly and practical, this book on face and facework in interpersonal communication is an excellent sourcebook.

Laura K. Guerrero, Peter A. Andersen, and Walid A. Afifi, *Close Encounters: Communication in Relationships*, 3rd ed. (Thousand Oaks, CA: Sage, 2011). An intermediate-level textbook, it provides an excellent and readable summary of current theory and research. Explains and explores the theories that dominate the study of interpersonal communication.

Terri A. Karis and Kyle D. Killian (eds.), *Intercultural Couples: Exploring Diversity in Intimate Relationships* (New York: Brunner-Routledge, 2008). Highlights both the struggles and the successes of close intercultural relationships. Contributions in this volume explore the positive and negative issues that emerge in such marriages.

John G. Oetzel and Stella Ting-Toomey (eds.), *The Sage Handbook of Conflict Communication: Integrating Theory, Research, and Practice* (Thousand Oaks, CA: Sage, 2006). Provides a comprehensive overview of theory and research on conflict in a variety of human communication settings. Includes a substantial section on conflict in intercultural communication.

Julia T. Wood, *Interpersonal Communication: Everyday Encounters*, 7th ed. (Belmont, CA: Wadsworth/Centage, 2011). A presentation of the fundamentals of interpersonal communication that parallels some of the major topics in this textbook.

CHAPTER

11

Episodes, Contexts, and Intercultural Interactions

KEY TERMS

There is a repetitiveness to everyday communication experiences that helps to make them understandable and predictable. The recurring features of these common events, which we call social episodes, allow you to anticipate what people may do, what will likely happen, and what variations from the expected sequence of events could mean.

SOCIAL EPISODES IN INTERCULTURAL RELATIONSHIPS

People undertake intercultural relationships in predictable ways. In this section, we describe how communication experiences are grouped into common events. Our point is that people's interactions are structured by their participation in events that are quite predictable and routine.

The Nature of Social Episodes

Think about how your daily life is structured. If you are like most people, there is a great deal of predictability in what you do each day and even with whom you do it. If you are attending a college or university, much of your life is taken up with such activities as attending class, studying, texting with a classmate in the cafeteria, working at a job, going shopping, meeting a friend after work, attending a party, and eating dinner. These are the kinds of structures in your life that we refer to as social episodes—that is, interaction sequences that are repeated over and over again. Not only do these social episodes recur, but their structure is also very predictable. The individuals who participate in these episodes generally know what to expect from others and what others expect from them. It is almost as if there were an unwritten script that tells you roughly what to say, whom to say it to, and how to say it.

Take the example of going to class. You probably attend class in a room filled with chairs, or tables with chairs, that face the front of the room. When you take a seat, you put your notebooks and other texts on the floor or under the desk. You keep your chair oriented the way all the other chairs are oriented. The room is arranged so that you can look at the teacher, and there is a clearly marked space in the front of the room for the teacher to stand or sit. When you enter the classroom, you never consider taking that spot. You expect the teacher, when she or he walks into the room, to do so. You do not expect the teacher to walk into the room and take the chair next to you.

If you get to class early enough, you might engage in small talk with another student. There are fairly predictable topics you might discuss, depending on how well you know each other. You probably talk about the class, whether you have done the reading, how your work is going, and the assignments. You might talk about the teacher and analyze his or her strengths and weaknesses. You might talk about the weather, the latest sports scores, or other common topics.

You expect the teacher to give a lecture or in some other way provide a sense of structure for the class. You take notes if the teacher gives a lecture, trying to summarize the key points. If you talk to a classmate while the teacher is lecturing, you whisper rather than talk in a loud voice. If the teacher did not enact the behaviors you expect from the person playing the part of "teacher," you might complain about it to others. Similarly, if one of your fellow classmates did not follow the expected behaviors for "being a student," you might think there was something wrong.

The purpose of this extended example is to underscore our point that much of what people do is made up of social episodes, which are repetitive, predictable, and routine behaviors that form the structure of their interactions with others. These social episodes provide information about how to interpret the verbal and nonverbal symbols of the interactants. The meanings of the symbols are understood because of the context in which they are given. Because those who participate in a social episode usually have the same understanding about what is to take place, they usually know how to behave, what to say, and how to interpret the actions and intentions of others.

CULTURE CONNECTIONS

Suppose that I am walking down my own street near sundown; I meet a neighbor whom I do not know well standing at her doorway, and we greet each other. Routinely, I will be invited in, and if I do enter, I will be offered food in addition to the obligatory cup of tea and perhaps urged to stay for a meal. Since I know that to accept would be an imposition, I probably will not enter, and if I do, I will refuse the continued urgings to accept more with profuse thanks, and go on my way. Again, if I admire something in a friend's house, he or she may offer it to me, but I will refuse to accept it, although the ritual of offer and refusal with thanks may last many minutes. Similar graceful byplay will surround a quarrel between diners about the paying of a check in a restaurant and even the sequence with which people pass through a doorway. A conversation with the driver while riding in an uncrowded taxicab may cause his sense of being a host to overbalance his sense of performing a commercial service, leading to his saying at fare time, "Be my guest" *(befarmaid)*. The passenger will of course insist on paying, adding a little extra in appreciation of the gift of hospitality, and go on his way cheered.

—Mary Catherine Bateson ∎

In social episodes that include intercultural interactions, however, those involved may—and in all likelihood will—have very different expectations and interpretations about people's behaviors and intentions. As the interaction becomes more and more ambiguous, the expected behaviors that pattern the social episode also become more unpredictable and problematic. Though your culture teaches you to interpret the meanings and behaviors in social episodes in particular ways, other cultures may provide their members with very different interpretations of these same experiences.

Components of Social Episodes

There are five components of social episodes, each of which influences intercultural communication: cultural patterns, social roles, rules of interaction, interaction scenes, and interaction contexts.

CULTURAL PATTERNS   Cultural patterns are shared judgments about what the world is and what it should be, and widely held expectations about how people should behave. The patterns of a culture's beliefs and values, described in Chapters 4 and 5, permeate the ways in which members of a culture think about their world.

Cultural patterns are like tinted glasses that color everything people see and to which they respond. The episodes that are used to structure people's lives—attending class, eating dinner, playing with a friend, going to work, talking with a salesperson—are certainly common to many cultures. But the interpretations that are imposed on these behaviors vary greatly, depending on the cultural patterns that serve as the lens through which the social episodes are viewed.

Joseph Forgas and Michael Bond found that Hong Kong Chinese and Australian students, although leading superficially very similar lives—going to classes, studying, and so on—perceived various social episodes very differently. The perceptions of the Chinese students reflected values and cultural patterns associated with that culture's emphasis on community, the collective good, and acceptance of authority. The Australian students saw the same episodes in terms of self-confidence, competition, and the pleasure they might receive from the interactions in which they participated.[1]

SOCIAL ROLES   A social role is a set of expected behaviors associated with people in a particular position. Common roles that exist in most cultures include student, mother, father, brother, sister, boss, friend, service person, employee, salesclerk, teacher, manager, soldier, woman, man, and mail carrier. The role that you take in a particular social episode strongly suggests to you the way in which you should act. If you are participating in an episode of a boss giving an employee a performance review, you would expect to behave very differently if you were the employee rather than the boss. If another person is upset about the comments of a coworker, your response would be influenced by the particular role you play in relationship to the upset person. Are you in the role of friend, relative, or employer? Your answer to the question will definitely affect how you respond to the person's concerns. In many episodes, you play clearly defined roles that give you guidance about what you should say to the other person and even how you should say it. Furthermore, the role you are playing is matched by the roles of others in the episode. You have expectations for yourself based on your roles, and you also have expectations for others based on their roles.

The expectations for appropriate behavior for the roles of student and teacher are quite apparent in the example at the beginning of this section. However, appropriate behaviors for these roles will vary greatly among cultures. In many Asian cultures, it is not acceptable to ask a teacher questions or to whisper to another student. Students are expected to stand up when the teacher enters the room and again when the teacher leaves

CULTURE CONNECTIONS

Many outsiders do not understand the simple differences in culture and lifestyle in France... or the French way of thinking. They invariably go home with stories of those quaint little Frenchies. Those who are able to adjust are changed forever and usually choose to stay, or to return when possible.

The pace of life in France is different. For me, used to a rather hectic American day, it was a relief when I slowed down and actually enjoyed my life in the French style. I had been the person who ate a peanut butter and jelly sandwich over the keyboard. I never had time to cook; my family survived on take-out, fast food, Boston Chicken and lamb chops on Sunday nights. Suddenly meals and social life were the most important parts of my day. Yes, more important than my work. In France they were the anchors to a day and to a lifetime. In the United States, we rarely made time for our friends and the people we cared most about— and they understood this, or said they did, because they were in the same jam. In France nothing was more important.

In France people sat at table for three hours (or longer), ate slowly and had real conversations. This contrasted with an eat-and-run American pattern or even the "independent dinners" we had in our family because everyone had different activities and different needs. Other than Thanksgiving, Jewish holiday feast dinners and a few dinners out with friends, I don't remember sitting at table for much more than an hour in the United States... or having discussions about politics and philosophy.

People in France made less money than those in the United States but still lived better—partly because of this slower pace of life, partly because of the cultural importance of a good meal (with good wine, *bien sûr*) and partly because, with less discretionary income, priorities were better defined. If a French person had to choose between new clothes or a concert ticket, the concert ticket usually won out.

—Suzy Gershman   ■

This Vietnamese wedding illustrates the components of social episodes: cultural patterns, social roles, rules of interaction, interaction scenes, and interaction contexts.

the room. The students would never call a teacher by his or her first name but only by a formal title.

The role of friends also varies greatly from culture to culture. European Americans have a tendency to call a lot of people "friends," and they often separate their friends into different categories based on where the friendship is established. They might have friends at work, friends from their neighborhoods, and friends from clubs or organizations to which they belong. Many of these friendships are fairly transitory and might last only as long as people work for the same organization or live in the same neighborhood. When the place in which the friendship is conducted is no longer shared in common, the friendship no longer exists in any active sense. In many other cultures, people may have fewer friends, but these friendships are often maintained for longer periods of time.

The importance of this discussion to your participation in intercultural communication should be obvious. Even though you may think you are fulfilling a particular role (such as that of student, friend, houseguest, or customer), the expectations of the role may vary widely between your culture and the culture in which you are interacting. There are also sets of rules that generally govern the interactions among people in an episode. Some of these rules are related to specific roles, but others are simply norms or guides to govern behavior.

RULES OF INTERACTION Rules of interaction provide a predictable pattern or structure to social episodes and give relationships a sense of coherence.[2] Rules of interaction are not written down somewhere, nor are they typically shared verbally. Instead, they operate at the level of unwritten, unspoken expectations. Most of the time, people are not even consciously aware of the rules that govern a social episode until the rules are broken. Think about the various kinds of rules, for example, that govern the interactions at a wedding. In addition to the various roles (bride, groom, parents, bridesmaids, groomsmen, and guests), there are many rules embedded in the different types of weddings that occur. A wedding invitation from a U.S. American couple that is engraved on heavy linen paper and announces a candlelight ceremony at dusk suggests something about the rules governing what a guest should wear and how a guest should act. In contrast, a photocopied invitation on colored paper announcing that pizza and beer will be served following the ceremony suggests a very different set of rules. Often, in weddings between people from different cultures, a portion of the "rules of interaction" from each culture is enacted. During such intercultural marriages, the rules that are typically unspoken may have been verbalized and negotiated, as the ceremony tries to honor and appreciate both cultural heritages.[3]

Rules of interaction include such diverse aspects as what to wear, what is acceptable to talk about, the sequence of events, and the artifacts that are part of the event. B. Aubrey Fisher stated:

> Virtually every social relationship has rules to determine what is appropriate and what is inappropriate for that relationship. For some relationships, the most important rules can be found in a larger social context. Meet someone at a church social,

and you will probably conform to rules appropriate to interpersonal communication in a church. For other relationships, the important rules are created during the process of interaction. After you get to know someone, you are more likely to be innovative and to do something "different."[4]

In France, for instance, you would never talk about your work at a dinner party, even if all of the people there were in some way connected to the same place of work. Among most U.S. American businesspeople, however, it is commonplace to expect talk about business at a dinner table. An invitation to a dinner party can mean that immediately upon arrival, you will be given the meal and only after you have eaten will you sit and talk leisurely with your hosts; or the invitation may mean that you must spend a substantial period of time before a meal is served in having drinks and talking with your hosts. Do you bring gifts such as flowers, wine, or candy? If so, are there particular artifacts that are taboo, such as wine or other forms of alcohol in Saudi Arabia or chrysanthemums in Italy (which are only given at a funeral)? If you are offered something to eat or drink, must it be accepted because to do otherwise would be considered an insult? (Yes, for Azerbaijanis, among others.)[5]

Even the definition of what constitutes a "meal" can vary from culture to culture. One such example is that of Doug, who is invited to dinner in the home of a Nepalese woman named Sangita and her husband, Lopsang.

> [Sangita] ushers Doug into a lower-level room of their two-story home and invites him to sit on a bench among several that surround a large coffee table. Soon, Sangita and her son, Rinji, begin bringing down food: buffalo, chicken, lentil pancakes. After about one and one-half hours of eating and talking, Doug thanks Sangita and rises to leave.
>
> In disbelief, Sangita asks, "Where are you going? We haven't had dinner yet." Doug is astounded until Sangita explains, "These were just snacks. From now on you should know that if you haven't had the rice, you haven't eaten yet!"[6]

The three ascend the stairs to the upper level of the house, where Sangita's husband greets them before a dinner table set with rice, vegetables, buffalo jelly, and dal (lentil stew). As Doug learned, the rules of interaction provide culture-specific instructions about what should and should not occur in particular social episodes.

CULTURE CONNECTIONS

"How you doin?" The young woman at the cash register smiled at him.

That was one thing Mustafa still had not gotten used to. In America everyone asked everyone how they were doing, even complete strangers. He understood that it was a way of greeting more than a real question. But then he didn't know how to greet back with the same graceless ease.

"I am fine, thank you," he said. "How are you?"

The girl smiled. "Where are you from?"

One day, Mustafa thought, I will speak in such a way that no one will ask this rude question because they will not believe, even for a minute, that they are talking to a foreigner. He picked up his plastic bag and walked outside.

—Elif Shafak ∎

INTERACTION SCENES   Interaction scenes are made up of the recurring, repetitive topics that people talk about in social conversations. Most conversations are organized around these ritualized and routinized scenes, which are the chunks of conversational behavior adapted to the particular circumstances.

Kathy Kellermann describes a standard set of interaction scenes that are commonly used by college students in U.S. universities when engaged in informal conversations. As Figure 11.1 indicates, Kellermann's research suggests that interaction scenes are organized into subsets, so that the scenes in subset 1 come before those in subset 2, and so on; however, within a particular subset, no specific order of scenes exists. Consequently, an informal conversation among acquaintances might include such topics as a ritualized greeting ("Hello!"), a reference to the other person's health ("How are you?"), a reference to the present situation, a discussion of the weather, a comment on people known in common, other common interests, a positive evaluation of the other person ("Nice to see you again!"), a reason for terminating the conversation ("I'm late for a meeting"), and finally a good-bye sequence.[7] Notice that certain scenes are part of more than one subset; these scenes function as a bridge to link the subsets together and thus help the conversation to flow from idea to idea.

Conversations among people from other cultures have a similar structure. That is, a standard set of scenes or topics is used to initiate and maintain conversations, and the conversations flow from beginning to end in a more or less predictable pattern, which is typically understood and followed by the interactants. However, there are important differences in the ways the conversations of people from other cultures are organized and sequenced, including the types of topics discussed and the amount of time given to each one.

The actual topics in an interaction scene can vary widely from one culture to another. In Hong Kong, for instance, conversations among males often include inquiries about the other person's health and business affairs. In Denmark or the French portion of Belgium, questions about people's incomes are to be avoided. In Algeria, topics such as the weather, health, or the latest news are acceptable, but one would almost never inquire about female family members. In Ecuador and Chile, it is appropriate, almost obligatory, to inquire politely about the other person's family. Among Africans, a person is expected to inquire first about a person's well-being before making a request.

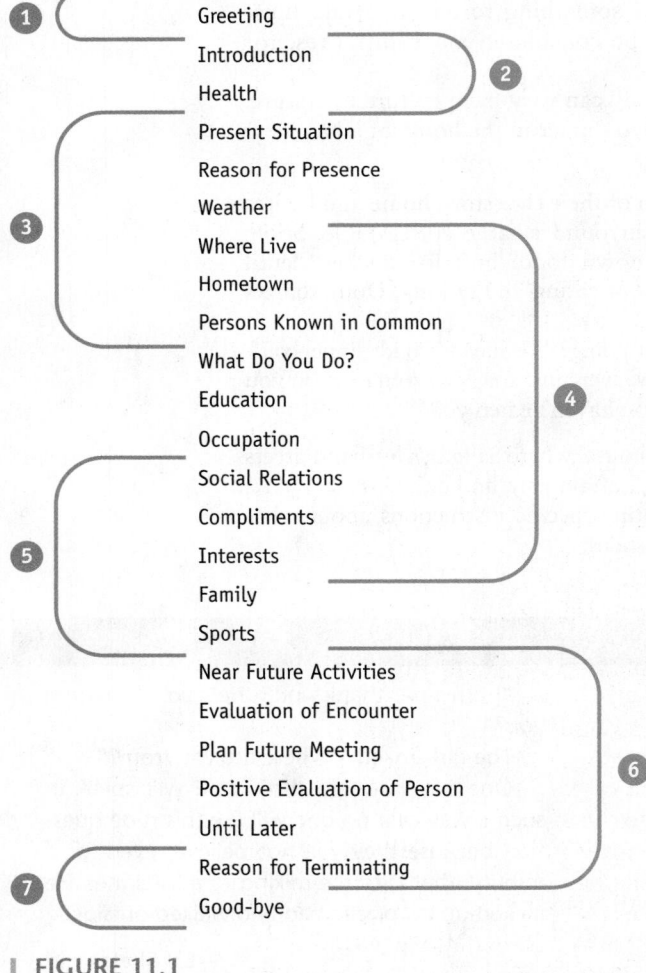

1. Greeting
   Introduction
   Health
2.
3. Present Situation
   Reason for Presence
   Weather
   Where Live
   Hometown
   Persons Known in Common
   What Do You Do?
   Education
   Occupation
4.
5. Social Relations
   Compliments
   Interests
   Family
   Sports
   Near Future Activities
   Evaluation of Encounter
   Plan Future Meeting
   Positive Evaluation of Person
6.
   Until Later
   Reason for Terminating
7. Good-bye

FIGURE 11.1

Typical interaction scenes.[8]

CULTURE CONNECTIONS

On the way, Susan gave me a quick course in Vietnamese table manners. She said, "Don't leave your chopsticks sticking up in the rice bowl. That's a sign of death, like the joss sticks in the bowls in cemeteries and family altars. Also, everything is passed on platters. You have to try everything that's passed to you. If you empty a glass of wine or beer, they automatically refill it. Leave half a glass if you don't want any more."

"Sounds like South Boston."

"Listen up. The Vietnamese don't belch like the Chinese do to show they enjoyed the meal. They consider that crude, as we do."

"I don't consider belching crude. But then, I don't belong to the Junior League."

—Nelson DeMille ■

The amount of time, and therefore the extent of detail, that is given to each conversational topic may also vary from one culture to another. For example, an Asian colleague who was working in the United States was asked one Monday morning, "What did you do over the weekend?" He began on Friday evening and listed every event that had taken place over the past two days. It was far more information than his polite U.S. American colleague wanted to know, and she spread the word around the office: "Never ask what he did on his time off!"[9] It is well known that European Americans like to get down to business in their conversations and will typically avoid elaborate sequences of small talk. Social and business conversations among the Saudis and Kuwaitis, on the other hand, will include far more elaborate greeting rituals, some phrases actually being repeated several times before the conversation moves on to subsequent sequences. Similarly, when Africans meet, they typically inquire extensively about the health and welfare of each other's parents, relatives, and family members. Although the Japanese do not typically repeat words or phrases, they also prefer to spend considerably more time in the "getting to know you" phase of social conversations.

Difficulties can arise in intercultural interactions when the participants differ in their expectations. At a predominantly African American university in Washington, D.C., for example, an encounter took place between an African American student; African students from Tanzania, Nigeria, and Kenya; and Caribbean students from Jamaica, Trinidad, and Tobago. The African American student, who did not share the others' expectations about the need for elaborate greeting rituals before making a simple request, walked into the graduate assistant's office to inquire about the time. "Hi! Does anyone know what time it is?" he asked, without any formal greetings. No one responded. After a few moments, he repeated his question, apparently frustrated. The African and Caribbean students looked up but continued with their work without responding. At this point the Nigerian student, who realized that the problems were due to incompatible expectations, responded to the first student's question. The student thanked him and left the room. When the African American had gone, the other students wondered aloud why the Nigerian had answered the question. "He has no respect," one of them remarked. "How could he walk into the room and ask about the time without greeting anyone?" another argued. Interestingly, both the African American and the other students were simply attempting to conform to their own expectations about the appropriate behaviors in an interaction scene involving strangers who are making requests.

CULTURE CONNECTIONS

The women of the groom's family tend to at least twenty simmering pots. They pour water into giant calabashes filled with dry rice and *swish-swish* rhythmically till the water goes cloudy and the dust settles. A marriage is hardly an invitation-only affair—everyone and anyone who can shows up. And everyone expects to eat. "There's no such thing as a real fête without a meal," the mayor has told me over and over. The groom is expected to provide a plate of food for an entire village of people. What would they think if they knew we throw rice *on the ground* at weddings in America?

—Sarah Erdman   ■

INTERACTION CONTEXTS   Interaction contexts are the settings or situations within which social episodes occur. Contexts impose a "frame" or reference point around communication experiences by helping people to determine what specific actions should mean, what behaviors are to be expected, and how to act appropriately and effectively in a particular interaction.

CONTEXTS FOR INTERCULTURAL COMMUNICATION

U.S. Americans are increasingly being asked to participate in social episodes within three specific contexts that we would like to highlight: health care, education, and business. Each provides an important and recurring meeting ground where people from many cultures converge and interact. We now describe in greater detail the particular importance and challenge of these three contexts.

The Health Care Context

In Chapter 1 we indicated that the need for intercultural competence arises, in part, because of the increased cultural mixing that has occurred across national boundaries and within the United States itself. The health care context affects doctors, nurses, counselors, and health care workers as well as patients, families, communities, and cultural groups. Within the health care context, there are often multiple cultures represented among those who are the medical providers, others who work in the setting, and still others who benefit from the services provided.[10] The consequences to human life and suffering from a lack of intercultural competence in the health care context should be obvious.

Communication scholars have been studying the specific characteristics of the intercultural health care context in an effort to improve communication competence.[11] Similarly, health care professionals have responded to the intercultural imperative by including courses that are designed to increase intercultural communication skills within their professional training and development programs. In fact, health care professionals are increasingly educated and trained with the goal of improving their intercultural competence. The elements of intercultural competence on which we focus— one's knowledge, motivations, skills, and social practices—are repeatedly identified as essential to the health care professional seeking to provide the best health care.[12] The nursing profession, for instance, has developed a specialization in "transcultural nursing" and has a well-established professional organization, the Transcultural Nursing Society. Similarly, the American Academy of Nursing has developed specific

CULTURE CONNECTIONS

Santana had attended Hmong funerals before and understood how important rituals were to the family of the deceased and their clan. Families often had long waits before they could bury their dead because only three Hmong funeral chapels existed in the city.

Hmong funerals normally lasted three days except for extreme cases where a family may not have enough financial resources. They believed it took that long to give a soul instruction for its journey to the world of her ancestors.

—Christopher Valen ■

recommendations on the development of intercultural competence for the nursing profession.[13] Textbooks, training materials, and studies of the competence of students and faculty alike are now common within nursing education settings, and courses in transcultural nursing are standard offerings in many undergraduate and graduate nursing degree programs.[14] Resource materials are now available to assist all health care providers as they interact with people representing a range of cultural backgrounds.[15] Indeed, as a prerequisite to their certification, many health care providers are asked to demonstrate their competence in interacting with diverse cultural groups.[16] This increased emphasis on intercultural competence extends to college health officials,[17] speech language therapists,[18] occupational therapists,[19] mental health professionals,[20] and even child psychiatric practitioners.[21]

CULTURE'S INFLUENCE ON THE HEALTH CARE CONTEXT   Cultural patterns provide the lenses through which people come to understand their world. All participants in the health care context—the providers, the patients, their families, and the larger social world—draw from their own cultural patterns and expectations about what constitutes appropriate and effective medical care. Cultural patterns often lead to very clear expectations about the right and wrong ways to treat illnesses and help people—expectations that are not necessarily shared by those from other cultures.[22] Cultures high in power distance, for example, often have rigid role expectations that result in very brief patient consultations with their medical providers.[23]

While scholars have offered several ways to conceptualize the systematic relationship of cultural patterns to health care, there is a remarkable similarity among their presentations. Three general approaches characterize beliefs about health that cultures might adopt to explain issues of illness and wellness: magico-religious or personalistic, holistic or naturalistic, and biomedical or Western.[24] These three approaches bear a strong resemblance to elements in the cultural patterns we described in Chapters 4 and 5.

In the magico-religious or personalistic approach, health and illness are closely linked to supernatural forces. Mystical powers, typically outside of human control, cause health and illness. A person's health is therefore at the mercy of these powerful forces for good or evil. Sometimes illnesses occur because of transgressions or improper actions; the restoration of health is thus a gift, or even a reward, for proper conduct. Within this approach, health and illness are usually seen as anchored in or related to the whole community rather than to a specific individual. The actions of one person, then, can dramatically affect others. Treatments for illnesses within this framework are directed toward soothing or removing problematic supernatural forces rather than toward changing something organic within the individual. Healers, who are best equipped to deal with both the spiritual and the physical worlds, perform such treatments. Some African cultures, for example, believe that demons and evil spirits cause illness.[25] Many Asian cultures also believe in the supernatural as an important source of illness.[26] Hmong people, for example, believe that

> every thing, living and nonliving, has a spirit that must not be disrespected.... A transgression against a spirit, whether it be the spirit of a living or inanimate object, will result in retribution if corrective action is not taken.... [Retributions] could be anything the person perceives to be negative, but typically, they are such things as illness, disease, injury, or death.[27]

Within the United States, cultural groups with many members who share such beliefs include various Latino and African American cultures.

In the holistic or naturalistic approach, humans desire to maintain a sense of harmony with the forces of nature. Illness is explained in systemic terms and occurs when organs in the body (such as the heart, spleen, lungs, liver, and kidneys) are out of balance with some aspect of nature. There is thus a great emphasis on the prevention of illness by maintaining a sense of balance and good health. Good health, however, means more than just an individual's biological functioning. Rather, it includes her or his relationship to the larger social, political, and environmental circumstances. Some diseases, for instance, are thought to be caused by external climatic elements such as wind, cold, heat, dampness, and dryness. Native Americans, for example, often define health in terms of a person's relationship to nature; health occurs if a person is in harmony with nature, whereas sickness occurs because a principle of nature has been violated. As Richard Dana suggests, "Healing the cultural self for American Indians and Alaska Natives must be holistic to encompass mind, body, and spirit."[28]

A common distinction within the holistic approach is contrasting both foods and diseases as either hot or cold. The classification of a disease as hot or as cold links it both to a diagnosis and to a treatment.[29] The ancient Chinese principle of yin and yang captures the essence of this distinction; everything in the universe is either positive or negative, cold or hot, light or dark, male or female, plus or minus, and so on, and people should have a harmonious balance between these opposing forces in their approach to all of life's issues.

In the biomedical or Western approach, people are thought to be controlled by biochemical forces. Consequently, objective, physical data are sought. Good health

CULTURE CONNECTIONS

I came here with three months of training under my belt. I was packed off to this village with only a collection of health-education books and a head full of vague ideas. I wanted no direction, no preconceived mission, and that's what I've gotten. I am here to see what I can make starting from scratch, and the tiny village of Nambonkaha is my ready canvas. But where do you start when health is vast and elusive at the same time? How do you promote behavior change so that people have more control over the state of their bodies but stop at the threshold where important traditions get destroyed? And how can you presume to change anything at all as an outsider with a two-year contract?

—Sarah Erdman   ■

is achieved by knowing which biochemical reactions to set in motion. Disease occurs when a part of the body breaks down, resulting in illness or injury. Doctors and nurses provide treatments by fixing the biochemical problem affecting the "broken part," thus making the body healthy again. This approach is closely linked to European American cultural patterns and has had a major influence on the development of the health care system in the United States. Indeed, the biomedical approach is so dominant within the United States that it is sometimes very difficult for individuals—providers and patients alike—to act competently in and adapt themselves to alternative cultural patterns.[30]

Often these approaches to health and health care will collide in an intercultural encounter. An example of the collision of the magico-religious approach with the Western approach occurred in a sixth-grade classroom when the teacher saw red marks on the neck and forehead of a Vietnamese student. The teacher suspected abuse, but the marks were caused by the student's parents, who had treated the student's cold by dipping a coin into oil and rubbing very hard until the coin turned the skin red. The parents believed that internal bad winds caused illness, and, by bringing the winds to the surface, a person can be healed of colds and upper respiratory problems. People from many Asian cultures hold similar beliefs.

Such health care clashes, often due to differing cultural patterns, highlight the impact of cultural differences on health care practices. An emergency room nurse, for example, took a call from a teenager whose father cut himself with an electric hedge trimmer and was bleeding heavily. Following a typical emergency room triage protocol, the nurse asked what the family was doing to treat the wound and learned that this Iranian family was treating it with honey, which they believed had the power to heal. After the family arrived at the emergency room, the nurse was surprised to see that the wound had already begun to close, and the bleeding had stopped.[31]

FAMILY AND GENDER ROLES IN THE HEALTH CARE CONTEXT   The role of the individual patient, in contrast to the role of the family, is an important difference in the functioning of health care systems. The health care system in the United States typically focuses solely on the individual patient as the source of the medical problems in need of a cure. Yet many cultures in the United States are more collectivist and group-oriented, and this difference can be the basis for serious problems and misunderstandings. Cultures that value the community or the extended family, for instance, may influence people's willingness to keep important health care appointments. Navajo women, for example, who often give priority to family members' needs, have been known to forgo clinic appointments when someone from the extended family stops in to visit and ask for help.[32] Likewise, competent treatment for Latino patients may require the involvement and agreement of other family members, not just the patient.[33]

The responsibilities of family members in the health care context differ widely across cultures. Among the Amish communities in the United States, for example, the family includes a large, extended group, with adult members of the extended family having obligations and responsibilities to children other than their own biological ones. Hospital rules that give rights and responsibilities only to members of the immediate family pose challenges when an Amish child is hospitalized. The large number of people who expect to make lengthy visits to the child may prove difficult for the medical staff.[34] Similarly, when suggesting health care intervention strategies for Pacific Islanders and Hawaiians, experts recommend focusing on the entire family, rather than on just the identified patient, in order to be effective.[35]

In many cultures, health care providers are expected to talk about the nature of the illness and its prognosis with family members but not with the patient. It is the family members, not the patient, who are expected to make decisions about the nature of treatment.[36] Of course, intercultural difficulties may occur when the family's ideas about the appropriate course of treatment differ from those of the medical staff. A Latino teenager, for example, was hospitalized at an oncology unit. Problems occurred when his family took him home for a day but did not follow the medical rules for such visits. He ate forbidden foods, did not return to the hospital at the specified time, and generally did not follow other aspects of his treatment. The medical team was upset with the family because the patient suffered a setback. The parents, however, knew that their son had only a limited time to live and wanted him to be with his family and enjoy what little time he had.[37]

Many cultures have strong expectations about modesty, and expectations about bodily displays for women can make the medical examination itself a source of intercultural difficulties. In some cultures, for instance, role requirements governing appropriate behaviors for women do not permit undressing for an examination by male physicians or nurses. Somali, Hmong, and Latina women, for example, value modesty and have strong social taboos against showing too much of their bodies to others; disrobing for a medical examination by male practitioners may be embarrassing and difficult.[38] Similarly, many women are uncomfortable revealing personal information in the presence of sons and daughters who may accompany them to the medical appointment.

Cultural differences in the role requirements that restrict interaction between women and men may also require that the medical caregivers be sensitive to important differences in needs and expectations. For example, ten-year-old Ahmed was hospitalized with complications from an appendectomy, and his mother had planned to stay all night with him in his hospital room. She became very distressed when Patrick, another boy, was brought into the room as a patient; Patrick's father also wanted to stay with his son, but the appropriate role behavior for Ahmed's mother precludes interactions with men outside of her family. Fortunately, the hospital staff was sensitive to the cultural issues; they recognized the importance of the problem and moved Ahmed to a private room.[39]

CONVERSATIONAL STRUCTURES, LANGUAGE, AND NONVERBAL COMMUNICATION   Because of different interaction rules, the medical interview between caregiver and patient can be another source of intercultural communication problems. Latinos and Arabs, for example, may engage in extensive small talk before indicating their reasons for the medical visit. Interviews with Native Americans may be punctuated with extensive periods of silence. Medical interviewers may consider such small talk or silence as a "waste of time" rather than a vital component of the person's cultural pattern that affects his or her comfort level and willingness to proceed with the interview.[40] Similarly, direct and explicit discussions with many Asians and Asian Americans may pose serious threats to their face, and the use of indirection or other face-saving strategies may be preferred.[41]

In many cultures, doctors are perceived as authority figures with whom one must agree in the face-to-face medical interview. A patient may know that he or she will not be able to follow a proposed treatment plan but may be reluctant to respond to the doctor in a way that might appear to be a challenge to the doctor's authority.[42] Similarly, individuals from cultures that see health care workers as authority figures will be reluctant to initiate interaction and ask questions.[43] Patients from individualistic and low-context cultures, for instance, often feel that it is very important to communicate verbally with their physicians, and they are therefore very motivated to do so. Conversely, patients from collectivistic and high-context cultures may be much more apprehensive about participating as a patient in their medical care, and they may therefore avoid conversing with their physicians during medical interviews.[44] Latinos, for example, may not want to provide direct answers to questions posed by the health care provider.[45]

A major challenge arises when patients and medical care workers do not even speak the same language.[46] Large urban hospitals and health care offices reflect the increasing multilingual characteristics of the United States. Interpreters often play a key role in allowing health care workers to communicate with patients.[47] Health care education programs offer nurses, doctors, and other health care professionals many pathways to improve their ability to interact with patients who primarily speak other languages.[48]

Ambiguities in the use of language can present additional difficulties in diagnosing and treating illnesses. Idiomatic language in the health care context can create misunderstandings, such as when a nurse indicated to a Chinese-born physician that a patient had "cold feet" about an upcoming surgery and the physician, seeking to rule out circulatory problems, ordered vascular tests.[49] Or consider the case of a pharmacist who instructs a patient to "take three pills a day at mealtime" and expects that the patient will take one pill at each of three meals. Patients who come from cultures that do not separate their day into three major mealtimes may instead take all three pills simultaneously at the one large meal that is eaten every day.[50] Or consider the patient with limited English-language ability who is told that the results of his medical tests are positive. Is this good news or bad?[51] A misunderstanding with possible very serious ramifications happened to a nonnative speaker of English who had signed a consent form for a tubal ligation, or "having her tubes tied." The woman thought that "tied" meant that the procedure could be reversed should she later decide to have more children. Only the skilled intervention of an interculturally competent health care worker, who understood the ambiguity in the language and clearly explained to the woman the consequences of her decision, prevented a medical procedure that was not wanted.[52]

It is not just ambiguities in understanding and in translating the English language that challenge health care workers. The Spanish language, for example, which is used by several U.S. cultures, has many words and grammatical constructions that vary from one cultural group to another.[53] Such cases highlight problems of intercultural misunderstandings and the implications for "informed consent" in the intercultural health care context.

Nonverbal communication can also pose unique challenges in the health care context. Eye contact, for example, can have a multitude of meanings. Consider an interaction between a European American health care professional and a Nigerian patient. To the European American, direct eye contact is expected; when not given, individuals are often regarded as untrustworthy and disrespectful. To the Nigerian, who comes from a large power-distance culture, direct eye contact may be avoided to show respect. Similarly, the appropriate use of touch varies widely across cultures, and, therefore, health care professionals must adjust their use of touch as healing to be appropriate to the cultures of their patients. For Haitians, touching during conversations would be expected.[54] European Americans will touch the head of a child as a sign of friendliness,

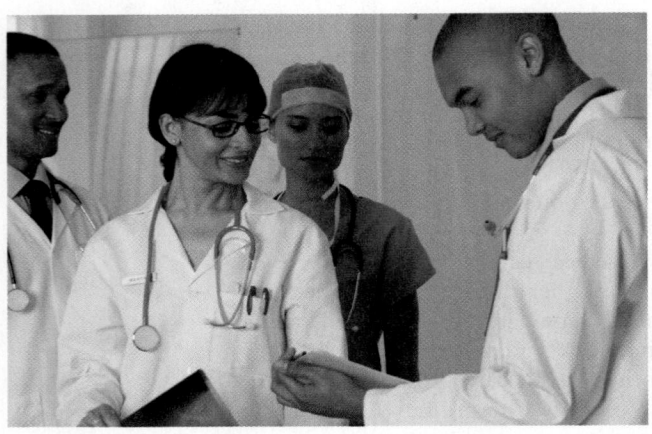

A multicultural medical team meets to discuss their patients' needs.

yet for some Southeast Asians such a gesture would be understood as an insult, since the head is the locus of one's soul.[55] Similar difficulties in the health care setting can arise for other aspects of nonverbal communication. As you reflect on these differences, you will understand the importance of intercultural competence.

The effective treatment of a patient's pain by the health care professional requires an ability to interpret the patient's nonverbal and verbal codes so that a culturally appropriate medical intervezntion can occur. In many cultures, for instance, individuals are taught to be more stoic and circumspect in verbally identifying the severity of their pain. In other cultures, there is an expectation that one will use very emotional and dramatic terms to describe one's experience of pain.[56]

INTERCULTURAL COMPETENCE IN THE HEALTH CARE CONTEXT Health care professionals assume a special responsibility in ensuring that they understand their patients in order to treat them effectively. This responsibility requires a willingness to attempt to understand the cultural patterns—the beliefs, values, interaction norms, and social practices—of their patients. There are excellent reference books now available to health care professionals in which the general characteristics of various cultures are presented. However, in the health care context, as in all others, you must remember that each individual may or may not share the preferences of her or his cultural group.

The Educational Context

The U.S. educational system—from pre-kindergarten to college and on through graduate or professional school—increasingly requires competent intercultural communication skills from all of its participants. Because of the culturally diverse student populations throughout the U.S. educational system, people must give increased attention to the factors that students, parents, teachers, administrators, other educational professionals, and ordinary citizens face when challenged to communicate in the educational context.[57]

CULTURE'S INFLUENCE ON THE EDUCATIONAL CONTEXT All participants in the educational context—teachers, students, parents, school administrators, and other staff—bring their cultures' beliefs, values, norms, and social practices with them. Differences in cultural backgrounds may produce developmental variations in children's cognitive, physical, and motor abilities, as well as in their language, social skills, and emotional maturity.[58]

Communication within a classroom, on the playground of an elementary school, or within a college dormitory is typically governed by a set of rules based within one cultural group. In the United States, the dominance of the patterns associated with the European American culture pervade the educational system, and they set the expectations for teachers, administrators, and students about how to behave and learn effectively and appropriately.[59] Yet, for most teachers in U.S. schools, it is an everyday occurrence to have students who come from cultural backgrounds other than their

CULTURE CONNECTIONS

Similarly, and perhaps not surprisingly, both my graduate students and many undergraduates in Rwanda and in Graz also struggled to find a voice in my classes, since most had encountered only traditional lectures in the past. I succeeded in getting them to express opinions contrary to the textbooks or ones even simply questioning my own assertions only by breaking them into small groups and assigning them a sort of "devil's advocacy" role that they found unusual and at first very uncomfortable (though one they soon approached with enthusiasm). Upon returning to the States, I noted again how our students are far more active participants in the classroom and are far less reticent at challenging the perspectives of their teachers and peers. While our students' knowledge base vis-à-vis world events may be thin in comparison to their European and African counterparts, their ability to engage critically with each other, with popular culture, and with the latest trends in technology and identity-political issues is much sharper.

—Donald E. Holl ■

own. The demographic profiles of students in U.S. schools routinely identify many who speak different languages at home and who come from a large range of cultures. As a student in high school, or now in college, we anticipate that you may have already experienced classrooms of people from diverse cultural backgrounds. Alternatively, you may be a parent interacting within your children's school system. Students in intercultural communication classes may be preparing to teach in elementary or high schools. As the statistics presented in Chapter 1 demonstrate, many current teachers, and certainly most future teachers, will work in a setting that demands the knowledge, motivations, and skills of a competent intercultural communicator. In the words of Janet Bennett and Riikka Salonen, the U.S. campus today is "culturally complicated."[60]

For many students, attending school can itself be an intercultural experience. Elvira, for example, is a junior Filipina American student who, on a daily basis, crosses the cultural boundary from her Filipino home to her U.S. American high school. Although she attends regularly and receives very high grades, she is concerned that the school experiences cut her off from her sister and friends. Sonia, similarly, is a Mexican American high school student who is very popular with her Latina friends but consistently feels like an outsider at school. This makes it very difficult for her to be academically and socially successful in the educational context.[61]

Scholars in communication and education have begun to document the many ways that cultural differences can lead to dissimilarities in interpretations and expectations about competent behaviors

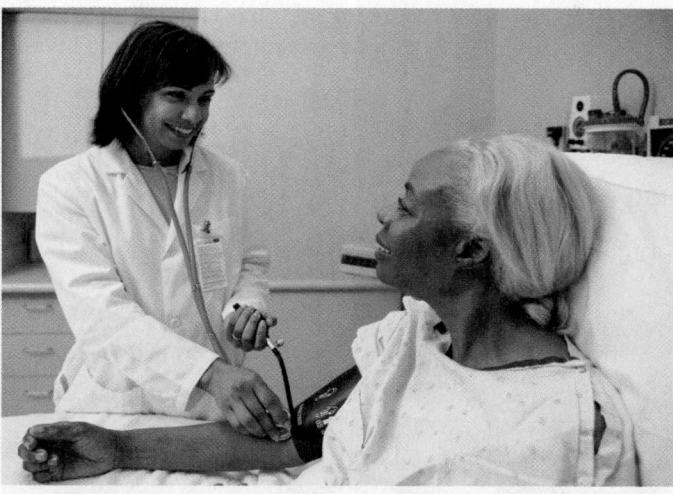

Because health care may require the display of and access to one's body in ways that are private and personal, cultural expectations about what behaviors are permitted or prohibited are particularly important in such settings.

for students and teachers. Problematic issues include differences in expectations concerning such classroom behaviors as the rules for participation and turn taking, discipline and control, and even pedagogical approaches such as lectures, group learning, and self-paced work. Intercultural problems also arise when parents and other family members attempt to communicate with various officials representing the school.

THE ROLE OF THE TEACHER   Recall from Chapter 4 that cultures differ in the ways they choose to define activities, social relations, the self, the world, and the passage of time. All of these choices can influence preferences for how students and teachers relate to each other in the classroom.[62]

Teachers have a unique and powerful influence on student interactions in the classroom and beyond. Consequently, classroom teachers have a potent effect on how well students from different cultures learn. Teachers come to the classroom carrying with them both their unique personality characteristics and the influences of their culture. Increasingly, scholars and practitioners in the United States and throughout the world are recognizing the unique challenges of multicultural classrooms, and substantial attention is now being given to ways that teachers can be more effective and appropriate—that is, more competent—in their adaptations to the learning styles that characterize the range of their students' cultures.

Teachers, then, have a particular responsibility to demonstrate intercultural communication competence. Often the first step for teachers is to become more aware of cultural differences and then adapt when and how they approach their students. For example, an African American teacher of French, in what had been a predominantly European American East Coast suburban high school, described his experiences in teaching a changing, more multicultural student body:

> When I first found myself teaching classes of mostly black kids, I went home frustrated every night because I knew I wasn't getting through to them, and they were giving me a hard time. It only started getting better when I finally figured out that I had to reexamine everything I was doing.[63]

The intercultural nature of classrooms, from elementary school through college, requires educators to adapt their approaches to teaching and learning.

This teacher began by being aware of cultural differences and then changing his approach to teaching.[64] Students training to become teachers in elementary and secondary schools in the United States now are routinely required to take courses to learn how to adapt their classroom teaching in ways that maximize learning for culturally diverse students.[65]

Those who prefer a more hierarchical relationship between individuals will structure the relationship between student and teacher with greater status differences. German instructors, for example, tend to be more formal, aloof, and socially distant than their U.S. counterparts.[66] Similarly, within many Asian and Asian American cultures, teachers are highly revered and respected. Students and parents would not openly and directly question the authority and statements of a teacher. Consider, for example, the types of communication messages and the proper role behaviors of students and teachers in Chinese classrooms. If you are familiar with U.S. classrooms, compare your experiences with the following:

> Students who are late for class should get the teacher's permission to enter the classroom. Even in college, students have to sit quietly in rows that face the teacher, listen attentively, and take careful notes. Students must also raise their hands and stand at attention when they answer or want to ask questions. Not raising a hand is a violation of classroom rules, and not standing up is a violation of the reverence rule.[67]

As this description conveys, the Chinese classroom is characterized by a high degree of formality. Many people from cultures with similar preferences for formality are shocked to find European American teachers' penchant for informality. Many U.S. professors, for instance, encourage students to call them by their first names; while many students prefer such informality, some feel uncomfortable because it suggests disrespect.

Even a teacher's seemingly inconsequential personal preference has the potential to create discomfort among students and their parents, often without the teacher even knowing about it. One such example is of a woman who was considered one of the best English teachers in her school but who used red ink to address her notes of encouragement to the students. For Koreans, particularly Buddhists, the teacher's "insignificant" preference to write the students' names in red ink created enormous distress, because a Buddhist only writes people's names in red at their death or at the anniversary of their death.[68]

CLASSROOM INTERACTION Cultural characteristics also influence what is appropriate and effective communication within the classroom. That is, culture shapes what is considered to be desirable and undesirable classroom behaviors. From the expectations for students interacting with their teachers, to the manner in which they relate with one another, to the language and topics considered appropriate for teachers and students to discuss, to the overall structure of interaction within a classroom—culture affects perceptions of competent classroom interaction.

Students from collectivistic cultures are generally more accepting of messages about appropriate classroom behaviors and will comply with teachers' requests about classroom management.[69] Even the nature of teachers' persuasive messages differs across cultures. Chinese college teachers, for instance, appeal to the group in gaining student compliance, whereas European American teachers, with a cultural preference for individualism, stress the benefit to the specific student.[70] Within the classroom, the treatment of personal property is also influenced by the culture of the students, with "personal" items such as toys, books, and clothing perceived very differently in individualistic and in collectivistic cultures.[71]

This typical preschool classroom depicts the intercultural character of many schools across the United States.

CULTURE CONNECTIONS

Karen Randolph had been teaching high school English in the United States before she accepted a teaching job at a teacher's college in China. She found her new environment and her new teaching assignment exciting. Both her students and her colleagues seemed a bit shy of her, but Karen was sure that in time they would all come to be friends.

In the classroom, however, Karen was very frustrated. When she asked a question, the class was silent. Only if she called on a particular student would she get an answer, often a very good one. She could not understand why they wouldn't volunteer when they obviously knew the answers. They were very quiet when she was speaking in front of the class, and never asked questions, let alone interrupt with an opinion, but as soon as the class ended, they would cluster around her desk to ask their questions one-by-one. They would also offer their suggestions about the lesson at this time.

Karen often asked her students to work in small groups during class, especially when they were editing each other's writing. They were slow to move into groups and when they did, they often simply formed a group with the people sitting next to them. Finally she devised her own system of forming groups to get them to interact with students sitting in another section of the classroom.

Most frustrating of all, after she taught her class how to edit essays, she found that the students were likely to write vague and not very helpful remarks on their classmates' papers. They would say nice things about the essays and correct small grammatical errors, but seemed unwilling to criticize them in a way that would help another student revise the essay. They usually accepted her criticism of their writing with good spirits and promises to improve. In fact they frequently asked for more correction of their English from her than she wanted to give. She felt that one hundred per cent grammatical correctness was not as important as learning how to correct what they had written on their own and with the help of others. After all, they would not always have a teacher to tell them what was good and not so good about their English writing.

—Linell Davis   ∎

Classroom discussion and participation also vary greatly across cultures. Donal Carbaugh, who studies how culture is displayed in people's conversations, makes a comparison between European American expectations for classroom interaction and those of the Blackfeet, one of the Native American tribes in the United States. These differences are displayed in Table 11.1. It is easy to imagine a classroom with a European American teacher and students from both cultures; in a typical classroom with European American norms, the European American students would feel comfortable with their learning experience, whereas the Native American students would be hesitant to volunteer, speak out, or raise concerns unless the teacher specifically calls on them by name.[72] Similarly, many Asian and Asian American students rarely ask the teacher directly for clarification; to do so might be regarded as a challenge to the teacher's authority and could threaten his or her face should the answer not be known. Korean students are often unwilling to talk with their teachers even when the teachers have incorrectly calculated the students' scores on an exam.[73] Like Native Americans and Korean Americans, Cambodian American students often display classroom behaviors that are consistent with their cultural backgrounds, rather than with the dominant European American expectations. The contrasting expectations for these behaviors are displayed in Table 11.2.

TABLE 11.1

BLACKFEET AND EUROPEAN AMERICAN COMMUNICATION IN A CLASSROOM SETTING[74]

	Blackfeet	European American
Primary Mode	Silence	Speaking
Cultural Premise	Listener-active, interconnected	Speaker-active, constructive
Secondary Mode	Verbal speaking	Silence
Social Position	Differences by gender and age	Commonality, equality
Typical Speaker	Elder male	Citizen
Cultural Persona	Relational connection	Unique individual
Values	Nature, heritage, modesty, stability	Upward mobility, change, progress

Students from many cultures who go to school in the United States may find it difficult to adapt to the verbal style expected of them. Conversely, when U.S. students study overseas, they often experience difficulties in understanding the cultural expectations related to the educational context.[75] Yet a willingness to speak in class

TABLE 11.2

CONTRASTING CULTURAL BELIEFS ABOUT EDUCATION[76]

Cultural Beliefs of European Americans	Cultural Beliefs of Cambodian Americans
Children should develop a personal point of view about the world. This prepares them for a society in which individuality is prized and democracy is the ideal.	Personal views are less important than the culture's accepted views. Children's personal views are not valued until the children reach maturity.
Questioning (and even challenging) the teacher's ideas plays a vital role in learning.	Children should not question what adults tell them, even if they do not understand it well.
Learning proceeds *whole* to *part*. Children first need to see the big picture; then they can analyze the whole to discover the relevant parts and make connections.	Learning proceeds *part* to *whole*. Children first need to learn the parts; then they can synthesize them into a coherent whole as they mature.
Children are naturally curious and want to learn; motivation is inherent in the individual.	Children are naturally curious and want to learn; a child's motivation is subordinate to cultural expectations and adult authority.
Mistakes are a natural part of learning and should be integrated into the classroom experience.	Mistakes that are made in front of others cause one to lose face.
Students are expected to be active, independent learners by questioning, making connections, and articulating their evolving understandings.	Students are expected to conform to the pace and structure of learning that is orchestrated by adults.

is a communication characteristic highly valued by European American teachers and students, whose cultural framework celebrates individual achievement and responsibility. To students from cultures that emphasize the collective good and the maintenance of face, however, such behaviors in the classroom are too competitive, as they disrupt the group's harmony and separate people from one another.[77] Native American fifth and sixth graders, for example, perceived their high-verbal teachers to have less competence in their oral delivery of messages.[78] Similarly, African American children, whose culture emphasizes the development of verbal skills and expressiveness, are often affected in their classroom interactions with their European American teachers:

> In both verbal and nonverbal language, they [African American children] are more theatrical, show greater emotion, and demonstrate faster responses and higher energy.... African-American speakers are more animated, more persuasive, and more active in the communication process. They often are perceived as confrontational because of this style. On the other hand, the school, and most Anglo-American teachers, are more oriented toward a passive style, which gives the impression that the communicator is somewhat detached, literal, and legalistic in use of the language. Most African-American students find this style distancing and dissuasive.[79]

Another example of cultural consequences on the learning environment can be seen in the research of Steven T. Mortenson, who found that, while both Chinese and U.S. American students responded to academic failure in similar ways, the Chinese students were less likely to express their concerns about academic failure to others and instead were more likely to experience physical illnesses.[80]

Turn taking within the classroom is also governed by cultural expectations. Watch how teachers in your various classes regulate the flow of conversations and contributions. A teacher has a particular set of expectations about who speaks in the classroom as well as when and how to speak. Is it acceptable for students to talk among themselves? How loudly can they talk to each other? How long can private conversations continue before the teacher asks for them to stop? How do students get permission to speak in class? All of these classroom behaviors, which are crucial to how teachers evaluate their students and how students evaluate teachers and classroom environments, are grounded in cultural expectations.

Cultural patterns directly affect preferred ways to learn in the classroom. Think for a moment about the classroom experiences you have had. Did they encourage students to work cooperatively in groups? Or were classroom activities designed to encourage students to work alone, succeeding or failing on their individual merits? Latino children, whose culture teaches the importance of family and group identities, are more likely to value cooperativeness than competitiveness.[81] Because Native American cultural patterns emphasize the group, harmony with nature, and circularity, children from that culture often respond better to learning approaches that are noncompetitive, holistic, and cooperative.[82] European American children, in contrast, often prefer learning approaches that emphasize competition, discrete categories for information, and individual achievement.

FAMILIES AND THE EDUCATIONAL SYSTEM   Another key set of relationships in which competent intercultural communication is essential is in the interaction of parents, and sometimes other family members, with school personnel, including teachers, administrators, and others. Consider the following examples of how parents from different cultural backgrounds with children in kindergarten classrooms interacted with their children and the classroom:

> In Ms. Nelson's kindergarten classroom, some of the Chinese-American mothers come to school every day at lunch time. They bring hot lunches and hand-feed their five year old children. In another kindergarten classroom, Mexican American mothers walk their children to school. When the bell rings they enter the classrooms with their children. They walk the children to their tables and help them take off their jackets. They hang their children's jackets and book bags on the hooks, generally located in the back of the room, before leaving the classroom. When parent-teacher conferences were held in early October, a European-American mother proudly told her child's teacher that Elizabeth could tie her own shoes when she was four years old.[83]

The range of differences in parental involvement with their children at this grade level indicates the wide range of meanings and expectations that parents from different cultures can have of teachers in just one school and one classroom.

Teachers' evaluations of their students' classroom behaviors are also heavily influenced by cultural expectations. Judgments that Asian Americans are "too quiet," or that African Americans are "too animated," are anchored in teachers' cultural preferences. If teachers make such judgments, it will decrease their ability to work with parents to improve their students' overall learning.[84]

Because the value of education differs from one culture to another, the importance of a student's success in school will also vary. For Thais and Filipinos, for instance, education affects the entire family's status and social standing. By excelling in school, therefore, children bring honor to their families while preparing for future successes that will further enhance the family stature. Education is thus a family concern, rather than an individual achievement.[85]

Even the need for the customary parent–teacher conferences may not make much sense to parents from cultures in which there is no expectation that parents will play an active role in decisions about their children's education.[86] Many Middle Eastern parents, for instance, expect their children to do well in school. Thus, when the children actually do well, there is generally less overt praise or material reward than is common in the United States; the children are doing what is expected of them. However, when the children do not do well, parents may present a variety of attitudes including denial, blaming the school, blaming the child, and feeling ashamed.[87]

Similar expectations exist among many Asian and Asian American parents. A teacher's request for a routine conference, for example, may be met with a sense of skepticism or a deep concern that a disobedient child may have dishonored the family. Because of face-saving needs, the parents may even assume that the exact nature of this problem will not be stated explicitly but must be discerned through a clever analysis and interpretation of the teacher's subtle clues. The teacher's bland statements that their child behaves well are therefore regarded as merely a social politeness. Not wanting to heap unlimited praise on the child for fear of setting false expectations, the teacher may unwittingly provide the parents with just the sort of high-context hints and generalities about the child's faults and weaknesses that they will interpret as an indication of a deeper and more difficult problem in need of correction.

A poignant example of the consequences of differing cultural expectations, complicated by linguistic difficulties, is the story of Magdalena, a Mexican immigrant mother, and her son Fabian. Because Fabian was not behaving appropriately in school, the

CULTURE CONNECTIONS

...it has become apparent that teaching difficulties are not simply a question of language but are rooted in profound cultural differences.

For example,... "here in the northern part of Europe, we place great emphasis on autonomous learning, and expect students to work independently and critically present the information they are presented."

Elsewhere in Europe, teaching methods emphasize students' listening to lectures, taking notes, demonstrating their learning through written tests, and being able to repeat what they have learned from professors' lectures.

Moreover, information that some students might grasp immediately could leave those from another culture befuddled. Students'

ways of handling that kind of confusion also vary widely.

"In a lot of Asian cultures, there is a big thing about not losing face," Ms. Lauridsen notes, "and students don't want to admit that they don't know what they need to know."

In such cases, even if students do end up trying to confront the issue head-on, meeting with their professor might not solve the problem. Because many Asian students tend to "always nod and smile like they have understood," she says, "it takes a while to decode some of these students, for those of us who are used to more direct ways of interacting."

—Aisha Labi ∎

school officials asked Magdalena to have Fabian evaluated by a professional. Concerned about Fabian and wanting to be responsive to the school's request, Magdalena took Fabian to see their family doctor. As the situation at the school became more negative, the teachers and administrators believed that Magdalena was ignoring the seriousness of the problem and was not responding to their request. Ultimately, Fabian was expelled from school.[88] As Jerry McClelland and Chen Chen conclude:

> The combination of the school's instructions, the interpreter's translation, and her comprehension of the message resulted in Magdalena not understanding that a counselor's report, rather than a physician's report, was being requested. Given the Mexican culture in which she grew up, Magdalena was puzzled by the message to have Fabian checked. Magdalena said that in Mexico, if there is a problem with a child, the teacher and parent talk to each other and do not bring in a third person to give an opinion.[89]

INTERCULTURAL COMPETENCE IN THE EDUCATIONAL CONTEXT   The challenge to develop one's intercultural competence and fulfill the promise that cultural diversity brings to the educational context is aptly summarized by Josina Macau. She suggests that creating a constructive learning environment in an age of cultural diversity requires that people be sensitive to different and sometimes competing experiences.[90]

The intercultural challenges of communicating in a different language, which we described in Chapters 7 and 9, are also prominent within the educational context. Increasingly, schools in the United States are finding it advantageous to translate the letters that are sent home to parents so that immigrant parents are better able to learn about key school issues and events. Even then, the unique features of each language, the presence of colloquialisms, and the use of specialized terminology still present barriers for parents whose primary language is not U.S. English.[91]

The starting point for developing intercultural competence in the educational context is to understand one's own cultural background. It is particularly important that teachers and administrators recognize their culture's influence on expectations about how class-rooms should operate and how students should behave. The stakes for developing inter-cultural competence in education are very high. Although the following example focuses on Native Americans, it is equally true of students from a variety of cultural backgrounds. It illustrates the importance of the educational context and the potential for both perma-nent and harmful consequences as a result of interactions within that context.

> When many young Native American children enter the classroom, they frequently find themselves in foreign environments where familiar words, values, and life-styles are absent. As the classroom activities and language become increasingly different from the familiar home environment, the students suffer a loss of self-confidence and self-esteem, a loss that is sometimes irreparable.[92]

The Business Context

Like the health care and educational contexts, the business context is now intercul-tural. From customers, to employees, to markets, to financing, and even to organi-zational structures, businesspeople now work in a world in which success requires intercultural competence.[93] Commerce and trade are global and affect us daily. Books, training programs, Web sites, and courses abound with information, advice, and skills necessary to navigate our global world with intercultural communication competence skills.[94] Just look at your possessions and you will see ample evidence of the products that have crossed national and cultural boundaries. People, however, are the key ingre-dients in the intercultural business world.

Throughout most of your working life, you will likely be within an intercultural business context; some of your customers, coworkers, supervisors, and subordinates will come from cultures that differ from your own. Many organizations today include people who were born in one country, educated in another, and work in yet another. Many projects—by architects, engineers, businesspeople, medical personnel, financial managers, and others—are done in one country during their daytime and electronically transmitted across the globe for work during the daytime of another culture's workforce. Intercultural communication in the business context is also increased by the availability of easy, fast, and inexpensive communication. As a recent analysis suggests, by the year 2020 "our office will be everywhere; our team members will live halfway around the world."[95] Employees of multinational business organizations are now able to conduct their work using email, chat boards, and verbal interaction over various Internet services. Similarly, most of the major cities in the world can be reached within a day or two; these technological innovations make communication quick and sometimes inexpensive, but they also can be potentially problematic because of intercultural differences.

Many businesspeople are inadequately prepared to take an intercultural assignment or to work in an intercultural team. The critical need to coach individuals to work in intercultural settings has spawned a large industry of professionals who provide training to companies and to individuals who are about to have an intercultural assignment.[96] A Google search for "intercultural training" or "intercultural preparation" will provide evidence of the prevalence of this instruction. Many corporations now hire training firms to improve the productivity of their operations with their intercultural workforce.[97]

CULTURE'S INFLUENCE ON THE BUSINESS CONTEXT The discussion of cultural patterns in Chapters 4 and 5 is particularly useful in understanding culture's influence on the business context. Differences in cultural patterns create widely dissimilar expectations

Work settings are increasingly culturally diverse, providing opportunities to improve intercultural competence.

for how a business is structured and what is considered appropriate and effective—and therefore successful—communication within that business. Chapter 5 described several taxonomies that depicted the dimensions of cultural patterns, which directly apply to business organizations around the world. While all of the dimensions described are useful for understanding issues in the intercultural business context, our discussion will highlight differences in business practices that are related to the individualism–collectivism dimension.

Cultural variations in people's relationships to their organizations are important in understanding the intercultural business context. Is the critical unit of analysis and of human action the individual or the group? Specific areas of intercultural business that are associated with variations in individualism–collectivism include the following:

■ *Who speaks for the organization?* In organizations within individualistic cultures, a single person may represent a company in its negotiations. In collectivistic cultures, a group of representatives would likely be involved in negotiations.

■ *Who decides for the organization?* Organizations within individualistic cultures likely empower their negotiators to make decisions that are binding on the company. Such decisions are often made rapidly and without consultation from the home office once the negotiations have begun. Organizations within collectivistic cultures may require extensive consultations among the delegation members and with the home office at each step in the negotiation process, as no single individual has the exclusive power for decision making.

■ *What motivates people to work?* Do people work because they are motivated by the possibility of individual rewards, as is common in individualistic cultures, or is group support and solidarity with one's colleagues a primary motivator? Reward systems to encourage employees' best efforts vary widely. In Mexico, for instance, though the individual is valued, rewards for independent actions and individual achievements that are successful with U.S. Americans may not be strong motivators. Thus, production contests and "employee of the month" designations to encourage Mexican employees are often unsuccessful.[98]

■ *What is the basis for business relationships?* In collectivistic cultures, it is vital that businesspeople establish cordial interpersonal relationships and maintain them over time. The assumption that it is possible to have a brief social exchange that will produce the degree of understanding necessary to establish a business agreement is simply incorrect. In many African cultures, for example, friendship takes precedence over business. Similarly, most Middle Easterners extend their preference for sociability to business meetings, where schedules are looser and the first encounter is only for getting acquainted and not for business.[99] A similar regard for establishing social relationships as a prelude to doing business is also common in China, Korea, and Japan.[100] Indeed, the very notion of trustworthiness differs across intercultural business relationships, with individualistic cultures often emphasizing personal integrity in judging another's trustworthiness while collectivistic cultures emphasize one's commitment to the group or the organization.[101]

A scholar-practitioner who has investigated the impact of cultural patterns on communication competence in the intercultural workforce is Fons Trompenaars.[102] After many years of studying companies around the globe, Trompenaars identified the cultural dimension of universalism–particularism as especially useful in understanding how business practices vary because of culture. Universalistic cultures prefer to make business decisions based on a consistent application of rules, whereas particularistic cultures choose instead to adapt the rules to specific circumstances and relationships. William B. Gudykunst and Yuko Matsumoto indicate that universalism–particularism is related to the individualism–collectivism dimension. They suggest that businesspeople from individualistic cultures tend to be universalistic and apply the same value standards to all, whereas those from collectivistic cultures tend to be particularistic and apply different value standards to ingroups and outgroups.[103] Some features of the impact of this variation on the conduct of international business include the following:

- *What is the meaning of a contract?* Someone from a universalistic culture may view the signed contract as binding on all, whereas someone from a particularistic culture may view the contract as valid only if the circumstances remain unchanged, which may include whether the person who signed the contract is still part of his or her company. For example, the Chinese concept of legal or contractual agreements differs from the U.S. concept. In the United States, of course, a business contract is binding and should be implemented precisely as agreed. In China, however, contracts are sometimes regarded more as statements of intent rather than as promises of performance. Therefore, they are binding only if the circumstances and conditions that were in effect when the contract was signed are still present when the contract should be implemented.[104]
- *Are job evaluations conducted uniformly or adapted to specific individuals?* Within universalistic cultures, all individuals in similar jobs are evaluated using standardized criteria. Within particularistic cultures, the performance criteria depend upon people's relationships with others and their standing within the organization.
- *Are corporate office directives typically heeded or circumvented?* In universalistic cultures, directives from corporate headquarters are valued and are heeded throughout the organization. In particularistic cultures, such directives are often ignored or circumvented because they don't apply to the particular circumstances of a specific subsidiary or branch office.

Just as in the educational and health care contexts, every element of the business context can be influenced by culture. In the remainder of this section, we first consider the important functions of business negotiations and conducting business deals. We next focus on the interpersonal work environment of intercultural teams. Finally, we look at cultural differences in conversational structure, in language, and in nonverbal communication that influence the business context. Though we separate these topics for ease of discussion, we do want to remind you that, in actual communication in the business context, their impact on culture forms an interconnected whole that cannot be understood as a set of unrelated parts.

BUSINESS NEGOTIATION AND DEAL MAKING   In the business context, the core activity of much intercultural communication involves negotiations to make a business deal. These interactions range from large multinational corporations that are discussing

CULTURE CONNECTIONS

From the U.K. about South Korea:

I work for a company that has a subsidiary in Korea, so I go there quite frequently on business. I have a good relationship with a highly esteemed Korean manager who has worked for us for many years. We now need to find a local firm to supply a component, and this manager has strongly recommended his brother's company. I am rather wary about this and feel that this manager has put me in an awkward position. I don't want to lay myself open to accusations of favoritism or even corruption. Anyway, I am going to put the job out to bid, but feel that by doing so I am risking creating ill feeling within my own company.

In South Korea loyalty to one's family is a duty, and your manager would be failing in his duty if he did not try to help his brother win the order. On the other hand, he appears to be a loyal member of your company too, and may genuinely feel that his brother's company is likely to be your best supplier. Certainly personal relationships can facilitate business wherever in the world you find yourself, and in Eastern Asia knowledge of someone's background and family is seen as providing a form of guarantee of their personal commitment to your business. Naturally you should listen to what other companies have to offer, but be prepared to spend time in discussions with the company your manager recommended. It would be silly to exclude the best contender because of an exaggerated sense of fairness, and certainly most Indians, East Asians, and Latin Americans would find such a decision totally incomprehensible.

—From a letter to Gwyneth Olofsson ■

multibillion-dollar contracts, to the small business owners who want something manufactured in another country, to salespeople who are selling products to people who are culturally different from themselves.

Cultures differ in the preferred flow or pacing of business negotiations.[105] In the initial stages of a negotiation, for example, German business managers may ask numerous questions about technical details. In Scandinavia, there is a great deal of initial frankness and a desire to get right down to business. Among the French, however, the early emphasis is on laying out all aspects of the potential deal. In contrast, many Italian and Asian managers may use these same initial stages to get to know the other person by talking about subjects other than the business deal. Likewise, preliminaries in Spain may take several days.[106] Similarly, problems often characterize Mexican and U.S. American negotiations, which arise from a greater emphasis on relational concerns by Mexican negotiators and on task behaviors by U.S. negotiators.[107] Many Africans also want a friendship to be established before doing business.[108]

In the United States and in other Western countries, where individual achievement is valued, advancements occur because of one's accomplishments, there is a shorter-term and results-oriented approach to negotiating, and a high priority is placed on getting the job done and accomplishing task-related objectives. Interpersonal communication is typically direct, confrontational, face-to-face, and informal. Negotiating teams are willing to make decisions and concessions in the public negotiation setting, where individuals within the team may disagree publicly with one another. One individual is usually given the authority to make decisions that are binding on all.[109] Russians, though, prefer a negotiation style that is more indirect and that may not address issues straightforwardly. To Russians, U.S. Americans often appear too confrontational and aggressive.[110] Among the Japanese, however, who value group loyalty and age, advancement is based on seniority, there is a long-term approach to negotiating, and the formation and nurturance of long-lasting business relationships are extremely important. Interpersonal communication is likely to be indirect, conciliatory in tone, and formal. Often an intermediary is used, the real decision making occurs privately and away from the actual negotiations, the negotiating teams make group decisions, and all team members are expected to present a united front.

Such differences may lead to difficulties and to failure for the unwary. Consider the attempts of a U.S. businessman trying to negotiate an important deal in India:

> Joel, in his frustration, tried to speed up matters, as a lot more issues had to be addressed, but the Indians felt that he was only interested in finalizing and implementing the deal. They also began to question his intelligence, abilities, and sincerity. His informal way of addressing them also made them uncomfortable and not respected. All in all, they didn't really trust him or the deal he was proposing.[111]

Table 11.3 provides additional examples of some of the key differences in ways of thinking and in expectations for business negotiations across five cultures.

Cultural differences in business practices are also evident in the use of interpersonal relationships for strategic purposes. In Colombia and other Latin American countries, for example, achieving objectives by using interpersonal connections to obtain jobs, contracts, supplies, and other contacts—that is, giving and receiving personal favors to create an interdependent network of relationships—is regarded very positively.[112] Similar customs exist in India and elsewhere. While not as widespread throughout the multicultural U.S. workplace, such practices as providing emotional support to fellow workers, and thereby building informal social networks that can be used strategically to circumvent the bureaucracy, are common.[113]

The importance and value of social hierarchy and face maintenance are illustrated by many Chinese businesses. Chinese businesspeople will likely have to check with their superiors before making any real decisions. In Chinese organizations, superiors are expected to participate in many decisions that U.S. managers might routinely delegate to subordinates. The Chinese process of consulting the next higher level in the hierarchy often continues up the bureaucratic ladder to the very top of the organization. Thus, the autonomy that is expected and rewarded in the United States may be regarded as insubordination in China.[114] Likewise, while decisions to approve or reject specific requests or proposals may be communicated clearly by Chinese managers, justifications for such decisions are often vague or omitted in an effort to protect the face of the employees. In a business negotiation involving Chinese and U.S. Americans, therefore, attempts by the U.S. team to insist on explanations for Chinese decisions may communicate a lack of respect and a failure to acknowledge the Chinese attempts at face maintenance.

Even when negotiations have been successful, cultural variations in common social practices can cause problems. Consider the experience of Richard, who works for a U.S. company that ships refrigerated containers to Asia. Richard's company uses a yellow marker to indicate that the product has passed inspection. This makes Chinese customers suspicious of the quality of the products, since Chinese manufacturers use yellow to signify a defective product.[115]

Intercultural competence is required for successful business transactions within the intercultural United States. Consider the circumstances of Ms. Youngson, the head of a corporate sales division, who sent her sales representatives to the Chinese and Korean

TABLE 11.3

BUSINESS NEGOTIATION IN FIVE COUNTRIES[116]

Dimension	Finland	India	Mexico	Turkey	United States
Goal: contract or relationship		Business is personal; establish relationships	Seek long-term relationships	Establish relationships before negotiating	Establish rapport quickly, then move to negotiating
Attitude: win-lose or win-win			Have a win-win attitude		Look for mutual gains, whenever possible
Personal style: informal or formal		Negotiations follow formal procedures, but the atmosphere is friendly and relaxed	Established etiquette must be followed		Do not use formality or rituals in business interaction
Communication: direct or indirect	Direct	"No" is harsh. Evasive refusals are common and more polite	Negotiators may seem indirect and avoid saying "no"	Politeness is important	Be direct and to the point
Time sensitivity: high or low	Begin business right away, without small talk. It is not appropriate to be late	Conduct business at a leisurely pace. "Time-is-money" is an alien concept	The business atmosphere is easy going	Do not expect to get right down to business. The pace of meetings and negotiations is slow	Expect quick decisions and solutions
Emotionalism: high or low	Use objective facts, rather than subjective feelings. Serious and reserved	Facts are less persuasive than feelings	Truth is based on feelings. Emotional arguments are more effective than logic	Show emotion. Feelings carry more weight than objective facts	Subjective feelings are not considered "facts." Points are made by accumulating facts
Team organization: one leader or consensus	Individuals are responsible for decisions	Decisions will be made at the top	Authority is vested in a few at the top. Seek consensus		Individuals with relevant knowledge and skills make decisions
Risk taking: high or low		Take risks	Avoid risk	Take risks	

merchants in her city. She had a competent sales crew, she thought, but they were remarkably unsuccessful. Finally, out of desperation, she sent her team to a training workshop designed to help them understand the needs of Asian customers. The team discovered numerous cultural errors in their sales approach, including the most grievous error of focusing on the business of the sale too quickly, rather than going through steps to establish a trusting relationship. The sales representatives also learned not to sit unless they were invited to do so, and they were trained not to put their materials on the desk or even to lean on it.[117]

CONVERSATIONAL STRUCTURE, LANGUAGE, AND NONVERBAL COMMUNICATION We began this chapter by describing the many social episodes that form much of the social interaction of our lives. The business context has many social episodes enacted with rules derived from various aspects of culture. Something seemingly as simple as the exchange of business cards can set the tone for subsequent business relationships. Many U.S. businesspeople simply take the business cards offered to them and, after a perfunctory glance, tuck them away; in most Asian cultures, however, the exchange of business cards requires a more involved ritual in which the cards are examined carefully upon their receipt.

The ease of international telecommunications brings businesspeople from around the globe into interactions using the common communication episode of "making an introductory telephone call." Yet something as straightforward as the protocol for a common telephone call is shaped by the many differences that one's culture creates. Variations in the purposes of telephone calls, their degree of formality, the expectations about appropriate opening and closing remarks, and the anticipated length of the conversation all present intricate choices for achieving intercultural communication competence.[118] In India, for example, call center workers have had to receive extensive intercultural training to deal with irate U.S. callers. The U.S. Americans

> often wanted a better deal or an impossibly swift resolution, and were aggressive and sometimes abrasive about saying so. The Indians responded according to their deepest natures: They were silent when they didn't understand, and they often committed to more than they could deliver. They would tell the Americans that someone would get back to them tomorrow to check on their problems, and no one would.[119]

As part of their training, the Indian workers watched *Friends* and *Ally McBeal* to get an initial grounding in U.S. interaction patterns. Then they were taught how to begin conversations, how to end them, how to express empathy, and—however unnatural it might feel—how to be assertive. Many books and Internet resources are now available to help people navigate a world in which intercultural business contacts are increasing because of modern communication technologies such as fax machines, emails, telephone conversations, and, of course, the Internet itself. These resources provide specific tips on conducting business in the intercultural business context.[120]

How language is used in business contexts is also highly influenced by culture. There is, for example, much more formality used by French-speaking Canadians than by English-speaking Canadians.[121] Mexican businesspeople are likely to use persuasive arguments that, from the perspective of many U.S. Americans, may seem "overly dramatic."[122] Differences in role expectations and in the rules for interactions between Japanese and U.S. American businesspeople are not confined to meetings that take place in Japan, nor are they limited to negotiations among teams from different organizations. Young Yun Kim and Sheryl Paulk found that communication problems and misunderstandings occurred within a Japanese-owned company in the United States because of the Japanese preference for indirectness and the U.S. American preference for directness.[123] Indonesians similarly prefer a high level of indirectness in business contexts.[124]

Specific aspects of differences in nonverbal communication also pose challenges in the business context. The United States, England, China, France, Japan, and Germany all operate with an expectation of punctuality. In Germany, being even two or three minutes late is considered insulting.[125] Nonverbal behaviors such as smiling, head nodding, and silence also differ across cultures, and competent business practices require an understanding of these distinctions. As Roong Sriussadapron describes,

> When an expatriate supervisor assigned a task to a Thai local employee, the Thai employee smiled, nodded his head, and said nothing. The expatriate supervisor thought that his assignment would be accomplished by his Thai subordinate without any problem while his Thai subordinate had made no commitment. In fact, he only acknowledged that he would try his best and keep working with no deadline unless he was clearly notified.... Culturally speaking, Thais usually do not refuse someone immediately when they are asked to do something. Thai employees rarely voice refusal to work, especially with their supervisors, even though they feel unwilling, unable, or unavailable.[126]

Touching, conversational distances, eye contact, and the use of silence also vary in their appropriate use. In India, men routinely hold hands as a sign of friendship and not of sexual interest; this challenges U.S. American and European managers, who interpret the nonverbal behavior quite differently.[127] Eye contact with Saudi Arabians can also become problematic for those from the United States or from some European cultures, because the Saudis do not engage in direct eye contact but rather sit much closer to those with whom they are interacting.[128] In Finland, silence is meant to encourage a person to keep talking, while Japanese negotiators employ silence as a tactic to maintain their control in the negotiation.[129]

Even the seating arrangements and protocol during many business negotiations are highly prescribed. Among the Japanese, tables are never round in such business settings, and the expression "head of the table" is meaningless. Contrary to the usual practice

CULTURE CONNECTIONS

Bernard asked Wai Ting, one of the staff members on his team, to create a report that he urgently needed for an upcoming meeting with his boss. Wai Ting listened to the request, and said "Yes." Hearing the "yes" Bernard assumed the report would be on his desk the next day.

After three days, Bernard had not received the report. He approached Wai Ting asking for the report. Wai Ting responded, "Yes, I am working on it." She then began to tell Bernard about the difficulty she was having with a different project.

Bernard was frustrated that the report wasn't ready and further baffled by Wai Ting's reply concerning another unrelated project. He knew he had been clear in his request. As a result of his frustration he cut her off and said to her, "Which part of the phrase 'The report is urgent.' didn't you understand?'" Wai Ting became silent and looked down.

—Beth Fisher-Yoshida and Kathy D. Geller   ∎

in the United States, the power seat is not necessarily occupied by the most senior person present. Rather, whoever is most knowledgeable about the specific discussion topic takes the power seat and is designated as the company's official spokesperson for this aspect of the negotiations. At the conclusion of the business meeting, ritualistic thank-you's are uttered while all are still seated, both sides arise simultaneously and begin bowing, and the power person from the host company is expected to stay with the "guests" until they are outside the premises and are able to depart.[130]

Conducting business among intercultural teams can be both challenging and rewarding.

INTERCULTURAL TEAMS The reality of the workforce and how business is conducted globally in today's world means that many individuals now work in intercultural teams. This puts a premium on those individuals who have the knowledge, motivation, and skills to become interculturally competent. Because individuals are now expected to work competently in intercultural teams, there are a spate of "how to" books and articles for these prospective team members,[131] along with simulations intended to improve the team's intercultural competence.[132] Sometimes those teams are within their own business organization and located within one office or geographic space, while at other times the members of the work teams are spread across the globe in different countries and different time zones. Supervisors increasingly have teams of people who are from different cultures and who work in different countries. One of the challenges for intercultural communication in today's business context is developing intercultural competence in multicultural teams.[133] Consider, for example, managing differences seemingly as small as preferences for email versus chat and discussion boards within intercultural teams; U.S. businesses preferred email, while the German businesses preferred discussion boards.[134]

Intercultural teams must accomplish their tasks and their work objectives, at times despite the presence of a common language with which to communicate. A recent study of intercultural teams within European organizations documents the importance of intercultural competence in language use. The findings suggest that team members often switch from one language to another in their attempts to be understood, that they pay extra attention to clarifying their ideas and explaining their thoughts as a means of reducing linguistic misunderstandings, and that English was the language most often used by these intercultural teams.[135]

Not all intercultural teams, of course, are successful. Corinne Rosenberg describes "culturally challenged" work teams of a U.S. multinational corporation that was located in Europe, serviced a geographic area that included parts of Africa and the Middle East, and reported to supervisors and colleagues in the United States. These work teams were replete with cultural misunderstandings and ineffective intercultural communication. Managers of these teams who were involved in email exchanges and conference calls ignored the reality of time zone differences. Cultural differences in communication style also contributed to the teams' ineffectiveness. British members, whose style was less direct and more reticent, felt undervalued. To adjust to the multicultural team environment, these British team members had to learn to adapt their communication style without feeling that they had given up their own cultural mannerisms.[136]

Work roles also differ across cultures. Among the Japanese, work roles are an extension of the family hierarchy. That is,

> presidents are "family heads," executives "wise uncles," managers "hard-working big brothers," workers "obedient and loyal children." American workers employed in Japanese-managed companies do not see themselves as "loyal and obedient children" and instead hold traditional American values of individualism, competitiveness, and social mobility.[137]

Another area in which cultural differences affect the business context is in gender expectations. As we suggested in Chapter 5, cultures differ in their prescriptive roles for men and women; in many cultures women are unlikely to have managerial or supervisory positions in business. Women from the United States may have to make careful adjustments to be interculturally competent in the business setting.[138]

Expectations about the "proper" way to conduct employee performance appraisals and provide a rationale for judgments and actions are another source of cultural differences. For example, Chinese managers do not provide their subordinates with the detailed performance appraisals that are customary in many U.S. firms. Feedback on failures and mistakes is often withheld, which allows subordinates to save face and maintain their sense of esteem for future tasks within the organization.[139] Similarly, what constitutes competent performance appraisals is culturally anchored. In the United States, performance appraisals are meant to be objective and are given in a direct style, typically in writing, and the employee is often expected to provide a written rebuttal. In Saudi Arabia, the interpersonal relationship is the basis for the appraisal; the appraisal is given indirectly and is not put in writing, but a Saudi will retreat if there is negativity implied. In Japan, the basis of the appraisal is a combination of objective and subjective; it is presented in writing, includes a high degree of formality but with criticisms very subtly presented, and it is rare for the employee to disagree.[140]

INTERCULTURAL COMPETENCE IN THE BUSINESS CONTEXT What kinds of knowledge, motivations, and skills constitute "competence" in the business context? The very nature of competence itself may differ across cultures. That is, cultures often can hold fundamentally different expectations about how competence ought to be displayed. Compare, for example, organizations with which you are familiar to the typical Thai organization. In Thai companies, people are perceived as communicatively competent only if they know how to avoid conflict with others, can control their emotional displays (both positive and negative), can use polite forms of address when talking to others, and can demonstrate respect, tactfulness, and modesty in their behaviors.[141]

As the workforce has become more culturally diverse, scholars and practitioners have tried to provide managers and their employees with the tools to work together successfully. Many managers now receive ongoing training about diversity issues, and company employees are often given similar opportunities to improve their knowledge, motivation, and skills.[142] Most people recognize that the cultural heterogeneity of the workforce brings with it special challenges and opportunities, both for companies and

CULTURE CONNECTIONS

When you are walking down the road in Bali and you pass a stranger, the very first question he or she will ask you is, "Where are you going? "The second question is, "Where are you coming from?" To a Westerner, this can seem like a rather invasive inquiry from a perfect stranger, but they're just trying to get an orientation on you, trying to insert you into the grid for the purposes of security and comfort. If you tell them that you don't know where you're going, or that you're just wandering about randomly, you might instigate a bit of distress in the heart of your new Balinese friend. It's far better to pick some kind of specific direction—*anywhere*—just so everybody feels better.

The third question a Balinese will almost certainly ask you is, "Are you married?" Again, it's a positioning and orienting inquiry. It's necessary for them to know this, to make sure that you are completely in order in your life. They really want you to say yes. It's such a relief to them when you say yes. If you're single, it's better not to say so directly. And I really recommend that you not mention your divorce at all, if you happen to have had one. It just makes the Balinese so worried. The only thing your solitude proves to them is your perilous dislocation from the grid. If you are a single woman traveling through Bali and somebody asks you, "Are you married?" the best possible answer is: "Not yet." This is a polite way of saying, "No," while indicating your optimistic intentions to get that taken care of just as soon as you can.

Even if you are eighty years old, or a lesbian, or a strident feminist, or a nun, or an eighty-year-old strident feminist lesbian nun who has never been married and never intends to get married, the politest possible answer is still: "Not yet."

—Elizabeth Gilbert ∎

for the individuals who work in them.[143] Work teams that are culturally diverse, for example, are often more innovative and creative than are culturally homogeneous work groups,[144] but only if the team can use its differences to its advantage.[145] Percy W. Thomas bluntly summarizes the challenge to us all:

> Twenty years of studying, teaching, and seeking to understand human reactions to differences of all sorts has led me to three conclusions: (1) People lack the communication skills, sensitivity, understanding, flexibility, and trust necessary to establish effective relationships; (2) many reactions to people who are culturally, racially, ethnically, and sexually different are based on irrational fears and nonsensical stereotypes; and (3) people do not know how to deal with their irrational fears, attitudes, beliefs, and behaviors as they relate to inappropriate and counterproductive responses to diversity.[146]

The stakes for businesses are very high. Companies can lose the valuable talent of good employees when cultural differences affect their work negatively. Employees themselves experience their work environments in such a way as to affect their own mental well-being.

Charles R. Bantz summarizes the lessons he learned from working in a multicultural research team that was engaged in a long-term project spanning several years and several continents. Bantz recommends that increased attention and effort in four key areas would be most useful: gathering information about the multiple perspectives that will inevitably be present; maintaining flexibility and a willingness to adapt to differing situations, issues, and needs; building social relationships as well as task cohesion; and clearly identifying and emphasizing mutual long-term goals.[147]

CULTURE CONNECTIONS

During the evenings, Majed and his family often entertained guests, who almost always dropped by unannounced, as is the Kurdish custom. Guests often stopped by during the days as well, and at least one woman of the house was always expected to be home to receive them, with tea, fruit, and candy at the ready.

—Christiane Bird ■

EPISODES, CONTEXTS, AND INTERCULTURAL COMPETENCE

Recall from Chapter 3 that interaction contexts are a component of intercultural competence related to the associations between two people interacting in specific settings. The discussion in this chapter on social episodes and interaction contexts elaborates on these important ideas. Just as a picture hung on the wall has a frame around it, each intercultural encounter is surrounded or defined by a cultural frame. Competence in intercultural communication requires understanding the nature of this cultural frame.

People frame their intercultural encounters by the definitions or labels they give to particular social episodes. The activities in which you interact are chunked or grouped into social episodes that are influenced by your cultural patterns, roles, rules, interaction scenes, and interaction contexts. Someone else may take a social episode that to you is "small talk with a classmate" as "an offer of friendship." What is to you a businesslike episode of "letting off steam with a coworker about one of her mildly irritating habits" may be viewed as "public humiliation." Do not assume that what you regard as appropriate social roles and sensible rules of interaction will necessarily be comfortable or even acceptable to another.

SUMMARY

Social episodes are the repetitive, predictable, and routine behaviors that form the structure of one's interactions with others. Social episodes are made up of cultural patterns, social roles, rules of interaction, interaction scenes, and interaction contexts. People frame intercultural interactions by the expectations they have for particular social episodes.

Three specific social contexts—health care, education, and business—have become prominent meeting grounds where people from many cultures converge and interact. Each context was described in some detail to illustrate the importance of intercultural competence in everyday experiences.

FOR DISCUSSION

1. What are social episodes? When, if ever, are people affected by them?
2. What are interaction contexts? How does culture affect interaction contexts?
3. Describe an intercultural encounter you have had in the health care, education, or business context. What issues or concerns surfaced as a result of this intercultural encounter? How did you deal with these concerns?
4. What actions can people take to be more interculturally competent in everyday contexts?

FOR FURTHER READING

Richard Brislin, *Working with Cultural Differences: Dealing Effectively with Diversity in the Workplace* (Westport, CT: Praeger, 2008). An up-to-date presentation of ideas and information. Provides an interdisciplinary summary of theories and their application.

Bonnie M. Davis, *How to Coach Teachers Who Don't Think Like You: Using Literacy Strategies to Coach across Content Areas* (Thousand Oaks, CA: Corwin/Sage, 2008). A practical guide to working with teachers. Contains specific suggestions for effective and appropriate intercultural coaching and teaching.

Geri-Ann Galanti, *Caring for Patients from Different Cultures*, 4th ed. (Philadelphia: University of Pennsylvania Press, 2008). Explores many of the salient differences in expectations of competent health care across cultures.

Robert T. Moran, Philip R. Harris, and Sarah V. Moran, *Managing Cultural Differences: Global Leadership Strategies for the 21st Century*, 7th ed. (Boston: Elsevier, 2007). A compendium of how various business functions are influenced by cultural differences.

Larry D. Purnell, *Guide to Culturally Competent Health Care*, 2nd ed. (Philadelphia: F. A. Davis, 2009). An excellent guide to the effects of beliefs, values, norms, and social practices on expectations of appropriate and effective communication within the health care setting.

CHAPTER

12

The Potential for Intercultural Competence

KEY TERMS

It should be clear by now that we are personally committed to understanding the dynamics of culture and its effects on interpersonal communication. William Shakespeare suggested that the world is a stage filled with actors and actresses, but they come from different cultures and they need to coordinate their scripts and actions to accomplish their collective purposes. The image of a multicultural society is one that we firmly believe will characterize most people's lives in the twenty-first century. Intercultural communication will become far more commonplace in people's day-to-day activities, and the communication skills that lead to the development of intercultural competence will be a necessary part of people's personal and professional lives.

It should also be clear that intercultural communication is a complex and challenging activity. Intercultural competence, although certainly attainable in varying degrees, will elude everyone in at least some intercultural interactions. Nevertheless, we hope that, in addition to the challenges of intercultural interaction, this book also reminds you of the joys of discovery that can occur when interacting with people whose culture differs from your own.

In this closing chapter, we turn our attention to some final thoughts about enhancing your intercultural competence. First we look at intercultural contacts and explore what makes them more likely to be beneficial. Next we discuss some critical ethical issues that affect intercultural interactions. Following this, we offer a point of view about certain events that have been particularly newsworthy. By focusing on these events, we offer a glimpse into the ways that enormously powerful events and experiences can shape an entire generation's intercultural interactions—that is, how members of that generation are likely to perceive and engage people from other cultures. We also look at the apparent dichotomies that seem to shape individuals and nations in today's world. We conclude with an expression of optimism about the future of intercultural communication and with a renewed awareness of the need for a lifelong commitment to improving our multicultural world.

INTERCULTURAL CONTACT

Many people believe that creating the opportunity for personal contact fosters positive attitudes toward members of other groups. Indeed, this assumption provides the rationale for numerous international exchange programs for high school and college students. There are also international "sister city" programs, wherein a U.S. city pairs itself with a city in another country and encourages the residents of both cities to visit with and stay in one another's homes. Sometimes, of course, intercultural contact does overcome the obstacles of cultural distance, and positive attitudes between those involved do result.

Unfortunately, there is a great deal of historical and contemporary evidence to suggest that contact between members of different cultures does not always lead to good feelings. In fact, under many circumstances, such contact only reinforces negative attitudes or may even change a neutral attitude into a negative one. For instance, tourists in other countries are sometimes repelled by the local inhabitants, and immigrants to the United States have not always been accepted by the communities into which they have settled. In some communities and among some people, there is still much prejudice and negative feeling between European Americans and African Americans. The factors that lead to cordial and courteous interactions among people from different cultural groups are very complicated. One factor, that of access to and control of institutional and economic power, strongly influences attitudes between members of different cultures.

Power Differences Between Groups

Not all groups within a nation or region have equal access to sources of institutional and economic power. When cultures share the same political, geographic, and economic landscapes, some form of a status hierarchy often develops. Groups of people who are distinguished by their religious, political, cultural, or ethnic identity often struggle

Intercultural contacts do not always lead to positive feelings among the participants. The sign in this store window, which protests U.S. immigration policies, illustrates the tensions that can occur in some intercultural encounters.

among themselves for dominance and control of the available economic and political resources. The cultural group that has primary access to institutional and economic power is often characterized as the dominant culture.

Internationally, there have been numerous instances of genocide, "ethnic cleansing," civil wars that pit one cultural group against another, and numerous outbreaks of violence between members of cultural groups sharing the same territory. In the United States, racial tensions between African Americans and European Americans have resulted in numerous incidents. Immigrants from various parts of the world have experienced open hostility, and sometimes violent reactions, from people who live in areas where they have settled. When these kinds of competitive tensions characterize the political and economic setting in which individuals from differing cultures interact, intercultural communication is obviously affected.

Scholars have given considerable attention to the influence of dominant and subordinate group membership on interpersonal and intercultural communication processes.[1] The results of their investigations suggest that there is a very interesting set of relationships among the factors that affect these interactions. For instance, members of dominant cultures will often devalue the language styles of subordinate cultural members and judge the "correctness" of their use of preferred speech patterns. In some cases, members of subordinate cultures will try to accommodate or adapt their speech to that of the dominant culture. In other circumstances, they will very deliberately emphasize their group's unique speech characteristics when they are in the presence of people from the dominant culture.

Special forms of language are often used to signal identification among members of the subordinate group and to indicate a lack of submission to the dominant group. Similarly, members of the dominant group are likely to retain the special characteristics of their language, including preferences for certain words, accents, and linguistic patterns, and may therefore devalue the linguistic patterns of others. For example, there are instances in which European Americans have devalued the use of Black Standard English.[2]

Members of the dominant group also have much greater access to public and mass communication channels. They may be excessively influential in determining the conversational topics that are regarded as socially relevant, the societal issues that are deemed important enough to be worthy of public attention, and the "proper" language for expressing one's views in social discussions. As muted group theory suggests, individuals who do not belong to the dominant group are often silenced by a lack of opportunities to express their experiences, perceptions, and worldviews. Essentially, the power of the dominant group's communication may function to silence or "mute" the voices of subordinate group members. To have their concerns recognized publicly, subordinate group members may be obliged to use the language and communication styles of the dominant group. Although muted group theory was initially applied to women's marginalized voices,[3] it has also been applied to cultural groups.[4] Mark Orbe, for example, describes African American males as a muted group, since their talking patterns and worldview are not part of the dominant group's norms in the United States.[5] By addressing the ways in which groups are marginalized, muted group theory allows us to understand the basic concerns of nondominant voices and encourages a more equitable world where no voices are silenced.

CULTURE CONNECTIONS

The Chens had been living in the UK for four years, which was long enough to have lost their place in the society from which they had emigrated but not long enough to feel comfortable in the new. They were no longer missed; Lily had no living relatives anyway, apart from her sister Mui, and Chen had lost his claim to clan land in his ancestral village. He was remembered there in the shape of the money order he remitted to his father every month, and would truly have been remembered only if that order had failed to arrive.

But in the UK, land of promise, Chen was still an interloper. He regarded himself as such. True, he paid reasonable rent to Brent Council for warm and comfortable accommodation, quarters which were positively palatial compared to those which his wife Lily had known

in Hong Kong. That English people had competed for the flat which he now occupied made Chen feel more rather than less of a foreigner; it made him feel like a gatecrasher who had stayed too long and been identified. He had no tangible reason to feel like this. No one had yet assaulted, insulted, so much as looked twice at him. But Chen knew, felt it in his bones, could sense it between his shoulder-blades as he walked past emptying public houses on his day off; in the shrinking of his scalp as he heard bottles rolling in the gutter; in a descending silence at a dark bus stop and its subsequent lifting; in an unspoken complicity between himself and others like him, not necessarily of his race.

—Timothy Mo ∎

Attitudes Among Cultural Members

The naïve view of intercultural contact—that all intercultural contacts are likely to be beneficial—has been proven repeatedly to be incorrect. While it is true that, in general, increased intercultural contact often leads to increased liking and appreciation of others, there are innumerable examples of increased contacts leading to increased prejudice, hatred, and interpersonal violence. What, then, are the conditions under which intercultural contacts are most likely to result in favorable outcomes? What do we know about the attitudes that form when people have frequent intercultural contacts with one another?

In his classic study on the contact hypothesis, Gordon Allport described four interrelated conditions that function together to reduce the likelihood of prejudice and increase positive attitudes as a result of intercultural communication. Interestingly, each of these conditions affects the motivational component of intercultural competence. The first condition is that there must be support from the top. That is, if the high-status individuals—those who are in charge or who are recognized as authority figures—support the intercultural contact, it is more likely to lead to a positive outcome.

The second condition for positive intercultural interactions is that those involved have a personal stake in the outcome. This means that the individuals involved have something to gain if they are successful—and something to lose if they are unsuccessful—that makes them regard the interactions as personal. If people treat one another as equals and are equally invested in the outcome of the interaction, there is increased motivation to do well and to make the interaction thrive.

The third condition affecting the likelihood that intercultural interactions will be positive is that the actual intercultural contacts are viewed as pleasing, constructive, and enjoyable. Interactions in which people cooperate and feel good about their experiences will increase the prospects for further intercultural contacts.

The fourth condition that Allport said will likely lead to favorable attitudes as a consequence of intercultural interactions is related to the perceived outcome of the interaction. When all parties have the potential to be effective—that is, when the members of both cultures either have common goals or view the interaction as allowing them to achieve their own individual goals—then successful cooperation is possible, and the interactants are very likely to perceive the intercultural contact as having the potential to be beneficial.[6]

Recent investigations suggest an additional factor that also affects attitudes and outcomes of intercultural contacts. Recall from Chapter 10, where we discussed anxiety/uncertainty management theory, that learning about people from another culture increases the ability to predict what they are likely to do. That is, intercultural contact can reduce the *perceived threat* that occurs when strangers meet. If people feel threatened by culturally different others, they are likely to increase their identification with their ingroup and increase their prejudices toward their outgroups. When that happens, intercultural contacts are less likely to be favorable.[7] Even groups that are in the majority sometimes see the presence of people from other cultures as threatening. For example, consider the perceived threat and consequent reactions of U.S. Americans to immigrants who are viewed as willing to work for a lower wage.

Outcomes of Intercultural Contact

Both fictional and nonfictional accounts of intercultural contacts are replete with references to individual and cultural changes. References are made to people who "go native" and who seem to adjust or adapt to life in the new culture. References are also made to those who retain their own cultural identity by using only their original language and by living in cultural ghettos. During the height of the British Empire in India, for example, many British officials and their families tried to re-create the British lifestyle in India, in a climate not conducive to tuxedos and fancy dresses, with layers and layers of slips and decorative fabrics.

It is generally accepted that intercultural communication creates stress for most individuals. In intercultural communication, the certainty of one's own cultural framework is gone, and there is a great deal of uncertainty about what other code systems mean. Individuals who engage in intercultural contacts for extended periods of time will respond to the stress in different ways. Most will find themselves incorporating at least some behaviors from the new culture into their own repertoire. Some take on the characteristics, the norms, and even the values and beliefs of another culture willingly and easily. Others resist the new culture and retain their old ways, sometimes choosing to spend time in enclaves populated only by others like themselves. Still others simply find the problems of adjusting to a new culture to be intolerable, and they leave if they can.

People's reactions also change over time. That is, the initial reactions of acceptance or rejection often shift as increased intercultural contacts produce different kinds of outcomes. Such changes in the way people react to intercultural contacts are called *adaptation.*

ADAPTATION   Words such as *assimilation, adjustment, acculturation,* and even *coping* are used to describe how individuals respond to their experiences in other cultures. Many of these terms refer to how people from one culture react to prolonged contact with those from another. Over the years, different emotional overtones have been attached to these terms. To some people, for instance, *assimilation* is a negative outcome; to others, it is positive. Some consider *adjustment* to be "good," whereas for others it is "bad."

CULTURE CONNECTIONS

The encounter of worlds and worldviews is the shared experience of our times. We see it in the great movements of modern history, in colonialism and the rejection of colonialism, in the late twentieth-century "politics of identity"—ethnic, racial, and religious. We experience our own personal versions of this encounter, all of us, whether Christian, Hindu, Jewish, or Muslim; whether Buddhist, Apache, or Kikuyu; whether religious, secular, or atheist. What do we make of the encounter with a different world, a different worldview? How will we think about the heterogeneity of our immediate world and our wider world? This is our question, our human question, at the end of the twentieth century.

—Diana L. Eck ∎

We offer an approach that allows you to make your own value judgment about what constitutes the right kind of outcome. We believe that competent adjustment to another culture will vary greatly from situation to situation and from person to person. We have used the broader term of *adaptation* to characterize these adjustments because it subsumes various forms of cultural or individual adaptation.

Adaptation is the process by which people establish and maintain relatively stable, helpful, and mutually shared relationships with others upon relocating to an unfamiliar cultural setting.[8]

Note that this definition suggests that, when individuals adapt to another culture, they must learn how to "fit" themselves into it. Again, remember that different individuals and different groups will make the fit in different ways.

Adaptation includes physical, biological, and social changes. Physical changes occur because people are confronted with new physical stimuli—they eat different food, drink different water, live in different climates, and reside in different kinds of housing. When people are exposed to a new culture, they may undergo actual physical or biological changes. People deal with new viruses and bacteria; new foods cause new reactions and perhaps even new allergies. Prolonged contact between groups results in intermarriage, and the children of these marriages are born with a mixture of the genetic features of the people involved. Social relationships change with the introduction of new people. Outgroups may become bonded with the ingroups, for example, in opposition to the new outgroup members. Such changes may also cause individuals to define themselves in new and different ways.[9]

Alternatively, the culture itself might change because of the influences of people from other cultures. The French, for example, have raised concerns about the effects of the English language on their own language and culture. Traditional societies have sometimes expressed this distress about the Westernization or urbanization of their cultures.

Intercultural adaptations are made both by the host culture and by the sojourners. As this photo illustrates, many Irish grocery stores now carry traditional Polish food items, since sojourners from other cultures often long for the familiar tastes from home.

CULTURE SHOCK VERSUS ADAPTATION   Sustained intercultural contact that requires to-tal immersion in another culture may produce a phenomenon that has sometimes been called culture shock. Anthropologist Kalvero Oberg, who provided an early elabora-tion of the term, describes some of the reasons it occurs:

> Culture shock is precipitated by the anxiety that results from losing all our fa-miliar signs and symbols of social intercourse. These signs or cues include the thousand and one ways in which we orient ourselves to the situations of daily life: when to shake hands and what to say when we meet people, when and how to give tips, how to give orders to servants, how to make purchases, when to ac-cept and when to refuse invitations, when to take statements seriously and when not. Now these cues, which may be words, gestures, facial expressions, customs, or norms, are acquired by all of us in the course of growing up and are as much a part of our culture as the language we speak or the beliefs we accept. All of us depend for our peace of mind and our efficiency on hundreds of these cues, most of which we are not consciously aware.[10]

That is, culture shock is said to occur when people must deal with a barrage of new perceptual stimuli that are difficult to interpret because the cultural context has changed. Things taken for granted at home require virtually constant monitor-ing in the new culture to ensure some degree of understanding. The loss of pre-dictability, coupled with the fatigue that results from the need to stay consciously focused on what would normally be taken for granted, produces the negative re-sponses associated with culture shock. These can include

> excessive washing of the hands; excessive concern over drinking water, food, dishes, and bedding; fear of physical contact with attendants or servants; the absent-minded, far-away stare (sometimes called the "tropical stare"); a feeling of helplessness and a desire for dependence on long-term residents of one's own nationality; fits of anger over delays and other minor frustrations; delay and outright refusal to learn the language of the host country; excessive fear of being cheated, robbed, or injured; great concern over minor pains and eruptions of the skin; and finally, that terrible longing to be home, to be able to have a good cup of coffee and a piece of apple pie, to walk into that corner drugstore, to visit one's relatives, and in general, to talk to people who really make sense.[11]

An interesting consequence of the many new information technologies is that those residing far from their homeland, in an unfamiliar cultural environment, can now communicate more easily and more often with friends and family back home. Similarly, the new media provide the ability to have virtual interactions with others from one's culture—via email, listservs, chat groups, blogs, Internet-based video and phone conversations, and instant messages—who are also living in the "foreign" land and are experiencing comparable difficulties and a lack of predictability that living in a new culture brings. Such exchanges, for example, can provide an opportunity to describe one's feelings of alienation and homesickness while adjusting to a new culture, thus reducing the stresses and strains of the adaptation.

Often associated with culture shock are the U-curve and W-curve hypotheses of cultural adaptation. In the U-curve hypothesis, the initial intercultural contacts are characterized by a positive, almost euphoric, emotional response. As fatigue mounts and culture shock sets in, however, the individual's responses are more and more negative, until finally a low point is reached. Then, gradually, the individual develops a more positive attitude and the new culture seems less foreign, until a positive emotional response once again occurs.

The U-curve hypothesis has been extended to the W-curve, which includes the person's responses to her or his own culture upon return.[12] It posits that a second wave of culture shock, which is similar to the first and has been called re-entry shock, may occur when the individual returns home and must readapt to the once taken-for-granted practices that can no longer be followed without question. Some returnees to the United States, for instance, have difficulties with the pace of life, the relative affluence around them, and the seemingly superficial values espoused by the mass media. Others are frustrated when their colleagues and friends seem uninterested in their intercultural experiences, which may have changed them profoundly, but instead want simply to fill them in on "what they missed." Such re-entry problems, of course, are not confined to U.S. Americans who have been to another culture.[13] Japanese school-age children who returned from living in English-speaking countries, for instance, have identified readjustment problems because of their differences from their peers, the precise expectations for their behaviors in school, their reduced proficiency in the Japanese language, and their interpersonal styles.[14] One girl had to dye her hair black because it had lightened from the sun. Another had to remind herself continually, "I shouldn't be different from others; I should do the same as others in doing anything."[15]

Though initially regarded as plausible, the U-curve and the W-curve hypotheses do not provide sufficiently accurate descriptions of the adaptation process. They do not account, for instance, for those whose experiences remain favorable, for those who fail to adapt and return home prematurely, or for those whose level of discomfort changes little during the adaptation period. Rather, there seem to be a variety of possible adaptation patterns that individuals could experience, depending on their particular circumstances. The pattern of adaptation varies widely from one individual to the next, and therefore no single pattern can be said to characterize the typical adaptation process.[16]

CULTURE CONNECTIONS

The trauma of dislocation varied, of course, by generation and gender. Young husbands felt the pain of not being able to provide, with great wounds to their male dignity and self-respect. Older people like my grandparents missed the comforts of retirement in a familiar milieu, with old friends and trusted servants; they felt vulnerable in a strange country, with a language they couldn't speak. But the loss everyone felt together, among the most acute, was the loss of gardens. Trees, flowers, the garden courtyard occupy a hallowed space in Iranian culture. Just look through the photo albums of an old Iranian family. You'll find faded images of parents seated outside on a raised divan covered with Persian rugs, with children playing by a fountain, or amidst a grove of trees, in the background. In one of my favorite stories that Maman would tell me as a child, my great-grandmother, in a fit of wounded rage at my great-grandfather, taking a second wife, ordered the leveling of one of the oldest mulberry orchards—tall, proud trees that had grown for decades, destroyed in revenge for his betrayal. She had found no better metaphor for the death of her love than the destruction of trees. In California, the absence of gardens seemed the bitterest part of our reconstructed lives.

—Azadeh Moaveni ∎

The term *culture shock* can now be seen to describe a pattern in which the individual has severe negative reactions on contact with another culture. Such extreme responses, however, in which the person's knowledge, motivation, and skills are initially insufficient to cope with the strangeness of a new culture, are among many likely reactions. We therefore prefer the more general term adaptation to refer to the pattern of accommodation and acculturation that results from people's contact experiences with another culture. As many theorists have suggested, it is through adaptation that personal transformation from cultural contact takes place.[17]

THE ADAPTATION PROCESS   Efforts to describe the adaptation process suggest a more complex set of patterns than the U-curve and W-curve hypotheses provide. Daniel J. Kealey found that the U-curve was an accurate description of the adaptation process for only about 10 percent of the individuals he studied; the majority experienced little change (30 percent remained highly satisfied, 10 percent stayed moderately satisfied, and 15 percent maintained a low level of satisfaction throughout); and another 35 percent had an extremely low level of satisfaction initially but improved continuously for the duration of their intercultural assignment.[18] Interestingly, many in this latter group, who experienced the most severe adjustment stress, eventually became the most competent in their ability to function in another culture. Additional research confirms that, for many individuals, the initial experience when adapting to an unfamiliar culture is not excitement and joy but rather stress and anxiety due to the unfamiliarity of the situation.[19] Similarly, there are multiple patterns of adaptation when returning "home" after an extended sojourn in another culture, which affect the psychological health, social adjustment, and cultural identity of the returning individuals.[20]

There is also ample evidence to suggest that the adaptation process has multiple dimensions or factors associated with it.[21] For example, Mitchell R. Hammer, William B. Gudykunst, and Richard L. Wiseman suggested that intercultural effectiveness consists of three such dimensions: the ability to deal with psychological stress, skill in communicating with others both effectively and appropriately, and proficiency in establishing interpersonal relationships.[22] Recent research by Colleen Ward and her colleagues has identified two primary dimensions of adaptation: psychological and sociocultural. The former is similar to Hammer and his colleagues' first dimension, and the latter seems to combine the remaining two. Psychological adaptation refers to one's personal well-being and good mental health while in the intercultural setting. Sociocultural adaptation refers to one's competence in managing the everyday social interactions that occur in daily life.[23]

Despite such distinctions, the adaptation process is often viewed, inaccurately, as a single "package" of related features that all follow the same trajectory of change for a given individual. However, distinct patterns of change likely characterize each dimension of adaptation. Thus, for instance, the time it takes to make a psychological adjustment to the pace of life in an unfamiliar culture may be very different from the rate of sociocultural adaptation to the culture's expectations regarding the use of indirection in language.

TYPES OF ADAPTATION   Answers to two important questions shape the response of individuals and groups to prolonged intercultural contact, thus producing different outcomes. The first concern is whether it is considered important to maintain one's cultural identity and to display its characteristics. The second concern involves whether people believe it is important to maintain relationships with their outgroups.[24]

Assimilation occurs when it is deemed relatively unimportant to maintain one's original cultural identity but it is important to establish and maintain relationships with other cultures. The metaphors of the United States as a melting pot and as a tributary stream, which are both described in Chapter 3, illustrate the choice described in Figure 12.1 as *assimilation*. The melting pot metaphor envisions many cultures giving up their individual characteristics to build the new, homogenized cultural identity of the United States. Similarly, the tributaries metaphor suggests that all cultural groups will eventually assimilate into the mainstream or dominant cultural group. Thus, assimilation means taking on the beliefs, values, norms, and social practices of the dominant cultural group.

When an individual or group retains its original cultural identity while seeking to maintain harmonious relationships with other cultures, integration occurs. Countries such as Switzerland, Belgium, and Canada, with their multilingual and multicultural populations, are good examples. The metaphors from Chapter 3 that describe the United States as a tapestry and as a garden salad both describe *integration* as the preferred style of adaptation. Integration produces distinguishable cultural groups that work cooperatively to ensure that the society and the individuals continue to function well. Both integration and assimilation promote harmony and result in an appropriate fit of individuals and groups to the larger culture.

When individuals or groups do not want to maintain positive relationships with members of other groups, the outcomes are starkly different. If a culture does not want positive relationships with another culture and if it also wishes to retain its cultural characteristics, separation may result. The metaphor from Chapter 3 of the United States as a rainbow illustrates the choice of *separation* as the desired form of acculturation. If the separation occurs because the more politically and economically powerful culture does not want the intercultural contact, the result of the forced separation is called segregation. The history of the United States provides numerous examples of segregation in its treatment of African Americans. If, however, a nondominant group chooses not to participate in the larger society in order to retain its own way of life, the separation is called seclusion. The Amish are a good example of this choice.

When individuals or groups neither retain their cultural heritage nor maintain positive contacts with the other cultural groups, marginalization occurs. This form of adaptation is characterized by confusion and alienation. None of the metaphors described in Chapter 3 characterize marginalization, as this form of adaptation is widely regarded as dysfunctional and ineffective. The choices of both marginalization and separation are reactions against other cultures. The fit these outcomes achieve in the adaptation process is based on battling against, rather than working with, the other cultures in the social environment.[25]

For purposes of simplification, Figure 12.1 suggests that each of the questions must be answered as wholly "yes" or "no." In reality, however, people could choose a variety of points between these two extremes. The French, for example, while certainly not isolationists, have raised concerns about the effects of the English language on their own language and culture. Similarly, traditional societies have sometimes been distressed about the Westernization or urbanization of their cultures while simultaneously expressing a desire for increased contact and trade. To a large extent, the types of adaptation also represent the diverse and often divergent views on culture, identity, and intercultural relations that are often expressed in the United States and beyond.[26]

Obviously, not all individuals adapt and acculturate similarly. A study of nearly 8,000 immigrant teenagers now living in one of thirteen countries showed that the *integration* profile was most common (36.4 percent), followed by *separation* (22.5 percent), *marginalization* (22.4 percent), and *assimilation* (18.7 percent). More importantly, there was a close association between the adaptation profiles and the quality of the

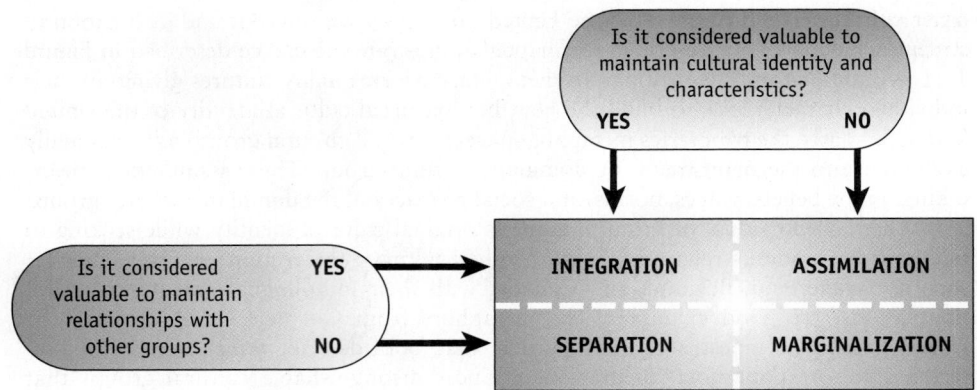

FIGURE 12.1
Forms of acculturation.[27]

adaptation outcomes. Those with an *integration* profile did the best on both psychological and sociocultural adaptation outcomes; those with a *separation* profile had, as one might expect, moderately good psychological adaptation but poor sociocultural adaptation; those with an *assimilation* profile had moderately poor psychological adaptation and slightly negative sociocultural adaptation; and those with a *marginalization* profile did the worst on both psychological and sociocultural adaptation outcomes.[28]

The adaptation process is perhaps best seen as an ongoing set of choices about intercultural communication, rather than as a single global selection among alternatives such as assimilation or integration.[29] Some people find these daily challenges of responding to another culture to be too stressful and overwhelming. If possible, such individuals will choose to return to their culture of origin; if they cannot do so, various kinds of maladaptive adjustments, or even mental illnesses, can occur. At the opposite extreme, and of particular interest to us, are those individuals who move easily among many cultures. Such people generally have a profound respect for many varied points of view and are able to understand others and to communicate appropriately and effectively with people from a variety of cultures. Such individuals are able to project a sense of self that transcends

CULTURE CONNECTIONS

Over the centuries the bravery of the Maasai had enabled them to acquire large tracts of land and the highest number of stock per person among tribes in Africa. Now, however, things have changed drastically for the worse. Our huge landholdings are shrinking at a frightening rate and our livestock is diminishing as a result of the encroachment on our land by other tribes and the creation of national parks by the governments of Kenya and Tanzania. Without the land and cattle, there will be no Maasai. But my people are still holding on and continue to celebrate our culture despite the urgent demands that we change our ways and assimilate to contemporary modes of living. If change must come, as seems inevitable, it must be gradual, not abrupt. We will adapt, we will survive.

—Tepilit Ole Saitoti ∎

any particular cultural group. Young Yun Kim uses the term intercultural personhood to describe the progression by which individuals move beyond the thoughts, feelings, and behaviors of their initial cultural framework to incorporate other cultural realities.[30]

Interculturally competent communicators integrate a wide array of culture-general knowledge into their behavioral repertoires, and they are able to apply that knowledge to the specific cultures with which they interact. They are also able to respond emotionally and behaviorally with a wide range of choices in order to act appropriately and effectively within the constraints of each situation. They have typically had extensive intercultural communication experiences, and they have learned to adjust to alternative patterns of thinking and behaving.

Tourism as a Special Form of Intercultural Contact

While sojourners, immigrants, businesspeople, refugees, students, and many others cross cultural boundaries and engage in intercultural interactions, tourism is perhaps the most common way that face-to-face intercultural contacts occur.[31] Annually there are about 64 million international tourists visiting the United States,[32] and each year almost 26 million U.S. tourists travel to international destinations.[33] Consequently, international tourism represents one of the largest means of intercultural contact.

Tourists, of course, have a variety of motives for visiting other places. For many, the goal is to experience the human-made cultural and historical sites that are unique to a locale: the Opera House in Sydney, the Prado in Madrid, Borobudur in Indonesia, the Anthropology Museum in Mexico City, the terracotta warriors in Xi'an, Olduvai Gorge in Tanzania, the Statue of Liberty in New York, the Dome of the Rock in Jerusalem, or countless other attractions that are of interest to tourists. For others, the goal of tourism is to experience the unique natural wonders of a place: the Barrier Reefs, the Galapagos Islands, Victoria Falls, Sequoia National Park, the Alps, or Patagonia. For still others, the goal is recreational adventure: to climb, surf, pedal, paddle, hike, swim, sail, cruise, or ride in unfamiliar lands and waters. Some people become international tourists for reasons that involve medical, religious, ecological, or humanitarian outcomes; for others, the desired outcome is shopping or escapist relaxation. Often, a combination of these motives is involved in tourists' decisions to travel. As you might expect, we fervently hope that an overriding motive for your future international travel will be to experience firsthand, and to appreciate more fully, the everyday lives of other people and cultures.

Although tourists have many reasons for crossing cultural boundaries, there are three reasons why the tourist experience creates a special and unique form of intercultural contact. First, the motives for tourists' intercultural contacts are predominantly short-term and voluntary. Tourists' experiences in a specific culture are typically brief, and their choices about where, when, and how to travel are voluntary.[34] Moreover, many of these choices are changed along the way; tourists often have substantial discretion to lengthen or shorten their time in a specific place, and they also have considerable flexibility to alter their daily arrangements and adapt their experiences to their individual preferences.

Second, the nature of the tourist–host relationship is distinctive and not necessarily perceived in the same way by the tourist and the host. For tourists, encounters with host-culture members are often viewed as desirable and interesting. After all, a primary reason that many tourists travel internationally is to have the interpersonal experiences that these intercultural encounters provide. One U.S. American family touring Costa Rica found, for example, that a flat tire in their rented automobile resulted in the serendipitous outcome of a local man helping them to fix it and then inviting them to his home for coffee; this encounter allowed the U.S. American family to visit a Costa Rican family in their own home, which was a highlight of their trip. Similarly, tourists to Europe often have delightful experiences when meeting tourists from other cultures, and many tourists enthusiastically recall their stories of bargaining with members of the host culture to purchase their souvenirs and gifts. But for some members of the host culture, tourist–host encounters may be viewed through a very different lens. Those whose jobs require them to cater to the needs of tourists—tour guides, taxi drivers, and all sorts of people working in so-called "tourist ghettos" where there are high densities of tourists' accommodations, attractions, and amenities—may view their interactions as just another superficial encounter with yet another tourist. Sometimes host-culture members experience tourists as impatient and unwilling to accept local customs, and sometimes tourists' inability to speak the local language and understand the local expectations of appropriateness and effectiveness can be an irritant and a barrier to host-culture relationships. Additionally, the tourist–host relationship is often asymmetric and commercialized; tourists have economic power, host nationals have information power, and the commercial goal is to exchange one form of power for the other. That is, tourists have money to spend, and host-culture members want that money to be used to purchase their services or products. Conversely, host-culture members know the local customs, costs, and cultural experiences, and they may be expected to share that knowledge as a precondition to receiving economic benefits from the tourist.

Finally, tourism also exists within a social and political context. Although it often provides economic benefits for those living in the tourist destination and allows people from one culture to learn about another, tourism can also produce serious negative consequences. At some popular tourist destinations, for instance, the tourists actually outnumber the native population, and tourists may consume natural resources at a greater rate than they can be replaced. At the centuries-old archeological site of Cambodia's Angkor Wat, tourists use so much water from underground wells that the foundations of the famous temples are sinking; this is literally undermining the attractions those tourists came to experience.[35] Sometimes, the prospect of increased economic development, and the tourist revenues they can generate, will alter the traditional relationships of a group of people to their land and property. About a generation ago, on Indonesia's island of Bali, Dutch multinational corporations began buying from the locals on Kuta Beach their prime but undeveloped land. The corporations then built huge tourist hotels and a massive tourist industry, which now rivals that of Hawai'i. Unfortunately, the former locals had to move to less desirable places, and

CULTURE CONNECTIONS

Diversity, of course, is not pluralism. Diversity is simply a fact, but what will we make of that fact, individually and as a culture? Will it arouse new forms of ethnic and religious chauvinism and isolation? Or might it lead to a genuine pluralism, a positive and interactive interpretation of plurality? These are critical questions for the future, as people decide whether they value a sense of identity that isolates and sets them apart from one another or whether they value a broader identity that brings them into real relationship with one another.

—Diana L. Eck ∎

Rapid transportation allows many tourists to experience the wonders of previous civilizations, such as those at Cambodia's Angkor Wat (left) and Peru's Machu Picchu (right). Such World Heritage locations dramatically illustrate the ethical dilemma of balancing the positive outcomes of tourism with the practical realities of site destruction.

they now use the revenues from the sale of their land to buy motorbikes, on which they commute to the hotels and tourist venues in order to work as low-paid employees.

Yet the short-term and often scripted intercultural communication of the tourist can also lead to meaningful, long-term, and deeply personal commitments that are stimulated by an intense desire to learn more about another culture. An unexpected encounter between a tourist and a restaurant owner, for example, can be the start of a long-term relationship that is sustained by letters, emails, phone calls, Internet social networks, and even visits from various family members. Many tourists also experience the intellectual excitement that comes from learning about a particular culture or place; they then choose to study that culture more formally and make repeated, in-depth visits to it. Thus tourism is a frequently used avenue by which individuals can sustain a long-term interest in another culture and gain the motivation, acquire the knowledge, and develop the skills to function within that culture in an interculturally competent way.

THE ETHICS OF INTERCULTURAL COMPETENCE

Those who attempt to achieve intercultural competence must face a number of ethical dilemmas. It is imperative to explore the following issues to become aware of the choices that are made all too often without due consideration and reflection.

There are three key ethical dilemmas. The first is summarized in the adage "When in Rome, do as the Romans do." The second asks if it is possible to judge a particular belief, value, norm, or social practice as morally reprehensible. If so, when and under what circumstances? Stated in a slightly different way, if all cultures have differing beliefs, values, norms, and social practices, does that mean there are no true rights and wrongs? The third dilemma relates to the consequences of intercultural contacts. Are they necessarily positive for individuals and their societies? In other words, should all intercultural contacts be encouraged?

When in Rome...

A fundamental issue confronting those who are in the midst of another culture is a decision about how much they should change their behaviors to fit the beliefs, values, norms, and social practices of those with whom they interact. Whose responsibility is it to attempt to take into account cultural differences in communication? Is it the responsibility of the visitors, newcomers, or sojourners to adjust their behaviors to the cultural framework of the host culture, or should members of the host culture adjust their communication and make allowances for the newcomers and strangers? Because English predominates in the United States, are all those who live in the United States required to use English? To what extent must individuals adapt their cultural beliefs, values, norms, and social practices to the dominant cultural patterns?

The old saying "When in Rome, do as the Romans do," which clearly places the responsibility for change on the newcomer, offers a great deal of wisdom, but it cannot be followed in all circumstances. In most cases, behaviors that conform to cultural expectations show respect for the other culture and its ways. Conformity with common cultural practices also allows the newcomer to interact with and to meet people from the host culture on some kind of genuine basis. Respecting differences in nonverbal and verbal codes means that the ethical intercultural communicator takes responsibility for learning as much about these codes as is possible and reasonable. Naturally, what is possible and reasonable will vary, depending on a range of circumstances. Sometimes, wholesale adoption of new cultural practices by a group of newcomers may be seen as disrespectful and can upset those from the host culture. In the past, for example, U.S. and European students visiting India wore Indian clothes, didn't wear shoes, and lived in very poor circumstances. Many Indians regarded this "conforming" behavior as insulting and disrespectful of their cultures. The visitor to a culture cannot simply adopt the beliefs and practices of a new culture without also risking being perceived as insincere and superficial.

Sometimes it is difficult for people to change their behaviors to match cultural patterns that contradict their own beliefs and values. For example, many U.S. American women, who were taught to value freedom and equality, may find it difficult to respond positively to cultural practices that require women to wear veils in public and to use

---

CULTURE CONNECTIONS

Every time I return to India—about once a year—it is different yet the same. Yes, there's now a McDonald's at my favorite market in Delhi, but around the corner is the shop that sells statues of Ganesh. Yes, there are cellphones and ATMs and Internet cafés, but none has made a dent in the bedrock of Indian culture. These latest foreign intruders are no different from the Mughals or the British or any of the other interlopers who over the centuries tried to subdue the subcontinent. India always emerged victorious, not by repelling these invaders but by subsuming them.

The Taj Mahal is today considered the quintessential Indian icon, yet it was built by a seventeenth-century Mughal emperor who at the time wasn't Indian at all. He is now. Likewise, McDonald's caved to the Indian palate and, for the first time, dropped Big Macs and all hamburgers from its menu, since Hindus don't eat beef. Instead, it serves McAloo Tikki and the McVeggie and a culinary hybrid, the Paneer Salsa Wrap. McDonald's didn't change India, as some feared. India changed McDonald's.

—Eric Weiner ∎

male drivers or chaperones. The ethical dilemma that intercultural communicators face is the decision about how far to go in adapting their behaviors to another culture. Should people engage in behaviors that they regard as personally wrong or difficult? At what point do people lose their own sense of self, their cultural identities, and their moral integrity? At what point does the adoption of new cultural behaviors offend and insult others? One of the challenges and delights of intercultural communication is in discovering the boundaries and touchstones of one's own moral perspective while simultaneously learning to display respect for other ways of dealing with human problems.

Another perspective from which to explore the ethical issues embedded in the adage is that of the "Romans." A common point of view, often expressed by U.S. Americans about those who have recently immigrated to the United States or who still retain many of the underlying patterns of their own culture, is that since these people now live in the United States, they should adjust to its cultural ways. The same comments are often made about students from other countries who come to the United States to study.

We ask you to consider the experiences of those people who immigrate to or study in another country. Perhaps you are such a person. Or perhaps your parents or grandparents did so. Not all immigrants or students have freely chosen the country where they now reside. Large numbers of people migrate from one country to another because political, military, and economic upheavals in their own country make living and learning there nearly impossible. For many, the choice to leave is juxtaposed against a choice to die, to starve, or to be politically censored. We also ask you to consider how difficult it must be for people to give up their culture. Remember how fundamental your cultural framework is, how it provides the logic for your behavior and your view of the world. How easy would it be for you if you were forced into new modes of behavior? Adjustment to another culture is difficult.

Are Cultural Values Relative or Universal?

A second ethical issue confronting the intercultural communicator is whether it is ever acceptable to judge the people of a culture when their behaviors are based on a radically different set of beliefs, values, norms, and social practices. Are there any values that transcend the boundaries of cultural differences? Are there any universally right or wrong values?

A culturally relativistic point of view suggests that every culture has its own set of values and that judgments can be made only within the context of the particular culture. Most people do not completely subscribe to this view, partly because it would lead to a lack of any firm beliefs and values on which to build a sense of self-identity.

David Kale argues that there are two values that transcend all cultures. First, the human spirit requires that all people must struggle to improve their world and to maintain their own sense of dignity, always within the context of their own particular culture. Thus, Kale suggests that "the guiding principle of any universal code of intercultural communication, therefore, should be to protect the worth and dignity of the human spirit."[36] The second universal value is a world at peace. Thus, all ethical codes must recognize the importance of working toward a world in which people can live at peace with themselves and one another.[37]

Ethical intercultural communicators continually struggle with the dilemmas presented by differences in cultural values. The tensions inherent in seeking to be tolerant of differences while holding firmly to one's own critical cultural values must always be reconciled. Kale's suggestions for responding ethically to cultural differences in values are excellent starting points for the internal dialogue that all competent intercultural communicators must conduct.

CULTURE CONNECTIONS

Mr. Malchode moved then to sit beside me. Quietly he said, "I don't mean to insult, but for your own sake you should know as a white you're intruding here. This is *our* place. It's not a zoo for tourists to see how 'natives' live. Even now we can't drink in a Messina hotel bar—the prices are trebled to keep us out. But you take it for granted you can come and drink here—you're white, so you can drink wherever you choose. Do you know enough about South Africa to understand what I say?"

—Dervla Murphy ■

Do the Ends Justify the Means?

The final ethical dilemma we wish to raise concerns these questions: Should all intercultural contacts be encouraged? Are the outcomes of intercultural contacts positive? Are all circumstances appropriate for intercultural contact? In short, do the ends justify the means?

We have been shamelessly enthusiastic about the potential benefits and delights of intercultural interaction. Nevertheless, certain outcomes may not necessarily be justified by the means used to obtain them.

As an ethical intercultural communicator, some of the following questions must be confronted:

- Is it ethical to go to another country, for whatever reason, if you are naïve and unprepared for cultural contact?
- Should intercultural contacts be encouraged for those who speak no language but their own?
- Should those who are prejudiced seek out intercultural contacts?
- Is it ethical to send missionaries to other countries?
- Is it acceptable to provide medical assistance to help a culture resist a disease, when in providing the assistance you may destroy the very infrastructure and nature of the indigenous culture?
- Is it justifiable for the sojourner from one culture to encourage a person from another culture to disregard his or her own cultural values?

There are no simple answers to any of these questions, but the competent intercultural communicator must confront these ethical dilemmas.

Ethics—Your Choices

We have offered few specific answers to these ethical dilemmas because every person must provide his or her own response. In the context of your own experiences and your own intercultural interactions, you must resolve the ethical dilemmas that will inevitably occur in your life. Kale provides four principles to guide you as you develop your own personal code of ethics. Ethical communicators should do the following:

- Address people of other cultures with the same respect that the communicators would like to receive themselves.
- Try to describe the world as they perceive it as accurately as possible.
- Encourage people of other cultures to express themselves in their unique natures.
- Strive for identification with people of other cultures.

THE PERILS AND PROSPECTS FOR INTERCULTURAL COMPETENCE

Today's world is buffeted by an enormously powerful set of forces. Some of these forces are not unique to this era but have existed at other times throughout the history of the world. Some, however, are wholly new, and they cause profound and, to some extent at least, unpredictable changes. The changes are set in motion by the speed with which global capital, information, goods, and the people who would trade in them can move across borders and throughout the world. Indeed, we are living through, yet do not fully understand, an unprecedented series of revolutions in communications, transportation, and technologies, that impose instantaneous interconnectedness upon most of the world's nations, their cultures, and their economies.

Impact of National and International Events on Intercultural Communication

Consider the following examples, each of which is drawn from the history of the United States and has had profound consequences throughout the world. As you reflect upon them, identify for yourself how these events were first shaped by global forces and, in turn, helped to shape subsequent global events.

October 24, 1929: A Thursday in early fall, and the day when the Roaring Twenties abruptly ended. The U.S. stock market crashed, a decade of unbelievable prosperity ended, and the Great Depression rapidly followed. Within three years, U.S. stock prices lost nearly 90 percent of their value, banks and other financial institutions collapsed, factories and businesses failed, unemployment soared by 700 percent, and ordinary working people were destitute, having lost both their jobs and their accumulated savings. Hunger was widespread, breadlines were commonplace, and medical care and durable goods were unaffordable. U.S. Americans of a certain age have been forever seared by these events, and their collective experiences have figured prominently in the subsequent financial, political, social, and vocational choices that were made.

December 7, 1941: A Sunday that, for many elderly U.S. Americans, will always be remembered as a day of infamy. The surprise was complete. The assault came that morning in two waves—the first at 7:53, the second an hour later. By ten o'clock it was over; more than 350 airplanes broke off the attack and left Oahu for their carriers, which returned them to Japan. Pearl Harbor, home of the U.S. Pacific Fleet, had been crippled: more than 2,000 dead, thousands more injured, eight battleships damaged or destroyed, and nearly 200 planes ruined. In the wake of this attack, both patriotism and fear followed; as the United States was drawn into World War II, courageous men and women marched off to defend their homeland and their loved ones, bravely facing death and the unfathomable horrors of war. Many who remained in the United States shared a newfound pride; U.S. flags were prominently displayed, slogans and posters to encourage patriotism were seemingly everywhere, and a desire to contribute to the national effort was palpable. Sadly, however, some U.S. Americans became the targets of unbridled fear, including those loyal U.S. citizens of Japanese ancestry who were required to abandon most of their belongings and were forced into detention camps for several years.

November 22, 1963: A warm Friday afternoon in Dallas. With an enthusiastic crowd cheering, President John F. Kennedy's motorcade passed through the streets in an open car. Suddenly at least three shots rang out, and the president was hit twice: in the base of the neck and squarely in the head. By 1:00 p.m.—less than an hour later—he was declared dead at the age of forty-six. In remarks prepared for delivery later that day, Kennedy had intended to say,

> In a world of complex and continuing problems, in a world full of frustrations and irritations, America's leadership must be guided by the lights of learning and reason or else those who confuse rhetoric with reality and the plausible with the possible will gain the popular ascendancy with their seemingly swift and simple solutions to every world problem.[38]

Three days later, leaders from more than ninety nations attended Kennedy's funeral; a million people lined the route as a horse-drawn caisson bore his body to St. Matthew's Cathedral for a requiem mass. Then, as more than a hundred million people watched on television, the president was buried in Arlington National Cemetery, where an eternal flame still marks his grave. Now, some fifty years later, U.S. Americans of a certain age remember vividly the tragedy of that assassination.

January 28, 1986: A cold Tuesday morning in Florida. At 11:38, a rocket left its launching pad for a seven-day mission. The countdown hadn't exactly gone smoothly, with weather and equipment problems plaguing the launch, but after seven delays spanning five days, liftoff finally occurred. From classrooms across the nation, excited schoolchildren watched the live coverage of the shuttle's launch. They were eagerly following the successes of Christa McAuliffe, the first "teacher in space," who was to speak to them the following day in a live telecast. Just seventy-three seconds into the flight, the unimaginable happened: the space shuttle *Challenger* exploded in a fiery blast, instantly killing all seven crew members. The United States was stunned and shaken. The State of the Union address, scheduled for delivery by President Ronald Reagan that evening, was postponed. In its place, the president delivered a short but moving tribute to the fallen astronauts; three days later, a sorrowful nation participated via television in a memorial service in Houston, as the United States mourned the "*Challenger* Seven." Ask U.S. Americans of a certain age—particularly those who were schoolchildren then—and they will tell you that they vividly remember the day the *Challenger* exploded; nearly three decades later, many can recall precisely where they were and what they were doing when they learned of the disaster.

September 11, 2001: A clear Tuesday morning, with a hint of fall in the air. Abruptly, at 8:50 a.m., a hijacked commercial airplane smashed into the northern tower of New York's World Trade Center. Twenty-four minutes later, another hijacked plane struck

CULTURE CONNECTIONS

Our efforts to counter hatred, intolerance, and indifference must continue simultaneously at individual and structural levels. We must try to influence for good the minds and hearts of individual people through dialogue and confidence building. These efforts must be reinforced by our efforts to create just structures in society to support the ongoing work of negotiations in the human community. Only then will we have a chance to negate the terrible consequences of the tremendous conflicts facing humankind today.

—Nelson Mandela ■

the southern tower. Twenty-four minutes after that, at 9:38, a third hijacked plane struck the Pentagon, a portion of which collapsed. Pandemonium ensued. Within the hour, these jet-fueled fireballs had caused both of the 110-story towers of the World Trade Center to collapse. Fears of additional attacks were widespread: the Sears Tower in Chicago was evacuated, an antiterrorism division was mobilized in Los Angeles, and Seattle's Space Needle was closed. Air traffic was halted, financial markets were shut down, troops were mobilized, life was more uncertain, U.S. flags were everywhere, patriotism was admired, and donations soared. And Arab Americans—indeed, U.S. Americans from many cultures, as well as countless peace-loving Muslims throughout the world—were a bit more fearful for the safety of their children. More than ten years later, many U.S. Americans are still haunted by their experiences on that day.

What sense should we make of these experiences, particularly the most recent of them? Each of these experiences, and many others we could provide from around the world, creates an indelible memory for certain U.S. Americans, who then pass on the lessons learned from them. Each of these events fundamentally alters the basic and often unquestioned understandings that people have of their social world, and it raises issues such as the following:

- What does my culture and nation represent to others? That is, from the perspective of others who view us differently than we view ourselves, what does my culture and nation stand for?
- What are the beliefs, values, norms, and social practices that seem to guide my culture's actions?
- In what ways am I interconnected with other cultures and economies in the world?
- To what extent should I trust people who seem different from me? To what extent must I trust people who seem different from me, as a prerequisite for our mutual survival?

Events such as those described above have profound effects on many individuals, who subsequently shape the understandings of others in the generations that follow. The September 11 terrorist attacks, for example, continue to test the very fabric of the United States as a multicultural nation. Fears and uncertainties have encouraged people to evaluate others negatively based solely on such attributes as their physical appearance, choice of religious observance, culture of origin, and the like. But they have also created a healthy reevaluation of national priorities, the values inherent in a multicultural nation, and the means to achieve these goals.

Forces That Pull Us Together and Apart

There are two powerful yet opposing forces that are tugging on the United States and its many cultures. Indeed, these forces are not exclusive to the United States; they affect every nation and culture on earth, often in significant ways.

The opposing forces could variously be described as engagement versus isolationism, globalism versus nationalism, secularism versus spiritualism, consumerism versus fundamentalism, or capitalism versus tribalism. That is, intercultural relationships among cultural groups throughout the world are simultaneously being pushed together and wrenched apart. Though the terminology to describe these potent forces may vary, and their influential consequences may fluctuate widely across cultures and regions of the world, they nevertheless provide us with powerful lenses through which to view the changing interrelationships among the world's cultures.

One such force—promoting engagement, globalism, secularism, and capitalism—is nurtured and sustained by the economic interdependence of today's world. Economic interdependence, in turn, is linked to the rapid communications systems that now connect people virtually in real time, as events are displayed instantaneously through a variety of powerful technological innovations—television, film, videos, music, and the Internet. Transportation systems, as well, can quickly take people from one part of the globe to another, and tourism creates many opportunities to interact. Almost anywhere one travels, there will be familiar signs of the interdependent global economy: television shows such as *Star Search*, *CSI*, and *American Idol* have burgeoning international audiences; film and musical performers, from rap to salsa to classical, from Kanye West to Avril Lavigne to the St. Petersburg Opera, are known internationally; and MTV is seen globally, with local shows adapting the U.S. format by tailoring it to their audiences' sensibilities. One can hear Peruvian musicians on a corner in Brussels, Beatles tunes in an elevator in Malaysia, reggae music on the streets of Guatemala City, and African rhythms at a park in San Francisco. In many parts of the world, the music played on the radio stations could be described, at least in part, as global and not representative of that country's musical traditions.

Closely related to mass media's impact is the speed of communications that now link much of the world. Events that occur in one country are displayed, within minutes, to people thousands of miles away. As a consequence, events in one part of the world have dramatic consequences in others. International telephone usage is on the rise; Skype, texting, and the Internet have drastically reduced the cost for such calls, which has made it much easier to communicate over long distances.

Added to all of these forces is the stark reality of global economic interdependence. There are obvious signs of this "sharing" of the world's economy. The now-familiar KFC, Pizza Hut, and McDonald's fast-food outlets are seemingly everywhere; Fords, Toyotas, and Volkswagens are driven the world over; and consumer products by Coca Cola, Sony, Nestle, and Bayer are marketed internationally. The world traveler could easily assume, incorrectly, that similarities in consumer products and media messages result from or will lead to a homogenization of world cultures.

A counterpoint to these forces for globalization is another, and equally powerful, set of constraints. These alternative influences—for isolationism, nationalism, spiritualism, and tribalism—derive from a desire to preserve, protect, and defend what is seen as unique but threatened: the culture's language, religion, values, or way of life. As an example of these forces, consider the frequent desires expressed among members of a culture to protect its language from the intrusions of other languages; France, for example, is very vigilant about keeping non-French words from the national language. Similarly, people may elect to safeguard their economies from foreign products; they may do this formally—with protectionist tariffs on goods from other nations, particularly if the foreign goods compete favorably with those locally grown or manufactured—or informally, by common consent—witness the dearth of

CULTURE CONNECTIONS

It is always interesting to see something from another person's point of view, particularly if that person is very different from oneself. It broadens the horizons.

—Nury Vittachi   ■

Japanese-manufactured automobiles in Detroit, where major U.S. automakers are located. Cultures and nations may also attempt to protect their people from the deleterious effects of the beliefs, values, norms, and social practices imposed on them from the "outside," which might negatively influence people's behaviors. Prohibitions of certain imported films, or videos of artistic performances, frequently occur in many places.

There is no doubt that these two sets of forces are powerful and dynamic, and they will likely shape much of the human experience in the twenty-first century. Discussions about these countervailing forces often come down to asking which forces are stronger: those promoting globalization and homogenization or those that encourage cultures to maintain their distinctiveness and unique ways of living?

We believe that what is missing from most discussions about the relative strengths of these forces is an understanding of the effects of culture on the human communication process. While these forces simultaneously push us together and pull us apart, what hasn't been acknowledged is that humans still bring their cultural backgrounds to their interpretations of these global events and symbols, which then shape the ways they make sense of them. That is, McDonald's arches, Jackie Chan's films, Jay-Z's music videos, Internet chatroom messages, and acts of "humanitarianism" are all interpreted and analyzed through individuals' differing cultural and national structures.

The patina of familiarity and commonality does not necessarily produce a shared understanding of the nature of everyday events. In Chapter 1, we discussed the important distinction between understanding and agreement in communication outcomes. The goal of living in a multicultural world, therefore, may sometimes mean that we must attempt to achieve understanding while recognizing that agreement may not always be likely, or even possible. Perhaps, however, we can sometimes "agree to disagree," with respect, civility, and caring. Intergroup tensions have characterized human interaction since the beginning of time, and they are not likely to abate soon. Stereotyping, we have suggested, is a natural and inevitable human tendency to categorize groups of others and thereby make the world more predictable. Our challenge is to assess individuals on their own merits, rather than merely as members of groups or nations, while simultaneously recognizing that humans typically identify, and often react to their worlds, as members of a culture.

Cultures and their symbolic systems also can change over time. No culture is static. Even cultures that have minimal contact with the outside world are affected by changing ecological conditions and events, which in turn change how cultures experience and understand their own familiar world.

We suggest that both the forces promoting globalization and those encouraging individuation are mediated by the cultural patterns—the beliefs, values, norms, and social practices—of all peoples. Even identical messages, therefore, are interpreted differently by those whose codes and cultures differ. Even identical media, such as Internet discussion groups, can encourage an understanding of oneself and others or they can promote alienation and foster hate. In short,

These U.S. American high school graduates must live in a multicultural world that demands competent intercultural skills.

as Charles Ess and Fay Sudweeks suggest, a genuinely intercultural global village is an alternative to the polarizing options of "Jihad" and "McWorld."[39] Thus, while we recognize the far-reaching effects of technological, societal, and economic forces, we must also remember that one's culture provides the meaning systems by which all messages are experienced and interpreted.

CONCLUDING REMARKS

We began this book with a sense of optimism but also with a deep concern about the pressing need for intercultural communication competence. Here in the twenty-first century, such competence is an essential attribute for personal survival, professional success, national harmony, and international peace. The challenge of living in a multicultural world is the need to transcend the unpredictability of intercultural interactions, to cope with the accompanying fears that such interactions often engender, and to feel joy and comfort in the discovery of cultural variability.

Our focus has been on the interpersonal hurdles—the person-to-person problems—that arise in coping with the realities of cultural diversity. We commend and encourage all who have struggled to adjust to the multicultural nature of the human landscape. Inclusion of others is the means to a better future, so we should be "widening our circle"[40] by acknowledging and celebrating cultural differences in all aspects of our lives.

The need for an intercultural mentality to match our multicultural world, the difficulties inherent in the quest of such a goal, the excitement of the challenges, and the rewards of the successes are summarized in the words of Troy Duster:

> There is no longer a single racial or ethnic group with an overwhelming numerical and political majority. Pluralism is the reality, with no one group a dominant force. This is completely new.[41]

We urge you to view this book and each intercultural experience as steps in a lifelong commitment to competence in intercultural communication. Intercultural competence is, in many ways, an art rather than a science. Our hope is that you will use your artistic talents to make the world a better place in which people from all cultures can live and thrive.

CULTURE CONNECTIONS

As we embrace difference, we come to realize that while we are all "human beings," our way of *being* is in fact influenced by the cultural experiences that create our frames of reference and our beliefs. While the external trappings of this global world (dress, transport, homes, entertainment) are increasingly the same, it becomes ever more important to value different points of view and ways of experiencing the world. Now, more than ever before, to be successful in our lives and our work as transnational leaders, we need to recognize the importance of diversity as a key factor for our success. Each of us must develop an awareness of the paradoxes that thwart our best intentions in communicating and working effectively with others. For it is in this awareness that trust and respect will flourish.

—Beth Fisher-Yoshida and Kathy D. Geller ■

SUMMARY

When one cultural group lives near other cultural groups, various forms of adaptation occur. The desire to maintain both an identification with the culture of origin and positive relationships with other cultures influences the type of adaptation that is experienced. An intercultural transformation occurs when people are able to move beyond the limits of their own cultural experiences to incorporate the perspectives of other cultures into their own interpersonal interactions.

Ethical issues in the development of intercultural competence concern questions about whose responsibility it is to adjust to a different culture, issues about right and wrong, and the degree to which all intercultural contacts should be encouraged. Global forces, which bring people together, are met with equally strong forces that pull people apart. Always, however, people bring to every communicative interaction understandings that are filtered and framed by their own cultural patterns. Thus, one must interpret intercultural experiences within the context of national and international events.

FOR DISCUSSION

1. Are there historical examples in the United States of some groups dominating and subordinating other cultural groups? Are there contemporary examples?
2. What responsibility does a visitor from another culture have to the host culture's ways of living, thinking, and communicating? For example, should people visiting from another culture accept or engage in behaviors they find ethically wrong but that the host culture sanctions as ethically correct?
3. What are some of the advantages and problems with cultural relativism?
4. Are there universal values that you believe are found in every culture? Explain.
5. What can we do—what can you do—to make the world a place where many cultures thrive?

FOR FURTHER READING

Clifford G. Christians and John C. Merrill (eds.), *Ethical Communication: Moral Stances in Human Dialogue* (Columbia, MO: University of Missouri Press, 2009). What does it mean to be an ethical communicator? This book explores the lives of well-known individuals (predominantly from Western cultures) such as Plato, Ghandi, and Mother Theresa, as a means of understanding the components of ethical communication.

Cathy N. Davidson, *36 Views of Mount Fuji: On Finding Myself in Japan*, 2nd ed. (Durham, NC: Duke University Press, 2006). A well-written narrative that documents the adaptation of a European American woman to life in Japan, as well as a description of her readjustments upon returning to her culture of origin.

Jane Jackson, *Language, Identity and Study Abroad: Sociocultural Perspectives* (Oakville, CT: Equinox, 2008). Both a guide to international study experiences and an analysis of how these cultural experiences, if done well, can help one to develop an appreciation for people from other cultures.

Ann Kelleher and Laura Klein, *Global Perspectives: A Handbook for Understanding Global Issues*, 4th ed. (Boston: Longman, 2011). A broad presentation of salient ways to understand global issues. Focuses on ecological, economic, political, psychological, and sociological perspectives.

Thomas F. Pettigrew and Linda R. Tropp, *When Groups Meet: The Dynamics of Intergroup Contact* (New York: Psychology Press, 2011). Under what circumstances does intercultural contact reduce prejudice? This book explores some of the issues that affect one's ability to engage in competent intercultural interactions.

Resources

C hapter 5 discusses several conceptual taxonomies to explain variations in cultural patterns. Here we provide specific cultural-level information on three of the taxonomies—those by Geert Hofstede, Shalom Schwartz, and the GLOBE team. Each taxonomy is organized and arranged by geographic region.

As we have suggested previously, no single dimension in any of the taxonomies is sufficient to describe a culture. Every culture has an intricate and interrelated pattern of preferences, which means that one must use multiple dimensions to understand it.

Before you look at the information provided here, we offer an important caution. The information that follows describes group-level or *cultural* tendencies, rather than the specific propensities of any given individual. Indeed, as we hope you already realize, no one—not you, not anyone you know, not anyone you will ever meet—is perfectly and exactly "typical" of these cultural tendencies. Though the data provided on the various taxonomies give insights into cultural trends, you should understand and expect that people within a culture will vary—in some cases quite substantially—from the cultural norm.

GEERT HOFSTEDE'S CULTURAL TAXONOMY[1]

Culture	Power Distance	Uncertainty Avoidance	Individualism– Collectivism	Masculinity– Femininity	Time Orientation	Indulgence– Restraint
Anglo						
Australia	−101	−71	191	63	−105	117
Canada	−97	−84	149	16	−43	104
England	−115	−141	186	89	20	108
Ireland	−148	−141	107	100	−92	90
New Zealand	−177	−80	145	47	−55	135
South Africa	−50	−80	86	74	−51	
USA	−92	−93	195	68	−84	104
Latin Europe						
Belgium (French)	35	111	115	58		
Canada (French)	−26	−32	119	−21		
France	39	81	111	−31	69	14
Israel	−219	59	40	−10	−34	
Italy	−45	33	132	110	61	−67
Malta	−17	124	61	−10	3	95

310

Culture	Power Distance	Uncertainty Avoidance	Individualism–Collectivism	Masculinity–Femininity	Time Orientation	Indulgence–Restraint
Portugal	16	159	−73	−94	−76	−54
Spain	−12	81	27	−36	7	−4
Switzerland (French)	49	11	82	47		
Nordic Europe						
Denmark	−195	−193	124	−172	−47	113
Finland	−125	−37	78	−120	−34	54
Norway	−134	−76	103	−214	−47	45
Sweden	−134	−167	111	−230	28	149
Germanic Europe						
Austria	−228	11	44	157	57	81
Belgium	7	128	140	−31	148	54
Germany	−115	−10	94	89	152	−22
Luxembourg	−92	11	65	6	73	50
Netherlands	−101	−63	149	−183	86	104
Switzerland	−158	−50	103	121	115	95
Eastern Europe						
Bulgaria	49	76	−61	−47	94	−130
Croatia	63	55	−48	−47	49	−54
Czech Republic	−12	29	57	42	98	−72
Estonia	−92	−32	65	−99	148	−130
Greece	2	193	−40	42	−5	23
Hungary	−64	63	149	205	49	−63
Latvia	−73	−19	107	−209	94	−144
Lithuania	−83	−10	65	−157	148	−130
Poland	39	111	65	79	−34	−72
Romania	143	98	−61	−36	24	−112
Russia	157	120	−23	−68	144	−112
Serbia	124	107	−82	−31	24	−76
Slovakia	209	−71	32	320	127	−76
Slovenia	54	89	−73	−157	11	14
Latin America						
Argentina	−50	81	6	37	−109	77
Brazil	44	37	−27	0	−9	63
Chile	16	81	−90	−110	−63	
Colombia	35	55	−132	79	−138	171
Costa Rica	−115	81	−123	−146		
Ecuador	86	−2	−153	74		
El Salvador	30	115	−107	−47	−109	198

(Continued)

Culture	Power Distance	Uncertainty Avoidance	Individualism–Collectivism	Masculinity–Femininity	Time Orientation	Indulgence–Restraint
Guatemala	166	146	−161	−62		
Jamaica	−69	−236	−23	100		
Mexico	101	63	−61	105	−92	234
Panama	166	81	−140	−26		
Peru	21	85	−119	−36	−88	5
Surinam	119	107	11	−62		
Trinidad	−59	−54	−119	47	−138	158
Uruguay	7	141	−35	−57	−84	36
Venezuela	101	37	−136	126	−125	248
Sub-Saharan Africa						
East Africa	21	−67	−73	−42	−91*	−25*
West Africa	82	−58	−102	−15	−126*	42*
Middle East						
Arab Countries	96	3	−27	21	−103*	−62*
Morocco	49	3	6	21	−134	−90
Turkey	30	76	−31	−21	−1	18
Southern Asia						
Bangladesh	96	−32	−102	32	3	−112
India	82	−119	15	37	20	−85
Indonesia	86	−84	−128	−15	65	−31
Iran	−7	−37	−15	−31	−134	−22
Malaysia	209	−136	−77	6	−22	54
Pakistan	−22	11	−128	6	15	−202
Philippines	162	−102	−52	79	−80	−13
Thailand	21	−15	−102	−78	−59	0
Vietnam	49	−162	−102	−47	44	−45
Confucian Asia						
China	96	−162	−102	89	169	−94
Hong Kong	39	−167	−82	42	61	−126
Japan	−26	107	6	241	173	−13
Singapore	68	−258	−102	−5	107	5
South Korea	2	76	−111	−52	222	−72
Taiwan	−7	7	−115	−21	194	18

A large positive score means that the culture is high on that dimension. A large negative score means that the culture is low on that dimension. The average score is zero. Ratings are in standardized scores, with the decimal point omitted.

* East African data are from Rwanda, Tanzania, Uganda, Zambia, and Zimbabwe; West African data are from Burkina Faso, Ghana, Mali, and Nigeria; Arab data are from Algeria, Egypt, Iran, Iraq, Jordan, and Saudi Arabia.

SHALOM SCHWARTZ'S CULTURAL TAXONOMY[2]

Culture	Intellectual Autonomy	Affective Autonomy	Embeddedness	Egalitarianism	Hierarchy	Harmony	Mastery
Anglo							
Australia	5	77	−48	39	−11	−11	21
Canada (English)	44	103	−80	42	−57	−64	116
England	75	153	−110	87	−2	−37	46
Ireland	54	113	−93	80	−57	−84	65
New Zealand	82	143	−127	95	−16	2	97
United States	−36	78	−28	−3	7	−187	97
Latin Europe							
Canada (French)	123	162	−157	109	−107	39	8
France	205	177	−145	136	−29	62	−139
Israel (Jewish)	54	63	−43	31	39	−246	52
Italy	149	−30	−80	218	−169	198	−81
Portugal	51	31	−88	195	−102	82	110
Spain	169	40	−117	203	−114	148	−88
Switzerland (French)	254	166	−184	139	−64	125	−126
Nordic Europe							
Denmark	113	160	−147	128	−109	46	−18
Finland	154	96	−103	80	−123	105	−177
Norway	90	44	−83	162	−194	125	−56
Sweden	195	149	−164	80	−116	145	−81
Germanic Europe							
Austria	146	158	−167	76	−134	95	−11
Belgium	80	92	−132	192	−148	109	−62
Germany (East)	90	156	−154	98	−130	145	40
Germany (West)	169	124	−187	143	−107	198	−49
Netherlands	134	128	−147	128	−98	9	21
Switzerland (German)	85	149	−110	87	18	−27	21
Eastern Europe							
Bosnia Herzegovina	−38	−30	56	−10	−139	−47	−94

(Continued)

Culture	Intellectual Autonomy	Affective Autonomy	Embeddedness	Egalitarianism	Hierarchy	Harmony	Mastery
Bulgaria	−10	2	21	−208	78	36	52
Croatia	5	88	53	−32	48	−1	72
Cyprus	−128	−47	63	61	−86	−4	8
Czech Republic	75	6	−48	−88	−27	82	−120
Estonia	−25	−19	6	−40	−68	95	−94
Georgia	−84	2	83	−10	28	22	−132
Greece	16	88	−93	57	−116	125	199
Hungary	62	33	−46	−66	−91	105	−132
Latvia	−28	4	11	−137	−123	145	−120
Macedonia	−23	−85	31	−107	87	2	40
Poland	−5	−26	18	−77	39	−54	−62
Romania	72	−1	−1	−77	−77	29	78
Russia	−8	10	6	−115	87	−41	14
Serbia	100	46	−53	−92	−166	−21	59
Slovakia	−10	−89	9	−40	−77	148	−69
Slovenia	141	50	−19	−47	−164	142	−145
Ukraine	−64	6	36	−141	50	−51	33
Latin America							
Argentina	3	52	−66	102	−55	−14	−11
Bolivia	3	−142	70	20	73	29	−43
Brazil	−15	12	−41	76	7	2	−5
Chile	−2	−81	−36	139	−20	102	−100
Colombia	−8	29	18	1	128	−120	59
Costa Rica	10	6	−73	61	−11	36	46
Mexico	8	−119	28	16	−48	158	−24
Peru	−8	−91	33	57	96	−104	91
Venezuela	28	−38	−11	31	−57	−11	46
Sub-Sahara Africa							
Cameroon	−192	−253	167	−3	28	55	−215
Ethiopia	−100	−161	186	−107	−2	55	−94
Ghana	−113	−184	120	16	78	−163	116
Namibia	−77	−32	63	−77	44	−94	78
Nigeria	−172	−175	154	39	87	−90	−24
Senegal	−113	−203	164	87	66	−147	−126
South Africa	−123	4	60	−62	57	−54	−30
Uganda	−136	−148	110	−111	149	−17	52
Zimbabwe	−136	27	63	−145	76	−134	161

Culture	Intellectual Autonomy	Affective Autonomy	Embeddedness	Egalitarianism	Hierarchy	Harmony	Mastery
Middle East							
Egypt	−110	−182	164	−100	−32	−14	−177
Israel (Arab)	−13	−17	75	−32	60	−150	103
Jordan	−72	−19	102	−107	37	−117	167
Oman	−154	−112	177	−74	−43	−104	−69
Turkey	31	−17	−4	31	144	69	27
Yemen	−167	−194	209	16	−13	−107	−94
Southern Asia							
India	−79	4	46	−88	162	−34	218
Indonesia	−100	−9	120	−137	50	−67	−62
Iran	−95	−93	97	−59	203	−110	−18
Malaysia	−46	−91	139	−103	−20	−124	−18
Nepal	−67	−89	97	−21	158	105	123
Pakistan	−146	−66	130	−14	23	−11	40
Philippines	−97	−87	60	−36	78	6	−113
Thailand	−79	33	58	−148	203	−61	−37
Confucian Asia							
China	−38	−30	−11	−171	263	−81	301
Hong Kong	−13	−49	−6	−70	130	−173	91
Japan	116	57	−73	−122	71	62	78
Singapore	−120	−30	53	−32	110	−87	−37
South Korea	−28	0	−26	−100	128	−150	174
Taiwan	8	−36	9	−141	80	32	40
Pacific Islands							
Fiji	−123	−49	135	−6	55	−64	−107

A large positive score means that the culture is high on that dimension. A large negative score means that the culture is low on that dimension. The average score is zero. Ratings are standardized scores, with the decimal point omitted.

The GLOBE Team's Cultural Taxonomy[3]

Culture	Power Distance	Uncertainty Avoidance	In-Group Collectivism	Institutional Collectivism	Gender Egalitarianism	Assertiveness	Future Orientation	Performance Orientation	Humane Orientation
Anglo									
Australia	-102	38	-132	10	-1.63	38	53	64	42
Canada (English)	-83	69	-119	31	-0.81	-24	129	96	86
England	-4	81	-144	5	-0.90	3	94	-4	-78
Ireland	-4	23	1	90	-2.14	-59	29	64	187
New Zealand	-66	97	-200	133	-2.12	-193	-81	152	50
South Africa (white)	-1	-12	-86	88	-1.98	124	61	3	-128
USA	-68	-2	-121	-11	-1.79	111	66	96	18
Latin Europe									
France	27	44	-104	-75	-0.98	-2	-79	3	-147
Israel	-104	-25	-59	50	-2.20	25	1	-4	3
Italy	63	-62	-26	-134	-2.06	-18	-129	-127	-98
Portugal	66	-42	52	-77	-0.92	-131	-29	-122	-38
Spain	85	-32	44	-94	-2.69	76	-73	-21	-164
Switzerland (French)	-73	135	-175	-7	-1.57	-180	92	37	-33
Nordic Europe									
Denmark	-305	175	-219	130	-0.19	-91	129	30	76
Finland	-66	142	-145	90	-1.76	-88	85	-70	-27
Sweden	-75	192	-201	229	-0.43	-204	118	-92	3
Germanic Europe									
Austria	-51	165	-38	12	-2.47	130	133	84	-78
Germany (prev. East)	90	165	-84	-162	-2.55	159	23	-2	-147

Culture	Power Distance	Uncertainty Avoidance	In-Group Collectivism	Institutional Collectivism	Gender Egalitarianism	Assertiveness	Future Orientation	Performance Orientation	Humane Orientation
Germany (prev. West)	20	175	-152	-108	-2.44	111	92	37	-194
Netherlands	-252	89	-196	50	-1.36	49	165	54	-48
Switzerland	-63	200	-159	-44	-2.80	100	191	206	-104
Eastern Europe									
Albania	-130	67	84	69	-0.79	202	3	174	119
Georgia	13	-110	145	-51	-1.22	11	-94	-53	20
Greece	56	-128	19	-235	-1.41	119	-96	-220	-160
Hungary	94	-172	16	-169	0.22	175	-137	-163	-158
Kazakhstan	35	-83	18	10	-0.43	87	-60	-129	-21
Poland	-16	-90	53	67	0.05	-21	-159	-51	-102
Russia	85	-212	68	59	0.19	-123	-209	-173	-31
Slovenia	39	-63	41	-28	-0.11	-37	-55	-107	-63
Latin America									
Argentina	114	-85	52	-139	-1.38	22	-166	-110	-21
Bolivia	-157	-134	47	-49	-1.22	-93	-51	-119	-8
Brazil	39	-93	7	-99	-1.87	17	-8	-14	-91
Colombia	94	-98	82	-103	-0.90	17	-125	-39	-78
Costa Rica	-102	-57	26	-75	-1.19	-104	-53	5	65
Ecuador	104	-80	93	-82	-2.52	-13	-23	25	121
El Salvador	123	-90	30	-127	-2.28	130	-10	-92	-81
Guatemala	104	-143	68	-129	-2.66	-67	-131	-70	-42
Mexico	13	3	79	-44	-0.98	84	5	1	-23
Venezuela	56	-119	55	-68	-1.03	52	-107	-190	35
Sub-Sahara Africa									
Namibia	30	6	-84	-28	-0.33	-61	-77	-105	-27
Nigeria	152	21	58	-25	-2.69	175	53	-44	3
South Africa (Black)	-305	71	-5	33	-0.92	60	172	137	54

(Continued)

Culture	Power Distance	Uncertainty Avoidance	In-Group Collectivism	Institutional Collectivism	Gender Egalitarianism	Assertiveness	Future Orientation	Performance Orientation	Humane Orientation
Zambia	35	-10	97	85	-3.09	-18	-49	15	245
Zimbabwe	121	-2	60	-30	-2.61	-21	-16	35	78
Middle East									
Egypt	-59	-17	70	59	-3.23	-61	3	42	138
Kuwait	-11	8	92	57	-3.85	-137	-127	-36	93
Morocco	152	-85	101	-89	-3.15	103	-127	-26	22
Qatar	-104	-28	-58	59	-1.00	-7	-14	-158	72
Turkey	97	-88	103	-51	-3.01	106	-23	-66	-31
Southern Asia									
India	73	-2	108	31	-2.99	-110	74	37	104
Indonesia	4	1	75	69	-2.01	-75	3	76	129
Iran	63	-81	123	-87	-2.74	-26	-32	118	31
Malaysia	1	102	52	85	-1.33	-72	159	59	168
Philippines	66	-45	168	95	-0.98	-34	66	91	221
Thailand	111	-38	78	-51	-1.76	-134	-90	-41	155
Confucian Asia									
China	-30	129	92	123	-2.58	-102	-21	86	59
Hong Kong	-49	26	26	-28	-1.44	143	40	172	-40
Japan	-13	-15	-68	222	-2.20	-147	96	30	46
Singapore	-42	190	70	154	-0.80	9	265	196	-128
South Korea	106	-101	56	225	-4.07	71	27	111	-59
Taiwan	4	29	63	81	-2.23	-59	25	113	5

A large positive score means that the culture is high on that dimension. A large negative score means that the culture is low on that dimension. The average score is zero. Ratings are standardized scores, with the decimal point omitted. For the Gender Egalitarianism dimension, a large positive score means that the cultural value is feminine. A large negative score means that the cultural value is masculine. A score of zero indicates egalitarianism.

NOTES

CHAPTER 1

1. U.S. Census Bureau, "2010 Census Data," accessed October 5, 2011, from http://2010.census.gov/2010census/data/; U.S. Census Bureau, "ACS Demographic and Housing Estimates: 2010 American Community Survey 1-Year Estimates," accessed October 9, 2011, from http://factfinder2.census.gov/faces/tableservices/jsf/pages/productview.xhtml?pid=ACS_10_1YR_DP05&prodType=table; U.S. Census Bureau, "Hispanic or Latino Origin by Race," accessed October 10, 2011, from http://factfinder2.census.gov/faces/tableservices/jsf/pages/productview.xhtml?pid=ACS_10_1YR_B03002&prodType=table. We have included Native Hawaiians, Pacific Islanders, and Alaska Natives in the Native American demographics.

2. Richard Fry, "Latinos Account for Half of U.S. Population Growth since 2000," Pew Hispanic Center, October 23, 2008. Accessed November 9, 2011, from http://pewresearch.org/pubs/1002/latino-population-growth; Richard Fry, "Latino Children: A Majority Are U.S.-Born Offspring of Immigrants," Pew Hispanic Center, May 28, 2009. Accessed November 9, 2011, from http://pewresearch.org/pubs/1235/latino-children-immigrants-american-born; Karen R. Humes, Nicholas A. Jones, and Roberto R. Ramirez, "Overview of Race and Hispanic Origin: 2010," U.S. Census Bureau, March 2011. Accessed November 9, 2011, from http://www.census.gov/prod/cen2010/briefs/c2010br-02.pdf; Paul Mackun and Steven Wilson, "Population Distribution and Change: 2000 to 2010," U.S. Census Bureau, March 2011. Accessed November 9, 2011, from http://www.census.gov/prod/cen2010/briefs/c2010br-01.pdf.

3. Mark Mather, "2010 Census Data Highlight Shifting Racial Makeup of U.S.," Population Reference Bureau, March 25, 2011. Accessed November 9, 2011, from http://prbblog.org/index.php/2011/03/25/2010-census-us-race/.

4. Jeffrey S. Passel and D'Vera Cohn, "U.S. Population Projections: 2005–2050," Pew Hispanic Center, February 11, 2008. Accessed November 9, 2011, from http://pewhispanic.org/reports/report.php?ReportID=85; U.S. Census Bureau, "An Older and More Diverse Nation by Midcentury," August 4, 2008. Accessed November 9, 2011, from http://www.census.gov/newsroom/releases/archives/population/cb08-123.html. We have included Native Hawaiians, Pacific Islanders, and Alaska Natives in the Native American demographics.

5. William A. Henry III, "Beyond the Melting Pot," *Time*, April 9, 1990, 28.

6. D'Vera Cohn, "Multi-Race and the 2010 Census," Pew Research Center, April 6, 2011. Accessed November 10, 2011, from http://pewresearch.org/pubs/1953/multi-race-2010-census-obama; Humes, Jones, and Ramirez; Gregory Rodriguez, "President Obama: At Odds with Clear Demographic Trends toward Multiracial Pride," *Los Angeles Times*, April 4, 2011. Accessed November 9, 2011, from http://www.latimes.com/news/opinion/commentary/la-oe-rodriguez-column-obama-race-20110404,0,3716973.column; U.S. Census Bureau, "2010 Census Data"; Hope Yen, "Multiracial people fastest growing U.S. demographic," *Los Angeles Daily News*, May 29, 2009, A19.

7. Yen.

8. Sharon R. Ennis, Merarys Ríos-Vargas, and Nora G. Albert, "The Hispanic Population: 2010," U.S. Census Bureau, May 2011. Accessed October 10, 2011, from http://www.census.gov/prod/cen2010/briefs/c2010br-04.pdf.

9. Sonya Rastogi, Tallese D. Johnson, Elizabeth M. Hoeffel, and Malcolm P. Drewery, Jr., "Black or African American Population for the United States, Regions, States, and for Puerto Rico: 2010," U.S. Census Bureau, September 2011. Accessed October 10, 2011, from http://www.census.gov/prod/cen2010/briefs/c2010br-04.pdf, http://www.census.gov/prod/cen2010/briefs/c2010br-06.pdf.

10. Rastogi, Johnson, Hoeffel, and Drewery, Jr.

11. Humes, Jones, and Ramirez.

12. Stephen Ceasar, "U.S. Hispanic population tops 50 million," *Los Angeles Times*, March 25, 2011, A15; Passel and Cohn.

13. Elizabeth M. Grieco and Edward N. Trevelyan, "Place of Birth of the Foreign-Born Population: 2009," U.S. Census Bureau, October 2010. Accessed November 10, 2011, from http://www.census.gov/prod/2010pubs/acsbr09-15.pdf.

14. Grieco and Trevelyan.

15. U.S. Census Bureau, "Language Spoken at Home: 2010 American Community Survey 1-Year Estimates," accessed October 10, 2011, from http://factfinder2.census.gov/faces/tableservices/jsf/pages/productview.xhtml?pid=ACS_10_1YR_S1603&prodType=table; U.S. Census Bureau, "Selected Social Characteristics in the United States: 2010 American Community Survey 1-Year Estimates," accessed October 10, 2011, from http://factfinder2.census.gov/faces/tableservices/jsf/pages/productview.xhtml?pid=ACS_10_1YR_DP02&prodType=table; U.S. Census Bureau, "Language Spoken at Home by Ability to Speak English for the Population 5 Years and Over," accessed October 10, 2011, from http://factfinder2.census.gov/faces/tableservices/jsf/pages/productview.xhtml?pid=ACS_10_1YR_B16001&prodType=table.

16. *The Chronicle of Higher Education*, Almanac Issue 2011–12, August 26, 2011, 4.

17. U.S. Census Bureau, "Selected Characteristics of the Native and Foreign-Born Populations: 2010 American Community Survey 1-Year Estimates," accessed October 10, 2011, from http://factfinder2.census.gov/faces/tableservices/jsf/pages/productview.xhtml?pid=ACS_10_1YR_S0501&prodType=table; U.S. Census Bureau, "Place of Birth by Educational Attainment in the United States," accessed October 10, 2011, from http://factfinder2.census.gov/faces/tableservices/jsf/pages/productview.xhtml?pid=ACS_10_1YR_B06009&prodType=table; U.S. Census Bureau, "Place of Birth by Marital Status in the United States," accessed October 10, 2011, from http://factfinder2.census.gov/faces/tableservices/jsf/pages/productview.xhtml?pid=ACS_10_1YR_B06008&prodType=table; U.S. Census Bureau, "Marital Status by Nativity," accessed October 10, 2011, from http://factfinder2.census.gov/faces/tableservices/jsf/pages/productview.xhtml?pid=ACS_10_1YR_B12005&prodType=table.

18. Antonia Pantoja and Wilhelmina Perry, "Cultural Pluralism: A Goal to Be Realized," *Voices from the Battlefront*, ed. Marta Moreno Vega and Cheryll Y. Greene (Trenton, NJ: Africa World Press, 1993), 136.

19. Susan Abram, "U.S. Finds Flaws in Minority Schooling," *Los Angeles Daily News*, October 12, 2011, A1, A5.

20. U.S. Department of Education, National Center for Education Statistics, 2008, "Digest of Education Statistics, 2007 (NCES 2008-022)," Tables 179, 196, 197, 198. Accessed July 3, 2008, from http://nces.ed.gov/fastfacts/display.asp?id=98.

21. *The Chronicle of Higher Education*, Almanac Issue 2011–12.

22. Institute of International Education, "Open Doors 2011 Fast Facts." Accessed November 14, 2011, from http://www.iie.org/en/Research-and-Publications/~/media/Files/Corporate/Open-Doors/Fast-Facts/Fast-Facts-2011.ashx.

23. *The Chronicle of Higher Education*, Almanac Issue 2011–12, 32.

24. Institute of International Education, "Open Doors 2011 Fast Facts.

25. Karin Fischer, "Internationally, the Business of Education is Booming," *The Chronicle of Higher Education*, May 29, 2009, A1, A31–32.

26. Marshall McLuhan, *The Gutenberg Galaxy: The Making of Typographic Man* (Toronto: University of Toronto Press, 1962).

27. http://www.wolframalpha.com/input/?i=total+cell+phones+in+use; http://www.wolframalpha.com/input/?i=total+Internet+users; http://www.wolframalpha.com/input/?i=total+world+population.

28. Richard W. Fisher, "Globalization's Impact on U.S. Growth and Inflation." Remarks before the Dallas, TX State Assembly, May 22, 2006. Accessed December 17, 2007, from http://www.dallasfed.org/news/speeches/fisher/2006/fs060522.cfm.

29. Vinnie Mirchandani, "Globalization and Technology," *Deal Architect*, July 8, 2007. Accessed December 17, 2007, from http://dealarchitect.typepad.com/deal_architect/globalization_and_technology/index.html.

30. Rob Baedeker, "Where Americans Visit Most," Forbes Traveler.com, November 1, 2007.

Accessed December 19, 2007, from http://www
.forbestraveler.com/best-lists/countries-americans-
visit-story.html.

31. Thomas L. Friedman, *The World Is Flat: A Brief
History of the Twenty-First Century*, further up-
dated and expanded ed. (New York: Picador, 2007).
To trace the development of Friedman's ideas, see
also: Thomas L. Friedman, *The Lexus and the Olive
Tree* (New York: Farrar, Straus, & Giroux, 1999).

32. Rana Foroohar, "Uncertainty? Don't Be So Sure,"
Time, October 31, 2011, 28.

33. Foroohar.

34. U.S. Census Bureau and U.S. Bureau of Economic
Analysis, "U.S. International Trade in Goods and
Services: July 2011," accessed October 12, 2011,
from http://www.census.gov/foreign-trade/Press-
Release/current_press_release/ft900.pdf.

35. International Trade Administration, "New Travel
and Tourism Forecast Projects Strong Growth for
International Arrivals through 2016," accessed
November 5, 2011, from http://tinet.ita.doc.gov/
tinews/archive/tinews2011/20111027.html.

36. Bill Saporito, "A Great Leap Forward," *Time*,
October 31, 2011, 34–39.

37. Roger Dow, "Tourism in America: Removing
Barriers and Promoting Growth," Testimony
before the United States Senate Committee On
Commerce, Science & Transportation, April 5,
2011. Accessed October 13, 2011, from http://
images.magnetmail.net/images/clients/USTRAVEL/
attach/USTravel_RogerDowTestimony_
SenCommerceSubcmteExports_Final_411.pdf.

38. Institute of International Education, "Open
Doors 2011: International Student Enrollment
Increased by 5 Percent in 2010/11," November
14, 2011. Accessed November 14, 2011, from
http://www.iie.org/Who-We-Are/News-and-
Events/Press-Center/Press-Releases/2011/2011-
11-14-Open-Doors-International-Students.

39. U.S. Census Bureau, *Asian Owned Firms: 2002*
(Washington, DC: August 2006).

40. Jim Hopkins, "African American Women Step Up
in Business World," *USAToday*, August 24, 2006.
Accessed December 18, 2007, from http://www
.usatoday.com/money/smallbusiness/2006-08-24-
women-biz-usat_x.htm.

41. HispanicBusiness.com. Accessed October 13,
2011, from http://www.hispanicbusiness.com/
research/hispanicbusinessprojections.asp.

42. Andrew Romano, "How Dumb Are We?"
Newsweek, March 28 and April 4, 2011, 56–60.

43. Robert Shuter, "The Centrality of Culture,"
Southern Communication Journal 55 (1990): 241.

44. President's Commission on Foreign Language and
International Studies (Washington, DC: Department
of Health, Education and Welfare, 1979), 1.

45. Southern Poverty Law Center, "Hate Map,"
accessed October 12, 2011, from http://www
.splcenter.org/get-informed/hate-map.

46. U.S. Census Bureau, "Table 322. Hate Crimes—
Number of Incidents, Offenses, Victims, and Known
Offenders by Bias Motivation: 2000 to 2008,"
accessed October 12, 2011, from http://www.census
.gov/compendia/statab/2012/tables/12s0322.pdf.

47. Catharine R. Stimpson, "A Conversation, Not
a Monologue," *Chronicle of Higher Education*,
March 16, 1994, B1.

48. See Frank E. X. Dance, "The 'Concept' of
Communication," *Journal of Communication*
20 (1970): 201–10; Frank E. X. Dance and
Carl E. Larson, *The Functions of Human
Communication* (New York: Holt, Rinehart and
Winston, 1976), Appendix A.

49. Dance.

50. Dance and Larson.

51. Claude E. Shannon and Warren Weaver, *The
Mathematical Theory of Communication*
(Urbana: University of Illinois Press, 1949).

52. Interactional models include David K. Berlo, *The
Process of Communication* (New York: Holt,
Rinehart and Winston, 1960); Wilbur Schramm,
The Process and Effects of Mass Communication
(Urbana: University of Illinois Press, 1954); and
Bruce H. Westley and Malcolm S. MacLean, Jr., "A
Conceptual Model for Communication Research,"
Journalism Quarterly 34 (1957): 31–38.

53. Dean Barnlund, *Interpersonal Communication:
Survey and Studies* (Boston: Houghton Mifflin,
1968), 512.

54. Donald W. Klopf, *Intercultural Encounters: The
Fundamentals of Intercultural Communications*
(Englewood, CO: Morton, 1987), 23–24.

55. John C. Condon, *Good Neighbors:
Communicating with the Mexicans* (Yarmouth,
ME: Intercultural Press, 1985), 34.

56. Edward T. Hall, *Beyond Culture* (Garden City,
NY: Anchor, 1977).

57. Quoted in Frank Bures, "A New Accent on
Diversity," *Christian Science Monitor*, November
6, 2002, pp. 11–15. See also: Richard Rodriguez,
Brown: The Last Discovery of America (New
York: Viking, 2002).

CHAPTER 2

1. Alfred L. Kroeber and Clyde Kluckhohn, *Culture: A Critical Review of Concepts and Definitions* (Cambridge, MA: Harvard University Press, 1952).

2. John R. Baldwin, Sandra L. Faulkner, Michael L. Hecht, and Sheryl L. Lindsley, (eds.), *Redefining Culture: Perspectives across the Disciplines* (Mahwah, NJ: Lawrence Erlbaum Associates, 2006).

3. Mary Jane Collier and Milt Thomas, "Cultural Identity: An Interpretive Approach," *Theories in Intercultural Communication*, ed. Young Yun Kim and William B. Gudykunst (Newbury Park, CA: Sage, 1988), 103.

4. See, for example, Maurice Berger, *White Lies: Race and the Myths of Whiteness*, 1st ed. (New York: Farrar, Straus, Giroux, 1999); Karen Brodkin, *How Jews Became White Folks and What That Says About Race in America* (New Brunswick, NJ: Rutgers University Press, 2000); Paul D. Buchanan, *Race Relations in the United States: A Chronology, 1896–2005* (Jefferson, NC: McFarland & Co., 2005); Eric L. Goldstein, *The Price of Whiteness: Jews, Race, and American Identity* (Princeton: Princeton University Press, 2006); Laura E. Gómez, *Manifest Destinies: The Making of the Mexican American Race* (New York: New York University, 2007); Noel Ignatiev, *How the Irish Became White* (New York: Routledge, 1996); David R. Roediger, *Working toward Whiteness: How America's Immigrants Became White: The Strange Journey from Ellis Island to the Suburbs* (New York: Basic Books, 2005); David R. Roediger, *The Wages of Whiteness: Race and the Making of the American Working Class,* Haymarket Series Rev. ed. (New York: Verso, 2007); Karen Elaine Rosenblum and Toni-Michelle Travis, *The Meaning of Difference: American Constructions of Race, Sex and Gender, Social Class, Sexual Orientation, and Disability*, 5th ed. (New York: McGraw-Hill, 2008).

5. Brodkin, 74. See also Gómez, *Manifest Destinies: The Making of the Mexican American Race.*

6. For an elaboration of this idea, see Brodkin; Richard Delgado and Jeanne Stefancic (eds.), *Critical White Studies: Looking Behind the Mirror* (Philadelphia: Temple University Press, 1997); Marc Edelman, "Devil, Not-Quite-White, Rootless Cosmopolitan: Tsuris in Latin America, the Bronx, and the USSR," *Composing Ethnography: Alternative Forms of Qualitative Writing*, ed. Carolyn Ellis and Arthur Bochner (Walnut Creek, CA: Altamira Press, 1996); Ruth Frankenberg, *White Women, Race Matters: The Social Construction of Whiteness* (Minneapolis: University of Minnesota Press, 1993); Ian F. Haney Lopez, *White by Law: The Legal Construction of Race* (New York: New York University Press, 1996); Noel Ignatiev, *How the Irish Became White* (Cambridge, MA: Harvard University Press, 1995); Rosenblum and Travis; David Stowe, "Uncolored People: The Rise of Whiteness Studies," *Lingua Franca* 6 (1996): 68–77.

7. Thierry Devos and Mahzarin R. Banaji, "American = White?" *Journal of Personality and Social Psychology* 88 (2005): 447–466.

8. Myron W. Lustig, "WSCA 2005 Presidential Address: Toward a Well-Functioning Intercultural Nation," *Western Journal of Communication* 69 (2005): 377–379.

9. Gustav Ichheiser, *Appearances and Realities: Misunderstanding in Human Relations* (San Francisco: Jossey-Bass, 1970), 8.

10. David McCullough, 1994 Commencement Address at the University of Pittsburgh, quoted in *Chronicle of Higher Education*, June 8, 1994, B2.

11. Geert Hofstede, *Culture's Consequences: International Differences in Work-Related Values* (Beverly Hills, CA: Sage, 1980).

12. Peter A. Andersen, "Cues of Culture: The Basis of Intercultural Differences in Nonverbal Communication," *Intercultural Communication: A Reader*, 8th ed., ed. Larry A. Samovar and Richard E. Porter (Belmont, CA: Wadsworth, 1997), 244–256; Edward T. Hall, *The Hidden Dimension* (New York: Doubleday, 1966); Miles L. Patterson, *Nonverbal Behavior: A Functional Perspective* (New York: Springer-Verlag, 1983); Carol Zinner Dolphin, "Variables in the Use of Personal Space in Intercultural Transactions," *Intercultural Communication: A Reader*, 8th ed., ed. Samovar and Porter, 266–276.

13. Peter A. Andersen, Myron W. Lustig, and Janis F. Andersen, "Changes in Latitude, Changes in Attitude: The Relationship Between Climate and Interpersonal Communication Predispositions," *Communication Quarterly* 38 (1990): 291–311.

14. Charlotte Evans, "Barbed Wire, the Cutting Edge in Fencing," *Smithsonian* 22(4) (July 1991): 72–78, 80, 82–83.

15. Henry Chu, "Ancient Hindu Blessings a Mouse Click Away," *Los Angeles Times*, December 24, 2007, A3.

16. See Clayton Jones, "Cultural Crosscurrents Buffet the Orient," *Christian Science Monitor*, December 8, 1993, 11, 13; Sheila Tefft, "Satellite Broadcasts Create Stir Among Asian Regimes," *Christian Science Monitor*, December 8, 1993, 12–13.

17. George A. Barnett and Meihua Lee, "Issues in Intercultural Communication Research," *Cross-Cultural and Intercultural Communication*, ed. William B. Gudykunst (Thousand Oaks, CA: Sage, 2003), 268.

18. James W. Chesebro, "Communication, Values, and Popular Television Series—a Twenty-Five Year Assessment and Final Conclusions," *Communication Quarterly* 51 (2003): 367–418.

19. Chesebro, 395–396, 398.

20. Daniel L. Hartl and Andrew G. Clark, *Principles of Population Genetics*, 4th ed. (Sunderland, MA: Sinauer Associates, 2007).

21. Sandra Scarr, A. J. Pakstis, S. H. Katz, and W. B. Barker, "Absence of Relationship between Degree of White Ancestry and Intellectual Skills within a Black Population," *Human Genetics* 39 (1977): 69–86; Sandra Scarr and Richard A. Weinberg, "I.Q. Test Performance of Black Children Adopted by White Families," *American Psychologist* 31 (1976): 726–739; Richard A. Weinberg, Sandra Scarr, and Irwin D. Waldman, "The Minnesota Transracial Adoption Study: A Follow-up of IQ Test Performance at Adolescence," *Intelligence* 16 (1992): 117–135. See also: Thomas J. Bouchard, Jr., David T. Lykken, Matthew McGue, Nancy Segal, and Auke Tellegen, "Sources of Human Physiological Differences: The Minnesota Study of Twins Reared Apart," *Science* 250 (1990): 223–228.

22. Michael Winkelman, *Ethnic Relations in the U.S.: A Sociohistorical Cultural Systems Approach* (Minneapolis: West, 1993), 67–68.

23. Cary Quan Gelernter, "Racial Realities," *Seattle Times/Post-Intelligencer*, January 15, 1989, K1. Reprinted in *Ethnic Groups*, vol. 4., ed. Eleanor Goldstein (Boca Raton, FL: Social Issues Resources Ser., 1994), art. no. 83.

24. Alan R. Rogers and Lynn B. Jorde, "Genetic Evidence on Modern Human Origins," *Human Biology* 67 (1995): 1–36; Linda Vigilant, Mark Stoneking, Henry Harpending, Kristen Hawkes, and Allan C. Wilson, "African Populations and the Evolution of Human Mitochondrial DNA," *Science* 253 (1991): 1503–1507; Nicholas Wade, *Before the Dawn: Recovering the Lost History of Our Ancestors* (New York: Penguin, 2006).

25. Wade.

26. Lundy Braun, "Race and Genetics—Reifying Human Difference: The Debate on Genetics, Race, and Health," *International Journal of Health Services* 36 (2006): 557–574; Eddie L. Hoover, "There Is No Scientific Rationale for Race-Based Research," *Journal of the National Medical Association* 99 (2007): 690–692; Race, Ethnicity, and Genetics Working Group, "The Use of Racial, Ethnic, and Ancestral Categories in Human Genetics Research," *American Journal of Human Genetics* 77 (2005): 519–532.

27. Troy Duster, "The Molecular Reinscription of Race: Unanticipated Issues in Biotechnology and Forensic Science," *Patterns of Prejudice* 40 (2006): 427–441; Reanne Frank, "What to Make of It? The (Re)Emergence of a Biological Conceptualization of Race in Health Disparities Research," *Social Science & Medicine* 64 (2007): 1977–1983.

28. Luigi Luca Cavalli-Sforza, "Human Evolution and Its Relevance for Genetic Epidemiology," *Annual Review of Genomics and Human Genetics* 8 (2007): 1–15.

29. Cavalli-Sforza, 13–14.

30. Charmaine D. M. Royal and Georgia M. Dunston, "Changing the Paradigm from 'Race' to Human Genome Variation," *Nature Genetics* 36 (2004): S5–S7.

31. For discussions of race and ethnicity as social and political distinctions rather than as a genetic one, see Linda Martín Alcoff and Eduardo Mendieta (eds.), *Identities: Race, Class, Gender, and Nationality* (Malden, MA: Blackwell, 2003); John D. Buenker and Lorman A. Ratner (eds.), *Multiculturalism in the United States: A Comparative Guide to Acculturation and Ethnicity* (Westport, CT: Greenwood, 2005); Rodney D. Coates (ed.), *Race and Ethnicity: Across Time, Space, and Discipline* (Boston: Brill, 2004); Stephanie Cole and Alison M. Parker (eds.), *Beyond Black & White: Race, Ethnicity, and Gender in the U.S. South and Southwest* (Arlington, TX: Texas A&M University Press, 2004); Steve Fenton, *Ethnicity* (Cambridge: Polity Press, 2003); Joan Ferrante and Prince Brown

(eds.), *The Social Construction of Race and Ethnicity in the United States*, 2nd ed. (Upper Saddle River, NJ: Prentice Hall, 2001); Nancy Foner and George M. Fredrickson (eds.), *Not Just Black and White: Historical and Contemporary Perspectives on Immigration, Race, and Ethnicity in the United States* (New York: Russell Sage Foundation, 2004); David Theo Goldberg and John Solomos (eds.), *A Companion to Racial and Ethnic Studies* (Malden, MA: Blackwell, 2002); Harry Goulbourne (ed.), *Race and Ethnicity: Critical Concepts in Sociology* (New York: Routledge, 2001); Yasmin Gunaratnam, *Researching Race and Ethnicity: Methods, Knowledge, and Power* (Thousand Oaks, CA: Sage, 2003); Joseph F. Healey, *Diversity and Society: Race, Ethnicity and Gender* (Thousand Oaks, CA: Pine Forge Press, 2004); Joseph F. Healey and Eileen O'Brien (eds.), *Race, Ethnicity, and Gender: Selected Readings* (Thousand Oaks, CA: Pine Forge Press, 2004); Caroline Knowles, *Race and Social Analysis* (Thousand Oaks, CA: Sage, 2003); Maria Krysan and Amanda Lewis (eds.), *The Changing Terrain of Race and Ethnicity* (New York: Russell Sage Foundation, 2004); Jennifer Lee and Min Zhou (eds.), *Asian American Youth: Culture, Identity, and Ethnicity* (New York: Routledge, 2004); Peter Osborne and Stella Sandford (eds.), *Philosophies of Race and Ethnicity* (New York: Continuum, 2002); Raymond Scupin (ed.), *Race and Ethnicity: An Anthropological Focus on the United States and the World* (Upper Saddle River, NJ: Prentice Hall, 2003); Miri Song, *Choosing Ethnic Identity* (Malden, MA: Blackwell, 2003); Paul Spickard (ed.), *Race and Nation: Ethnic Systems in the Modern World* (New York: Routledge, 2004).

32. Audrey Smedley and Brian D. Smedley, "Race as Biology Is Fiction, Racism as a Social Problem Is Real: Anthropological and Historical Perspectives on the Social Construction of Race," *American Psychologist* 60 (2005): 16–26.

33. Marshall R. Singer, *Intercultural Communication: A Perceptual Approach* (Englewood Cliffs, NJ: Prentice Hall, 1987).

34. Jared Diamond, *Guns, Germs, and Steel: The Fates of Human Societies* (New York: W. W. Norton, 1999).

35. Diamond, 92.

36. The idea that the degree of heterogeneity among participants distinguishes intercultural from intracultural communication, and thereby results in a continuum of "interculturalness" of the communication, was first introduced by L. E. Sarbaugh, *Intercultural Communication* (Rochelle Park, NJ: Hayden, 1979).

37. For a more detailed discussion of the relationships among these terms, see Molefi Kete Asante and William B. Gudykunst (eds.), *Handbook of Intercultural Communication* (Newbury Park, CA: Sage, 1989), 7–13.

CHAPTER 3

1. The melting pot metaphor for U.S. cultural diversity was popularized by Israel Zangwell's play of 1908, *The Melting Pot*. The idea was anticipated more than 100 years earlier in Crèvecoeur's description of America: "Here individuals of all nations are melted into a new race of men, whose labors and posterity will one day cause great changes in the world." See J. Hector St. John Crèvecoeur, *Letter from an American Farmer* (New York: Albert and Charles Boni, 1782/1925), 55; Israel Zangwell, *The Melting Pot. Drama in Four Acts* (New York: Arno Press, 1908/1975). For an updated view of the melting pot metaphor, see Tamar Jacoby (ed.), *Reinventing the Melting Pot: The New Immigrants and What It Means to Be American* (New York: Basic Books, 2004).

2. See also Myron W. Lustig, "WSCA 2005 Presidential Address: Toward a Well-Functioning Intercultural Nation," *Western Journal of Communication* 69 (2005): 377–379.

3. Judith N. Martin, Robert L. Krizek, Thomas K. Nakayama, and Lisa Bradford, "Exploring Whiteness: A Study of Self Labels for White Americans," *Communication Quarterly* 44 (1996): 125–144; Thomas K. Nakayama and Robert L. Krizek, "Whiteness: A Strategic Rhetoric," *Quarterly Journal of Speech* 81 (1995): 291–309.

4. See Rodolfo O. de la Garza, Louis DeSipio, F. Chris Garcia, John Garcia, and Angelo Falcon, *Latino Voices: Mexican, Puerto Rican, & Cuban Perspectives on American Politics* (Boulder, CO: Westview Press, 1992).

5. James Diego Vigil, *From Indians to Chicanos: The Dynamics of Mexican American Culture* (Prospect Heights, IL: Waveland Press, 1980), 1.

6. Mark Z. Barabak, "Differences Found Among U.S. Hispanics," *San Diego Union*, August 30, 1991, A2.

7. Juan L. Gonzales, Jr., *Racial and Ethnic Groups in America* (Dubuque, IA: Kendall/Hunt, 1990), 199.

8. Earl Shorris, "The Latino vs. Hispanic Controversy," *San Diego Union-Tribune*, October 29, 1992, B11. See also Earl Shorris, *Latinos: A Biography of the People* (New York: Norton, 1992); Earl Shorris, "Latinos: The Complexity of Identity," *Report on the Americas* 26 (1992): 19–26.

9. See, for example, Michael L. Hecht and Sidney A. Ribeau, "Sociocultural Roots of Ethnic Identity: A Look at Black America," *Journal of Black Studies* 21 (1991): 501–513; George B. Ray, *Language and Interracial Communication in the United States: Speaking in Black and White* (New York: Peter Lang, 2009).

10. Kimberly Rios Morrison and Adrienne H. Chung, "'White' or 'European American'? Self-Identifying Labels Influence Majority Group Members' Interethnic Attitudes," *Journal of Experimental Social Psychology* 47 (2011): 165–170.

11. For a discussion of the concept of communication competence, see Lily A. Arasaratnam, "Empirical Research in Intercultural Communication Competence," *Australian Journal of Communication* 34.1 (2007): 105–117; Michael Byram, Adam Nichols, and David Stevens, *Developing Intercultural Competence in Practice* (New York: Multilingual Matters, 2001); Daniel J. Canary and Brian H. Spitzberg, "A Model of Competence Perceptions of Conflict Strategies," *Human Communication Research* 15 (1989): 630–649; Daniel J. Canary and Brian H. Spitzberg, "Attribution Biases and Associations between Conflict Strategies and Competence Outcomes," *Communication Monographs* 57 (1990): 139–151; Darla K. Deardorff, "Identification and Assessment of Intercultural Competence as a Student Outcome of Internationalization," *Journal of Studies in Intercultural Education* 10 (2006): 241–266; Darla K. Deardorff (ed.), *The Sage Handbook of Intercultural Competence* (Thousand Oaks, CA: Sage, 2009); Robert L. Duran and Brian H. Spitzberg, "Toward the Development and Validation of a Measure of Cognitive Communication Competence," *Communication Quarterly* 43 (1995): 259–275; DeWan Gibson and Mei Zhong, "Intercultural Communication Competence in the Healthcare Context," *International Journal of Intercultural Relations* 29 (2005): 621–634; David A. Griffith, "The Role of Communication Competencies in International Business Relationship Development," *Journal of World Business* 37 (2002): 256–265; Christopher Hajek and Howard Giles, "New Directions in Intercultural Communication Competence: The Process Model," *Handbook of Communication and Social Interaction Skills*, ed. John O. Greene and Brant R. Burleson (Mahwah, NJ: Erlbaum, 2003), 935–957; Mitchell R. Hammer, "The Intercultural Conflict Style Inventory: A Conceptual Framework and Measure of Intercultural Conflict Resolution Approaches," *International Journal of Intercultural Relations* 2 (2005): 675–695; Mitchell R. Hammer, Milton J. Bennett, and Richard L. Wiseman, "Measuring Intercultural Sensitivity: The Intercultural Development Inventory," *International Journal of Intercultural Relations* 27 (2003): 421–443; Paige Johnson, A. Elizabeth Lindsey, and Walter R. Zakahi, "Anglo American, Hispanic American, Chilean, Mexican, and Spanish Perceptions of Competent Communication in Initial Interaction," *Communication Research Reports* 18 (2001): 36–43; Shelley D. Lane, *Interpersonal Communication: Competence and Contexts* (Boston: Allyn & Bacon, 2008); Ildikó Lázár, Martina Huber-Kriegler, Denise Lussier, Gabriela S. Matai, and Christiane Peck (eds.), *Developing and Assessing Intercultural Communicative Competence: A Guide for Language Teachers and Teacher Educators* (Graz, Austria: Council of Europe, 2007); Myron W. Lustig and Brian H. Spitzberg, "Methodological Issues in the Study of Intercultural Communication Competence," *Intercultural Communication Competence*, ed. Richard L. Wiseman and Jolene Koester (Newbury Park, CA: Sage, 1993), 153–167; Alexei V. Matveev, "Describing Intercultural Communication Competence: In-Depth Interviews with American and Russian Managers," *Qualitative Research Reports in Communication* 5 (2004): 55–62; Robyn Moloney, "Forty Per Cent French: Intercultural Competence and Identity in an Australian Language Classroom," *Intercultural Education* 20 (2009): 71–81; Michal A. Moodian (ed.), *Contemporary Leadership and Intercultural Competence: Exploring Cross-Cultural Dynamics within Organizations* (Thousand Oaks, CA: Sage, 2009); Charles Pavitt, "The

Ideal Communicator as the Basis for Competence Judgments of Self and Friend," *Communication Reports* 3 (1990): 9–14; Stefanie Rathje, "Intercultural Competence: The Status and Future of a Controversial Concept," *Language and Intercultural Communication* 7 (2007): 254–266; Gert Rickheit and Hans Strohner (eds.), *Handbook of Communication Competence* (New York: Mouton de Gruyter, 2008); Brian H. Spitzberg, "Issues in the Development of a Theory of Interpersonal Competence in the Intercultural Context," *International Journal of Intercultural Relations* 13 (1989): 241–268; Brian H. Spitzberg, "An Examination of Trait Measures of Interpersonal Competence," *Communication Reports* 4 (1991): 22–29; Brian H. Spitzberg, "The Dialectics of (In)competence," *Journal of Social and Personal Relationships* 10 (1993): 137–158; Brian H. Spitzberg, "The Dark Side of (In)Competence," *The Dark Side of Interpersonal Communication*, ed. William R. Cupach and Brian H. Spitzberg (Hillsdale, NJ: Erlbaum, 1994), 25–49; Brian H. Spitzberg, "Intimate Violence," *Competence in Interpersonal Conflict*, ed. William R. Cupach and Daniel J. Canary (New York: McGraw-Hill, 1997), 174–201; Brian H. Spitzberg and Claire C. Brunner, "Toward a Theoretical Integration of Context and Competence Research," *Western Journal of Speech Communication* 56 (1991): 28–46; Brian H. Spitzberg and Gabrielle Chagnon, "Conceptualizing Intercultural Communication Competence," *The Sage Handbook of Intercultural Competence*, ed. Deardorff, 2–52; Brian H. Spitzberg and William R. Cupach, *Interpersonal Communication Competence* (Beverly Hills, CA: Sage, 1984); Brian H. Spitzberg and William R. Cupach, *Handbook of Interpersonal Competence Research* (New York: Springer-Verlag, 1989); Brian H. Spitzberg and William R. Cupach, "Interpersonal Skills," *Handbook of Interpersonal Communication*, 3rd ed., ed. Mark L. Knapp and John R. Daly (Newbury Park, CA: Sage, 2002), 564–611; Brian H. Spitzberg and William R. Cupach (eds.), *The Dark Side of Interpersonal Communication*, 2nd ed. (Mahwah, NJ: Erlbaum, 2007); Stella Ting-Toomey, "Researching Intercultural Conflict Competence," *Journal of International Communication* 13 (2007): 7–30; Wen-Shing Tseng and Jon Mark Streltzer (eds.), *Cultural*

Competence in Health Care (New York: Springer, 2008); Yu-Wen Ying, "Variation in Acculturative Stressors over Time: A Study of Taiwanese Students in the United States," *International Journal of Intercultural Relations* 29 (2005): 59–71.

12. Brian H. Spitzberg, "Communication Competence: Measures of Perceived Effectiveness," *A Handbook for the Study of Human Communication*, ed. Charles H. Tardy (Norwood, NJ: Ablex, 1988), 67–105.

13. William R. Cupach and T. Todd Imahori, "Identity Management Theory: Communication Competence in Intercultural Episodes and Relationships," *Intercultural Communication Competence*, ed. Wiseman and Koester, 112–131; T. Todd Imahori and Mary L. Lanigan, "Relational Model of Intercultural Communication Competence," *International Journal of Intercultural Relations* 13 (1989): 269–286.

14. Fathi S. Yousef, "Human Resource Management: Aspects of Intercultural Relations in U.S. Organizations," *Intercultural Communication: A Reader*, 5th ed., ed. Larry A. Samovar and Richard E. Porter (Belmont, CA: Wadsworth, 1988), 175–182.

15. Darla K. Deardorff, *The Identification and Assessment of Intercultural Competence as a Student Outcome of Internationalization at Institutions of Higher Education in the United States*, Unpublished doctoral dissertation, North Carolina State University, Raleigh, 2004; Darla K. Deardorff, "Identification and Assessment of Intercultural Competence as a Student Outcome of Internationalization," *Journal of Studies in Intercultural Education* 10 (2006): 241–266; Margaret D. Pusch, "The Interculturally Competent Global Leader," *The Sage Handbook of Intercultural Competence*, ed. Deardorff, 66–84.

16. Ann Neville Miller, "An Exploration of Kenyan Public Speaking Patterns with Implications for the American Introductory Public Speaking Course," *Communication Education* 51 (2002): 168–182.

17. Jolene Koester and Margaret Olebe, "The Behavioral Assessment Scale for Intercultural Communication Effectiveness," *International Journal of Intercultural Relations* 12 (1988): 233–246; Margaret Olebe and Jolene Koester,

"Exploring the Cross-Cultural Equivalence of the Behavioral Assessment Scale for Intercultural Communication," *International Journal of Intercultural Relations* 13 (1989): 333–347.

18. Brent D. Ruben, "Assessing Communication Competency for Intercultural Adaptation," *Group and Organization Studies* 1 (1976): 334–354; Brent D. Ruben, Lawrence R. Askling, and Daniel J. Kealey, "Cross-Cultural Effectiveness," *Overview of Intercultural Training, Education, and Research, Vol. I: Theory*, ed. David S. Hoopes, Paul B. Pedersen, and George W. Renwick (Washington, DC: Society for Intercultural Education, Training and Research, 1977), 92–105; Brent D. Ruben and Daniel J. Kealey, "Behavioral Assessment of Communication Competency and the Prediction of Cross-Cultural Adaptation," *International Journal of Intercultural Relations* 3 (1979): 15–48.

CHAPTER 4

1. Milton Rokeach, *Beliefs, Attitudes, and Values: A Theory of Organization and Change* (San Francisco: Jossey-Bass, 1969).

2. Elisabeth Bumiller, *May You Be the Mother of a Hundred Sons* (New York: Fawcett Columbine, 1990), 11.

3. Milton Rokeach, *The Nature of Human Values* (New York: Free Press, 1973); Milton Rokeach, "Value Theory and Communication Research: Review and Commentary," *Communication Yearbook* 3, ed. Dan Nimmo (New Brunswick, NJ: Transaction, 1979), 7–28.

4. Shalom H. Schwartz, "Universals in the Content and Structure of Values: Theoretical Advances and Empirical Tests in 20 Countries," *Advances in Experimental Social Psychology*, vol. 25, ed. Mark P. Zanna (San Diego: Academic Press, 1992), 1–65; Shalom H. Schwartz, "Are There Universal Aspects in the Structure and Content of Values?" *Journal of Social Issues* 50 (1994): 19–45; Shalom H. Schwartz and Anat Bardi, "Value Hierarchies across Cultures: Taking a Similarities Perspective," *Journal of Cross-Cultural Psychology* 32 (2001): 268–290; Shalom H. Schwartz, Gila Melech, Arielle Lehmann, Steven Burgess, Mari Harris, and Vicki Owens, "Extending the Cross-Cultural Validity of the Theory of Basic Human Values with a Different Method of Measurement," *Journal of Cross-Cultural Psychology* 32 (2001): 519–542; Shalom H. Schwartz and Tammy Rubel, "Sex Differences in Value Priorities: Cross-Cultural and Multimethod Studies," *Journal of Personality and Social Psychology* 89 (2005): 1010–1028; Shalom H. Schwartz and Lilach Sagiv, "Identifying Culture-Specifics in the Content and Structure of Values," *Journal of Cross-Cultural Psychology* 26 (1995): 92–116; Shalom H. Schwartz, Markku Verkasalo, Avishai Antonovsky, and Lilach Sagiv, "Value Priorities and Social Desirability: Much Substance, Some Style," *British Journal of Social Psychology* 36 (1997): 3–18; Naomi Struch, Shalom H. Schwartz, and Willem A. van der Kloot, "Meanings of Basic Values for Women and Men: A Cross-Cultural Analysis," *Personality and Social Psychology Bulletin* 28 (2002): 16–28.

5. Norine Dresser, *Multicultural Manners: Essential Rules of Etiquette for the 21st Century*, rev. ed. (Hoboken, NJ: Wiley, 2005) 30.

6. Dresser, 91.

7. William B. Gudykunst and Carmen M. Lee, "Cross-Cultural Theories," *Cross-Cultural and Intercultural Communication*, ed. William B. Gudykunst (Thousand Oaks, CA: Sage, 2003), 12.

8. Florence Rockwood Kluckhohn and Fred L. Strodtbeck, *Variations in Value Orientations* (Evanston, IL: Row, Peterson, 1960).

9. See Edgar H. Schein, *Organizational Culture and Leadership*, 2nd ed. (San Francisco: Jossey-Bass, 1992). Based on the work of Kluckhohn and Strodtbeck and all subsequent scholars of cultural patterns, we have added the "time" dimension to Schein's typology.

10. Our primary sources for this section include John C. Condon and Fathi Yousef, *An Introduction to Intercultural Communication* (Indianapolis: Bobbs-Merrill, 1975); Edward C. Stewart, *American Cultural Patterns: A Cross-Cultural Perspective* (Pittsburgh: Regional Council for International Education, 1971); Edward C. Stewart and Milton J. Bennett, *American Cultural Patterns: A Cross-Cultural Perspective*, rev. ed. (Yarmouth, ME: Intercultural Press, 1991). Excellent resources with thorough descriptions of the cultural patterns of U.S. American cultural groups include Don C. Locke, *Increasing Multicultural Understanding: A Comprehensive Model* (Newbury Park, CA: Sage, 1992); Eleanor W. Lynch and Marci J. Hanson, *Developing*

Cross-Cultural Competence: A Guide for Working with Young Children and Their Families (Baltimore: Paul Brookes, 1992); and Esther Wanning, Culture Shock: USA (Singapore: Times Books International, 1991).

11. Rajesh Kumar and Anand Kumar Sethi, Doing Business in India: A Guide for Western Managers (New York: Palgrave Macmillan, 2005).

12. Yale Richmond, From Da to Yes: Understanding the East Europeans (Yarmouth, ME: Intercultural Press, 1995), 110.

13. See, for example, Ringo Ma, "The Role of Unofficial Intermediaries in Interpersonal Conflicts in the Chinese Culture," Communication Quarterly 40 (1992): 269–278.

14. Mary Jane Collier, Sidney A. Ribeau, and Michael L. Hecht, "Intracultural Communication Rules and Outcomes Within Three Domestic Cultures," International Journal of Intercultural Relations 10 (1986): 452. Also see Mary Jane Collier, "A Comparison of Conversations Among and Between Domestic Cultural Groups: How Intra- and Intercultural Competencies Vary," Communication Quarterly 36 (1988): 122–144.

15. Jack L. Daniel and Geneva Smitherman, "How I Got Over: Communication Dynamics in the Black Community," Quarterly Journal of Speech 62 (1976): 29.

16. Daniel and Smitherman, 31.

17. Jamake Highwater, The Primal Mind (New York: New American Library, 1981).

18. Melvin Delgado, "Hispanic Cultural Values: Implications for Groups," Small Group Behavior 12 (1981): 75.

19. Kumar and Sethi, 60.

20. Daniel and Smitherman, 29.

21. Daniel and Smitherman, 32.

CHAPTER 5

1. Edward T. Hall, Beyond Culture (Garden City, NY: Anchor, 1977).

2. Geert Hofstede, Culture's Consequences: Comparing Values, Behaviors, Institutions, and Organizations across Nations, 2nd ed. (Thousand Oaks, CA: Sage, 2001); Geert Hofstede, Cultures and Organizations: Software of the Mind (London: McGraw-Hill, 1991); Geert Hofstede, Gert Jan Hofstede, and Michael Minkov, Cultures and Organizations: Software of the Mind:

Intercultural Cooperation and Its Importance for Survival, 3rd ed. (New York: McGraw-Hill, 2010).

3. Denise Rotondo Fernandez, Dawn S. Carlson, Lee P. Stepina, and Joel D. Nicholson, "Hofstede's Country Classification 25 Years Later," Journal of Social Psychology 137 (1997): 43–54; Bradley L. Kirkman, Kevin B. Lowe, and Cristina B. Gibson, "A Quarter Century of Culture's Consequences: A Review of Empirical Research Incorporating Hofstede's Cultural Values Framework," Journal of International Business Studies 37 (2006): 285–320.

4. Data from twenty-two countries were originally reported in: Chinese Culture Connection, "Chinese Values and the Search for Culture-Free Dimensions of Culture," Journal of Cross-Cultural Psychology 18 (1987): 143–164. See also Hofstede, Hofstede, and Minkov (2010).

5. Hofstede, Cultures and Organizations, 164–166; Culture's Consequences, 360.

6. Hofstede, Hofstede, and Minkov, Cultures and Organizations; Geert Hofstede, Gert Jan Hofstede, Michael Minkov, and Henk Vinken, "VSM 08: Values Survey Module 2008 Manual," 2008. Accessed May 5, 2011, from http://geerthofstede.com/research—vsm/vsm-08.aspx; Ronald Inglehart (ed.), Human Values and Social Change: Findings from the Values Surveys (Boston: Brill, 2003); Ronald Inglehart, Miguel Basañez, Jaime Díez-Medrano, Loek Halman, and Ruud Luijkx, Human Beliefs and Values: A Cross-Cultural Sourcebook Based on the 1999–2002 Values Surveys (Mexico: Siglo XXI Editores, 2004); Ronald Inglehart, Miguel Basañez, and Alejandro Moreno, Human Values and Beliefs: A Cross-Cultural Sourcebook. Findings from the 1990–1993 World Values Survey (Ann Arbor: University of Michigan Press, 1998); Ronald Inglehart and Pippa Norris, Rising Tide: Gender Equality and Cultural Change around the World (Cambridge: Cambridge University Press, 2003); Michael Minkov, What Makes Us Different and Similar: A New Interpretation of the World Values Survey and Other Cross-Cultural Data (Bulgaria: Klasika y Stil, 2007); Peter B. Smith, "Book Review: Michael Minkov, What Makes Us Different and Similar: A New Interpretation of the World Values Survey and Other Cross-Cultural Data," International Journal of Cross Cultural Management 8 (2008): 110–112.

7. Our description of Schwartz's ideas is based on the following: Ronald Fischer, C.-Melanie Vauclair,

Johnny R. J. Fontaine, and Shalom H. Schwartz, "Are Individual-Level and Country-Level Value Structures Different? Testing Hofstede's Legacy with the Schwartz Value Survey," *Journal of Cross-Cultural Psychology* 41 (2010): 135–151; Johnny R. J. Fontaine, Ype H. Poortinga, Luc Delbeke, and Shalom H. Schwartz, "Structural Equivalence of the Values Domain across Cultures: Distinguishing Sampling Fluctuations from Meaningful Variation," *Journal of Cross-Cultural Psychology* 39 (2008): 345–365; Sonia Roccas, Lilach Sagiv, Shalom H. Schwartz, and Ariel Knafo, "The Big Five Personality Factors and Personal Values," *Personality and Social Psychology Bulletin* 28.6 (2002): 789–801; Lilach Sagiv and Shalom H. Schwartz, "Value Priorities and Readiness for Out-Group Social Contact," *Journal of Personality and Social Psychology* 69 (1995): 437–448; Shalom H. Schwartz, "Individualism-Collectivism: Critique and Proposed Refinements," *Journal of Cross-Cultural Psychology* 21 (1990): 139–157; Shalom H. Schwartz, "Universals in the Content and Structure of Values: Theoretical Advances and Empirical Tests in 20 Countries," *Advances in Experimental Social Psychology*, ed. Mark P. Zanna, Vol. 25 (San Diego: Academic Press, 1992), 1–65; Shalom H. Schwartz, "Are There Universal Aspects in the Structure and Content of Values?" *Journal of Social Issues* 50 (1994): 19–45; Shalom H. Schwartz, "Beyond Individualism-Collectivism: New Cultural Dimensions of Values," *Individualism and Collectivism: Theory, Method and Applications*, ed. Uichol Kim, Harry C. Triandis, Cigdem Kagitçibasi, Sang-Chin Choi, and Gene Yoon (Newbury Park, CA: Sage, 1994), 85–119; Shalom H. Schwartz, "Value Priorities and Behavior: Applying a Theory of Integrated Value Systems," *The Psychology of Values*, ed. Clive Seligman, James M. Olson, and Mark P. Zanna (Mahwah, NJ: Erlbaum, 1996), 1–24; Shalom H. Schwartz, "Values and Culture," *Motivation and Culture*, ed. Donald Munro, John F. Schumaker, and Stuart C. Carr (New York: Routledge, 1997), 69–84; Shalom H. Schwartz, "Values and Behavior: Strength and Structure of Relations," *Personality and Social Psychology Bulletin* 29 (2003): 1207–1220; Shalom H. Schwartz, "Evaluating the Structure of Human Values with Confirmatory Factor Analysis," *Journal of Research in Personality* 38 (2004): 230–255;

Shalom H. Schwartz, "Mapping and Interpreting Cultural Differences around the World," *Comparing Cultures: Dimensions of Culture in a Comparative Perspective*, ed. Henk Vinken, Joseph Soeters, and Peter Ester (Boston: Brill, 2004), 43–73; Shalom H. Schwartz, "A Theory of Cultural Value Orientations: Explication and Applications," *Comparative Sociology* 5 (2006): 137–182; Shalom H. Schwartz, "A Theory of Cultural Value Orientations: Explication and Applications," *Measuring and Mapping Cultures: 25 Years of Comparative Value Surveys*, ed. Yilmaz Esmer and Thorleif Pettersson (Boston: Brill, 2007), 33–78; Shalom H. Schwartz, "Studying Values: Personal Adventure, Future Directions," *Journal of Cross-Cultural Psychology* 42 (2011): 307–319; Shalom H. Schwartz and Anat Bardi, "Influences of Adaptation to Communist Rule on Value Priorities in Eastern Europe," *Political Psychology* 18 (1997): 385–410; Shalom H. Schwartz and Anat Bardi, "Moral Dialogue across Cultures: An Empirical Perspective," *Autonomy and Order: A Communitarian Anthology*, ed. Edward W. Lehman (Lanham, England: Rowman & Littlefield, 2000); Shalom H. Schwartz and Anat Bardi, "Value Hierarchies across Cultures: Taking a Similarities Perspective," *Journal of Cross-Cultural Psychology* 32 (2001): 268–290; Shalom H. Schwartz and Klaus Boehnke, "Evaluating the Structure of Human Values with Confirmatory Factor Analysis," *Journal of Research in Personality* 38 (2004): 230–255; Shalom H. Schwartz, Gila Melech, Arielle Lehmann, Steven Burgess, Mari Harris, and Vicki Owens, "Extending the Cross-Cultural Validity of the Theory of Basic Human Values with a Different Method of Measurement," *Journal of Cross-Cultural Psychology* 32 (2001): 519–542; Shalom H. Schwartz and Tammy Rubel, "Sex Differences in Value Priorities: Cross-Cultural and Multimethod Studies," *Journal of Personality and Social Psychology* 89 (2005): 1010–1028; Shalom H. Schwartz and Lilach Sagiv, "Identifying Culture-Specifics in the Content and Structure of Values," *Journal of Cross-Cultural Psychology* 26 (1995): 92–116; Shalom H. Schwartz, Markku Verkasalo, Avishai Antonovsky, and Avishai Sagiv, "Value Priorities and Social Desirability: Much Substance, Some Style," *British Journal of Social Psychology* 36 (1997): 3–18; Peter B. Smith, Mark F. Peterson, and Shalom H. Schwartz,

"Cultural Values, Sources of Guidance, and Their Relevance to Managerial Behavior: A 47-Nation Study," *Journal of Cross-Cultural Psychology* 34 (2002): 297–303; Peter B. Smith and Shalom H. Schwartz, "Values," *Handbook of Cross-Cultural Psychology*, ed. John W. Berry, Marshall H. Segall, and Cigdem Kagitçibasi, 2nd ed. Vol. 3. Social behavior and applications (Boston: Allyn and Bacon, 1997); Naomi Struch, Shalom H. Schwartz, and Willem A. van der Kloot, "Meanings of Basic Values for Women and Men: A Cross-Cultural Analysis," *Personality and Social Psychology Bulletin* 28 (2002): 16–28; Christin-Melanie Vauclair, Katja Hanke, Ronald Fischer, and Johnny Fontaine, "The Structure of Human Values at the Culture Level: A Meta-Analytical Replication of Schwartz's Value Orientations Using the Rokeach Value Survey," *Journal of Cross-Cultural Psychology* 42 (2011): 186–205; Evert van de Vliert, Shalom H. Schwartz, Sipke E. Huismans, Geert Hofstede, and Serge Daan, "Temperature, Cultural Masculinity, and Domestic Political Violence: A Cross-National Study," *Journal of Cross-Cultural Psychology* 30 (1999): 291–314.

8. Our discussion of the GLOBE research is based on the following: Neal Ashkanasy, Vipin Gupta, Melinda S. Mayfield, and Edwin Trevor-Roberts, "Future Orientation," *Culture, Leadership, and Organizations: The GLOBE Study of 62 Societies*, ed. Robert J. House, Paul J. Hanges, Mansour Javidan, Peter W. Dorfman, and Vipin Gupta (Thousand Oaks, CA: Sage, 2004), 282–342; Neal M. Ashkanasy, Edwin Trevor-Roberts, and Louise Earnshaw, "The Anglo Cluster: Legacy of the British Empire," *Journal of World Business* 37 (2002): 28–39; Gyula Bakacsi, Takács Sándor, Karácsonyi András, and Imrek Viktor, "Eastern European Cluster: Tradition and Transition," *Journal of World Business* 37 (2002): 69–80; Dale Carl, Vipin Gupta, and Mansour Javidan, "Power Distance," *Culture, Leadership, and Organizations: The GLOBE Study of 62 Societies*, ed. House, Hanges, Javidan, Dorfman, and Gupta, 513–563; Mary Sully De Luque and Mansour Javidan, "Uncertainty Avoidance," *Culture, Leadership, and Organizations: The GLOBE Study of 62 Societies*, ed. House, Hanges, Javidan, Dorfman, and Gupta, 602–653; Deanne N. Den Hartog, "Assertiveness," *Culture, Leadership, and Organizations: The GLOBE Study of 62 Societies*, ed. House, Hanges, Javidan, Dorfman, and Gupta,

395–436; P. Christopher Earley, "Leading Cultural Research in the Future: A Matter of Paradigms and Taste," *Journal of International Business Studies* 37 (2006): 922–931; Cynthia G. Emrich, Florence L. Denmark, and Deanne N. Den Hartog, "Cross-Cultural Differences in Gender Egalitarianism," *Culture, Leadership, and Organizations: The GLOBE Study of 62 Societies*, ed. House, Hanges, Javidan, Dorfman, and Gupta, 343–394; Michele J. Gelfand, Dharm P. S. Bhawuk, Lisa Hisae Nishii, and David J. Bechtold, "Individualism and Collectivism," *Culture, Leadership, and Organizations: The GLOBE Study of 62 Societies*, ed. House, Hanges, Javidan, Dorfman, and Gupta, 437–512; Vipin Gupta and Paul J. Hanges, "Regional and Climate Clustering of Societal Cultures," *Culture, Leadership, and Organizations: The GLOBE Study of 62 Societies*, ed. House, Hanges, Javidan, Dorfman, and Gupta, (178–218; Vipin Gupta, Paul J. Hanges, and Peter Dorfman, "Cultural Clusters: Methodology and Findings," *Journal of World Business* 37 (2002): 11–15; Vipin Gupta, Gita Surie, Mansour Javidan, and Jagdeep Chhokar, "Southern Asia Cluster: Where the Old Meets the New?" *Journal of World Business* 37 (2002): 16–27; Paul J. Hanges and Marcus W. Dickson, "Agitation over Aggregation: Clarifying the Development of and the Nature of the GLOBE Scales," *Leadership Quarterly* 17 (2006): 522–536; Paul J. Hanges, Marcus W. Dickson, and Mina T. Sipe, "Rationale for GLOBE Statistical Analyses," *Culture, Leadership, and Organizations: The GLOBE Study of 62 Societies*, ed. Robert J. House, Paul J. Hanges, Mansour Javidan, Peter W. Dorfman, and Vipin Gupta (Thousand Oaks, CA: Sage, 2004), 219–233; Paul J. Hanges, Julie S. Lyon, and Peter W. Dorfman, "Managing a Multinational Team: Lessons from Project Globe," *Advances in International Management* 18 (2005): 337–360; Geert Hofstede, "What Did GLOBE Really Measure? Researchers' Minds Versus Respondents' Minds," *Journal of International Business Studies* 37 (2006): 882–896; Robert J. House, Mansour Javidan, Paul Hanges, and Peter Dorfman, "Understanding Cultures and Implicit Leadership Theories across the Globe: An Introduction to Project GLOBE," *Journal of World Business* 37 (2002): 3–10; Robert J. House, Paul J. Hanges, Mansour Javidan, Peter W. Dorfman, and Vipin Gupta (eds.), *Culture, Leadership, and Organizations: The GLOBE Study of 62 Societies*

(Thousand Oaks, CA: Sage, 2004); Robert J. House and Mansour Javidan, "Overview of GLOBE," *Culture, Leadership, and Organizations: The GLOBE Study of 62 Societies*, ed. House, Hanges, Javidan, Dorfman, and Gupta, 9–28; Jon P. Howell, José DelaCerda, Sandra M. Martínez, Leonel Prieto, J. Arnoldo Bautista, Juan Ortiz, Peter Dorfman, and Maria J. Méndez, "Leadership and Culture in Mexico," *Journal of World Business* 42 (2007): 449–462; Mansour Javidan, "Performance Orientation," *Culture, Leadership, and Organizations: The GLOBE Study of 62 Societies*, ed. House, Hanges, Javidan, Dorfman, and Gupta, 239–281; Mansour Javidan and Markus Hauser, "The Linkage between GLOBE Findings and Other Cross-Cultural Information," *Culture, Leadership, and Organizations: The GLOBE Study of 62 Societies*, ed. House, Hanges, Javidan, Dorfman, and Gupta, 102–121; Mansour Javidan and Robert J. House, "Leadership and Cultures around the World: Findings from GLOBE: An Introduction to the Special Issue," *Journal of World Business* 37 (2002): 1–2; Mansour Javidan, Robert J. House, and Peter W. Dorfman, "A Nontechnical Summary of GLOBE Findings," *Culture, Leadership, and Organizations: The GLOBE Study of 62 Societies*, ed. House, Hanges, Javidan, Dorfman, and Gupta, 29–48; Mansour Javidan, Robert J. House, Peter W. Dorfman, Paul J. Hange, and Mary Sully de Luque, "Conceptualizing and Measuring Cultures and Their Consequences: A Comparative Review of Globe's and Hofstede's Approaches," *Journal of International Business Studies* 37 (2006): 897–914; Jorge Correia Jesuino, "Latin Europe Cluster: From South to North," *Journal of World Business* 37 (2002): 81–89; Hayat Kabasakal and Muzaffer Bodur, "Arabic Cluster: A Bridge between East and West," *Journal of World Business* 37 (2002): 40–54; Hayat Kabasakal and Muzaffer Bodur, "Humane Orientation in Societies, Organizations, and Leader Attributes," *Culture, Leadership, and Organizations: The GLOBE Study of 62 Societies*, ed. House, Hanges, Javidan, Dorfman, and Gupta, (Thousand Oaks, CA: Sage, 2004), 564–601; Mark F. Peterson and Stephanie L. Castro, "Measurement Metrics at Aggregate Levels of Analysis: Implications for Organization Culture Research and the GLOBE Project," *Leadership Quarterly* 17 (2006): 506–521; Peter B. Smith, "When Elephants Fight, the Grass Gets Trampled: The GLOBE and Hofstede Projects," *Journal of*

International Business Studies 37 (2006): 915–921; Erna Szabo, Felix C. Brodbeck, Deanne N. Den Hartog, Gerhard Reber, Jürgen Weiblere, and Rolf Wunderer, "The Germanic Europe Cluster: Where Employees Have a Voice," *Journal of World Business* 37 (2002): 55–68.

9. See C. Harry Hui and Harry C. Triandis, "Individualism-Collectivism: A Study of Cross-Cultural Researchers," *Journal of Cross-Cultural Psychology* 17 (1986): 225–248; Harry C. Triandis, *The Analysis of Subjective Culture* (New York: Wiley, 1972); Harry C. Triandis, *Culture and Social Behavior* (New York: McGraw-Hill, 1994); Harry C. Triandis, *Individualism & Collectivism* (Boulder, CO: Westview, 1995); Harry C. Triandis and S. Arzu Wasti, "Culture," *The Influence of Culture on Human Resource Management Processes and Practices*, ed. Dianna Stone and Eugene Stone-Romero (New York: Earlbaum, 2008), 1–24.

10. Adapted from: Schwartz, "A Theory of Cultural Value Orientations: Explication and Applications," 142; Shalom H. Schwartz, *Cultural Values Influence and Constrain Economic and Social Change*, Unpublished Manuscript; Shalom H. Schwartz, *Cultural Value Orientations: Nature & Implications of National Differences*, Unpublished Manuscript; Shalom H. Schwartz, "Values: Cultural and Individual," *Fundamental Questions in Cross-Cultural Psychology*, ed. Fons J. R. van de Vijver, Athanasios Chasiotis, and Seger M. Breugelmans (New York: Cambridge University Press, 2011), 473; Christin-Melanie Vauclair, Katja Hanke, Ronald Fischer, and Johnny Fontaine, "The Structure of Human Values at the Culture Level: A Meta-Analytical Replication of Schwartz's Value Orientations Using the Rokeach Value Survey," *Journal of Cross-Cultural Psychology* 42 (2011): 188.

11. Adapted from: House, Hanges, Javidan, Dorfman, and Gupta.

CHAPTER 6

1. Marilynn Brewer and Donald T. Campbell, *Ethnocentrism and Intergroup Attitudes* (New York: Wiley, 1976).

2. Osei Appiah, "Effects of Ethnic Identification on Web Browsers' Attitudes toward and Navigational Patterns on Race-Targeted Sites," *Communication Research* 31 (2004): 312–337; Danette Ifert Johnson, "Music Videos and National Identity on

Post-Soviet Kazakhstan," *Qualitative Research Reports in Communication* 7 (2006): 9–14; Dana E. Mastro, "A Social Identity Approach to Understanding the Impact of Television Messages," *Communication Monographs* 70 (2003): 98–113; Andrew F. Wood and Matthew J. Smith, *Online Communication: Linking Technology, Identity, and Culture*, 2nd ed. (Mahwah, NJ: Lawrence Erlbaum, 2005).

3. Our labels are analogous to Triandis's tripartite distinction among one's collective self, public self, and private self. See: Harry C. Triandis, "The Self and Social Behavior in Differing Cultural Contexts," *Psychological Review* 96 (1989): 506–520. See also: Henri Tajfel, *Differentiation between Social Groups* (London: Academic Press, 1978); Henri Tajfel, *Human Groups and Social Categories: Studies in Social Psychology* (Cambridge: Cambridge University Press, 1981).

4. Julia T. Wood uses the term "social diversity" to refer to differences among people because of age, gender, sexual orientation, and the like. See: Julia T. Wood, "Celebrating Diversity in the Communication Field," *Communication Studies* 49 (1998): 172–178.

5. See, for example, Amanda J. Godley, "Literacy Learning as Gendered Identity Work," *Communication Education* 52 (2003): 273–285; Radha S. Hegde and Barbara Dicicco-Bloom, "Working Identities: South Asian Nurses and the Transnational Negotiations of Race and Gender," *Communication Quarterly* 50 (2002): 90–95; Patricia S. Parker, "Negotiating Identity in Raced and Gendered Workplace Interactions: The Use of Strategic Communication by African American Women Senior Executives within Dominant Culture Organizations," *Communication Quarterly* 50 (2002): 251–268; Saskia Witteborn, "Of Being an Arab Woman before and after September 11: The Enactment of Communal Identities in Talk," *Howard Journal of Communications* 15 (2004): 83–98.

6. Our discussion of the stages of cultural identity draws heavily upon the works of Jean S. Phinney, particularly Jean S. Phinney, "Ethnic Identity in Adolescents and Adults: Review of Research," *Psychological Bulletin* 108 (1990): 499–514; Jean S. Phinney, "Ethnic Identity and Self-Esteem: A Review and Integration," *Hispanic Journal of Behavioral Sciences* 13 (1991): 193–208; Jean S. Phinney, "A Three-Stage Model of Ethnic Identity Development in Adolescence," *Ethnic Identity: Formation and Transmission among Hispanics and Other Minorities*, ed. Martha E. Bernal and George P. Knight (Albany: State University of New York Press, 1993), 61–79; Jean S. Phinney, "Ethic Identity and Acculturation," *Acculturation: Advances in Theory, Measurement, and Applied Research*, ed. Kevin M. Chun, Pamela Balls Organista, and Gerardo Marín (Washington, DC: American Psychological Association, 2003), 63–81; Jean S. Phinney, "Ethnic Identity Exploration in Emerging Adulthood," *Emerging Adults in America: Coming of Age in the 21st Century*, ed. Jeffery Jensen Arnett and Jennifer Lynn Tanner (Washington, DC: American Psychological Association, 2006), 117–134; Jean S. Phinney and Anthony D. Ong, "Conceptualization and Measurement of Ethnic Identity: Current Status and Future Directions," *Journal of Counseling Psychology* 54 (2007): 271–281.

7. See, for example, Daniel Bernardi (ed.), *The Persistence of Whiteness: Race and Contemporary Hollywood Cinema* (New York: Routledge, 2007); Nyla R. Branscombe, Michael T. Schmitt, and Kristin Schiffhauer, "Racial Attitudes in Response to Thoughts of White Privilege," *European Journal of Social Psychology* 37 (2007): 203–215; Michael K. Brown, Martin Carnoy, Elliott Currie, Troy Duster, David B. Oppenheimer, Marjorie M. Shultz, and David Wellman, *Whitewashing Race: The Myth of a Color-Blind Society* (Berkeley: University of California Press, 2003); Leda Cooks, "Pedagogy, Performance, and Positionality: Teaching About Whiteness in Interracial Communication," *Communication Education* 52 (2003): 245–257; Thierry Devos and Mahzarin R. Banaji, "American = White?" *Journal of Personality and Social Psychology* 88 (2005): 447–466; Steve Garner, *Whiteness: An Introduction* (New York: Routledge, 2007); Robert Jensen, *The Heart of Whiteness: Confronting Race, Racism, and White Privilege* (San Francisco: City Lights, 2005); Frances E. Kendall, *Understanding White Privilege: Creating Pathways to Authentic Relationships across Race* (New York: Routledge, 2006); Eric D. Knowles and Kaiping Peng, "White Selves: Conceptualizing and Measuring a Dominant-Group Identity," *Journal of Personality and Social Psychology* 89 (2005): 223–241; George Lipsitz, *The*

Possessive Investment in Whiteness: How White People Profit from Identity Politics, Revised and Expanded ed. (Philadelphia: Temple University Press, 2006); Judith N. Martin and Olga Idriss Davis, "Conceptual Foundations for Teaching About Whiteness in Intercultural Communication Courses," *Communication Education* 50 (2001): 298–313; Ann Neville Miller and Tina M. Harris, "Communicating to Develop White Racial Identity in an Interracial Communication Class," *Communication Education* 54 (2005): 223–242; Nelson M. Rodriguez and Leila E. Villaverde (eds.), *Dismantling White Privilege: Pedagogy, Politics, and Whiteness* (New York: P. Lang, 2000); David R. Roediger, *Working toward Whiteness: How America's Immigrants Became White: The Strange Journey from Ellis Island to the Suburbs* (New York: Basic Books, 2005); David R. Roediger, *The Wages of Whiteness: Race and the Making of the American Working Class*, Revised ed. (New York: Verso, 2007); Karen Elaine Rosenblum and Toni-Michelle Travis (eds.), *The Meaning of Difference: American Constructions of Race, Sex and Gender, Social Class, Sexual Orientation, and Disability*, 5th ed. (New York: McGraw-Hill, 2008); Shelly Tochluk, *Witnessing Whiteness: First Steps toward an Antiracist Practice and Culture* (Lanham, MD: Rowman & Littlefield Education, 2008); John T. Warren, "Doing Whiteness: On the Performative Dimensions of Race in the Classroom," *Communication Education* 50 (2001): 91–108; John T. Warren and Kathy Hytten, "The Faces of Whiteness: Pitfalls and the Critical Democrat," *Communication Education* 53 (2004): 321–339.

8. Judith N. Martin, Robert L. Krizek, Thomas K. Nakayama, and Lisa Bradford, "Exploring Whiteness: A Study of Self Labels for White Americans," *Communication Quarterly* 44 (1996): 125–144. See also: Thomas K. Nakayama and Robert L. Krizek, "Whiteness: A Strategic Rhetoric," *Quarterly Journal of Speech* 81 (1995): 291–309.

9. Fernando P. Delgado, "Chicano Ideology Revisited: Rap Music and the (Re)Articulation of Chicanismo," *Western Journal of Communication* 62 (1998): 95–113; Fernando P. Delgado, "All Along the Border: Kid Frost and the Performance of Brown Masculinity," *Text and Performance Quarterly* 20 (2000): 388–401; Danette Ifert Johnson, "Music Videos and National Identity

on Post-Soviet Kazakhstan"; Dana E. Mastro, "A Social Identity Approach to Understanding the Impact of Television Messages."

10. For interesting research on identity change among Asian Indians and Chinese Americans, see, respectively, Jean Bacon, "Constructing Collective Ethnic Identities: The Case of Second Generation Asian Indians," *Qualitative Sociology* 22 (1999): 141–160; Zhuojun Joyce Chen, "Chinese-American Children's Ethnic Identity: Measurement and Implications," *Communication Studies* 51 (2000): 74–95. For an analysis of the shifting identities of European Americans, see Ronald L. Jackson, II, and Katherine Simpson, "White Positionalities and Cultural Contracts: Critiquing Entitlement, Theorizing, and Exploring the Negotiation of White Identities," *Ferment in the Intercultural Field*, ed. William J. Starosta and Guo-Ming Chen (Thousand Oaks, CA: Sage, 2003), 177–210.

11. Radhika Gajjala, "Interrogating Identities: Composing Other Cyberspaces," *Intercultural Alliances: Critical Transformation*, ed. Mary Jane Collier (Thousand Oaks, CA: Sage, 2003), 167–188.

12. Young Yun Kim, "Identity Development: From Cultural to Intercultural," *Interaction and Identity*, ed. Hartmut B. Mokros (New Brunswick, NJ: Transaction, 1996), 350.

13. Katherine Grace Hendrix, Ronald L. Jackson, and Jennifer R. Warren, "Shifting Academic Landscapes: Exploring Co-Identities, Identity Negotiation, and Critical Progressive Pedagogy," *Communication Education* 52 (2003): 177–190; S. Lily Mendoza, Rona T. Halualani, and Jolanta A. Drzewiecka, "Moving the Discourse on Identities in Intercultural Communication: Structure, Culture, and Resignifications," *Communication Quarterly* 50 (2002): 312–327; Kristin Moss and William V. Faux II, "The Enactment of Cultural Identity in Student Conversations on Intercultural Topics," *Howard Journal of Communications* 17 (2006): 21–37; Patricia S. Parker, "Negotiating Identity in Raced and Gendered Workplace Interactions"; Rosenblum and Travis.

14. Theodor Gomperz, *Greek Thinkers: A History of Ancient Philosophy*, Vol. 1 (trans. L. Magnus) (New York: Scribner's, 1901), 403–404.

15. William G. Sumner, *Folkways* (Boston: Ginn, 1940), 27.

16. Sumner.

17. Walter Lippmann, *Public Opinion* (New York: Harcourt, Brace, 1922), 25.

18. Marilynn B. Brewer, "When Stereotypes Lead to Stereotyping: The Use of Stereotypes in Person Perception," *Stereotypes and Stereotyping*, ed. C. Neil Macrae, Charles Stangor, and Miles Hewstone (New York: Guilford, 1996), 254–275; Diane M. Mackie, David L. Hamilton, Joshua Susskind, and Francine Rosselli, "Social Psychological Foundations of Stereotype Formation," *Stereotypes and Stereotyping*, ed. Macrae, Stangor, and Hewstone, 41–78; Charles Stangor and Mark Schaller, "Stereotypes as Individual and Collective Representations," *Stereotypes and Stereotyping*, ed. Macrae, Stangor, and Hewstone, 3–37.

19. Micah S. Thompson, Charles M. Judd, and Bernadette Park, "The Consequences of Communicating Social Stereotypes," *Journal of Experimental Social Psychology* 36 (2000): 567–599; Vincent Y. Yzerbyt, Alastair Coull, and Steve J. Rocher, "Fencing Off the Deviant: The Role of Cognitive Resources in the Maintenance of Stereotypes," *Journal of Personality and Social Psychology* 77(3) (1999): 449–462.

20. See Charles M. Judd and Bernadette Park, "Definition and Assessment of Accuracy in Social Stereotypes," *Psychological Review* 100 (1993): 109–128; Carey S. Ryan, Bernadette Park, and Charles M. Judd, "Assessing Stereotype Accuracy: Implications for Understanding the Stereotyping Process," *Stereotypes and Stereotyping*, ed. Macrae, Stangor, and Hewstone, 121–157.

21. Marilynn B. Brewer, "Social Identity, Distinctiveness, and In-Group Homogeneity," *Social Cognition* 11 (1993): 150–164; Klaus Fiedler and Eva Walther, *Stereotyping as Inductive Hypothesis* (New York: Psychology Press, 2004); Bonnie L. Haines, *Bigger Than the Box: The Effects of Labeling* (Oacoma, SD: Unlimited Achievement Books, 2004); E. E. Jones, G. C. Wood, and G. A. Quattrone, "Perceived Variability of Personal Characteristics in In-Groups and Out-Groups: The Role of Knowledge and Evaluation," *Personality and Social Psychology Bulletin* 7 (1981): 523–528; John T. Jost and Brenda Major (eds.), *The Psychology of Legitimacy: Emerging Perspectives on Ideology, Justice, and Intergroup Relations* (New York: Cambridge University Press, 2001);

Charles M. Judd and Bernadette Park, "Out-Group Homogeneity: Judgments of Variability at the Individual and Group Levels," *Journal of Personality and Social Psychology* 54 (1988): 778–788; Paul Martin Lester and Susan Dente Ross (eds.), *Images that Injure: Pictorial Stereotypes in the Media*, 2nd ed. (Westport, CT: Praeger, 2003); Toni Lester (ed.), *Gender Nonconformity, Race, and Sexuality: Charting the Connections* (Madison: University of Wisconsin Press, 2002); P. W. Linville and E. E. Jones, "Polarized Appraisals of Out-Group Members," *Journal of Personality and Social Psychology* 38 (1980): 689–703; Craig McGarty, Vincent Y. Yzerbyt, and Russell Spears (eds.), *Stereotypes as Explanations: The Formation of Meaningful Beliefs about Social Groups* (New York: Cambridge University Press, 2002); Brian Mullen and L. Hu, "Perceptions of Ingroup and Outgroup Variability: A Meta-Analytic Integration," *Basic and Applied Social Psychology* 10 (1989): 233–252; Thomas M. Ostrom, Sandra L. Carpenter, Constantine Sedikides, and Fan Li, "Differential Processing of In-Group and Out-Group Information," *Journal of Personality and Social Psychology* 64 (1993): 21–34; Michael Pickering, *Stereotyping: The Politics of Representation* (New York: Palgrave, 2001); David J. Schneider, *The Psychology of Stereotyping* (New York: Guilford, 2004).

22. David Barsamian, "Albert Mokhiber: Cultural Images, Politics, and Arab Americans," *Z Magazine*, May 1993, 46–50. Reprinted in *Ethnic Groups*, vol. 4, ed. Eleanor Goldstein (Boca Raton, FL: Social Issues Resources Ser., 1994), art. no. 73.

23. Ziva Kunda and Bonnie Sherman-Williams, "Stereotypes and the Construal of Individuating Information," *Personality and Social Psychology Bulletin* 19 (1993): 97.

24. John J. Seta and Catherine E. Seta, "Stereotypes and the Generation of Compensatory and Noncompensatory Expectancies of Group Members," *Personality and Social Psychology Bulletin* 19 (1993): 722–731.

25. C. Neil Macrae, Alan B. Milne, and Galen V. Bodenhausen, "Stereotypes as Energy-Saving Devices: A Peek Inside the Cognitive Toolbox," *Journal of Personality and Social Psychology* 66 (1994): 37–47.

26. Judee K. Burgoon, Charles R. Berger, and Vincent Waldron, "Mindfulness and Interpersonal

Communication," *Journal of Social Issues* 56 (2000): 105–127.

27. Gordon W. Allport, *The Nature of Prejudice* (New York: Macmillan, 1954).

28. John F. Dovidio, John C. Brigham, Blair T. Johnson, and Samuel L. Gaertner, "Stereotyping, Prejudice, and Discrimination: A Closer Look," *Stereotypes and Stereotyping*, ed. C. Neil Macrae, Charles Stangor, and Miles Hewstone (New York: Guilford, 1996), 276–319.

29. Richard W. Brislin, *Cross-Cultural Encounters: Face-to-Face Interaction* (New York: Pergamon Press, 1981), 42–49.

30. For a recent test of the ego-defensive functions of attitudes, see Maria Knight Lapinski and Franklin Boster, "Modeling the Ego-Defensive Function of Attitudes," *Communication Monographs* 68 (2001): 314–324.

31. Steven Fein and Steven J. Spencer, "Prejudice as Self-Image Maintenance: Affirming the Self Through Derogating Others," *Journal of Personality and Social Psychology* 73 (1997): 31–44.

32. Teun A. van Dijk, *Communicating Racism: Ethnic Prejudice in Thought and Talk* (Newbury Park, CA: Sage, 1987).

33. Marilynn B. Brewer, "The Psychology of Prejudice: Ingroup Love or Outgroup Hate?" *Journal of Social Issues* 55 (1999): 429–444. See also: Marilynn B. Brewer and Wendi L. Gardner, "Who Is This 'We'? Levels of Collective Identity and Self Representations," *Journal of Personality and Social Psychology* 71 (1996): 83–93.

34. For a discussion of the causes and consequences of some forms of racism, see Joe R. Feagin and Hernán Vera, *White Racism: The Basics* (New York: Routledge, 1995).

35. Robert Blauner, *Racial Oppression in America* (New York: Harper and Row, 1972), 112.

36. Dalmas A. Taylor, "Race Prejudice, Discrimination, and Racism," *Social Psychology*, ed. A. Kahn, E. Donnerstein, and M. Donnerstein (Dubuque, IA: Wm. C. Brown, 1984); cited in Phyllis A. Katz and Dalmas A. Taylor, "Introduction," *Eliminating Racism: Profiles in Controversy*, ed. Phyllis A. Katz and Dalmas A. Taylor (New York: Plenum, 1988), 6.

37. Katz and Taylor, 7.

38. Blauner.

39. James M. Jones, "Racism in Black and White: A Bicultural Model of Reaction and Evolution," *Eliminating Racism: Profiles in Controversy*, ed. Katz and Taylor, 130–131.

40. S. Elizabeth Bird, "Gendered Construction of the American Indian in Popular Media," *Journal of Communication* 49 (1999): 78.

41. Bird, 80. See also: Richard Morris, "Educating Savages," *Quarterly Journal of Speech* 83 (1997): 152–171.

42. Jones, 118–126.

43. Katz and Taylor, 7.

44. Jenny Yamoto, "Something about the Subject Makes It Hard to Name," *Race, Class, and Gender in the United States: An Integrated Study*, 2nd ed., ed. Paula S. Rothenberg (New York: St. Martin's Press, 1992), 58.

45. For discussions of racism and prejudice, see Linda Jacobs Altman, *Racism and Ethnic Bias: Everybody's Problem* (Berkeley Heights, NJ: Enslow, 2001); Joseph F. Aponte and Laura R. Johnson, "The Impact of Culture on the Intervention and Treatment of Ethnic Populations," *Psychological Intervention and Cultural Diversity*, 2nd ed., ed. Joseph F. Aponte and Julian Wohl (Boston: Allyn & Bacon, 2000), 18–39; Martha Augoustinos and Katherine J. Reynolds (eds.), *Understanding Prejudice, Racism, and Social Conflict* (Thousand Oaks, CA: Sage, 2001); Michael D. Barber, *Equality and Diversity: Phenomenological Investigations of Prejudice and Discrimination* (Amherst, NY: Humanity Books, 2001); Benjamin P. Bowser, Gale S. Auletta, and Terry Jones, *Confronting Diversity Issues on Campus* (Newbury Park, CA: Sage, 1994); Bernard Boxill (ed.), *Race and Racism* (New York: Oxford University Press, 2001); John C. Brigham, "College Students' Racial Attitudes," *Journal of Applied Social Psychology* 23 (1993): 1933–1967; Richard W. Brislin, "Prejudice and Intergroup Communication," *Intergroup Communication*, ed. William B. Gudykunst (London: Arnold, 1986), 74–85; Brislin (1981): 42–49; Charles E. Case, Andrew M. Greeley, and Stephan Fuchs, "Social Determinants of Racial Prejudice," *Sociological Perspectives* 32 (1989): 469–483; Farhad Dalal, "Insides and Outsides: A Review of Psychoanalytic Renderings of Difference, Racism, and Prejudice," *Psychoanalytic Studies* 3 (2001): 43–66; Samuel L. Gaertner and John F. Dovidio, "The Aversive Form of Racism," *Prejudice, Discrimination and Racism: Theory*

and Research, ed. John F. Dovidio and Samuel L. Gaertner (New York: Academic Press, 1986), 61–89; Ellen J. Goldner and Safiya Henderson-Holmes (eds.), *Racing and (E)Racing Language: Living with the Color of Our Words* (Syracuse, NY: Syracuse University Press, 2001); Harry Goulbourne (ed.), *Race and Ethnicity: Critical Concepts in Sociology* (New York: Routledge, 2001); Edgar Jones, "Prejudicial Beliefs: Their Nature and Expression," *Racism and Mental Health: Prejudice and Suffering*, ed. Kamaldeep Bhui (Philadelphia: Jessica Kingsley, 2002), 26–34; David Milner, "Racial Prejudice," *Intergroup Behavior*, ed. John C. Turner and Howard Giles (Chicago: University of Chicago Press, 1981), 102–143; Scott Plous (ed.), *Understanding Prejudice and Discrimination* (Boston: McGraw-Hill, 2003); Peter Ratcliffe, *The Politics of Social Science Research: Race, Ethnicity, and Social Change* (New York: Palgrave, 2001); Albert Ramirez, "Racism toward Hispanics: The Culturally Monolithic Society," *Eliminating Racism: Profiles in Controversy*, ed. Katz and Taylor, 137–157; Paula S. Rothenberg (ed.), *Race, Class, and Gender in the United States: An Integrated Study*, 5th ed. (New York: W. H. Freeman, 2001); James R. Samuel, *The Roots of Racism* (New York: Vantage, 2001); David O. Sears, "Symbolic Racism," *Eliminating Racism: Profiles in Controversy*, ed. Katz and Taylor, 53–84; Key Sun, "Two Types of Prejudice and Their Causes," *American Psychologist* 48 (1993): 1152–1153; Nicholas J. Ucci, *The Structure of Racism: Insights into Developing a New Language for Socio-Historical Inquiry* (Centereach, NY: Cybergraphic Fine Art, 2001); Ian Vine, "Inclusive Fitness and the Self-System: The Roles of Human Nature and Sociocultural Processes in Intergroup Discrimination," *The Sociobiology of Ethnocentrism: Evolutionary Dimensions of Xenophobia, Discrimination, Racism, and Nationalism*, ed. Vernon Reynolds, Vincent Falger, and Ian Vine (London: Croom Helm, 1987), 60–80; Bernd Wittenbrink, Charles M. Judd, and Bernadette Park, "Evidence for Racial Prejudice at the Implicit Level and Its Relationship with Questionnaire Measures," *Journal of Personality and Social Psychology* 72 (1997): 262–274.

46. Jacqueline N. Sawires and M. Jean Peacock, "Symbolic Racism and Voting Behavior on Proposition 209," *Journal of Applied Social Psychology* 30 (2000): 2092–2099.

47. Brigham, 1934.

48. Cheryl R. Kaiser and Carol T. Miller, "Stop Complaining! The Social Costs of Making Attributions to Discrimination," *Personality and Social Psychology Bulletin* 27 (2001): 254–263.

CHAPTER 7

1. Charles F. Hockett, "The Origin of Speech," *Human Communication: Language and Its Psychobiological Bases* (Readings from Scientific American) (San Francisco: Freeman, 1982), 5–12.

2. No, it isn't just that *tka* begins with two consonant sounds. *Spring* begins with three such sounds. For an interesting discussion of the rules of language, see Steven Pinker, *The Language Instinct: How the Mind Creates Language* (New York: HarperCollins, 1994).

3. Roger Brown, *Social Psychology* (New York: Free Press, 1965); quoted in Donald W. Klopf, *Intercultural Encounters: The Fundamentals of Intercultural Communication* (Englewood, CO: Morton, 1987), 137.

4. Wen Shu Lee, "In the Names of Chinese Women," *Quarterly Journal of Speech* 84 (1998): 283–302.

5. Jeanette S. Martin and Lillian H. Chaney, *Global Business Etiquette: A Guide to International Communication and Customs* (Westport, CT: Praeger, 2006), 25.

6. Wen Shu Lee, "On Not Missing the Boat: A Processual Method for Intercultural Understanding of Idioms and Lifeworld," *Journal of Applied Communication Research* 22 (1994): 141–161.

7. Christiane F. Gonzalez, "Translation," *Handbook of International and Intercultural Communication*, ed. Molefi Kete Asante and William B. Gudykunst (Newbury Park, CA: Sage, 1989), 484.

8. For an example of the kinds of translation problems that occur in organizations with a multilingual workforce, see Stephen E. Banks and Anna Banks, "Translation as Problematic Discourse in Organizations," *Journal of Applied Communication Research* 19 (1991): 223–241. See also: Henriette W. Langdon, *The Interpreter Translator Process in the Educational Setting* (Sacramento, CA: Resources in Special Education, California State University with the California Department of Education, 1994).

9. Eugene A. Nida, *Toward a Science of Translating* (Leiden, Netherlands: E. J. Brill, 1964).

10. Lee Sechrest, Todd L. Fay, and S. M. Zaidi, "Problems of Translation in Cross-Cultural Communication," *Intercultural Communication: A Reader*, 5th ed., ed. Larry A. Samovar and Richard E. Porter (Belmont, CA: Wadsworth, 1988), 253–262.

11. There is some evidence that, as early as the fifteenth century, an Asian scholar named Bhartvhari, in a work titled *Vahyapidan*, argued that speech patterns are determined by social contexts.

12. Edward Sapir, *Language: An Introduction to the Study of Speech* (New York: Harcourt Brace, 1921).

13. Edward Sapir; quoted in Benjamin Lee Whorf, "The Relation of Habitual Thought and Behavior to Language," *Language, Thought, and Reality: Selected Writings of Benjamin Lee Whorf*, ed. J. B. Carroll (Cambridge, MA: MIT Press, 1939/1956), 134.

14. For a thorough summary and discussion of the experimental research in psychology investigating the validity of the linguistic determinism hypothesis, see John J. Gumperz and Stephen C. Levison (eds.), *Rethinking Linguistic Relativity* (Cambridge: Cambridge University Press, 1996); Curtis Hardin and Mahzarin R. Banaji, "The Influence of Language on Thought," *Social Cognition* 11 (1993): 277–308; Earl Hunt and Franca Agnoli, "The Whorfian Hypothesis: A Cognitive Psychology Perspective," *Psychological Review* 98 (1991): 377–389.

15. Whorf suggested there may be about 7 words for *snow*, though the actual number is closer to 12. Over time and numerous retellings of this example, however, the number of words claimed to represent forms of snow has increased dramatically, typically to the 17 to 23 range; the *New York Times* once cavalierly referred to 100 different words. See Geoffrey K. Pullum, *The Great Eskimo Vocabulary Hoax, and Other Irreverent Essays on the Study of Language* (Chicago: University of Chicago Press, 1991); "The Melting of a Mighty Myth," *Newsweek*, July 22, 1991, 63.

16. Richard W. Brislin, Kenneth Cushner, Craig Cherrie, and Mahealani Yong, *Intercultural Interactions: A Practical Guide* (Beverly Hills, CA: Sage, 1986), 276.

17. John C. Condon and Fathi S. Yousef, *An Introduction to Intercultural Communication* (Yarmouth, ME: Intercultural Press, 1975), 182.

18. Michael Cole and Sylvia Scribner, *Culture and Thought: A Psychological Introduction* (New York: Wiley, 1974), 2.

19. Eleanor Rosch Heider and Donald C. Olivier, "The Structure of the Color Space in Naming and Memory for Two Languages," *Cognitive Psychology* 3 (1972): 337–354.

20. For a more complete discussion of the evidence on variations in vocabulary of the color spectrum, see Cole and Scribner, 45–50; Thomas M. Steinfatt, "Linguistic Relativity: Toward a Broader View," *Language, Communication, and Culture: Current Directions*, ed. Stella Ting-Toomey and Felipe Korzenny (Newbury Park, CA: Sage, 1989), 35–75.

21. Stephen W. Littlejohn, *Theories of Human Communication*, 4th ed. (Belmont, CA: Wadsworth, 1992), 209.

22. Li-Rong Lilly Cheng, *Assessing Asian Language Performance: Guidelines for Evaluating Limited-English-Language Proficient Students* (Rockville, MD: Aspen, 1987), 8.

23. For an interesting discussion of the difficulties that Mandarin speakers might have in using English pronouns appropriately, see Stephen P. Banks, "Power Pronouns and Intercultural Understanding," *Language, Communication, and Culture: Current Directions*, ed. Stella Ting-Toomey and Felipe Korzenny (Newbury Park, CA: Sage, 1989), 180–198.

24. Michael Dorris, *The Broken Cord* (New York: Harper and Row, 1989), 2.

25. See Earl Hunt and Franca Agnoli, "The Whorfian Hypothesis: A Cognitive Psychology Perspective," *Psychological Review* 98 (1991): 377–389.

26. Wilma M. Roger, *National Foreign Language Center Occasional Papers* (Washington, DC: Johns Hopkins University Press, February 1989).

27. Studies of the relationship of language and intercultural communication are often conducted under the rubric of "intergroup behavior" or "intergroup communication." See, for example, John C. Turner and Howard Giles (eds.), *Intergroup Behavior* (Chicago: University of Chicago Press, 1981).

28. Henri Tajfel, "Social Categorization, Social Identity, and Social Comparison," *Differentiation Between Social Groups*, ed. Henri Tajfel (London: Academic Press, 1978).

29. Aaron Castelan Cargile and Howard Giles, "Language Attitudes Toward Varieties of English: An American-Japanese Context," *Journal of Applied Communication Research* 26 (1998): 336–356; Howard Giles and Patricia Johnson, "The Role of Language in Ethnic Group Relations," *Intergroup Behavior*, ed. Turner and Giles, 199–243.

30. Joshua A. Fishman, "Language and Ethnicity," *Language, Ethnicity, and Intergroup Relations*, ed. Howard Giles (London: Academic Press, 1977), 15–58.

31. William B. Gudykunst, "Cultural Variability in Ethnolinguistic Identity," *Language, Communication, and Culture: Current Directions*, ed. Stella Ting-Toomey and Felipe Korzenney (Newbury Park, CA: Sage, 1989), 223.

32. Règl Allard and Rodrique Landry, "Subjective Ethnolinguistic Vitality Viewed as a Belief System," *Journal of Multilingual and Multicultural Development* 7 (1986): 1–12. For a review of the vitality framework, see Jake Harwood, Howard Giles, and Richard Y. Bourhis, "The Genesis of Vitality Theory: Historical Patterns and Discourse Dimensions," *International Journal of the Sociology of Language* 108 (1994): 167–206.

33. Howard Giles and Arlene Franklyn-Stokes, "Communicator Characteristics," *Handbook of International and Intercultural Communication*, ed. Molefi Kete Asante and William B. Gudykunst (Newbury Park, CA: Sage, 1989), 117–144.

34. Nancy F. Burroughs and Vicki Marie, "Communication Orientations of Micronesian and American Students," *Communication Research Reports* 7 (1990): 139–146.

35. Jennifer Fortman, "Adolescent Language and Communication from an Intergroup Perspective," *Journal of Language & Social Psychology* 22 (2003): 104–111; Cynthia Gallois, Howard Giles, Elizabeth Jones, Aaron C. Cargile, and Hiroshi Ota, "Accommodating Intercultural Encounters: Elaborations and Extensions," *Intercultural Communication Theory*, ed. Richard L. Wiseman (Thousand Oaks, CA: Sage, 1995), 115–147; Howard Giles and Nikolas Coupland, *Language: Contexts and Consequences* (Pacific Grove, CA: Brooks/Cole, 1991); Howard Giles and Kimberly A. Noels, "Communication Accommodation in Intercultural Encounters," *Readings in Cultural Contexts*, ed. Judith N. Martin, Thomas K.

Nakayama, and Lisa A. Flores (Mountain View, CA: Mayfield, 1998), 139–149; Howard Giles and Patricia Johnson, "Ethnolinguistic Identity Theory: A Social Psychological Approach to Language Maintenance," *International Journal of the Sociology of Language* 68 (1987): 66–99; Howard Giles, Anthony Mulac, James J. Bradac, and Patricia Johnson, "Speech Accommodation Theory: The Next Decade and Beyond," *Communication Yearbook 10*, ed. Margaret McLaughlin (Newbury Park, CA: Sage, 1987), 13–48; Bettina Heinz, "Backchannel Responses as Strategic Responses in Bilingual Speakers' Conversations," *Journal of Pragmatics* 35 (2003): 1113–1142; Han Z. Li, "Cooperative and Intrusive Interruptions in Inter- and Intracultural Dyadic Discourse," *Journal of Language & Social Psychology* 20 (2001): 259–284; Hung Ng Sik and John Anping He, "Code-switching in Tri-generational Family Conversations among Chinese Immigrants in New Zealand," *Journal of Language & Social Psychology* 23 (2004): 28–48.

36. Ellen Bouchard Ryan, Howard Giles, and Richard J. Sebastian, "An Integrative Perspective for the Study of Attitudes Toward Language Variation," *Attitudes toward Language Variation: Social and Applied Contexts*, ed. Ellen Bouchard Ryan and Howard Giles (London: Arnold, 1982), 1.

37. Michael L. Hecht, Mary Jane Collier, and Sidney Ribeau, *African American Communication: Ethnic Identity and Cultural Interpretation* (Newbury Park, CA: Sage, 1993), 84–89.

38. John R. Edwards, *Language Attitudes and Their Implications Among English Speakers*, ed. Ellen Bouchard Ryan and Howard Giles (London: Arnold, 1982), 22.

39. Hope Bock and James H. Pitts, "The Effect of Three Levels of Black Dialect on Perceived Speaker Image," *Speech Teacher* 24 (1975): 218–225; James J. Bradac, "Language Attitudes and Impression Formation," *Handbook of Language and Social Psychology*, ed. Howard Giles and W. Peter Robinson (Chichester, England: Wiley, 1990), 387–412.

40. Geneva Smitherman, *Talkin That Talk: Language, Culture, and Education in African America* (New York: Routledge, 2000), 21–22.

41. For an excellent summary of research on the effect of accent and dialect variations among ethnic groups, see Giles and Franklyn-Stokes. See also: Diane M. Badzinski, "The Impact of Accent and

Status on Information Recall and Perception Information," *Communication* 5 (1992): 99–106.

42. Edwards, 22–27.

43. Richard Rodriguez, *Hunger of Memory: The Education of Richard Rodriguez* (Toronto: Bantam Books, 1982), 14–16. Italics in original.

44. Hecht, Collier, and Ribeau 90; U. Dagmar Scheu, "Cultural Constraints in Bilinguals' Codeswitching," *International Journal of Intercultural Relations* 24 (2000): 131–150.

45. Abdelala Bentahila, *Language Attitudes among Arabic-French Bilinguals in Morocco* (London: Multilingual Matters, 1983), 27–65.

CHAPTER 8

1. Albert E. Scheflen, "On Communication Processes," *Nonverbal Behavior: Applications and Cross-Cultural Implications*, ed. Aaron Wolfgang (New York: Academic Press, 1979), 1–16.

2. Sheila J. Ramsey, "Nonverbal Behavior: An Intercultural Perspective," *Handbook of Intercultural Communication*, ed. Molefi Kete Asante, Eileen Newmark, and Cecil A. Blake (Beverly Hills, CA: Sage, 1979), 111.

3. Edward T. Hall, *The Silent Language* (Garden City, NY: Doubleday, 1959).

4. Charles Darwin, *The Expression of Emotions in Man and Animals* (New York: Appleton, 1872).

5. For an examination of hand gestures, see Marcel Kinsbourne, "Gestures as Embodied Cognition: A Neurodevelopmental Interpretation," *Gesture* 6 (2006): 205–214. For an examination of vocal characteristics, see Robert W. Frick, "Communicating Emotion: The Role of Prosidic Features," *Psychological Bulletin* 97 (1985): 412–429; Marcel Kinsbourne, "Gestures as Embodied Cognition: A Neurodevelopmental Interpretation," *Gesture* 6 (2006): 205–214.

6. Research on gestures includes: Dedre Gentner and Susan Goldin-Meadow (eds.), *Language in Mind: Advances in the Study of Language and Thought* (Cambridge, MA: MIT Press, 2003); Susan Goldin-Meadow, *Hearing Gesture: How Our Hands Help Us Think* (Cambridge, MA: Harvard University Press, 2003); Susan Goldin-Meadow, *The Resilience of Language: What Gesture Creation in Deaf Children Can Tell Us about How All Children Learn Language* (New York: Psychology Press, 2003); Jana M. Iverson and Susan Goldin-Meadow, "Why People Gesture When They Speak," *Nature* 396 (1998): 228; Jana M. Iverson and Susan Goldin-Meadow, "The Resilience of Gesture in Talk: Gestures in Blind Speakers and Listeners," *Developmental Science* 4 (2001): 416–422. Research on emotional expressions includes: Irenäus Eibl-Eibesfeldt, *Love and Hate: The Natural History of Behavior Patterns* (New York: Holt, 1974); Irenäus Eibl-Eibesfeldt, "Universals in Human Expressive Behavior," *Nonverbal Behavior: Applications and Cultural Implications*, ed. Aaron Wolfgang (New York: Academic Press, 1979), 17–30.

7. Robert Ardrey, *The Territorial Imperative: A Personal Inquiry into the Animal Origins of Property and Nations* (New York: Atheneum, 1966).

8. Paul Ekman, *Emotion in the Human Face* (Cambridge, England: Cambridge University Press, 1982); Paul Ekman, "Facial Expression," *Nonverbal Behavior and Communication*, 2nd ed., ed. Aron W. Siegman and Stanley Feldstein (Hillsdale, NJ: Erlbaum, 1987), 97–116; Paul Ekman, "Facial Expression and Emotion," *American Psychologist* 48 (1993): 384–392; Paul Ekman, *Emotions Revealed: Recognizing Faces and Feelings to Improve Communication and Emotional Life* (New York: Times Books, 2003); Paul Ekman and Wallace V. Friesen, "Constants across Cultures in the Face and Emotion," *Journal of Personality and Social Psychology* 17 (1971): 124–129; Paul Ekman and Wallace V. Friesen, *Unmasking the Face: A Field Guide to Recognizing Emotions from Facial Clues* (Englewood Cliffs, NJ: Prentice-Hall, 1975); Paul Ekman and Wallace V. Friesen, "A New Pan-Cultural Expression of Emotion," *Motivation and Emotion*, 10, 159–168; Paul Ekman and Dacher Keltner, "Universal Facial Expressions of Emotion: An Old Controversy and New Findings," *Nonverbal Communication: Where Nature Meets Culture*, ed. Ullica Segerstråle and Peter Molnár (Mahwah, NJ: Erlbaum, 1997), 27–46; Alan J. Fridlund, Paul Ekman, and Harriet Oster, "Facial Expressions of Emotion: Review of Literature, 1970–1983," *Nonverbal Behavior and Communication*, 2nd ed., ed. Aron W. Siegman and Stanley Feldstein (Hillsdale, NJ: Erlbaum, 1987), 143–224.

9. Ross Buck, *The Communication of Emotion* (New York: Guilford, 1984); Ross Buck, "Nonverbal Communication: Spontaneous and Symbolic Aspects," *Communication*

Theory, 5 (1995): 393–396; Ross Buck and C. Arthur VanLear, "Verbal and Nonverbal Communication: Distinguishing Symbolic, Spontaneous, and Pseudo-Spontaneous Nonverbal Behavior," *Journal of Communication* 52 (2002): 522–541.

10. David Matsumoto, "Cultural Influences on Facial Expressions of Emotion," *Southern Communication Journal* 56 (1991): 128–137; David Matsumoto, Andres Olide, Joanna Schug, Bob Willingham, and Mike Callan, "Cross-Cultural Judgments of Spontaneous Facial Expressions of Emotion," *Journal of Nonverbal Behavior* 33 (2009): 213–238.

11. Marc D. Pell, Laura Monetta, Silke Paulmann, and Sonja A. Kotz, "Recognizing Emotions in a Foreign Language," *Journal of Nonverbal Behavior* 33 (2009): 107–120.

12. Michael Argyle, *Bodily Communication* (New York: International Universities Press, 1975), 95.

13. David Matsumoto, "Cultural Similarities and Differences in Display Rules," *Motivation and Emotion* 14 (1990): 195–214; David Matsumoto, Seung Hee Yoo, Johnny Fontaine, Ana Maria Anguas-Wong, Monica Arriola, Bilge Ataca,...Elvair Grossi, "Mapping Expressive Differences around the World: The Relationship between Emotional Display Rules and Individualism Versus Collectivism," *Journal of Cross-Cultural Psychology* 39 (2008): 55–74.

14. Thomas J. Schofield, Ross D. Parke, Erica K. Castañeda, and Scott Coltrane, "Patterns of Gaze between Parents and Children in European American and Mexican American Families," *Journal of Nonverbal Behavior* 32 (2008): 171–186.

15. Linda C. Halgunseth, Jean M. Ispa, and Duane Rudy, "Parental Control in Latino Families: An Integrated Review of the Literature," *Child Development* 77 (2006): 1282–1297.

16. Michael Argyle, "Rules for Social Relationships in Four Cultures," *Australian Journal of Psychology* 38 (1986): 309–318; Michael Argyle, Monika Henderson, Michael Harris Bond, Yuichi Iizuka, and Alberta Contarello, "Cross-Cultural Variations in Relationship Rules," *International Journal of Psychology* 21 (1986): 287–315.

17. Miles L. Patterson, Yuichi Iizuka, Mark E. Tubbs, Jennifer Ansel, Masao Tsutsumi, and Jackie Anson, "Passing Encounters East and West: Comparing Japanese and American Pedestrian Interactions,"

Journal of Nonverbal Behavior 31 (2007): 155–166; Miles L. Patterson and Joann Montepare, "Nonverbal Behavior in a Global Context Dialogue Questions and Responses," *Journal of Nonverbal Behavior* 31 (2007): 167–168.

18. Piotr Szarota, "The Mystery of the European Smile: A Comparison Based on Individual Photographs Provided by Internet Users," *Journal of Nonverbal Behavior* 34 (2010): 249–256.

19. Norine Dresser, *Multicultural Manners: Essential Rules of Etiquette for the 21st Century*, rev ed. (Hoboken, NJ: Wiley, 2005), 15.

20. Robert G. Harper, Arthur N. Wiens, and Joseph D. Matarazzo, *Nonverbal Communication: The State of the Art* (New York: Wiley, 1978).

21. Sharon Ruhly, *Intercultural Communication*, 2nd ed. (Chicago: Science Research Associates, 1982), 23–26.

22. John C. Condon and Fathi S. Yousef, *Intercultural Communication* (Indianapolis: Bobbs-Merrill, 1975), 122.

23. Dresser, 56.

24. Ray Birdwhistell, *Kinesics and Context: Essays on Body Motion Communication* (Philadelphia: University of Pennsylvania, 1970), 34.

25. Joseph A. DeVito, *Messages: Building Interpersonal Communication Skills* (New York: Harper and Row, 1990), 218.

26. Miles L. Patterson, *More than Words: The Power of Nonverbal Communication* (Barcelona, Spain: Aresta, 2011), 10.

27. Our description and analysis of the functions of nonverbal communication represent a synthesis of those that have been described and researched widely. See, for example: Peter A. Andersen, *Nonverbal Communication: Forms and Functions*, 2nd ed. (Long Grove, IL: Waveland Press, 2008); Judee K. Burgoon, Laura K. Guerrero, and Cory Floyd, *Nonverbal Communication* (Boston: Allyn & Bacon, 2010); Miles L. Patterson, *Nonverbal Behavior: A Functional Perspective* (New York: Springer-Verlag, 1983); Patterson, *More than Words*.

28. David Matsumoto, Andres Olide, Joanna Schug, Bob Willingham, and Mike Callan, "Cross-Cultural Judgments of Spontaneous Facial Expressions of Emotion," *Journal of Nonverbal Behavior* 33 (2009): 213–238; Jessica L. Tracy and Richard W. Robins, "The Nonverbal Expression of Pride: Evidence for Cross-Cultural Recognition," *Journal of Personality and Social*

Psychology 94 (2008): 516–530; Sowan Wong, Michael Harris Bond, and Patricia M. Rodriguez Mosquera, "The Influence of Cultural Value Orientations on Self-Reported Emotional Expression across Cultures," *Journal of Cross-Cultural Psychology* 39 (2008): 224–229.

29. Sally Planalp, *Communicating Emotion: Social, Moral, and Cultural Processes* (New York: Cambridge University Press, 1999); Sally Planalp, "Varieties of Emotional Cues in Everyday Life," *The Nonverbal Communication Reader: Classic and Contemporary Readings*, ed. Laura K. Guerrero and Michael L. Hecht (Long Grove, IL: Waveland, 2008), 397–401; Sally Planalp, Victoria Leto DeFrancisco, and Diane Rutherford, "Varieties of Cues to Emotion in Naturally Occurring Situations," *Cognition and Emotion* 10 (1996): 137–153.

30. Marc D. Pell, Laura Monetta, Silke Paulmann, and Sonja A. Kotz, "Recognizing Emotions in a Foreign Language," *Journal of Nonverbal Behavior* 33 (2009): 107–120.

31. Sheena Iyengar, *The Art of Choosing* (New York: Twelve, 2010).

32. Jonah Lehrer, *How We Decide* (New York: Houghton Mifflin Harcourt, 2009), 13, 26.

33. Mark L. Knapp and Judith A. Hall, *Nonverbal Communication in Human Interaction*, 7th ed. (Boston: Wadsworth, Cengage Learning, 2010); Mark Knapp and Anita L. Vangelisti, *Interpersonal Communication and Human Relationships* 6th ed. (Boston: Pearson, 2009).

34. The term *body language* is inaccurate because "nonverbal communication involves more than the body; it is not a language; and the meaning is usually not specific, but conditional on the larger social context." Patterson, *More than Words*, 9.

35. Paul Ekman and Wallace V. Friesen, "The Repertoire of Nonverbal Behavior: Categories, Origins, Usage, and Coding," *Semiotica* 1 (1969): 49–98.

36. Dresser, 11–12.

37. Tom Brosnahan and Pat Yale, *Turkey: A Lonely Planet Travel Survival Kit*, 5th ed. (Victoria, Australia: Lonely Planet, 1996), 27.

38. Simone Pika, Elena Nicoladis, and Paula Marentette, "How to Order a Beer: Cultural Differences in the Use of Conventional Gestures for Numbers," *Journal of Cross-Cultural Psychology* 40 (2009): 70–80.

39. Jeanette S. Martin and Lillian H. Chaney, *Global Business Etiquette: A Guide to International Communication and Customs* (Westport, CT: Praeger, 2006), 52.

40. Sana Reynolds and Deborah Valentine, *Guide to Cross-Cultural Communication*, 2nd ed. (Boston: Prentice Hall, 2011), 84.

41. Martin and Chaney; Reynolds and Valentine.

42. Paul Ekman, Wallace V. Friesen, and Phoebe Ellsworth, *Emotion in the Human Face: Guidelines for Research and an Integration of Findings* (New York: Pergamon Press, 1972).

43. Michael Harris Bond, "Emotions and Their Expressions in Chinese Culture," *Journal of Nonverbal Behavior* 17 (1993): 245–262.

44. Marianne LaFrance and Clara Mayo, "Racial Differences in Gaze Behavior During Conversations: Two Systematic Observational Studies," *Journal of Personality and Social Psychology* 33 (1976): 547–552.

45. Edward T. Hall, *The Hidden Dimension* (Garden City, NY: Doubleday, 1966).

46. Edward T. Hall and Mildred Reed Hall, *Understanding Cultural Differences* (Yarmouth, ME: Intercultural Press, 1990), 12.

47. Hall and Hall, 180.

48. Hall and Hall, 10.

49. Stanley E. Jones and A. Elaine Yarbrough, "A Naturalistic Study of the Meanings of Touch," *Communication Monographs* 52 (1985): 19–56.

50. Nancy M. Henley, *Body Politics: Power, Sex, and Nonverbal Communication* (Englewood Cliffs, NJ: Prentice-Hall, 1977).

51. Hall and Hall, 11.

52. Dean Barnlund, "Communication Styles in Two Cultures: Japan and the United States," *Organizational Behavior in Face-to-Face Interaction*, ed. Adam Kendon, Richard M. Harris, and Mary Ritchie Key (The Hague: Mouton, 1975), 427–456.

53. Sidney M. Jourard, "An Exploratory Study of Body Accessibility," *British Journal of Social and Clinical Psychology* 5 (1966): 221–231.

54. Rosita Daskel Albert and Gayle L. Nelson, "Hispanic/Anglo-American Differences in Attributions to Paralinguistic Behavior," *International Journal of Intercultural Relations* 17 (1993): 19–40.

55. Mara B. Adelman and Myron W. Lustig, "Intercultural Communication Problems as Perceived by Saudi Arabian and American

Managers," *International Journal of Intercultural Relations* 5 (1981): 349–364; Myron W. Lustig, "Cultural and Communication Patterns of Saudi Arabians," *Intercultural Communication: A Reader*, 5th ed., ed. Larry A. Samovar and Richard E. Porter (Belmont, CA: Wadsworth, 1988), 101–103.

56. John Reader, *Man on Earth* (New York: Harper and Row, 1988), 91. Reader's ideas are based on Paul Spencer, *The Samburu: A Study in Gerontocracy in a Nomadic Tribe* (London: Routledge and Kegan Paul, 1968).

57. Reader, 163.

58. Edward T. Hall, "The Hidden Dimensions of Time and Space in Today's World," *Cross-Cultural Perspectives in Nonverbal Communication*, ed. Fernando Poyatos (Toronto: C. J. Hogrefe, 1988), 151.

59. Young-Ok Lee, "Perceptions of Time in Korean and English," *Human Communication* 12 (2009): 119–138.

60. Hall, *The Silent Language*.

61. Jane Engle, "Punctuality: Some Cultures Are Wound Tighter than Others," *Los Angeles Times*, December 11, 2005, I3.

62. Hall, *The Silent Language*, 178.

63. Alexander Gonzalez and Philip G. Zimbardo, "Time in Perspective," *Psychology Today* 19 (March 1985): 20–26.

64. Trudy Milburn, "Enacting 'Puerto Rican Time' in the United States," *Constituting Cultural Difference through Discourse*, ed. Mary Jane Collier (Thousand Oaks, CA: Sage, 2001), 47–76.

65. Martin and Chaney, 37.

66. William S. Condon, "Cultural Microrhythms," *Interaction Rhythms: Periodicity in Communicative Behavior*, ed. Martha Davis (New York: Human Sciences Press, 1982), 66.

67. Befu 1975, as quoted in Ramsey, "Nonverbal Behavior," 118.

68. Peter A. Andersen, *Nonverbal Communication*, 78.

69. Holley S. Hodgins and Richard Koestner, "The Origins of Nonverbal Sensitivity," *Personality and Social Psychology Bulletin* 19 (1993): 466–473.

70. Anna-Marie Dew and Colleen Ward, "The Effects of Ethnicity and Culturally Congruent and Incongruent Nonverbal Behaviors on Interpersonal Attraction," *Journal of Applied Social Psychology* 23 (1993): 1376–1389.

71. David Matsumoto and Tsutomu Kudoh, "American-Japanese Cultural Differences in Attributions of Personality Based on Smiles," *Journal of Nonverbal Behavior* 17 (1993): 231–243. See also Ann Bainbridge Frymier, Donald W. Klopf, and Satoshi Ishii, "Affect Orientation: Japanese Compared to Americans," *Communication Research Reports* 7 (1990): 63–66; Donald W. Klopf, "Japanese Communication Practices: Recent Comparative Research," *Communication Quarterly* 39 (1991): 130–143.

CHAPTER 9

1. Laurie G. Kirszer and Stephen R. Mandell, *The Brief Wadsworth Handbook*, 5th ed. (Boston: Wadsworth, 2006); Laurie G. Kirszer and Stephen R. Mandell, *The Concise Wadsworth Handbook*, 2nd ed. (Boston: Wadsworth, 2008); Laurie G. Kirszer and Stephen R. Mandell, *The Pocket Wadsworth Handbook*, 4th ed. (Boston: Wadsworth, 2008); Laurie G. Kirszer and Stephen R. Mandell, *The Wadsworth Handbook*, 8th ed. (Boston: Wadsworth, 2008).

2. Kaplan's original name for this specialization was "contrastive rhetoric." See Robert B. Kaplan, "Cultural Thought Patterns in Inter-Cultural Education," *Language Learning: A Journal of Applied Linguistics* 16 (1966): 1–20.

3. See, for example, Ulla Connor, "New Directions in Contrastive Rhetoric," *TESOL Quarterly* 36 (2002): 493–510; Ulla Connor, *Contrastive Rhetoric: Cross-Cultural Aspects of Second Language Writing* (Cambridge: Cambridge University Press, 1996); Ulla Connor, "Introduction," *Journal of English for Academic Purposes* 3.4 (2004): 271–276; Ulla Connor, "Intercultural Rhetoric Research: Beyond Texts," *Journal of English for Academic Purposes* 3.4 (2004): 291–304; Ulla Connor, "Mapping Multidimensional Aspects of Research: Reaching to Intercultural Rhetoric," *Contrastive Rhetoric: Reaching to Intercultural Rhetoric*, ed. Ulla Connor, Ed Nagelhout, and William V. Rozycki (Philadelphia: John Benjamins, 2008), 299–315; Xiaoming Li, "From Contrastive Rhetoric to Intercultural Rhetoric: A Search for Collective Identity," *Contrastive Rhetoric: Reaching to Intercultural Rhetoric*, ed. Ulla Connor, Ed Nagelhout, and William V. Rozycki (Philadelphia: John Benjamins, 2008), 11–24.

4. John Hinds, "Reader versus Writer Responsibility: A New Typology," *Writing Across Languages: Analysis of L2 Written Text*, ed. Ulla Connor and Robert B. Kaplan (Reading, MA: Addison-Wesley, 1987), 141–152.

5. Connor, *Contrastive Rhetoric*.

6. Satoshi Ishii, "Thought Patterns as Modes of Rhetoric: The United States and Japan," *Intercultural Communication: A Reader*, 4th ed., ed. Larry A. Samovar and Richard E. Porter (Belmont, CA: Wadsworth, 1985), 97–102.

7. David Cahill, "The Myth of the 'Turn' in Contrastive Rhetoric," *Written Communication* 20 (2003): 170–194.

8. Yamuna Kachru, "Writers in Hindi and English," *Writing Across Languages and Cultures: Issues in Contrastive Rhetoric*, ed. Alan C. Purvis (Newbury Park, CA: Sage, 1988), 109–137.

9. Arpita Misra, "Discovering Connections," *Language and Social Identity*, ed. John L. Gumperz (Cambridge: Cambridge University Press, 1982), 57–71.

10. Linda Wai Ling Young, "Inscrutability Revisited," *Language and Social Identity*, ed. John L. Gumperz (Cambridge: Cambridge University Press, 1982), 72–84.

11. Keiko Hirose, "Pursuing the Complexity of the Relationship between L1 and L2 Writing," *Journal of Second Language Writing* 15 (2006): 142–146; Yunxia Zhu, "Understanding Sociocognitive Space of Written Discourse: Implications for Teaching Business Writing to Chinese Students," *IRAL: International Review of Applied Linguistics in Language Teaching* 44 (2006): 265–285.

12. Donald G. Ellis and Ifat Maoz, "Cross-Cultural Argument Interactions between Israeli-Jews and Palestinians," *Journal of Applied Communication Research* 30 (2002): 181–194; M. Sean Limon and Betty H. La France, "Communication Traits and Leadership Emergence: Examining the Impact of Argumentativeness, Communication Apprehension, and Verbal Aggressiveness in Work Groups," *Southern Communication Journal* 70 (2005): 123–133; Ringo Ma, "The Role of Unofficial Intermediaries in Interpersonal Conflicts in the Chinese Culture," *Communication Quarterly* 40 (1992): 269–278; Alicia M. Prunty, Donald W. Klopf, and Satoshi Ishii, "Argumentativeness: Japanese and American Tendencies to Approach and Avoid Conflict," *Communication Research*

Reports 7 (1990): 75–79; Judith Sanders, Robert Gass, Richard Wiseman, and Jon Bruschke, "Ethnic Comparison and Measurement of Argumentativeness, Verbal Aggressiveness, and Need for Cognition," *Communication Reports* 5 (1992): 50–56; Sunwolf, "The Pedagogical and Persuasive Effects of Native American Lesson Stories, Sufi Wisdom Tales, and African Dilemma Tales," *Howard Journal of Communications* 10 (1999): 47–71; Shinobu Suzuki and Andrew S. Rancer, "Argumentativeness and Verbal Aggressiveness: Testing for Conceptual and Measurement Equivalence across Cultures," *Communication Monographs* 61 (1994): 256–279; Lynn H. Turner and Robert Shuter, "African American and European American Women's Visions of Workplace Conflict: A Metaphorical Analysis," *Howard Journal of Communications* 15 (2004): 169–183.

13. Barry L. Thatcher, "Writing Policies and Procedures in a U.S./South American Context," *Technical Communication Quarterly* 9.4 (2000): 365–399. See also Barry Thatcher, "Intercultural Rhetoric, Technology Transfer, and Writing in U.S.-Mexico Border Maquilas," *Technical Communication Quarterly* 15.3 (2006): 383–405.

14. James F. Hamill, *Ethno-Logic: The Anthropology of Human Reasoning* (Urbana: University of Illinois Press, 1990), 23.

15. Stephen Toulmin, *Human Understanding, Volume I: The Collective Use and Evolution of Concepts* (Princeton, NJ: Princeton University Press, 1972).

16. For a thorough discussion about the relationship between cultural patterns and argumentation, see Andrea Rocci, "Pragmatic Inference and Argumentation in Intercultural Communication," *Intercultural Pragmatics* 3–4 (2006): 409–422.

17. Sunwolf.

18. Ann Neville Miller, "An Exploration of Kenyan Public Speaking Patterns with Implications for the American Introductory Public Speaking Course," *Communication Education* 51 (2002): 168–182.

19. Xiaosui Xiao, "From the Hierarchical *Ren* to Egalitarianism: A Case of Cross-Cultural Rhetorical Mediation," *Quarterly Journal of Speech* 82 (1996): 38–54.

20. Yanrong Chang, "Courtroom Questioning as a Culturally Situated Persuasive Genre of Talk," *Discourse & Society* 15 (2004): 705–722.

21. Robert B. Kaplan, "Foreword: What in the World Is Contrastive Rhetoric?" *Contrastive Rhetoric*

Revisited and Redefined, ed. Clayann Gillima Panetta (Mahwah, NJ: Erlbaum, 2001), vii–xx.

22. Christina Janik, "As Academics We Are Not Disposed to Say 'I Know the World Is Round—': Marking of Evidentiality in Russian and German Historiographic Articles," *Cross-Linguistic and Cross-Cultural Perspectives on Academic Discourse*, ed. Eija Suomela-Salmi and Fred Dervin (Philadelphia: John Benjamins, 2009), 19–32.

23. For a detailed discussion of cultural variability in argumentation, see Rocci.

24. Barbara Johnstone, "Linguistic Strategies for Persuasive Discourse," *Language, Communication, and Culture: Current Directions*, ed. Stella Ting-Toomey and Felipe Korzenny (Newbury Park, CA: Sage, 1989), 139–156.

25. Robert Shuter, "The Culture of Rhetoric," *Rhetoric in Intercultural Contexts*, ed. Alberto Gonzalez and Dolores V. Tanno (Thousands Oaks, CA: Sage, 2000), 12–13.

26. Ora-Ong Chakorn, "Persuasive and Politeness Strategies in Cross-Cultural Letters of Request in the Thai Business Context," *Journal of Asian Pacific Communication* 16 (2006): 3–46.

27. Abdulrahman M. Alhudhaif, *A Speech Act Approach to Persuasion in American and Arabic Editorials*, Diss. Purdue University, 2006. Dissertation Abstracts International, 66A, 10, Apr, 3623.

28. Angela Eagan and Rebecca Weiner, *Culture Shock! China: A Survival Guide to Customs and Etiquette* (Tarrytown, NY: Marshall Cavendish, 2007); Kathy Flower, *Culture Smart! China: A Quick Guide to Customs and Etiquette* (Portland, OR: Graphic Arts Books, 2003); Stanley B. Lubman, "Negotiations in China: Observations of a Lawyer Communicating with China," *Communicating with China*, ed. Robert A. Kapp (Chicago: Intercultural Press, 1983); Stanley B. Lubman (ed.), *China's Legal Reforms* (Oxford: Oxford University Press, 1996); Stuart Strother and Barbara Strother, *Living Abroad in China* (Berkeley, CA: Avalon Travel Publishing, 2006); Hu Wenzhong and Cornelius L. Grove, *Encountering the Chinese: A Guide for Americans* (Yarmouth, ME: Intercultural Press, 1991).

29. Anne Bliss, "Rhetorical Structures for Multilingual and Multicultural Students," *Contrastive Rhetoric Revisited and Redefined*, ed. Clayann Gilliam Panetta (Mahwah, NJ: Erlbaum, 2001).

30. Bliss.

31. John C. Condon, *Good Neighbors: Communicating with the Mexicans* (Yarmouth, ME: Intercultural Press, 1985).

32. Daniel Dolan, "Conditional Respect and Criminal Identity: The Use of Personal Address Terms in Japanese Mass Media," *Western Journal of Communication* 64 (1998): 459–473.

33. Johnstone.

34. Connor, *Contrastive Rhetoric*, 167.

35. Donal Carbaugh, *Cultures in Conversation* (Mahwah, NJ: Erlbaum, 2005), 19.

36. For another example that documents the difference between European American and Russian preferences in persuasive styles, see Maria Loukianenko Wolfe, "Different Cultures—Different Discourses? Rhetorical Patterns of Business Letters by English and Russian Speakers," *Contrastive Rhetoric: Reaching to Intercultural Rhetoric*, ed. Connor, Nagelhout, and Rozycki, 87–121.

37. Howard Giles, Nikolas Coupland, and John Wiemann, "'Talk Is Cheap…' but 'My Word Is My Bond': Beliefs about Talk," *Sociolinguistics Today: International Perspectives*, ed. Kingsley Bolton and Helen Kwok (New York: Routledge, 1992), 218–243.

38. D. Lawrence Kincaid, "Communication East and West: Points of Departure," *Communication Theory: Eastern and Western Perspectives*, ed. D. Lawrence Kincaid (San Diego: Academic Press, 1987), 337.

39. Giles, Coupland, and Wiemann.

40. Donald W. Klopf, *Intercultural Encounters: The Fundamentals of Intercultural Communication*, 2nd ed. (Englewood Cliffs, NJ: Morgan, 1991), 181.

41. June Ock Yum, "Korean Philosophy and Communication," *Communication Theory: Eastern and Western Perspectives*, ed. D. Lawrence Kincaid (San Diego: Academic Press, 1987), 79.

42. Yum, 83.

43. Peter Nwosu, "Negotiating with the Swazis," *Howard Journal of Communication* 1 (1988): 148.

44. Aino Sallinen-Kuparinen, James C. McCroskey, and Virginia P. Richmond, "Willingness to Communicate, Communication Apprehension, Introversion, and Self-Reported Communication Competence: Finnish and American Comparisons," *Communication Research Reports* 8 (1991): 55–64.

45. Carbaugh, xxi.

46. Keith H. Basso, "'To Give Up on Words': Silence in Western Apache Culture," *Cultural Communication and Intercultural Contact*, ed. Donal L. Carbaugh (Hillsdale, NJ: Erlbaum, 1990), 303–320.

47. Basso, 308.

48. Klopf.

49. William B. Gudykunst and Stella Ting-Toomey, *Culture and Interpersonal Communication* (Newbury Park, CA: Sage, 1988), 99–116.

50. Sherry L. Beaumont and Shannon L. Wagner, "Adolescent-Parent Verbal Conflict: The Roles of Conversational Styles and Disgust Emotions," *Journal of Language & Social Psychology* 23 (2004): 338–368; Deborah A. Cai, Steven R. Wilson, and Laura E. Drake, "Culture in the Context of Intercultural Negotiation: Individualism–Collectivism and Paths to Integrative Agreements," *Human Communication Research* 26 (2000): 591–607; Min-Sun Kim, "Culture-Based Interactive Constraints in Explaining Intercultural Strategic Competence," *Intercultural Communication Competence*, ed. Richard L. Wiseman and Jolene Koester (Newbury Park, CA: Sage, 1993), 132–150; Min-Sun Kim, "Cross-Cultural Comparisons of the Perceived Importance of Conversational Constraints," *Human Communication Research* 21 (1994): 128–151; Min-Sun Kim, "Toward a Theory of Conversational Constraints: Focusing on Individual-Level Dimensions of Culture," *Intercultural Communication Theory,* ed. Richard L. Wiseman (Thousand Oaks, CA: Sage, 1995), 148–169; Min-Sun Kim and Mary Bresnahan, "Cognitive Basis of Gender Communication: A Cross-Cultural Investigation of Perceived Constraints in Requesting," *Communication Quarterly* 44 (1996): 53–69; Min-Sun Kim, John E. Hunter, Akira Miyahara, Ann-Marie Horvath, Mary Bresnahan, and Hye-Jin Yoon, "Individual- vs. Culture-Level Dimensions of Individualism and Collectivism: Effects on Preferred Conversational Styles," *Communication Monographs* 63 (1996): 29–49; Min-Sun Kim, Renee Storm Klingle, William F. Sharkey, Hee Sun Park, David H. Smith, and Deborah Cai, "A Test of a Cultural Model of Patients' Motivation for Verbal Communication in Patient-Doctor Interactions," *Communication Monographs* 67 (2000): 262–283; Min-Sun Kim and William F. Sharkey, "Independent and Interdependent Construals of Self: Explaining Cultural Patterns of Interpersonal Communication in Multi-Cultural Organizational Settings," *Communication Quarterly* 43 (1995): 20–38; Min-Sun Kim, William F. Sharkey, and Theodore M. Singelis, "The Relationship of Individuals' Self-Construals and Perceived Importance of Interactive Constraints," *International Journal of Intercultural Relations* 18 (1994): 117–140; Min-Sun Kim, Ho-Chang Shin, and Deborah Cai, "Cultural Influences on the Preferred Forms of Requesting and Re-Requesting," *Communication Monographs* 65 (1998): 47–66; Min-Sun Kim and Steven R. Wilson, "A Cross-Cultural Comparison of Implicit Theories of Requesting," *Communication Monographs* 61 (1994): 210–235; Akira Miyahara, Min-Sun Kim, Ho-Chang Shin, and Kak Yoon, "Conflict Resolution Styles Among Collectivist Cultures: A Comparison between Japanese and Koreans," *International Journal of Intercultural Relations* 22 (1998): 505–525.

51. Thomas Kochman, "Force Fields in Black and White," *Cultural Communication and Intercultural Contact*, ed. Donal Carbaugh (Hillsdale, NJ: Erlbaum, 1990), 193–194.

52. Melanie Booth-Butterfield and Felecia Jordan, "Communication Adaptation Among Racially Homogeneous and Heterogeneous Groups," *Southern Communication Journal* 54 (1989): 265.

53. Yale Richmond, *From Da to Yes: Understanding the East Europeans* (Yarmouth, ME: Intercultural Press, 1995), 118.

54. Ronald Scollon and Suzanne Wong-Scollon, "Athabaskan-English Interethnic Communication," *Cultural Communication and Intercultural Contact*, ed. Carbaugh, 270.

55. Scollon and Wong-Scollon, 273.

CHAPTER 10

1. John Paul Feig, *A Common Core: Thais and Americans,* rev. Elizabeth Mortlock (Yarmouth, ME: Intercultural Press, 1989), 50.

2. Hu Wenzhong and Cornelius L. Grove, *Encountering the Chinese: A Guide for Americans,* 2nd ed. (Yarmouth, ME: Intercultural Press, 1991).

3. John C. Condon, *Good Neighbors: Communicating with the Mexicans* (Yarmouth, ME: Intercultural Press, 1985).

4. Mary Jane Collier and Elirea Bornman, "Core Symbols in South African Intercultural Friendships," *International Journal of Intercultural Relations* 23 (1999): 133–156; Daniel Perlman and Beverley Fehr, "The Development of Intimate Relationships," *Intimate Relationships: Development, Dynamics, and Deterioration,* ed. Daniel Perlman and Steven Duck (Newbury Park, CA: Sage, 1987), 13–42.

5. Michael L. Hecht, Mary Jane Collier, and Sidney A. Ribeau, *African American Communication: Ethnic Identity and Cultural Interpretation* (Newbury Park, CA: Sage, 1993).

6. Mary Jane Collier, "Cultural Background and the Culture of Friendships: Normative Patterns," paper presented at the annual conference of the International Communication Association, San Francisco, May 1989. See also Mary Jane Collier, "Conflict Competence within African, Mexican, and Anglo American Friendships," *Cross-Cultural Interpersonal Communication,* ed. Stella Ting-Toomey and Felipe Korzenny (Newbury Park, CA: Sage, 1991), 132–154.

7. Stanley O. Gaines, Jr., "Communalism and the Reciprocity of Affection and Respect among Interethnic Married Couples," *Journal of Black Studies* 27 (1997): 352–364; Stanley O. Gaines, Jr., et al., "Patterns of Attachment and Responses to Accommodative Dilemmas Among Interethnic/Interracial Couples," *Journal of Social and Personal Relationships* 16 (1999): 275–285; Stanley O. Gaines, Jr., et al., "Links between Race/Ethnicity and Cultural Values as Mediated by Racial/Ethnic Identity and Moderated by Gender," *Journal of Personality and Social Psychology* 72 (1997): 1460–1476; Stanley O. Gaines, Jr., with Raymond Buriel, James H. Liu, and Diana I. Ríos, *Culture, Ethnicity, and Personal Relationship Processes* (New York: Routledge, 1997). See also: Stella D. Garcia and Semilla M. Rivera, "Perceptions of Hispanic and African-American Couples at the Friendship or Engagement Stage of a Relationship," *Journal of Social and Personal Relationships* 16 (1999): 65–86; Regan A. R. Gurung and Tenor Duong, "Mixing and Matching: Assessing the Concomitants of Mixed-Ethnic Relationships," *Journal of Social and Personal Relationships* 16 (1999): 639–657; John McFadden, "Intercultural Marriage and Family: Beyond the Racial Divide," *Family Journal* 9 (2001): 39–42.

8. Irene I. Blea, *Toward a Chicano Social Science* (New York: Praeger, 1988); Richard Lewis, Jr., George Yancy, and Siri S. Bletzer, "Racial and Nonracial Factors That Influence Spouse Choice in Black/White Marriages," *Journal of Black Studies* 28 (1997): 60–78.

9. Don C. Locke, *Increasing Multicultural Understanding: A Comprehensive Model* (Newbury Park, CA: Sage, 1992), 55.

10. Lin Yutang, *The Chinese Way of Life* (New York: World, 1972), 78.

11. We have synthesized a variety of sources to provide this generalization.

12. William D. Wilder, *Communication, Social Structure and Development in Rural Malaysia: A Study of Kampung Kuala Bera* (London: Athlone Press, 1982), 107.

13. Joe Cummings, Susan Forsyth, John Noble, Alan Samagalski, and Tony Wheelan, *Indonesia: A Travel Survival Guide* (Berkeley, CA: Lonely Planet Publications, 1990), 321.

14. Albert Mehrabian, *Silent Messages* (Belmont, CA: Wadsworth, 1971).

15. Edward T. Hall, *The Hidden Dimension* (Garden City, NY: Doubleday, 1966).

16. Peter A. Andersen, Myron W. Lustig, and Janis F. Andersen, "Changes in Latitude, Changes in Attitude: The Relationship Between Climate and Interpersonal Communication Predispositions," *Communication Quarterly* 38 (1990): 291–311.

17. Wilder, 105.

18. Feig, 41.

19. William O. Beeman, *Language, Status, and Power in Iran* (Bloomington: Indiana University Press, 1986), 86.

20. Harvey Taylor, "Misunderstood Japanese Nonverbal Communication," *Gengo Seikatsu* (Language Life), 1974; quoted in Helmut Morsbach, "The Importance of Silence and Stillness in Japanese Nonverbal Communication: A Cross-Cultural Approach," *Cross-Cultural Perspectives in Nonverbal Communication,* ed. Fernando Poyatos (Toronto: C. J. Hogrefe, 1988), 206.

21. We rely primarily on Leslie A. Baxter, "A Dialectical Perspective on Communication Strategies in Relationship Development," *Handbook of Personal Relationships: Theory, Research, and Interventions,* ed. Stephen W. Duck (New York: Wiley, 1988), 257–273; Leslie A. Baxter, "Dialectical Contradictions in

Relationship Development," *Journal of Social and Personal Relationships* 7 (1990): 69–88; Leslie A. Baxter and Barbara M. Montgomery, *Relating: Dialogues and Dialectics* (New York: Guilford Press, 1996). See also: Irwin Altman, "Dialectics, Physical Environments, and Personal Relationships," *Communication Monographs* 60 (1993): 26–34; Irwin Altman, Anne Vinsel, and Barbara B. Brown, "Dialectic Conceptions in Social Psychology: An Application to Social Penetration and Privacy Regulation," *Advances in Experimental Social Psychology,* vol. 14, ed. Leonard Berkowitz (New York: Academic Press, 1981), 107–160; Carl W. Backman, "The Self: A Dialectical Approach," *Advances in Experimental Social Psychology,* vol. 21, ed. Leonard Berkowitz (New York: Academic Press, 1988), 229–260; Daena Goldsmith, "A Dialectic Perspective on the Expression of Autonomy and Connection in Romantic Relationships," *Western Journal of Speech Communication* 54 (1990): 537–556; Angela Hoppe-Nagao and Stella Ting-Toomey, "Relational Dialectics and Management Strategies in Marital Couples," *The Southern Communication Journal* 67 (2002): 142–159.

22. See Erving Goffman, *Interaction Ritual: Essays on Face-to-Face Behavior* (Garden City, NY: Anchor Books, 1967).

23. David Yau-fai Ho, "On the Concept of Face," *American Journal of Sociology* 81 (1976): 867–884.

24. Penelope Brown and Stephen Levinson, "Universals in Language Use: Politeness Phenomena," *Questions and Politeness: Strategies in Social Interaction,* ed. Esther N. Goody (Cambridge: Cambridge University Press, 1978), 56–289; Penelope Brown and Stephen Levinson, *Politeness: Some Universals in Language Use* (Cambridge: Cambridge University Press, 1987). Though Brown and Levinson's ideas have been criticized on several points, the portions of their ideas expressed here are generally accepted. For a summary of the criticisms, see Karen Tracy and Sheryl Baratz, "The Case for Case Studies of Facework," *The Challenge of Facework: Cross-Cultural and Interpersonal Issues,* ed. Stella Ting-Toomey (Albany: State University of New York Press, 1994), 287–305.

25. Greg Leichty and James L. Applegate, "Social-Cognitive and Situational Influences on the Use of Face-Saving Persuasive Strategies," *Human Communication Research* 17 (1991): 451–484.

26. We have modified Lim's terminology and concepts somewhat but draw on his overall conception. See Tae-Seop Lim, "Politeness Behavior in Social Influence Situations," *Seeking Compliance: The Production of Interpersonal Influence Messages,* ed. James Price Dillard (Scottsdale, AZ: Gorsuch Scarisbrick, 1990), 75–86; Tae-Seop Lim, "Facework and Interpersonal Relationships," *The Challenge of Facework: Cross-Cultural and Interpersonal Issues,* ed. Ting-Toomey, 209–229; Tae-Seop Lim and John Waite Bowers, "Facework: Solidarity, Approbation, and Tact," *Human Communication Research* 17 (1991): 415–450.

27. Lim, 211.

28. The Wade–Giles system for the Romanization of Chinese words is used, rather than the newer *pin-yin* system, in order to maintain consistency with the terms used in the quotes by Hu and by Hsu. Terms that the Wade–Giles system would render as *lien* and *mien-tzu* are written in the *pin-yin* system as, respectively, *lian* and *mian zi.*

29. Hsien Chin Hu, "The Chinese Concepts of 'Face,'" *American Anthropologist* 46 (1944): 45–64.

30. Francis L. K. Hsu, "The Self in Cross-Cultural Perspective," *Culture and Self: Asian and Western Perspectives,* ed. Anthony J. Marsella, George DeVos, and Francis L. K. Hsu (New York: Tavistock, 1985), 33.

31. Leo Rosten, *The Joys of Yiddish* (New York: McGraw-Hill, 1968), 234.

32. Hu, 61–62.

33. Brown and Levinson, 66.

34. Brown and Levinson; see also Robert T. Craig, Karen Tracy, and Frances Spisak, "The Discourse of Requests: Assessment of a Politeness Approach," *Human Communication Research* 12 (1986): 437–468.

35. Ron Scollon and Suzie Wong-Scollon, "Face Parameters in East-West Discourse," *The Challenge of Facework: Cross-Cultural and Interpersonal Issues,* ed. Ting-Toomey, 133–157.

36. Scollon and Wong-Scollon, 137.

37. Lijuan Stahl, "Face-Negotiation," unpublished manuscript (San Diego: San Diego State University, 1993), 12–13.

38. See Jared R. Curhan, Margaret A. Neale, and Lee Ross, "Dynamic Valuation: Preference Changes

in the Context of Face-to-Face Negotiation," *Journal of Experimental Social Psychology* 40 (2004): 142–151; John G. Oetzel, "The Effects of Ethnicity and Self-Construals on Self-Reported Conflict Styles," *Communication Reports,* 11 (1998): 133–144; John G. Oetzel, "The Influence of Situational Features on Perceived Conflict Styles and Self-Construals in Work Groups," *International Journal of Intercultural Relations,* 23 (1999): 679–695; John G. Oetzel and Stella Ting-Toomey, "Face Concerns in Interpersonal Conflict: A Cross-Cultural Empirical Test of the Face Negotiation Theory," *Communication Research* 30 (2003): 599–624; John G. Oetzel, Stella Ting-Toomey, Tomoko Masumoto, Yumiko Yokochi, Xiaohui Pan, Jiro Takai, and Richard Wilcox, "Face and Facework in Conflict: A Cross-Cultural Comparison of China, Germany, Japan, and the United States," *Communication Monographs* 68 (2001): 235–258; Stella Ting-Toomey, "Toward a Theory of Conflict and Culture," *Communication, Culture, and Organizational Processes,* ed. William B. Gudykunst, Lea P. Stewart, and Stella Ting-Toomey (Beverly Hills, CA: Sage, 1985), 71–86; Stella Ting-Toomey, "Intercultural Conflict Styles: A Face-Negotiation Theory," *Theories in Intercultural Communication,* ed. Young Yun Kim and William B. Gudykunst (Newbury Park, CA: Sage, 1988), 213–235; Stella Ting-Toomey, "Intergroup Diplomatic Communication: A Face-Negotiation Perspective," *Communicating for Peace,* ed. Felipe Korzenny and Stella Ting-Toomey (Newbury Park, CA: Sage, 1990), 75–95; Stella Ting-Toomey, "Intercultural Conflict Competence," *Competence in Interpersonal Conflict,* ed. William R. Cupach and Daniel J. Canary (New York: McGraw-Hill, 1997), 120–147; Stella Ting-Toomey, and Beth-Ann Cocroft, "Face and Facework: Theoretical and Research Issues," *The Challenge of Facework: Cross-Cultural and Interpersonal Issues,* ed. Ting-Toomey, 307–340; Stella Ting-Toomey and John G. Oetzel, *Managing Intercultural Conflict Effectively* (Thousand Oaks, CA: Sage, 2001).

39. Min-Sun Kim, "Culture-Based Interactive Constraints in Explaining Intercultural Strategic Competence," *Intercultural Communication Competence,* ed. Richard L. Wiseman and Jolene Koester (Newbury Park, CA: Sage, 1993), 132–150; Min-Sun Kim, "Cross-Cultural Comparisons of the Perceived Importance of Conversational Constraints," *Human Communication Research* 21 (1994): 128–151; Min-Sun Kim, "Toward a Theory of Conversational Constraints: Focusing on Individual-Level Dimensions of Culture," *Intercultural Communication Theory,* ed. Richard L. Wiseman (Thousand Oaks, CA: Sage, 1995), 148–169; Min-Sun Kim, John E. Hunter, Akira Miyahara, Ann-Marie Horvath, Mary Bresnahan, and Hye-Jin Yoon, "Individual-vs.-Culture-Level Dimensions of Individualism and Collectivism: Effects on Preferred Conversational Styles," *Communication Monographs* 63 (1996): 29–49; Min-Sun Kim and William F. Sharkey, "Independent and Interdependent Construals of Self: Explaining Cultural Patterns of Interpersonal Communication in Multi-Cultural Organizational Settings," *Communication Quarterly* 43 (1995): 20–38; Min-Sun Kim, William F. Sharkey, and Theodore M. Singelis, "The Relationship of Individuals' Self-Construals and Perceived Importance of Interactive Constraints," *International Journal of Intercultural Relations* 18 (1994): 117–140; Min-Sun Kim and Steven R. Wilson, "A Cross-Cultural Comparison of Implicit Theories of Requesting," *Communication Monographs* 61 (1994): 210–235; Akira Miyahara, Min-Sun Kim, Ho-Chang Shin, and Kak Yoon, "Conflict Resolution Styles Among 'Collectivist' Cultures: A Comparison between Japanese and Koreans," *International Journal of Intercultural Relations* 22 (1998): 505–525.

40. Dean C. Barnlund, "Apologies: Japanese and American Styles," *International Journal of Intercultural Relations* 14 (1990): 193–206; William R. Cupach and T. Todd Imahori, "Managing Social Predicaments Created by Others: A Comparison of Japanese and American Facework," *Western Journal of Communication* 57 (1993): 431–444; William R. Cupach and T. Todd Imahori, "A Cross-Cultural Comparison of the Interpretation and Management of Face: U.S. American and Japanese Responses to Embarrassing Predicaments," *International Journal of Intercultural Relations* 18 (1994): 193–219; Naoki Nomura and Dean Barnlund, "Patterns of Interpersonal Criticism in Japan and the United States," *International Journal of Intercultural Relations* 7 (1983): 1–18; Kiyoko Sueda and Richard L. Wiseman, "Embarrassment Remediation in Japan and the United States,"

International Journal of Intercultural Relations 16 (1992): 159–173.

41. Stahl, 14.

42. Robyn Penman, "Facework in Communication: Conceptual and Moral Challenges," *The Challenge of Facework: Cross-Cultural and Interpersonal Issues,* ed. Stella Ting-Toomey (Albany: State University of New York Press, 1994), 21.

43. For an elaboration of uncertainty reduction theory, see Walid A. Afifi and Laura K. Guerrero, "Motivations Underlying Topic Avoidance in Close Relationships," *Balancing the Secrets of Private Disclosures,* ed. Sandra Petronio (Mahwah, NJ: Erlbaum, 2000), 165–179; Charles R. Berger, "Communicating Under Uncertainty," *Interpersonal Processes: New Directions in Communication Research,* ed. Michael E. Roloff and Gerald R. Miller (Newbury Park, CA: Sage, 1987), 39–62; Charles R. Berger and James J. Bradac, *Language and Social Knowledge: Uncertainty in Interpersonal Relations* (London: Arnold, 1982); Charles R. Berger and Richard J. Calabrese, "Some Explorations in Initial Interaction and Beyond: Toward a Developmental Theory of Interpersonal Communication," *Human Communication Research* 1 (1975): 99–112; James J. Bradac, "Theory Comparison: Uncertainty Reduction, Problematic Integration, Uncertainty Management, and Other Curious Constructs," *Journal of Communication* 51 (2001): 456–476; Glen W. Clatterbuck, "Attributional Confidence and Uncertainty in Initial Interaction," *Human Communication Research* 5 (1979): 147–157; William Douglas, "Uncertainty, Information-Seeking, and Liking During Initial Interaction," *Western Journal of Speech Communication* 54 (1990): 66–81; Daena J. Goldsmith, "A Normative Approach to the Study of Uncertainty and Communication," *Journal of Communication* 51 (2001): 514–533; William B. Gudykunst, "The Influence of Cultural Similarity, Type of Relationship, and Self-Monitoring on Uncertainty Reduction Processes," *Communication Monographs* 52 (1985): 203–217; William B. Gudykunst, Elizabeth Chua, and Alisa J. Gray, "Cultural Dissimilarities and Uncertainty Reduction Processes," *Communication Yearbook* 10, ed. Margaret McLaughlin (Beverly Hills, CA: Sage, 1984), 456–469; William B. Gudykunst and Tsukasa Nishida, "Individual and Cultural Influences on Uncertainty Reduction," *Communication Monographs* 51 (1984): 23–36; William B. Gudykunst, Seung-Mock Yang, and Tsukasa Nishida, "A Cross-Cultural Test of Uncertainty Reduction Theory: Comparisons of Acquaintances, Friends, and Dating Relationships in Japan, Korea, and the United States," *Human Communication Research* 11 (1985): 407–455; Jolanda Jetten, Michael A. Hogg, and Barbara-Ann Mullin, "In-Group Variability and Motivation to Reduce Subjective Uncertainty," *Group Dynamics* 4 (2000): 184–198; Kathy Kellermann and Rodney Reynolds, "When Ignorance Is Bliss: The Role of Motivation to Reduce Uncertainty in Uncertainty Reduction Theory," *Human Communication Research* 17 (1990): 5–75; Leanne K. Knobloch and Denise Haunani Solomon, "Information Seeking beyond Initial Interaction: Negotiating Relational Uncertainty within Close Relationships," *Human Communication Research* 28 (2002): 243–257; Angela Y. Lee, "The Mere Exposure Effect: An Uncertainty Reduction Explanation Revisited," *Personality & Social Psychology Bulletin* 27 (2001): 1255–1266; James W. Neuliep and Erica L. Grohskopf, "Uncertainty Reduction and Communication Satisfaction during Initial Interaction: An Initial Test and Replication of a New Axiom," *Communication Reports* 13 (2000): 67–77; Sally Planalp and James M. Honeycutt, "Events That Increase Uncertainty in Personal Relationships," *Human Communication Research* 11 (1985): 593–604; Sally Planalp, Diane K. Rutherford, and James M. Honeycutt, "Events That Increase Uncertainty in Personal Relationships II: Replication and Extension," *Human Communication Research* 14 (1988): 516–547; Michael Sunnafrank, "Predicted Outcome Value during Initial Interactions: A Reformulation of Uncertainty Reduction Theory," *Human Communication Research* 13 (1986): 3–33; Thomas C. Taveggia and Lourdes Santos Nieves Gibboney, "Cross Cultural Adjustment: A Test of Uncertainty Reduction Principle," *International Journal of Cross Cultural Management* 1 (2001): 153–171.

44. William B. Gudykunst, "Toward a Theory of Effective Interpersonal and Intergroup Communication: An Anxiety/Uncertainty Management (AUM) Perspective," *Intercultural*

Communication Competence, ed. Wiseman and Jolene Koester, 33–71; William B. Gudykunst, "Anxiety/Uncertainty Management Theory: Current Status," *Intercultural Communication Theory,* ed. Wiseman, 8–58; William B. Gudykunst, "Intercultural Communication Theories," *Handbook of International and Intercultural Communication,* 2nd ed., ed. William B. Gudykunst and Bella Mody (Thousand Oaks, CA: Sage, 2002), 183–205; William B. Gudykunst and Carmen M. Lee, "Cross-Cultural Communication Theories," *Handbook of International and Intercultural Communication,* 2nd ed., ed. Gudykunst and Mody, 25–50; William B. Gudykunst and Tsukasa Nishida, "Anxiety, Uncertainty, and Perceived Effectiveness of Communication across Relationships and Cultures," *International Journal of Intercultural Relations* 25 (2001): 55–71; Kimberly N. Hubbert, William B. Gudykunst, and Sherrie L. Guerrero, "Intergroup Communication Over Time," *International Journal of Intercultural Relations* 23 (1999): 13–46; Craig R. Hullett and Kim Witte, "Predicting Intercultural Adaptation and Isolation: Using the Extended Parallel Process Model to Test Anxiety/Uncertainty Management Theory," *International Journal of Intercultural Relations* 25 (2001): 125–139; Walter G. Stephan, Cookie White Stephan, and William B. Gudykunst, "Anxiety in Intergroup Relations: A Comparison of Anxiety/Uncertainty Management Theory and Integrated Threat Theory," *International Journal of Intercultural Relations* 23 (1999): 613–628.

45. Edna B. Foa and Uriel G. Foa, "Resource Theory: Interpersonal Behavior as Exchange," *Social Exchange: Advances in Theory and Research,* ed. Kenneth Gergen, Martin S. Greenberg, and Richard H. Willis (New York: Plenum Press, 1980), 77–101.

46. Judee K. Burgoon, "Nonverbal Violation of Expectations," *Nonverbal Interaction,* ed. John M. Wiemann and Randall P. Harrison (Beverly Hills, CA: Sage, 1983), 77–111; Judee K. Burgoon, "Interpersonal Expectations, Expectancy Violations, and Emotional Communication," *Journal of Language and Social Psychology* 12 (1993): 30–48; Judee K. Burgoon, "Cross-Cultural and Intercultural Applications of Expectancy Violations Theory," *Intercultural Communication Theory,* ed. Wiseman, 194–214.

47. Berger, 41.

48. William B. Gudykunst and Tsukasa Nishida, "Individual and Cultural Influences on Uncertainty Reduction," *Communication Monographs* 51 (1984): 23–36. See also: Kevin Avruch, "Type I and Type II Errors in Culturally Sensitive Conflict Resolution Practice," *Conflict Resolution Quarterly* 20 (2003): 351–371; Michelle LeBaron, *Bridging Cultural Conflicts: A New Approach for a Changing World* (San Francisco, CA: Jossey-Bass, 2003); Dieter Senghaas, *The Clash within Civilizations: Coming to Terms with Cultural Conflicts* (New York: Routledge, 2002); Stella Ting-Toomey and John G. Oetzel, *Managing Intercultural Conflict Effectively* (Thousand Oaks, CA: Sage, 2001); Catherine H. Tinsley, "How Negotiators Get to Yes: Predicting the Constellation of Strategies Used across Cultures to Negotiate Conflict," *Journal of Applied Psychology* 86 (2001): 583–593.

49. Tsukasa Nishida, "Sequence Patterns of Self-Disclosure Among Japanese and North American Students," paper presented at the Conference on Communication in Japan and the United States, Fullerton, CA, March 1991.

50. Judith A. Sanders, Richard L. Wiseman, and S. Irene Matz, "A Cross-Cultural Comparison of Uncertainty Reduction Theory: The Cases of Ghana and the United States," paper presented at the annual conference of the International Communication Association, San Francisco, May 1989.

51. Changsheng Xi, "Individualism and Collectivism in American and Chinese Societies," *Our Voices: Essays in Culture, Ethnicity, and Communication,* ed. Alberto González, Marsha Houston, and Victoria Chen (Los Angeles: Roxbury, 1994), 155.

52. Robert Littlefield, "Self-Disclosure Among Some Negro, White, and Mexican-American Adolescents," *Journal of Counseling Psychology* 21 (1974): 133–136.

53. Sidney Jourard, "Self-Disclosure Patterns in British and American College Females," *Journal of Social Psychology* 54 (1961): 315–320.

54. Stella Ting-Toomey, "Intimacy Expressions in Three Cultures: France, Japan, and the United States," *International Journal of Intercultural Relations* 15 (1991): 29–46.

55. Dean Barnlund, *Public and Private Self in Japan and the United States: Communicative Styles of*

Two Cultures (Tokyo: Simul Press, 1975); Ting-Toomey, "Intimacy Expressions."

56. Sidney Jourard, *Self-Disclosure: An Experimental Analysis of the Transparent Self* (New York: Wiley, 1971).

57. George Levinger and David J. Senn, "Disclosure of Feelings in Marriage," *Merrill Palmer Quarterly* 13 (1987): 237–249.

58. See Ting-Toomey and Oetzel, *Managing Intercultural Conflict Effectively*; Stella Ting-Toomey, and John G. Oetzel, "Cross-Cultural Face Concerns and Conflict Styles: Current Status and Future Directions," *Handbook of International and Intercultural Communication*, 2nd ed., ed. Gudykunst and Mody, 143–163. Our discussion of managing conflict also draws on a number of works, including John G. Oetzel, "The Effects of Self-Construals and Ethnicity on Self-Reported Conflict Styles," *Communication Reports* 11 (1998): 133–144; John G. Oetzel, "The Influence of Situational Features on Perceived Conflict Styles and Self-Construals in Work Groups," *International Journal of Intercultural Relations* 23 (1999): 679–695; John Oetzel, Stella Ting-Toomey, Tomoko Masumoto, Yumiko Yokochi, Xiaohui Pan, Jiro Takai, and Richard Wilcox, "Face and Facework in Conflict: A Cross-Cultural Comparison of China, Germany, Japan, and the United States," *Communication Monographs* 68 (2001): 235–258; Stella Ting-Toomey, "Managing Conflict in Intimate Intercultural Relationships," *Conflict in Personal Relationships*, ed. Dudley D. Cahn (Hillsdale, NJ: Erlbaum, 1994), 47–77; Stella Ting-Toomey, Kimberlie K. Yee-Jung, Robin B. Shapiro, Wintilo Garcia, Trina J. Wright, and John G. Oetzel, "Ethnic/Cultural Identity Salience and Conflict Styles in Four US Ethnic Groups," *International Journal of Intercultural Relations* 24 (2000): 47–81; Ting-Toomey, "Conflict and Culture"; Stella Ting-Toomey, "Conflict Styles in Black and White Subjective Cultures," *Current Research in Interethnic Communication*, ed. Young Yun Kim (Beverly Hills, CA: Sage, 1986); Ting-Toomey, "Face Negotiation Theory" (see note 38).

59. Deborah A. Cai, Steven R. Wilson, and Laura E. Drake, "Culture in the Context of Intercultural Negotiation: Individualism-Collectivism and Paths to Integrative Agreements," *Human Communication Research* 26 (2000): 591–617;

Ge Gao, "An Initial Analysis of the Effects of Face and Concern for 'Other' in Chinese Interpersonal Communication," *International Journal of Intercultural Relations* 22 (1998): 467–482.

60. Ringo Ma, "The Role of Unofficial Intermediaries in Interpersonal Conflicts in the Chinese Culture," *Communication Quarterly* 40 (1992): 269–278.

61. Ting-Toomey, "Face Negotiation Theory" (see note 38).

62. Susan Cross and Robert Rosenthal, "Three Models of Conflict Resolution: Effects on Intergroup Expectancies and Attitudes," *Journal of Social Issues* 55(3) (1999): 561–580.

CHAPTER 11

1. Joseph P. Forgas and Michael H. Bond, "Cultural Influences on the Perception of Interaction Episodes," *Journal of Cross-Cultural Psychology* 11 (1985): 75–88.

2. Robert T. Craig and Karen Tracy, *Conversational Coherence: Form, Structure, and Strategy* (Beverly Hills, CA: Sage, 1983); Susan B. Shimanoff, *Communication Rules: Theory and Research* (Beverly Hills, CA: Sage, 1980).

3. Wendy Leeds-Hurwitz, "Intercultural Weddings and the Simultaneous Display of Multiple Identities," *Communicating Ethnic & Cultural Identity*, ed. Mary Fong and Rueyling Chuang (Lanham, MD: Rowman & Littlefield, 2004), 135–148.

4. B. Aubrey Fisher, *Interpersonal Communication: Pragmatics of Human Relationships* (New York: Random House, 1987), 59.

5. Norine Dresser, *Multicultural Manners: Essential Rules of Etiquette for the 21st Century*, revised ed. (Hoboken, NJ: Wiley, 2005), 72.

6. Dresser, 76.

7. Kellermann calls these interaction scenes "Memory Organization Packets" (MOPs). Our description of her research is based on the following: Kathy Kellermann, "The Conversation MOP II: Progression Through Scenes in Discourse," *Human Communication Research* 17 (1991): 385–414; Kathy Kellermann, "The Conversation MOP: A Model of Pliable Behavior," *The Cognitive Bases of Interpersonal Communication*, ed. Dean E. Hewes (Hillsdale, NJ: Erlbaum, 1995); Kathy Kellermann and Tae-Seop Lim, "The Conversation MOP III: Timing of Scenes in

Discourse," *Journal of Personality and Psychology* 54 (1990): 1163–1179.

8. Kellermann, "The Conversation MOP II, 388.

9. Jeanette S. Martin and Lillian H. Chaney, *Global Business Etiquette: A Guide to International Communication and Customs*, 2nd ed. (Santa Barbara, CA: Praeger, 2012).

10. See, for example, David T. Cowan and Ian Norman, "Cultural Competence in Nursing: New Meanings," *Journal of Transcultural Nursing* 17 (2006): 82–88.

11. See Linda C. Lederman (ed.), *Beyond These Walls: Readings in Health Communication* (New York: Oxford University Press, 2008); Bernard Moss, *Communication Skills in Health and Social Care*, 2nd ed. (Thousand Oaks, CA: Sage, 2012); Renata Schiavo, *Health Communication: From Theory to Practice* (San Francisco: Jossey-Bass, 2007); Kevin B. Wright, Lisa Sparks, and H. Dan O'Hair, *Health Communication in the 21st Century* (Malden, MA: Blackwell, 2008); Heather M. Zoller and Mohan J. Dutta, *Emerging Perspectives in Health Communication: Meaning, Culture, and Power* (New York: Routledge, 2008).

12. See, for example, Barbara Broome, "Culture 101," *Urologic Nursing* 26 (2006): 486–489; Cathleen A. Collins, Shawn I. Decker, and Karen A. Esquibel,"Definitions of Health: Comparison of Hispanic and African-American Elders," *Journal of Multicultural Nursing & Health* 12 (2006): 14–19; Linda Dayer-Berenson, *Cultural Competencies for Nurses: Impact on Health and Illness* (Sudbury, MA: Jones and Bartlett, 2011); Y. S. Kim-Godwin, P. N. Clarke, and L. Barton, "A Model for the Delivery of Culturally Competent Community Care," *Journal of Advanced Nursing* 35 (2001): 918–925; Susan Kleiman, "Discovering Cultural Aspects of Nurse-Patient Relationships," *Journal of Cultural Diversity* 13 (2006): 83–86; Tsveti Markova and Barbara Broome, "Effective Communication and Delivery of Culturally Competent Health Care," *Urologic Nursing* 27 (2007): 239–241; Mary Sobralske and Janet Katz, "Culturally Competent Care of Patients with Acute Chest Pain," *Journal of the American Academy of Nurse Practitioners* 17 (2005): 324–329; Vasso Vydelingum, "Nurses' Experiences of Caring for South Asian Minority Ethnic Patients in a General Hospital in England," *Nursing Inquiry* 13 (2006): 23–32; Yu Xu and Ruth Davidhizar, "Intercultural Communication in Nursing Education: When Asian Students and American Faculty Converge," *Journal of Nursing Education* 44 (2005): 209–215.

13. Linda Dayer-Berensen, *Cultural Competencies for Nurses: Impact on Health and Illness* (Sudbury, MA: Jones and Bartlett, 2011).

14. See, for example, Joyceen S. Boyle and Margaret M. Andrews (eds.), *Transcultural Concepts in Nursing Care*, 6th ed. (Philadelphia: Lippincott, 2012); Barbara Broome and Teena McGuinness, "A CRASH Course in Cultural Competence for Nurses," *Urologic Nursing* 27 (2007): 292–295; Nancy Campbell-Heider, Karol Pohlman Rejman, Tammy Austin-Ketch, Kay Sackett, Thomas Hugh Feeley, and Nancy C. Wilk, "Measuring Cultural Competence in a Family Nurse Practitioner Curriculum," *Journal of Multicultural Nursing & Health* 12 (2006): 23–24; Maureen Campesino, "Beyond Transculturalism: Critiques of Cultural Education in Nursing," *Journal of Nursing Education* 47 (2008): 298–304; Geri-Ann Galanti, *Caring for Patients from Different Cultures*, 4th ed. (Philadelphia: University of Pennsylvania Press, 2008); Joyce Newman Giger and Ruth Elaine Davidhizar (eds.), *Transcultural Nursing: Assessment and Intervention*, 5th ed. (St. Louis: Mosby/Elsevier, 2008); Suzan Kardong-Edgren, "Cultural Competence of Baccalaureate Nursing Faculty," *Journal of Nursing Education* 46 (2007): 360–366; Madeleine Leininger and Marilyn McFarland (eds.), *Cultural Care Diversity and Universality: A Worldwide Nursing Theory*, 2nd ed. (Sudbury, MA: Jones and Bartlett, 2006); Chato Rasoal, Tomas Jungert, Stephan Hau, Elinor Edvardsson Stiwne, and Gerhard Andersson, "Ethnocultural Empathy among Students in Health Care Education," *Evaluation & the Health Professions* 32 (2009): 300–313; Roberta Waite and Christina J. Calamaro, "Cultural Competence: A Systemic Challenge to Nursing Education, Knowledge Exchange, and the Knowledge Development Process," *Perspectives in Psychiatric Care* 46 (2010): 74–80; Barbara Jones Warren, "The Fluid Process of Cultural Competence at the Graduate Level: A Constructionist Approach," *Transforming Nursing Education: The Culturally Inclusive Environment*, ed. Susan Dandridge Bosher and Margaret Dexheimer Pharris (New York:

Springer, 2009), 179–206; Megan J. Wood and Marsha Atkins, "Immersion in Another Culture: One Strategy for Increasing Cultural Competency," *Journal of Cultural Diversity* 13 (2006): 50–54.

15. See, for example, Larry D. Purnell, *Guide to Culturally Competent Health Care*, 2nd ed. (Philadelphia: F. A. Davis, 2009); Larry D. Purnell and Betty J. Paulanka, *Transcultural Health Care: A Culturally Competent Approach*, 4th ed. (Philadelphia: F. A. Davis, 2013).

16. Kathryn Hopkins Kavanagh and Virginia Knowlden (eds.), *Many Voices: Toward Caring Culture in Healthcare and* Healing (Madison, WI: University of Wisconsin Press, 2004).

17. "Quick Takes," *Inside Higher Education*, February 24, 2011. Online edition.

18. Sharon E. Kummerer and Norma A. Lopez-Reyna, "The Role of Mexican Immigrant Mothers' Beliefs on Parental Involvement in Speech–Language Therapy," *Communication Disorders Quarterly* 27 (2006): 83–94.

19. See, for example, Lucinda Dale, Raeanne Albin, Shelley Kapolka-Ullom, Annjanette Lange, Megan McCan, Kacey Quaderer, and Nikki Shaffer, "The Meaning of Work in the U.S. for Two Latino Immigrants from Colombia and Mexico," *Work* 25 (2005): 187–196.

20. See, for example, Madonna G. Constantine and Derald Wing Sue (eds.), *Strategies for Building Multicultural Competence in Mental Health and Educational Settings* (Hoboken, NJ: Wiley, 2005); George Henderson, Dorscine Spigner-Littles, and Virginia Hall Milhouse, *A Practitioner's Guide to Understanding Indigenous and Foreign Cultures: An Analysis of Relationships between Ethnicity, Social Class and Therapeutic Intervention Strategies*, 3rd ed. (Springfield, IL: Charles C. Thomas, 2006); June L. Leishman, "Culturally Sensitive Mental Health Care: A Module for 21st Century Education and Practice," *International Journal of Psychiatric Nursing Research* 11 (2006): 1310–1321; Jane S. Mahoney, Elizabeth Carlson, and Joan C. Engebretson, "A Framework for Cultural Competence in Advanced Practice Psychiatric and Mental Health Education," *Perspectives in Psychiatric Care* 42 (2006): 227–233.

21. Andres J. Pumariega, Eugenio Rothe, and Kenneth Rogers, "Cultural Competence in Child Psychiatric Practice," *Journal of the American Academy of Child and Adolescent Psychiatry* 48 (2009): 362–366.

22. Halime Celik, Tineke A. Abma, Guy A. Widdershoven, Frans C. B. van Wijmen, and Ineke Klinge, "Implementation of Diversity in Healthcare Practices: Barriers and Opportunities," *Patient Education and Counseling* 71 (2008): 65–71; Hanzade Dogan, Verena Tschudin, İnci Hot, and İbrahim Özkan, "Patients' Transcultural Needs and Carers' Ethical Responses," *Nursing Ethics* 16 (2009): 683–696.

23. Ludwien Meeuwesen, Atie van den Brink-Muinen, and Geert Hofstede, "Can Dimensions of National Culture Predict Cross-National Differences in Medical Communication?" *Patient Education and Counseling* 75 (2009): 58–66.

24. Rohini Anand and Indra Lahiri, "Intercultural Competence in Health Care—Developing Skills for Interculturally Competent Care," *The Sage Handbook of Intercultural Competence*, ed. Darla K. Deardorff (Thousand Oaks, CA: Sage, 2009), 387–402; Boyle and Andrews; Wen-Shing Tseng and Jon Streltzer, *Cultural Competence in Health Care* (New York: Springer, 2008); Kim Witte and Kelly Morrison, "Intercultural and Cross-Cultural Health Communication: Understanding People and Motivating Healthy Behaviors," *Intercultural Communication Theory*, ed. Richard L. Wiseman (Thousand Oaks, CA: Sage, 1995), 216–246.

25. Rachel E. Spector, *Cultural Diversity in Health and Illness*, 7th ed. (Upper Saddle River, NJ: Pearson Prentice Hall, 2009).

26. Sam Chan and Deborah Chen, "Families with Asian Roots," *Developing Cross-Cultural Competence: A Guide for Working with Young Children and Their Families*, 4th ed., ed. Eleanor W. Lynch and Marci J. Hanson (Baltimore: Paul Brookes, 2011), Chapter 8.

27. Avonne Yang, "Coming Home to Nursing Education for a Hmong Student, Hmong Nurse, and Hmong Nurse Educator," *Transforming Nursing Education: The Culturally Inclusive Environment*, ed. Susan Dandridge Bosher and Margaret Dexheimer Pharris (New York: Springer, 2009), 117.

28. Richard H. Dana, "The Cultural Self as Locus for Assessment and Intervention with American Indians/Alaska Natives," *Journal of Multicultural Counseling and Development* 28 (2000): 66–82.

29. Chan and Chen.

30. For a discussion of these issues, see D. Patricia Gray and Debra J. Thomas, "Critical Reflections on Culture in Nursing," *Journal of Cultural Diversity* 13 (2006): 76–82; Kavanagh and Knowlden.

31. The examples here and in the preceding paragraph are adapted from Dresser.

32. Ursula M. Wilson, "Nursing Care of American Indian Patients," *Ethnic Nursing Care: A Multicultural Approach*, ed. Modesta Soberano Orque, Bobbie Bloch, and Lidia S. Ahumada Monroy (St. Louis: Mosby, 1983), 277.

33. Joan Kuipers, "Mexican Americans," *Transcultural Nursing: Assessment and Intervention*, 5th ed., ed. Joyce Newman Giger and Ruth Elaine Davidhizar (St. Louis: Mosby Year Book, 2008), Chapter 9.

34. Boyle and Andrews.

35. Noreen Mokuau and Pemerika Tauili'ili, "Families with Native Hawaiian and Samoan Roots," *Developing Cross-Cultural Competence: A Guide for Working with Young Children and Their Families*, 4th ed., ed. Eleanor W. Lynch and Marci J. Hanson (Baltimore: Paul Brookes, 2011), Chapter 10.

36. Witte and Morrison.

37. Kavanagh and Knowlden. See also Kuipers.

38. See Jennifer Carroll, Ronald Epstein, Kevin Fiscella, Teresa Gipson, Ellen Volpe, and Pascal Jean-Pierre, "Caring for Somali Women: Implications for Clinician–Patient Communication," *Patient Education & Counseling* 66 (2007): 337–345; Kuipers; Yang.

39. Adapted from Dresser.

40. Pamela J. Kalbfleisch, "Effective Health Communication in Native Populations in North America," *Journal of Language and Social Psychology* 28 (2009): 158–173. See also Geri-Ann Galanti, *Caring for Patients from Different Cultures*, 4th ed. (Philadelphia: University of Pennsylvania Press, 2008); Kavanagh and Knowlden; Kuipers; Wilson.

41. Chan and Chen.

42. Galanti; Witte and Morrison.

43. See, for example, Cora C. Muñoz and Joan Luckmann, *Transcultural Communication in Nursing*, 2nd ed. (Clifton Park, NY: Thomson/ Delmar Learning, 2005).

44. Min-Sun Kim, Renee Storm Klingle, William F. Sharkey, Hee Sun Park, David H. Smith, and Deborah Cai, "A Test of a Cultural Model of Patients' Motivation for Verbal Communication in Patient-Doctor Interactions," *Communication Monographs* 67 (2000): 262–283.

45. Nilda Chong, *The Latino Patient: A Cultural Guide for Health Care Providers* (Yarmouth, ME: Intercultural Press, 2002).

46. Hanzade Dogan, Verena Tschudin, İnci Hot, and İbrahim Özkan, "Patients' Transcultural Needs and Carers' Ethical Responses," *Nursing Ethics* 16 (2009): 683–696.

47. See, for example, R. E. Nailon, "Nurses Concerns and Practices with Using Interpreters in the Care of Latino Patients in the Emergency Department," *Journal of Transcultural Nursing* 17 (2006): 119–128.

48. See, for example, Roxanne Amerson and Shelley Burgins, "Hablamos Español: Crossing Communication Barriers with the Latino Population," *Journal of Nursing* 44 (2005): 241–243.

49. Susan Raye Moffitt and Mimi Jenko, "Transcultural Nursing Principles: An Application to Hospice Care," *Journal of Hospice and Palliative Nursing* 8 (2006): 172–180.

50. Boyle and Andrews.

51. Galanti.

52. Haffner, "Translation Is Not Enough: Interpreting in a Medical Setting," *Western Journal of Medicine* 157 (1992): 255–260.

53. Muñoz and Luckmann.

54. Joyce Newman Giger and Ruth Elaine Davidhizar, *Transcultural Nursing: Assessment and Intervention*, 5th ed. (St. Louis, MO: Mosby Elsevier, 2008).

55. Susan Raye Moffitt and Mimi Jenko, "Transcultural Nursing Principles: An Application to Hospice Care," *Journal of Hospice and Palliative Nursing* 8 (2006): 172–180.

56. For an excellent summary of cultural variations in what is regarded as the appropriate verbal and nonverbal means to express pain, see Mary Sobralske and Janet Katz, "Culturally Competent Care of Patients with Acute Chest Pain," *Journal of the American Academy of Nurse Practitioners*, 17 (2005): 324–329.

57. For an informative discussion of curricular challenges to be more sensitive to diversity across all segments of education, see Regan A. R. Gurung and Tenor Duong, "Mixing and Matching: Assessing the Concomitants of Mixed-Ethnic Relationships," *Journal of Social and Personal Relationships* 16 (1999): 639–657.

58. Jeffrey Trawick-Smith, *Early Childhood Development: A Multicultural Perspective*, 5th ed. (Upper Saddle River, NJ: Pearson Merrill Prentice Hall, 2010).

59. Lisa Delpit, *Other People's Children: Cultural Conflict in the Classroom* (New York: New Press, 2006); Lori Mestre, *Librarians Serving Diverse Populations: Challenges and Opportunities* (Chicago: American Library Association, 2010); C. Raeff, Patricia M. Greenfield, and Blanca Quiroz, "Conceptualizing Interpersonal Relationships in the Cultural Contexts of Individualism and Collectivism," *Variability in the Social Construction of the Child*, ed. S. Harkness, C. Raeff, and C. Super, *New Directions in Child Development*, no. 87 (San Francisco: Jossey-Bass, 2000); Rosa Hernandez Sheets, *Diversity Pedagogy: Examining the Role of Culture in the Teaching-Learning Process* (Boston: Pearson, 2005).

60. Janet Bennett and Riikka Salonen, "Intercultural Communication and the New American Campus," *Change* March/April (2007): 46. See also Kenneth Cushner and Jennifer Mahon, "Intercultural Competence in Teacher Education—Developing the Intercultural Competence of Educators and Their Students: Creating the Blueprints," *The Sage Handbook of Intercultural Competence*, ed. Deardorff, 304–320.

61. Patricia Phelan, Ann Locke Davidson, and Hanh Cao Yu, "Students' Multiple Worlds: Navigating the Borders of Family, Peer, and School Cultures," *Renegotiating Cultural Diversity in American Schools*, ed. Patricia Phelan and Ann Locke Davidson (New York: College Press, 1993), 52–88.

62. For an interesting ethnographic description of how Navajo philosophy permeates curriculum and classroom behaviors in a Navajo community college, see Charles A. Braithwaite, "Sa'ah Naagháí Bik'eh Hózhóón: An Ethnography of Navajo Educational Communication Practices," *Communication Education* 46 (1997): 219–233.

63. Gary R. Howard, "As Diversity Grows, So Must We," *Educational Leadership* 64 (2007): 16–22.

64. In an interesting study of teachers from rural, urban, and suburban schools across the Midwestern geographic region of the United States, Jennifer Mahon found that teachers had a tendency to minimize the cultural differences that were present among their students. See Jennifer Mahon, "Under the Invisibility Cloak? Teacher Understanding of Cultural Difference," *Intercultural Education* 17 (2006): 391–405.

65. Bonnie M. Davis, *How to Teach Students Who Don't Look Like You: Culturally Relevant Teaching Strategies* (Thousand Oaks, CA: Corwin Press, 2006); Bonnie M. Davis, *How to Coach Teachers Who Don't Think Like You: Using Literacy Strategies to Coach across Content Areas* (Thousand Oaks, CA: Corwin Press, 2008); Carlos J. Ovando and Mary Carol Combs, *Bilingual and ESL Classrooms: Teaching in Multicultural Contexts*, 5th ed. (Boston: McGraw Hill, 2012); Kikanza J. Nuri-Robins, Delores B. Lindsey, Randall B. Lindsey, and Raymond D. Terrell, *Culturally Proficient Instruction: A Guide for People Who Teach*, 3rd ed. (Thousand Oaks, CA: Corwin Press, 2012); Teresa A. Wasonga, "Multicultural Education Knowledgebase, Attitudes and Preparedness for Diversity," *International Journal of Educational Management* 29 (2005): 67–74.

66. K. David Roach and Paul R. Byrne, "A Cross-Cultural Comparison of Instructor Communication in American and German Classrooms," *Communication Education* 50 (2001): 1–14.

67. Shuming Lu, "Culture and Compliance Gaining in the Classroom: A Preliminary Investigation of Chinese College Teachers' Use of Behavior Alteration Techniques," *Communication Education* 46 (1997): 20–21.

68. Dresser, 64.

69. Cristy Lee, Timothy Levine, and Ronald Cambra, "Resisting Compliance in the Multicultural Classroom," *Communication Education* 46 (1997): 29–43; Lu, 10–28.

70. Lu, 24–25.

71. Adapted from Donal Carbaugh, *Cultures in Conversation* (Mahwah, NJ: Erlbaum, 2005), 92.

72. Carbaugh.

73. Eunkyong L. Yook and Rosita Albert, "Perceptions of the Appropriateness of Negotiation in Education Settings: A Cross-Cultural Comparison among Koreans and Americans," *Communication Education* 47 (1998): 18–29.

74. Carrie Rothstein-Fisch and Elise Trumbull, *Managing Diverse Classrooms: How to Build on Students' Cultural Strengths* (Alexandria, VA: Association for Supervision and Curriculum

Development, 2008); Elise Trumbull, Carrie Rothstein-Fisch, Patricia M. Greenfield, and Blanca Quiroz, *Bridging Cultures between Home and School: A Guide for Teachers* (Mahwah, NJ: Erlbaum, 2001).

75. Jolene Koester and Myron W. Lustig, "Communication Curricula in the Multicultural University," *Communication Education* 40 (1991): 250–254.

76. Adapted from Ken Pransky, *Beneath the Surface: The Hidden Realities of Teaching Culturally and Linguistically Diverse Young Learners, K-6* (Portsmouth, NH: Heinemann, 2008), 29.

77. For an extended discussion of some of the ways in which individualistic and collectivist cultures shape behavior in the classroom, see Rothstein-Fisch and Trumbull.

78. Paul David Bolls, Alex Tan, and Erica Austin, "An Exploratory Comparison of Native American and Caucasian Students' Attitudes toward Teacher Communicative Behavior and toward School," *Communication Education* 46 (1997): 198–202. See also Paul David Bolls and Alex Tan, "Communication Anxiety and Teacher Communication Competence Among Native American and Caucasian Students," *Communication Research Reports* 13 (1996): 205–213.

79. Barbara J. Shade and Clara A. New, "Cultural Influences on Learning: Teaching Implications," *Multicultural Education: Issues and Perspectives*, 2nd ed., ed. James A. Banks and Cherry A. McGee Banks (Boston: Allyn & Bacon, 1993), 320. See also James A. Banks and Cherry A. McGee Banks (eds.), *Multicultural Education: Issues and Perspectives*, 7th ed., (Hoboken, NJ: Wiley, 2010).

80. Steven T. Mortenson, "Cultural Differences and Similarities in Seeking Social Support as a Response to Academic Failure: A Comparison of American and Chinese College Students," *Communication Education* 55 (2006): 127–146.

81. Maria E. Zuniga, "Families with Latino Roots," *Developing Cross-Cultural Competence: A Guide for Working with Young Children and Their Families*, 4th ed., ed. Eleanor W. Lynch and Marci J. Hanson (Baltimore: Paul Brookes, 2011), Chapter 7.

82. Jennie R. Joe and Randi Suzanne Malach, "Families with American Indian Roots,"

Developing Cross-Cultural Competence: A Guide for Working with Young Children and Their Families, 4th ed., ed. Lynch and Hanson, Chapter 5; Cornel Pewewardy, "Toward Defining a Culturally Responsive Pedagogy for American Indian Children: The American Indian Magnet School," *Multicultural Education for the Twenty-First Century, Proceedings of the Second Annual Meeting, National Association for Multicultural Education, February 13–16th, 1992*, ed. Carl A. Grant (Morristown, NJ: Paramount, 1992), 218.

83. Rosa Hernández Sheets, *Diversity Pedagogy: Examining the Role of Culture in the Teaching-Learning Process* (Boston: Pearson, 2005), 4.

84. Rothstein-Fisch and Trumbull.

85. See Chan and Chen.

86. Leigh Chiarelott, Leonard Davidman, and Kevin Ryan, *Lenses on Teaching: Developing Perspectives on Classroom Life*, 4th ed. (Belmont, CA: Thomson Wadsworth, 2006); Trumbull, Rothstein-Fisch, Greenfield, and Quiroz.

87. Virginia-Shirin Sharifzadeh, "Families with Middle Eastern Roots," *Developing Cross-Cultural Competence: A Guide for Working with Young Children and Their Families*, 4th ed., ed. Lynch and Hanson, Chapter 11.

88. Jerry McClelland and Chen Chen, "Standing up for a Son: School Experiences of a Mexican Immigrant Mother," *Hispanic Journal of Behavioral Science* 19 (1997): 281–300.

89. McClelland and Chen, 291. See also Delores B. Hunt, Linda D. Jungwirth, Jarvis V. N. C. Pahl, and Randall. B. Lindsey, *Culturally Proficient Learning Communities: Confronting Inequities through Collaborative Curiosity* (Thousand Oaks, CA: Sage, 2009).

90. Josina Macau, *Embracing Diversity in the Classroom: Communication Ethics in an Age of Diversity*, ed. Josina Makau and Ronald C. Arnett (Urbana: University of Illinois Press, 1997), 48–67. See also Josina M. Makau and Debian L. Marty, *Cooperative Argumentation: A Model for Deliberative Community* (Prospect Heights, IL: Waveland 2001).

91. David Farbman, "A New Day for Kids," *Educational Leadership*, May 2007, 62–65.

92. Joe and Malach, 109.

93. See, for example, James Calvert Scott, "Developing Cultural Fluency: The Goal of International Business Communication

Instruction in the 21st Century," *Journal of Education for Business* 74 (1999): 140–143.

94. See, for example, Abel Adekola and Bruno S. Sergi, *Global Business Management: A Cross-Cultural Perspective* (Burlington, VT: Ashgate, 2007); Daniel Altman, *Connected: 24 Hours in the Global Economy* (New York: Farrar, Straus and Giroux, 2007); Elizabeth Kathleen Briody and Robert T. Trotter, *Partnering for Organizational Performance: Collaboration and Culture in the Global Workplace* (Lanham, MA: Rowman & Littlefield, 2008); George Cairns, *International Business* (Thousand Oaks, CA: Sage, 2008); Penny Carté and Chris J. Fox, *Bridging the Culture Gap: A Practical Guide to International Business Communication*, 2nd ed. (Philadelphia: Kogan Page, 2008); John D. Daniels and Jeffrey A. Krug, *International Business and Globalization* (Thousand Oaks, CA: Sage, 2007); John D. Daniels, Lee H. Radebaugh, and Daniel P. Sullivan, *International Business: Environments and Operations*, 12th ed. (Upper Saddle River, NJ: Pearson Prentice Hall, 2009); Kamal Fatehi, *Managing Internationally: Succeeding in a Culturally Diverse World* (Thousand Oaks, CA: Sage, 2008); Jeremy Haft, *All the Tea in China: How to Buy, Sell, and Make Money on the Mainland* (New York: Portfolio, 2007); Charles W. L. Hill, *International Business: Competing in the Global Marketplace*, 6th ed. (Boston: McGraw-Hill/Irwin, 2007); Jorma Larimo, *Contemporary Euromarketing: Entry and Operational Decision Making* (Binghamton, NY: International Business Press, 2007); Andrea Mandel-Campbell, *Why Mexicans Don't Drink Molson: Rescuing Canadian Business from the Suds of Global Obscurity* (Vancouver: Douglas & McIntyre, 2007); Dorothy McCormick, Patrick O. Alila, and Mary Omosa, *Business in Kenya: Institutions and Interactions* (Nairobi: University of Nairobi Press, 2007); William Hernández Requejo and John L. Graham, *Global Negotiation: The New Rules* (New York: Palgrave Macmillan, 2008); Carl Rodrigues, *International Management: A Cultural Approach*, 3rd ed. (Thousand Oaks, CA: Sage, 2008); Alan M. Rugman, *The Oxford Handbook of International Business*, 2nd ed. (New York: Oxford University Press, 2008); Oded Shenkar and Yadong Luo, *International Business*, 2nd ed. (Thousand Oaks, CA: Sage, 2008); Gabriele Suder, *International Business* (Thousand Oaks, CA: Sage, 2008); Xiaowen Tian, *Managing International Business in China* (New York: Cambridge University Press, 2007); Yonggui Wang and Richard Li-Hua, *Marketing Competences and Strategic Flexibility in China* (New York: Palgrave Macmillan, 2007); Frederick F. Wherry, *Global Markets and Local Crafts: Thailand and Costa Rica Compared* (Baltimore: Johns Hopkins University Press, 2008); Stephen White, *Media, Culture and Society in Putin's Russia* (New York: Palgrave Macmillan, 2008); John Yabs, *International Business Operations in Kenya*, 2nd ed. (Nairobi: Lelax Global Ltd., 2007); George S. Yip and Audrey J. M. Bink, *Managing Global Customers: An Integrated Approach* (New York: Oxford University Press, 2007).

95. Jeanne C. Meister and Karie Willyerd, *The 2020 Workplace: How Innovative Companies Attract, Develop, and Keep Tomorrow's Employees Today* (New York: HarperCollins, 2010).

96. See, for example, Dennis Briscoe, Randall Schuler, and Ibraiz Tarique, *International Human Resource Management: Policies and Practices for Multinational Enterprises*, 4th ed. (New York: Routledge, 2011); Pawan Budhwar, Randall S. Schuler, and Paul R. Sparrow (eds.), *International Human Resource Management* (Thousand Oaks, CA: Sage, 2009); Paula Caligiuri, David Lepak, and Jaime Bonache, *Managing the Global Workforce* (Hoboken, NJ: Wiley, 2010); Jeremy Comfort and Peter Franklin, *The Mindful International Manager: How to Work Effectively across Cultures* (Philadelphia: Kogan Page, 2010); Michael Dickmann and Yehuda Baruch, *Global Careers* (New York: Routledge, 2011); Paul Evans, Vladimir Pucik, and Ingmar Björkman, *The Global Challenge: International Human Resource Management*, 2nd ed. (New York: McGraw-Hill Irwin, 2011); Roger Herod, *Managing the International Assignment Process: From Selection through Repatriation* (Alexandria, VA: Society for Human Resource Management, 2009); Mijnd Huijser, *The Cultural Advantage: A New Model for Succeeding with Global Teams* (Boston: Intercultural Press, 2006); Hugh Scullion and David G. Collings (eds.), *Global Talent Management* (New York: Routledge, 2011); Paul R. Sparrow (ed.), *Handbook of International Human Resource Management: Integrating People, Process, and Context* (Chichester, UK: Wiley, 2009); Charles M. Vance and Yongsun

Paik, *Managing a Global Workforce: Challenges and Opportunities in International Human Resource Management*, 2nd ed. (Armonk, NY: M. E. Sharpe, 2011); Mijnd Huijser, *The Cultural Advantage: A New Model for Succeeding with Global Teams* (Boston: Intercultural Press, 2006).

97. See, for example, Tanya Mohn, "Going Global, Stateside" *New York Times*, March 9, 2010, B8; Roger Yu, "Cultural Training Has Global Appeal; Working as a Team or Getting a Deal Done Takes Understanding," *USA Today*, December 22, 2009, B3.

98. Robert T. Moran, Philip R. Harris, and Sarah V. Moran, *Managing Cultural Differences: Global Leadership Strategies for the 21st Century*, 7th ed. (Boston: Elsevier, 2007).

99. Moran, Harris, and Moran.

100. Tim Ambler, Morgen Witzel and Chao Xi, *Doing Business in China*, 3rd ed. (New York: Routledge, 2009); Thomas L. Coyner with Song-Hyon Jang, *Doing Business in Korea: An Expanded Guide* (Seoul, Korea: Seoul Selection, 2010); Boye Lafayette De Mente, *Business Guide to Japan: A Quick Guide to Opening Doors & Closing Deals* (Rutland, VT: Tuttle, 2006).

101. Masami Nishishiba and David Ritchie, "The Concept of Trustworthiness: A Cross-Cultural Comparison Between Japanese and U.S. Business People," *Journal of Applied Communication Research* 28 (2000): 347–367.

102. Fons Trompenaars and Charles Hampden-Turner, *Managing People across Cultures* (Chichester, England: Capstone, 2004); Fons Trompenaars and Charles Hampden-Turner, *Innovating in a Global Crisis: Riding the Whirlwind of Recession* (Oxford: Infinite Ideas, 2009); Fons Trompenaars and Charles Hampden-Turner, *Riding the Waves of Innovation: Harness the Power of Global Culture to Drive Creativity and Growth* (New York: McGraw-Hill, 2010); Fons Trompenaars and Charles Hampden-Turner, *Riding the Waves of Culture: Understanding Diversity in Global Business*, 3rd ed. (New York: McGraw-Hill, 2012); Fons Trompenaars and Peter Prud'Homme, *Managing Change across Corporate Cultures* (Chichester, England: Capstone, 2004); Fons Trompenaars and Peter Woolliams, *Business across Cultures* (Chichester, England: Capstone, 2003); Fons Trompenaars and Peter Woolliams, *Marketing across Cultures* (Chichester, England: Capstone, 2004).

103. William B. Gudykunst and Yuko Matsumoto, "Cross-Cultural Variability of Communication in Personal Relationships," *Communication in Personal Relationships Across Cultures*, ed. William B. Gudykunst, Stella Ting-Toomey, and Tsukasa Nishida (Thousand Oaks: Sage, 1996), 19–56.

104. Cindy P. Lindsay and Bobby L. Dempsey, "Ten Painfully Learned Lessons about Working in China: The Insights of Two American Behavioral Scientists," *Journal of Applied Behavioral Science* 19 (1983): 265–276.

105. See, for example, Lawrence W. Tuller, *Doing Business beyond America's Borders: The Do's, Don'ts and Other Details of Conducting Business in 40 Countries* (Irvine, CA: Entrepreneur Press, 2008), 289–305.

106. Alex Blackwell, "Negotiating in Europe," *Hemispheres*, July 1994, 43.

107. Jo Ann G. Heydenfeldt, "The Influence of Individualism/Collectivism on Mexican and US Business Negotiation," *International Journal of Intercultural Relations*, 24 (2000): 383–407.

108. Moran, Harris, and Moran.

109. Alan Goldman, "Communication in Japanese Multinational Organizations," *Communicating in Multinational Organizations*, ed. Richard L. Wiseman and Robert Shuter (Thousand Oaks, CA: Sage, 1994), 49–59.

110. Yale Richmond, *From Nyet to Da: Understanding the New Russia* (Boston: Intercultural Press, 2009).

111. Rajesh Kumar and Anand Kumar Sethi, *Doing Business in India: A Guide for Western Managers* (New York: Palgrave Macmillan, 2005), 104–105.

112. Lecia Archer and Kristine L. Fitch, "Communication in Latin American Multinational Organizations," *Communicating in Multinational Organizations*, ed. Richard L. Wiseman and Robert Shuter (Thousand Oaks, CA: Sage, 1994), 75–93.

113. Myria Watkins Allen, Patricia Amason, and Susan Holmes, "Social Support, Hispanic Emotional Acculturative Stress and Gender," *Communication Studies* 49 (1998): 139–157; Patricia Amason, Myria Watkins Allen, and Susan A. Holmes, "Social Support and Acculturative Stress in

the Multicultural Workplace," *Journal of Applied Communication Research* 27 (1999): 310–334.

114. Lindsay and Dempsey.

115. Dresser, 68.

116. Adapted from Lynn E. Metcalf, Allan Bird, Mahesh Shankarmahesh, Zeynep Aycan, Jorma Larimo, and Di' dimo Dewar Valdelamar, "Cultural Tendencies in Negotiation: A Comparison of Finland, India, Mexico, Turkey, and the United States," *Journal of World Business* 41 (2006): 382–394.

117. Dresser, 156.

118. Farid Elashmawi and Philip R. Harris, *Multicultural Management 2000: Essential Cultural Insights for Global Business Success* (Houston: Gulf, 1998), 118–125; Moran, Harris, and Moran.

119. David Streitfeld, "A Crash Course on Irate Calls," *Los Angeles Times*, August 2, 2004, A1, A8. See also Paul Davies, *What's This India Business? Offshore Outsourcing, and the Global Services Revolution* (Yarmouth, ME: Nicholas Brealey International, 2004).

120. See, for example, Martin and Chaney.

121. Moran, Harris, and Moran.

122. Moran, Harris, and Moran.

123. Young Yun Kim and Sheryl Paulk, "Interpersonal Challenges and Personal Adjustments: A Qualitative Analysis of the Experiences of American and Japanese Co-Workers," *Communicating in Multinational Organizations*, ed. Wiseman and Shuter, 117–140. See also Alan E. Omens, Stephen R. Jenner, and James R. Beatty, "Intercultural Perceptions in United States Subsidiaries of Japanese Companies," *International Journal of Intercultural Relations* 11 (1987): 249–264; David W. Shwalb, Barbara J. Shwalb, Delwyn L. Harnisch, Martin L. Maehr, and Kiyoshi Akabane, "Personal Investment in Japan and the U.S.A.: A Study of Worker Motivation," *International Journal of Intercultural Relations* 16 (1992): 107–124.

124. Moran, Harris, and Moran.

125. Lillian H. Chaney and Jeanette S. Martin, *Intercultural Business Communication*, 4th ed. (Upper Saddle River, NJ: Pearson Prentice Hall, 2007).

126. Roong Sriussadapron, "Managing International Business Communication Problems at Work: A Pilot Study in Foreign Companies in Thailand," *Cross Cultural Management: An International Journal* 13 (2006): 330–344.

127. Kumar and Kumar Sethi, 110.

128. Moran, Harris, and Moran.

129. David C. Thomas, *Cross-Cultural Management: Essential Concepts*, 2nd ed. (Thousand Oaks, CA: Sage, 2008).

130. Richard H. Reeves-Ellington, "Using Cultural Skills for Cooperative Advantage in Japan," *Human Organization* 52 (1993): 203–215.

131. See, for example, Mijnd Huijser, *The Cultural Advantage: A New Model for Succeeding with Global Teams* (Boston: Intercultural Press, 2006); Robert T. Moran, William E. Youngdahl, and Sarah V. Moran, "Intercultural Competence in Business— Leading Global Projects: Bridging the Cultural and Functional Divide," *The Sage Handbook of Intercultural Competence* ed. Deardorff, 287–303.

132. See, for example, Jessica Hirshorn, *Rocket: A Simulation on Intercultural Teamwork* (Boston: Nicholas Brealey, 2010).

133. Aparna Nancherla, "Gaining Control of the Remote Workforce," *T+D*, August 2010, 20.

134. Elizabeth Gareis, "Virtual Teams: A Comparison of Online Communication Channels," *Journal of Language for International Business* 17 (2006): 6–21.

135. Evelyne Glaser and Manuela Guilherme, "Intercultural Competence for Multicultural Teams: A Qualitative Study," *Intercultural Competence for Professional Mobility [CD]*, ed. Evelyne Glaser, Manuela Guilherme, María del Carmen Méndez García, and Terry Mughan (Strasbourg: Council of Europe Publishing, 2007), 1–19.

136. Corinne Rosenberg, "EMEA-US Culture Clash: Resolving Diversity Issues through Reflective Evaluated Action Learning," *Industrial and Commercial Training* 37 (2005): 304–308.

137. William I. Gordon, "Organizational Imperatives and Cultural Modifiers," *Business Horizons* 27 (1984): 81.

138. See, for example, Christalyn Branner and Tracey Wilson, *Doing Business with Japanese Men: A Woman's Handbook* (Berkeley, CA: Stone Bridge Press, 1993).

139. Moran, Harris, and Moran.

140. Moran, Harris, and Moran.

141. Nongluck Sriussadaporn-Charoenngam and Fredric M. Jablin, "An Exploratory Study of Communication Competence in Thai Organizations," *Journal of Business Communication* 36 (1999): 382–418.

142. For an assessment of the conditions under which training outcomes are maximized, see Herman Aguinis and Kurt Kraiger, "Benefits of Training and Development for Individuals and Teams, Organizations, and Society," *Annual Review of Psychology* 60 (2009): 451–474.

143. See, for example, Maureen Guirdham, *Communicating across Cultures at Work*, 3rd ed. (New York: Palgrave Macmillan, 2011); Geraldine Healy and Franklin Oikelome, *Diversity, Ethnicity, Migration, and Work: International Perspectives* (New York: Palgrave, 2011); Chaunda L. Scott and Marilyn Y. Byrd (eds.), *Workforce Diversity: Current and Emerging Issues and Cases* (Thousand Oaks, CA: Sage, 2012); Monir H. Tayeb, *International Human Resource Management: A Multinational Company Perspective* (Oxford: Oxford University Press, 2005); R. Roosevelt Thomas, Jr., *World Class Diversity Management: A Strategic Approach* (San Francisco: Berrett-Koehler, 2010).

144. Steven H. Cady and Joanie Valentine, "Team Innovation and Perceptions of Consideration: What Difference Does Diversity Make?" *Small Group Research* 30 (1999): 730–750. See also: Georges Buzaglo and Susan A. Wheelan, "Facilitating Work Team Effectiveness: Case Studies from Central America," *Small Group Research* 30 (1999): 108–129; Graeme L Harrison, Jill L. McKinnon, Anne Wu, and Chee W. Chow, "Cultural Influences on Adaptation to Fluid Workgroups and Teams," *Journal of International Business Studies* 31 (2000): 489–505.

145. Jolanta Aritz and Robyn C. Walker, "Group Composition and Communication Styles: An Analysis of Multicultural Teams in Decision-Making Meetings," *Journal of Intercultural Communication Research* 38 (2009): 99–114; Norbert L. Kerr and R. Scott Tindale, "Group Performance and Decision Making," *Annual Review of Psychology* 55 (2004): 623–655; Lisa Millhous, "The Experience of Culture in Multicultural Groups: Case Studies of Russian-American Collaboration in Business," *Small Group Research* 30(3) (1999): 280–308.

146. Percy W. Thomas, "A Cultural Rapport Model," *Valuing Diversity*, ed. Lewis Brown Griggs and Lente Louise Louw (New York: McGraw-Hill, 1995), 136–137.

147. Charles R. Bantz, "Cultural Diversity and Group Cross-Cultural Team Research," *Journal of Applied Communication Research* 21 (1993): 1–20.

CHAPTER 12

1. For summaries of these studies, see William B. Gudykunst (ed.), *Intergroup Communication* (London: Arnold, 1986); Ellen Bouchard Ryan and Howard Giles (eds.), *Attitudes toward Language Variation: Social and Applied Contexts* (London: Arnold, 1982).

2. Cynthia Gallois, Arlene Franklyn-Stokes, Howard Giles, and Nikolas Coupland, "Communication Accommodation in Intercultural Encounters," *Theories in Intercultural Communication*, ed. Young Yun Kim and William B. Gudykunst (Newbury Park, CA: Sage, 1988), 157–188.

3. Marsha Houston and Cheris Kramarae, "Speaking from Silence: Methods of Silencing and Resistance," *Discourse & Society* 2 (1991): 387–399; Cheris Kramarae, *Women and Men Speaking* (Rowley, MA: Newbury House, 1981).

4. Mary M. Meares, John G. Oetzel, Annette Torres, Denise Derkacs, and Tamar Ginossar, "Employee Mistreatment and Muted Voices in the Culturally Diverse Workplace," *Journal of Applied Communication Research* 32 (2004): 4–27.

5. Mark P. Orbe, "'Remember, It's Always Whites' Ball': Descriptions of African American Male Communication," *Communication Quarterly* 42 (1994): 287–300.

6. Gordon W. Allport, *The Nature of Prejudice* (New York: Macmillan, 1954). For a review of the research on the contact hypothesis, see: Todd D. Nelson, *The Psychology of Prejudice*, 2nd ed. (Boston: Allyn & Bacon, 2006); Thomas F. Pettigrew, "Future Directions for Intergroup Contact Theory and Research," *International Journal of Intercultural Relations* 32 (2008): 187–199; Thomas F. Pettigrew and Linda R. Tropp, "A Meta-Analytic Test of Intergroup Contact Theory," *Journal of Personality and Social Psychology* 90 (2006): 751–783; Thomas F. Pettigrew and Linda R.

Tropp, *When Groups Meet: The Dynamics of Intergroup Contact* (New York: Psychology Press, 2011); Thomas F. Pettigrew, Linda R. Tropp, Ulrich Wagner, and Oliver Christ, "Recent Advances in Intergroup Contact Theory," *International Journal of Intercultural Relations* 35 (2011): 271–280; Linda R. Tropp and Robyn K. Mallett (eds.), *Moving Beyond Prejudice Reduction: Pathways to Positive Intergroup Relations* (Washington, DC: American Psychological Association, 2011).

7. Jim Blascovich, Wendy Berry Mendes, Sarah B. Hunter, and Brian Lickel, "Stigma, Threat, and Social Interactions," *The Social Psychology of Stigma*, ed. Todd F. Heatherton, Robert E. Kleck, Michelle R. Hebl, and Jay G. Hull (New York: Guilford Press, 2003), 307–333; Jim Blascovich, Wendy Berry Mendes, Sarah B. Hunter, Brian Lickel, and Neneh Kowai-Bell, "Perceiver Threat in Social Interactions with Stigmatized Others," *Journal of Personality and Social Psychology* 80 (2001): 253–267; Jim Blascovich, Wendy Berry Mendes, Sarah B. Hunter, Brian Lickel, and Sarah Hunter, "Challenge and Threat during Social Interactions with White and Black Men," *Personality and Social Psychology Bulletin* 28 (2002): 939–952; Oliver Christ, Miles Hewstone, Nicole Tausch, Ulrich Wagner, Alberto Voci, Joanne Hughes, and Ed Cairns, "Direct Contact as a Moderator of Extended Contact Effects: Cross-Sectional and Longitudinal Impact on Outgroup Attitudes, Behavioral Intentions, and Attitude Certainty," *Personality and Social Psychology Bulletin* 36 (2010): 1662–1674; Kristin Davies, Linda R. Tropp, Arthur Aron, Thomas F. Pettigrew, and Stephen C. Wright, "Cross-Group Friendships and Intergroup Attitudes: A Meta-Analytic Review," *Personality and Social Psychology Review* 15 (2011): 332–351; Kristof Dhont, Arne Roets, and Alain Van Hiel, "Opening Closed Minds: The Combined Effects of Intergroup Contact and Need for Closure on Prejudice," *Personality and Social Psychology Bulletin* 37 (2011): 514–528; Angel Gómez, Linda R. Tropp, and Saulo Fernández, "When Extended Contact Opens the Door to Future Contact: Testing the Effects of Extended Contact on Attitudes and Intergroup Expectancies in Majority and Minority Groups," *Group Processes & Intergroup Relations* 14 (2011): 161–173; P. J. Henry and Curtis D. Hardin, "The Contact Hypothesis Revisited: Status Bias in the Reduction of Implicit Prejudice in the United States and Lebanon," *Psychological Science* 17 (2006): 862–868; Gordon Hodson, "Do Ideologically Intolerant People Benefit from Intergroup Contact?" *Current Directions in Psychological Science* 20 (2011): 154–159; Stefania Paolini, Jake Harwood, and Mark Rubin, "Negative Intergroup Contact Makes Group Memberships Salient: Explaining Why Intergroup Conflict Endures," *Personality and Social Psychology Bulletin* 36 (2010): 1723–1738; Stefania Paolini, Miles Hewstone, Ed Cairns, and Alberto Voci, "Effects of Direct and Indirect Cross-Group Friendships on Judgments of Catholics and Protestants in Northern Ireland: The Mediating Role of an Anxiety-Reduction Mechanism," *Personality and Social Psychology Bulletin* 30 (2004): 770–786; Katharina Schmid, Miles Hewstone, Nicole Tausch, Ed Cairns, and Joanne Hughes, "Antecedents and Consequences of Social Identity Complexity: Intergroup Contact, Distinctiveness Threat, and Outgroup Attitudes," *Personality and Social Psychology Bulletin* 35 (2009): 1085–1098; Linda R. Tropp and Thomas F. Pettigrew, "Differential Relationships between Intergroup Contact and Affective and Cognitive Dimensions of Prejudice," *Personality and Social Psychology Bulletin* 31 (2005): 1145–1158.

8. The definition is modified from one proposed by Young Yun Kim. See Young Yun Kim, "Adapting to an Unfamiliar Culture," *Handbook of International and Intercultural Communication*, 2nd ed., ed. William B. Gudykunst and Bella Mody (Thousand Oaks, CA: Sage, 2002), 260; Young Yun Kim, *Becoming Intercultural: An Integrative Theory of Communication and Cross-Cultural Adaptation* (Thousand Oaks, CA: Sage, 2001), 31; Young Yun Kim, "Adapting to a New Culture: An Integrative Communication Theory," *Theorizing about Intercultural Communication*, ed. William B. Gudykunst (Thousand Oaks, CA: Sage, 2005), 375–400.

9. John W. Berry, Uichol Kim, and Pawel Boski, "Psychological Acculturation of Immigrants," *Cross-Cultural Adaptation: Current Approaches*, ed. Young Yun Kim and William B. Gudykunst (Newbury Park, CA: Sage, 1988).

10. Kalvero Oberg, "Cultural Shock: Adjustment to New Cultural Environments," *Practical Anthropology* 7 (1960): 176.

11. Oberg.

12. D. Bhugra, "Migration and Depression," *Acta Psychiatrica Scandinavica* 108 (2003): 67–72; Michael Brein and Kenneth H. David, "Intercultural Communication and the Adjustment of the Sojourner," *Psychological Bulletin* 76 (1971): 215–230; Kevin F. Gaw, "Reverse Culture Shock in Students Returning from Overseas," *International Journal of Intercultural Relations* 24 (2000): 83–104; John T. Gullahorn and Jeanne E. Gullahorn, "An Extension of the U-Curve Hypothesis," *Journal of Social Issues* 14 (1963): 33–47; Daniel J. Kealey, "A Study of Cross-Cultural Effectiveness: Theoretical Issues, Practical Applications," *International Journal of Intercultural Relations* 13 (1989): 387–428; Otto Klineberg and W. Frank Hull, *At a Foreign University: An International Study of Adaptation and Coping* (New York: Praeger, 1979); Jolene Koester, "Communication and the Intercultural Reentry: A Course Proposal," *Communication Education* 23 (1984): 251–256; Judith N. Martin, "The Intercultural Reentry: Conceptualizations and Suggestions for Future Research," *International Journal of Intercultural Relations* 8 (1984): 115–134; Craig Storti, *The Art of Coming Home* (Yarmouth, ME: Intercultural Press, 2001); Ching Wan, "The Psychology of Culture Shock," *Asian Journal of Social Psychology* 7 (2004): 233–234; Colleen Ward, Stephen Bochner, and Adrian Furnham, *The Psychology of Culture Shock*, 2nd ed. (New York: Routledge, 2001).

13. See, for instance, Sarah Brabant, C. Eddie Palmer, and Robert Gramling, "Returning Home: An Empirical Investigation of Cross-Cultural Reentry," *International Journal of Intercultural Relations* 14 (1990): 387–404.

14. Walter Enloe and Philip Lewin, "Issues of Integration Abroad and Readjustment to Japan of Japanese Returnees," *International Journal of Intercultural Relations* 11 (1987): 223–248; Louise H. Kidder, "Requirements for Being 'Japanese': Stories of Returnees," *International Journal of Intercultural Relations* 16 (1992): 383–393.

15. Enloe and Lewin, 235.

16. Nancy Adler, "Re-Entry: Managing Cross-Cultural Transitions," *Group and Organization Studies* 6 (1981): 341–356; Austin Church, "Sojourner Adjustment," *Psychological Bulletin* 91 (1982): 540–572; Dennison Nash, "The Course of Sojourner Adaptation: A New Test of the U-Curve Hypothesis," *Human Organization* 50 (1991): 283–286.

17. See, for example, Young Yun Kim and Brent D. Ruben, "Intercultural Transformation: A Systems Theory," *Theories in Intercultural Communication*, ed. Kim and Gudykunst, 299–321.

18. Daniel J. Kealey, "A Study of Cross-Cultural Effectiveness: Theoretical Issues, Practical Applications," *International Journal of Intercultural Relations* 13 (1989): 387–428.

19. Lorraine Brown and Immy Holloway, "The Initial Stage of the International Sojourn: Excitement or Culture Shock?" *British Journal of Guidance & Counselling* 36.1 (2008): 33–49.

20. Betina Szkudlarek, "Reentry—A Review of the Literature," *International Journal of Intercultural Relations* 34.1 (2010): 1–21.

21. Andrew G. Ryder, Lynn E. Alden, and Delroy L. Paulhus, "Is Acculturation Unidimensional or Bidimensional? A Head-to-Head Comparison in the Prediction of Personality, Self-Identity, and Adjustment," *Journal of Personality and Social Psychology* 79 (2000): 49–65.

22. Mitchell R. Hammer, William B. Gudykunst, and Richard L. Wiseman, "Dimensions of Intercultural Effectiveness: An Exploratory Study," *International Journal of Intercultural Relations* 2 (1978): 382–393.

23. See Anne-Marie Masgoret and Colleen Ward, "Culture Learning Approach to Acculturation," *The Cambridge Handbook of Acculturation Psychology*, ed. David L. Sam and John W. Berry (Cambridge: Cambridge University Press, 2006), 58–77; Andy S. J. Ong and Colleen Ward, "The Construction and Validation of a Social Support Measure for Sojourners: The Index of Sojourner Social Support (ISSS) Scale," *Journal of Cross-Cultural Psychology* 36 (2005): 637–661; Colleen Ward, "Thinking Outside the Berry Boxes: New Perspectives on Identity, Acculturation and Intercultural Relations," *International Journal of Intercultural Relations* 32 (2008): 105–114; Colleen Ward and Antony Kennedy, "Locus of Control, Mood Disturbance, and Social Difficulty During Cross-Cultural Transitions," *International Journal of Intercultural Relations* 2 (1992): 175–194; Colleen Ward and Antony Kennedy, "Where's the 'Culture' in Cross-Cultural Transition? Comparative Studies of Sojourner Adjustment,"

Journal of Cross-Cultural Psychology 24 (1993): 221–249; Colleen Ward and Antony Kennedy, "Acculturation and Cross-Cultural Adaptation of British Residents in Hong Kong," *Journal of Social Psychology* 133 (1993): 395–397; Colleen Ward and Antony Kennedy, "Acculturation Strategies, Psychological Adjustment, and Sociocultural Competence During Cross-Cultural Transactions," *International Journal of Intercultural Relations* 18 (1994): 329–343; Colleen Ward and Antony Kennedy, "The Measurement of Sociocultural Adaptation," *International Journal of Intercultural Relations* 23 (1999): 659–677; Colleen Ward and Chan-Hoong Leong, "Intercultural Relations in Plural Societies," *The Cambridge Handbook of Acculturation Psychology*, ed. David L. Sam and John W. Berry (Cambridge: Cambridge University Press, 2006), 484–503; Colleen Ward, Chan-Hoong Leong, and Meilin Low, "Personality and Sojourner Adjustment: An Exploration of the Big Five and the Cultural Fit Proposition," *Journal of Cross-Cultural Psychology* 35 (2004): 137–151; C. Ward and A. Rana-Deuba, "Home and Host Culture Infuences on Sojourner Adjustment," *International Journal of Intercultural Relations* 24 (2000): 291–306; Colleen Ward and Wendy Searle, "The Impact of Value Discrepancies and Cultural Identity on Psychological and Sociological Adjustment of Sojourners," *International Journal of Intercultural Relations* 15 (1991): 209–225.

24. See John W. Berry, "Immigration, Acculturation, and Adaptation," *Applied Psychology: An International Review* 46 (1997): 5–68; John W. Berry, "Conceptual Approaches to Acculturation," *Acculturation: Advances in Theory, Measurement, and Applied Research*, ed. Kevin M. Chun, Pamela Balls Organista, and Gerardo Marín (Washington, DC: American Psychological Association, 2003), 17–37; John W. Berry, "Acculturation: Living Successfully in Two Cultures," *International Journal of Intercultural Relations* 29 (2005): 697–712; John W. Berry, "Contexts of Acculturation," *The Cambridge Handbook of Acculturation Psychology*, ed. Sam and Berry, 27–42; John W. Berry, "Stress Perspectives on Acculturation," *The Cambridge Handbook of Acculturation Psychology*, ed. Sam and Berry, 43–57; John W. Berry, "Acculturation," *Handbook of Socialization: Theory and Research*, ed. Joan E. Grusec and Paul D. Hastings (New York: Guilford, 2007), 543–558; John W. Berry, "Globalisation and Acculturation," *International Journal of Intercultural Relations* 32 (2008): 328–336; Berry, Kim, and Boski; John W. Berry, Jean S. Phinney, David L. Sam, and Paul Vedder, "Immigrant Youth: Acculturation, Identity, and Adaptation," *Applied Psychology: An International Review* 55 (2006): 303–332; Seth J. Schwartz and Byron L. Zamboanga, "Testing Berry's Model of Acculturation: A Confirmatory Latent Class Approach," *Cultural Diversity and Ethnic Minority Psychology* 14 (2008): 275–285; Colleen Ward, "Thinking Outside the Berry Boxes: New Perspectives on Identity, Acculturation and Intercultural Relations," *International Journal of Intercultural Relations* 32 (2008): 105–114.

25. Berry, Kim, and Boski, 71.

26. See Young Yun Kim, "Ideology, Identity, and Intercultural Communication: An Analysis of Differing Academic Conceptions of Cultural Identity," *Journal of Intercultural Communication Research* 36 (2007): 237–253. See also: Young Yun Kim, "Unum and Pluribus: Ideological Underpinnings of Interethnic Communication in the United States," *International Journal of Intercultural Relations* 23 (1999): 591–611; Young Yun Kim, "Unum vs. Pluribus: Ideology and Differing Academic Conceptions of Ethnic Identity," *Communication Yearbook*, 26 ed. William B. Gudykunst (Mahwah, NJ: Erlbaum, 2003), 298–325; Young Yun Kim, "Adapting to a New Culture: An Integrative Communication Theory," *Theorizing about Intercultural Communication* (Thousand Oaks, CA: Sage, 2005), 375–400.

27. Adapted from Berry, "Globalisation and Acculturation," 328–336.

28. John W. Berry, Jean S. Phinney, David Sam, and Paul Vedder (eds.), *Immigrant Youth in Cultural Transition: Acculturation, Identity, and Adaptation across National Contexts* (Mahwah, NJ: Erlbaum, 2006); John W. Berry, Jean S. Phinney, David L. Sam, and Paul Vedder, "Immigrant Youth: Acculturation, Identity, and Adaptation," *Applied Psychology: An International Review* 55 (2006): 303–332.

29. Robin Goodwin, *Personal Relationships across Cultures* (New York: Routledge, 1999); Inga Jasinskaja-Lahti, Karmela Liebkind, Magdalena

Jaakkola, and Anni Reuter, "Perceived Discrimination, Social Support Networks, and Psychological Well-Being among Three Immigrant Groups," *Journal of Cross-Cultural Psychology* 37 (2006): 293–311; Karmela Liebkind, "Acculturation and Stress: Vietnamese Refugees in Finland," *Journal of Cross-Cultural Psychology* 27 (1996): 161–180.

30. Young Yun Kim, "Intercultural Personhood: Globalization and a Way of Being," *International Journal of Intercultural Relations* 34 (2008): 359–368.

31. Our discussion of tourism is informed by the following sources: Fleura Bardhi, Jacob Ostberg, and Anders Bengtsson, "Negotiating Cultural Boundaries: Food, Travel and Consumer Identities," *Consumption Markets & Culture* 13.2 (2010): 133–157; Tracy Berno and Colleen Ward, "Innocence Abroad: A Pocket Guide to Psychological Research on Tourism," *American Psychologist* 60.6 (2005): 593–600; Bodil Stilling Blichfeldt and Inès Kessler, "Interpretive Consumer Research: Uncovering the 'Whys' Underlying Tourist Behavior," *Handbook of Tourist Behavior: Theory & Practice*, ed. Metin Kozak and Alain Decrop (New York: Routledge, 2009), 3–15; Antónia Correia and Miguel Moital, "Antecedents and Consequences of Prestige Motivation in Tourism: An Expectancy-Value Motivation," *Handbook of Tourist Behavior: Theory & Practice*, ed. Metin Kozak and Alain Decrop (New York: Routledge, 2009), 16–32; Sara Dolnicar and Byron Kemp, "Tourism Segmentation by Consumer-Based Variables," *Handbook of Tourist Behavior: Theory & Practice*, ed. Kozak and Decrop, 177–194; Elfriede Fursich, "Packaging Culture: The Potential and Limitations of Travel Programs on Global Television," *Communication Quarterly* 50 (2002): 204–226; Juergen Gnoth and Andreas H. Zins, "Emotions and Affective States in Tourism Behavior," *Handbook of Tourist Behavior: Theory & Practice*, ed. Kozak and Decrop, 195–207; Robert Govers and Frank M. Go, "Tourism Destination Image Formation," *Handbook of Tourist Behavior: Theory & Practice*, ed. Kozak and Decrop, 35–49; Petri Hottola, "Culture Confusion: Intercultural Adaptation in Tourism," *Annals of Tourism Research* 31 (2004): 447–466; Maria Månsson, "The Role of Media Products on Consumer Behavior in Tourism," *Handbook of Tourist Behavior: Theory & Practice*, ed. Kozak and Decrop, 226–236; Gianna Moscardo, "Understanding Tourist Experience through Mindfulness Theory," *Handbook of Tourist Behavior: Theory & Practice*, ed. Kozak and Decrop, 99–115; Siew Imm Ng, Julie Anne Lee, and Geoffrey N. Soutar, "Tourists' Intention to Visit a Country: The Impact of Cultural Distance," *Tourism Management* 28 (2007): 1497–1506; Pettigrew and Tropp, "A Meta-Analytic Test of Intergroup Contact Theory", Margaret J. Pitts, "Identity and the Role of Expectations, Stress, and Talk in Short-Term Student Sojourner Adjustment: An Application of the Integrative Theory of Communication and Cross-Cultural Adaptation," *International Journal of Intercultural Relations* 33 (2009): 450–462; Jan Rath, *Tourism, Ethnic Diversity, and the City* (New York: Routledge, 2007); Yvette Reisinger, "Cross-Cultural Differences in Tourist Behavior," *Handbook of Tourist Behavior: Theory & Practice*, ed. Kozak and Decrop, 235–255; Kristin Bervig Valentine, "Yaqui Easter Ceremonies and the Ethics of Intense Spectatorship," *Text and Performance Quarterly* 22 (2002): 280–296; Colleen Ward and Tracy Berno, "Beyond Social Exchange Theory: Attitudes toward Tourists," *Annals of Tourism Research* 38 (2011): 1556–1569.

32. Data are based on 2011 estimates. See International Trade Administration, "New Travel and Tourism Forecast Projects Strong Growth for International Arrivals through 2016," Accessed November 5, 2011, from http://tinet.ita.doc.gov/tinews/archive/tinews2011/20111027.html.

33. Data are based on 2009 estimates. See U.S. Department of Commerce, International Trade Administration, "Profile of U.S. Resident Travelers Visiting Overseas Destinations: 2009 Outbound," Accessed November 5, 2011, from http://www.tinet.ita.doc.gov/outreachpages/download_data_table/2009_Outbound_Profile.pdf.

34. For an analysis of the utility of the "short-term–long-term" and "voluntary–involuntary" dimensions of intercultural communication encounters, see Adrian Furnham and Stephen Bochner, *Culture Shock: Psychological Reactions to Unfamiliar Environments* (London: Mathuen,

1986), 56. See also Adrian Furnham, "Tourism and Culture Shock," *Annals of Tourism Research* 11 (1984): 41–57.

35. Ker Munthit, "Cambodia Tourism Boom a Mixed Blessing," *Desert News* (Salt Lake City), December 3, 2006, online edition; Dante Ramos, "Touring the Tragic Kingdom," *Boston Globe*, October 28, 2007, online edition.

36. David W. Kale, "Ethics in Intercultural Communication," *Intercultural Communication: A Reader*, 6th ed., ed. Larry A. Samovar and Richard E. Porter (Belmont, CA: Wadsworth, 1991), 423.

37. Kale.

38. John F. Kennedy, "Remarks Prepared for Delivery at the Trade Mart in Dallas," November 22, 1963. Accessed online July 6, 2008, from http://www.jfklibrary.org/Historical+Resources/Archives/Reference+Desk/Speeches/JFK/003POF03TradeMart11221963.htm.

39. Charles Ess and Fay Sudweeks, *Culture, Technology, and Communication: Towards an Intercultural Global Village* (Albany: State University of New York Press, 2001). See also Radhika Gajjala, "Interrogating Identities: Composing Other Cyberspaces," *Intercultural Alliances: Critical Transformation*, ed. Mary Jane Collier (Thousand Oaks, CA: Sage, 2003), 167–188.

40. Myron W. Lustig (Primary Program Planner), "Widening Our Circle," Conference theme for the annual convention of the Western States Communication Association (Albuquerque, 2004).

41. Troy Duster, "Understanding Self-Segregation on the Campus," *Chronicle of Higher Education*, September 25, 1991, B2.

RESOURCES

1. Cultural data are adapted from information reported in Geert Hofstede, Gert Jan Hofstede, and Michael Minkov, *Cultures and Organizations: Software of the Mind: Intercultural Cooperation and Its Importance for Survival*, 3rd ed. (New York: McGraw-Hill, 2010). Cultural groupings are based on information reported in Vipin Gupta and Paul J. Hanges, "Regional and Climate Clustering of Societal Cultures," *Culture, Leadership, and Organizations: The GLOBE Study of 62 Societies*, ed. Robert J. House, Paul J. Hanges, Mansour Javidan, Peter W. Dorfman, and Vipin Gupta (Thousand Oaks, CA: Sage, 2004), 178–218.

2. Cultural data are adapted from information provided by Shalom Schwartz. Cultural groupings are based on information reported in Vipin Gupta and Paul J. Hanges, "Regional and Climate Clustering of Societal Cultures," *Culture, Leadership, and Organizations: The GLOBE Study of 62 Societies*, ed. House, Hanges, Javidan, Dorfman, and Gupta, 178–218.

3. Cultural data are adapted from information on actual cultural practices, as reported in Robert J. House, Paul J. Hanges, Mansour Javidan, Peter W. Dorfman, and Vipin Gupta (eds.), *Culture, Leadership, and Organizations: The GLOBE Study of 62 Societies* (Thousand Oaks, CA: Sage, 2004). Cultural groupings are based on information reported in Vipin Gupta and Paul J. Hanges, "Regional and Climate Clustering of Societal Cultures," *Culture, Leadership, and Organizations: The GLOBE Study of 62 Societies*, ed. House, Hanges, Javidan, Dorfman, and Gupta, 178–218.

CREDITS

PHOTO CREDITS

Pages 2, 7, 27, 32, 34, 36, 39, 80, 87, 89, 96, 106, 109, 119, 140, 143, 145, 155, 156, 163, 170, 172, 199, 208, 215(l), 215(r), 227, 254, 267, 288, 291, 299(l), 299(r), 307 © Myron W. Lustig and Jolene Koester

Pages 4, 11 © Jim West/Alamy

Page 6, © Somos Images/Alamy

Pages 8, 274 © Corbis Cusp/Alamy

Page 10, © PCN Photography/Alamy

Page 11, © Janine Wiedel Photolibrary/Alamy

Pages 28(l), 113, © Corbis Flirt/Alamy

Pages 28(r), 264, 265 © Blend Images/Alamy

Page 35, © Philip Scalia/Alamy

Page 51, © John Robertson/Alamy

Pages 56, 266 © Jon Parker Lee/Alamy

Page 62, © David Muenker/Alamy

Page 65, © Presselect/Alamy

Page 71, © Gordon Sinclair/Alamy

Page 75, © Pep Roig/Alamy

Page 78(l), © RubberBall/Alamy

Page 78(r), © GoGo Images Corporation/Alamy

Page 102, © Jon Arnold Images Ltd/Alamy

Pages 120, 131 © Purestock/Alamy

Page 132, © Jonathan Goldberg/Alamy

Page 139, © Jeff Greenberg/Alamy

Page 160, © Ted Pink/Alamy

Page 177, © Jenny Matthews/Alamy

Page 183, © Beyond Fotomedia GmbH/Alamy

Page 190, © Imagestate Media Partners Limited—Impact Photos/Alamy

Page 191, © Paul Doyle/Alamy

Page 196, © Idealink Photography/Alamy

Page 203, © Arcaid Images/Alamy

Page 212, © Bill Bachmann/Alamy

Page 220, © Image Source/Alamy

Page 229, © Corbis/SuperStock

Page 231, © Myrleen Pearson/Alamy

Page 242, © Fancy/Alamy

Page 246, © Morgan Lane Photography/Alamy

Page 281, © moodboard/Alamy

TEXT CREDITS

p. 3: "Beyond the Melting Pot" by William A. Henry II from TIME, April 9, 1990.

p. 5: "Cultural Pluralism: A Goal to Be Realized," by Antonia Pantoja and Wilhelmina Perry from VOICES FROM THE BATTLEFRONT, edited by Marta Moreno Vega and Cheryll Y. Greene.

p. 6: "Internationally, the Business of Education is Booming," by Karin Fisher from THE CHRONICLE OF HIGHER EDUCATION, May 29, 2009.

p. 6: "Globalization and Technology," by Vinnie Mirchandani from DEAL ARCHITECT, July 8, 2007.

p. 7: TRANSLATIONS OF BEAUTY by Mia Yun. Copyright © 2004 Mia Yun. Reprinted with permission of Atria Books, a Division of Simon & Schuster, Inc.

p. 9: "Tourism in America: Removing Barriers and Promoting Growth," Testimony by Roger Dow before the United States Senate Committee On Commerce, Science & Transportation, April 5, 2011.

p. 9: "How Dumb Are We?" by Andrew Roman from NEWSWEEK MAGAZINE, March 20, 2011.

p. 10: President's Commission on Foreign Language and International Studies (Washington, DC: Department of Health, Education and Welfare, 1979).

p. 10: "A Conversation, Not a Monologue," by Catharine Stimpson from THE CHRONICLE OF HIGHER EDUCATION, March 16, 1994.

p. 11: "Together in world of differences and delights" by Al Martinez from LOS ANGELES DAILY NEWS, December 20, 2010. Copyright © 2010 Al Martinez. Reprinted by permission of the author.

p. 14: Copyright © 2012. Reprinted with permission of Hispanic Outlook Magazine; www.HispanicOutlook.com.

p. 17: INTERPERSONAL COMMUNICATION: SURVEY AND STUDIES by Dean Barnlund (Houghton Mifflin:1968).

p. 17: INTERCULTURAL ENCOUNTERS: THE FUNDAMENTALS OF INTERCULTURAL COMMUNICATIONS by Donald W. Klopf (Morton Publishing Company: 1987).

p. 18: "How to be Invisible" by Seth Stevenson from NEWSWEEK, April 19, 2010.

p. 18: GOOD NEIGHBORS: COMMUNICATING WITH THE MEXICANS by John C. Condon (Intercultural Press: 1985).

p. 19: "International Education: Broadening the Base of Participation," by Johnetta R. Cole, 43rd International Conference on Educational Exchange, Charleston, South Carolina, November 1990 (Council on International Exchange).

p. 20: DEAD AT DAYBREAK by Deon Meyer (Little, Brown and Company: 2005).

p. 21: "A New Accent on Diversity" by Frank Bures from THE CHRISTIAN SCIENCE MONITOR, November 6, 2002.

p. 21: THE PRIMAL MIND: VISION AND REALITY IN INDIAN AMERICA by Jamake Highwater (Plume: 1995).